S0-AZQ-247

DESIGNING STEEL STRUCTURES

Methods and Cases

PRENTICE-HALL INTERNATIONAL SERIES
IN CIVIL ENGINEERING AND ENGINEERING MECHANICS

N. M. Newmark and W. J. Hall, editors

Designing Steel Structures

METHODS AND CASES

SOL E. COOPER
Structural Engineer
Adjunct Professor, San Francisco State University

with

Andrew C. Chen
AIA

Prentice-Hall, Inc., Englewood Cliffs, New Jersey 07632

Library of Congress Cataloging in Publication Data

COOPER, SOL E.
Designing steel structures.

(Prentice-Hall international series in civil engineering and engineering mechanics)
Bibliography: p.
Includes index.
1. Building, Iron and steel. 2. Structural design.
I. Chen, Andrew C. II. Title. III. Series.
TA684.C748 1985 624.1'821 84-11686
ISBN 0-13-201385-1

Printed in the United States of America

10 9 8 7 6 5 4 3 2 1

Editorial/production supervision and interior design: Joan L. Stone
Cover design: Judith Winthrop
Manufacturing buyer: Anthony Caruso
Cover photograph and interior line work: Andrew C. Chen

ISBN 0-13-201385-1 01

PRENTICE-HALL INTERNATIONAL, INC., *London*
PRENTICE-HALL OF AUSTRALIA PTY. LIMITED, *Sydney*
EDITORA PRENTICE-HALL DO BRASIL, LTDA., *Rio de Janeiro*
PRENTICE-HALL CANADA INC., *Toronto*
PRENTICE-HALL OF INDIA PRIVATE LIMITED, *New Delhi*
PRENTICE-HALL OF JAPAN, INC., *Tokyo*
PRENTICE-HALL OF SOUTHEAST ASIA PTE. LTD., *Singapore*
WHITEHALL BOOKS LIMITED, *Wellington, New Zealand*

To Shirley Cooper, my dearest and most severe critic,
whose considerable expertise, thank goodness,
is not in engineering.

Contents

Preface

There is no shortage of books on structural steel design. However, when, in 1977, I started to teach the subject to engineering undergraduates, I had difficulty finding the "right" text. I had spent the previous 38 years in the field of structural design and construction, with about 30 of those years principally focused on the design of structures, many of them using steel. I was anxious to introduce to eager young minds that process which I had found useful in translating engineering theory into effective designs. I hoped at the same time to reveal the excitement involved in conceiving the "right" solution of a structural system addressed to specific needs and able to resist gracefully the battering of all the force systems it may have to experience in a long and useful life. This for students too often bored with the process of finding "answers" to individual textbook problems with only remote connection to the world they seek to conquer, or for young practitioners whose early assignments often seem tedious and unimaginative.

We hope this work will be useful to students and their teachers and, equally, to practitioners of civil and structural engineering, particularly in the early stages of their careers. We address it as well to other engineers who must deal with structural systems within their designs, and to architects for whom understanding of the structural skeleton is crucial in making decisions about form and function, which are at the heart of their design creations. The supporting author, an architect, has felt that need in his work.

This is intended as a *practical* book, with problems addressed at the level of mathematical sophistication that is usual for the majority of structural engineering designers. This approach arises from the authors' perception of the needs of a large number of readers; it in no sense springs from disdain for more sophisticated mathematical approaches or their power in extending the boundaries of our understanding of the performance of structures. The engineering profession has and needs both approaches. The practical is used here, because it most easily combines theoretical engineering with the nonquantifiable, nonautomatic phases of design, what engineers call their *art*. However, we have attempted to be theoretically valid in our discussion, to clarify the theoretical background behind some standardized and codified methods, to point out the limitations of such methods, and to point to the need for more sophisticated mathematical methods when they are necessary for the completeness or safety of a design.

The practical engineer who is fully aware of his responsibility to client, public, and his own self-esteem draws on his own background and talents and, at the boundary of these, recognizes the need to seek and consult others whose special talents can complement his own. The engineered product is not the consequence of brilliant insights alone, but also that of a technology involving a large number of people, incorporating the inherited wisdom of earlier generations and a constantly improving base in the systematic understanding of physical and social processes.

The calculation methods presented are *hand* methods. The problems in analysis are relatively simple and nonrepetitive and hand methods work. They also maintain a close link between the designer's thinking and intuition and the structural systems and choices he must deal with. However, the logical systems are presented so as to be easily convertible to digital computer programs, a process ultimately necessary to get the most utility from many design solutions. In presenting the logic, we attempt to avoid the pitfalls of the completely mechanical. Creative design requires that the designer be in control. At key points, the logical program, hand or computerized, must have choices and escapes, so the designer can lead the results, taking into account the quantifiable as well as the nonquantifiable.

The quantitative work uses the English system of units. American practice is still, for the most part, in these units. Structural shapes produced by the American steel industry are most conveniently used with this system, and published data on cross sections and strength are based on them. Conversion to SI (metric) is possible in each case, but there are pitfalls in arithmetic manipulation and basic inefficiency in arithmetical solutions that require conversion. The SI system is most effectively used when it is incorporated directly into the total technology. It would be a disservice to the reader to complicate the process of conceptual learning by using units that do not take best advantage of the available data in the form in which it is generally used.

THE THREE-PART STRUCTURE

The three-part structure of the book is an attempt by the authors to introduce the thinking and information-retrieval process that is effective for designers. Look at Part I as crucial general background. Add to that the information in Part II about selection of specific types of members, joints, and systems. Add what you already know, or may have to find out, about the physical and other principles involved in the design. Apply all the collected background plus the designer's ability to select, integrate, and create to the solution of the specific problems in the design systems discussed in Part III, following those worked out by the writer and working out those suggested for development by the reader.

It is neither necessary nor useful, and probably not possible, to master Parts I and II before venturing into Part III. Rather, learn the structure of the information in the first two parts and develop mastery by application to the case studies and problems in Part III. In our experience, this is the learning process that most young designers follow in fact as their abilities mature; this is the way designs evolve. It is the approach we recommend to the reader.

The material herein represents the views of the writer, or more precisely those of Cooper, sharpened by the help, insights, and advice of Chen and others. The approaches used are those that have been useful to me in the many years during which I have had the opportunity to help execute a great variety of structural and related designs, supplemented by what was learned by additional years as iron worker, construction engineer, consultant to designers and construction companies, and recently as teacher of engineering students. In recent years, Cooper and Chen have collaborated on a number of designs. Humility has been added by observation and sometimes investigation of engineering failures. Insight has been gained by the opportunity to work under and with some great and creative engineering designers and teachers. Engineering designs represent the further development by an individual or a group of people of a great deal of knowledge and information handed down through the engineering profession or others. But in the final analysis, the work, if it is not just copied, is individual.

The reader should not expect that the formats, procedures, and approaches presented here are universally accepted dogma. They work for us. They seem to work, but always with some variation, for many whom we have learned from, worked with, supervised, and helped to train. They work directly for some problems, but need adaptation for others. The engineering mind needs past solutions and tried formats as anchors. But it needs to be free to break new ground or to differ with the way others see it or the precepts of *conventional wisdom*.

We do not claim to have created new science herein. The scientific background of structural engineering, at least in its simpler applications, is the least controversial part of a structural design. It is in the other parts, what are often called the *art* of design engineering, that the greatest differences arise. The science is given its very important due throughout this work. Our own approach to the art is recommended for careful reading and consideration so that the reader can draw from it what suits, and adapt and change it to suit his or her own style of thinking, understanding of the design context, or the special nature of the problem.

The problem of gender is one that is constantly with us. The authors are convinced that the introduction of women into engineering is not only just but makes available much untapped talent that is needed at all levels of responsibility. However, the language gets in our way. The reader soon tires of the very clumsy "his or her" form, and we have used it only sparingly. We have not found a satisfactory substitute in general for the use of "his," or in some instances "hers," in both cases without prejudice as to the real sex of the individuals. In this instance, the authors themselves are "he's," but take no credit for that accident.

THE BOOK AS TEACHING MATERIAL

Most of the material in the book was developed in the course of teaching the subject of structural steel design in a three-unit, one-semester introductory course. There is, I believe, more than can be covered in one semester. I assume others, as I, would select some material to cover and omit or lightly touch on other matters. I have presented in single semesters, partly with the help of other texts when this work was not yet complete, all or most of Part I; the material on member selection in Part II, including Chapters 7 through 10, and parts of Chapters 11 through 14; and two and sometimes three of the case studies in Part III, as follows: mill building and plate girder bridge; or stiffleg derrick, tier building, and plate girder bridge; or mill building, tier building, and plate girder based on Chapter 14.

In structuring the course, I use the approach suggested to readers above. After setting the stage with the material in Part I, we examine tension and compression members (Chapters 8 and 9). At that point, I find it useful to start the first case study. From that point, for eight to nine weeks, we look at case studies while, in separate lectures of the same weeks, I continue to present material from Part II. To the extent possible, the methods of Part II are presented slightly before or at the same time as they are used in the case studies. This is not always completely possible, since the needs of the design force us at times to go beyond the selection procedures already covered. I try to fill in the gaps in discussion. A possible schedule, similar to those I have used, is shown on the next page, based on two 1½-hour meetings per week in a 15-week semester. There are two exams and three design reports. I use a meeting during the final exam week to discuss and synthesize results of student reports. With the luxury of a two-semester course or a design lab, I would, of course, use a different schedule. However, I would not abandon the idea that the case studies and member selection procedures support each other and are best studied together.

Assignments based on the case studies are system design problems. They are assigned in the first week of the case study and due after it has been completed. I encourage peer discussion among students and group designs.

In our curriculum, we require senior students to do independent design projects singly or in a group under the supervision of a faculty member. Many have used problems relating to the case studies: changing the context, extending the level of detail, using computer solutions for refinement, or seeking different solutions to the same or similar design problems. They have less trouble reaching for new design solutions after following the design process in system applications. Suggestions for such projects are given after each case study.

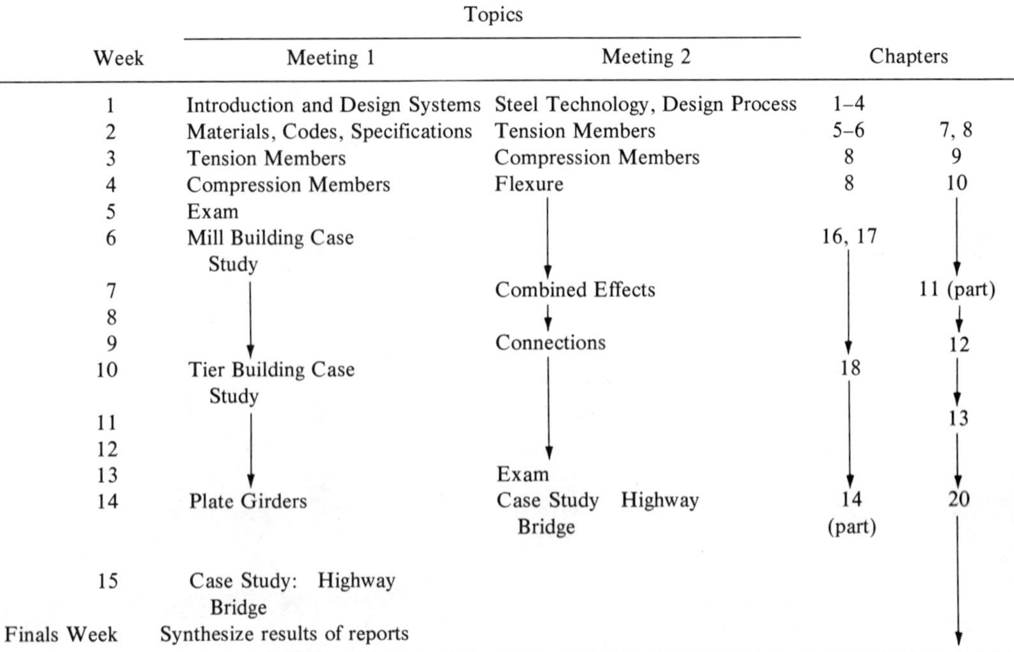

Week	Topics: Meeting 1	Topics: Meeting 2	Chapters	
1	Introduction and Design Systems	Steel Technology, Design Process	1–4	
2	Materials, Codes, Specifications	Tension Members	5–6	7, 8
3	Tension Members	Compression Members	8	9
4	Compression Members	Flexure	8	10
5	Exam			
6	Mill Building Case Study		16, 17	
7		Combined Effects		11 (part)
8				
9		Connections		12
10	Tier Building Case Study		18	
11				13
12				
13		Exam		
14	Plate Girders	Case Study Highway Bridge	14 (part)	20
15	Case Study: Highway Bridge			
Finals Week	Synthesize results of reports			

IN GRATEFUL ACKNOWLEDGMENT

Much thanks is owed and gratefully extended to many individuals who helped in this work: Alvaro Collin, Steve Johnston, Jim Naftzger, Bob Preece, Bob Ray, and Herman Zutraun, friends and fellow engineers for many years, who brought their seasoned judgment to bear on critical review and suggestions for many of the chapters; Mehmet Celebi and Peter Pfaelzer, fellow teachers of engineering at San Francisco State University; Richard W. Golden, Gail McGovern, and especially Linda Force, former students now in practice who devoted many hours and great diligence to review, correction, and that special criticism needed to guarantee that the work suited the needs of the student and young engineer. All of these and many others helped to steer me away from error in either calculation or concept and to guide my thinking on the makeup of the book. They are, of course, not to blame for mistakes that may have crept in in spite of their best efforts and mine. I hope readers will point these out for future correction.

Without Andy Chen, who illustrated the work and also brought the different understanding of an experienced architect to this engineering work, there would be no book today.

The manuscript was prepared with the help, much appreciated, of Roxi Berlin, and also Shelley Dizon, Yvonne Kanis, Donna Henderson, George W. T. Lew, Shuen Yuh Lo, and Mark Smith.

DESIGNING STEEL STRUCTURES

Methods and Cases

Part I

Background to Structural Steel Design

1

Introduction: Field of Structural Design

1.1 INTRODUCTION

"Find a need and fill it" is an advertising slogan for a concrete supplier. In slightly revised form, "Given a need, fill it," it could be a definition of the task of a (structural) designer. Just that broad, just that unspecified! Write on a blank page. Create a solution. A good basic credo, but one which omits the step that is most frustrating to novitiates—how to begin.

Perhaps the most common frustration for a beginning student in structural design arises when he or she is asked to start a solution. An example has been given in a text. Having defined a problem requiring the choice of an appropriate member for a structure, the author suggests that we "try a W14 × 145"* (or some other), and then proceeds to test the adequacy of the trial section, if necessary adjusting the choice slightly.

The student (or engineer) is then asked to solve a different problem where a W14 × 145 is clearly not appropriate. What now? The number of possible choices stretches out through page after unfamiliar page of available steel shapes, as well as another infinity of choices that do not use standard shapes and possibly not even steel. How then to begin?

The author of that text has come to the problem with a degree of engineering judgment that the student does not yet possess. He may have already tried and

*An industry standard designation, to be clarified later.

rejected five other possible approaches to the solution, which, in the interest of brevity, he spares the student. In an analogous situation, a senior engineer, conscious of time and budget, may apply his* trained judgment to consider and reject a multitude of possible solutions; only then does he assign one or two approaches, likely winners, for in-depth investigation and development by a junior engineer.

We will attempt to chart the sea of structural design, converting it from a vast undifferentiated space to a finite number of design lanes, which, since they do not all lead to the same end, can be the subject of informed choice. The process, then, starts with a series of questions.

What is the problem?
What possible solutions suggest themselves? (How to begin?)
Which seems most likely to be the best solution(s)?
Which shall I try first?
Does it work?
How does it compare to other likely candidates?
Is this the solution of choice?
If not, which is?

Each of these questions could be followed by "Why?"

The assumption is made that the designer brings to each question or can get from informed consultants the necessary tools of mathematics, engineering mechanics, economics, social science, and so on, that make it possible to investigate the question adequately.

We hope to provide some of the additional necessary tools that apply to structures that are to be built of structural steel. A good deal of specific information about that material is necessary if it is to be used appropriately. We will examine examples of structural steel designs, trying to clarify how design thinking develops. In the process, we hope to lay out design methods that can be adapted to other structures, regardless of the material.

Since our audience will include beginners, we will use relatively unsophisticated methods of analysis, which are adequate for a large class of structures, but not all. From time to time we will point to the need for more sophisticated investigation of part or all of a problem. Our mathematical methods will be primarily "hand" as distinguished from "computer," since we believe design thinking can be most easily clarified for the student, and often for the seasoned designer, with "hand" methods. However, when formal logical systems apply, they will usually be presented in forms easily adaptable to digital computer programming.

*The terms "he," "his," "him," and the like, are used in the neuter sense and to avoid awkward circumlocutions.

1.2 THE FIELD OF STRUCTURAL STEEL DESIGN

At the start of a structural design, an engineer must make a number of basic choices, among them being the materials to be used in construction and the structural system(s) in which they are to be used. For the past century, the choice of material has often been structural steel. This book will examine the conditions that make the selection of structural steel appropriate and some of the ways in which it can be effectively used. To do that, it will be helpful to discuss the forms in which steel is usually provided, the organization of the industries that produce, fabricate, and build with it, and the way the engineer interacts with that industry. We will have to recall some of the fundamentals of mechanics on which our understanding of structural systems is based and apply them to the special characteristics of this very strong, highly versatile material. We will need to examine the crystallized engineering thinking that exists in the form of codes and specifications, whose intelligent use may simplify the designer's problem, but which, inappropriately used, may lead to inadequate or overly expensive designs or, even worse, structural failures.

The number of materials commonly used in primary structural systems is limited. A list that includes steel, reinforced and unreinforced masonry (concrete and other), wood, and aluminum would be almost exhaustive for most definitions of primary structures. Strictly considered, the list should also include the supporting soil. Table 1.1 lists the materials often used in a number of different types of structures. The design engineer usually has to choose among the alternatives shown. Very often the appropriate choice is structural steel.

Our attention will for the most part be directed toward primary structures. The term *primary structures* cannot be precisely defined; but it can be understood by example. Consider the main supporting system of a building: floor slabs, beams, girders, columns, bracing; omit nonbearing walls and partitions, utility distribution lines, sculptural embossments, and the like. Similarly, consider a bridge, superstructure to foundation, without lighting or utility lines that may be carried on it, possibly with sign supports but not signs. Consider the framework of a derrick without driving machinery, the hull of a ship without machinery, navigating gear, and so on. Similar examples will no doubt occur to the reader.

The structural action of many other systems follows engineering principles similar to those of primary structures. The only significance of the separation is that most of the attention of structural engineers is directed to primary structures, and much of what we will say about the organization of industry and engineering applies specifically to work done on primary structures. Steel shapes and plates are supplied to this work as *structurals* and sometimes as *light gage steel* preformed into relatively stiff *decking* and *siding*. The characteristics of the steels used are selected to be useful in primary structures; steels formulated for other uses may not be acceptable.

Specifications for steel manufacture and for the design of structures are made compatible. The structural designer, having chosen a type of steel and a structural system, has also chosen a predictable and controllable series of events from mill to

TABLE 1.1 THE FIELD OF STRUCTURAL DESIGN: PARTIAL LISTING OF THE SPHERES IN WHICH STRUCTURAL ENGINEERING PLAYS A PART, SOME TYPES OF STRUCTURES INVOLVED, AND MATERIALS COMMONLY USED

Field	Structures	Materials
Civil Engineering		
Roads and railroads	Pavements	Processed earth, asphalt, AC, PC concrete
	Bridges	*Steel,* reinforced concrete (R/C), prestressed concrete, wood, hybrid
	Retaining structures	R/C, *steel,* wood, stone, . . .
	Service buildings	See Architecture
	Vehicles	*Steel,* aluminum
Water supply and sewerage	Dams	Earth, concrete
	Power houses	*Steel,* R/C
	Reservoirs, tanks	Earth, concrete, *steel,* plastic
	Pipelines	*Steel,* iron, R/C, PVC
Ports	Piers, wharves, drydocks, docks	*Steel,* concrete, wood
Architecture	Buildings	*Steel,* R/C, wood, masonry
	Building accessories	Same plus plastics, processed materials
Naval architecture	Hulls	*Steel,* wood, fiberglass, ferrocement, R/C
Machinery, industrial processing, mining	Buildings	Similar to Architecture
	Tanks	*Steel,* concrete, plastics
	Pressure vessels	*Steel,* prestressed concrete
	Foundations	R/C, *steel*
	Pipes	*Steel,* cast iron, concrete, clay, plastics

jobsite, which, properly supervised, can lead to a structure that will be "as designed" in form and quality.

Such steel *structures* as auto bodies, pipelines, pressure vessels, signs, and the like, which have principal uses other than primary structural support, are usually designed under different conditions and with different characteristics emphasized. The structural engineer who has mastered the use of "structural steel" sometimes must cross the vague line to deal with these other uses. In doing so, he should investigate not only the similarities but also the differences between *structures* and the structural action of other systems.

Primary structures may include steel in the form of reinforcing bars combined with other materials to make reinforced concrete, reinforced unit masonry, or reinforced earth. This usage differs markedly in character from that of structural steel. We will exclude reinforcing bars from the field of steel structures as we define it. However, we will include the use of structural steel with reinforced concrete in what are called *composite structures*, such as bridge girders where a reinforced concrete

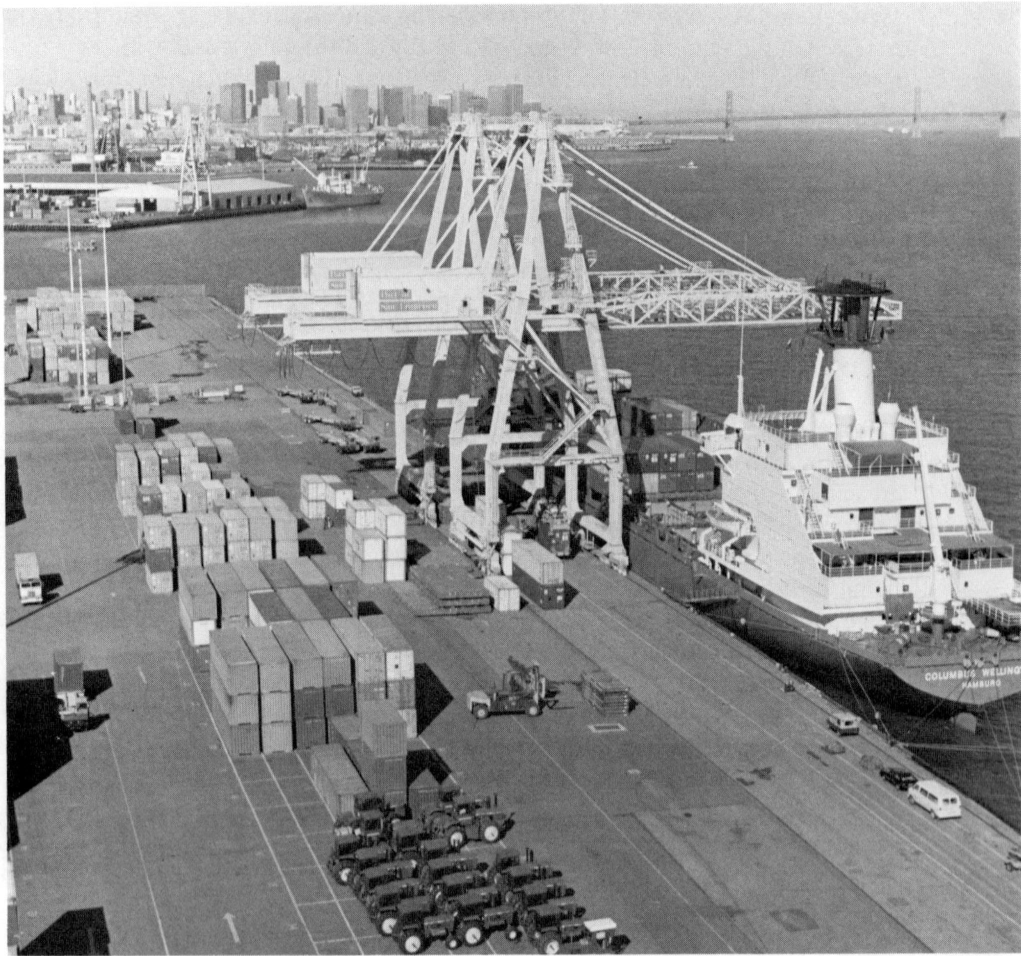

A modern port and the city and regions it serves includes many examples of engineered structures. (Photo courtesy of the Port of San Francisco.)

deck is designed to act structurally in combination with steel plates or shapes to make a combined resisting member.

1.3 TOTAL DESIGN: INTEGRATED SYSTEMS, CODES, SAFETY

This book deals with structural steel design from the stance of the design engineer and the way the design problem is approached by him or her. The engineer is responsible to a client to provide the most effective design for the client's needs and budget that is consistent with overriding public policy. The structural engineer usually provides designs that affect both the success of a client and the safety and well-being of the public. Often, the public is the client. The engineer works within a team that includes other engineers and design professionals whose systems need to be comfortably meshed with his own. The success of a structural design is dependent on the success of the total design and construction project: not the least steel tonnage for the job, but the least steel tonnage for the job consistent with the requirements of time, accommodation to energy constraints, and so on; not the least construction cost, but the least lifetime costs or most advantageous investment cost; often not the fastest construction schedule, but the greatest adaptability. The choices are as many as client needs are diverse. The engineer's bag of tricks, his approaches to solutions, must be just as varied.

At rock bottom, the structural design must be safe—a basic article of faith among engineers. But "How safe?" is often the subject of bitter controversy. Codes and specifications are often adopted by governments or private associations that attempt to define minimum acceptable levels of safety and translate them into design constraints. The designer must have the key to the code's translation, that is, be able to understand the conditions envisioned by the minimum requirements of the code. He must also understand the real conditions that his structure will experience and define structural safety in terms of those conditions, exceeding code "minimums" where engineering analysis and judgment tell him they are insufficient. Similarly, a client's operating needs may dictate a design of quality beyond that required by the code. We will follow the way of design practice, studying the codes and specifications as both constraints on our design freedom and aids to our design thinking, in both instances the crystallized judgment of highly qualified professionals. But we will emphasize the unique nature of each design problem and examine the special conditions that require special treatment.

Similar things can be said about standardized designs. Some design problems occur often enough and under similar enough conditions to invite standard designs, which can be offered *off-the-shelf* by specialized manufacturers. The standardized designs and the manufacturing processes that they make possible are often the least-cost solution to all or part of the engineer's problem. The designer's responsibility is to be aware of them and to understand their virtues and their limitations. When they fit the problem best, they can and should be incorporated as "purchase" items into the total design provided for the client. Where they fall short, the designer may be able to arrange modifications by the manufacturer, or may reject them and provide a different solution.

1.4 STRUCTURE OF THIS BOOK

The book is arranged in three parts, which, while separate, complement each other. It can be read sequentially, page by page. However, we feel it will be mastered best if the reader takes a somewhat different approach, more closely paralleling that followed by designers in practice.

Part I is intended to provide essential background information about steel, structures, and the design process. It touches a number of subjects, none of them in great depth, but can, we believe, provide a context within which later work can be better understood. Read it quickly for the essence of the information. Refer to it again when the need for parts of its content becomes clear to you in dealing with the ideas of Parts II and III. When greater depth is needed (for example in understanding metallurgy or construction management), go to specialized materials listed in the References or elsewhere.

Part II provides the tools needed to select or design members to be used in structural systems such as are designed in Part III. One can, as is more traditional in texts, master Part II on its own before attempting the design of structural systems. We consider that approach both less instructive and less fun, and suggest a more interactive one.

Part III draws selectively from chapters of Part II. After the first three chapters (7, 8, and 9) of Part II, a reader should be ready to venture into the first (Mill Building) case study (Chapter 17), in the meantime learning those parts of Chapters 10 and 11 that are used in that study. Some may prefer to go to the Stiffleg Derrick Case Study of Chapter 19, which draws on the same earlier chapters in a different context. The Tier Building Case Study (Chapter 18) will be easier to follow if the reader starts from a good base in the analysis of indeterminate structures, as well as the material of Chapters 8 through 11. On the other hand, it may be read simultaneously with Chapters 10 and 11, each part helping one to understand the other.

Chapter 18 also clarifies some issues about the design process originally addressed in Chapters 2 and 4, on which the reader should refresh himself when starting Chapter 18. If the reader has not previously mastered indeterminate analysis, Chapter 18 may be read in a different way, absorbing some of the concepts and power of indeterminacy, and following later with mastery of the procedures. At the same time, ideas about the purpose of preliminary design, illustrated in that chapter, will become clear. The Highway Bridge Case Study (Chapter 20) and Chapter 14 are directly complementary.

Chapters 12 and 13 cover the indispensable subject of design of connections. They may be read separately or "as needed" in connection with both the earlier chapters of Part II and the system designs in Part III. Interconnections between these chapters and others are suggested at various points in the book.

The case studies are illustrative of problems met by engineers in designing total structures. There is no pretense that the solutions presented are the only or even the best ones. In each case, they illustrate the consequences of one design approach by one designer to a design problem. The reader is encouraged to try different approaches, compare the results, and trace the consequences of each design decision through the system being designed. That, in our mind, is the road to design mastery.

Much of the material of this book is based on the eighth edition (1978) of the *Steel Construction Manual*[7] and the 1982 edition of the *Uniform Building Code*.[33]

Inevitably, these will become out of date before many years, making the material here harder to follow. We did not feel we had a choice. These materials, or similar equally time limited alternatives, are indispensable parts of the design process in U.S. practice. We limited the amount of such material to the minimum we considered necessary. The most important chapter, for our purposes, of the *Uniform Building Code* is included as Appendix A. We hope to update the material of the book when it becomes essential to understanding.

1.5 ANALYSIS AND DESIGN

In recent years the engineering profession has tended to divide itself into analysts and designers, a trend that can easily be carried too far. Our purpose here is to address issues of design, but, as will be seen first in Chapter 2, and again very sharply in all of Part III, design decisions must be based on analysis. More profoundly, to make any but the simplest design choices, the designer must understand the analytical consequences of the systems being considered. Again, we propose an interactive approach to learning, with design and analysis each buttressing and clarifying the other, with no single prescribed sequence in which they are learned.

2 Structural Systems

2.1 INCONVENIENCE, FAILURE, AND THE PURPOSE OF DESIGN

On December 22, 1982, according to the *San Francisco Chronicle*, "Embarcadero One," a modern office building in San Francisco, was swaying so violently that workers on the upper floors left early. "It was swaying so bad," one office worker on the thirty-ninth floor was quoted as remarking, "that I couldn't stand up."

This was not a replay of the 1906 earthquake. It was a wind and rain storm whose intensity, unusual but not unprecedented by San Francisco standards, would hardly have fazed veterans of Florida hurricanes or Hong Kong typhoons. The highest wind speeds reported in the *Chronicle* that day for the San Francisco Bay Area were 90 miles per hour at the top of two of the highest local peaks.

Many of the early departing workers from Embarcadero One that afternoon were forced to delay their homeward trip to Marin County until the Golden Gate Bridge, swaying too violently for traffic safety, was reopened to traffic. At no time that day was either Embarcadero One or the Golden Gate Bridge in danger of either collapse or even, as far as we know, significant distress in its parts.

If, on the other hand, the 1906 earthquake had been replayed that day, the probability is high that Embarcadero One and the Golden Gate Bridge would have escaped damage, but, in the opinions of many structural engineers, some other San Francisco buildings would not. This despite dedicated attention to earthquake-resistant design by California structural engineers since 1906, UBC-mandated seismic design requirements since 1927, California legislation since 1933, and considerable success by West Coast structural engineers in their continuous search for structural systems suitable to resist earthquake effects both safely and economically.

Discomfort in the upper floors of Embarcadero One that windy day in 1982 and even the temporary closing of the bridge were not indicators of design deficiencies. But the collapse of the Kansas City Hyatt Regency "skyways" on July 17, 1981 has been attributed to defects in design, as has the collapse of a new hospital and a number of freeway overpasses in the Southern California earthquake of February 9, 1971. The 1982 events resulted in a temporary loss of function and minor annoyance that could only have been avoided, if at all, at great dollar cost which few would propose. Those of 1971 and 1981 involved collapse, loss of use and investment, and loss of life.

We are led by these events to attempt a definition of a structure, although we realize (see Chapter 1) that a fully satisfactory definition is not possible. A *structure* is an assemblage of members into a supporting system on which the day-to-day activities of people may be conducted. The success of a structural design may then be measured by:

- How well and how reliably the structure provides support.
- To what degree its form allows for and even encourages the activity intended.
- How well current technology is used in the structural design so that the preceding can be done well at minimum justifiable cost.

An *engineered structure* as a physical thing must be analyzable by the principles of mechanics for its ability to stand safely against attacking loads. It must also, although usually with less mathematical precision, be analyzable for the degree to which it carries out its other functions.

The process of structural design must be more than analysis. It must create structural systems to be analyzed. However, as with the chicken and the egg, it is difficult, and even useless, to say which came first. Structural engineers seek from among those systems whose methods of analysis are already understood alternative ones that can solve the underlying functional design problems. Sometimes the search leads to new combinations of analyzable systems or even new concepts of systems that require new methods of analysis. These may flow by intuitive leaps from the known mathematical and physical properties of materials and systems, but must ultimately be tested by analysis. Actual new structural systems often are used before complete analysis is possible and thus rely on extrapolation from previous knowledge. In that case, tests in use may be necessary before a fully analyzable structure is possible.

For example, the aerodynamic theory of bridge response to wind forces came after the first Tacoma Narrows Bridge collapsed in 1940. The resulting mathematical theory by Bleich and others[21,65] explained that failure and a number of other historic

Steel arches carry a high-level highway bridge across the New River Gorge in West Virginia. (Photo courtesy of U.S. Steel Corp.)

failures of suspension bridges. By alteration of existing bridges and application to new ones, the theory has prevented other failures from this cause. The advantages of suspension bridge designs were so great that many, mostly successful such bridges were built before a satisfactory answer to the aerodynamic problem was found, even before the nature of the problem was perceived.

Designs are typically conceived in conceptual terms based on predictions of the results of loads and are refined or sometimes abandoned in favor of others as a result of analysis. The process is iterative, sometimes requiring several iterations. In Chapter 4, we will look further into the design process. At this point we turn to some matters relating to structural systems.

2.2 STRUCTURAL SYSTEMS AND SUBSYSTEMS

A structural system is an assemblage of members that interact to perform a more comprehensive function than those of the individual members. Each member, or sometimes groups of members, may be considered as a subsystem. The nature of interaction between subsystems and systems determines the needs of each. A structural design starts with the definition of the larger system. Design of a subsystem can only be successful if its interaction with the larger system at their juncture is

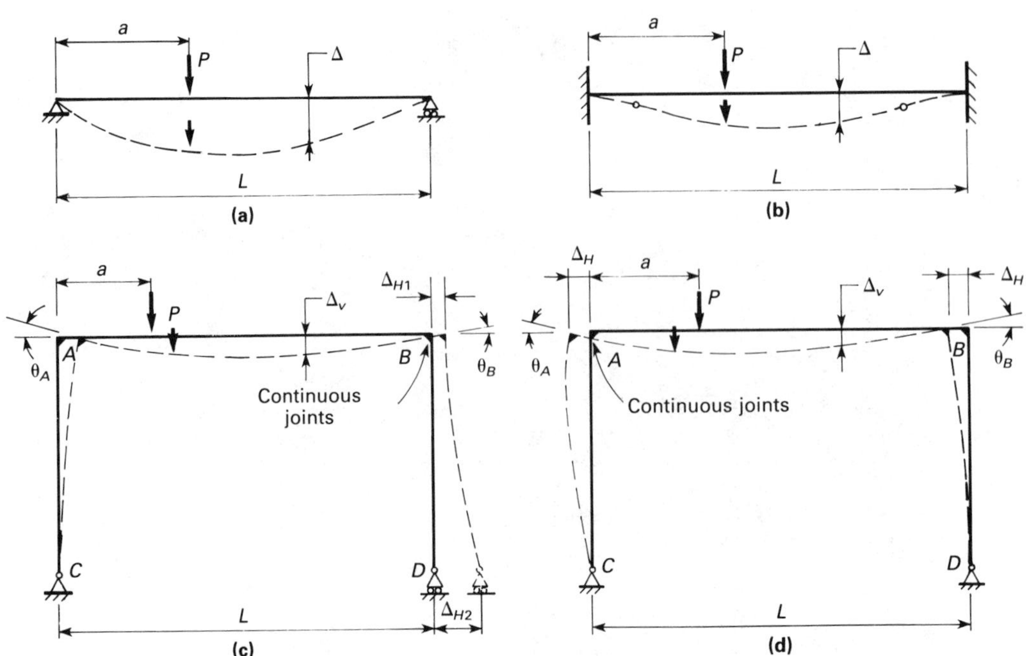

Figure 2.1 Actions of members depend on actions of system: (a) simple span can be analyzed completely if P, a, and L are known; (b) same applies to fixed end beam; (c) statically determinate frame resembles (a); (d) design of member AB in statically indeterminate frame cannot be attempted unless the relative stiffnesses of all members are known.

understood. For example, in Fig. 2.1, it will be seen immediately that the design demands of the simple span and fixed-ended beam are different. Either is easy to analyze, since their end (boundary) conditions are fully prescribed. The primary analysis of the frame of Fig. 2.1(c) also seems straightforward, since it is statically determinate. However, complete analysis requires examination of the deflected shape, which depends on the stiffnesses* of the members. The analysis of the indeterminate frame of Fig. 2.1(d) must start from the evaluation of the relative stiffnesses of the members.

In the procedures in Part II for selection and/or design of members, it will be necessary to prescribe the conditions at the boundaries arbitrarily. In the case studies of Part III, we will see how, for those structural systems, the nature of the system determines the boundary conditions for the member and therefore the design requirements of the member. We will also see that the selections of the members will often change their boundary conditions.

What follows in this chapter is intended to reveal some general characteristics of structural systems and subsystems as illustrated by a limited number of them closely

*Stiffness is defined in Section 2.3.

related to the material covered in other parts of this book. Many more systems exist in a rich heritage shared by builders and engineers. We recommend references 37, 43, 52, and 61 for insights into structural philosophy, a field too often neglected in engineering education.

2.3 DEFINITIONS

If words are to be a medium of communication, the user and the receiver must react to them in the same way. Without needing to define a word anew each time, both must have internalized the same definition. However, language is a living medium within which meanings evolve, often by analogy drawn from experience, which is different for different people. So words grow to have different meanings for different people, a fact understood well by some advertisers. Scientists seeking precision define terms precisely so that fellow scientists understand them in the same way. In the process, they reject alternative definitions of the same terms that continue in general use. Engineers as scientists seek and need precision of meaning. However, engineers serve public needs directly and cannot (should not) avoid rubbing elbows with a public who have different definitions than we do for our terms.

A case in point are some words that we must attempt to define in this chapter. The first definition in our *American College Dictionary*,* 1961 edition, of the word *rigid* is *stiff*. Yet we will define *stiffness*, as used in common structural engineering usage, in a way that contrasts it with *rigidity*. *Rigid*, in that definition, is *not flexible*, and *flexible* is, among other things, *elastic*. But we expect most of our rigid structures to be elastic by a definition of elasticity in common engineering use: elastic action is such that any load-induced strains return to zero when the load is removed. The litany of contradictions is a long one, many of them inevitable, and accounts for many misunderstandings between engineers and their public, sometimes even among engineers.

We will define a few special terms as used in this chapter. For the most part, the definitions reflect those in use in the "structural engineering community" and the usage of the same terms in other parts of this book. If and when that is not true, we hope the difference will be clear from the context. Some other terms are, we believe, unambiguous; others are defined in Appendix B.

Compliant As a characteristic of a structural system, the ability to change form in response to load without loss of function. May involve changes in macrogeometry, as in suspension cables, or in microgeometry, as in ductile flow in members or joints. Small deformations from elastic strains do not represent compliant response.

Diaphragm A resisting surface rigid enough to transfer force effects to other locations with almost no internal change in geometry. *Examples*: floor slab of building, shear wall, each subjected to loads in its plane.

*It doesn't matter here that this is not the most complete, unabridged, or even the most current, dictionary available.

Instability As applied to a structural system, the inability to tolerate low magnitudes of at least one type or direction of force or moment without collapse or loss of function. In stiff or rigid systems, a small disturbing effect on an unstable system changes the basic geometry of the system, and therefore the system itself (see Fig. 2.3). Compliant systems seek stable equilibrium by changes in system geometry.

Rigidity As a characteristic of a structural system, the ratio of a force to the displacement it causes in the direction of the force. May be equated with translational stiffness.

Rigid joint A joint that is internally rigid, so that it can be translated or rotated without changing the angle between the tangents to the elastic curves of members connected at the joint.

Shear wall A rigid wall designed to resist horizontal forces in its plane. May also resist gravity forces.

Stability A requirement of a structural resisting system, best understood as the opposite of *instability*. Equilibrium principles can only be applied to stable structures.

Stiffness (rotational) The moment on an end of an elastic flexural member that will cause a rotation of one radian around a transverse axis if the other end is fixed against rotation (resistance of a beam to rotation).

Stiffness (translational) See *rigidity*.

2.4 DISPLACEMENT* IN STRUCTURAL DESIGN

The emphasis in most early engineering education is primarily on the consequence of force in terms of stress. *Displacements* of members or *deformation* of materials are seen as derivative consequences of stress, although stress and strain are ineluctably tied and neither is really derivative of the other. Elastic displacements of structures arise as a consequence of accumulations of strain in the structural members. However, deflection itself may be no more an indication of danger in a building than it is in a diving board. Members cannot resist load without stress or its partner, deflection. The process of design must start from that as a given and proceed to consider displacement in all of its aspects, both helpful and harmful. To list a few:

1. Only the simplest structures, the statically determinate ones, can be analyzed without reference to the deformation characteristic of the members and the system. Even for these, excessive sag or vibration of the members may make the design nonfunctional.

**Displacement* and *deflection* are sometimes used interchangeably. Displacement is the more general term.

2. *Indeterminate* structural *systems* can only be analyzed by methods based on displacements. Sources of deformation must be identified and the members selected at least as to relative magnitudes of the significant size parameters like *area, moment of inertia*, and *length* before an analysis is possible (e.g., analysis based on *slope deflection* relationships must start with the relative rotational stiffnesses of all members known in advance). Conversely, by careful selection of the size parameters of different members, the designer can lead the effects of force to the members he has selected to resist them. Given a small and large coil spring in parallel, both deflecting the same distance under the action of a load, the large spring supports most of the load.
3. *Small deflection theory* of structures starts with the assumption that deflections are small enough so that the simplified equations usually used in analysis ($f_b = Mc/I$, etc.) are applicable. Diagrams of deformed structures must usually use exaggerated deflection scales to show the deflected shapes. Usually (barring unusual occurrences such as San Francisco's December 22, 1982), the deflections themselves are not perceptible. However, deflections of structures may not be as small as the term "small deflection" might imply. Table 2.1 cites some commonly accepted criteria and their implications for elastic movement. These are not by any means the largest deflections tolerated. In the tier building of Chapter 18, a combined system is used, part of which is intended to minimize perceptible movements in high winds; the other part allows more movement, but maximizes energy absorption in an extreme earthquake.
4. Deflection becomes objectionable when it detracts from the function of a structure. Sagging of a structural floor may cause partition doors to stick and plaster in ceilings and partitions to crack. Large-amplitude movements cause instruments to malfunction. Windows are too rigid to permit with impunity the amount of distortion that structural frames tolerate. Yet they may occupy the same planes and must be prevented, by joint separations filled with soft resilient materials, from resisting the structural loads they are not equipped to handle. Movements involving high acceleration, which may or may not be associated with large periodic amplitudes, upset people and equipment to the point, in extremes, of inhabitability. Excessive acceleration of bridge decks leads to both riding discomfort and loss of driving safety, even though member stresses may be low.
5. Small changes in the geometry of structures may interfere with the function of nonstructural elements. *Sagging* floors make door and window openings nonrectangular, causing them to jam. Windows break from building *drift*, as do nonstructural wall panels. Even axial elastic deformation of building columns may cause undesirable compression on nonstructural veneers attached to walls, dislodging bricks or pieces of stone.
6. The utility of a building or its contents may be severely compromised by acceleration of the floors or walls in response to wind loads. Recall again the events of December 22.

TABLE 2.1 DEFLECTIONS PERMITTED BY COMMON CRITERIA

Phenomenon	Sketch	Criterion	Example
Deflection of beam	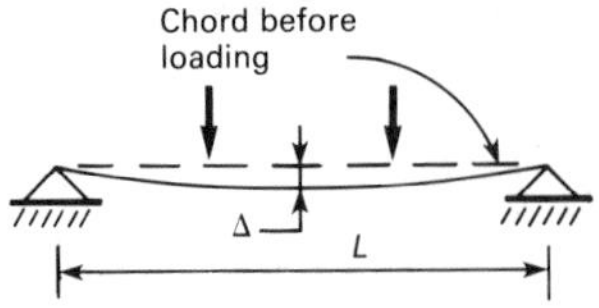 Chord before loading; Δ; L	$\Delta \le \frac{L}{240}$	$\Delta \le 2$ in. for span = 40 ft
Drift of building	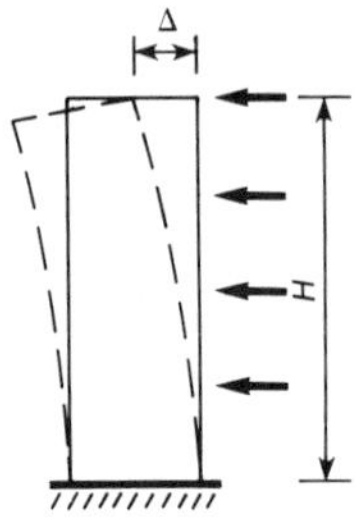 Δ; H	$\Delta \le \frac{H}{500}$	$\Delta = 1$ ft for $H = 500$ ft
Deflection of highway bridge from traffic	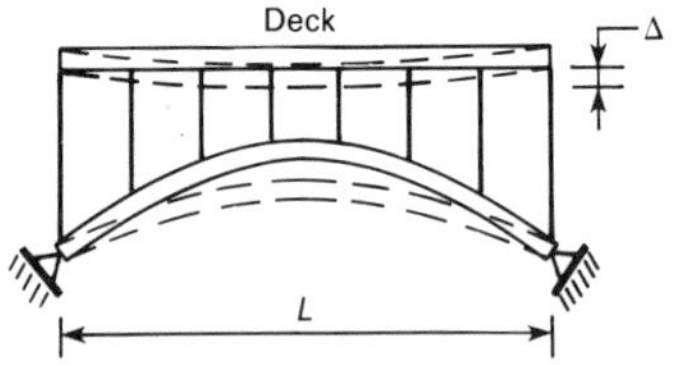 Deck; Δ; L	$\Delta \le \frac{L}{800}$	$\Delta \le 1.25$ ft for span = 1000 ft

None of the effects listed imply structural distress or loss of safety, although these are both possible if the stresses experienced are excessive for the members.

2.5 STABILITY

For a structure to stand, it must be stable. This sounds like a truism, not worthy of repetition. Yet, by the evidence of engineering failures, too many of the structural engineers weaned on stress fail to grasp it. Among them were the designers of two structures whose buckling failures are described at the beginning of Chapter 9. A significant thing about instability is that it does not imply high stress, but the inability to tolerate small disturbances. The ball at the crest of the hill in Fig. 2.2 is in unstable equilibrium. The slightest disturbance will cause it to roll downhill. The one in the trough is in stable equilibrium. If displaced, it will roll back to its original position.

The compression members whose buckling is described in Chapter 9 did so at low stress because the shape of their cross section and their length made them

Unstable
equilibrium
Stable
equilibrium

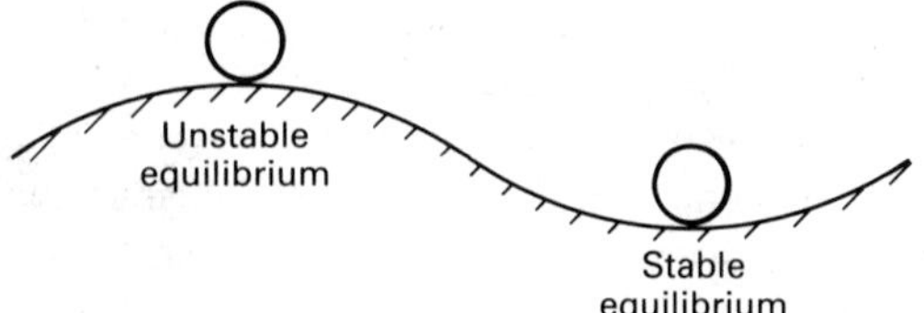

Figure 2.2 Stable and unstable equilibrium.

P
P
P
(a)
(b)
(c)
Two intersecting braced planes cannot resist torque. *ABCD* and *EFGH* have constant shape. *EFGH* rotates. Stiff faces *AEHD* and *DHGC* warp. Faces *AEFB* and *BFGC* warp and change angles between members.
F
G
E
H
B
C
A
D
(d)
(e)

Figure 2.3 Instability in structures: (a) slender column is not stiff enough to support large load without buckling; (b) narrow compression flange of beam is not stiff enough to prevent buckling with large load; (c) rectangular system of pinned-end members has no lateral resistance; (d) torsional stability requires shear couples in parallel and opposite resisting planes; (e) stable frame cannot stand on unstable soil.

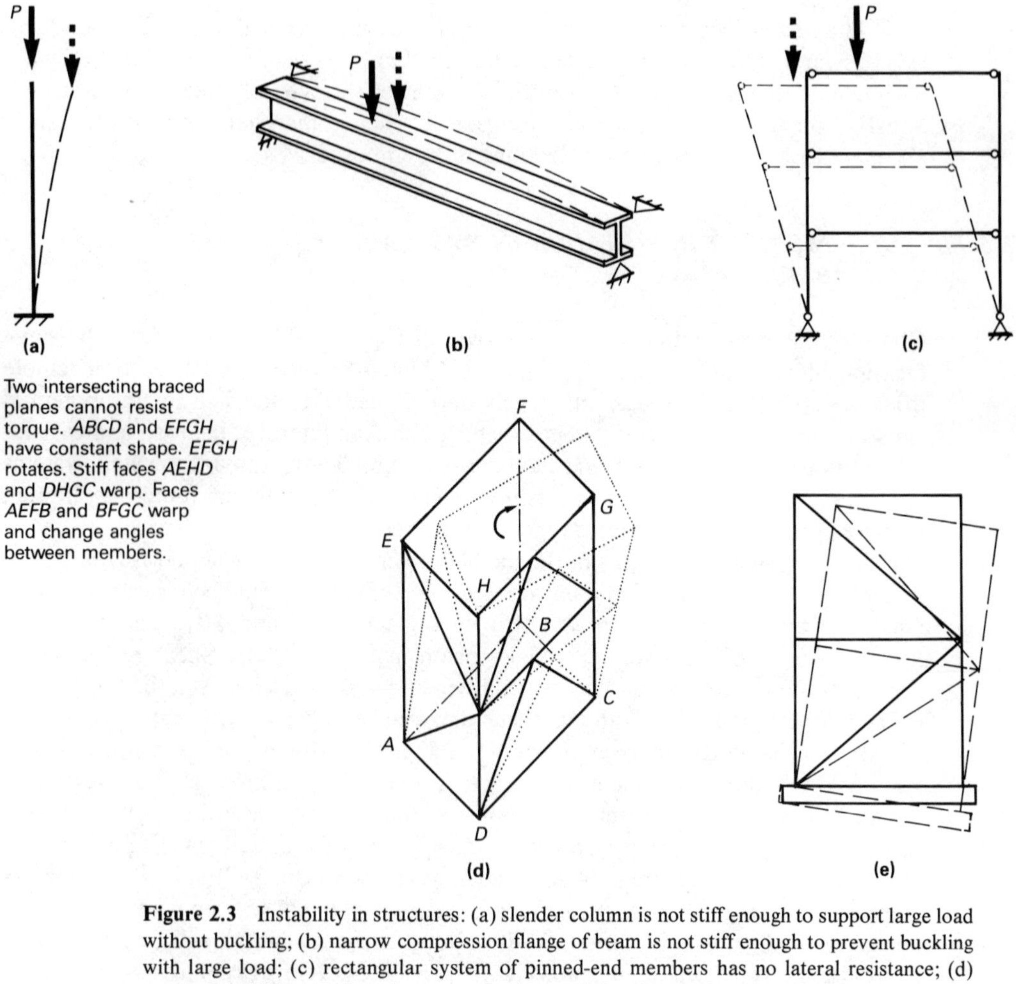

unstable at that stress, although the steel of which they were made could resist much more. Children building houses of cards recognize very quickly how small a disturbance is necessary to destroy their equilibrium. A complete, stable structural system is one that has elements able to resist small disturbances trying to cause movement in any direction. An adequate system is a stable one that has enough strength and stiffness to resist the loads to which it will be subjected within acceptable limits of stress and deflection. The buckling failure of the column in Fig. 2.3(a) and that of the beam in Fig. 2.3(b) occur perpendicular to the direction of both primary stress and load due to insufficient stiffness. The collapse of the frame in Fig. 2.3(c) is inevitable since there is nothing in the system to resist horizontal forces, no matter how small. This is a house of cards. Torsional instability or instability of supporting soils also lead to collapse.

The systems we examine must first be stable for forces and moments around any axis. Given that, we can look at their deflected shapes on the assumption that they are adequate, studying the adequacy of the system for the type of loads to be assigned. Satisfied on that account, we can proceed to choose materials and member sizes adequate for the stresses and deflections to be imposed.

2.6 FLEXURE: MOMENT RESISTANCE IN SYSTEMS AND MEMBERS

Consider the three simple structures shown in Fig. 2.4. They all perform the same function illustrated in Fig. 2.4(a). A load, P, must be resisted at a location remote from possible points of support; it acts perpendicular to the line joining potential support points. A structural system is needed that can intercept the load and transfer its effects to supports. Shear and moment diagrams for the three illustrated systems are identical. The system characteristics are significantly different, requiring different design approaches and different types of members.

In the system of Fig. 2.4(b) the member spanning between supports A and B is longer than the distance AB and has no internal stiffness. It could be a cable, which is capable of being deformed into two straight segments, AC and AB, without internal distress. The load is resisted by axial tension in the members. Shear resistance is offered by the vertical component of axial tension and varies between the portions to left and right of the load as in the shear diagram of part (a). Moment resistance is provided by the horizontal component of axial tension, which must be coupled with a horizontal reaction component at a lever arm that is maximum at the position of maximum sag, S. The moment at any point on the system is described in the moment diagram of Fig. 2.4(a). The deflected shape is that of the moment diagram turned upside down. The low point of the sag is at the load and will shift if distance a changes. Changing a also changes S. It is a *compliant* system, one whose geometry changes with the position of the load. It requires supports at both ends, each capable of providing both horizontal and vertical resistance. It also requires considerable space below line AB. The disadvantages that result from these characteristics in many

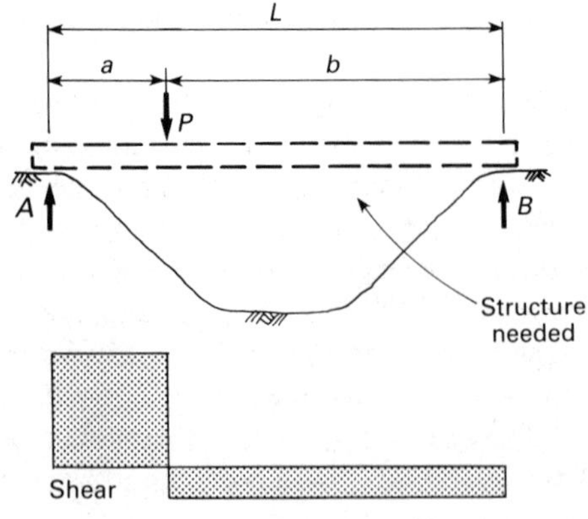

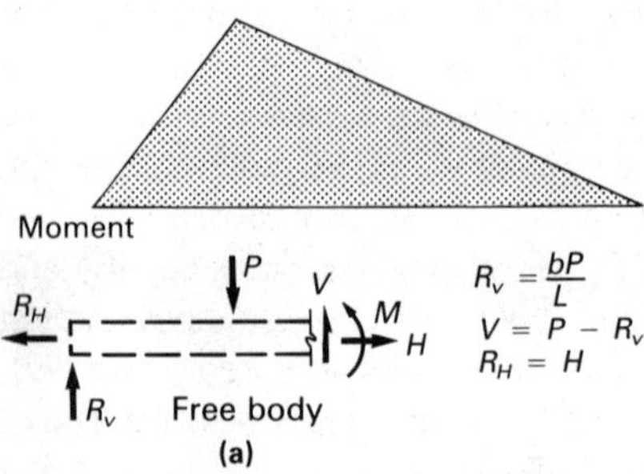

(a)

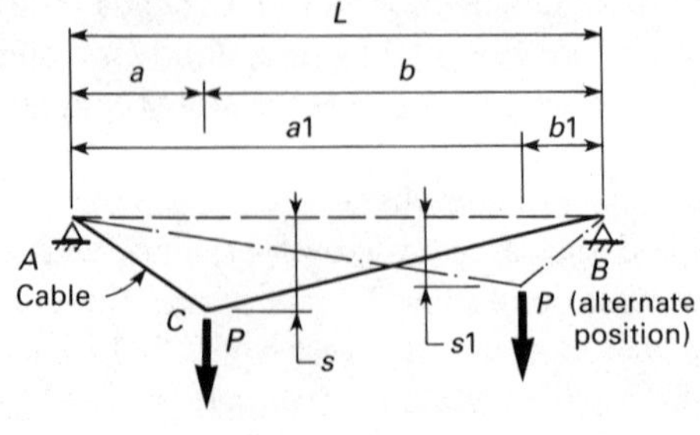

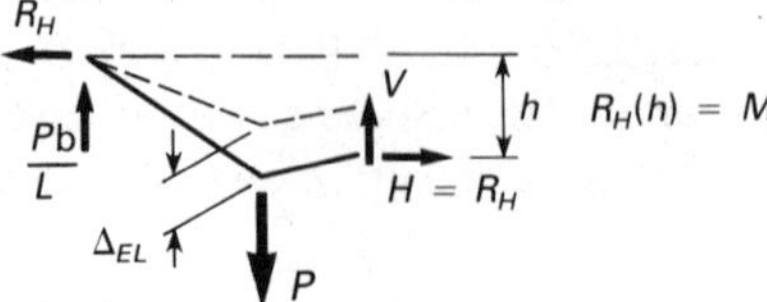

(b) Tension solution (axial)

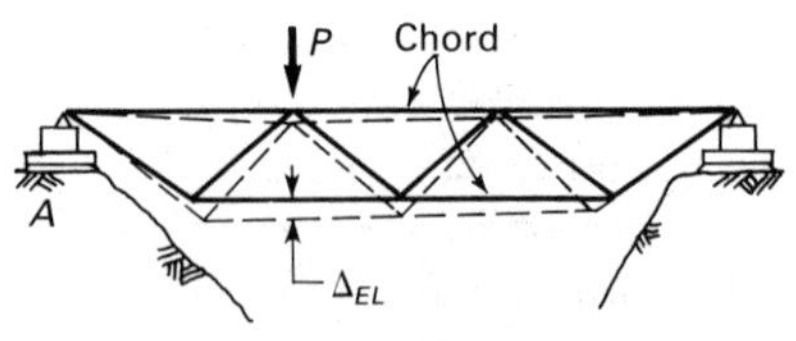

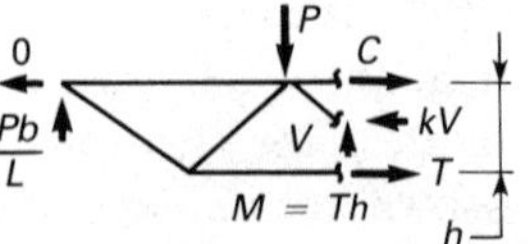

(c) Truss solution (axial member loads)

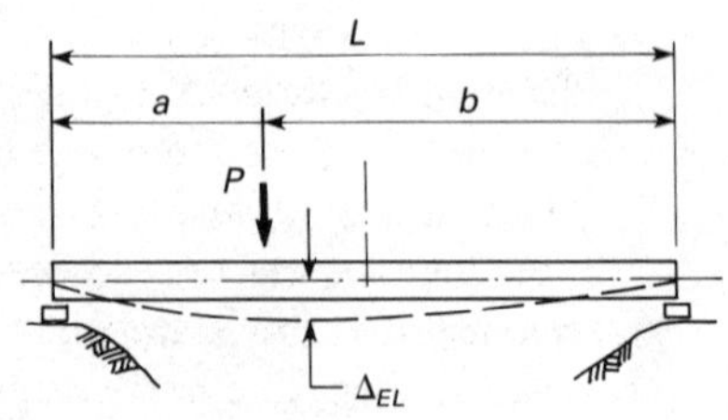

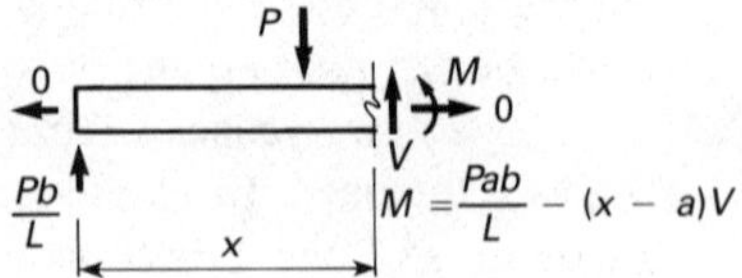

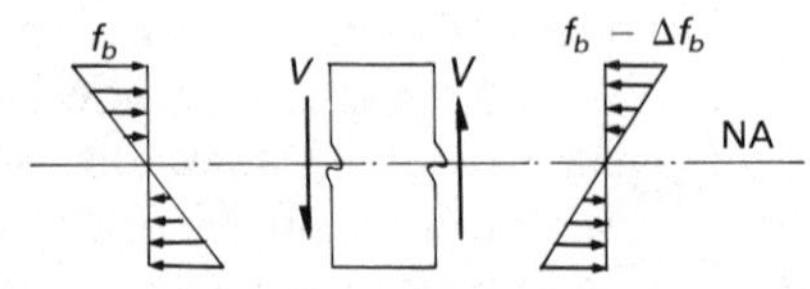

(d) Beam solution (flexural). f_b varies with moment and distance from neutral axis

Δ_{EL} = Elastic deflection

Figure 2.4 Alternative structural systems.

situations should be obvious. However, many magnificent bridges have been built on variants of this system.

The system of Fig. 2.4(c) uses a truss to do the same job. In an *ideal* pin-jointed truss, all members are axially loaded, some in tension, others in compression. If chords are parallel, shear is resisted by the vertical component of the force in a diagonal, and moment by coupling of horizontal chord forces at the distance h. The structure is unchanged by the application of load P. Slopes of members and distance h are constant. Vertical support is clearly necessary at each end. Horizontal support is necessary at one end at least to keep the truss from sliding off its supports. Horizontal support at the other end is possible, but may be undesirable, particularly if the temperature of the members varies over time. The system requires considerable depth h. As shown, the required space must exist below line AG. However, the truss could be turned upside down if space is more available above AB.

The beam in the system of Fig. 2.4(d) also does the same job. It is rigid in the same sense defined previously. A single member resists both shear and moment with almost invariant geometry. Both shear and normal stress vary continuously as different functions of distance above and below the neutral axis. Moment resistance may be considered to be the integral of differential couples, each the product of stress, differential area, and lever arm integrated over the cross section. Put another way, the resisting moment arises because bending stress, f_b, varies over the cross section and changes sign above and below the neutral axis. This may be defined as flexural action in a member. The beam requires almost no space other than its own depth. As with the truss, it requires vertical support at both ends and horizontal support at one end at least; horizontal support at the other end may be undesirable.

In all three cases we must assume that the materials are capable of resisting the stresses imposed. In all cases, there is some elastic deflection, which depends on the material and the cross-sectional geometry of members. The magnitude of such deflection may be an important consideration in design. All these systems may and often do work well with steel members.

Turned upside down and made stiff and noncompliant as is necessary for compressive systems, the suspension cable becomes an arch. However, since an arch is stiff, it cannot change shape to accommodate load effects axially, and therefore must have capacity for bending moment as well as axial load. See Fig. 2.5(a). Early masonry arches minimized this problem since the proportion of transitory loads to the permanent (dead) loads was low and the profile of the arch could be idealized to minimize bending [Fig. 2.5(a)]. Steel arches need not be bound by that restriction, since they can be made flexurally resistant. The rigid frame of Fig. 2.5(c) is an evolution of the arch, which can only be made with materials resistant to both tensile and compressive stress and stiff cross sections. Structural steel is one of several materials that lend themselves to this form.

Examine now the consequences if the vertical load, P, does not act in the plane of resistance of the systems in Fig. 2.4. We are forced to abandon the snug harbor of two-dimensional approaches and think in three dimensions. We also find it helpful to think of subsystems. In Fig. 2.6(a) and (b), we see two cables in place of the one

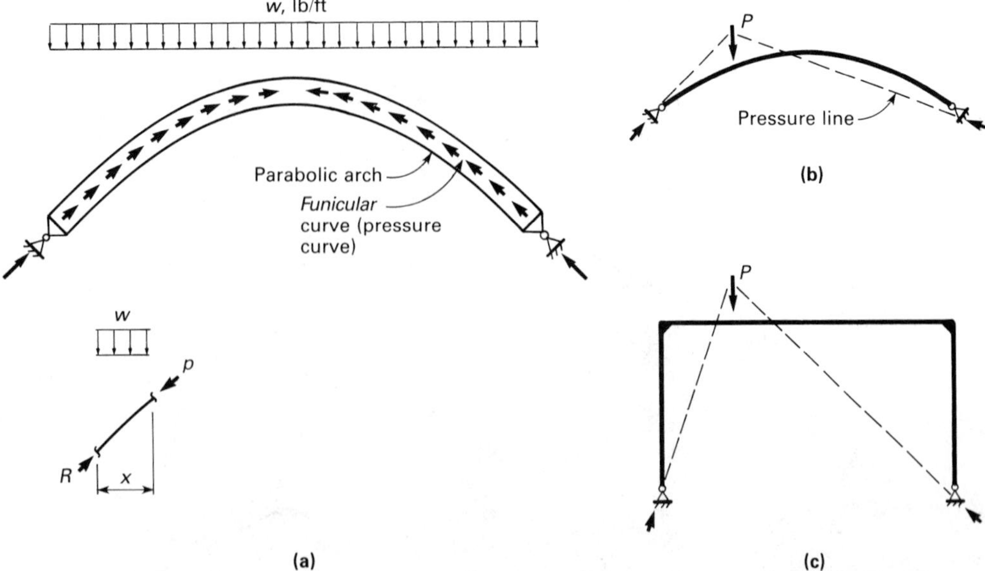

Figure 2.5 Arches and rigid frames: (a) "Ideal" uniformly loaded arch is parabolic to match the shape of the *funicular curve,* the curve representing the resultant vector of all internal pressures (as p on partial free body). The parabolic shape also matches that of a moment diagram for a straight beam, uniformly loaded. (b) The pressure line caused by a concentrated load deviates from the curve of the arch, causing bending moment in the arch as well as axial pressure. The shape of the pressure line is the same as that on a similarly loaded beam. The slopes must be consistent with the *moment area* relations used in indeterminate analysis. (c) A rigid frame may be treated as an arch as in (b).

in Fig. (2.4(b). The load is between the cables, but its effects must be transferred to them, which can be done in any of the three ways of Fig. 2.4. The fully compliant solution of Fig. 2.6(a) must rotate the planes of the individual cables before it can reach equilibrium. It might be used, for example, to hang a suspended warning light. Figure 2.6(b) introduces a secondary beam, allowing the planes of the cables to remain vertical. A primitive suspension bridge would have a series of such beams (planks), acting as a walking surface.

The parallel trusses of Fig. 2.6(c) have spanning between them a secondary truss; it could just as easily be a secondary beam. However, the stiffness of the members makes translation in the y direction undesirable. Unlike the fully compliant cable system, which can be stable in any available equilibrium position, systems of stiff members become unstable if they are not stiff enough between constraints against movement in any direction. Horizontal bracing members are introduced to prevent such translation, creating a third truss in the horizontal xy plane. A similar solution is shown in Fig. 2.6(d), this time with beams.

Figure 2.6(e) introduces a new mode of resistance. A single beam is sufficient to span between the two supports provided two conditions are fulfilled. The force and

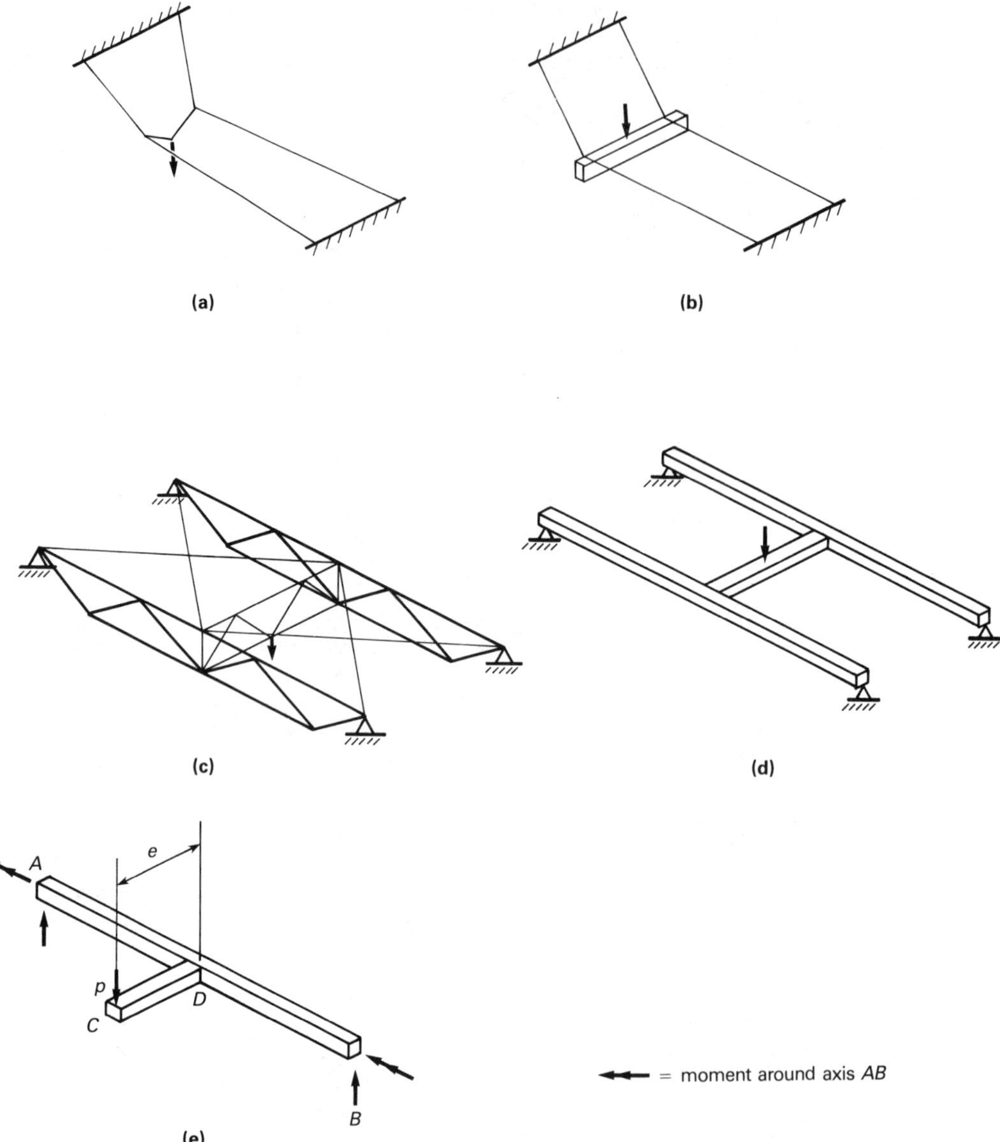

Figure 2.6 Subsystems for loads not in plane of primary system: (a) Load on cable between cables. Compliant system seeks equilibrium condition. (b) Load on beam between cables. Cables in vertical planes. (c) and (d) Stiff trusses and beams with truss or beam transverse member. Lateral bracing required. (e) Load on outrigger from single beam. *AB* must have both bending and torsional strength. Reactions must resist shear and torque.

moment *Pe* must be transferred to the beam *AB* by a cantilever beam *CD*, after which not only must the effect of *P* be transferred to the supports by shear and bending in *AB*, but the beam must be able to resist the twisting effect of the moment *Pe* around its longitudinal axis. We will find in later chapters that the beam characteristics that make for efficient resistance to bending and torsion are not the same, but the same beam can be made to incorporate effective resistance to both. At this point, we should note that the last few pages have introduced all of the force and moment effects that are experienced by simple systems: axial load, shear, bending, and torsion. The reader has no doubt met all these in studies of strength of materials and has examined the stresses that result on members. He will recognize one of the central consequences of the laws of equilibrium. If every action requires an equal and opposite resisting action, there must be a continuous path within a member or a system between the point of application of a force or moment and its ultimate supports.

2.7 A MULTISTORY BUILDING AS A STRUCTURAL SYSTEM

Now let us look at the multistory building shown schematically in Fig. 2.7. The building may be considered [Figs. 2.7(b) and (c)] as a beam cantilevering from its foundation, subjected to a general system of forces and moments represented by force vectors applied at the floor levels. The building must deflect in response to the loads,* just as the analogous beam would. The reactions at the foundation are the negative of all the original input loads, plus the reaction to the cantilever bending moment, which may be considered on a building as an *overturning* moment. Between input and reaction, the building must resist on a continuous path the shear effects, the bending moments that result from shear acting over a distance, and the twisting effects.

A common breakdown of the loads on such a structure may be seen in Table 2.2. In response to the loads, the building shortens, bends, and twists.

Thus far, the building acts like a torsionally resistant beam. However, it has certain special demands as a building. A building is designed to be occupied. Most of the volume must be voids between structural parts, some of which are occupied by the activities of people, and some by the various nonstructural systems (heating, ventilation, lighting, windows, stairs, elevators, partitions, surface finishes, etc.), all the myriad things that convert this cantilever beam into a "machine for living."

Each floor requires a structural subsystem that can, with only minor, acceptable distortion, support both horizontal and vertical loads, transferring their immediate effects to vertical elements, which must ultimately transfer the load effects to the supporting earth at the foundation. Within a floor, vertical resistance is offered by a

*The word *loads* will be taken to include both forces and moments. The effect of a moment may be to twist or to bend or both.

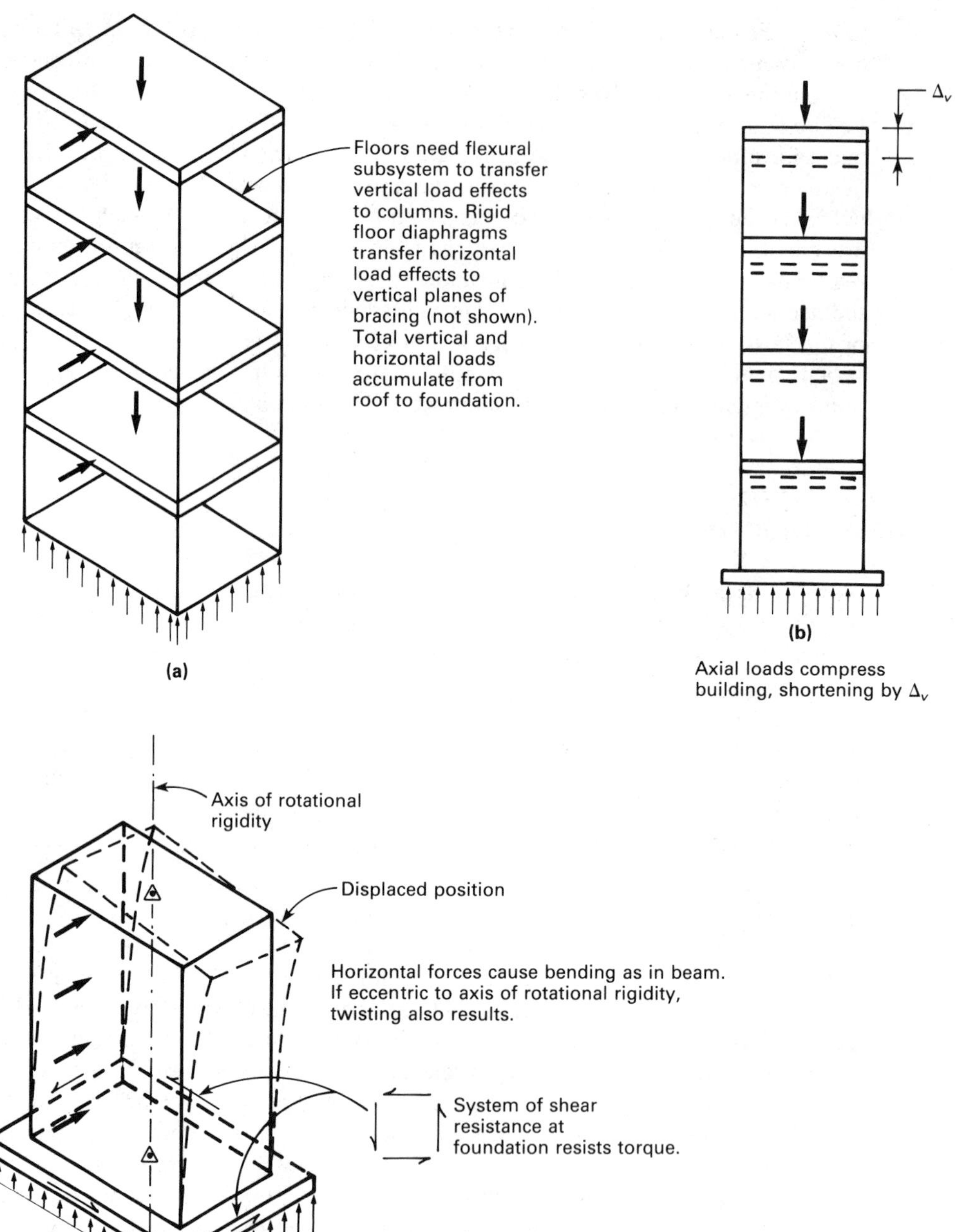
Floors need flexural subsystem to transfer vertical load effects to columns. Rigid floor diaphragms transfer horizontal load effects to vertical planes of bracing (not shown). Total vertical and horizontal loads accumulate from roof to foundation.
(a)
Δv
(b)
Axial loads compress building, shortening by Δv
Axis of rotational rigidity
Displaced position
Horizontal forces cause bending as in beam. If eccentric to axis of rotational rigidity, twisting also results.
System of shear resistance at foundation resists torque.
Foundation pressure is result of axial load plus overturning from horizontal load.
(c)

Figure 2.7 Response to loads, multistory building.

TABLE 2.2 LOADS ON BUILDING

Direction[a]	Source	
F_z	Gravity effects	
	Dead load	Permanent weight of structure and contents
	Live load	Weight of contents that change over time
F_x, F_y	Wind load	Pressure of wind on exposed surface
	Seismic load	Acceleration of mass of structure in response to ground acceleration due to an earthquake; function of mass, varying frequency, and intensity of earthquake input and natural frequencies of the structure
M_z		Effect of eccentricity of F_x and/or F_y with respect to the center of resistance to torsion
M_x, M_y		Effect of eccentricity of F_z

[a]z direction is vertical. x, y and z are mutually perpendicular.

hierarchy of flexural members (slabs, beams, and girders) supported on columns and/or load-bearing walls. Horizontal loads are resisted by the floor slabs, which act as deep (small ratio of span to depth) girders in their horizontal resisting plane, transferring the effects of horizontal loads to the vertical subsystems, which must transfer their shear, moment, and twisting effects to the foundation.

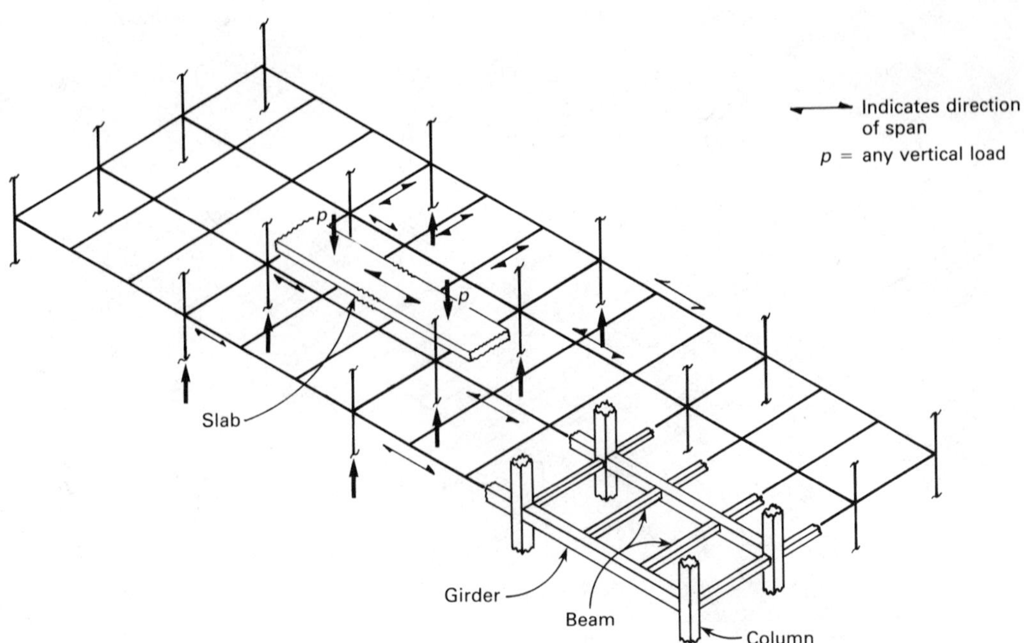

Figure 2.8 Hierarchy of vertical-load-resisting elements, building floor.

2.8 HIERARCHIES OF VERTICAL-LOAD-RESISTING MEMBERS

Illustrated in Fig. 2.8 is a simple hierarchy of vertical load-resisting members, similar to one used in the Tier Building Case Study of Chapter 18. A load at any point on a floor is supported first by a slab that spans its effects to supporting beams. The reactions at the beams become loads on the beams, which span their effects in the perpendicular direction to supporting girders, which in turn span the beam reactions to the columns and through them to the foundations. Hierarchical systems of this sort are very common. They will also be seen in the Mill Building Case Study of Chapter 17 for both vertical and horizontal loads and, in truncated form, in the Plate Girder Bridge Case Study of Chapter 20. In the mill building, the parts of the vertical load system are renamed "roof decks," "purlins," "trusses," and "columns," although their functions do not change. The bridge of Chapter 20 uses only slabs and girders, with the girders supported directly on the foundations. Other, larger bridges, as illustrated in Fig. 2.9, have slabs, stringers (longitudinal beams), floor beams (transverse beams supporting the stringers), girders or trusses (longitudinal), and piers or towers in lieu of columns supported on the foundations.

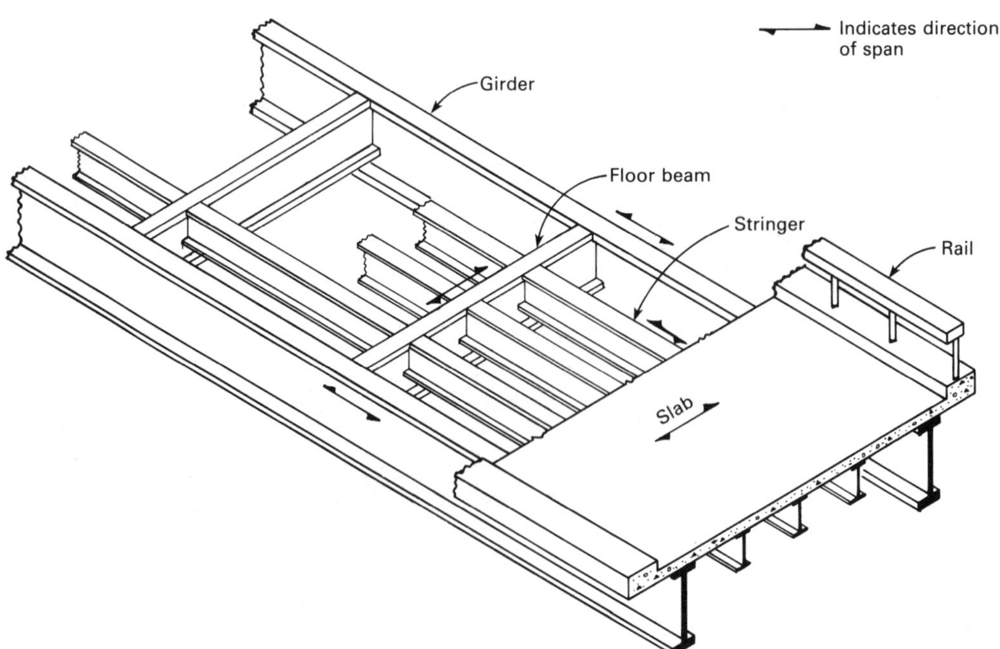

Figure 2.9 Hierarchy of vertical-load-resisting elements, bridge deck.

2.9 HIERARCHIES OF HORIZONTAL-LOAD-RESISTING SYSTEMS

To trace the effects of horizontal loads on a structure, we will return to the example of a multistory building. Again, it must be possible to trace a path for the effects of the load from their origin through the resisting system to the foundation. A simple example is shown in Fig. 2.10(a). Although the forces shown are parallel to the transverse building walls, the reader will recognize from earlier discussion that stability requires shear-resisting vertical planes in both principal directions.

The applied loads shown may actually have originated from wind pressure on the outside walls whose effects were spanned to the floor levels by the walls. They may similarly represent earthquake-induced inertia forces acting on the masses of the walls and floors in response to acceleration of the supporting ground. The floors act like horizontal diaphragms or deep, rigid girders, transferring shear to the vertical resisting planes in the side walls. The, in this case, braced side walls continue the

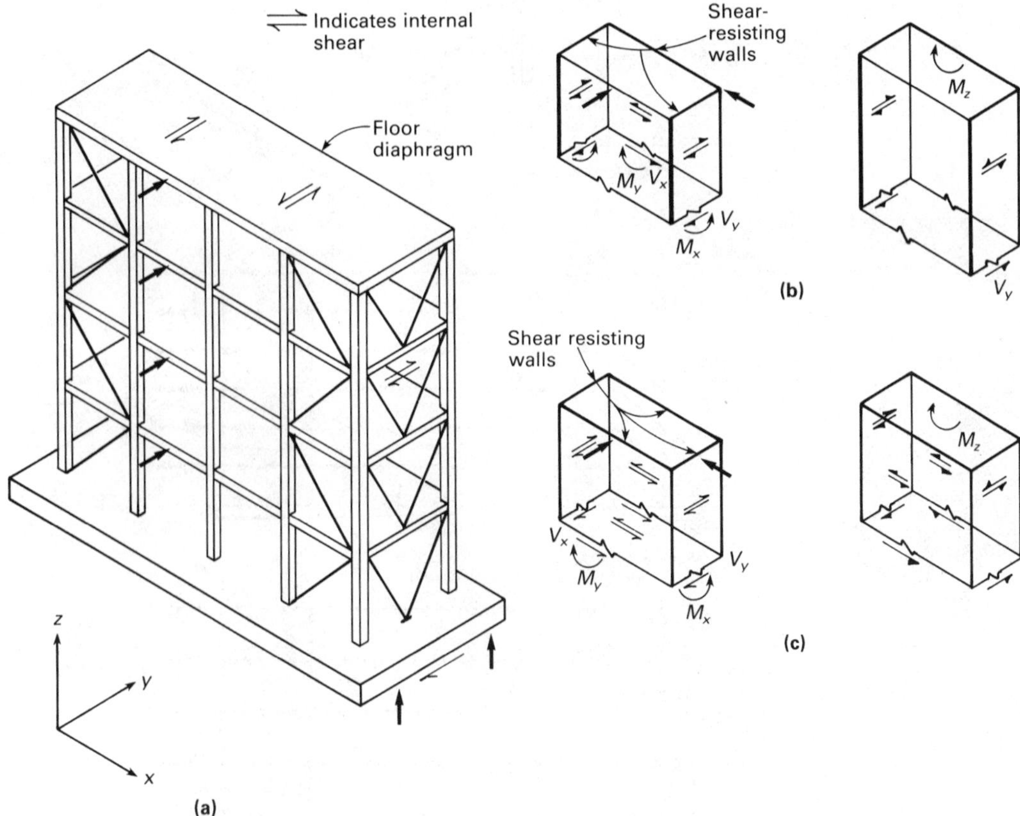

Figure 2.10 Shear paths for horizontal loads on a building.

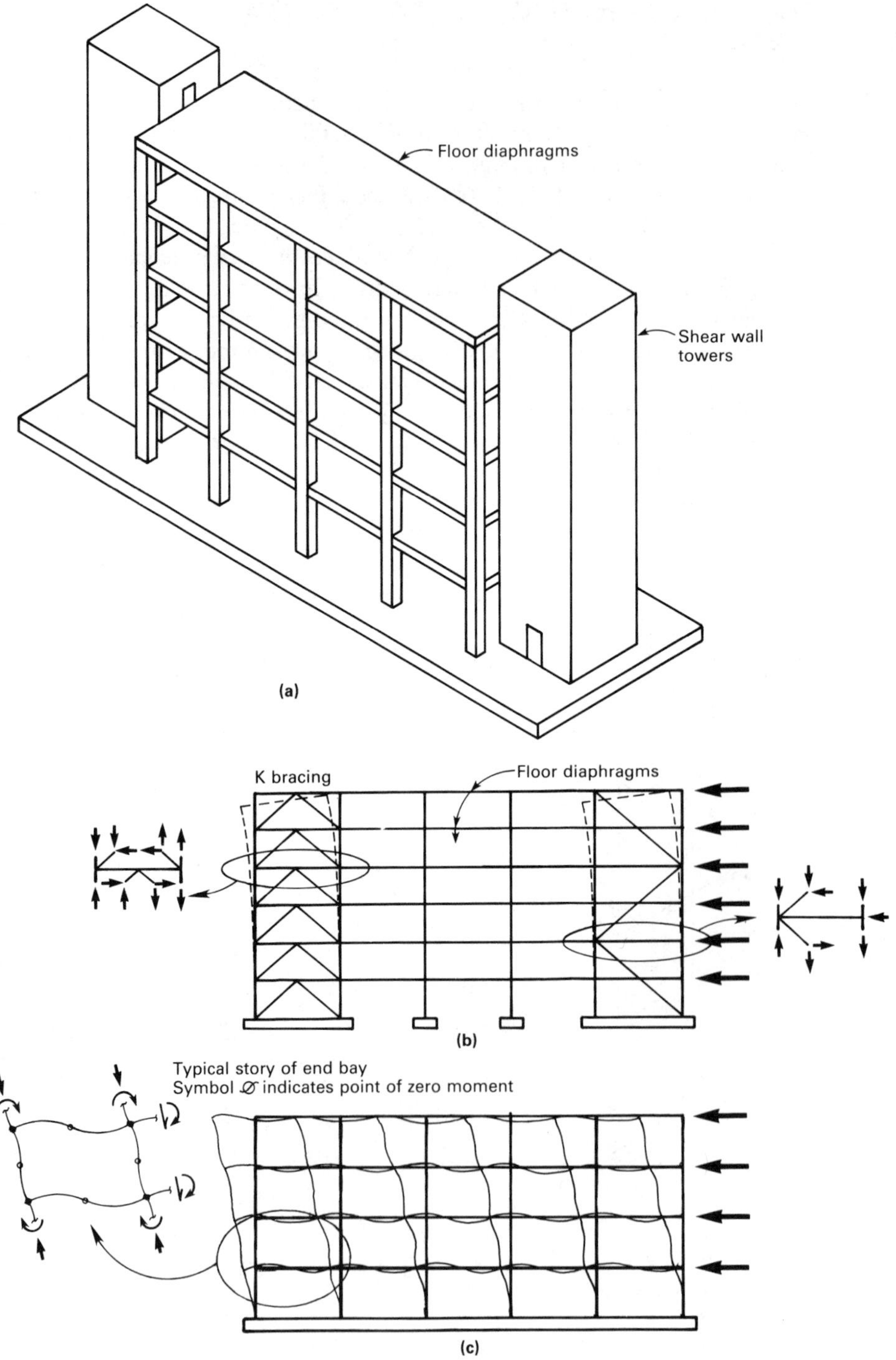
Floor diaphragms
Shear wall towers
(a)
K bracing
Floor diaphragms
(b)
Typical story of end bay
Symbol ∅ indicates point of zero moment
(c)

shear transfer along with their overturning effects to the base of the building. The foundation must supply all the required equilibrating resistance.

The actual shear path depends on the nature of the resisting system. Figure 2.10(b) shows a minimum system required for stability. A path exists for shearing effects of both X and Y loads and the twisting moment, M_z, which results if the resultant loads do not go through the center of resistance. All torque must be resisted by opposing shears in the side walls. In the system of Fig. 2.10(c), all four outside walls are shear resistant. The force effects go to each pair, causing shear in the direction of each force. Torque is resisted by both pairs of walls with pairs of equal and opposite shear vectors. Figures 2.10(b) and (c) are analogous to *open* channel members and *closed* tubes, which will be met in a later chapter. The closed tube is more efficient in limiting both the shear stresses and the twisting rotation resulting from torsion.

The vertical resisting planes may transfer the horizontal shear down to the foundation in one of a number of ways, several of which are shown in Fig. 2.11. The two shear wall towers of Fig. 2.11(a) consist of rigid vertical diaphragms much like the floor slabs, except that they cantilever from the foundations. The floors between shear walls tie the two shear wall towers together, making it possible for them to share the total horizontal load no matter where it originated. In the tier building of Chapter 18, we will use only one such tower at the center of the building, collecting loads from the outer reaches of the building via the floor diaphragms. Even in a steel frame building, shear walls of the sort shown are likely to be made of reinforced concrete or other masonry, doing double duty by resisting horizontal shear and enclosing space at the same time.

Either of the two types of bracing shown in Fig. 2.11(b) resist shear by the characteristic action of axially loaded members, as in a truss. The small free bodies extracted from the system demonstrate the role of the diagonals in resisting the shear, while the resistance to overturning moment develops progressively from the top down as the diagonals transfer the vertical components of their loads to the columns acting as chords of the vertical trusses. Braced systems tend to be rigid.

The moment-resisting frame of Fig. 2.11(c) relies on the flexural strength and stiffness of its columns and girders to transfer the shear from floor to floor, and ultimately to the base. It is the least rigid of the systems shown, although its joints are rigid, accounting for their common title of rigid frames (also called *Vierendeel trusses*"). The columns bend and, as a consequence, translate horizontally. Perhaps more significantly, the rigidity of the joints causes the girders to bend. They are constrained by the columns against vertical translation. But their bending allows the girder–column joints to rotate, causing further translation of the entire limber system. Even with reasonable flexural design stress, movement of a tall moment-resisting frame may become objectionable. Various devices are used to mitigate this effect,

---

Figure 2.11 Shear-resisting systems for horizontal loads: (a) shear wall system (shear wall towers may house stairs, elevators, utilities); (b) braced walls; (c) moment-resisting frame (Vierendeel truss).

while retaining the advantages this system offers for seismic resistance. Perhaps the least efficient approach is to increase member sizes far beyond those necessary for reasonable stress. The dual system of Chapter 18 is one of several such adaptations that have been used. Unfortunately, we will have to omit others from the scope of this book.

3

Structural Steel Technology

3.1 THE DECISION TO USE STEEL

None of the discussion in the previous chapter has been limited to steel structures, although structural steel is one possible material of choice for any of the structures discussed. One of the earliest decisions to be made by a structural designer is which material to use from a number of possible choices suitable for the structural system being considered. The basis of choice is sometimes clear-cut, but often debatable. Few would question the choice of structural steel for the Golden Gate or Verrazano Narrows Bridge, although in recent years precast, prestressed concrete has been used, with steel tie cables, for longer and longer bridges. Early work in offshore drilling structures was almost exclusively done with structural steel; more recently reinforced concrete towers have found a share of that market.

We will see in the case study of Chapter 18 that the designer cannot "prove" that steel is to be preferred over reinforced concrete for that structure, except on the basis of his company's judgment about the reliability of the two materials for ductile performance, a judgment not shared by all qualified people. The author designed a steel bridge over a North Slope river in Alaska because of the need to transport it to the site after assembly in the "*lower 48*" states. The choice in California would probably have been concrete.

However, some general statements are possible. Steel is a manufactured material that must be made from iron by carefully controlled chemical and physical alteration at high temperatures and is more expensive than most structural materials on the basis of price per pound. However, in strength per unit of weight, it exceeds all traditional structural materials except aluminum. In completed structures, it is sometimes clearly the least expensive in total dollars of structural frame and often competitive with alternative choices. Its range of application is very broad, from buildings even smaller and lighter than the mill building we will examine in Chapter 18, to the longest bridges. It often displays advantages in time of erection, ease of transport, and ease of alteration and dismantling, an important criterion for the stiffleg derrick of Chapter 19. It lends itself to mixed use with other materials, as will be seen in Chapters 14 and 20. The selection of steel for most structures should not be an automatic decision; however, it often turns out to be the "right" one.

3.2 STEEL TECHNOLOGY: FROM MINE TO CONSTRUCTION

When the structural designer chooses steel, he is mobilizing a piece of a total technology. How does this technology work? How does the engineer and/or client know what to expect from it?

Figure 3.1 shows the principal actors in the technology and their relationships. The primary path is from materials through steel mills to erection site, either directly or, more commonly, via fabrication shops or specialty manufacturers. The mill supplies steel in the form of large flat plates, thin sheets, or long pieces of varying cross section called *shapes* (see Fig. 7.1), specially formulated to supply the structural market (more on that in Chapter 5). Quantities produced in a *rolling* of a particular cross section are usually large, supplying a number of orders, and require scheduling several months in advance. Lengths are cut to order, varying up to 75 feet; longer lengths may be possible by special arrangement. Prices include a *base* steel price, plus *add-ons* for specific shapes, required qualities beyond the minimums in the governing specifications, and other sources of adjustment. Usually, the cost of steel is finally known only when a specific order is placed.

Fabricating shops convert the primary steel from the mill-delivered form into forms suitable for erection at the construction site. This requires cutting to size, punching holes, and the joining of pieces into relatively large *assemblies* by bolting, welding, or riveting of joints. The practical limit to the size of shop assemblies may be set by the capacity of the shop, the capacity of the transport system between shop and field site (roads, railroads, and barges having different limits), or the limits of the equipment at the construction site. If performed under shop conditions, most operations are usually cheaper and often more dependable than the same operations done in the field. An effective designer usually tries to tailor the design to make maximum use of the advantages of the fabricating shops.

Ultimately, either as single pieces or as subassemblies, the steel must be brought to the construction site and erected, that is, joined into the total steel structure as

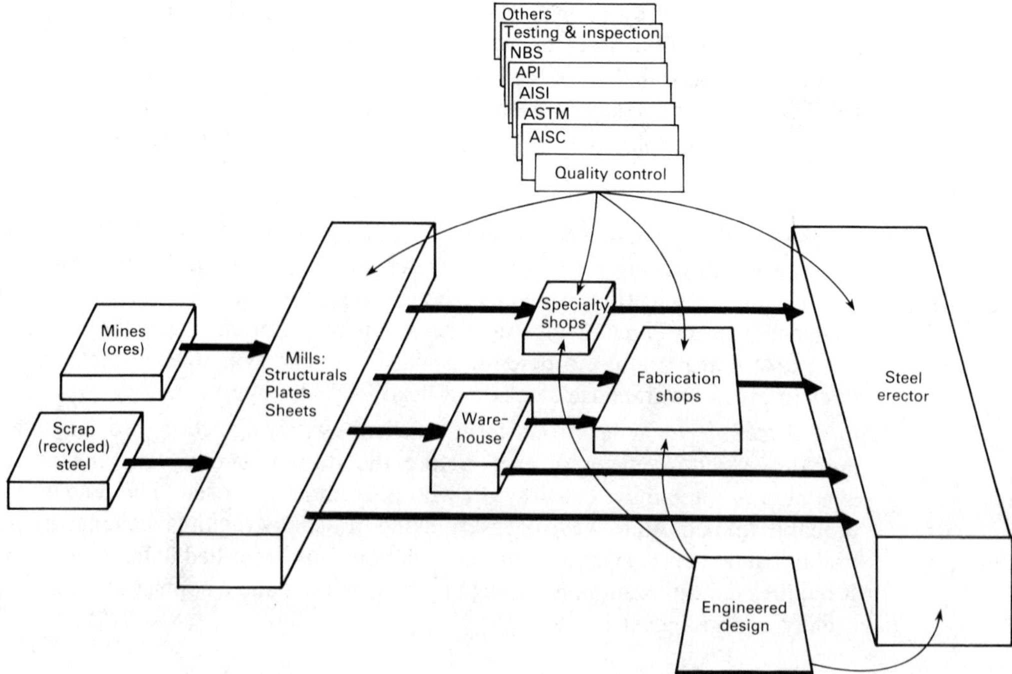

Figure 3.1 Components of structural steel technology.

designed. The steel structure itself becomes a part of a larger system of civil, structural, mechanical, electrical, architectural, and other special components supplied in accordance with the designs of a coordinated group of design professionals.

3.3 CONTROL OF TIME

Examine Fig. 3.1 from the point of view of time. The process of materials → mills → fabrication → erection put on a time scale represents the lead time a designer must expect between the agreement to build and the time of availability of the completed structural frame. The duration of each part and the total time are functions of the specific job, subject to variables such as the total manufacturing commitment of the mills, competition for space in the fabricating shop, and the capacity of local shops. The designer's choices may affect the required time. For example, does the design require unusually long and complicated fabrication? If so, are there compensating advantages?

Why worry about time? The client has at least two concerns, which must become the concerns of the designer. What is the date of *beneficial use* for (1) earning revenues or (2) being available for the intended function? What is the investment cost of delay? These concerns lead to various attempts to short-circuit the normal process, usually at added cost, but often with compensating advantages. Common shortcuts are:

1. *Buying steel from warehouse stocks*. This is usually possible for small quantities from the limited number of plates and shapes stocked in warehouses, a much smaller selection than can be bought from the mills, and usually sold at higher unit cost. Time savings may be a number of months, but the designer must take extra trouble to determine what is really available. Using general rules to predict warehouse stocks may lead to disappointment.
2. *Fast tracking*. In at least one form, a client (with engineering advice) may purchase steel in advance, even before the design is complete, making it available to the builder (contractor) who is ultimately chosen. The reader can probably think of other ways of fast tracking. It usually requires a change in the traditional owner–designer–contractor relationships described in later chapters. It requires careful preliminary design by the engineer and a sophisticated client, usually a large industrial one.

3.4 CONTROL OF QUALITY

Standing alongside each of the primary actors in Fig. 3.1 are the elements of a quality-control system, represented by bodies with several functions.

1. To translate agreed qualities of materials and manufacture into standards. Agreement to supply under these standards obligates the supplier to produce an item (material or completed product) that meets defined levels of quality. In this category, for example, are ASTM standards for supply of steel materials (see Chapter 4) and the AISC *Code of Standard Practice for Steel Buildings and Bridges*.[3]
2. To provide and execute standard tests measuring conformance with the quality standards. For example, a *Charpy V-notch impact test* may be used as an indicator of performance of steel under repetitive or low-temperature loading.
3. To inspect the manufacturing and erection process and certify that the materials and processes used have produced a structure conforming to the requirements of the engineer's design.

A partial list of organizations in the quality-control system is included with Fig. 3.1. Standards are usually adopted by agreement of trade associations or official and semiofficial standards-setting bodies organized under the National Bureau of Standards, the Department of Commerce, and a network of engineering testing laboratories and inspection services spread throughout the country. For each commercially adopted standard there are usually equivalent standards set by the federal government for federal purchasing and sometimes special standards adopted by local regulatory bodies or groups of building officials.

We will be mostly concerned here with a much smaller number.

ASTM American Society for Testing and Materials. Probably the most widely used standards-setting body in American commercial practice. Among a very much larger list, the ASTM has specifications governing the properties and quality of almost all steels used in structural practice and standard tests to be used in verifying steel properties.

AISC American Institute of Steel Construction. A trade association representing the fabricated steel industry, the AISC adopts industry standards and publishes the *Manual of Steel Construction*, a useful source of engineering and fabrication information on structural steel. The manual includes in Part 5 what is probably the most widely used specification for the design of steel structures, the "AISC Specifications for Design, Fabrication, and Erection of Structural Steel for Buildings."

AWS The American Welding Society includes industry and public bodies and professional engineers. Among its publications is the *Structural Welding Code—Steel*, covering quality standards for welding used in the design and fabrication of steel structures. The AWS Code is now adopted almost in its entirety as part of the AISC Specification.

ICBO The International Conference of Building Officials publishes standards for building construction, often adopting ASTM standards by reference. ICBO also publishes the *Uniform Building Code* (UBC), which many public bodies adopt as local ordinances governing the requirements to be met to ensure orderly development and public safety. The UBC prescribes, among other things, minimum design loads for structures and design requirements for structures made of various materials. The chapter on structural steel design usually follows very closely the AISC Specification.

AISI The American Iron and Steel Institute is a trade association representing steel suppliers. Among their publications is the *Specification for Design of Structures from Cold Formed, Light Gage Steel*, representing the state-of-the-art engineering wisdom in the use of such materials.

Other organizations with functions paralleling some of the above-named, include the following:

AASHTO American Association of State Highway and Transportation Officials

ANSI American National Standards Institute

AREA American Railway Engineering Association

API American Petroleum Institute

There are many others.

For the student or relative novice, it is probably best to concentrate on the small number listed here, master the system in which they operate, and adapt to special industries and special conditions when necessary.

3.5 COSTS OF STRUCTURAL STEEL

Engineers have long recognized that their design responsibility is not only to provide physically adequate structures; they must do so at costs to their clients at least competitive with other equally adequate design alternatives. Those who take this part of their responsibility seriously have found themselves very frustrated in recent decades of highly volatile fabrication and construction costs and almost equally volatile costs of primary steel.

Cost analysis becomes very difficult for engineers who expect the same degree of dependability in cost data as they find in formulas relating stress to strain or those describing other aspects of structural mechanics. Still, the client demands and deserves cost information. Even if this were not so, intelligent design choice requires it. Although an adequate treatment of methods of calculating and predicting costs would take much more time and space than are available in this book, it is not possible to ignore it completely.

The cost of primary concern to the investor or owner of a structure is the final cost of the structure. Engineering procedures must recognize and be able to manipulate those parts that make up the final cost. We have already pointed to some of the sources of costs earlier in this chapter. Steel brought to the job site already contains congealed costs of ores, primary steel manufacture, and fabrication. At the construction site, additional costs are incurred in erection of the steel. Time becomes a separate cost dimension when it is recognized that the investor incurs interest costs from the date at which money must be advanced, while the return, in money or use, begins only when the structure is ready for use.

Engineers wishing to evaluate design approaches seek rules of thumb, simplified general rules by which they can predict costs in dollars per unit weight of steel [usually by historic usage, dollars per ton of 2000 pounds ($/ton)]. As with their clients, they are primarily interested in $/ton *in place* (i.e., after erection). The designers can themselves, from their calculation of steel requirements, convert figures in $/ton to other useful forms, such as $/square foot of building or $/linear foot of bridge.

There are no adequate rules of thumb. There are a number of useful publications available published annually by people who follow trends of construction costs. A few are listed in the references. Tables 3.1 and 3.2 are examples from one of them. Used carefully, these can be very helpful in predicting costs. The engineer using them should be aware that:

- They represent the results of averaged samples.

- They are based on last year's market conditions, possibly imperfectly adjusted.
- They represent common, but not specific mixes of fabrication and erection costs.

The careful user checks the basis of the reported figures.

- Do they include the overhead costs incurred by the steel erector? Usually yes.
- If the steel erector is a subcontractor, do they include the overhead and administration costs incurred by the general contractor? Often no.
- Is anticipated profit to each of these included? These are costs to the investor.
- Are quality-control costs included? Usually not, although some may be buried in unrecognizable form in costs listed as "materials," "labor," or "overhead."
- Are engineering design costs included? Usually no.
- Are contract administration costs included? Usually no.

In spite of these caveats, the engineer must make an evaluation. The decision to choose steel for a structure in preference to other materials implies an opinion that steel is "better" for this structure. Since for many structures several choices can yield adequately safe designs, "better" often must be measured as either less expensive or more desirable by other considerations. Short of a vague measure of $/unit of use, we fall back on $/square foot or $/linear foot, and apply a "judgment factor" to nonquantifiable considerations.

We will risk a few observations that designers may find useful.

1. Costs of steel as a material are published from time to time in newspapers and technical publications (e.g., *Engineering News Record*) based on information supplied by mills. These are usually base prices for standard *structurals*. Prices actually charged may include many *adds* and *deducts*, which can only be ascertained by calling the supplier. For the most complete information, call the steel companies.
2. The base price of the usual structural steels supplied under ASTM standard numbers included in the AISC Specification is only slightly higher for *high-strength low-alloy* steels than for the lower-strength *carbon* steels (see Chapter 5).
3. Mill rollings are scheduled months in advance. Best prices require lead time between the date of placement of orders and the time the steel is needed at the site of fabrication or erection.
4. Fabrication costs, in $/ton, of *high-strength low-alloy* steels are almost identical to *carbon* steels.

TABLE 3.1 UNIT COSTS, STRUCTURAL METALS, BUILDING CONSTRUCTION

5.1	STRUCTURAL METALS	CREW	DAILY OUTPUT	UNIT	BARE COSTS MAT.	BARE COSTS INST.	BARE COSTS TOTAL	TOTAL INCL O&P
50	**STRUCTURAL STEEL PROJECTS** Bolted, unless mentioned otherwise							
020	Apts., nursing homes, etc., steel bearing, 1 to 2 stories	E-5	10.30	Ton	914	240	1,154	1,325
030	(71) 3 to 6 stories	"	10.10		946	245	1,191	1,375
040	7 to 15 stories	E-6	14.20		965	245	1,210	1,425
050	(72) Over 15 stories	"	13.90		995	250	1,245	1,450
070	Offices, hospitals, etc., steel bearing, 1 to 2 stories	E-5	10.30		890	240	1,130	1,300
080	3 to 6 stories	E-6	14.40		920	245	1,165	1,350
090	7 to 15 stories		14.20		938	245	1,183	1,375
100	Over 15 stories		13.90		966	250	1,216	1,425
110	For multi-story masonry wall bearing construction, add					30%		
130	Industrial bldgs., 1 story, beams & girders, steel bearing	E-5	12.90		910	190	1,100	1,275
140	Masonry bearing	"	10		910	245	1,155	1,350
150	Industrial bldgs., 1 story, under 10 tons,							
151	steel from warehouse, trucked	E-2	7.50	Ton	1,080	250	1,330	1,525
160	1 story with roof trusses, steel bearing	E-5	10.60		1,000	230	1,230	1,425
170	Masonry bearing	"	8.30		1,000	295	1,295	1,500
190	Monumental structures, banks, stores, etc., minimum	E-6	13		930	270	1,200	1,400
200	Maximum	"	9		1,600	390	1,990	2,325
220	Churches, minimum	E-5	11.60		850	210	1,060	1,225
230	Maximum	"	5.20		1,200	475	1,675	1,975
280	Power stations, fossil fuels, minimum	E-6	11		805	320	1,125	1,350
290	Maximum		5.70		1,300	615	1,915	2,300
295	Nuclear fuels, non-safety steel, minimum		7		950	500	1,450	1,750
300	Maximum		5.50		1,130	640	1,770	2,150
304	Safety steel, minimum		2.50		1,200	1,400	2,600	3,325
307	Maximum		1.50		1,430	2,350	3,780	4,900
310	Roof trusses, minimum	E-5	13		770	190	960	1,100
320	Maximum		8.30		1,280	295	1,575	1,825
321	Schools, minimum		14.50		805	170	975	1,125
322	Maximum		8.30		1,310	295	1,605	1,850
340	Welded construction, simple commercial bldgs., 1 to 2 stories	E-7	7.60		825	325	1,150	1,350
350	7 to 15 stories	E-9	8.30		1,015	440	1,455	1,725
370	Welded rigid frame, 1 story, minimum	E-7	15.80		850	155	1,005	1,150
380	Maximum	"	5.50		1,100	450	1,550	1,825
400	(73) High strength steels, add to A36 price, minimum				75			83M
410	Maximum				152			165M
430	Column base plates, light	2 Sswk	2,000	Lb.	.45	.16	.61	.75
440	Heavy plates	E-2	15,000	"	.39	.12	.51	.60
460	Castellated beams, light sections, to 50#/L.F., minimum		10.70	Ton	995	175	1,170	1,325
470	Maximum		7		1,205	265	1,470	1,675
490	Heavy sections over 50# per L.F., minimum		11.70		855	160	1,015	1,150
500	Maximum		7.80		1,065	240	1,305	1,500
520	(74) High strength bolts in place, light reaming, 3/4" bolts, average	2 Sswk	165	Ea.	.97	1.94	2.91	4.10
530	7/8" bolts, average	"	160	"	1.48	2	3.48	4.76
550	(120) Steel domes							
551								
590	(76) Galvanizing structural steel, under 1 ton, add to above			Ton	350			385M
600	Over 20 ton, add to above			"	250			275M
60	**WELD ROD** Steel, type E6010, 1/8" diameter, less than 500#			Lb.	.60			.66M
010	500# to 1000#				.58			.63M
020	1000# to 20,000#				.54			.59M
040	Steel, type E6011, 1/8" diameter, less than 500#				.61			.67M
050	500# to 1000#				.59			.64M
060	1000# to 20,000#				.55			.60M
065	Steel, type E7018 (low hydrogen) 1/8" diam., less than 500#				.58			.63M
066	500# to 1000#				.56			.61M
067	1000# to 20,000#				.52			.57M
070	Steel, type E7024 (jet weld) 1/8" diam., less than 500#				.55			.60M
071	500# to 1000#				.53			.58M
072	1000# to 20,000#				.49			.53M
080	Deduct for 5/32" diameter, type E6010 or type E6011				.05			.05M
081	Semi-automatic coils, 1/16" diameter, 3000# lots				.56			.61M

TABLE 3.1 Continued

5.1	STRUCTURAL METALS	CREW	DAILY OUTPUT	UNIT	BARE COSTS MAT.	BARE COSTS INST.	BARE COSTS TOTAL	TOTAL INCL O&P
65	**WELDING**, Field. Cost per welder, no operating engineer	E-14	8	Hr.	1.80	29	30.80	44
020	With 1/2 operating engineer	E-13	8		1.80	38	39.80	57
030	(77) With 1 operating engineer	E-12	8	↓	1.80	47	48.80	70
050	With no operating engineer, minimum	E-14	13.30	Ton	1.20	17.30	18.50	27
060	Maximum	"	2.50		4.80	92	96.80	140
080	With one operating engineer per welder, minimum	E-12	13.30		1.20	28	29.20	42
090	Maximum	"	2.50	↓	4.80	150	154.80	220
120	Continuous fillet, stick welding, incl. equipment,							
130	single pass, 1/8" thick, 0.1#/L.F.	E-14	240	L.F.	.06	.96	1.02	1.46
140	3-16" thick, 0.2#/L.F.		120		.12	1.92	2.04	2.93
150	1/4" thick, 0.3#/L.F.		80		.18	2.88	3.06	4.39
161	5/16" thick, 0.4#/L.F.		60		.24	3.84	4.08	5.85
180	3 passes, 3/8" thick, 0.5#/L.F.		48		.30	4.79	5.09	7.30
201	4 passes, 1/2" thick, 0.7#/L.F.		34		.42	6.75	7.17	10.30
220	5 to 6 passes, 3/4" thick, 1.3#/L.F.		19		.78	12.10	12.88	18.50
240	8 to 11 passes, 1" thick, 2.4#/L.F.	↓	10		1.44	23	24.44	35
260	For all position welding, add, minimum					20%		
270	Maximum					300%		
290	For semi-automatic welding, deduct, minimum					5%		
300	Maximum					15%		
400	Cleaning and welding plates, bars, or rods							
401	to existing beams, columns, or trusses	E-14	12	↓	.30	19.20	19.50	28

5.2	METAL JOISTS & DECKS	CREW	DAILY OUTPUT	UNIT	BARE COSTS MAT.	BARE COSTS INST.	BARE COSTS TOTAL	TOTAL INCL O&P
10	**BULB TEE** subpurlins, 40 psf L.L., painted, 5' span	E-1	5,900	S.F.	.16	.09	.25	.31
010	8' span		4,200		.22	.13	.35	.43
030	11' span	↓	2,700	↓	.39	.20	.59	.72
060	(78) For galvanizing, add			Lb.	.20			.22M
20	**LIGHTGAGE JOISTS** punched, double nailable 10" deep, 14 ga.	E-1	1,000	L.F.	2.43	.53	2.96	3.46
070	12 gauge		1,000		3.23	.53	3.76	4.34
090	(81) 12" deep, 14 gauge		880		2.70	.61	3.31	3.87
100	12 gauge	↓	880	↓	3.57	.61	4.18	4.83
110	(76) For galvanizing, add			Lb.	.20			.22M
30	**METAL DECKING** Steel floor panels, over 15,000 S.F.							
020	Cellular units, galvanized, 2" deep, 20-20 gauge	E-4	1,460	S.F.	2	.49	2.49	2.94
030	18-18 gauge		1,390		2.33	.51	2.84	3.34
040	3" deep galvanized 20-20 gauge		1,375		2.10	.52	2.62	3.10
050	18-20 gauge		1,350		2.30	.53	2.83	3.33
060	18-18 gauge		1,290		2.55	.55	3.10	3.65
070	16-18 gauge		1,230		2.75	.58	3.33	3.91
080	16-16 gauge		1,150		2.90	.62	3.52	4.13
100	4-1/2" deep, galvanized, 20-18 gauge		1,100		3.50	.65	4.15	4.84
110	18-18 gauge		1,040		3.70	.68	4.38	5.10
120	16-18 gauge		980		3.85	.72	4.57	5.35
130	16-16 gauge	↓	935		4.15	.76	4.91	5.75
150	For acoustical deck, add				.75			.82M
170	For cells used for ventilation, add				.25			.27M
190	For multi-story or congested site, add			↓		50%		
200								
210	Open type, galv., 1-1/2" deep, 22 ga. under 50 square	E-4	4,500	S.F.	.63	.16	.79	.93
240	Over 50 square		4,900		.53	.14	.67	.80
260	20 ga., under 50 square		3,865		.77	.18	.95	1.13
270	Over 50 square		4,170		.65	.17	.82	.98
290	18 ga., under 50 square		3,800		.95	.19	1.14	1.33
300	Over 50 square		4,100		.81	.17	.98	1.16
320	3" deep, over 50 sq., 22 gauge		3,600		.76	.20	.96	1.14
330	20 gauge		3,400		.90	.21	1.11	1.31
340	18 gauge		3,200		1.12	.22	1.34	1.57
350	16 gauge		3,000		1.35	.24	1.59	1.85
370	4-1/2" deep, long span roof, 20 gauge		2,700		1.70	.26	1.96	2.27
380	18 gauge		2,460		1.85	.29	2.14	2.48
390	16 gauge		2,350		2.20	.30	2.50	2.88
410	6" deep, long span, 18 gauge		2,000		2	.36	2.36	2.74

TABLE 3.2 COST INDEXES

19.1 CITY COST INDEXES

DIVISION		ALABAMA												ALASKA			ARIZONA		
		BIRMINGHAM			HUNTSVILLE			MOBILE			MONTGOMERY			ANCHORAGE			PHOENIX		
		MAT.	INST.	TOTAL	MAT.	INST.	TOTAL	MAT.	INST.	TOTAL	MAT.	INST.	TOTAL	MAT.	INST.	TOTAL	MAT.	INST.	TOTAL
2	SITE WORK	83.2	92.4	87.4	115.7	89.4	104.0	118.7	91.4	106.5	85.8	90.3	87.8	106.4	131.4	117.5	95.2	92.3	93.9
3.1	FORMWORK	91.8	70.3	75.2	88.7	64.6	70.1	108.1	75.9	83.3	102.3	76.5	82.4	114.1	148.4	140.5	109.2	92.7	96.5
3.2	REINFORCING	98.4	76.4	90.2	95.9	72.4	87.1	83.0	74.6	79.9	83.0	76.4	80.6	117.9	137.7	125.3	114.9	103.0	110.5
3.3	CAST IN PLACE CONC.	92.7	90.5	91.4	102.1	91.3	95.7	100.1	92.1	95.4	101.5	92.6	96.2	114.5	116.1	115.5	104.1	88.6	95.0
3	CONCRETE	93.9	80.9	85.9	97.9	78.6	86.1	97.5	83.9	89.1	97.1	84.5	89.4	115.3	131.3	125.1	107.8	91.6	97.8
4	MASONRY	87.3	73.9	77.0	93.3	77.4	81.1	90.5	80.0	82.4	93.6	71.4	76.4	132.3	145.7	142.7	84.8	90.8	89.4
5	METALS	96.0	81.4	91.2	99.7	79.4	93.0	93.9	80.8	89.6	94.8	81.3	90.4	116.6	129.3	120.8	98.2	97.6	98.0
6	WOOD & PLASTICS	88.9	71.8	79.4	100.9	66.5	81.8	102.5	78.3	89.0	101.2	77.0	87.7	121.8	144.1	134.2	94.7	92.7	93.6
7	MOISTURE PROTECTION	86.6	69.6	81.0	92.0	75.8	86.7	87.7	73.2	82.9	88.5	76.1	84.4	95.6	146.9	112.5	95.0	97.1	95.7
8	DOORS, WINDOWS, GLASS	91.5	72.9	81.6	98.7	70.6	83.7	95.5	79.1	86.7	103.4	76.2	88.8	119.8	144.2	132.9	103.5	96.7	99.9
9.1	LATH & PLASTER	90.3	72.1	75.9	95.1	71.8	76.7	92.0	83.8	85.5	109.8	77.5	84.3	123.3	145.0	140.4	87.4	99.4	96.8
9.2	DRYWALL	80.6	74.2	77.4	96.5	62.8	79.8	96.5	79.9	88.3	97.1	76.0	86.6	121.1	145.8	133.3	101.5	92.6	97.1
9.5	ACOUSTICAL WORK	98.6	71.3	82.9	101.0	65.3	80.4	94.6	77.5	84.8	94.6	76.2	84.0	113.0	145.8	131.9	103.4	91.9	96.8
9.6	FLOORING	109.1	77.2	99.8	108.9	75.7	99.2	113.1	84.5	104.7	104.8	76.2	96.4	117.0	145.8	125.4	96.1	94.8	95.7
9.8	PAINTING	121.8	76.7	86.1	104.2	78.2	83.6	121.7	83.6	91.5	120.0	64.6	76.2	137.0	159.7	154.9	102.6	88.1	91.1
9	FINISHES	93.1	74.9	82.9	100.4	69.8	83.3	102.8	81.5	90.9	101.5	72.2	85.0	121.5	150.5	137.8	100.3	91.5	95.4
10-14	TOTAL DIV. 10-14	100.0	80.5	94.5	100.0	80.3	94.4	100.0	82.2	94.9	100.0	79.7	94.2	100.0	145.7	112.9	100.0	103.2	100.9
15	MECHANICAL	97.0	77.4	87.2	99.4	79.1	89.3	97.5	80.4	89.0	99.0	73.8	86.4	108.0	146.0	126.9	98.0	104.0	101.0
16	ELECTRICAL	94.2	78.6	83.3	91.7	80.2	83.6	91.6	91.2	91.3	103.7	76.1	84.3	101.7	145.2	132.3	101.4	99.6	100.1
1-16	WEIGHTED AVERAGE	93.8	77.9	85.3	99.1	77.7	87.6	97.8	82.8	89.7	97.4	77.4	86.7	110.9	141.7	127.5	99.4	96.0	97.6

DIVISION		ARIZONA			ARKANSAS						CALIFORNIA								
		TUCSON			FORT SMITH			LITTLE ROCK			ANAHEIM			BAKERSFIELD			FRESNO		
		MAT.	INST.	TOTAL	MAT.	INST.	TOTAL	MAT.	INST.	TOTAL	MAT.	INST.	TOTAL	MAT.	INST.	TOTAL	MAT.	INST.	TOTAL
2	SITE WORK	107.1	98.1	103.1	99.2	92.1	96.0	113.3	94.8	105.0	100.6	111.5	105.5	94.7	108.9	101.1	90.3	121.5	104.2
3.1	FORMWORK	99.0	95.5	96.3	102.4	68.7	76.4	105.3	72.6	80.1	95.2	127.3	120.0	119.0	127.1	125.3	99.9	131.4	124.2
3.2	REINFORCING	117.9	103.0	112.3	116.0	75.7	101.0	117.9	75.2	101.9	99.4	124.5	108.8	96.2	116.4	103.7	106.6	122.9	112.7
3.2	CAST IN PLACE CONC.	105.8	98.4	101.4	90.7	91.8	91.4	107.3	92.6	98.7	109.6	109.2	109.3	102.6	109.4	106.6	93.3	108.1	102.0
3	CONCRETE	107.4	97.7	101.4	99.2	80.9	87.9	109.5	82.8	93.1	104.3	118.0	112.7	104.2	117.3	112.3	97.9	119.0	110.9
4	MASONRY	93.1	95.3	94.8	97.6	75.3	80.4	100.6	78.0	83.1	118.3	124.6	123.2	108.5	117.0	115.0	95.9	110.3	107.0
5	METALS	91.9	101.0	94.9	89.8	80.9	86.9	106.4	80.4	97.9	99.3	119.1	105.8	96.6	114.3	102.4	95.8	118.0	103.1
6	WOOD & PLASTICS	103.2	95.4	98.9	108.3	69.3	86.6	98.5	74.1	84.9	95.8	124.8	112.0	104.5	124.8	115.8	95.4	131.1	115.3
7	MOISTURE PROTECTION	106.5	95.3	102.8	84.5	71.6	80.3	85.4	73.0	81.3	108.4	127.2	114.6	86.4	103.9	92.2	107.7	112.4	109.3
8	DOORS, WINDOWS, GLASS	88.7	96.6	92.9	94.4	72.3	82.6	89.1	75.3	81.7	99.6	122.0	111.6	101.1	120.7	111.6	105.1	127.2	116.9
9.1	LATH & PLASTER	110.4	95.4	98.6	91.6	74.2	77.9	97.6	77.7	81.9	98.4	123.9	118.5	98.3	110.1	107.6	106.7	115.7	113.8
9.2	DRYWALL	95.4	95.3	95.3	96.5	68.7	82.7	107.1	73.9	90.6	98.5	125.3	111.8	106.2	122.1	114.1	97.8	129.6	113.6
9.5	ACOUSTICAL WORK	115.1	95.3	103.7	84.4	68.1	75.0	84.4	73.2	77.9	82.1	125.8	107.2	96.8	125.8	113.5	96.7	132.3	117.2
9.6	FLOORING	108.0	95.3	104.2	87.9	74.3	83.9	86.9	77.3	84.1	121.0	124.8	122.1	112.6	104.1	110.1	86.9	124.8	98.0
9.8	PAINTING	99.4	95.3	96.1	110.8	75.8	83.1	114.4	64.2	74.7	115.8	124.6	122.7	127.2	116.5	118.7	108.5	123.0	119.9
9	FINISHES	100.4	95.3	97.5	95.3	71.8	82.1	101.6	71.0	84.4	104.1	125.0	115.8	109.2	118.4	114.4	96.8	126.3	113.4
10-14	TOTAL DIV. 10-14	100.0	95.2	98.6	100.0	75.7	93.1	100.0	78.2	93.8	100.0	118.4	105.2	100.0	115.6	104.4	100.0	137.8	110.7
15	MECHANICAL	99.3	95.5	97.4	97.7	75.9	86.8	97.2	79.1	88.2	96.8	125.3	111.0	95.9	116.0	105.9	97.0	123.3	110.1
16	ELECTRICAL	103.7	95.2	97.8	99.9	68.6	77.9	92.7	81.0	84.5	98.6	124.5	116.9	101.6	121.5	115.6	109.8	122.9	119.0
1-16	WEIGHTED AVERAGE	100.4	96.2	98.2	96.5	76.0	85.5	100.4	79.4	89.1	101.6	122.2	112.7	99.6	117.0	109.0	98.9	121.1	110.8

DIVISION		CALIFORNIA																	
		LOS ANGELES			OXNARD			RIVERSIDE			SACRAMENTO			SAN DIEGO			SAN FRANCISCO		
		MAT.	INST.	TOTAL	MAT.	INST.	TOTAL	MAT.	INST.	TOTAL	MAT.	INST.	TOTAL	MAT.	INST.	TOTAL	MAT.	INST.	TOTAL
2	SITE WORK	82.5	119.3	98.7	98.1	113.1	104.8	94.2	110.8	101.6	82.2	107.7	93.6	90.3	109.9	99.1	133.4	115.6	125.5
3.1	FORMWORK	96.3	127.0	120.0	90.2	127.3	118.8	102.9	127.3	121.7	112.3	131.7	127.3	102.6	118.2	114.6	114.1	135.3	130.4
3.2	REINFORCING	78.1	123.7	95.1	99.4	124.2	108.7	113.0	124.0	117.1	99.4	132.7	111.8	114.9	123.7	118.2	81.1	123.7	97.0
3.3	CAST IN PLACE CONC.	97.6	107.4	103.4	102.6	109.2	106.5	102.6	109.5	106.7	116.2	109.2	112.0	96.3	120.3	110.4	112.0	122.1	118.0
3	CONCRETE	92.6	116.8	107.4	99.4	118.0	110.8	105.2	118.1	113.2	111.3	120.5	117.0	102.1	119.7	113.0	104.8	127.6	118.9
4	MASONRY	102.1	121.1	116.7	101.2	124.3	119.0	108.7	124.1	120.6	101.2	120.1	115.8	110.2	117.4	115.7	111.8	132.1	127.5
5	METALS	106.1	117.7	109.9	105.4	118.9	109.8	99.4	118.8	105.7	106.9	124.9	112.8	97.3	123.2	105.8	98.0	124.0	106.5
6	WOOD & PLASTICS	94.0	124.3	110.9	93.0	124.8	110.8	98.7	124.8	113.3	95.4	131.1	115.3	100.7	115.1	108.7	116.4	135.5	127.1
7	MOISTURE PROTECTION	105.0	126.4	112.0	89.8	125.3	101.5	90.4	121.4	100.6	85.2	125.2	98.3	96.5	116.2	103.0	93.3	129.5	105.2
8	DOORS, WINDOWS, GLASS	100.3	121.7	111.8	104.2	122.0	113.7	104.2	121.9	113.7	92.1	132.6	113.7	110.0	113.9	112.1	108.1	134.1	122.0
9.1	LATH & PLASTER	97.6	124.6	118.9	97.4	128.4	121.9	97.4	129.6	122.8	99.7	133.4	126.3	110.2	123.9	121.0	113.0	140.3	134.5
9.2	DRYWALL	97.3	122.8	109.9	99.6	125.2	112.3	95.0	125.1	109.9	96.0	134.2	114.9	115.1	119.8	117.4	105.1	137.2	121.0
9.5	ACOUSTICAL WORK	103.4	125.7	116.2	88.4	125.8	109.9	88.4	125.8	109.9	86.7	132.3	112.9	105.9	116.1	111.8	103.4	137.3	122.9
9.6	FLOORING	98.7	125.0	106.4	93.2	124.5	102.4	93.2	124.4	102.3	84.7	132.7	98.7	110.0	124.8	114.3	91.8	136.9	105.0
9.8	PAINTING	93.5	116.2	111.4	92.6	111.6	107.7	99.7	124.1	119.0	101.4	144.4	135.4	94.4	128.7	121.5	121.0	149.7	143.7
9	FINISHES	101.3	121.3	112.4	96.5	120.7	110.0	94.7	125.0	111.7	93.6	137.4	118.2	110.8	123.2	117.7	104.1	141.7	125.2
10-14	TOTAL DIV. 10-14	100.0	118.3	105.1	100.0	118.3	105.1	100.0	118.2	105.1	100.0	141.2	111.6	100.0	117.9	105.0	100.0	143.6	112.3
15	MECHANICAL	101.4	124.2	112.7	99.5	122.8	111.1	96.6	121.9	109.2	98.2	126.2	112.1	101.2	124.8	112.9	101.2	157.5	129.2
16	ELECTRICAL	98.5	117.1	111.6	98.6	140.3	127.9	98.2	110.0	106.5	109.8	132.7	125.9	108.6	120.0	116.6	109.1	148.8	137.0
1-16	WEIGHTED AVERAGE	99.1	120.2	110.4	99.3	123.3	112.2	99.1	119.5	110.0	99.5	126.2	113.8	102.1	119.8	111.6	104.7	138.4	122.8

5. Building costs in \$/ton in place are based on net tonnage after fabrication, while mill prices are charged for total tonnage shipped to the fabricator. The difference may be as much as 3% to 5%.
6. Specifying special qualities, such as *guaranteed notch toughness* or corrosion resistance, incurs added cost in \$/ton.
7. For areas remote from the major steel mills, transportation may be a significant cost.
8. Modes of available transportation are significant. If railroad access is not available at either end, the maximum shipping length of members may be reduced and fabrication and erection costs increased. The availability of barge shipment offers greatest flexibility in the size of the assemblies shipped; cost savings result.
9. The designer who attempts, by complicated and unusual fabrication, to reduce the total tonnage of steel in place may actually increase the owner's cost in \$/square foot.
10. On the other hand, given a large enough number of similar designs or repetitive uses, a designer and/or manufacturer may make significant savings in ultimate costs by special fabrication or design techniques addressed to reducing steel tonnage (e.g., open-web joists, standard light industrial buildings).
11. For a particular structure, the minimum cost usually results from the use of a relatively small number of similar members connected by a relatively small number of types of joints. Even at the cost of some additional tonnage, simplicity often pays.

Since actual costs involve local conditions and time-related factors in the construction market, it is often useful to consult with local contractors and suppliers in making design evaluations of costs. Estimators for contractors follow cost trends closely. They often calculate materials costs from prices guaranteed by suppliers. They keep specific data on cost to them of labor, equipment, and overhead and are much more aware than most design engineers of what might be called the "hunger factor," the dips and rises in supply and demand that are highly time related and have a strong bearing on bid prices. In estimating the cost of work, they also factor in specific construction methods and the construction plant necessary to support them. These are very much dependent on the site and time. Imaginative, innovative designs may only be economically feasible if the designer ascertains that the proper mix of equipment and interested erector can be found to take advantage of the innovation.

Actual costs are only known after bid proposals have been received, and then only assuming well-prepared, clear contract documents. In the face of very poor predictability, it is sometimes worth investing in complete alternative designs in order to find out which will yield the best costs. This path, however, may only be feasible if both alternatives work equally well with the nonstructural parts of a total project.

Member selection procedures in Part II of this book focus mostly on the search for minimum acceptable weight/foot of suitable members. In the case studies in Part III, it is possible, to a limited extent, to consider other cost factors.

3.6 ENGINEERING IN STRUCTURAL STEEL TECHNOLOGY

Figure 3.1 shows the engineer affecting technology by providing engineered designs to be executed by fabricating shops and erectors. This is an oversimplification useful in a simplified presentation. It does not, however, clarify the total role of engineers and engineering in the process. Without attempting to untangle all the components of the highly complex interaction between engineers and the total technology, we will mention a few of the interactions.

1. Fabricating shops are not just blotters soaking up designs provided by engineering designers. They structure themselves and their equipment to serve the needs of the designs and design details that they are asked to follow in fabrication. On the other hand, they employ engineers, on staff or as consultants, whose function is to improve the ability of the shops to produce fabricated assemblies at least cost. This results in an active dialogue between design engineers and fabricators' engineers. Designers attain information about the limits of fabrication feasibility of their design concepts. The fabricators' improved capability offers opportunities that designers can use in improving their designs. The introduction of automatic and robotic welding processes by fabricators has had a profound effect on design, particularly in large structures.
2. Steel erectors employ or retain engineers whose function is to plan and optimize the construction process, seeking minimum cost and safe procedure. Their procedures must satisfy the designers that erection conditions will not adversely affect the ability of members to function according to design in the completed structure. The stresses and deformations of a steel structure during construction are often completely different from those analyzed by the designer for the completed structure. Each stage of erection brings a new system of stresses. Unless the erection procedures are supervised, the erection effects may be large and permanent and invalidate the dead-load calculations of the designer.

   An active dialogue often develops between the erectors' engineers and design engineers, particularly on major structures, each promoting their own client's interests, but each, if following professional precepts, anxious to protect the integrity of the structure.

   Designers aware of erection problems are able, in their design development, not only to avoid construction difficulties, but also to find design opportunities in the erection process. Figure 3.2 depicts such a case in connection with the construction of a subway station in San Francisco, part of

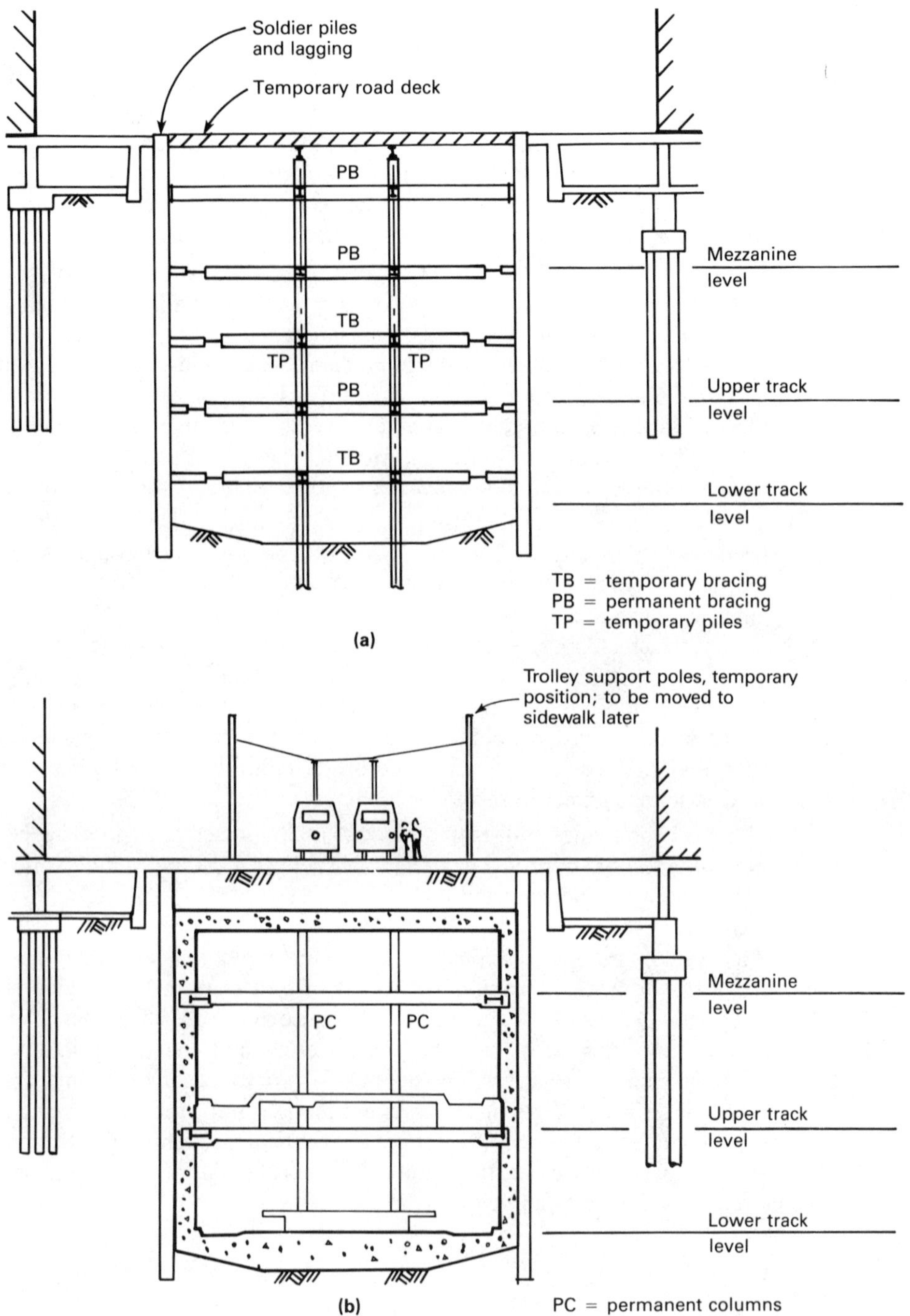

Figure 3.2 BART station construction bracing: (a) during construction; (b) after construction of station; street restoration still incomplete.

the Bay Area Rapid Transit System. The designers* designed the steel framing in such a way that it was convenient to use permanent members to double during construction as temporary bracing for the exposed sides of an excavation over 75 feet deep, 100 feet wide, and over 800 feet long. The author, who was involved as project engineer for the contractor's consultants, found the opportunity so provided to be quite useful for his client.

3. Engineers, on behalf of specialty shops, create designs suited to production methods that can take advantage of cost savings available through volume production. The cost of extensive research and testing, feasible because of volume production, often results in efficient design and cost savings. Information on the resulting products, available to design engineers, may establish that the standardized designs are suitable and less costly than the alternatives that might emerge from custom design. Modular components and standard building designs offered by manufacturers are a growing part of steel technology.
4. To a great extent, the nature of structural steel designs is controlled by the products (plates, shapes, bars, pipes, and others) available through the mills and the steel alloys that the mills are able and willing to produce. At any given time, current offerings act as constraints on the designer. On the other hand, albeit slowly, the steel industry is able to react to demand created by designers for new shapes or special steel properties. Demands of design for very cold Alaska Pipeline conditions led to the development of notch-tough steels suitable to avoid brittle fracture at cold temperatures.† Once available, the same steels are used for other conditions that invite brittle failure, such as highly restrained, welded joints subject to repetitive loading.
5. The steel companies and steel fabricators maintain active engineering programs both internally and through associations such as AISI and AISC, which act as liaison between industry and profession.

The system that leads to construction of steel structures is a highly engineered technology that utilizes the capitalized results of past engineering development and the continuing expansion of engineering knowledge in all its aspects. We have concentrated on the role of structural engineering and the direct interaction of industry engineers. The reader will no doubt be aware that each entity involved in the technology described utilizes in its own production and process planning a great variety of engineering talents. The processes existing throughout the technology of steel are engineered processes, the results of the special nature of engineer–industry–public relationships that characterize the modern era.

*Designers were PBQ&D, Inc.; the contractors, Winston-Drake-Early, a joint venture, retained Earl & Wright, Inc., as engineering consultant. The author was project engineer for Earl & Wright.

†Both research and design interest was sparked much earlier by brittle failure of a number of welded ships in Arctic waters during World War II.[54]

4

Design Process: Structural Design

4.1 HOW DESIGNS ARE BORN—AND TO WHOM

Unlike Athena, who sprang fully grown and fully armed from the brow of Zeus, the designs of mortal engineers must struggle to be born. In imitation of nature, they must evolve as survivors in a hostile environment, called upon to demonstrate their efficacy and survivability in competition with other solutions to the problems they address.

We will look at several stages of the evolution of a typical structural steel design and several of the arenas in which structural steel competes with other materials for consideration as the primary structural material. The scenario we present is not the only one found in structural practice. However, it is found often enough so that it will transplant to many design situations. The same basic process exists in most design development—from articulation of a need through the various stages of design to construction and ultimately to use. The specific relationships between people and institutions often differ.

4.2 STAGES OF A STRUCTURAL DESIGN

Flow Chart 4.1 capsulizes the various stages as they are often followed. The form of flow charts used in this book is described in Chapter 7. A very few comments should suffice here. Diamonds are decision points, which must be exited by at least two outgoing arrows representing possible decisions. Numbered bubbles are feedback

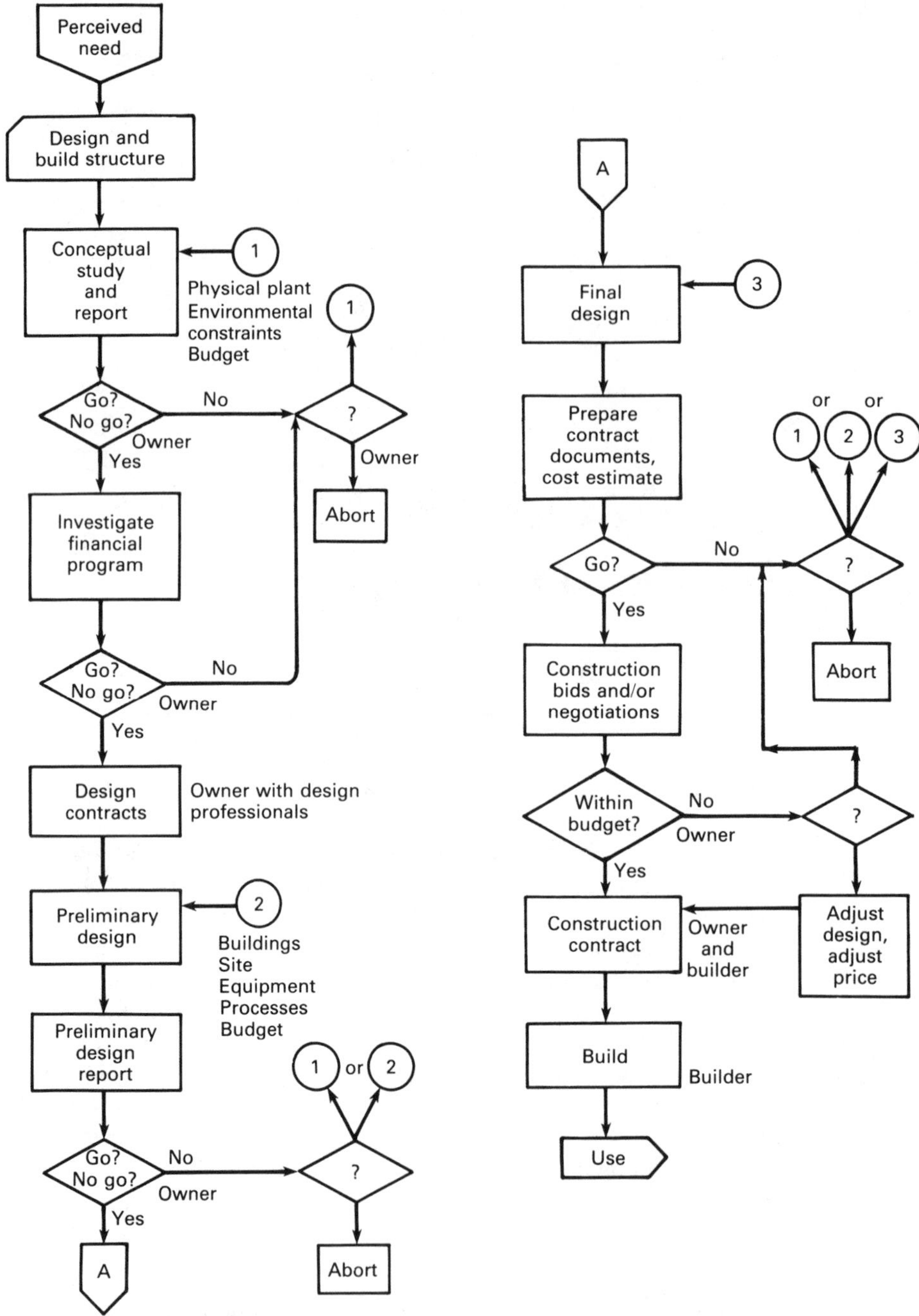

Owner = investor, land owner, developer, manufacturer, public body
Design professionals = architects, engineers, consultants, etc., depending on project
Builder = construction contractor

Flow Chart 4.1 Stages of a structural design decision path.

instructions entered from a decision diamond; a correspondingly numbered bubble with an arrow back into the flowstream tells the user where to reenter the flow. Large, open, lettered arrows are connectors from column to column (or, in other charts, page to page). What is shown is a process involving interaction between an owner, private or public, and a number of experts, among whom the structural designers are one part of a group of design professionals.

The specific actors and their interactions are different if the owner* is a private investor, from those involved in public works construction. They differ between commercial and industrial construction, between large and small industry, between local and national government, between U.S. and extraterritorial construction. Participants do not necessarily follow the stages set down here, step by formal step. The motives and actions of decision makers are often more complex than their imperfect representations in engineering models. However, the stages described are necessary for most successful design developments. Avoiding some stages or entering others prematurely courts disaster.

A structural design project starts when somebody perceives a need (or desire) for something to be built. The "somebody" may be a private entrepreneur, a representative of government, or an executive of a public service institution. The perceived structure is intended to satisfy the need. It may be seen as a unique structure or a part of some larger system to be started or extended. The process that follows includes a series of steps that answer some fundamental questions:

- Is the need real enough and urgent enough to command the financial and institutional support necessary to make the project "viable"?
- Is the design and construction of a structure an effective and desirable way to satisfy the need?
- What is the "right" structural solution?
- What is the necessary budget and can it be supported?

Ideally, the design process starts with a *conceptual study* done by experts equipped to answer broad questions of feasibility. The experts engaged may be the same team who will ultimately be used as designers if the project proves feasible, possibly augmented by program consultants knowledgeable about the financial markets, the needs of a specific industry or type of institution, and the like.

The questions to be addressed in a conceptual study are designed to test the efficacy of the proposed project; among them are the following:

- Is the need provable?
- Is a structure the preferred way to meet it?

**Owner* may be considered a generic term for the people or institutions that authorize and pay for the design and construction. Design professionals may have the owner as their direct client. Often there is one professional who deals with the owner as client and who retains others as consultants. The discussion that follows describes one such set of relationships.

- What type? Where? How big?
- What will the impacts be—social, environmental, financial—on the owner and others affected?
- Can others be expected to intervene? Adversely or favorably?
- How does this proposal compare with other possible proposals designed to serve the same need?
- Can capital be raised to finance the project? What is necessary to attract it?
- Can it be put in place in time?
- Is this the right time? the right place?

The design itself is seen in dim outline at this stage. But:

- Is there precedent for believing this design approach will work?
- Are the needs of the design well enough understood to establish a construction budget at this time?

The questions indicate the varied qualifications required of those who are to answer them.

There are few owners, public or private, who can finance a desired project without recourse to the financial market. There are fewer still who will choose to back even their own dreams without at least some encouraging information pointing to potential success. The decision to invest in an idea likely as not means the decision to forego another, possibly more feasible one.

At this stage, the design professionals must, on paper, put a physical shell around the plans of the owner and his team of advisers. They must define the required space and consider the environment in which users will be expected to live (e.g., air conditioning, lighting, amenities, facade).

What follows relates to designs when an *architectural* firm is the lead designers of, for example, a commercial or institutional building; the *structural* design group (or company) is one of a number of professional design groups contributing to the total design. In other situations, the lead may be taken by *structural* engineers (e.g., bridges), *process engineers* (e.g., chemical plants), or others appropriate to the type of facility and its location. With suitable substitutions of words, the process described applies to the general process of design in many settings.

If an *architectural* project, the architects will draw on their own knowledge of earlier successful projects, perhaps recombining elements, perhaps visualizing new approaches. They will call on consultants of various engineering disciplines to point out the possibilities and the limitations of what each discipline can supply to define the space and equipment needs for each engineered function. They will examine the needs of each and combine them all into a conceived physical design that can, at a minimal level of refinement, be converted into a report and/or a set of conceptual drawings. The level of detail is only sufficient so the owner and potential lenders (once again, private or public) can perceive the nature of the conceived design, and so that they and the design professionals, in their own spheres, can develop the potential for costs and return and effect on the surroundings. The product of each discipline has monetary elements that can be projected quantitatively and nonmonetary elements that may defy quantification, but are nevertheless the stuff of which human decisions are made.

The structural designers, in this phase, supply the element of the design concept that is represented by the structural supporting system. Based on familiarity with available alternatives, judgment as to their usefulness, and a minimum of necessary validating calculation, the engineers are called on to propose one or more structural solutions capable of housing the other elements that fit into the design concept.

Completion of the conceptual study brings the owner to the first *go–no go* decision, based now on systematically developed information. Did the study confirm the need? Did it point to a built solution as appropriate? Is there a potential market? Can financing be expected? Should the project be approached as a permanent one, or does it have a limited life? Can it live with its neighbors?

If, in the owner's mind and with the advice of his professional advisers, it still looks like a *go*, the next stage will provide another test. The search for funds may answer the following questions: Can the project be funded? Can it be funded at this time? If, once again, the result is a *go*, arrangements are made to proceed with the design. The design may proceed as conceived in the conceptual study or on a modified basis, based on considerations resulting from the previous two steps.

As indicated in Table 4.1, the resulting stage of design execution is itself divided into two stages, preliminary and final. Each serves a distinct function. Each is necessary. Each is followed by the opportunity for the owner to make progressively more informed decisions to *go, revise,* or *abort*. By custom, a design contract recognizes the right of the owner to decide, after the preliminary design stage, whether or not to proceed. A percentage of the total fee is assigned to this stage. A designer who proceeds to final design prematurely may be unable to bill for the work expended if the project is aborted.

The *preliminary design* is the next stage of refinement from conceptual design. Some of the same elements exist, but are investigated at a higher level of detail. Once again, if an architectural project, the architects will orchestrate the work of a number

TABLE 4.1 PARTS OF A TYPICAL BID PACKAGE

Invitation to bid
Bid form
Form of agreement
Specifications
General conditions
Special conditions
Technical specifications
Drawings

of engineering and other professionals whose designed systems must be blended into the total design.

We must now be dealing with a specific site, which may not have been true earlier. It is possible to determine means of access; locations on the site of structures and nonstructural components such as landscaping, drainage, and sculpture; area of buildings per story and number of stories; requirements of internal circulation via elevators, stairs, corridors; requirements set by regulatory bodies relating to fire safety, pollution emissions, energy usage, and so on; levels of comfort sought through ventilation, lighting, furnishings, surface textures; esthetic design and compatibility with surroundings.

Engineering and nonengineering disciplines will make basic choices among materials, types of equipment, and the systems in which they can be used, each to be filtered through the architects for compatibility with the needs of others and for their relative costs and effectiveness in the total design scheme. Story heights must accommodate the structural frame plus the concealed distribution systems for air conditioning, heating, electric power, and the like. Weights and operating characteristics of machinery become inputs to the problems of structure and acoustical control. The process points toward the ultimate design as an integration of a number of systems living together in harmony. The problem of each engineering discipline is to propose the system that best solves the problems of its discipline as a part of the integrated solution.

For the structural designer, the task is, in part, the definition of a structural system that can support the loads caused by occupancy and use, as well as those others, such as wind and earthquake loads, that are imposed by nature. This is to be done in a manner that facilitates the work of other designers. It involves the choice of construction materials and one among a number of possible framing configurations appropriate to the materials. It requires, with help from calculation, determination of the space required to accommodate the structural members, space which is then to be reserved for structure and not violated by other systems.

The Tier Building Case Study presented in Chapter 18 is limited to the preliminary design of a medium-sized tier building in which structural steel is determined to be the material of choice. The product of the preliminary structural design is the arrangement of members into a complete system developed in sufficient detail to determine that the system is workable, to claim the necessary space, and to determine budgeted costs. If accepted, such a preliminary design and corresponding designs by other disciplines are combined into the total project preliminary design and budget to be presented to the owner.

With a *go* decision after review of all preliminary designs, the engineers can proceed to final structural design confident that other needs will not intrude on the needs of the structure, and equally confident that the structural design will not bog down at a later point due to unanticipated system problems or the substitution of other competing structural systems.

The *final design stage* converts the preliminary design into a set of documents suitable for the technical portion of a construction contract. In the case of the structural designer, this involves the final choices of member sizes and connection details incorporated into drawings and specifications that define the structure physically and the level of quality required to realize the intention of the design, and therefore of the contract.

The structural design must itself be documented by a set of calculations intended to determine that the design is both safe and efficient. The calculations become more and more detailed and refined as the design process moves through its conceptual, preliminary, and final stages. The calculations define the loads anticipated on the structure in all reasonably expected combinations so that all potential weak points of the structural system are revealed. Load levels for design are set at no less than those required by the governing codes. On the other hand, if the nature of anticipated use indicates that higher load levels are to be expected, the higher levels are used in the calculations.

If there is to be a structural steel frame, it is usually a part of a larger structural design that includes other materials, the action of which must be recognized in the design of the steel frame. The nature of the foundation, often reinforced concrete, sometimes on stiff piles, sometimes on relatively deformable soil, determines a condition at the boundary of the structural frame that affects the action of the frame itself. Similarly, concrete or brick masonry used in bearing or shear walls may be other parts of a total resisting system of which the structural steel is one component. The structural drawings and specifications include all structural elements; the structural portion of the construction budget is derived from all structural parts.

Since there are gray areas that are subject to conflicting interpretation as to whether they are part of the structural design or of some other system, it becomes necessary for the structural designer to define carefully and in detail what is included in the structural design and the budget derived from it. The architect then has information that makes it possible to see that no items of work are neglected or duplicated.

If the final design survives the public regulatory process and final budget review, the focus shifts to the generation of construction contracts between owner and builder and then to construction itself. In this phase, a whole new set of relationships exists. The contractor has undertaken to supply for a stated price a structure that follows the geometry and details shown on the drawings and is of the quality required by the drawings and specifications. The designer's role may be limited to review and enforcement of design requirements as to quality and details. Sometimes, unanticipated conditions or new opportunities arise during construction, making it desirable or necessary to make design changes. This should be done by the design professionals and, in most situations, must by law be done by them or other licensed professionals.

At times, an owner will write a separate construction management contract with the designers or with specialists in construction management. A construction manager will act as the owner's agent in writing construction contracts with builders, monitoring their performance, and expediting completion. The designers' quality-control function may then become a part of the larger management function.

However, the control of quality is intended to guarantee that the structure will actually be the designed one, and should not be subordinated to other management functions with different purposes.

The process thus moves from articulation of need through various stages of conceptual, preliminary, and final design and through construction. Each stage involves a review of the adequacy of the design concept, as well as budget and financial feasibility. Each intermediate stage ends at a decision point, at which time the project may be continued, revised, or aborted. The end of the construction phase offers the next critical test, that of adequacy of the structure for its intended use. There may be other tests later, during earthquake, flood, high winds, or other circumstances that test whether the structure is really adequate to live in its environment and continue to serve its intended function. Each such test is a test of the structural designer and of the other professionals involved.

4.3 CONTRACT DOCUMENTS FOR CONSTRUCTION

Return at this time to Flow Chart 4.1. Immediately following "Final Design" is the stage of preparation of contract documents and cost estimates. The contract documents are the immediate product of the design process. The cost estimate prepared from the contract documents is the most reliable estimate available up to this point, until further information comes in from the results of the actual bidding—the lowest price for which one of a group of competing construction contractors will offer to build the structure described in the documents.

What are the contract documents? They are a group of documents intended to be the subject of a contract between the owner and a construction company (contractor), under which the contractor undertakes, for a stated price, to transform the design into physical being. As with the design itself, the contract may cover structures as well as other elements of the total project: utilities, communications, landscaping, decorating, and so on. As a legally binding agreement, a contract puts obligations on both parties to perform something of value to the other—the contractor to build, the owner to pay. It is enforceable* by the courts, but the basis for contract enforcement must be a "meeting of the minds," a set of documents describing the terms of the contract to which both parties ascribe the same meaning. Any uncertainty within the contract documents as to meaning clouds the ability of the courts to enforce and, therefore, of both parties to be sure of receiving the value due them. Uncertainty may arise, among other things, if the documents do not completely describe the work to be done, if different parts of the documents describe it differently or in contradictory terms, if practical limits of technology make it impossible to build the structure as designed, or if the wording or graphics used to describe the design are ambiguous, that is, subject to more than one reasonable

*The intention is not to resort to litigation, but to hold litigation in reserve as the ultimate guarantor.

interpretation. The burden is on the designer to make the technical portions of the contract complete, internally consistent, buildable, and unambiguous—in sum, enforceable. This is a separate burden from the responsibility also borne by the engineer–designer to design a structure that is adequate for its task and safe.

The collapse of two walkways at the Hyatt Regency Hotel in Kansas City in 1981 took over 100 lives and injured many others. In subsequent reports by the National Bureau of Standards, a number of questions were raised as to the adequacy of the design. Other questions went to alleged differences between the architectural and structural drawings in describing the connection that failed and incompleteness in the description of the connection on both sets of drawings. In this, as in most instances of the failure of structures, there were obviously many contributing factors, which were discussed in public and engineering reports and litigated. In cases of this sort, civil liability suits are often brought against all parties possibly at fault, including owners, designers, and builders. If fault is found, determination of who is liable may hinge on how well and how clearly the contract documents described the design and the obligations of the parties.

In the type of situation described in the previous section, an *Invitation to Bid* may be issued by the owner or his agent (architect, construction manager, or other), inviting qualified contractors to propose the price for which they would be willing to perform all the work described in the contract documents under conditions also described therein. As listed in Table 4.1, the package accompanying the invitation to bid includes a form on which the bid price is to be proposed, the form of agreement to be signed by the parties indicating consummation of the contract, and the detailed provisions, technical and nontechnical, that describe the required work and the conditions under which it is to be performed.

The form of agreement would provide that the work is to be done for the agreed price, possibly within a stipulated time, and in accordance with the requirements of the drawings and specifications.

The nontechnical portion of the specification* (see Table 4.2) includes the *General Conditions* and *Special Conditions*, both of which govern the relations between the parties relating to the conduct of the work. The General Conditions are usually standardized (e.g., on a standard form published by the American Institute of Architects) and are used identically in many contracts for a large class of construction. Special Conditions are specific to the particular project, arising from the nature of the design, the location, local ordinances, and the like.

The *technical specifications* and *drawings* result directly from the design and describe it in physical terms and in specified quality, the drawings covering the former, the technical specifications the latter. For example, the structural drawings of a steel-framed building would include:

Arrangement drawings, in several views, showing all members constituting the framing system at small scale.

*We refer here to *specifications* governing construction, not *design specifications*, which will be part of the content of the rest of this book.

TABLE 4.2 TYPICAL SUBJECTS IN GENERAL AND SPECIAL CONDITIONS

General conditions
Status of documents
Submission of samples
Conditions for changes in contracted work
Conditions for terminating the contract
Required insurance
Settlement of disputes
Special conditions
Supervision of the work
Responsibility to provide surveys
Maintenance of safety
Procedures for submitting shop drawings
Schedule

TABLE 4.3 TYPICAL DIVISIONS OF THE TECHNICAL SPECIFICATIONS[a]

General requirements
Site work
Concrete
Metals (including structural steel)
Moisture protection
Finishes
Mechanical work
Electrical work

[a]An incomplete list. See, for example, a complete list of divisions and subdivisions issued by the Construction Specifications Institute, Washington, D.C.

Detail drawings explaining, at larger scale, methods of joining members and of providing for connection to parts of the work that are not part of the steel frame.

The structural steel section of the specifications would set forth, among other things, provisions about the *quality* of the steel, *tests* required to establish its quality, and *shop drawings** to be provided by the fabricator of the steel.

Drafters of the design drawings may be engineers or technicians working under engineer instruction, augmented by design calculations or sketches. Technical specifications (see Table 4.3) are written by architects and engineers, sometimes the designers, and sometimes specialists in specification writing. The structural engineer would be responsible for the structural drawings and the structural parts of the technical specifications. Mechanical and electrical engineers and other consultants would do the same for their parts of the design. Architectural drawings describe the work in general, finishes in particular, and the arrangement of structures on the site. Work done in site preparation may be assigned to civil engineers and/or landscape architects. Coordination and cross-checking of the drawings falls on the architect with assistance from the consultants.

The relationships described here are specific to an architecturally led building project. However, the same functions must be carried out by the design team, no matter how organized. The challenge to the structural engineer, and to others in their spheres, is to make the drawings and specifications clear, complete, and unassailable. If not so, the minds may not have met in the contract and the owner may be faced with demands for additional compensation when the designer insists on the quality of work intended in the design.

*These are drawings used to instruct shop and field workers in the details needed for fabrication and erection. They are derived from and must be consistent with the design drawings, but should not be used to add engineering design requirements.

5

Steel as a Structural Material

5.1 GENERAL REMARKS

One of the most exciting episodes in an unforgettable educational TV series is the one where Jacob Bronowski describes the making of a Samurai sword. The exquisite blade consisted of over 30,000 layers of steel, precisely formulated, worked, and reworked under carefully controlled temperature to the condition best suited to its function. Combining the flexibility provided by large smooth crystals in the core with a cutting edge based on small, hard, jagged crystals, this weapon became the symbol of a class that dominated Japanese society for centuries.

Bronowski describes the process in his *The Ascent of Man*.[25] He points out, incidentally, that the knowledge of metallurgy that made the sword possible was unknown to the Samurai; it was instead the property of a different group of master artisans. The Samurai sword incident is one of several examples in a chapter "The Hidden Structure" in which Bronowski demonstrates the profound impact that the working of metals had on different civilizations.

Bronowski places the date of the earliest known smelting of iron before 2500 B.C. in Egypt, and the earliest widespread use that by the Hittites circa 1500 B.C. Steel was known in India in 1000 B.C. and by the Samurai days, circa A.D. 800, its properties were understood well enough to create what may still be the finest steel known. But it was not until the mid-nineteenth century that steel technology opened the doors to a new industrial age.

The ninth-century Japanese transferred their steel technology from generation to generation within a selected group by the spoken word, reinforced by highly elaborated rituals. Today, we do the same in mathematical formulas printed in widely available books. But the primary difference is not there. For steel to catalyze the Industrial Revolution, it had to be available in different forms, with properties different from those which suited a class that was a small part of the Japanese population of the time, had a limited need for this fine metallurgy, and had a great interest in preventing its wide dissemination.

The forms in which steel is used today are many and varied. Perhaps its most significant attribute, in addition to its very high strength, is the ability of modern steel technology to tease out of it a seemingly infinite number of properties and combinations of properties on demand: for auto fenders, soft and easily formable; for the chassis, tough, fatigue resistant, in stiff cross sections; for the springs, highly tempered, flexible; for a hammer head, very hard; for its handle, soft and tough. The reader can with little difficulty think of dozens of steel items from his or her own experience and match to each the properties that it should possess to best fulfill its function.

A large fabricated shop assembly is being shipped to the site of the bridge in Chapter 2 where it will be assembled into the structure. (Photo courtesy of U.S. Steel Corp.)

5.2 CONTROL OF STEEL PROPERTIES BY ALLOYING

5.2.1 General

Iron is extracted with relative ease from a number of different ores in which it occurs, the principal added requirements being carbon, limestone, air, and heat. Its technology is widely known and needs minimum sophistication in production plant or in chemical formulation. Practical forms of iron occur alloyed with up to 7% carbon, but are usually in the range below 2%. Steel is produced from iron–carbon alloys by removing most of the carbon, down to 1.7% and usually below, adding small amounts of other alloying materials, such as manganese, silicon, and copper, and removing, to the extent possible, impurities like sulfur and phosphorus. It is produced as a liquid at temperatures above 2100°F. The rate at which it is cooled and its chemical composition determine its microstructure and mechanical properties and the way in which it responds to stress and/or environmental changes. Its properties can again be altered by reheating in the range below 1300°F ± and controlled cooling. The result is a class of materials that requires very sophisticated control in manufacture, but has much greater strength than iron and physical and mechanical properties that are both variable and predictable. Strength, hardness, ductility, toughness, weldability, all are controllable by the manufacturing process.

Although steel products are sometimes cast into molds in liquid form, the more usual procedure is to reheat the large *ingots* in which they are originally cast to high temperature (> 2000°F). In this state the ingots are soft and malleable enough so they can be run through a series of rollers, which squeeze the steel first into *billets* or *slabs* and then into a desired shape, a very long piece of some predetermined, constant cross section. The cross section may be very thin, wide *sheets*; thin, narrow *strips* sometimes wound into coils; thicker, wide *plates* or narrow *bars*; or the variety of special cross-section *shapes* provided for structural and similar uses. The most common cross sections used in structures are shown in Fig. 7.1.

5.2.2 Chemistry

The chemistry of a steel determines the limit of the mechanical properties that can be achieved by the processes used in manufacture. Adding carbon increases strength, often at a sacrifice of ductility (see Fig. 5.1). High-carbon steels can be made very hard and brittle by rapid cooling or *quenching* from high temperature (above 1300°F). Hardness can be reduced and workability improved by *annealing*: heating to an intermediate temperature range, holding that temperature for a number of hours, and slowly cooling to room temperature. Properly combining carbon, manganese, and other elements can raise the yield and ultimate strength without the brittle properties associated with high carbon content; the result is high-strength easily

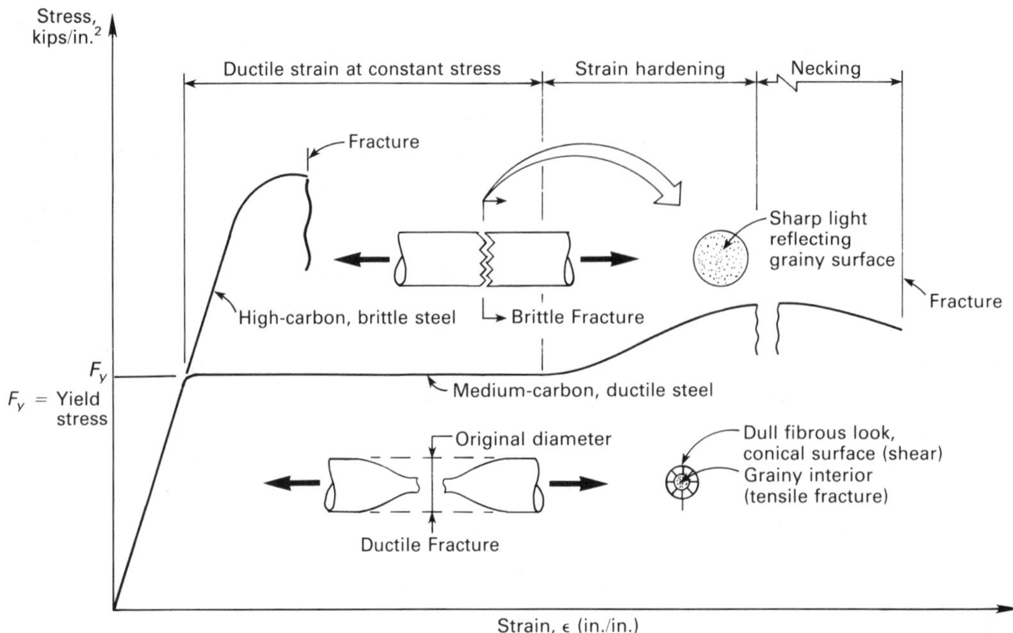

Figure 5.1 Ductile versus brittle performance in tension test (stress versus strain and nature of failure).

weldable steels, less sensitive to heating and cooling. Copper in small quantities reduces the rate and type of corrosion products (rust) associated with most iron and steel. Chrome–nickel steels are both strong and tough and unlike most steels neither corrosive nor magnetic.

The variations in properties are far too numerous to cover in this book. We will concentrate on those steels normally used in *structures* as defined in the introductory chapter. The *Metals Handbook* of ASM[15] and other specialized references[27,56] have much more complete information. Industry terminology distinguishes between *carbon steels* and *alloy steels*, with the latter arbitrarily defined as containing "more than 1.65% manganese, .60% silicon, and .60% copper." Table 5.1 shows, for the structural steels used with the design specification of the AISC, the amount of the principal alloying elements. A36* steel and the less common A529 are designated as *carbon* steels (carbon $< 0.30\%$), the *high-strength* steels are designated *low alloy*, and A514 a *heat-treated constructional alloy* steel. The *low alloy* term is used because the manganese, silicon, and copper, while significant, are not in high enough percentage for the *alloy steel* designation. In all cases, the total of all alloying elements is no more than 5% of the total composition of the steel.

*A36, and so on, are numbers identifying specifications issued by the American Society for Testing and Materials governing the properties of various steels. The full designation would include ASTM and the year of latest revision; e.g., ASTM A36-74. We have cropped the full designation for easier reading.

TABLE 5.1 MECHANICAL AND CHEMICAL REQUIREMENTS: STRUCTURAL STEELS[a]

Type	ASTM No.	F_y,[b] ksi	F_u,[b] ksi	Forms	Principal use	Chemistry,[b] % allowed						Relative Corrosion Resistance[d]
						C_{max}	Mn	Si	S_{max}	P_{max}	Other	
Carbon	A36	32 (>8″), 36 other	58–80	Plates Bars Shapes	Welded, bolted, riveted construction General structural	0.26–0.29[b]	0.60–1.20	0 and 0.15–0.30	0.05	0.04	Cu = 0 Cu = 0.20	1 2
	A53 Gr. B	35	60 min.	Pipe	Similar to A36 for applications with pipe	0.30	1.20	—	0.05	0.06		
	A500	33–46[c]	45–58[c]	Cold-formed tubing, round or shapes	Similar to A36							
	A501	36	58	Hot-formed tubing, round, square, other	Similar to A36	0.26	—	—	0.05	0.04	Cu = 0 Cu = 0.20	1 2
	A529	42	60–85	Group 1 shapes; plates to ½″	Similar to A36 Improved yield							
	A570	30–50[c]	49–65[c]	Sheet and strip	Cold-formed structural members for buildings; welded, bolted, screwed, riveted	0.25	0.90–1.35[c]	—	0.05	0.04	0.20 if Cu specified	

TABLE 5.1 *(continued)*

Type	ASTM No.	F_y,[b] ksi	F_u,[b] ksi	Forms	Principal use	Chemistry,[b] % allowed						Relative Corrosion Resistance[d]
						C_{max}	Mn	Si	S_{max}	P_{max}	Other	
High strength, low alloy	A242	42–50[b]	63–70[b]	Plates and bars to 4″; shapes	Riveted, bolted, general structural use; welding characteristics vary	0.15 or 0.20[c]	1.0 or 1.35[c]		0.05	0.15 or 0.04[c]	Cu = .20 or Cr = .05 + Si = .05	4 4
	A441	40–50[b]	60–70[b]	Plates and bars to 6″; shapes	Riveted, bolted, welded construction; general structural use	0.22	0.85–1.25	0.30 max	0.05	0.04	Min.: Cu = 0.20 V = 0.02	2
	A572	42–65[c]	60–80[c]	Plates and bars to 6″; shapes	Riveted, bolted, welded construction, except grades 55, 60 and 65 *not* for welded bridges; welding procedures depend on formulation; notch toughness subject to negotiation	0.21–0.26[b]	1.35–1.65[b]	—	0.05	0.04	Cb, V, N	
	A588	42–50[b]	63–70[b]	Plates and bars	Riveted and bolted	0.10–0.20[c]	0.98–1.35[c]	0.15–0.90[c]	0.04 or 0.05[c]	0.04	Cu, Cr, Ni, Mo,	4

				to 4″; shapes	construction; welded construction with suitable procedures						V, Zr, Cb, Ti[c]	
	A606	45–50[c]	65–70[c]	Sheet and strip	Cold-formed, light-gage structural applications; corrosion resistant	0.26	1.30		0.06			2–4[c]
	A607	45–70[c]	60–85[c]	Sheet and strip	Cold-formed, light-gage structural applications	0.26–0.30[c]	1.40–1.70[c]		0.06	0.05	Co, V, Cu = 0.20	1; or 2 if Cu specified
	A618	50	65–70[c]	Hot-formed tubing, round or shapes	Riveted and bolted construction; welded construction with proper procedure	0.22 or 0.23[c]	0.85–1.35[c]	0.0 or 0.30[c]	0.05	0.04	Cu = 0.20 for Gr. II, V, Co	1 if Cu is specified
Quenched and tempered alloy	A514	90 or 100[b]	100–130	Plates only, 4″ max	Welded structures *with great care*; dis-couraged if ductility important	0.32 or 0.21[c]	0.80–1.50[c]	0.15–0.90[c]	0.04	0.035	Various elements	

[a]Information incomplete and subject to change. Consult latest edition of specification.

[b]Varies by thickness and group; minimum unless noted.

[c]Varies by grade or type.

[d]Compared to carbon steel without copper.

Higher carbon content makes steels more hardenable, the degree of hardness being determined by the high temperature reached during heat treatment and rate of cooling. Steels high in carbon are very strong and can be hardened by very rapid cooling (*quenching*). Microcracks that form during quenching are usually eliminated by *tempering* (reheating for a sustained period and controlled cooling). The properties of *carbon* (or *medium carbon*) steels are much less sensitive to heating and quenching. In *high-strength, low-alloy steels*, additional strength comes from the additional alloying elements (particularly manganese), maintaining their insensitivity to the heating and cooling processes. The mechanical properties of the *heat-treated constructional alloy* steels are very strongly dependent on the fact that they are *quenched and tempered* as delivered and are altered by any subsequent reheating and cooling.

5.2.3 Mechanical Properties: Structural Steels

Strength. Strength is defined in units of stress (force/unit are)—in the usual usage of this book, kips/square inch (ksi), where a kip equals 1000 pounds. All structural steels (not all steels) have required minimum values of yield stress and specified ranges of ultimate stress. Table 5.1 gives these values, among other things, for the steels permitted under the AISC design specification (see next chapter). Minimum yield stress, F_y, varies from 36 ksi for ASTM A36 steel to 100 ksi for A514, with the corresponding values of minimum ultimate strength being 58 and 110 ksi. Figures 5.2 and 5.3 show stress–strain curves for several structural steels. The yield stress, F_y, is clear in the case of the two lower-strength steels, but gets less clear for A514, which has no pronounced yield stress. A nominal *yield stress* is defined at 0.2% offset as shown on the A514 curve of Fig. 5.3. Ultimate strength (see Fig. 5.3) is reached after strain hardening. The elastic design approach used in most of this book permits no nominal stresses as high as F_y. However, in some significant ways, local yield stresses and strains in the yield range are expected. In fact, the survivability of steel structures often relies on the ductility of the steel. Strain hardening and the corresponding stresses, $F_y \leq f \leq F_u$,* are not permitted. If allowed to occur, strain hardening alters the mechanical properties sufficiently to invalidate the underlying assumptions of both elastic and plastic design methods.

Ductility, brittleness, plasticity, elasticity. Consider the idealized stress–strain curves for a ductile steel and a brittle steel in Fig. 5.1. The ASM† defines *ductility* as the "ability of a material to deform plastically without fracturing," a measure of which is elongation in a tension test. *Brittleness* is defined as that "quality of a material that leads to crack propagation without plastic deformation." Ductile cracking requires nonelastic energy to be supplied and occurs slowly. Brittle cracking occurs suddenly, with little addition of energy to that stored elastically.

*For symbols, see Appendix B.

†Reference 15.

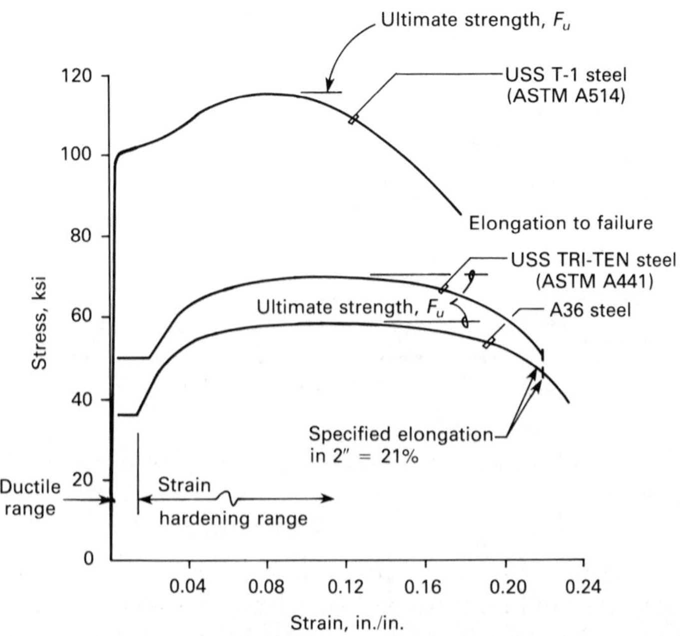

Figure 5.2 Typical stress–strain curves for structural steels. (Adapted with permission from U.S. Steel Co., *Steel Design Manual,* 1981)

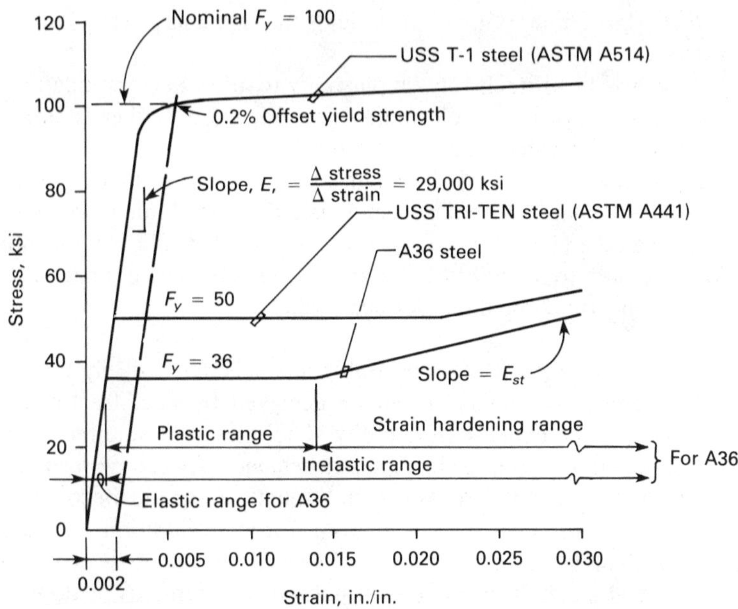

Figure 5.3 Typical stress–strain curves for structural steels: lower range of strains. (Adapted with permission from U.S. Steel Co., *Steel Design Manual,* 1981)

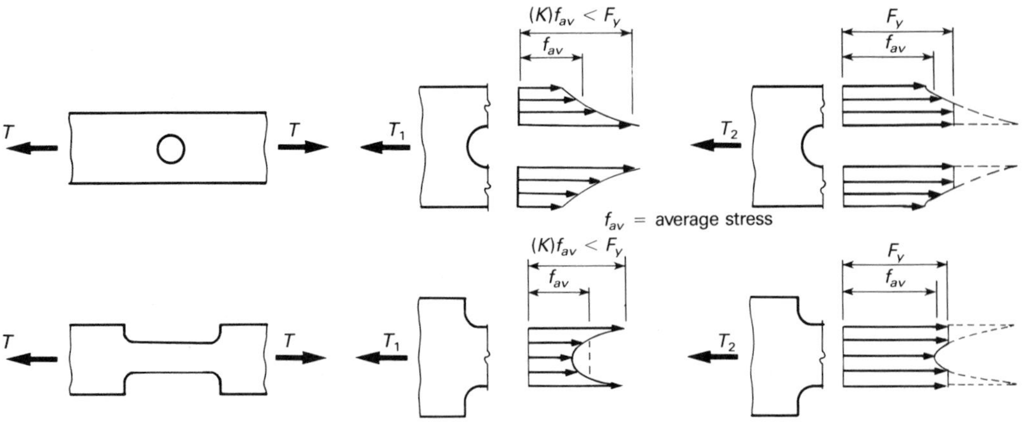

Figure 5.4 Capacity of elastoplastic steel to limit effect of stress magnification at holes and notches.

Readers who have witnessed or conducted standard tension tests in materials laboratories will recognize in Fig. 5.1 the "breaks" resulting from such tests. The amount of necking (reduction in area) before fracture and the appearance of the fracture places the specimen on a spectrum of ductility to brittleness. The energy absorbed before fracture is measured by the area under the stress–strain curve. The elastic portion represents a very small part of the energy that can be absorbed in an *elastoplastic* material, one that yields through a large range of strain at constant yield stress.

Structural steels rely on plasticity to allow accommodation to locally magnified stresses at holes and discontinuities. Figure 5.4 illustrates such accommodation. Since strain hardening changes the microstructure and related mechanical properties of the steel, structural designers seek to limit strains to levels below the strain-hardening range. Although usual specifications limit nominal allowable stresses to less than the yield stress, F_y, it is recognized that relief of magnified stress by yielding is necessary to avoid brittle fracture. Thus, the plastic range of strains in Fig. 5.2 and 5.3 is the range available for such relief.

Hardness is often desirable in steels, particularly for resistance to abrasion. With proper chemistry it can be achieved by heat treatment, but with increased strength at the expense of ductility. The search for suitable structural steels involves the attempt to preserve elastoplastic behavior. As can be seen in Fig. 5.2 and 5.3, this becomes more difficult with high strength. It would become even more difficult if additional strength were achieved through addition of carbon, with resulting embrittlement, rather than manganese, silicon, and other elements that are used in the structural steels. Fortunately, structural requirements seldom require high levels of hardness except in localized detail pieces, such as wear points associated with moving parts.

Fracture toughness and fatigue resistance are related phenomena. Failures due to insufficiency of either usually initiate as cracks at notches or discontinuities of section and result in fracture. Notches may be visible discontinuities, either planned or unplanned, or flaws in the materials resulting from the manufacturing or fabrication process. There are laboratory tests of materials that act as indicators of fracture toughness and are also of value in predicting fatigue resistance. However, observed failures do not correlate directly with any single test. Fractures observed under field conditions have usually been ascribed to combinations of conditions of which material properties are only one; design details, preexisting stress conditions, and loading rate are other variables. Fracture mechanics has not progressed to the point where simple quantitative procedures are available for design engineers. However, design techniques are available, which, added to control of notch toughness, reduce the likelihood of fracture. The literature on crack control is extensive and growing rapidly as the need becomes more obvious in today's complex structures. Reference 4 describes the state of the art of a rapidly changing field and offers an extensive bibliography.

Brittle fracture is abrupt, often occurs without warning, and can be catastrophic. Cracks, once started, propagate rapidly. Ductile fractures propagate more slowly and can often be identified by inspection as visible cracks or cracks that become indentifiable by inspection techniques using radiography (RT), magnetic particles (MT), dye penetrants (PT), and ultrasound (UT).

Fatigue failures usually start as shear cracks that develop slowly enough to be identifiable by programs of periodic inspection. Once identified, they are often repairable; left in place, the cracks themselves constitute notches or discontinuities that invite brittle failure. For this reason, the AWS specification permits no cracks in or adjacent to completed welds.

Fracture toughness is used as a measure of impact resistance, the ability to absorb sudden stress increases at a notch. Tests for toughness, such as the *Charpy test*, measure the energy required to fracture a standard specimen at a notch (see Fig. 5.5) with a suddenly applied blow. Increased toughness is associated with ductility. Toughness declines with lower temperature to a *nil ductility* level at which fracture energy is least. The study of toughness was greatly accelerated during World War II when many similar all-welded American ships split under Arctic low-temperature conditions.[54] Recent American construction in Alaska has led to the specification of minimum *Charpy* energy at low temperatures for work destined for low-temperature use. Similar specifications are used to cope with a similar problem in work with highly restrained welded joints at less severe temperatures.

Fatigue resistance is the ability of a material or a structure to resist repeated cycles of changing stress before fracture. *Fatigue limits* or *endurance limits* for steels are defined as the "maximum stress below which a material can presumably endure an infinite number of stress cycles." They are established in laboratory tests on highly polished regular specimens. Limits determined in this way can be very high. The

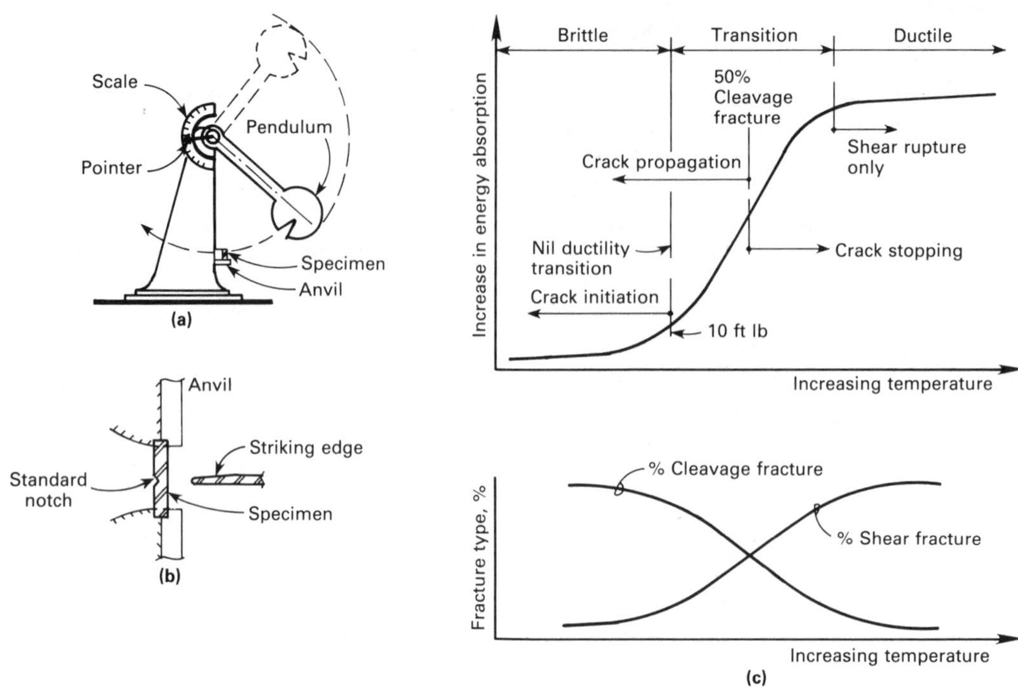

Figure 5.5 Temperature versus toughness, as measured by impact resistance in Charpy V-notch tests: (a) Charpy test apparatus; (b) detail of Charpy test; (c) typical Charpy results, variation with temperature, medium-carbon steel.

practical fatigue limit of a steel in a structure is affected by all the discontinuities involved in the actual construction, from surface scratches to deeply cut, sharp notches or attached welded details causing interruptions in strain continuity. Fatigue failures may take place at very low maximum stresses, often well below the yield strength of the material.

The procedures discussed in Chapter 8 establish allowable ranges of cycling stress, as functions of the number of cycles of varying stress anticipated during the life of a structure and the abruptness of discontinuities in details. Within the groups of structural steels acceptable under a design specification, fatigue resistance so established is independent of the steel used. If maximum stress for single load applications is higher than that caused by cycling loads, the use of high-strength steels may still be advantageous.

Fatigue failures are related primarily to the total number of cycles of large, repetitive stress changes during the life of a structure. In machinery and airplanes, vibrations tend to cause cycling stresses at constant input frequencies of rotating parts. In structures that experience highly variable loading, the changes are not so regular either in frequency or magnitude. Cycling forces come, for example, from highly variable ocean waves, loading cycles of cranes, and passage of trucks over bridges; stress ranges vary from cycle to cycle. In *practical* design procedures, fictitious regular stress cycles are used in proportioning members to resist fatigue.

Earthquake loads are cycling in nature, but occur infrequently enough that they do not represent fatigue problems. However, extreme seismic disturbances may cause large inelastic strain excursions beyond the plastic region, leading to fractures after a small number of cycles.

Weldability, although an imprecise, qualitative term, is very important in considering the suitability of a steel for structural use. ASTM-A7 steel, the standard structural steel as recently as 1963, was by 1969 retired from production, because its chemistry was not prescribed and its weldability not completely predictable,* this in spite of successful use in thousands of welded structures during the previous 20 years of active development of welding technology in structures. The modern standard structural steel, ASTM-A36, may be considered a fully weldable steel because, with the minimum care that can be easily applied and routinized, it can be successfully welded without loss in strength in either the parent or weld steel and with minimum change in other mechanical properties.

Successful welding requires that the steel at the weld be melted at temperatures above 3000°F, after which it cools to the temperature of the surrounding air. The rates of heating and cooling affect the grain structure and mechanical properties of the steel in ways similar to methods of heat treatment of the primary steel. Welding therefore involves a complex metallurgical process, taking place many times during the assembly of a welded steel structure. If, furthermore, a welded member is restrained against shortening while cooling, high tensile stresses may be *locked in* prior to application of design loads, reducing the additional stress from load that can be applied before yield. The local effect is analogous to stress magnification at notches as illustrated in Fig. 5.4. As in the notch case, ductility is essential to permit redistribution and averaging of stresses without brittle fracture.

Weldability is a function not only of the material, but also of the joint design and procedures used in welding. The American Welding Society's *Structural Welding Code* (AWS D1.1) lists requirements for preheating, interpass temperatures, and cooling rates for various thicknesses of each structural steel. In the case of A514 steel, which is heat treated during manufacture and which may be adversely affected by excessive preheat, maximum preheat is also specified. Specified welding procedures are also based on the type of welding process used.

*Welding involves localized melting of adjacent pieces of steel and fusion during recrystallization. If the chemistry is not known, it is not possible to predict the steel's sensitivity to hardening and embrittlement in cooling.

5.3 DESIRABLE STEEL PROPERTIES FOR FRAMED STRUCTURES

5.3.1 The Search for Structural Steels

Given the great variety of possible properties that can be derived from steels by control of chemistry and temperature, what is the right mix for structures? Clearly, it is not the same mix as was used in the Samurai sword, "ideal" though that may have been for absorbing the impact of a powerful blow while cutting through bone and sinew without dulling its hard cutting edge. If for no other reason, the sword steel would be rejected on grounds of cost. The work of building up the thickness in 30,000 separate layers, heating and cooling each time under controlled conditions over a period of days was necessary to produce one piece of steel that would not be too heavy for a strong man to handle nor so long that it would drag on the ground when sheathed. The reader may want to speculate on the cost in person hours per pound.

In addition, the extremely hard surface would make it impossible to punch holes for connectors and almost as difficult to drill. Welding of joints would quickly destroy the properties so carefully built up by heat treatment during manufacture. The extreme flexibility that augments its impact resistance, translated into deflection per unit load on a floor beam, would make a building floor so lively as to be unusable.

By 1939, the search for the right steel for structural use had led to the establishment of one basic steel for buildings and bridges (ASTM-A7, a "mild" or medium-carbon steel with $F_y = 33$ ksi), with the occasional use of "high-strength" steel, compounded with silicon, and some incidental steels for connecting materials. Contrast this with the multiplicity of structural steels listed in Table 5.1 and used today. In addition, special requirements, such as notch toughness and manufacturing process, may be negotiated with the mills.

The same period has seen several changes in the list of structural sections (shape of cross section and weight/foot) available from steel mills, with some being added, others eliminated, and many being slightly altered to reflect redesign of rolls and rolling procedures.

5.3.2 Desirable Criteria for Steel Selection for Structures

What criteria lead the steel industry to select steels to be manufactured for structural use? What criteria lead designers to select specific steels for their designs? The two questions are obviously related, although at any one time a designer may have to compromise between the most desirable properties and the properties of those steels available at competitive cost. The industry can supply a very large number of variants on steel in common use. However, it produces without price penalty only those steels for which there is widespread proven demand. Sometimes, as was the case for notch-

Multiple oxy-acetylene torches cut identically shaped pieces from a large steel plate. Movements of such torches may be controlled by a roller moving over a template (pattern) or by digital computer. Single pieces may be cut with a hand-guided torch. (Photo courtesy of U.S. Steel Corp.)

tough steel for Alaska Pipeline use, the demand predates the capacity of the industry.

The criteria for structural steels flow first from fabrication requirements. It must be possible, with relative ease, to cut the steel to length, to cut complicated shapes at ends and in the body of a member, to form flat plates into bent shapes, to punch or drill holes, and to weld into complex assemblies. See Fig. 5.6. These operations occur many times during the construction of a structure. Assemblies weighing in the tens of tons must be fabricated and joined in the field into structures weighing in the thousands of tons, and on a time schedule of a few months. Fabrication and erection must be carried out by workers whose skills are primarily manual and who should not be expected to understand the subtle effects of heat-treating processes on steel crystals and the mechanical properties of structures.

Structures require steels minimally sensitive to the changing temperatures of cutting and welding, soft enough to cut with tools of harder steel, corrosive enough to

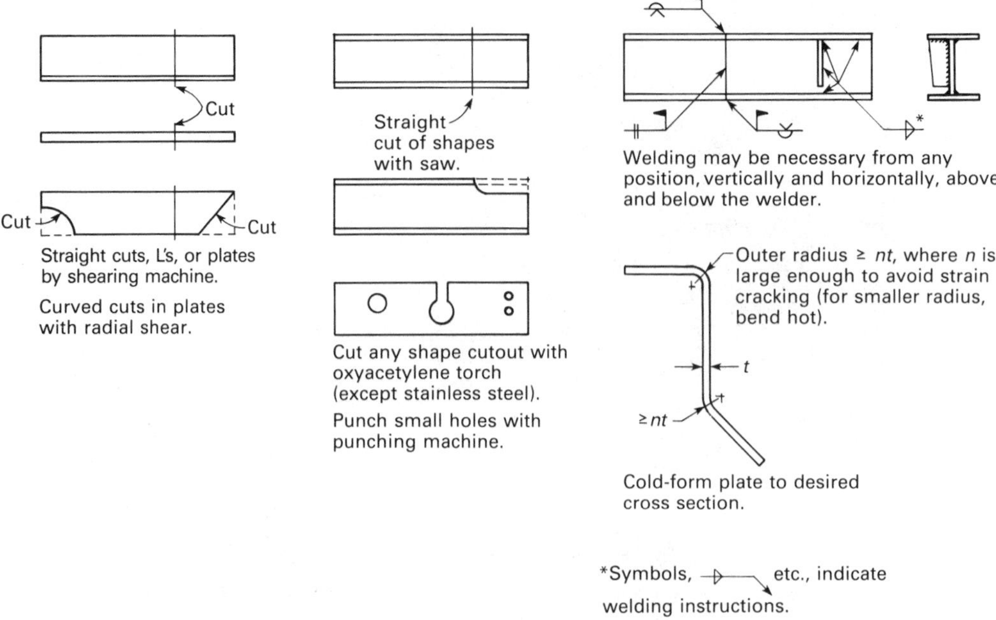

Figure 5.6 Some operations in shop fabrication and field assembly.

cut with oxyacetylene torches,* ductile enough to bend without cracking and without the need for prior heating, and ductile enough also that occasional local yielding at connections or elsewhere does not cause loss of useful life by strain hardening. The more complex the instructions necessary to maintain the integrity of the steel during fabrication, the more costly to achieve, and the more extensive the inspection necessary to achieve quality control. The structural steels are formulated with these requirements in mind, although sometimes some compromise of one quality is necessary to achieve another.

Different service conditions lead to different desirable criteria. Perhaps the least demanding structures are buildings in a temperate climate used where drastic load changes occur seldom and slowly, temperature changes are minimal, and, particularly for interior use, corrosion is a minor problem that can be prevented by a coat of corrosion-resistant paint or is avoided by application of fireproofing materials over the steel.

Exterior exposure accelerates corrosion. Anticorrosive coatings are helpful. Alloying with small percentages of copper causes formation of a coating of airtight oxides that act to inhibit further corrosion. Some recent bridge designs use copper-bearing steels and forego painting. In marine exposures, heavy in moisture-laden

*Steel is heated to visible red color and a jet of oxygen is supplied, which cuts by rapid oxidation.

salts, this may be insufficient and special paint systems may be required for protection.

Repetitive cyclic changes in loading require attention to fatigue resistance of the steel (also, as mentioned previously, to design details). A36 steel can be ordered with special requirements on uniformity of grain structure (excluding *rimming* or *capping* and requiring *silicon killing*, avoiding localized flaws that act as the site of fatigue cracks). The additional cost of high-strength steels is often not justified, since fatigue resistance is not enhanced; in some instances, it may be reduced.

Exposure to highly impacting loads (e.g., cranes, reciprocating machinery) or to extreme low temperature (Arctic or high mountain conditions), as well as fatiguing loads, requires enhanced notch toughness. Charpy impact test requirements can be specified for the service temperature of the steel. Heat treating of large structures after assembly is usually impossible. The preservation of notch toughness in welded structures requires attention to a number of factors:

- Choice of notch-tough steel.
- Avoidance of abrupt changes in cross section.
- Avoidance of localized flaws such as cracks or impurities within welds.
- Annealing of the heat-affected zone through control of the rate of temperature change before, during, and after welding (preheat and interpass temperatures and rate of cooling).

Table 5.1 includes for each steel authorized under the AISC Design Specification the type of use recommended and other pertinent information. Steels are divided into three basic groups by strength. All fulfill the preceding criteria, but in different degree. The carbon steels are the most common; usual shop procedures and quality control lead to reliable products. The higher strength of low-alloy steels is often purchased at small enough additional cost so that the ratio of cost to strength tends to favor them. They are easily weldable by customary shop procedures. However, their advantage of strength is limited. The lighter sections that might result from a direct comparison to carbon steels are sometimes unstable when considering column buckling (kl/r) or element stability in compression (width/thickness). In addition, smaller sections with higher allowable stresses lead to increased deflections. If, as is often the case, deflection criteria control design selections or fatigue is a design criterion, the higher strength of these steels may not be helpful.

The quenched and tempered steels supplied under A514 demand special care in use and should not be considered general-purpose steels. They are further limited in that they are furnished only as plates and bars and could only be used in the usual I, channel, or other shapes after a prior assembly stage; if welded into such shapes, additional heat treating is necessary after welding to restore the properties of the original steel and to avoid catastrophic brittle failure.

One obscure concern is worth considering. It is not easy to distinguish one type of steel from another visually. Nor can the different grain structures that result from differences in chemistry or heating history be seen without laboratory tests on pieces cut out of the structure. The chances of inadvertent substitution of improperly selected pieces grow the more a design uses multiple types of steel. Color coding and stamped identification help in keeping track of the steels. But the chance of mistakes is sufficient so that many designers specify only one type of steel for any specific structure; or if more than one type is used, the mixing is kept as simple and controllable as possible.

There are, however, instances where *hybrid* designs are efficient. For example, a plate girder might have an A36 web and A572 flanges (see Chapter 14). The two steels can be successfully welded to each other. The proportions of the thin, wide web are sufficiently different from the thicker, narrower flanges so that the chance of inadvertent substitution is minimal. Each represents a large enough proportion of total weight so that separate choice of steel may be justified. To a great extent, the decision is subject to the designer's judgment and general rules are not possible.

5.3.3 Steels Used in Connections

Creating structural systems from individual members requires that the members be connected, usually, today, by welding, bolting, or riveting. Special steels are used in these processes, which will be discussed in some detail in Chapters 12 and 13. Welding is done at high temperature, after which the steel is allowed to cool to room temperature. A temperature gradient is established from the location of the melted (and then recrystallized) weld zone and regions remote from it. Both the weld zone and a *heat-affected zone* adjacent to it are subjected to alterations in their microstructure. The demands on the weld metal are similar to those on the *base metal*, the steel being joined. Welding specifications require matching of the steels, as well as control of the rate of cooling after welding. Steel for bolts or rivets, on the other hand, sometimes has different demands from that of the connected steel, and its selection is based on the performance required of the joints rather than that of the members. In a long connection with multiple connectors at low load, some of the connectors are more highly loaded than others. Before failure, the connector loads must be equalized if the joint performance is to be predictable and the material used efficiently. This demands that connectors as well as connected members have the ability to redistribute stresses by ductile strain. Rivet steels (ASTM 502) and that of ordinary bolts (ASTM A307) are similar to the medium-carbon steels like A36. On the other hand, special steels (A325, A490) are used for *high-strength bolts* to make possible very high clamping forces between parts of a connection. As with other steels of very high strength, the properties of these steels change drastically if heat treated after manufacture. Specifications for their use do not permit this.

5.4 THE AISC MANUAL

5.4.1 Subject Matter

The *Manual of Steel Construction*,[7] published by AISC, is one of the primary tools available to designers as well as others involved in the technology of structural steel. It assembles in one place a great deal of information, which, for those who stop long enough to find out what it contains, is of great use. The subject matter is *structural* steels almost exclusively and the specifications used in structural steel design. The organization of the manual is divided into parts discussing:

1. The steels used for structures under the AISC Specification and the forms in which they can be purchased (Part 1).
2. Information and aids about design choices of particular types of members (Parts 2 through 4).
3. Specifications and codes used in the design of structural steel buildings and joints (Part 5).
4. Miscellaneous data and mathematical tables (Part 6).

The reader is urged to become familiar with the manual. The short discussion that follows is intended only as partial introduction.

5.4.2 Manual, Part 1

The bulk of Part 1 consists of the Shape Tables. These list all the standard shapes of cross section supplied by the industry specifically for structural use, with the dimensional properties necessary for designing and detailing. The kinds of available shapes are summarized in Fig. 7.1 of this book and discussed in other chapters where they are pertinent to the uses being considered.

The designations W, M, S, and HP are all sections in the shape of an I, the reason for each letter being historical. The W (earlier "wide flange") shapes are the most common in current use. C and MC are used for *channel* shapes, and L for *angles*. Tees (WT and ST), which are made by cutting W and S sections in half, are tabulated separately, as are the properties of double angles (⅃L) and several other sections made by combining primary sections. The pipes listed on page 1–89 of the Manual represent only a few of the many pipe sizes available. Note in that table that pipes of the same nominal size (e.g., 10 inch) are produced as *strong*, *extra-strong*, and *double-extra strong* by varying the wall thickness and the inside diameter while keeping the outside diameter constant. For pipes of over 12-inch diameter, this is no longer true: the nominal size is the inside diameter, and the outer diameter changes with wall thickness. Structural tubing is listed in available sizes, both square and rectangular. Round tubing, similar to pipes, is available, but seldom used in structures.

As will be discussed in detail in Part II of this book, design selections are usually based on the "best" combination of engineering properties listed. See, for example, AISC Manual, page 1–18, which shows the properties of the cross sections of several W sections. The shape is designated by type (W), nominal depth, and weight per lineal foot. Thus, a W18 × 119 is of I shape from the W series with nominal depth of 18 inches and weighs 119 pounds per lineal foot. A group of actual cross-section dimensions follows, in both decimal and fraction form, revealing among other things, that the actual depth is 18.97 inches. Note that the T dimension is constant for all W18 shapes. What is not quite so obvious is that the depth in-to-in of flanges ($d - 2t_f$) is also constant; the same inside rolls are used in rolling all W18's; variations in thickness of web and flanges are achieved by adjusting the spaces between rolls. This is true for many sizes of W section, a fact that is helpful in designing connections. *Compact section criteria* (see Chapter 10) are used in connection with Section 1.5.1.4 of the AISC Specification, which permits increased allowable stresses for sections able to mobilize their plastic moments (M_p) without buckling F_y' and F_y'''* columns show the maximum yield stress, F_y, for which the section is *compact* by criteria relating to the proportions of parts of the cross section. Parameters R_T and d/A_f are also dimensional criteria related to stability, which are discussed later. The remaining columns show strength and stiffness properties familiar from courses in mechanics of materials.

Both preceding and following the shape tables is other information useful for design and purchasing. The designer should be aware of the tolerances (deviations from published dimensions) permitted in the Standard Mill Practice Section. These act as a significant constraint on the precision of design.

5.4.3 Manual, Parts 2 through 4

Parts 2, 3, and 4 provide tables, curves, and other information useful in designing beams, columns, and connections. In large part, the information in these parts provides numerical solutions to formulas in the AISC Specification, tabulated in form to make member selection convenient. Tables are usually supplemented by explanatory notes.

5.4.4 Manual, Part 5

The AISC Design Specification and Commentary are discussed in detail in the next chapter. The balance of Part 5 includes a specification for joints using high-strength bolts, a code describing standard practices in contractual relations in the purchase of steel, and the conditions for quality certification by AISC.

*The listed values of F_y''' may be exceeded if there is little axial compressive stress. There are two formulas in the AISC Specification governing the d/t criterion (1.5-4a and 1.5-4b).

5.4.5 Manual, Part 6

Part 6 contains information on miscellaneous subjects available in many other places. The main value of this part is the grouping of this information into convenient form in the same volume that is used for many other aspects of steel design.

6
Anatomy of a Code and Specification

6.1 INTRODUCTION: BUILDING STANDARDS IN TWELFTH-CENTURY CHINA AND TWENTIETH-CENTURY AMERICA

In A.D. 1103 at the Imperial Capital of Kaifeng in Honan Province, the Emperor Hui Tsung ordered printed one of the oldest volumes of building standards known. The work was by Li Hsieh of the Department of Construction, and consisted of 1078 pages covering such things as arrangement of buildings, classification into eight classes, dimensions of individual members, and decorative artwork. Else Glahn[30] reproduces details in her report of a structural system that evolved over a period of 6000 years into a standardized system of nonrigid wood post and beam construction with bracket arms supporting heavy tile roofs (three to four times the weight of modern Western tile roofs). Although violating the preconceptions of many of today's earthquake engineers, Glahn points to 1000-year-old examples of this still-familiar type of construction as "proof that the standard plan gave protection against . . . great windstorms . . . and . . . earthquakes." She does not mention buildings that did not survive.

The standards of Li Hsieh illustrate both the value and the shortcomings of codification, which lead to extensive debate among building code writers today. By recording in a code the historic experience of successful building practice, the state preserves the wisdom of the past, simplifies the conduct of business and the training of builders, and applies standards of safe construction for the public benefit. On the other hand, to the extent that the code freezes construction practice to yesterday's state-of-the-art knowledge, it inhibits new ideas and new systems, perhaps more suited to the needs and possibilities of a new day. Scientific investigation of materials and mathematical modeling of building systems make it possible to evolve safe, new systems in far less time than 6000 years. Modern steel construction evolved since the nineteenth century; much of today's practice was unknown 50 years ago. Sound as they may have been, Li Hsieh's standards could not be used to design a skyscraper or a major arena today.

The authors believe that codification is necessary. Engineering science is not so fully developed nor known mathematical models so reflective of actual structures as to permit confident prediction of performance without laboratory testing and observation of actual structures in being. What is true for the collected knowledge of the engineering profession is even more true for the more limited knowledge of the individual engineer. Code limitations are necessary for safe structures and orderly construction practice. For the young engineer as well as the seasoned one, it is a comfort to rely on the crystallized professional knowledge reflected in design specifications rather than to pretend to be all-knowing and all-seeing. On the other hand, society is not served if its codes merely reflect the past and do not permit the careful application of new theory and new systems. Today's codes are usually a compromise between the past and the future of engineering knowledge, written by committees that properly contain both advocates of the preservation of historically preserved standards and those promoting new developments.

Codes and design specifications are a part of the professional and societal system for regulating construction. The engineer must understand them as both constraints and tools in design practice. We will, in this chapter, dissect a design specification in some detail and a building code to a more limited extent. We will not reproduce them, but point to what is there, how it is organized and how used. The specification will be the 1978 *Specification for the Design, Fabrication and Erection of Structural Steel for Buildings* of AISC. The code will be the 1982 *Uniform Building Code* (UBC) of the International Conference of Building Officials. Both are widely used. Neither is universally so. However, familiarity with the structure, contents, and purposes of these documents should make it relatively easy to see the similarities and differences in other codes and specifications that may govern specific design situations.

6.2 CODES AND SPECIFICATIONS

Engineering usage of the terms *code* and *specification* tends to be somewhat loose; it is difficult to find satisfactory definitions for each. The two terms are often used interchangeably, although we feel they should be separated. We will attempt to distinguish between them, but caution the reader to check the definition intended by the user in engineering reading or discussion.

Both codes and specifications are sets of requirements. However, we will treat codes as collections of laws, whereas specifications may be agreements by individuals or groups on quality or other standards.* For example, the *Uniform Building Code*[33] is written for adoption by governing bodies such as municipalities. When adopted by such a body, it becomes a set of legal requirements governing the various people involved in building. Violators are subject to legal and sometimes criminal sanction. A specification may, if contained in a contract, be enforced by civil courts, but penalties for noncompliance apply only to one of the individuals party to the contract.

The UBC is much broader in application than the AISC Specification. It contains within it one chapter (Chapter 27) that covers the same matters as the AISC Specification, and, in some editions of the UBC, is almost identical to it. If a part of a specification is written into the UBC, either in detail or by reference, the specification in this form attains the same legal status as the rest of the code.

Even without specific adoption, a widely used specification such as that of the AISC represents a standard of the engineering profession to which engineers are expected to adhere under usual circumstances. If the engineer's calculations claim conformance to this specification, such conformance is expected, and deviation may lead to professional or other penalties. If no specific specification is cited as a guide, the engineer's work is still expected to follow either accepted practice reflected in such a specification or to specifically prove the engineering soundness of the design approach used.

The reader should not confuse a design specification such as the one discussed here with a construction specification. A design specification is usually applied by an engineer in designing a structure. A construction specification, usually written by the engineer, is a set of requirements as to quality, detail, or performance set by an engineer on a supplier or builder.

6.3 THE UNIFORM BUILDING CODE†

The 1982 UBC consists of 48 chapters (1 through 60, but omitting 12 chapter numbers) plus 14 chapters of appendix; some of the latter are appended to chapters in the body of the code, others bear different numbers. The 60 chapters are contained within 11 parts plus the Appendix. The scope of the code is very broad and relates to engineered and nonengineered buildings and to matters of direct and indirect application to engineering work. A listing of the parts will indicate the scope. The headings are quoted from the UBC.

*By this definition, the AISC *Code of Standard Practice* and the *Code of Ethics* of the Engineers' Council for Professional Development are misnamed. Our dictionary lists the "collection of laws" definition first. The distinction made here is necessary to separate two types of documents, even though universal agreement on the words does not exist.

†Other general codes are not identical but cover similar topics.

Part I Administrative
II Definitions and Abbreviations
III Requirements Based on Occupancy
IV Requirements Based on Type of Construction
V Engineering Regulations—Quality and Design of the Materials of Construction
VI Detailed Regulations
VII Fire Resistive Standards for Fire Protection
VIII Regulations for Use of Public Streets and Projections over Public Property
IX Wall and Ceiling Coverings
X Special Subjects
XI Uniform Building Code Standards
Appendix

Most of the attention of the structural engineering designer is directed at Part V. Within that part, the initial chapter, Chapter 23*, covers "General Design Requirements," including methods of design and minimum prescribed loads for design. Loads[†] are divided into dead loads, live loads, wind loads, and seismic (earthquake) loads.

Dead loads are the weight of the structure and permanent attachments.

Live loads are roof and floor loads resulting from occupancy and use. Roof live loads vary from 12 to 20 pounds per square foot (psf) in snow-free regions to several hundred pounds per square foot where deep snow packs occur. Prescribed minimum floor live loads vary by occupancy from 40 psf in residences to 250 psf in some industrial and warehousing uses. Alternative concentrated loads are also shown. If actual loads, based on designed use, are greater than the listed loads, the larger loads are to be used.

Wind loads, based on a wind zone map of the United States, are to be applied as pressure in pounds per square foot of exposed surface. The pressures within any zone vary with height above the ground, type of terrain, and shape of the structure.

Seismic requirements are also covered in Chapter 23. Along with other seismic design requirements, formulas and procedures are presented that convert the dynamic accelerations due to earthquakes into pseudostatic horizontal forces varying with the mass of the structure. The amount of the horizontal force, expressed as a percentage of gravity loads on the mass, depends on location on an earthquake probability map, the type of framing system, and several other factors. Alternatively, the designer is permitted to submit data in support of a dynamic analysis.

Case studies in Part III of this book illustrate the use of the requirements of Chapter 23, as well as other parts of this code.

*See Appendix A, this book.

[†]Chapter 7 of this book contains further discussion of design loads.

The remaining chapters in Part V cover design requirements for common building materials: masonry, wood, concrete, steel, aluminum. Chapter 27 is the chapter on design in steel. It is similar to the design requirements of the AISC Specification.

Chapter 17 of Part IV classifies buildings into five types (I through V) based on degrees of fire resistance and requirements of public safety; type I is the highest classification. Chapters 18 through 22 describe the requirements of each type. Structural steel is permitted for all types, but in fire-resistive construction it must be protected against direct exposure to heat, which would weaken the steel. Requirements are shown in Table 17A (Chapter 17). Methods of fire protection are covered in Part VII. Fire protection requirements are expressed in *hours* from 0 to 4. These represent the minimum time for which the fireproofing is expected to protect the steel from weakening due to high temperature from a fire of standard intensity.

Thus, according to Table 17A, the structural frame of a type I building must be rated for three hours. Table 43A (Chapter 43) shows the minimum thicknesses of various insulating materials that must be used to protect the steel in order to be classified as "three hour" protection (e.g., two inches of grade B concrete or 1½ inches of perlite plaster on metal lath applied around steel columns or truss members).

Chapter 29 (Part VI) covers soil classifications and foundations and must be of interest in any structural design, since proper analysis of any structural frame requires understanding of the response to frame action at the foundation.

Chapter 60 (Part XI) lists, by number, standards for testing, materials, and miscellaneous items, much of which parallels similar standards published by ASTM and other standards bodies. A separate volume of *UBC Standards* gives detailed requirements of the standards. Standard 27-1 is referenced in Chapter 27, Section 2721(b), and adopts as permitted structural steels a group of steels listed by ASTM number. Section 2704 of that chapter refers to UBC Standard 27-3, which is based on the requirements for "stress variation or stress reversal design" of the AISC Specification.

The balance of the UBC is of general interest to the structural designer, as it is to anyone involved in building design, construction, or use. However, the degree of detailed attention required will vary with the engineer's responsibilities in a specific project.

6.4 AISC SPECIFICATION AND COMMENTARY

6.4.1 Purpose and Organization

> Structural steel design in accordance with requirements of the "Specification for the Design, Fabrication and Erection of Structural Steel for Buildings (year)."

This note, or one to similar effect, is included in the design notes of a very large proportion of steel-framed buildings and other steel structures designed in the United

States. Such a statement implies that the designer is thoroughly familiar with the requirements of the Specification and has observed all requirements pertinent to the design problem.

No advice is more pertinent to a new engineer preparing to design a steel structure than to "learn" the Specification. Learn what it is intended to do. Learn what matters it governs. Learn where specific requirements are to be found, those places which are obvious and those which are obscure or "hidden." Learn the organization of the Specification and the way to retrieve information from it. Learn how to interpret the sometimes confusing language in which words sometimes have specific historically determined meanings known best to those who have used them that way for many years.

Much of the learning will come from use. However, in this chapter we will attempt to clear up some of the early confusion that arises when faced with a complicated set of new rules to which one is expected to conform. Many of the new user's problems can be transcended if the awesome stack of closely printed pages is reduced to that smaller number which need consideration for the particular problem addressed. The organization of the Specification, which evolved through a series of revisions, makes this possible once it is understood.

The Specification and Commentary are published by AISC in a combined document. They are also included as a portion of Part 5 of the AISC Manual. It is important to distinguish various versions by the date of publication; we are discussing the 1978 edition.

To consider the organization of the Specification, it is convenient to divide it into six parts:

1. Nomenclature
2. Specification: Part 1, Elastic Design
3. Specification: Part 2, Plastic Design
4. Appendix A: Tables of Numerical Equivalents
5. Appendixes B through E
6. Commentary

It is also worth noting that the tables and curves found in Parts 2, 3, and 4 of the Manual are in large part based on direct numerical solutions of the formulas in the Specification.

The Specification itself consists of numbers 1 through 5 of the preceding list. The Commentary is helpful in interpreting the sometime obscure, legalistic language of the Specification and explaining the background from theory and tests of its various provisions.

The numbering system used, similar to the one used in this book, is one based on progressive subdivision set off by decimals. *Parts** are integers, *sections* decimals,

*There are only two Parts in the AISC Specification, but 20 chapters in this book. Sections of the Specification always start with 1 or 2. Sections of this book always start with the number of the chapter. Once this is understood, there will be little confusion as to which book is being referenced.

subsections subdecimals. Thus, 1.5.1.3.1 is the first subdivision of the third subdivision of subsection 1 of section 5 of part 1 of the Specification. This system is infinitely subdividible, convenient for cross-referencing both within the Specification and in discussion and correspondence about it.

The numbering of formulas is slightly different. Formulas (1.5-5a) and (1.5-5b) are found in Section 1.5, but one must hunt within that section for the subsections (1.5.1.4.2 and 1.5.1.4.3) in which they appear.

6.4.2 Nomenclature

A list of symbols, in Roman and Greek letters, is placed for convenient reference at the beginning of the Specification, with each symbol defined. This list includes all of the symbols used in formulas within the Specification. Where applicable, the definitions of symbols include dimensions. Many formulas contain constants that are only valid when the numerical substitutions are in the intended dimensions. For example, Formula (1.5-5a),

$$F_b = F_y \left[0.79 - 0.002 \left(\frac{b_f}{2t_f} \right) \sqrt{F_y} \right]$$

is only usable if F_b, F_y, and $\sqrt{F_y}$ are all expressed in kips per square inch.

Definitions of symbols are sometimes, but not always, repeated after formulas in the body of the Specification. Use of the nomenclature list will be found to be very convenient. This book uses many symbols from the AISC Nomenclature list, with the same meaning intended (see Appendix B).

6.4.3 Specification: Part 1, Elastic Design

Part 1 is the most commonly used portion of the AISC Specification. It covers the design requirements for *elastic* or *allowable stress* design, that approach which seeks a factor of safety in the ratio

$$\frac{\text{failure stress}}{\text{allowable stress}} = \text{factory of safety} > 1.0$$

Anticipated or required *service* loads are applied in formulas relating load to stress; the resulting stress must not exceed the allowable.

$$\text{service load stress} = f \leq F = \text{allowable stress} < F_y$$

Since service loads are applied to elastic structures, the required limits of deflection are checked by applying the design loads directly in formulas for deflection.

Part 1 consists of 26 sections, a list of which we have reproduced as Table 6.1, where we have grouped the 26 sections into 7 groups of related sections of which it will be noted that:

TABLE 6.1 TABLE OF CONTENTS—AISC SPECIFICATIONS

Table of Contents

Adapted with permission from AISC Specification, 1978.

I	Deals with a group of general items
II and III	Apply to selection of members and are needed very often
V	Will be needed in design of connections
IV and VI	Deal with special topics and are consulted when one or more of these topics is pertinent
VII	Deals with fabrication and erection.

The four sections, 1.1 through 1.4, which we group under I, merit some comments. Section 1.1 governs presentation of designs on drawings. Subsection 1.1.1 obligates the designer to provide some specific information. Subsection 1.1.2 obligates a fabricator to prepare drawings instructing shop workers in detail as to what is expected in fabricated pieces and assemblies. The design drawings are expected to provide information complete enough to present design requirements thoroughly and unambiguously. However, they are not usually in a form that is easily readable by workers trained in shop processes. Standard forms of shop drawings are available in an AISC publication,* but the practice of individual fabricators often deviates to some degree based on their own preferences. In usual contracts, fabricators are required to submit shop details for review by the design engineers prior to fabrication.

The *types of construction* defined in Section 1.2 are peculiar to steel structures. In flexural members, the shears and moments experienced by a member are a function not only of load and span, but also of the rigidity (resistance to rotation) of the joints. Depending on the types of end connections used, a structure is classified as *Type 1—Rigid Frame, Type 2—Simple Framing* (sometimes called *Simple Span Design*), or *Type 3—Semirigid Framing*. Connections discussed in Chapters 12 and 13 of this book are identified as applicable to one or the other.

Loads and forces specified for design in Section 1.3 are given somewhat less completely than in most building codes. Subsection 1.3.7 refers to the minimum requirements of the building code (A58.1) published by the American National Standards Institute, but does not inhibit application of other building codes where they apply. This book refers to other codes in specific cases, often to the *Uniform Building Code*.

On the other hand, since steel structures are often subjected to impact effects of crane elevator and machinery loads, both vertical and horizontal, Subsections 1.3.3 and 1.3.4 require that given percentages of the weights of moving masses be applied in addition to other loads. This critical requirement is sometimes omitted from building codes.

Section 1.4, "Materials," defines a specific list of steels permissible under the AISC Specification. They are identified by ASTM numbers or, in the case of welding materials, by AWS numbers. The writers of the Specification had the properties of the *approved* steels in mind when they wrote the Specification. Many provisions are

*American Institute of Steel Construction, *Detailing for Steel Construction* (Detroit: The Institute, 1983).

not necessarily valid for other steels. Designers who use other steels (e.g., see cables in Chapter 8) should recognize that they are acting outside the AISC Specification and take whatever precautions are necessary to assure sound design.

Manufacturers periodically develop new steels with properties they consider desirable and marketable. Use of new steels, based on manufacturers' information, often precedes standardization through ASTM or adoption in general specifications. Designers may require them in construction specifications, but in doing so they are not shielded by the AISC Specification.

The sections grouped under VII relate to requirements for fabrication and erection. They clearly apply to the operations of construction contractors. Designers may not immediately realize that they also apply to design. Although day-to-day reference to these sections is not often required during design development, the level of workmanship required is that expected in work designed under the AISC Specification. The designer may require other, higher standards, and execute designs based on them. If so, the special standards must be specifically set out in the contract documents. Failing that, contractors will base their price on the requirements of the AISC Specification.

Groups II through VI apply to the selection and design of members and systems and demand the most detailed attention of the designer. A longer list of "contents" would include the subsections and lesser divisions into which the sections are divided. In the process of choice of a structural system or member, it will often be necessary to comply with requirements of a number of subsections scattered through the Specification. In discussing such choices in the various chapters of Part II of this book, we have included lists of sections and subsections of the AISC Specification pertinent to specific subjects. For example, referring to Table 8.2, the first four subsections cited apply to all tension members, the next six to specific types, and Appendix B to tension members subjected to fatigue. There are also four other subsections cited governing "related considerations," which apply to connection details and which must be considered when tension members are selected.

6.4.4 Specification: Part 2, Plastic Design

The plastic approach to steel design starts by defining the strength of a member based on the limit set when all elements of a cross section are at yield stress, F_y. In a member subjected to bending moment, that condition exists at the plastic moment, M_p. M_p is larger than the yield moment, M_y, that moment where only the outer fibers are stressed to F_y, and stresses elsewhere vary linearly from zero at the neutral axis to F_y at the outer fiber (see Fig. 6.1). The strength of a continuous frame is usually calculated on the basis of the development of plastic moment at the most highly stressed points. Safety is maintained by limiting service loads to a decimal, less than one, times the loads that cause that moment.

The plastic design method is not covered in detail in this book. Readers are referred to reference 8 and others. It is most useful in rigid frame structures.

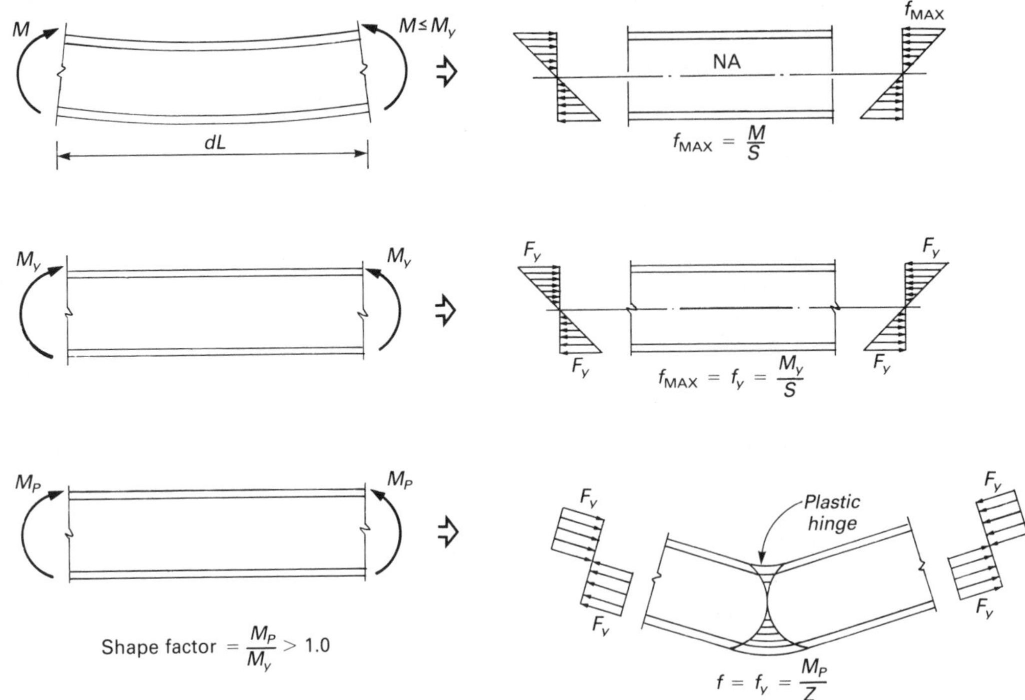

Figure 6.1 Elastic and plastic bending.

Part 2 of the AISC Specification establishes design and fabrication requirements applicable to plastic design. Load factors, (plastic load/service load), are set at constant values of 1.7 for dead and live loads, and 1.3 for conditions including design wind or earthquake.*

To guarantee that plastic moments can be developed without buckling of the elements of a cross section, requirements (Sections 2.5 through 2.7) are set on the proportions of such elements, similar to "compact section" requirements under Part 1. Compact sections are discussed in detail in our Chapter 10. In establishing the plastic strength of a frame, plastic *hinges* are established that, at M_p, are expected to allow rotation similar to a true pinned joint. For this to happen, fibers must be able to strain, plastically, through a much larger range than the elastic range, before strain hardening. Steels permitted under Section 2.2, a much smaller list than allowed under Section 1.4, are specifically selected for their desirable plastic properties. The list does not include any steels with minimum yield stress, F_y, greater than 65 ksi.

Frequent excursions into the range of plastic strains shorten the life of a steel structure even more drastically than repetitive cycling of elastic strains. Section 2.1

*The *load and resistance factor method*, a more recent probabilistic approach, is not included in the 1978 Specification. It appears to be gaining currency. An excellent discussion is found in Galambos.[28]

cautions against the use of plastic methods in designing crane runways. The author recommends that this caution extend to all structures designed for significant repetitive loadings.

In a sentence within Section 1.1, all "pertinent provisions" of Part 1 that are not modified by Part 2 are incorporated into that part.

The plastic method covered in Part 2 with the load factors it specifies is similar to, but not the same in concept as, the load factor approach, which has become familiar in recent years in the "strength" methods used in the design of reinforced concrete structures.[2] The load and resistance factor method of reference 28 is also in the same stream, but not the same.

6.4.5 Appendix A: Tables of Numerical Equivalents

Formulas in Part 1 of the Specification are often complex and may involve a number of independent variables. Experienced as well as new designers may be tempted to throw up their hands and wish for simpler, self-evident rules, questioning the value of such complexity. It is probably impossible to go back to the simpler rules that characterized earlier specifications when modern high-strength steels did not exist and designs were more standardized in forms that often accepted overdesign (i.e., the use of more steel than necessary) as desirable.

Appendix A, properly used, takes much of the onus off the complex formulas in the current code. The 12 tables, 3 of which are presented in separate parts for $F_y = 36$ and $F_y = 50$ ksi, represent solutions of equations in Part 1, showing values of dependent variables associated with various values of independent variables. Table 6.2, herein, is reproduced from a list of the tables that appears in the AISC Specification at the beginning of Appendix A. The pertinent formulas are associated with the tables that solve them.

The virtue of these tables lies in more than just the saving of time in solutions for single cases. It often takes more time to find the appropriate table and figure out how to use it than to find the necessary numerical solutions by calculation. The tendency for novices is to do just that. The chief virtue of these tables is as tools to facilitate design choices, making it possible to see quickly the trend of stress values and the implications of design choices. The solutions presented in Parts II and III of this book will often use the tables, as well as the formulas from which they derive. We suggest that students learn to do both. Use of the formulas directly reinforces some parts of basic learning and makes it possible to go beyond the bounds of the tables. Use of the tables will strengthen the ability of the engineer to think his way creatively through a design.

6.4.6 Appendixes B through E

Appendixes B through E address four different design matters that arise infrequently enough so that the Specification writers thought it would be convenient to deal with them outside the body of the Specification.

TABLE 6.2 TABLES, LIST OF APPENDIX A, AISC SPECIFICATION

		Reference Sect. No.
Table 1	Allowable Stress as a Function of F_y	1.5.1, 1.10.5
Table 2	Allowable Stress as a Function of F_u	1.5.1, 1.5.2
	Allowable Stress for Main and Secondary Compression Members:	1.5.1.3.1 (Formula 1.5-1)
Table 3-36	For 36 ksi Yield Stress Steel	1.5.1.3.2 (Formula 1.5-2)
Table 3-50	For 50 ksi Yield Stress Steel	1.5.1.3.3 (Formula 1.5-3)
	Values of C_a for Determining Allowable Stress for Main and Secondary Compression Members when $Kl/r \leq C_c$ (by equation $F_a = C_aF_y$):	1.5.1.3.1 (Formula 1.5-1) or C5 (Formula C5-1)
Table 4	For Steel of Any Yield Stress	
Table 5	Values of C_c for Use in Formulas (1.5-1) and (1.5-2), and in Table 4	1.5.1.3.1 1.5.1.3.2
Table 6	Slenderness Ratios of Elements as a Function of F_y	1.5.1.4.1 1.5.1.4.5 1.9.1.2 1.9.2.2 1.9.2.3 1.10.2
Table 7	Values of C_b for Use in Formulas (1.5-6a), (1.5-6b), and (1.5-7)	1.5.1.4.5
Table 8	Values of C_m for Use in Formula (1.6-1a)	1.6.1
Table 9	Values of F'_e for Steel of Any Yield Stress	1.6.1 (Formula 1.6-1a)
	Allowable Shear Stress in Webs of Plate Girders by Formula (1.10-1), Tension Field Action Not Included:	1.10.5.2
Table 10-36	For 36 ksi Yield Stress Steel	
Table 10-50	For 50 ksi Yield Stress Steel	
	Allowable Shear Stress in Webs of Plate Girders by Formula (1.10-2), Tension Field Action Included:	1.10.5.2
Table 11-36	For 36 ksi Yield Stress Steel	
Table 11-50	For 50 ksi Yield Stress Steel	
Table 12	Values of C_h for Determining Maximum Allowable Bending Stress in Hybrid Girders (by equation $F'_b = C_hF_b$)	1.10.6 (Formula 1.10-6)

Reproduced with permission.

Appendix B governs repetitive loading and is triggered when design live loads are expected to vary cyclically more than 20,000 times during the life of the structure. This is a basic condition of highway bridge design or structures exposed to ocean waves. It occurs in buildings in some special situations, such as use of cranes or machinery. Since the provisions are applicable for the most part to stress cycles that include tension, further consideration of fatigue loading is found with the discussion of tension members in Chapter 8.

The scope of *Appendix C* is stated in its opening Section, C1: "Axially loaded members and flexural members containing elements subject to compression and having a width–thickness ratio in excess of the applicable limits of Section 1.9. . . ." Section 1.9 sets limits for the ratio, width/thickness, for individual elements within a cross section, within which the allowable stresses found in Section 1.5 apply without any modification based on those ratios. In Section 1.9.2, ratios are in the form of a constant divided by $\sqrt{F_y}$, the constant varying for different cases. Section 1.9.3 gives a similar limit, applying to pipes and circular tubes of

$$\frac{\text{outer diameter}}{\text{wall thickness}} \leq \frac{3300}{F_y}$$

These limits are chosen to avoid loss of effectiveness by buckling of thin elements, which can occur at low compressive stress, a phenomenon similar to low stress buckling of the entire section when a column is long and slender.

Ordinarily, prudent and efficient design leads to slenderness ratios within the stated limits. In those cases where there is reason to exceed the limits, the adjustments required in Appendix C are designed to preserve the usual factor of safety.

Most structural steel designs, particularly where standard mill-produced shapes are used, utilize prismatic members (i.e., members of constant cross section). However, there are times when it is desirable to taper a member. Figure 6.2 illustrates some cases to which *Appendix D* applies. Section D1 limits the applicability of Appendix D to symmetrical tapered members with equal and constant flanges and depths varying linearly within prescribed limits. Where applicable, Appendix D gives requirements for allowable stresses that replace those in the body of Part 1.

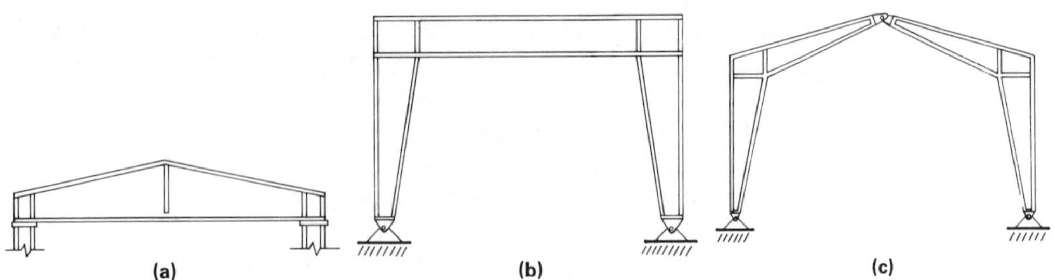

Figure 6.2 Tapered members: (a) tapered girder; (b) tapered columns in rigid frame; (c) tapered three-hinged *arch*.

Appendix E applies to high-strength bolted joints designed in "friction-type connections," connections whose resistance to shear depends on the available friction that results from the clamping effect of highly tightened bolts. Such connections will be discussed in Chapter 13. Allowable connector loads are prescribed in Specification Table 1.5.2.1. However, it is recognized that the degree of care used in preparing faying (overlapping) surfaces affects the real reliable limit of available friction. Appendix E defines nine classes of surface preparation and assigns allowable working stresses in shear that may be substituted for the usual values given in Table 1.5.2.1.

6.4.7 Commentary

The Commentary is not officially a part of the Specification. In theory, the Specification should be completely clear to an engineer reading it and the theoretical basis of its various provisions thoroughly understood by the reader. Neither of these assumptions is always true. The wording of the Specification is sometimes subject to more than one reasonable interpretation. The applicable theory may itself be evolving and be the subject of debate among the Specification writers themselves. Many formulas and other requirements are not directly derived from theory. They may be empirically derived from test results or field observations or be simplified "practical" forms of more complex relationships that are valid within specific limits.

The Commentary then becomes a useful tool for interpreting the Specification. It offers access to the background thinking of those who wrote it. Sometimes it includes the justification in theory, experience, and tests for provisions of the Specification. In some cases it includes useful tables and curves designed to reduce tedium in applying the Specification. In still other instances, there are comments that the Specification writers may consider to be of general interest or use.

An added use of the Commentary, from the point of view of experienced designers, is to offer quick insight into the changes that have taken place in a new edition of the Specification from the older one with which the designer is familiar and which they are often reluctant to give up. For the beginning as well as the seasoned engineer, the Commentary acts as a convenient and useful current supplement to whatever texts and theoretical works the engineer is accustomed to refer to in his or her work.

The Commentary is organized to directly parallel the Specification itself. Section and paragraph numbering on any particular subject is identical to that of the Specification provision the comment is intended to explain, except that figure and table numbers in the Commentary are preceded by the letter C.

Thus, Section 1.8 of the Specification covers requirements limiting the strength parameter, KL/r, of compression members, which is discussed in some detail herein in Chapter 9. The Commentary offers helpful discussion (Section 1.8), theoretical and recommended values of K for some standard cases (Table C1.8.1), and a nomographic solution for K useful in "rigid frame" structures (Fig. C1.8.2).

It is necessary to introduce a legal issue in connection with the Commentary. Although the Commentary is not officially part of the Specification and may not be specifically mentioned by a designer or legislative body that adopts the Specification, it may still have legal standing as part of the Specification. Courts often go beyond the apparent meaning of the wording of laws into evidence of what was in the minds of the legislators when the laws were adopted. The Commentary may in litigation be found to have both legal and professional standing; first as evidence of the thinking of the Specification writers; second, as evidence of generally accepted professional opinion at the time the Specification was written. If there is any question about the interpretation of the Specification, itself, it becomes not only useful but necessary for the engineer to know what the Commentary offers on the subject.

Part II
Fashioning of Tools

7
Introduction to Part II

7.1 FASHIONING OF TOOLS

The special purpose of Part II of this book may be called the fashioning of tools. The tools of the designer are functions of the mind. The reader has no doubt brought some tools to this reading: the principles of mechanics, an understanding of relations between stress and strain, the ability to analyze mathematically idealized systems responding to loads. These chapters will add some new tools. They will not make the other tools unnecessary. They will resemble some others that are specifically useful for other engineering materials. But these tools will be specific to structural steel—its power, its uses, its problems. We will address the various methods available to transform this very strong but very heavy and expensive man-made material, difficult to cut and shape, into structural systems adaptable to a multitude of uses, often able to span farther with less weight and at less cost than any other material available. Such methods must arise from understanding the nature of the material and the forms in which it can be found. They must include the ability to choose for each task the specific form most adapted to the task. They must be informed by knowledge of the methods available to connect steel members to each other and to other materials as at foundations. The nature of the connections often determines the choice of the members. The methods must flow from knowledge of the structural systems to which steel is specifically suited by its nature.

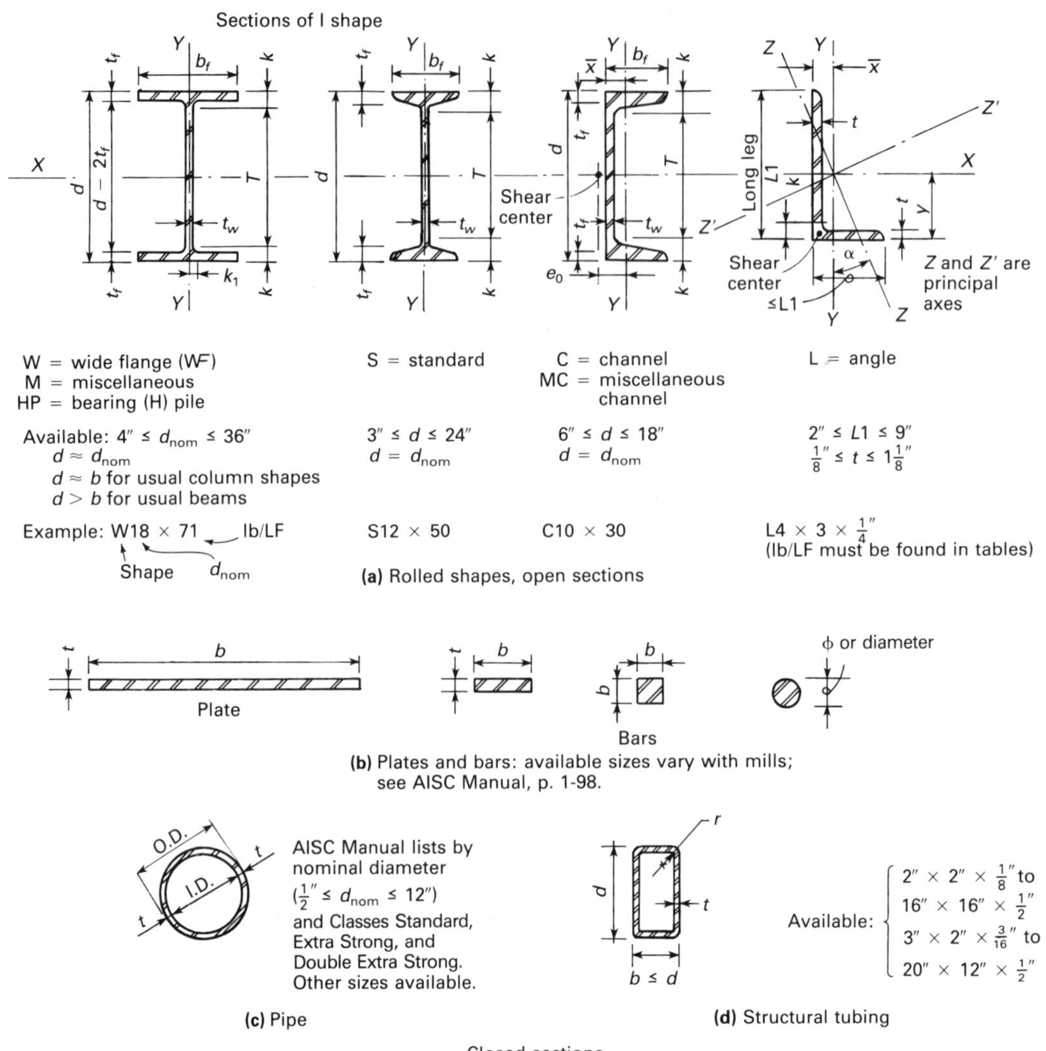

Figure 7.1 Basic structural sections available from steel mills.

7.2 PLATES, SHAPES: HEAVY STRUCTURAL SECTIONS

Examine Figs. 7.1 through 7.3, which show many of the forms of cross section in which steel is used. Consider how they evolved. Central to the problem is the very high strength of the material. Rupture strength of steels may be 20 to 100 times that of wood and, in tension, hundreds of times that of natural or manufactured masonry. Brought to a comparable task, the required cross-sectional area is very small. If all structural members were in tension, the task would be easy. Load/safe stress = area required. However, some members must resist compression, some flexure, some transverse and/or torsional shear.

The problems of compression involve not only area (A) but moment of inertia (I) of that area. A large ratio of effective length, L_e, to radius of gyration, r, will reduce the compressive capacity of a member. For ratios of $L_e/r > 120$ (approximately), the buckling stress is very far below the yield stress and cannot be improved by supplying stronger steel. Radius of gyration, r, equals $\sqrt{I/A}$. In flexure, a member must call for strength on its section modulus, S, where $S = I/c$, both I and c being determined by the distribution of the cross-sectional area. In the extreme, we may ask it to draw on its plastic reserve strength, the modulus becoming $Z = M_p/F_y$. The same flexing member will deflect under load in proportion to $1/I$. Transverse shear is usually associated with flexure. Although the ratio of bending moment/shear may be infinitely variable, it is usual that the area required to resist shear is a small part of that required in the same member to resist flexure. Resistance to torsion is most effective when the cross-sectional area is as far as possible from the center of twist and forms a continuous, closed path around that center. The controlling parameters for torsion are J and J/c, where J is again a (polar) moment of inertia of area and c a distance from the centroid. In all cases except tension, there is reason to spread out the area, with a large part of it as far from the geometric centroid as possible.

So if a member is to be used in tension only, we concern ourselves primarily with the area of cross section and the strength of the steel. Such a member is the cable, Fig. 7.2(d), made of a number of small wires of very high strength steel. Among its uses are the main and suspender cables (wire ropes) of suspension bridges and many types of hangers and stays used in hoisting systems. Any of the other cross sections in Figs. 7.1 through 7.3 can be and are used as tension members, but have other uses as well.

To achieve a large moment of inertia with relatively small area, the steel industry provides a variety of rolled sections called *shapes*. If used primarily for flexure about one axis, the W* and the now less common S shapes have most of their areas in the flanges, and large moment of inertia about one principal axis, small about the other. The thin web is intended to provide resistance to shear. A C section is similar but lacks symmetry. Used singly, it can develop problems associated with this lack. L sections are even more limited in flexure.

Where the use is primarily in compression, it is desirable to have large values of r and therefore large I about both principal axes. Optimally, $r_x = r_y$. In nominal sizes between 4 and 14 inches, some W sections are rolled with $b_f \approx d$, specifically for use as columns. These sections have r_y approximately equal to 60% of r_x, a far higher percentage than sections usually chosen for uniaxial bending. $r_x/r_y = 1.0$ in pipes or square tubes, making these sections ideal for compression members, although difficulties at connections and limits on the maximum normally available sizes prevent their universal adoption for this use.

*Industry designations W, S, HP, and M are all sections shaped like an I. The most commonly used sections for axial load and flexure are W's.

The process described results in cross sections with individual elements that are relatively wide and thin. For large ratios of width to thickness, individual elements in compression buckle out of plane at low stress (see Chapter 9), reducing the effective strength of the member. Most rolled shapes are deliberately proportioned to avoid this problem. Those that are not are flagged in the F_y' and F_y''' columns of the Shape Tables in Section 1 of the AISC Manual.

Where rolled sections do not have all the properties desired, special sections are often created by combining standard shapes or from plates. See Fig. 7.2. Several such in Fig. 7.2(a) have specific use for biaxial bending. C's and L's used symmetrically in pairs can be suitable for compression and/or bending and, in some systems, are easier to connect than W sections. An infinitely large number of sections can be created from plates joined by welding, bolting, or riveting into composite members tailored to the specific needs of the design problem. Reinforced concrete is often combined with structural steel in composite structures. *Orthotropic* systems[5] utilize steel plates, combinations of plates and shapes in two-way and three-way flexural systems. In all custom-designed sections, the designer will have to be particularly conscious of the limitations set by width/thickness ratios of individual elements and their potential effect in reducing the usable stresses in the steel. In the case of thin members in shear, as in plate girder webs, shear buckling can often be prevented by attaching stiffening plates or shapes.

An important difference between rolled and custom-tailored sections should be noted. Due to manufacturing methods, rolled and similar mill-supplied sections occur in long pieces of constant cross section. When used in any single member, a section chosen for a maximum requirement (shear, bending moment) at one point along the member's length may have much more strength and use much more material elsewhere than is necessary. For many, perhaps most, cases this seemingly wasteful use of steel is justified by the high cost of altering the mill-produced products. When the need for specially fabricated sections is established, the same constraints on the use of varying cross sections do not necessarily exist. Some of the cost of special fabrication of sections may be recaptured by providing material at each point along a member roughly as needed at that point. This very useful freedom should be used judiciously. Member stability must be considered, as must the effect of the member properties on the action of other members in the system. The saving in material cost from each change should be weighed against the cost of the additional fabrication. However, as a general rule, much more variation is reasonable when using special sections than with standard sections.

It is ordinarily possible to choose any one of a large variety of sections for a specific type of function. The young designer, perhaps frightened by so much freedom, may need advice on how to start. Actual choices are usually made from a more limited number of options, based on considerations of the structural system of which the member is a part and the connections that work most conveniently with the system. The chapters of Part II will clarify the methods of selection of members based on their uses and end connections. In the case studies of Part III, we will see how the

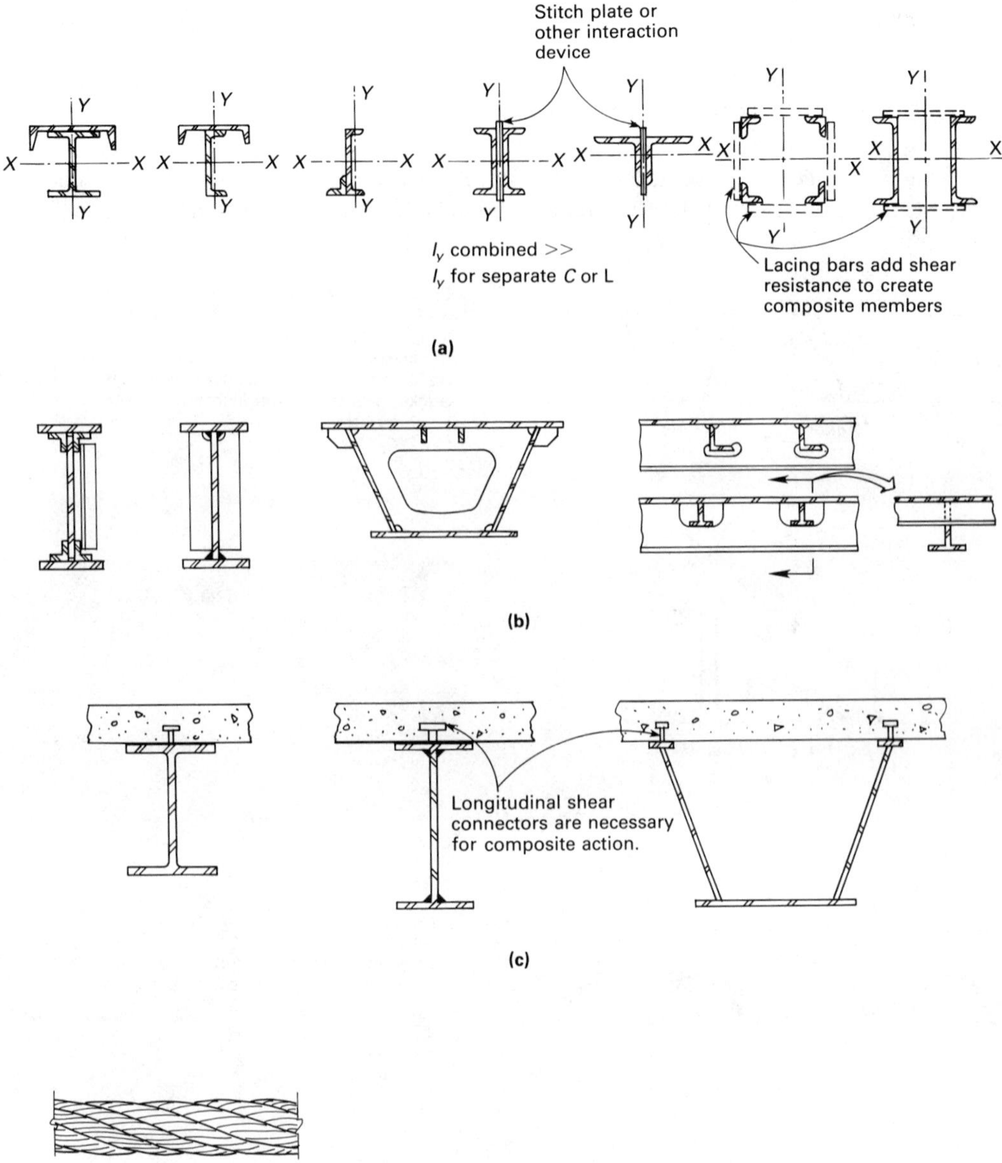

Figure 7.2 Derived sections: (a) composite (combined) sections; (b) plate girders and orthotropic sections (tailored from available plates and shapes); (c) composite sections of steel and reinforced concrete (using steel plates or shapes); (d) wire rope (cable).

type of system, its analysis, and the methods of Part II combine in creating structural frames.

7.3 COLD-FORMED STEEL MEMBERS

In light-gage steel construction, thin steel sheets are cold formed into structural sections, achieving structural strength and thickness through their form (see Fig. 7.3).

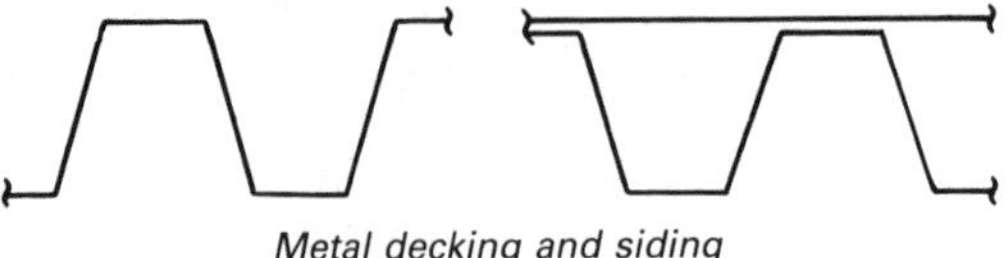

Metal decking and siding

Light-gage steel sheets achieve stiffness, flexural and shear strength when formed into structural shapes. Usual gages vary from #10 = 0.1345″ thickness to #28 = 0.049″. Upper limits may depend on forming equipment. Lower useful limits are set by buckling of wide, thin elements at low compressive or shear stress.

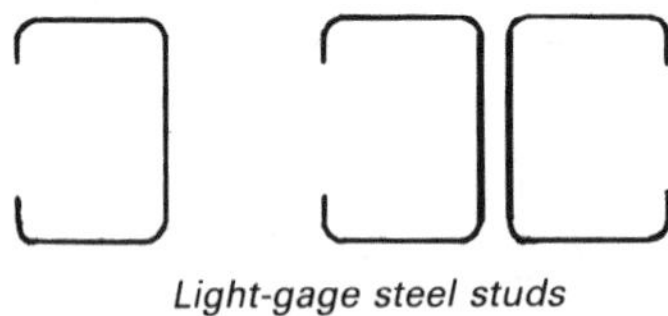

Light-gage steel studs

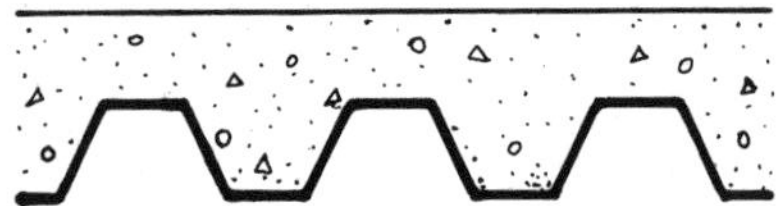

Light-gage steel decking may be used in composite action with concrete in slabs.

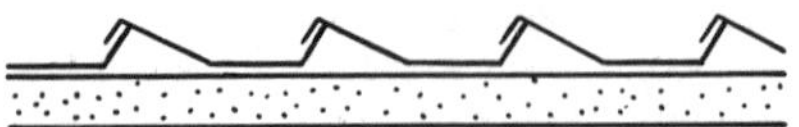

Light-gage steel used as part of insulated *sandwich* panels.

Figure 7.3 Some uses of light-gage steel in structures.

Use of light-gage steel is common in secondary structures such as partitions and ceiling framing, in some lightly loaded industrial buildings, and in roofs and floors of many structural steel buildings. Chapter 15 addresses some of the special design procedures necessary in dealing with this type of construction. Cold-formed deck and siding are used in the Mill Building and Tier Building Case Studies (Chapters 17 and 18).

7.4 LOADS

An engineered structure is expected to (1) stay in place and (2) maintain its geometric integrity so that (3) it can continuously support the functions for which it was built. In order for this to happen, the design engineer must attempt to predict the various loads and combinations of loads to which the structure will be exposed during its life, and provide:

1. Sufficient strength and stability to avoid collapse.
2. Sufficient stiffness so that the deflections that result from loads do not cause loss of function.
3. Ways to join the members so that their actions are consistent with the mathematical model used in design.

A search for guidance on load predictions reveals a great variety of sometimes confusing and contradictory information. It is possible to identify the sources of many loads, but sometimes much more difficult to quantify their effects.

The *Uniform Building Code* lists (Chapter 23, UCB)* four main categories of load: dead, live, wind, and seismic (earthquake). If gravity loads are applied or removed rapidly, the resulting acceleration of the masses is often calculated as a percentage increase of the load, sometimes called *impact*. The AISC Specification (1.3.3) specifies impact for moving live loads in buildings. The AASHTO and AREA specifications prescribe impact of vehicle loads on bridges.

Loads higher than minimum loads in the codes and specifications may be expected on specific structures or in specific locations. The designer is expected to provide for such increased loads if they can reasonably be anticipated. A general approach for a designer might be:

1. Seek guidance from standard sources, such as codes or specifications, on loads usually considered for the type of structure being designed (e.g., dead, live, wind).
2. Analyze, to the extent possible, the nature and magnitudes of unusual loads resulting from use and location (e.g., ice, vibrating machinery).

*See Appendix A.

TABLE 7.1 COMMON STRUCTURAL DESIGN LOADS

Type of Load		How Prescribed	Relative Predictability	Duration	Approaches to reserve of safety	
					Allow. stress[a]	Load Factor[b]
Vertical	Dead	Calculation of actual weight of structure and permanent attachments	*High*, but seldom perfect; either underestimate or overestimate can be dangerous	Usually life of structure	$F \times 1.0$	$L_c \times 1.4$
	Live (e.g., snow, traffic, floor loads)	Investigation of expected use of structure. Minimum may be standardized by codes. Designer may prescribe limits on user.	*Relatively poor*, but extreme values may be predictable for common structures	Varies, short or long duration but not constant	$F \times 1.0$	$L_p \times 1.7$
	Moving Live (e.g., vehicles)	Similar to live		Short duration, may be frequent	$F \times 1.0$	$L_p \times 1.7$
	Impact	Dynamic acceleration of moving live load, often a % to be added to live load	*Poor*, but can be improved by investigation	Short, but extreme value may be frequent	$F \times 1.0$	$L_c \times 1.7$

Horizontal (lateral or longitudinal)	Soil pressure	By soil mechanics and site investigation	*Poor*, but sampling and testing helps	Long term, may increase with time	$F \times 1.0$	$L_c \times 1.7$
	Hydrostatic pressure	Site investigation	Varies with site	Varies with structure and site	$F \times 1.0$	$L_c \times 1.7$
	Thermal	Investigate local environment and *structural* response. May be altered by details	*Good* if records sufficient; otherwise poor	Extremes usually short time	$F \times 1.25 \pm$ in combinations of loads	$L_c \times 1.5 \pm$
	Ice, streamflow	Local investigation	*Poor*; extremes may be very damaging	Extremes transitory	$F \times 1.25 \pm$	$L_c \times 1.5 \pm$
	Wind	Local investigation, ASCE or other standard reports based on fluid mechanics; field measurements	Extremes may be higher than usual code limits	Short and infrequent for extremes	$F \times 1.33$	$L_p \times 1.4 \pm$
	Seismic (earthquake)	Seismic mapping, dynamic modeling, equivalent static modeling by codes	Extreme effects very unpredictable	Short extremes, infrequent	$F \times 1.33 \pm$ for design level EQ.	$L_p \times 1.4 \pm$

[a] F = basic allowable stress: in AISC specifications may be F_t, F_a, F_b, F_v, F_p or other.

[b] L_c = loads calculated from actual or assumed conditions.

L_p = loads prescribed by specification or by designer based on conditions of use of structure. Load factor is multiplier shown.

3. For each, evaluate the degree of confidence you have, as designer, in defining the nature and magnitude of load.
4. Adjust your design factor of safety to recognize, among all other considerations, the degree of certainty about loads.

In the usual case, there is adequate guidance from standard sources, representing the results of a great deal of investigation and experience by the engineering profession. Step 1 is then sufficient and usually much better than any evaluation the individual designer could evolve independently. When usual standard procedures are not adequate, steps 2 through 4 become imperative. The designer falls back on research, investigation, and application of physical principals.

Table 7.1 lists a number of the loads that can be expected on many structures, with information pertinent to each. The basic division may be considered:

1. Dead loads: the weight of the structure and all permanent attachments.
2. Live loads: the effects of occupancy and use.
3. Environmentally related loads: those imposed by nature, usually independent of the use of the structure.

Temporary loads that occur during construction are not usually considered design loads on the completed structure. However, they are often considered by careful designers, particularly if their effects might compromise the structure. Among the loads listed, dead and live loads act vertically, being primarily due to gravity; environmental loads are indicated to be acting horizontally. Although a commonly used subdivision, this is not universally applicable.

Several of the loads represent pseudostatic surrogates for what are really forces resulting from dynamic acceleration of masses. The reader will recognize that wind and stream flow pressures result from interruption by the structure of fluid particles in a stream. Fluid mechanics reveals that the resulting pressures are equal to $p = KV^2$, where V is the extreme velocity of flow and K includes the effect of density, viscosity, and the form of the interfering structure.

Impact effects result from the acceleration of a moving mass. Vehicles on a bridge may be horizontally accelerated by starting and braking and vertically accelerated by the deflections of the supporting structure. Seismic loads result from the inertia effect of the structure's mass responding to the acceleration of the supporting soil as filtered through the resisting frame. They are often represented as static loads, calculated as a percentage of gravity acceleration acting on the mass affected. When applied to a vehicle, *impact* may appear as an added load equal to, say, 0.30 times the vehicle weight and applied whenever the vehicle (live) load is applied. In the seismic case, a horizontal force may be applied equal to, say, 0.15 times the weight of the structure (dead weight plus, in some cases, part of the live load associated with weight). For most structures, governing codes prescribe minimum seismic design forces and their distribution. This will be illustrated in some of the case studies in Part III.

Some structures warrant more complex dynamic analyses of the true effects of dynamic loads. However, we will limit our discussions to pseudostatic analyses. Structural design systems that emerge from such analyses are often considered satisfactory for the real dynamic loads.

With the exception of dead load, which acts at all times, all other loads will vary in magnitude from one time to another and may act at any one time on all or only some of the places where they apply. The intent of design is that the structure must resist all reasonably possible combinations of possible loads acting at maximum prescribed magnitudes or less. It must resist all such combinations without failure by collapse, overturning, sliding, tilting, or deformations that are injurious to function.

The *Uniform Building Code*, for example, requires consideration of at least five different combinations of loads at various percentages of their separate maxima [UBC 2303 (f)] and permits reductions in the intensity (pounds/square foot) of live loads when applied over large areas (UBC 2306 and Table 23C).

However, to identify some of the most dangerous effects, one may have to examine lesser combinations. Just as a hot air balloon rises when ballast is jettisoned, a water tank is more likely to turn over in a high wind if it is empty than if it is full. A light roof will lift off from wind suction where a heavy roof will not, requiring special pains to anchor the light roof to the building mass below.

For the most part, in learning to choose members in Part II of this book, we will define simple load cases, assuming they have emerged from prior analysis and evaluation. The case studies in Part III will offer the opportunity to examine more closely the origins of design loads and some of the ways they combine in testing structures. Section 16.3.5 extends the discussion here to explain the basis of loads in the case studies.

7.5 SAFETY FACTORS

7.5.1 Elastic Approach

Most of the work of later chapters is based on the approach in Part 1 of the AISC Specification, the historic *elastic* approach to steel design. Safety is sought by limiting the elastic stresses permitted to values considerably below the elastic limit of the steel. One form of allowable stress prescribed by the AISC Specification for tension members is set at $0.6F_y$ (AISC 1.5.1.1), giving a safety factor of $1/0.6 = 1.67$. Where phenomena such as buckling or fatigue cause loss of effective strength, allowable stresses are reduced in a search for a factor of safety close to 1.67. Higher allowable stresses ($F_{\text{basic}} \times 1.33$) are permitted for short-duration loads such as those from wind or earthquake. Due to delay in structure response to transitory loads, the apparent 25% reduction of safety factor does not actually represent a 25% loss of safety.

7.5.2 Plastic Approaches

Part 1 of the AISC Specification reflects the approach most generally used in current structural steel design. However, recognition of reserves of member strength beyond *first yield* has led to various moves to revise and one hopes refine the approach to safety. To some degree these are specific to steel structures. Some other changes parallel similar approaches to safety used with other materials. To account for the effects of extreme earthquakes, where "design" lateral force levels may be exceeded, designers often make their systems stable for strains far into the yield range of the ductile steel. This enhancement of safety is not covered in the AISC Specification.

A *plastic* approach is incorporated into the current AISC Specification as Part 2 of that document. A factor of safety is defined as the ratio of the load level that would cause collapse of a system to that which is permitted on the structure.

$$\text{factor of safety} = \frac{\text{collapse load}}{\text{permitted load}} > 1.0$$

This approach* starts from the recognition that, for flexure at least, the effective strength of a member is defined not by the moment, M_y, that causes yield stress in the outer fibers, but by the plastic moment, M_p, which may be 10% to 50% higher for different members and different bending axes. (As will be seen in later chapters, Part 1 of the Specification accounts for this phenomenon in part for particular conditions.) A safety factor of 1.7 is sought for usual loads and 1.3 for loads including wind or earthquake. All potential collapse mechanisms that could result from all possible load combinations must be identified. For axial loads, failure is defined similarly to Part 1.

The *load and resistance factor* approach is similar to the *plastic* approach, but replaces the constant factors of 1.7 and 1.3 with variable factors weighted for the degree of confidence in the magnitude of prescribed loads, the probability of their occurrence, and the evaluation of the frame resistance. Consistency in the real safety factor is sought by the approach described by Galambos[28] and has been used in a number of recent structures. The *load factor method*, which may be familiar to readers through study of reinforced concrete design methods, is somewhat similar to the *load and resistance factor* approach proposed for steel.

It is our conviction that the young designer, once having mastered one well-documented, reasonable approach to safety, can learn others with relative ease and can simultaneously consider the virtues of the alternatives. As of now, the most commonly used approach is that in Part 1 of the AISC Specification. However, the

*See references 8 and 11.

continuing attempts of the profession to improve design approaches point to exciting opportunities for young engineers in a dynamic profession.

7.6 COSTS

Chapter 3 included a limited discussion of costs of steel structures as they are seen by the designer. The reader planning to make design selections based on the methods of Part II would find it useful to review that discussion. For the most part, the procedures in Part II will seek to minimize costs by searching for the lowest weight in pounds/linear foot of members of constant cross section. In some instances, when variable cross sections are desirable, the effects of fabrication will be considered. Although this procedure does not always yield maximum economy, it seems the best compromise available when the total context is not known for each issue or problem addressed.

The case studies in Part III make it possible to restore context to some degree and extend the discussion of the effects of design choices on costs. As should be clear from Chapter 3, a fully adequate approach to costs is not possible in this or perhaps any single book.

7.7 FLOW CHARTS

The use of flow charts has been found by many to be a useful and systematic way of describing procedures in writing computer programs. We will use flow charts in a number of chapters of Parts II and III to describe systematic and, we believe, easily comprehensible ways to clarify complex problems of member selection or design development. The symbols and general approach resemble those found in a number of computer science texts. The charts used here are not presented in detailed enough form to translate directly into computer programs. Their use here is to assist the reader in digesting the ideas presented in the book.

The sequence of reasoning is followed by the direction of arrows connecting various boxes. Boxes used and their significance are as follows:

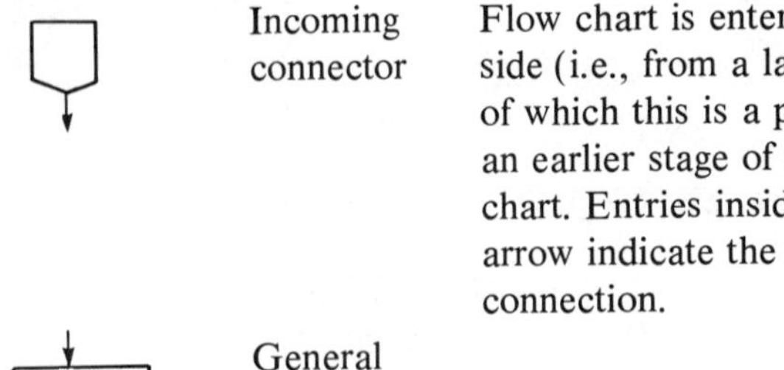

Incoming connector — Flow chart is entered from outside (i.e., from a larger problem of which this is a part) or from an earlier stage of this flow chart. Entries inside the open arrow indicate the nature of the connection.

General title or problem statement

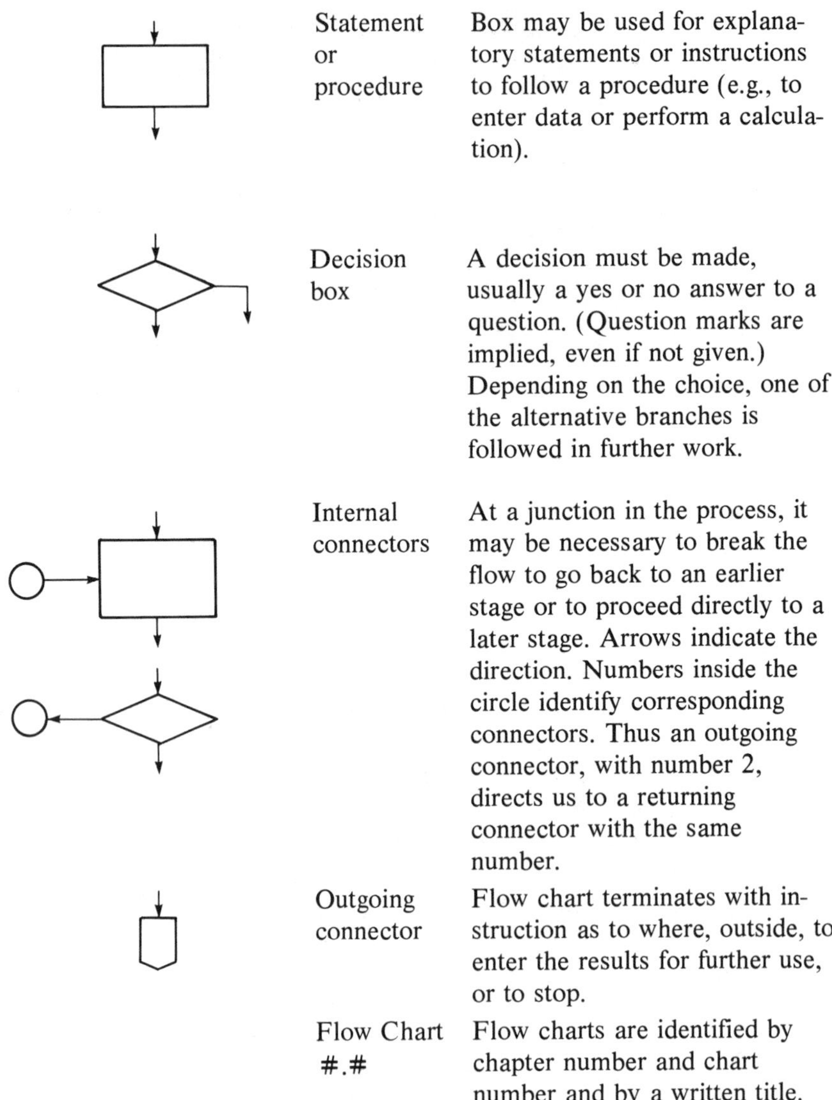

Symbol	Name	Description
	Statement or procedure	Box may be used for explanatory statements or instructions to follow a procedure (e.g., to enter data or perform a calculation).
	Decision box	A decision must be made, usually a yes or no answer to a question. (Question marks are implied, even if not given.) Depending on the choice, one of the alternative branches is followed in further work.
	Internal connectors	At a junction in the process, it may be necessary to break the flow to go back to an earlier stage or to proceed directly to a later stage. Arrows indicate the direction. Numbers inside the circle identify corresponding connectors. Thus an outgoing connector, with number 2, directs us to a returning connector with the same number.
	Outgoing connector	Flow chart terminates with instruction as to where, outside, to enter the results for further use, or to stop.
	Flow Chart #.#	Flow charts are identified by chapter number and chart number and by a written title.

7.8 REFERENCES TO AISC MANUAL AND SPECIFICATION

The chapters of Parts II and III include many references to the AISC Manual and Specification. We have, in many instances, shortened the references for less clumsy reading, and to avoid possible confusion between similar terms.

Where Manual appears by itself, the reference is to the eighth edition of the AISC Manual.

Where Specification or Spec. appears by itself, the reference is to the 1978 AISC Specification, which is included in the Manual in Part 5.

Where Commentary or AISC Commentary appears by itself, the reference is to the Commentary on the AISC Specification, 1978, which is also included in the Manual.

References to paragraphs or sections of the Specification are identified as "AISC 1.1.2," and so on, where the number is the identifying number within the Specification. All sections of Specification, Part 1, start with the number 1.

References to paragraphs or sections of the Commentary are similarly identified as "AISC Commentary 1.1.2," and so on.

If a reference is to "Manual, Part 2," and the like, it refers to one of the six parts of the Manual. References to parts of the Specification are given as "Specification, Part 1 or 2." Other references to "Parts" refer to the three parts of this book, unless otherwise qualified.

Numbers of formulas or equations in the Specification are given by section number and equation number. Where a number of a formula starts with a number other than 1 or 2, it identifies first the chapter number in this book, then the equation number. Thus, an equation in the Specification may be identified as AISC (1.3-2) or, simply (1.3-2). An equation discussed as part of the text may be (14-2), and so on.

In referring to steels whose properties are governed by ASTM Specifications, the complete reference requires the initials ASTM, the letter, and the number of the specification and the year of issue. We often omit both ASTM and the year. As in the usage of the AISC Specification, the most current issue is implied. Thus, "A441" is used to cover "ASTM A441-74."

8
Axially Loaded Tension Members

8.1 INTRODUCTION

We choose tension members as the first of the specific types to examine because they come closest to being "pure" in the sense that simple formulas can be used to relate load to stress and stress to allowable stress. It will be seen very quickly that even the tension problem is seldom pure. Procedures for design selection must recognize deviations from the ideal case.

8.2 AXIAL TENSION

Most of our readers will have witnessed the pure case. A highly machined, regular, and uniform specimen is carefully clamped into a laboratory testing device and subjected to successive increments of load and strain slowly applied until, following a stress–strain path specific to the type of steel, it ultimately displays, first, reduction in section and then rupture. This procedure produces a clearly defined yield stress, F_y, for most structural steels. Readers will recognize that axial tension implies a combination of shear and tension on planes at an angle to the direction of applied load. Rupture initiates at tensile stresses above yield, with cleavage on planes of high shear stress.

If we define the yield stress as the measure of member failure, which for many purposes* it truly is, then setting allowable tensile stresses F_t at $0.60F_y$,† we have a safety factor of $1/0.60 = 1.67$. If, on the other hand, we are concerned about what reserve we have against rupture, we must compare the member stress to the minimum guaranteed ultimate tensile stress, F_u. Table 8.1 lists, for usual structural steels, both the yield stress, F_y, and the minimum ultimate tensile stress, F_u. The ratio $(F_u)_{min}/F_y$ varies over a very wide range. For very thick bars of ASTM A36 with $F_y = 32$ ksi, $F_u/F_y = 1.81$; for ASTM A514 steel, it is 1.10.

The capacity of a tension member may be determined just as much by the nature of its end connections as by the cross-sectional area at points along its length. Holes punched or drilled for connecting bolts or rivets reduce the area locally, effectively reducing the strength at this weakened point. Welded attachments cause notch effects, which magnify local stresses. Sudden changes in cross section, whether increases or reductions, magnify local stresses in similar manner. If loads are static on ductile steel, stresses tend to equalize after yield without harm. However, if a member is subjected to many cycles of varying load, notch effects may significantly shorten its life.

When the center of resistance of a member's end connection is not on the centroidal axis of the member area, bending moments result from axial load.

$$(F_t)_{max} > (F_t)_{ave} = \frac{T}{A}$$

Temperature changes lead to shortening or lengthening of a member if it is free to change length. If relative movement between the ends is inhibited by other members in the system, the inhibited strain may result in extremely high tension with cold temperatures or buckling from increased temperatures.

The balance of this chapter will examine the uses of tension members, considerations and procedures in their selection, and the related provisions of the AISC Specification.

8.3 INSTANCES OF USE: TENSION MEMBERS

Examine Fig. 8.1. A number of common cases are shown, which may lead to selection of members for their capacity in tension. The simplest one, Fig. 8.1(a), illustrates the general problem. A tensile load is applied at one end of a straight member and resisted at the other. Since both ends have pin connections on the centroidal axis of the member, transverse shear cannot result from end loads. All forces, internal and external, must act on the line joining the pins. The member is axially loaded in tension. The member is free at the lower end to accommodate itself

*Elongation in the yield range may make a member unserviceable, even if the tensile stress is not enough to cause rupture.

†As in AISC Specification 1.5.1.

TABLE 8.1 YIELD AND MINIMUM ULTIMATE STRENGTH OF STRUCTURAL STEELS USED IN AISC SPECIFICATION, 1978

Steel type	ASTM designation	F_y^* Min. ksi	F_u^* Min. ksi	$\frac{F_u}{F_y}$	$F_{tg}^{\dagger} = 0.60F_y$	$F_{tn}^{\dagger} = 0.50F_u$	$\frac{F_y}{F_{tg}}$	$\frac{F_u}{F_{te}}$	$\frac{F_u}{F_{tg}}$
Carbon	A36	32	58	1.81	19.2	29.0	1.67	2.0	3.02
		36	58	1.61	21.6	29.0	1.67	2.0	2.69
	A529	42	60	1.43	25.2	30.0	1.67	2.0	2.39
High strength, low alloy	A441	40	60	1.50	24.0	30.0	1.67	2.0	2.50
		42	63	1.50	25.2	31.5	1.67	2.0	2.50
		46	67	1.50	27.6	33.5	1.67	2.0	2.50
		50	70	1.40	30.0	35.0	1.67	2.0	2.34
	A572 Gr. 42	42	60	1.43	25.2	30.0	1.67	2.0	2.93
	Gr. 50	50	65	1.30	30.0	32.5	1.67	2.0	2.17
	Gr. 60	60	75	1.25	36.0	37.5	1.67	2.0	2.09
	Gr. 65	65	80	1.23	39.0	40.0	1.67	2.0	2.05
Corrosion resistant, high strength, low alloy	A242	42	63	1.50	25.2	31.5	1.67	2.0	2.50
		46	67	1.50	27.6	33.5	1.67	2.0	2.50
		50	70	1.40	30.0	35.0	1.67	2.0	2.34
	A588	42	63	1.50	25.2	31.5	1.67	2.0	2.50
		46	67	1.50	27.6	33.5	1.67	2.0	2.50
		50	70	1.40	30.0	35.0	1.67	2.0	2.34
Quenched and tempered alloy	A514	90	100	1.11	54.0	50††	1.67	2.0	1.85††
		100	110	1.10	60.0	55††	1.67	2.0	1.84††

*For any ASTM designation, F_y and F_u vary with thickness or size grouping. See AISC Manual, Part 1, Table 2.

†The apparently constant ratios of F_y/F_{tg} and F_u/F_{tn} do not result in constant factors of safety since F_{tg} is applied to the gross area, ignoring the weakening caused by holes in bolted connections (AISC 1.5.1.1); F_{tn} is applied to a net cross-sectional area modified by a coefficient reflecting the nature of stress transfer at the end connection. $A_e = C_t A_n$ (AISC 1.14.2.2.)

††Not applicable in accordance with AISC 1.5.1.1 since $0.6\ F_y > 0.5F_u$.

Adapted from AISC Manual, Table 1.

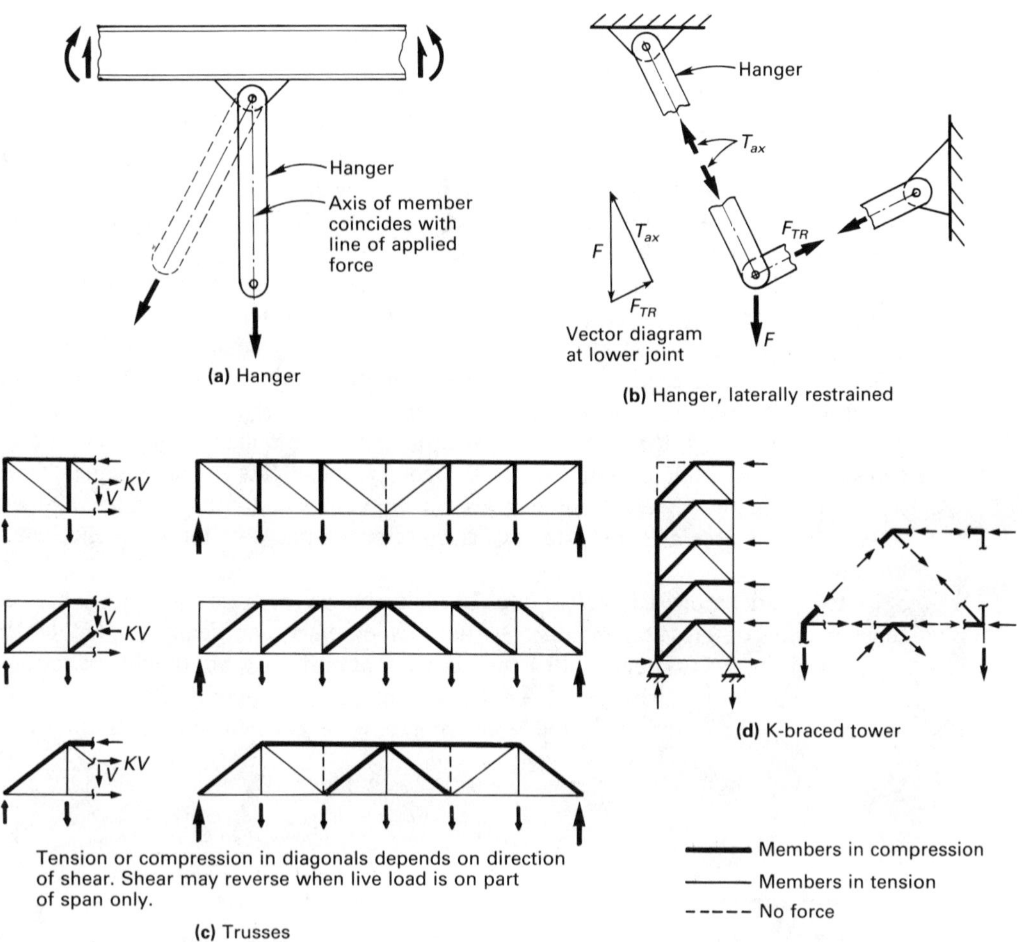

Figure 8.1 Use of tension and compression members in simple systems.

to any direction of force (i.e., to swing freely until the direction of its axis coincides with the direction of the resultant force). In Fig. 8.1(b), the force, F, is not axial with the member, and the member cannot rotate. However, F divides at the pin into two components, each of them causing an axial member force in a member. Figures 8.1(c) and (d) show a number of members, which are parts of larger structural systems. Tension members are easily identified for the load cases shown.

8.4 STRESS REVERSALS, MIXED USES, TEMPERATURE

Once again, examine Fig. 8.1. In some cases, it is clear that the member will always be in tension. If a floor or piece of equipment is suspended by hangers [Fig. 8.1(a)] and no other forces oppose gravity, the hanger will clearly be in tension. However, where tension arises as in Figs. 8.1(c) and (d), as a result of transverse loads on a truss or a system of bracing, there are often other combinations of loads that cause changes in the direction of system shear, and hence changes in the sign of the member force. A member subjected to tension from some load cases and compression from others must satisfy separately the criteria for tension members and the different criteria for compression members, in each case considering the maximum values of the load applicable to that set of criteria.

All the members shown in Fig. 8.1 are axially loaded from the effects of the force systems shown. Axially loaded members may also have to resist other forces causing shear and moment. Members under axial load and bending are discussed in Chapter 11. Examples of this chapter cover member selections for axial tension only. However, it should be recognized that all members resist vertical gravity forces acting on their own mass. If the member is not vertical, one component of gravity will cause transverse shear and bending. We will ignore these effects in this chapter on the assumption that self-weight bending effects are very small compared to axial load effects. For a design to be safe, an assumption of this sort must be checked by calculation.

Comparison of Figs. 8.2(a) and (b) highlights an important difference between tension and compression members. The axis of the eccentrically loaded tension member of Fig. 8.2(a) bends into a curved line that moves toward the line of action of the force, but cannot go beyond it. That of the compression member curves away from the line of action of the force. Deflection can only be stopped short of infinity if the

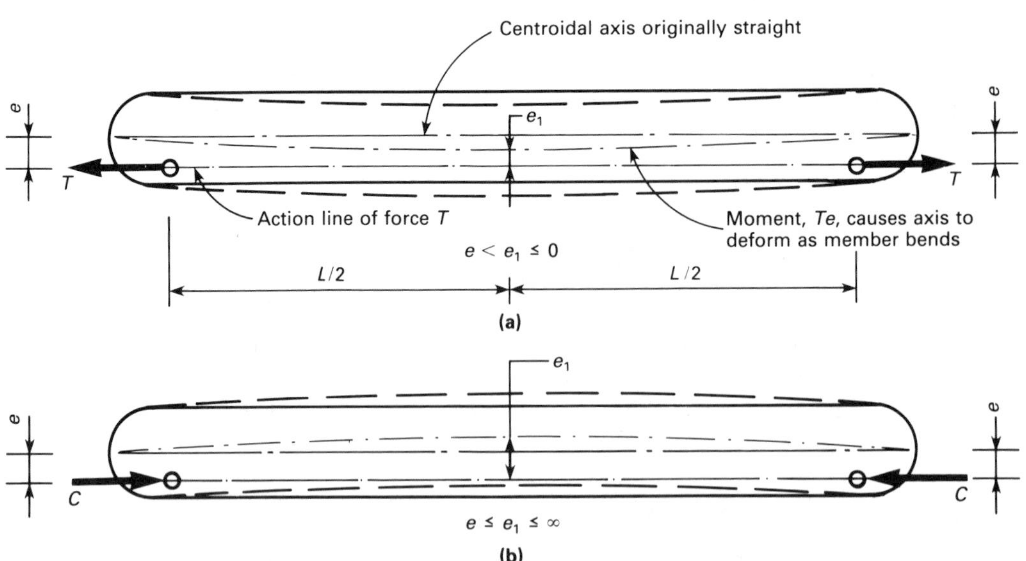

Figure 8.2 Effects of eccentric connections, tension and compression members.

member has sufficient bending resistance and stiffness. Tension members are self-stabilizing. Compression members tend to instability. A similar problem is illustrated in Fig. 8.3.

A small joint eccentricity is often neglected in tension members by relying on the self-stabilizing nature of tension members. A corollary effect, however, is that the maximum stress,

$$(f_t)_{\max} > \frac{T}{A} = (f_t)_{\text{ave}}$$

This may be considered acceptable for occasional or long-term loads. However, as will be seen later, it is not acceptable when loads are repetitive over many cycles, and the possibility of failure may hinge not only on the maximum stress but also on the range between minimum and maximum stress within a cycle. The AISC sets special requirements (e.g., AISC 1.15.3) for such cases.

Temperature change may create special problems in tension members with restrained ends. A member free to change length and subjected to a temperature change would change length by an amount $E_t \cdot L \cdot \Delta T$, where the coefficient of temperature expansion, E_t, is 6.5×10^{-6} in./in./°F. If the end joints prevent this change in length, the member experiences a change in stress, Δf, independent of length:

$$E_t \cdot \Delta T \cdot L - \Delta f \cdot \frac{L}{E} = 0$$

$$\Delta f = E_t \cdot \Delta T \cdot E = 0.1885 \text{ ksi/°F}$$

If change in length is fully prevented, a temperature change of 100°F causes stress of 18.85 ksi! Fortunately, the temperature range from ambient conditions is seldom ±100°F and full strain constraint at member ends seldom occurs. However, low temperature may cause significant increase in tensile stress. High temperature

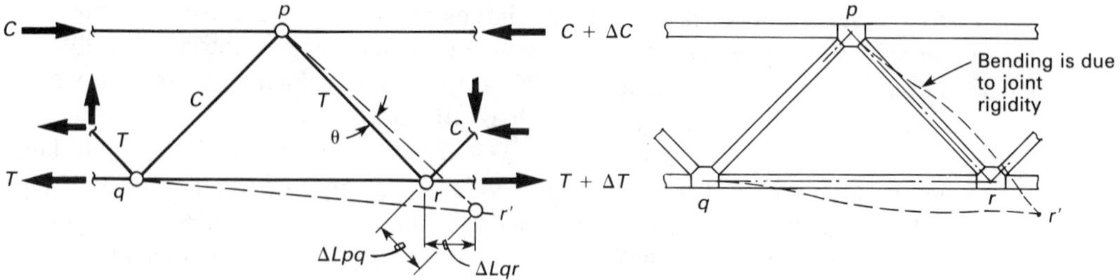

Figure 8.3 End moments in truss joints. Changes in length of members pq and qr require that r moves to r'. If ends are pinned, rotation through angle θ causes no stress. If joints resist rotation, end moments are *cranked in* at joints p, q, and r, forcing pr to bend. Magnitude of moments depends on stiffness of members.

reduces tensile stress, sometimes actually causing compression buckling in flexible members designed for tension. Many designers have abandoned the use of flexible tension members in cases where high-temperature buckling, even if not unsafe, is unsettling to users.

8.5 PROCEDURE FOR MEMBER SELECTION

Flow Chart 8.1 describes a logical procedure for the selection of a tension member. If axial tension loads control the selection, it should be sufficient. If there are other design loads on the same member, the selection must satisfy the axial tension requirement, either coincident with other loads or separately, depending on the results of analysis.

Note that there are two main branches: static loads and cycling loads. The static path applies when, as in most buildings, significant changes in stress due to variation of loads are expected to occur infrequently. The cycling path is used when significant stress changes will be frequent enough to raise the possibility of fatigue failure. The AISC Specification, which is followed in the flow chart, makes the division at 20,000 cycles. Other specifications use other numbers.

A stress cycle is defined as a change in stress through a range (f_{sr}). Stress cycles do not have to occur at regular frequencies as they do with many machines. The total number of excursions determines the branch to be considered.

The flow chart is usually entered from data developed in a larger system design and the analysis of that system. If there are cycling loads, data should include the number of cycles. If the static branch applies, three stages lead to the minimum required area, A_R, and minimum radius of gyration, r. Procedures and criteria for each stage are discussed later. At this point, the designer determines whether or not a section is available that provides $r \geq r_{\min}$ and $A \geq A_R$. Unless there are other criteria dictating otherwise, A should be as close to A_R as reasonably possible.

If A_R is available in the steel and shape assumed earlier, a tentative selection is made. If not, the choice of steel and shape must be reconsidered. Assuming there are no additional load conditions other than tension that may set more stringent or different controls on the same member, the selection holds and is used in further work. If there are other types of load, the selection must satisfy the criteria applicable to all load combinations. It should do so with the smallest satisfactory cross-sectional area reasonably available in the steel and shape that apply.

The member weight should be inspected for consistency with the input data or earlier assumptions. This check may lead to a change of input data and a new iteration. The decision should be based on (1) the magnitude and effect of the difference between the weight found and that assumed, and (2) an evaluation of this change in the context of the dead loads assumed for system design.

The cycling load branch appears only slightly more complicated. However, the detailed procedures will be found in the following to be considerably more complex. Figure 8.4 shows three cases of maximum and minimum tension and the requirements

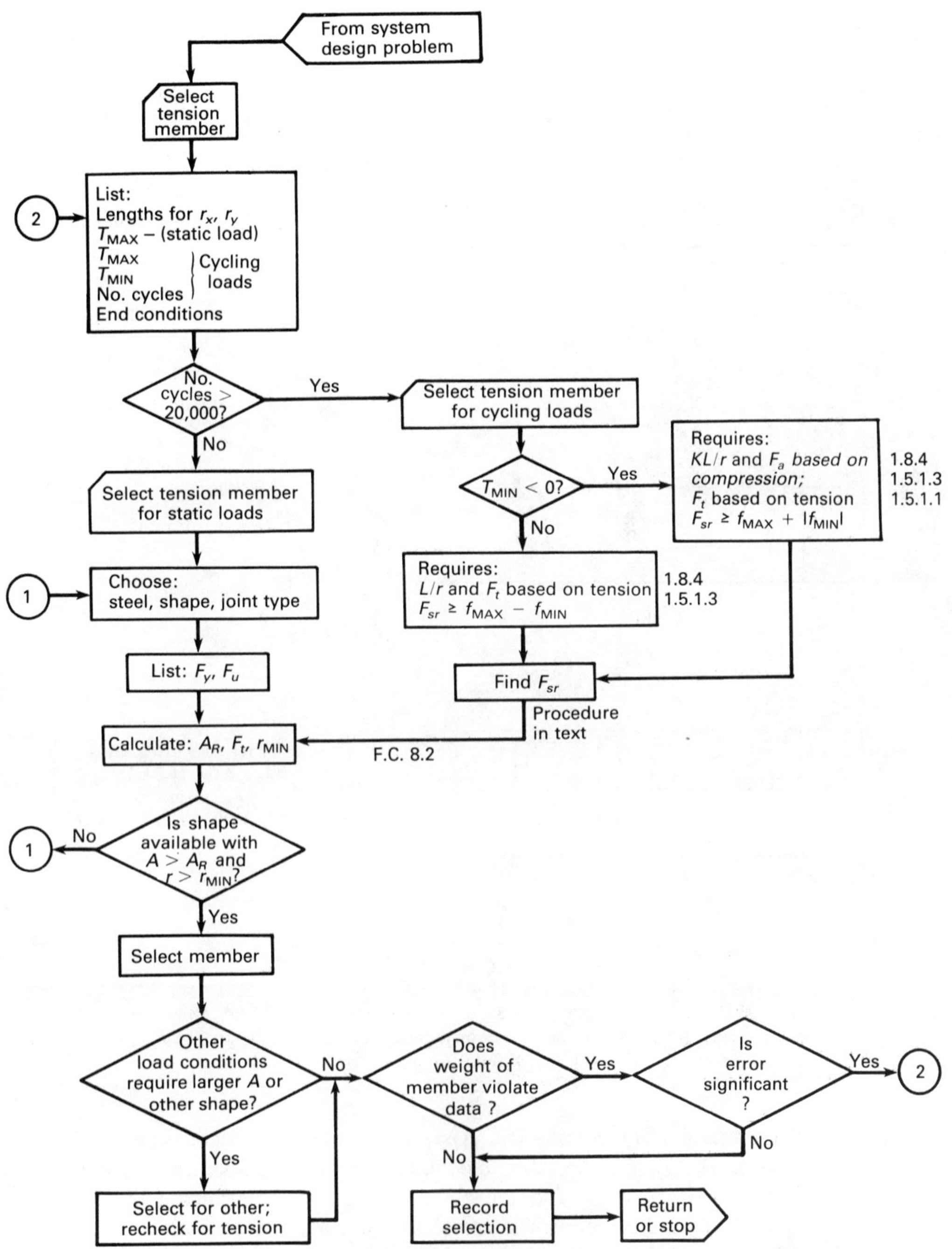

Flow Chart 8.1 Select tension member.

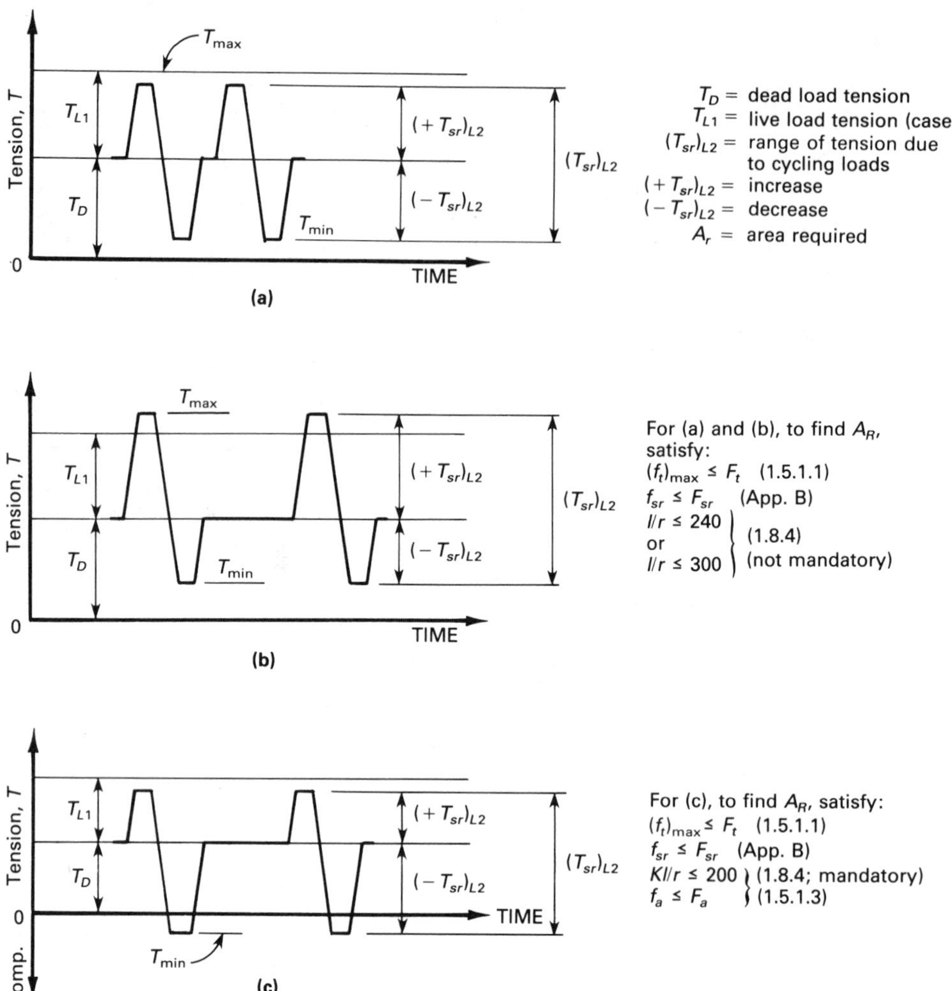

Figure 8.4 Static and cycling loads with dead load tension procedures per AISC Specification, 1978. $L1$ and $L2$ are different cases of live load. $L2$ occurs frequently; $L1$ occurs occasionally. (a) Static tension > maximum tension from cycling loads; (b) T_{max} results from cycling loads; $T_{min} > 0$; (c) $T_{min} < 0$ due to cycling loads.

that determine A_R. The use of AISC Appendix B for cycling loads is discussed later in this chapter. If $T_{min} < 0$ (i.e., T_{min} is compressive), the member must satisfy both the requirements set by T_{max} and those set by T_{min}. In determining the slenderness criterion, KL^*/r, the member is treated as a compression member even though $T_{max} \gg |T_{min}|$. Compression members are discussed in the next chapter.

*See Section 8.6.4 and Chapter 9.

8.6 AISC CRITERIA FOR SELECTION: STATIC OR MAXIMUM LOADS

8.6.1 Sections and Controls

Any type of steel cross section and any structural steel may be used for a tension member (see Figs. 7.1 and 7.2). The actual selection is usually determined by the total structural system, other members, and the nature of the end connections. The selection should satisfy several criteria:

1. Strength
2. Serviceability (deformation)
3. Stiffness

Table 8.2 lists sections of the AISC Specification giving specific requirements related to each of these criteria. The *Commentary* contains, under the same section headings, useful background explaining reasons for each. Some of these are also discussed here.

8.6.2 Allowable Stress, F_t, Area (A_R, A_g, A_n, A_e)*

AISC Section 1.5.1.1 governs allowable tensile stress, F_t, the limit of stress for an axially loaded member in tension. The allowable tension in such a member is

$$T_{\text{ALL}} = F_t A \tag{8-1}$$

A member to be selected for an axial tension, T, then requires a cross-sectional area,

$$A_R \geq \frac{T}{F_t} \tag{8-1a}$$

where the required area, A_R, may be either gross area or effective net area as applicable. For bolted or riveted members, selection of a member section requires consideration of the details of end connections. Connections are discussed in detail in Chapters 12 and 13. However, this chapter includes some discussion necessary to clarify the member selection process. Flow Chart 8.2 describes the selection procedure used in examples at the end of the chapter.

The first sentence of AISC 1.5.1.1, which applies to most tension members, sets two criteria for F_t, both of which must be met. When applied to the gross area of the cross section,

$$F_t = 0.60F_y \tag{8-2}$$

*When using symbols and formulas from the AISC Specification, use the dimensions in the specification; e.g., F_t is expressed in kips per square inch (ksi).

TABLE 8.2 SECTIONS OF AISC SPECIFICATION (1978) RELATING TO TENSION MEMBERS

			F_t		l/r	F_{sr}	
	Section	Subject	Strength criterion	Deformation criterion	Stiffness criterion	Cycling load criteria	Details
Primary selection criteria	1.5.1.1	Establish F_t	•	•			
	1.5.2	F_t, threaded rods	•	•			
	1.5.6	Adjust F_t, wind and seismic	•	•			
	1.8.4	Limit l/r			•		
	1.14.1 through 1.14.4	Define A_n, A_e	•				
	1.14.5	Pin-connected members (see also below)					•
	1.18.3	Built-up members			•		•
	App. B	Fatigue			•	•	•
Related considerations	1.5.1.5.1	Bearing stress F_p (pins)		•			•
	1.5.1.5.3	Bearing stress F_p (bolts and rivets)	•				•
	1.5.1.2.1	Shear stress F_v (pins)	•	•			•
	1.5.2	F_t and F_v (bolts and rivets)	•	•			•
	1.16	Rivets and bolts					•

leading to

$$A_g \geq \frac{T}{0.60F_y} \tag{8-3}$$

Alternatively, when applied to the effective net area,

$$F_t = 0.5F_u \tag{8-4}$$

$$A_e \geq \frac{T}{0.5F_u} \tag{8-5}$$

A_g is the gross area of the cross section and A_e the effective net area, both as defined in AISC 1.14. Formula (8-5), based on ultimate rupture, may be considered

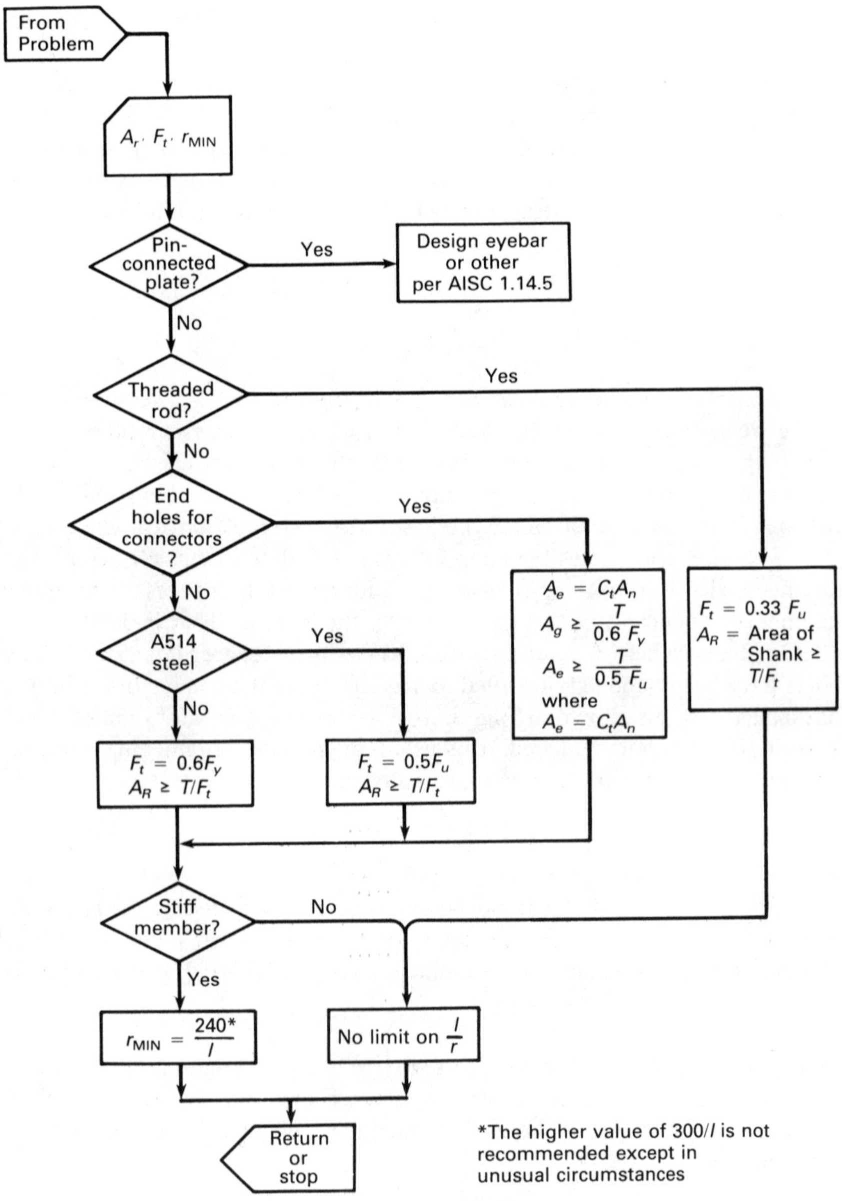

Flow Chart 8.2 Tension members: F_t, A_R, and r_{MIN} (does not include wire rope).

the criterion of strength, whereas formula (8-3) protects against the loss of serviceability that may arise from inelastic lengthening of the member. Inelastic stretching may endanger a system even if the member directly considered appears intact.

If end connections are welded and the area not otherwise reduced, $A_g = A_n = A_e$. Table 8.1 reveals that for all steels included under the AISC Specification except A514, $0.6F_y < 0.5F_u$. With that one exception, formula (8-3) governs the choice of minimum area for end-welded tension members. For example, with an end-welded tension member of A36 steel less than 8 in. thick, the smaller of the two choices controls.

$$F_t = 0.6F_y = 6(36) = 21.6 \text{ ksi} \quad \Leftarrow \text{controls}$$

or

$$F_t = 0.5F_u = 0.5(58) = 29.0.$$

For riveted and bolted members, application of Eq. (8-5) requires determination of the weakest cross section, which occurs where the area is reduced by drilling or punching holes for connectors. Figure 8.5 illustrates the search for the controlling section of reduced area, A_n, by an empirical method stipulated in AISC 1.14.2.1. The net area is the gross area minus the loss due to the most critical chain of n holes.

Figure 8.5(a) shows the simplest case of a plate in tension being connected to another plate with bolts or rivets. It will ultimately be necessary to determine the number of connectors necessary to transfer the load, T. That is delayed for Chapter 13. Our concern here is determination of the critical net area left to resist tension after holes have been punched or drilled to accommodate the connectors. For a connector of diameter, ϕ_c, as shown in Fig. 8.5(b), a hole is required of diameter, $\phi_H = \phi_c + 1/16$ in. If the hole is punched, as is usual, the process of punching removes a larger effective diameter than the nominal size of the hole.

$$\phi = \phi_c + \tfrac{1}{8} \text{ in.} \tag{8-6}$$

The loss of cross-sectional area per hole is ϕt.

Figures 8.5(c) and (d) show different possible chains of holes that could determine the critical net cross section for possible failure. When a chain of holes is straight and perpendicular to the member axis [Fig. 8.5(c)], it is easy to see that

$$A_n = A_g - \phi N t \tag{8-7}$$

In Fig. 8.5(d), the holes have been *staggered* by a longitudinal spacing, S, in an attempt to reduce the loss of cross section. There are now two possible chains of failure, one involving more holes than the other. On chain 2, the net area is, as in the previous case

$$A_n = A_g - \phi N_1 t \tag{8-7a}$$

Chain 3 includes N_2 holes, where $N_2 > N_1$, and is potentially weaker on that account. However, clearly the longer path along the chain replaces some of the loss. By an apparently arbitrary but long-tested criterion of AISC 1.14.2.1,

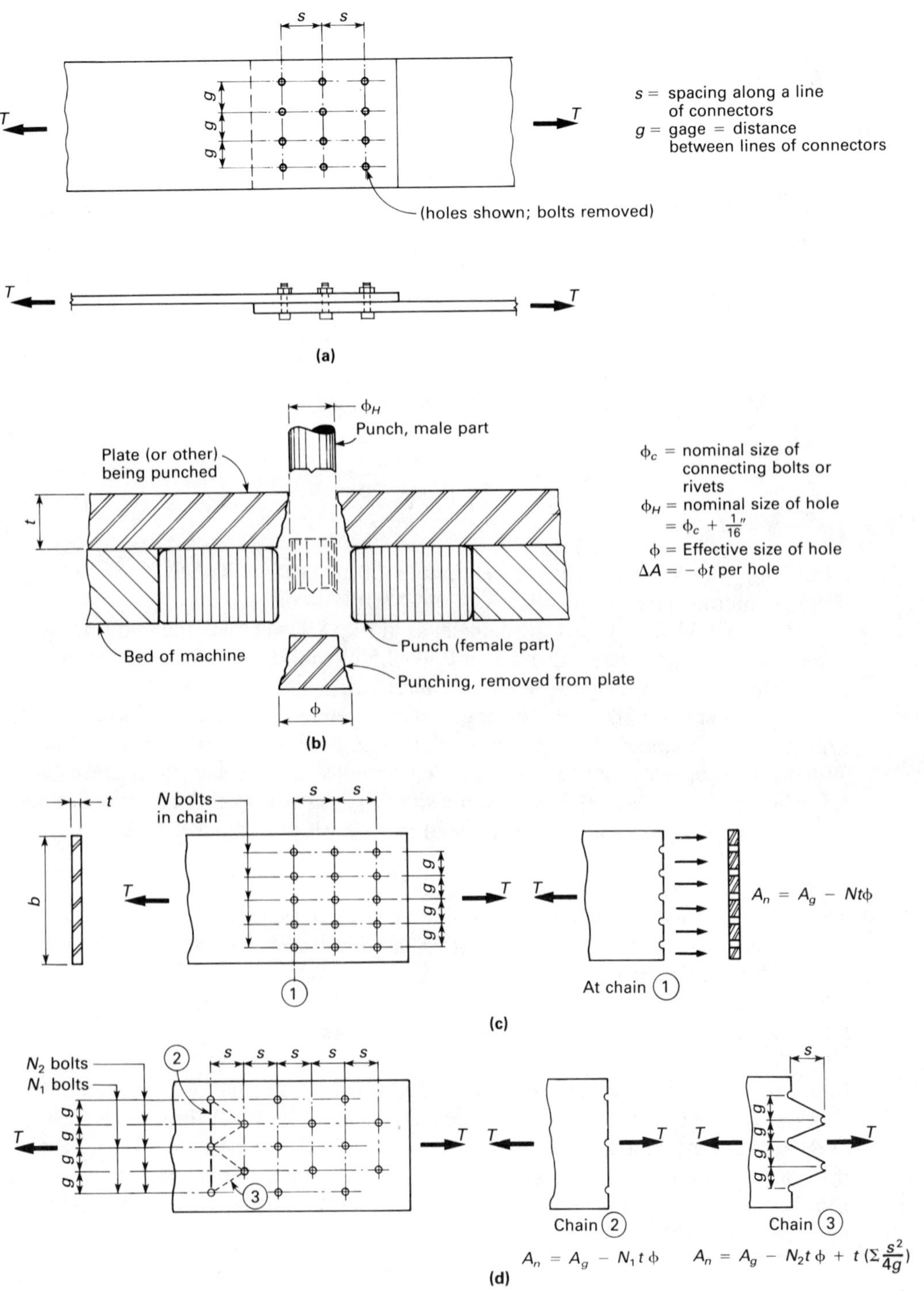

Figure 8.5 Net area A_n, plates in tension: (a) Connection of plates in tension; (b) loss of cross-section; (c) A_n (rows of bolts straight); (d) A_n (rows of bolts staggered).

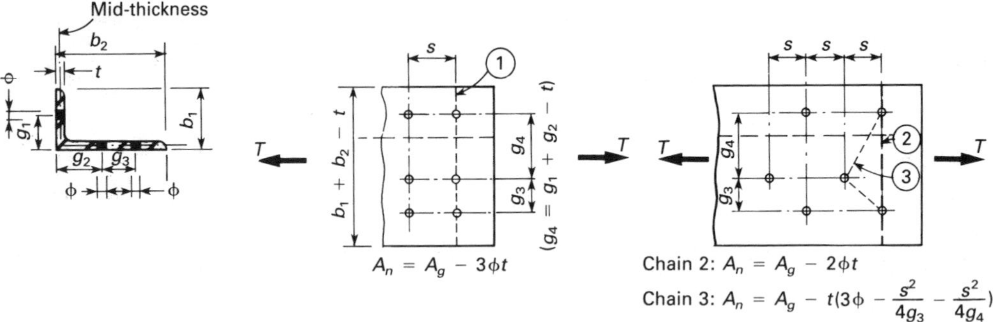

Figure 8.6 Determination of A_n for an angle.

$$A_n = A_g - t\left(\phi N_2 - \Sigma\frac{S^2}{4g}\right) \tag{8-7b}$$

If $(\Sigma S^2/4g > \phi(N_2 - N_1)$, chain 2 determines the net section. If $(\Sigma S^2/4g < \phi(N_2 - N_1)$, the control shifts to chain 3.

To apply AISC 1.14.2.1 to an angle, as in Fig. 8.6, "unfold" the mid-thickness of the angle to a flat surface: then, as in Figs. 8.5(c) and (d), A_n is determined by the applicable equation among (8-7), (8-7a), and (8-7b).

Determination of the net area does not completely measure the reduced tensile strength of the member. A further loss of strength occurs if the transfer of the load from the member to its connection takes place abruptly. Short connections cause poor stress flow and shear lag, effectively increasing $(f_t)_{max}$ as in a notch.* The effective net area, A_e, to be used in Eq. (8-5) is defined in AISC 1.14.2.2 and 1.14.2.3:

$$A_e = C_t A_n \tag{8-8}$$

where C_t is a reduction coefficient, varying between 0.9 and 0.75.

Procedures for determining A_n and A_e are illustrated numerically in Example 8.9.2, later in this chapter.

8.6.3 Threaded Parts and Pin-Connected Plates

In a threaded part in tension as at the threaded end of a round rod, a critical area exists at the root of the thread (see Fig. 8.7). Unlike Eq. (8-2) through (8-5), Eq. (8-9) applies a percentage of F_u to the gross area of cross section, $\pi d^2/4$, where d is the nominal diameter before cutting threads. A reduction of net area is assumed, based on the use of National Coarse threads. If deeper threads are cut, the equation must be modified.

*Readers have no doubt studied, in Strength of Materials, the magnifying effect of notches or holes on stresses.

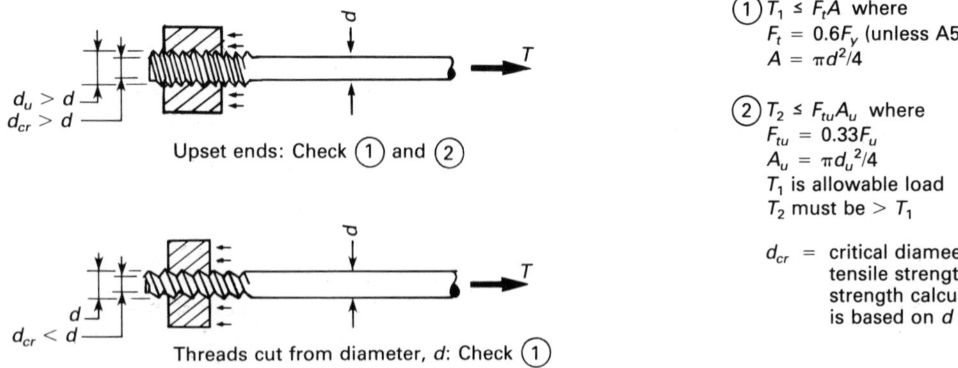

Figure 8.7 Threaded rods.

The allowable stress for this case may be found in AISC Table 1.5.2.1.

$$F_t = 0.33F_u \tag{8-9}$$

Figure 8.7 shows two cases of threaded rods, one where the threads are cut from a rod of constant diameter, the other where the threaded portion is *upset* to a larger diameter. Equation (8-9) applies in the former case. The latter case requires check of Eq. (8-9) and (8-3) on different cross sections. The requirement in Fig. 8.7(a), that $T_2 > T$ for upset rods, establishes a maximum load based on the serviceability criterion, F_y. To satisfy this requirement, it is necessary that

$$\frac{d_U}{d} \geq 1.35\sqrt{F_y/F_u} \tag{8-10}$$

Pin-connected members are treated separately (see Fig. 8.8). AISC 1.5.1.1 defines, for this case, a lower value of F_t than for bolted connections. Section 1.14.5, AISC, specifies special constraints on design. Pin connections avoid secondary stresses caused by rotational restraints in multiple-connector joints. However, there is only one connector and no fall-back if it fails. The large hole usually causes more drastic area reduction than small holes in bolted joints. The shear, normal, and principal stresses beyond the pin are quite complex. The design rules in Section 1.14.5 were derived empirically from tests and experience with the performance of structures.

There are two classes of pin-connected members, *eyebars* and others. The shape of the head in an eyebar is prescribed so as to promote optimum stress flow in the vicinity of the hole. The less favorable stress flow in the plates of Fig. 8.8(b) account for differences in required dimensions. Dimensional requirements are shown in Fig. 8.8 for both classes.

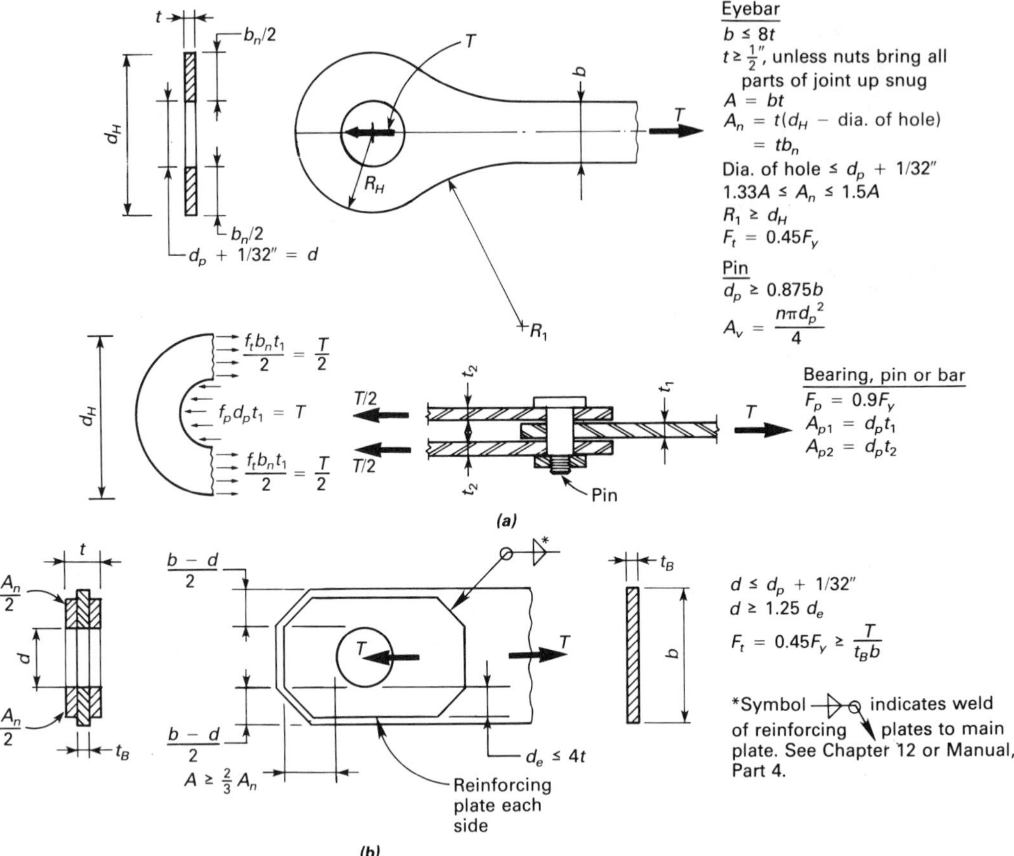

Figure 8.8 Pin-connected members in tension: $T \leq F_t A_n$, and $T \leq F_v A_v$, and $T \leq F_p A_p$. (a) Eyebar; (b) pin-connected plates.

In addition to tensile stress, selection requires consideration of the area in bearing between the pin and the bar, which is defined as A_p in Figure 8.8. To choose a pin diameter, one must also consider the shear stress in the pin, which is limited, by AISC 1.5.1.2.1 to

$$F_v = 0.4F_y \tag{8-11}$$

applied to either one or two times the cross-sectional area of the pin (see Figure 8.8). The allowable tension is the lowest of the values shown, limited by tensile stress in the bar, shear stress in the pin, or bearing stress between them.

In Chapter 13, we will find that bolt or rivet values are similarly governed by shear and bearing stresses.

8.6.4 Slenderness Ratio, l/r

Section 1.8.4, AISC, introduces the concept of slenderness ratio, which will be found to be very important for compression members. With respect to tension members, there is a nonmandatory but recommended maximum ratio:

$$\frac{l}{r} \leq 240 \text{ for main members}$$
$$\frac{l}{r} \leq 300 \text{ for bracing and secondary members} \tag{8-12}$$

Length, l, is the distance between points of lateral support (i.e., points along the member where it is prevented by external constraints from translating away from the axis joining the two ends). If there are no such constraints between the ends, l is equal to the length of the member, L. The use of l/r implies a significant value for the radius of gyration, r.

In our opinion, most bracing members are as important to the stability of structures as other *main* members. We are uncertain how to define *secondary* members, which concerns us little here, but very much in compression members. We will reserve further discussion of this problem for the next chapter. However, as of now, we recommend that, when in doubt, use criteria applying to main members.

Many satisfactory tension members, such as cables, square or round rods, or thin, wide bars, have trivial values of r. Such *flexible* members cannot be expected to fulfill Eq. (8–12) and are excluded from AISC 1.8.4. In Flow Chart 8.2, these are treated as nonstiff for l/r criteria.

Nonflexible members of significant r can be subject to objectionable vibration problems. Among other things, in joints not truly pinned this could lead to problems of fatigue. The authors recommend strict observance of the l/r criterion in such members.

A caution on l/r. A member may function primarily as a tension member, but under some unusual conditions experience small compressive forces. Even minor compression introduces the issue of possible failure by instability, and requires the more restrictive limit of $(KL/r) \leq 200$, which is found in AISC 1.8.4 and discussed more fully in the next chapter.

AISC 1.18.3 adds a requirement relating to the l/r of composite sections that will, as the slenderness matter itself, become more significant with respect to compression members. When a member is made up of two or more separate members acting together, the significant moment of inertia and radius of gyration is calculated as if they were a single member; they are calculated with respect to the principal axes of the combined section. For a member made up of n separate members,

$$I = \sum_{i=1}^{n} I_i + \Sigma A_i d_i^2$$

where I_i is the moment of inertia of the ith part about its axis parallel to the composite axis being considered, and d_i is the distance between the centroid of part i and that of the combined section. To enforce composite action, it is necessary to connect the members together so that longitudinal shear can be transferred between them. A limit of $l/r \leq 240$ is set for the individual members between such connections. In the next chapter we will see how such interconnections are often made. A more restrictive limit for interconnection of parts in contact assures that corrosion pockets do not develop due to small spacing between the parts.

8.7 CRITERIA FOR SELECTION, CYCLING LOADS, FATIGUE

When a member is subjected to a large number (>20,000 for the AISC Specification) of cycles of load change, protection against potential fatigue failure becomes a factor in member selection. The AISC Specification devotes Appendix B to the special design constraints that are set on cycling loads. Refer to Flow Chart 8.1 and Fig. 8.4. The required area A_R must satisfy Appendix B for the range of varying stress, in addition to the previously stated limits of maximum stress.

Investigation of the problem of cycling loads reveals the phenomenon of fatigue. When stress variations occur many times during the life of a member, failure may occur, even though the maximum stress from cycling loads is less than the failure stress for single instances of loading. The phenomenon is most important when the maximum stress is tensile. The actual stress variation may be between a negative (compression) stress and a positive (tension) stress or between two values of tensile stress. The number of cycles that lead to failure is more dependent on the stress range within the cycle than on the value of maximum stress.

It should be noted that seismic loads are of cycling nature, but occur through only a few cycles. In extreme earthquakes, if strains are high enough to cause strain hardening, a small number of cycles may be destructive.

Small stress variations below the endurance limit can be tolerated for an infinite number of cycles. Such small variations exist in all structures, causing no harmful effects. Large changes of stress occurring frequently require special consideration. The AISC Specification limits the allowable range of stress, F_{sr}, as a function of *loading condition* and *stress category*.

Figure 8.9 summarizes the procedure for determining F_{sr} in the formula

$$A_R \geq \frac{T_{\max} - T_{\min}}{F_{sr}} \tag{8-13}$$

The procedure is as follows:

1. Determine a *load condition* from 1 to 4 based on predicted number of cycles of varying load.
2. Choose a *stress category* from A to F based on the conditions and kind of stress listed in Table B2 and the illustrative examples in Fig. B1,* which are referenced in the table. Categories A through either E or F represent increasing

*Adapted here as Fig. 8.10.

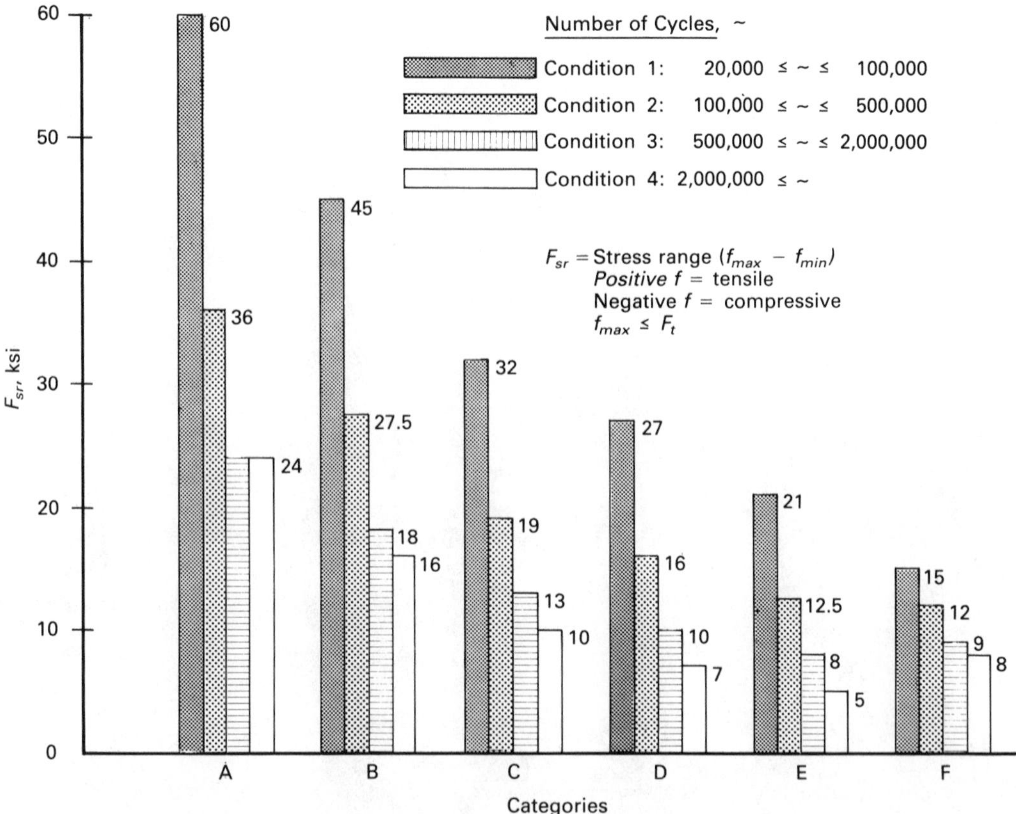

Figure 8.9 Allowable range of stress (F_{sr}) by loading condition and stress category, per AISC Specification, 1978, Table B3. Loading condition from Table B1; AISC stress categories from AISC Table B2 and AISC Fig. B1 (Fig. 8.10).

severity of notch effects, with category A applying to uniform or uniformly varying stress in ideally smooth, prismatic members.

3. Choose a limiting value of stress range, F_{sr}. Note that the specification does not intend that the designer apply a magnification factor to the stress, f_{sr}. The variation of stress magnification resulting from severity of notch is accounted for in the values of F_{sr}. The conditions shown in Fig. 8.10 represent levels of severity of notch effects.

$$F_{sr} = \frac{\text{range of load permitted for load condition and category}}{\text{area of member}} \tag{8-14}$$

Stress raisers may occur either at end connections or by attachments or section changes within the length of a member. The designer will soon discover that the 27

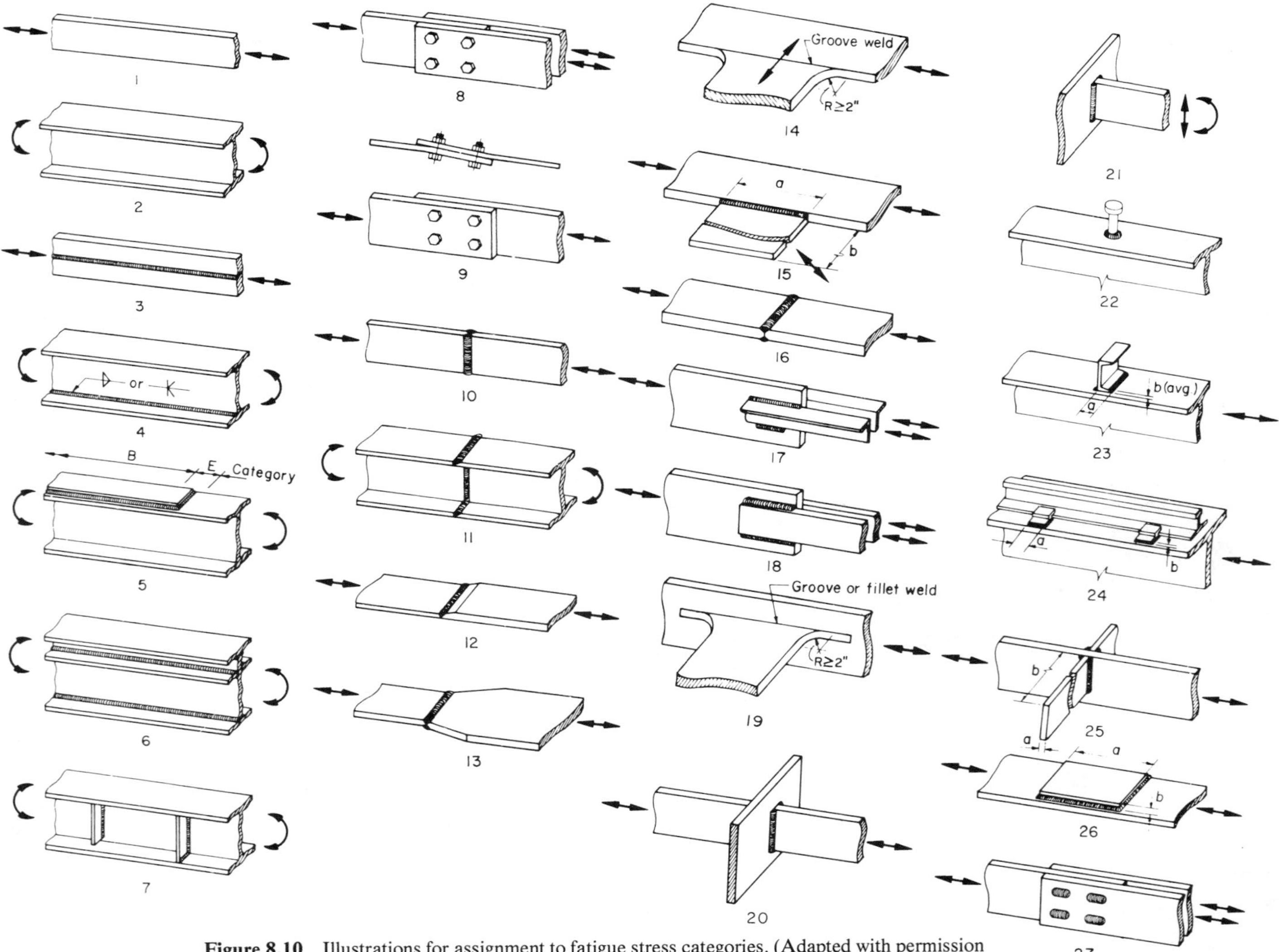

Figure 8.10 Illustrations for assignment to fatigue stress categories. (Adapted with permission from AISC Specification, Fig. B1)

cases shown in Fig. 8.10 do not cover all cases that arise. However, the effect of other cases may be approximated based on similarity of notch effects.

In Flow Chart 8.1, the calculation of the required area, A_R, requires finding the larger of that required for the maximum tension and for the range of stresses. The number of cycles of varying load may arise from factors of use that the designer cannot control. However, the stress category is often subject to designer control based on the care used in detailing. Particularly if, based on the stress category assumed, A_R is larger for cycling loads than for the maximum load, the designer may find it useful to improve the stress category by careful design of details.

8.8 WIRE ROPE

The cable, or wire rope, is suitable only for use as a tension member. Having essentially no stiffness ($I \to 0$; $r = \sqrt{I/A} \to 0$), a cable of any appreciable length is incapable of resisting compression. It can be bent around relatively small diameter sheaves or drums without encountering much flexural resistance or experiencing high bending stresses. This makes cables particularly suitable for hoisting equipment or suspension systems, in which the flexibility of the rope allows it to automatically assume the shape of an equilibrium system in tension. See Fig. 8.11. A chain has similar characteristics.

Wire ropes are used in many specialized structures such as suspension bridges (main suspension cable, hangers, and stays) and guyed towers (guys). They are sometimes used in buildings as in cable-suspended roofs. Several characteristics of cables are different from any of the other sections discussed in this book. We will point out some of these. However, we suggest that engineers using wire rope read further in cited references[14,64] and/or manufacturers' data.

A wire rope is made up of a number of small-diameter wires of very high strength steel ($F_u > 150$ ksi). Usually a group of wires is twisted together into a strand and a group of strands twisted together into a rope (Fig. 8.12). This results in a rope that can be handled, wound on reels, or otherwise bent into circular form without the individual wires separating. Manufacturers list a large variety of *constructions* of rope based on the number and size of wires in each strand and the number and arrangement of strands. Different constructions are desirable for different uses. Occasionally, as in the main cables of a large suspension bridge, the individual wires remain straight and parallel and are contained in place by wrapping.

The steels used in wire ropes are not among those identified as "approved for use" under the AISC Specification. They are, however, used commonly enough so that designers choose them with confidence. Since the details of the AISC Specification intended to provide rules for prudent use of steels do not apply, designers must apply their own rules. Information from manufacturers is helpful, as are precedents based on successful use and, where applicable, standards of ANSI.

There is no identifiable yield stress applicable to wire rope steels. Safe use is usually established by applying a factor of safety to the manufacturer's guaranteed ultimate strength of a cable of given construction, diameter, and steel. The safety

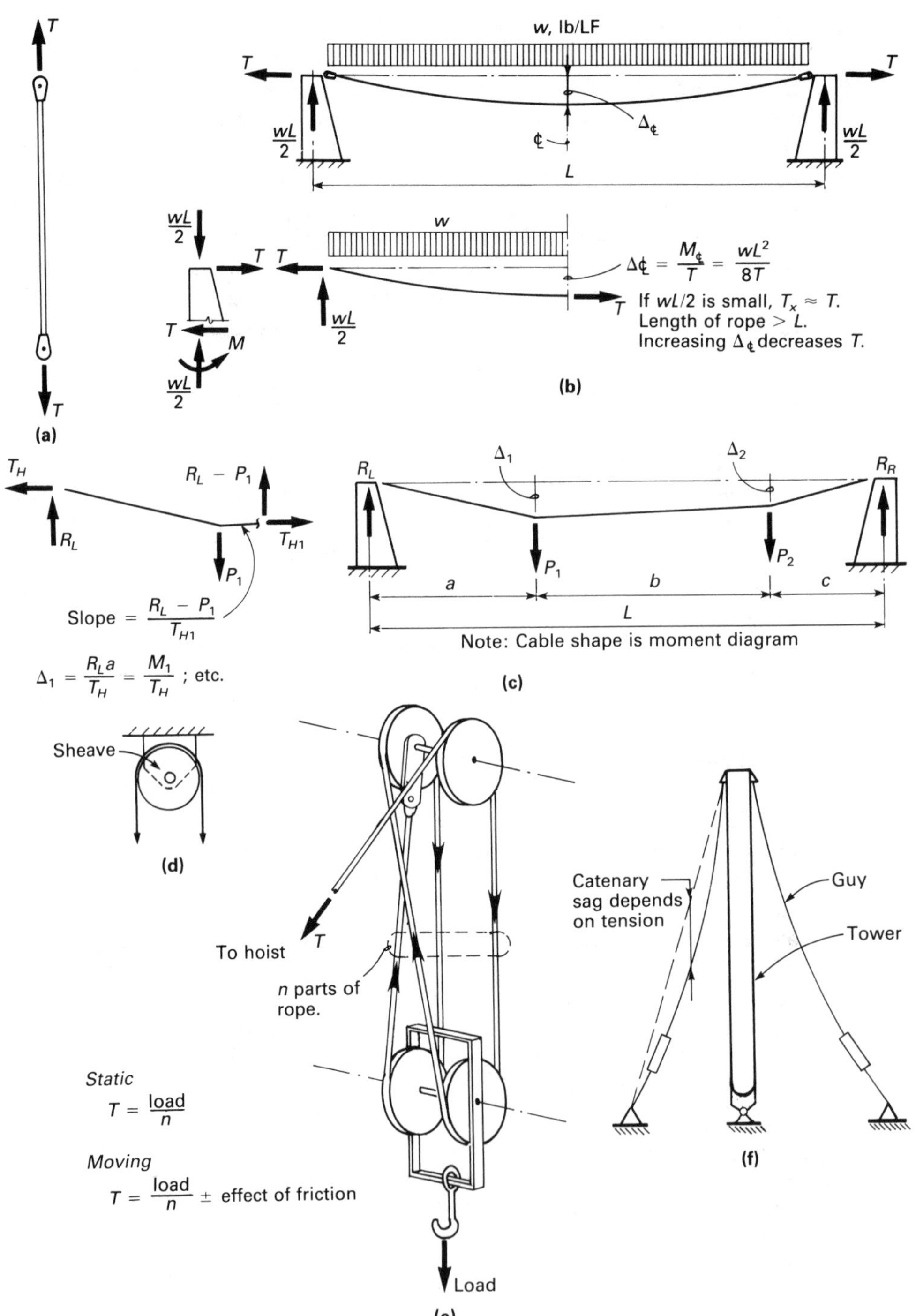

Figure 8.11 Cables in structural systems: (a) hanger; (b) horizontal rope under tension; (c) suspended system with concentrated loads; (d) sheave (pulley); (e) running rope for hoisting (schematic); (f) guyed tower.

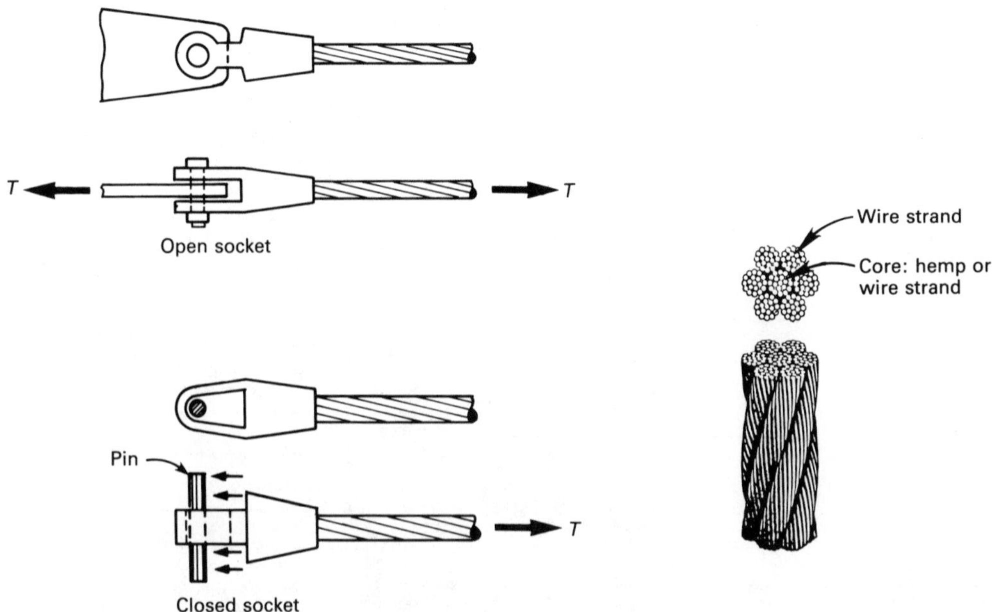

Figure 8.12 Wire rope (cable) with sockets.

factor will vary with the type of use, ranging up to 12, for example, on running ropes in elevators and as low as 3 in *standing* ropes, which are used straight, not bent around sheaves or drums. The actual strength of a rope will often decline over the period of use due to corrosion and, in the case of running ropes, due to mechanical working of wires when bending over sheaves. Actual factors of safety may thus vary over time and be somewhat less than those shown in the table for bright new rope. Corrosion protection is possible by galvanizing, but it may reduce strength as much as 10%. Because of the lack of pronounced yield stress or ductile performance in the stress–strain curve, stress magnification from kinks and notches can be particularly dangerous. However, a small number of individual wires can be broken with loss of only a small, proportionate part of the strength of a rope. When monitoring ropes used in construction equipment, the breaking of individual wires or kinking are seen as indicators of the need to replace a cable.

End fittings are supplied in many types for use with wire ropes. Attachment of the rope to adjacent members is usually by a pin passed through such an end fitting. The pin itself creates an ideal joint. However, only a limited number of types of end fittings can be attached in such a way as to *develop* 100% of the strength of rope (i.e., the breaking strength of the fitting and adjacent piece of rope $\geq$ the strength elsewhere in the rope). Sockets of types similar to those shown in Fig. 8.12, properly used, develop the full strength of the rope.

In twisted-strand construction, high tensile loads cause individual wires and strands to slip over adjacent ones, slightly realigning their positions in the cross

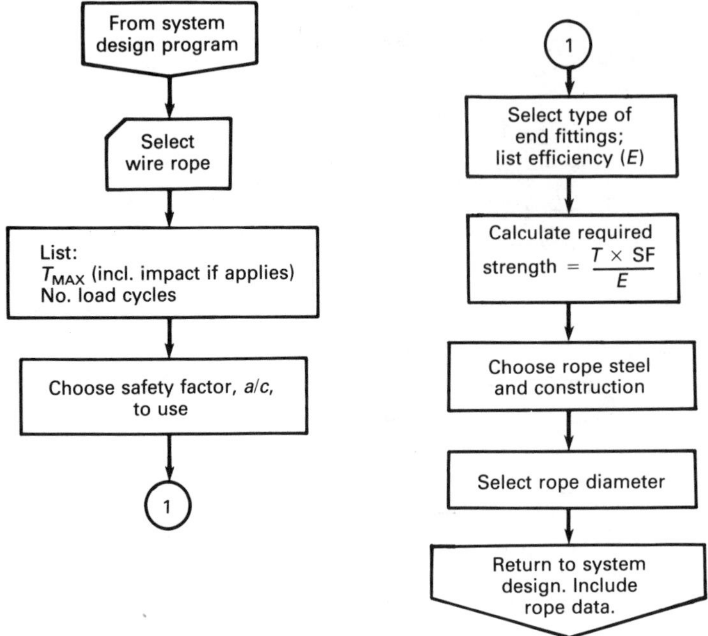

Flow Chart 8.3 Select wire rope.

section and causing the rope to lengthen. Thus a load–deflection curve will indicate an apparent "modulus" less than the Young's Modulus of the steel. The ratio

$$\frac{\text{load (ksi)}}{\text{stretch (in./in.)}}$$

for new wire rope may be as low as 16,000 ksi compared to Young's Modulus for steels of 29,000. The portion of stretching due to mechanical realignments is not recoverable after removal of load.

Flow Chart 8.3 for selection of a wire rope member differs somewhat from that for other tension members. For selection of the safety factor, the designer might seek guidance from authoritative bodies who commonly regulate the use of wire rope, such as the Army Corps of Engineers, ANSI, or a state division of industrial safety.

8.9 TENSION MEMBER EXAMPLES

8.9.1 Example: Tension Member Selection, Welded Ends

A tension diagonal in a truss such as illustrated in Fig. 8.1(c) has a maximum member load of 70 kips from dead + live loads, and 80 kips from dead + live + wind loads, and a length of 22′-0″. The members of the truss are being selected from commercially available angles, used in pairs and connected at ends with the help of ⅜″ gusset plates welded to each double angle. Steel being used is ASTM A36. Select a suitable section.

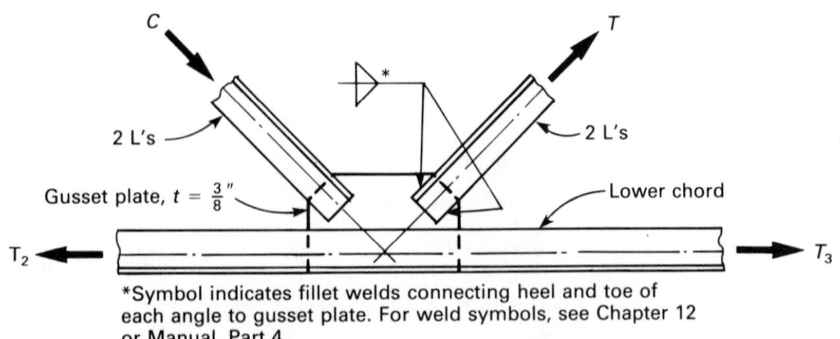

CALCULATIONS (Flow Chart 8.1)	*COMMENTS*
(1) Design conditions:	If not otherwise stated, the assumption of cycles < 20,000 is usual. The engineer responsible for analysis of the structure should know if this was a valid assumption.
$L = l = 22' = 0'' = 264''$	
cycles < 20,000	
$D + L = 70\ k$	AISC 1.5.6 permits F_t (1.33) for $D + L + W$, compared to F_t (1.0) for $D + L$. $D + L$ controls here.
$D + L + W = 80\ k < 1.33\ (70)$	
Steel ASTM A36: $F_y = 36$ ksi, $F_u \geq 58$ ksi	AISC Manual, Part 1, Table 1
⅃L section	
Ends welded to gusset plates.	
(2) Calculate A_R, F_t, $r_{\min}$:	Subroutine, Flow Chart 8.2
$F_t = 0.6F_y \rightarrow F_t = 0.6(36) = 21.6$ ksi	Use, since $A_g = A_n = A_e$
$F_t = 0.5F_u \rightarrow \quad 0.5(58) = 29.0$	
$A_g = A_n = A_e$	
$A_R = \dfrac{T}{F_t} = \dfrac{70}{21.6} = 3.24\ \text{in.}^2$	Welded ends; no holes Subscript R indicates "required"
$r_{\min} = \dfrac{l}{240} = \dfrac{264}{240} = 1.10\ \text{in.}$	

(3) Check available sizes in ⅃L 's.

Equal leg angles: ⅃L 3½ × 3½ × ¼

$A = 3.38$ in.$^2 > 3.24$

$r_x = 1.09 < 1.10$ NG

$r_y = 1.59$

AISC Manual, Part 1: Tables of Properties of Double Angle Sections. Selection for $A \geq A_R$ and both r_x and $r_y > r_{min}$.

Unequal leg angles:

⅂Г 4 × 3 × ¼: $A = 3.38$ in.^{2}

$r_x = 1.28$; $r_y = 1.29$ OK

Symbol ⅂Г indicates long legs back to back.

¬⌐ 4 × 3 × ¼: $A = 3.38$

$r_x = 0.896 < 1.10$ NG

$r_y = 1.92$

Symbol ¬⌐ indicates short legs back to back.

Use ⅂Г 4 × 3 × ¼.

⅂Г Selected for conditions given. Problem statement does not mention any other design loads.

The area requirement was satisfied by the three sections considered. Their weight per foot was identical. The r requirement was only satisfied by the 4 × 3 L's with long legs back to back. In general, for double angles, if $l_x = l_y$, the lightest section will consist of unequal legs, with long legs back to back, since r_y/r_x will be closest to 1.0 with that arrangement.

Section 1.18.3, AISC, requires that intermittent filler plates be connected to the two angles at maximum distances $l/r_z \leq 240$ for the individual angles. In this case, $r_z = 0.651$ in. for a single angle 4 × 3 × ¼. Fill plates are required at

$$\frac{240(0.651)}{12} = 13.02 \text{ ft max. spacing}$$

requiring one such plate in the 22-ft. length between gussets.

8.9.2 Example: Tension Member Selection, Bolted End Connections

The assembly of part (a) is to support a hanging (dead + live) load, P. Plate A is to be fabricated from ASTM A36 steel.

1. What is the permissible load, P, if holes are punched for ⅞-in.-diameter bolts according to the pattern of part (b) of the sketch?
2. What is the permissible load, P, if holes are punched for ⅞-in.-diameter bolts according to the pattern of part (c) of the sketch?

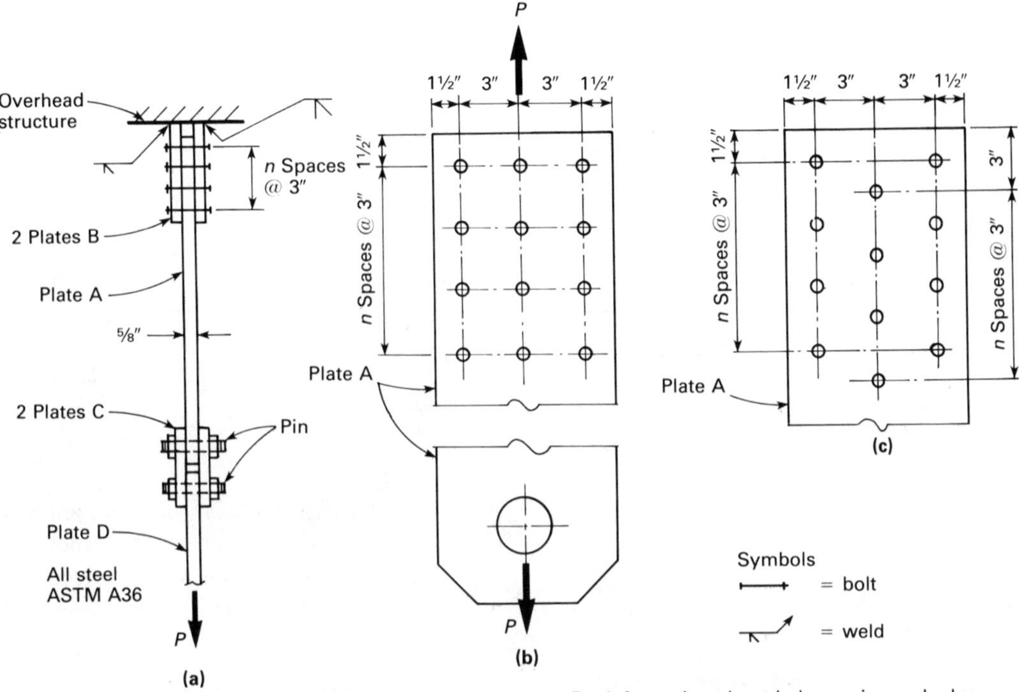

Notes: 1. For information about hole spacing and edge distances, see AISC Spec., Sec. 1.16
2. Connections are discussed in Chapters 12 and 13. Detailed knowledge about connectors is not necessary for this problem

It may be assumed that the other elements of the assembly can be made adequate for load P, as found here, and that the weight of the members of the assembly are included in load P.

$$n > 2 \text{ cycles of live load} \ll 20{,}000$$

CALCULATIONS	*COMMENTS*
(1) Steel: ASTM A36	
$F_u \geq 58$ ksi	
$F_y = 36$ ksi	
Hole dia. $= 1'' =$ connector $+ \frac{1}{8}'' = \phi$	See Fig. 8.5.

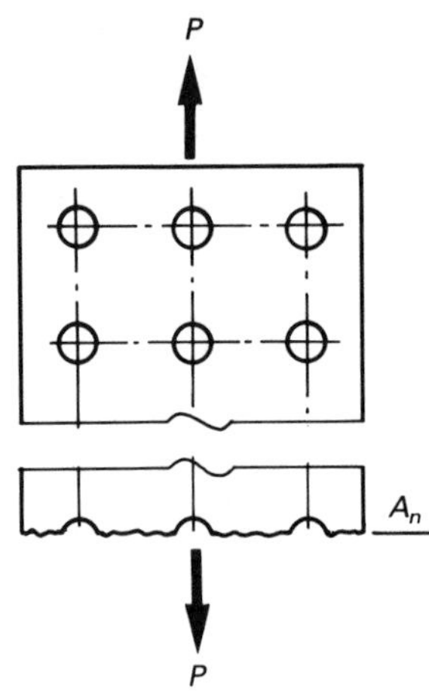

$A_n = t(b - 3\phi)$

$= 0.625[9 - 3(1)] = 3.75$ in.^{2}

$A_g = tb = 5.63$ in.^{2}

$A_e = 0.85A_n = 3.19$ in.^{2}

$P = 0.50F_u$

$A_e = 0.50(58)\ (3.19) = 92.5$ k

Critical areas, A_g and A_n, are cross-section areas. It is not necessary to know the value of n here.

AISC 1.14.2.2; $n > 2$

AISC 1.5.1.1

or

$P = 0.60F_y$,

$A_g = 0.60(36)\ (5.63) = 121.6$ k

Use $P_{ALL} = 92.5$.

Use smaller answer as allowable P.

(2)

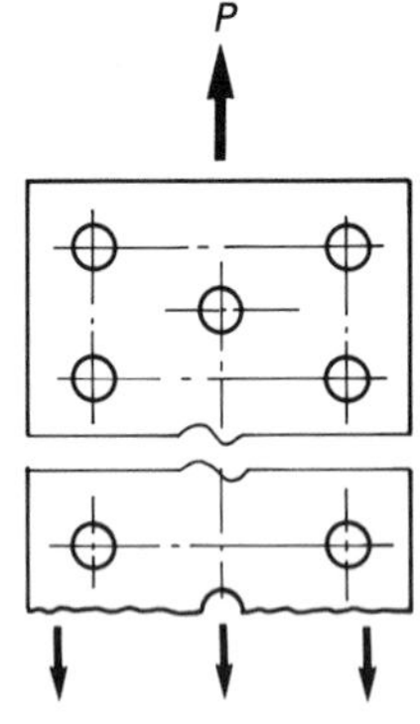

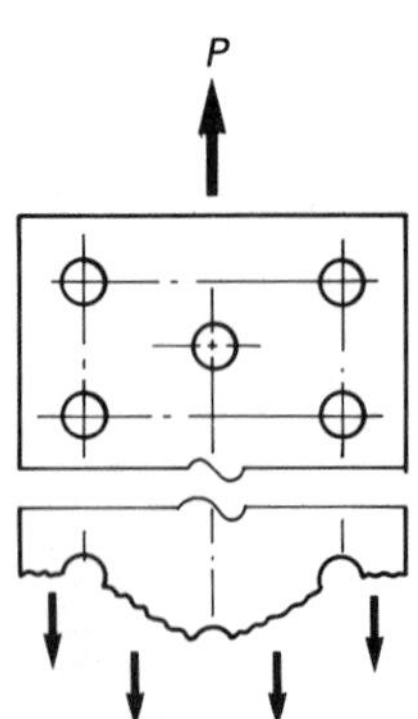

(2.1) $A_e = C_tA_n - C_t(b - N\phi)$

$= 0.85(0.625)[9 - 1(1)]$

$= 4.25 \text{ in.}^2$

$P = 0.50F_uA_e = 123.9$

Two possible chains of holes could determine possible failure. Use smaller of two A_e's. See Fig. 8.5(d).

(2.2) $A_e = C_tt\left(b - N\phi + 2\dfrac{s^2}{4g}\right)$

$= 0.85(0.625)\left[9 - 3(1) + (2)\dfrac{1.5^2}{4(3)}\right]$

$= 3.39 \text{ in.}^2 < 4.25$

$P = 0.50F_uA_e = 0.50(58)(3.39) = 98.3 \text{ k}$

$P = 0.60F_yA_g = 121.6$ — As before

Use $P_{ALL} = 98.3$ k. — (2.2) gives weakest chain.

Part (1) gave the smallest allowable load, which should be no surprise. In part (2), the capacity was increased by staggering the holes.

8.9.3 Example: Tension Member, Connected by Pin

For the hanger in Example 8.9.2, select a size and pin diameter for plate C. Show all dimensions at the end of the plate. Based on part (2.2), $P = 98$ kips. Do not use reinforcing plates. Do not use an eyebar. Steel in pin is A36.

CALCULATIONS	REMARKS
(1) Choose minimum diameter of pin, d_p, for shear:	
$P = 98 \text{ k} \leq 0.4F_yA_pn$	$n = 2$, since double shear
$98 \leq 0.4(36)\dfrac{(\pi)}{4}d_p^2(2);$	
$d_p = 2.08$ in.	

(2) Check edge distance:

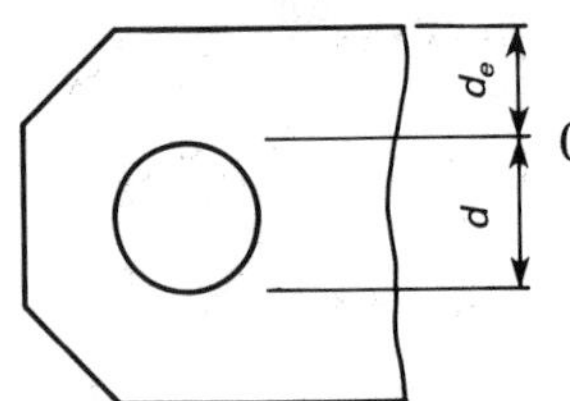

$d \geq 1.25d_e$

$0.5d + d_e = 4.5'' = 1.625d_e$ — AISC 1.14.5, para. 3

$d \geq 3.46''$; use 3.50

$d_p \geq 3.47'' > 2.08$ — $d_p = d - \frac{1}{32}''$

(3) Check tension, net section:

$$f_t = \frac{p}{A_n} = \frac{98}{0.625(9 - 3.50)}$$

AISC 1.5.1.1

$$A_n = 28.51 \text{ ksi} > 0.45F_y$$
$$= 16.20 \quad \text{NG}$$

Requires reinforcing plates; provide on each side for symmetry.

Add ΔA_n for $98 - 16.2(0.625)(5.5) = 42.3$ k:

$$\Delta A_n = \frac{42.3}{16.2} = 2.61 \text{ in.}^2$$

$$= 1.31 \text{ in.}^2/\text{side}$$

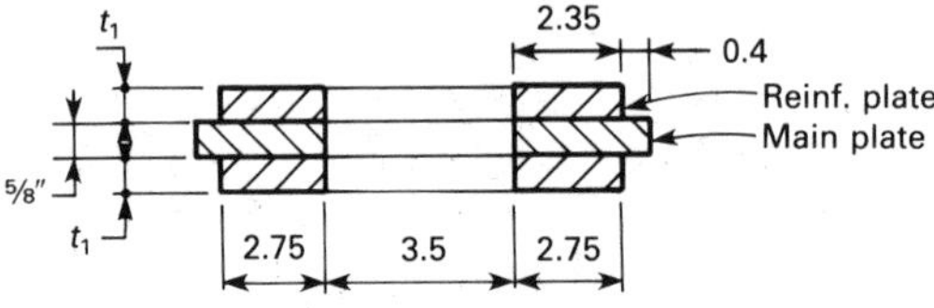

Reinforcing plate stops short of edge to allow room for welding (Chapter 12).

$$2.35t_1 = 1.31$$

$t_1 = 0.56$; try ⅝″

Thicknesses usually chosen in increments of $\frac{1}{16}''$.

(4) Check bearing stress:

$$f_p = \frac{P}{A_p} = \frac{98}{3(0.625)3.5}$$

$$= 14.93 \text{ ksi} \ll 0.9F_y = 32.4$$

AISC 1.5.1.5.1

(5) Details:

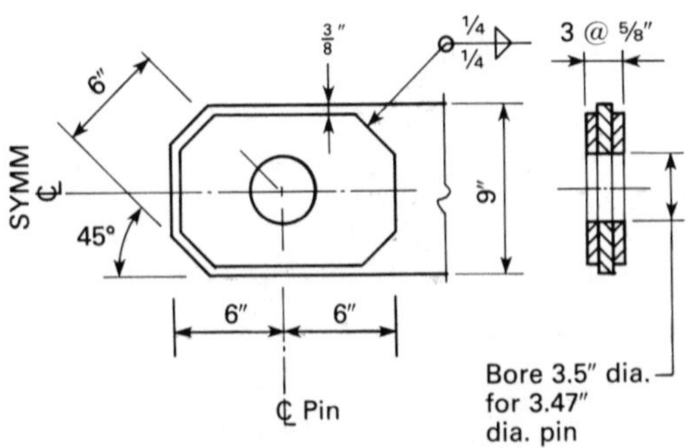

Note: A_n of section beyond pin $> \frac{2}{3}A_n$ of transverse section at pin. Size of fillet weld can be verified, when desired, by methods of Chapter 12. For other details, AISC 1.14.5.

The dimensions of the pin plate end emerge from "cook-book" rules of AISC 1.14.5, which have been validated by tests and usage.

A similar problem arises in the Stiffleg Derrick Case Study of Chapter 19. The solution can be found under the calculation for the *load bails* in that study.

8.9.4 Example: Eyebar Design

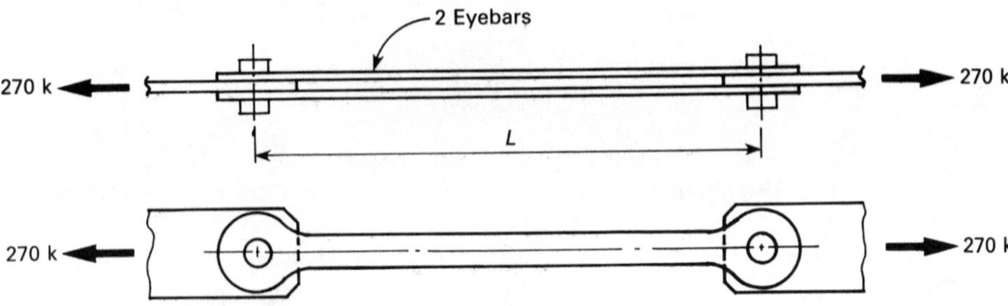

The eyebars shown are part of the *topping line* system of a derrick. They are to resist the total tension shown, which results from analysis of the worst design operating condition for the derrick. By design criteria, allowable stresses for operating conditions on the derrick are the same as those permitted by the AISC Design Specifications for Steel Buildings. Increases permitted by AISC 1.5.6 do not apply. Cycles of maximum load are predicted to be much less than 20,000. Design the eyebars, using steel with $F_y = 50{,}000$ psi.

The solution, based on the AISC Specification, is on Calculation Sheet R1 of the Stiffleg Derrick Case Study, Chapter 19. (Within Chapter 19, there is a Table of Contents listing calculation sheets by number.) Note the length, L, need not be known, since the l/r limitations of AISC 1.8.4 are not applicable; this is an example of a *flexible* member, as discussed in Section 8.6.4. Discussion of Calculation Sheet R1 is given in Section 19.6.4.

8.9.5 Example: Wire Rope in Tension

In place of the eyebars selected in Example 8.9.4, it is desired to consider the use of wire rope. Select the rope size and end fittings.

CALCULATIONS (See Flow Chart 8.3)	*REMARKS*
Use:	
Improved plow steel or better Galvanized rope, 6 × 19, construction safety factor ≥ 3.5 for "Standing Ropes" End socket efficiency = 100%	ANSI Standard 30.6 ANSI sets *minimum* safety factor = 3.0 based on manufacturer's guaranteed breaking strength of new rope. 6 × 19 rope has 6 twisted strands of 19 wires each.
required breaking strength = $\dfrac{\text{load (SF)}}{\text{efficiency (kips/ton)}} = \dfrac{270(3.5)}{1.0(2)} = 472.5$ tons	Manufacturers quote strength in tons of 2000 lb.
Select:	
Two ropes, 2⅜″ diameter Extra improved plow steel Independent wire rope core Galvanized Breaking strength = 0.9(2)(274) = 493 T > 472	Data are from information published by one manufacturer. May change from time to time. IWRC rope has an extra strand at the center. Galvanizing adds corrosion protection, but reduces rope strength by 10%; permitted by ANSI for standing ropes only.
Standard or *swaged* wire rope, closed sockets each end, each rope. Weight of two ropes = 2(10.4) = 20.8 lb/ft, plus end fittings	See Fig. 8.12. *Swaged* sockets are more compact than *standard*. Either is considered to have 100% efficiency; i.e., the socket and connection have strength at least equal to the guaranteed strength of the rope.

The ropes of this problem do the same job as the eyebars of the previous problem. The two solutions can be compared by weight. The two eyebars weigh

$$2\left(6 \times \frac{3}{4}\right) \frac{490}{144} = 30.6 \text{ lb/ft} + \text{eye ends}$$

The wire ropes would probably cost more per pound than eyebars. A cost per foot comparison would require price requests for the sections selected.

8.9.6 Example: Tension Member, Fatigue Possible

In Example 8.9.1, the 70-kip design load consists of 40 kips from dead load conditions plus 30 kips from live load. It is decided during the process of design to suspend from the truss a hoist that will cause 40 kips of tension in the member about 175,000 times during the *design life* of the structure. The designer considers it reasonable to reduce other live load effects to 20 kips when the hoist loads are applied. Size the member.

CALCULATIONS	*REMARKS*
(Similar to Flow Chart 8.1) Steel ASTM A36:	
$F_y = 36$ ksi	
$F_t = 21.6$ ksi	Example 8.9.1
$l/r \leq 240$	

New design loads:

	Case 1	Case 2	
D	40	40	
$L1$	30	20	
$L2$	0	40	← Controls fatigue
	70	100	← Control of T_{max}

Case 2 will control design for both T_{max} and fatigue.

$T_{min} \gg 0$

Select for T_{max}:

Not a fatigue criterion.

$$A_R = \frac{100}{21.6} = 4.63 \text{ in.}^2$$

$$r \geq \frac{22(12)}{240} = 1.10$$

	A	r_{min}	
]⌈ $4 \times 3 \times \frac{3}{8}$	4.97	1.26	
]⌈ $5 \times 3 \times \frac{5}{16}$	4.80	1.22	← Try

Double-angle shape tables, AISC Manual, Part 1; long legs back to back.

Check for fatigue, AISC App. B:

Table B1, AISC Spec. (or Fig. 8.9)

cycles = 175,000 → loading condition 2

End condition 17, Fig. B1, AISC → stress category E

Fig. B1, AISC (or Fig. 8.10)
Table B2, AISC Spec.

$$F_{sr} = 12.5 \text{ ksi}$$

Table B3, AISC Spec.

$$f_{sr} = \frac{40}{4.80} = 8.33 \text{ ksi} < 12.5$$

Use]⌈ $5 \times 3 \times \frac{5}{16}$.

Satisfies all criteria.

PROBLEMS

8.1. The conditions of Example 8.9.1 are the same except that the member is to be connected to the gusset plate with one line of 3/4-in.-diameter A325 bolts, each capable of transferring 15.5 kips in a double shear connection similar to that in the example. Size the member. (*Note:* It is not necessary to know how many bolts are required in order to select the member.)

8.2. Do Example 8.9.1, with the following changes:

(a) Use steel of ASTM A588.

(b) Use length of 15 ft and A36 steel.

(c) Use length of 15 ft and A588 steel.

(d) Original steel and length. Add 35-kip load from seismic effects.

8.3. In example 8.9.2, how could you arrange a staggered pattern of holes so that the allowable load would be 107.9 kips?

8.4. If the plate of Example 8.9.2, part (1), were steel conforming to ASTM A514, what would be the allowable load?

8.5. In the problem of Example 8.9.6, further investigation has led to the conclusion that the cycling load can be expected to occur at least 100 times per working day for the 50-year design life of the plant. Size the member.

8.6. Tension member *A* is connected to member *B* by a full penetration weld capable of transmitting the full strength of the smaller plate through the joint. The industrial building has been built and is in use. However, a reviewing agency responsible for industrial safety has found that a cycling load of ±40 kips can be expected at least 2.5×10^6 times during the life of the plant. They have ordered the plant shut down within a year unless remedial measures are taken. As a consulting engineer engaged by the plant owner for advice:

(a) Do you agree with the findings of the reviewing agency? Is the danger real? Support your conclusion.

(b) Can you offer a method to ameliorate the problem without changing the sizes of the members?

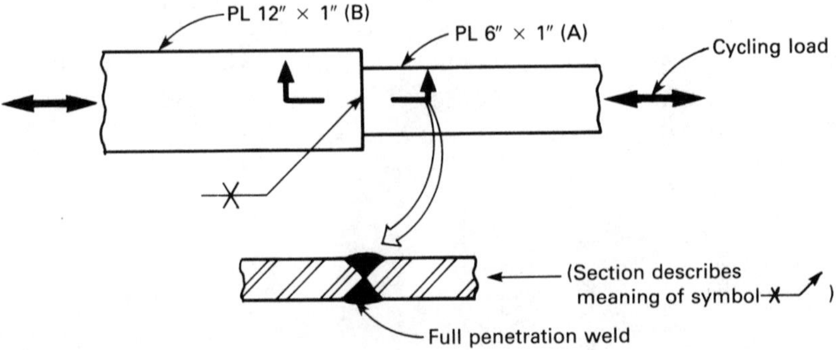

8.7. The Pratt Truss has been analyzed for the loads shown. Member loads are as shown on the truss. The truss is to be built with A36 steel of W shapes connected as shown. (The number of bolts is not sought at this time.) Member sizes should be selected from W6 or W8 series. A typical joint uses 3/4-in.-diameter bolts, minimum two per gage line.

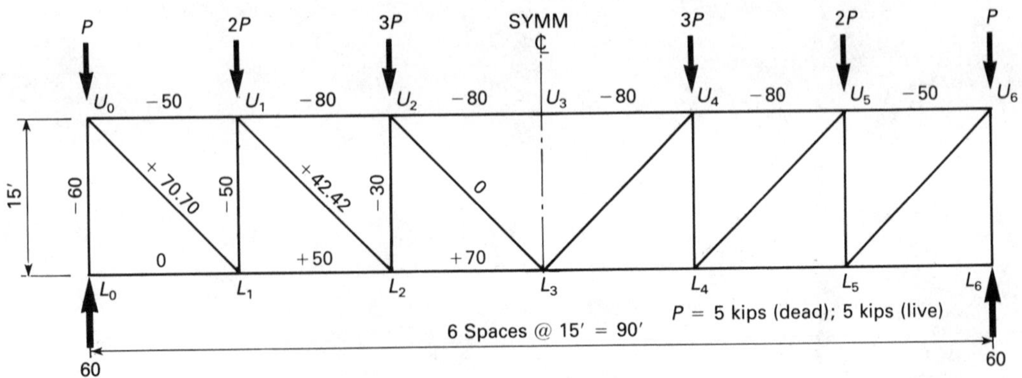

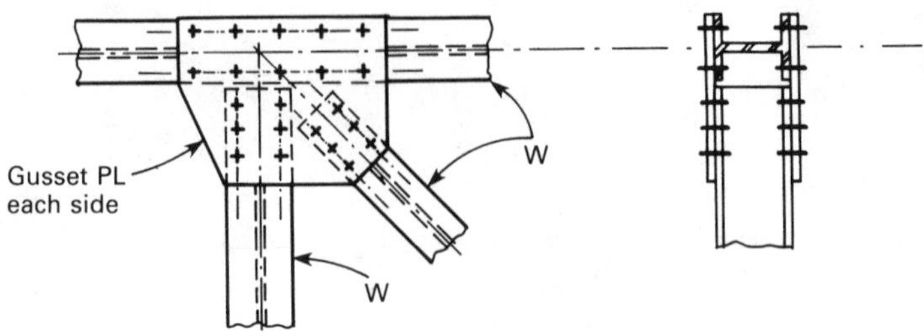

(a) Check the analysis.

(b) For the member loads as shown, choose member sizes for U_0L_1 and U_1L_2.

(c) Select a member size for U_2L_3. (*Hint:* Can you visualize a condition when the force in U_2L_3 would not be zero? In that case, would force be tensile or compressive? See AISC 1.8.4 and assume $K = 1.0$.)

(d) Select the size of a member continuous from L_0 to L_3 and L_3 to L_6. You may assume lateral (i.e., horizontal perpendicular to the truss) support at L_0, L_2, L_4, and L_6. Although continuous, the member may be assumed to be "pinned" at all points of vertical and lateral restraint. l/r may be based on length between "pins."

(e) May members L_0L_1 and L_5L_6 be removed? Why?

(f) May L_0L_1, be removed if L_5L_6 is retained? Why?

8.8. Do Problem 8.7 with welded end connections.

9 Axially Loaded Compression Members

9.1 INTRODUCTION: NATURE OF COMPRESSION FAILURE

9.1.1 Buckling Failures in the Real World

Perhaps the best way to illustrate the difference between compression and tension members is to describe briefly two construction failures that the author was called on to investigate. Leonhardt Euler, the Swiss physicist, who described compression buckling mathematically in the eighteenth century, would have had no difficulty predicting both.

The first instance was a building being built by the *lift slab* method, a construction technique by which floor slabs poured at ground level are suspended from the tops of the columns and raised into their final position by jacking in progressive stages. The design of the completed structure can be represented schematically by Fig. 9.1(a); the resistance to rotation and potential buckling mode are shown in Fig. 9.1(c). The case during lifting is shown diagrammatically in Figs. 9.1(b) and (d). According to observers, the slab drifted sideways about 3 inches when it was a few feet short of its final planned position. The columns were pulled back to plumb and lifting proceeded. A short time later, the slab drifted in the other direction, this time continuing to total collapse of the system.

Our calculations indicated that during lifting the columns were asked to sustain almost exactly the *Euler critical load* for a column pinned at its ends:

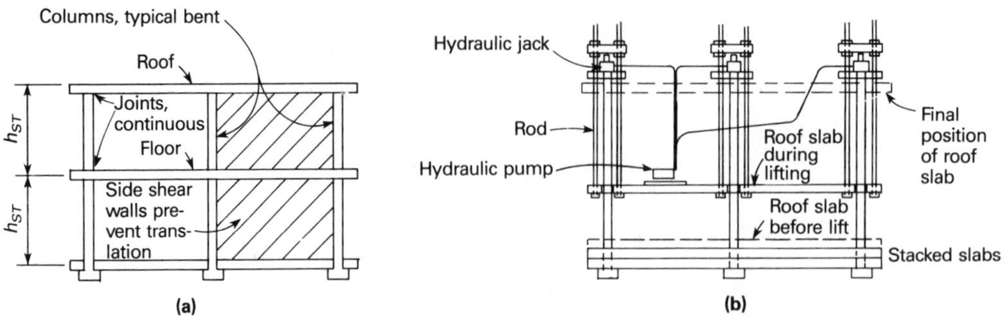

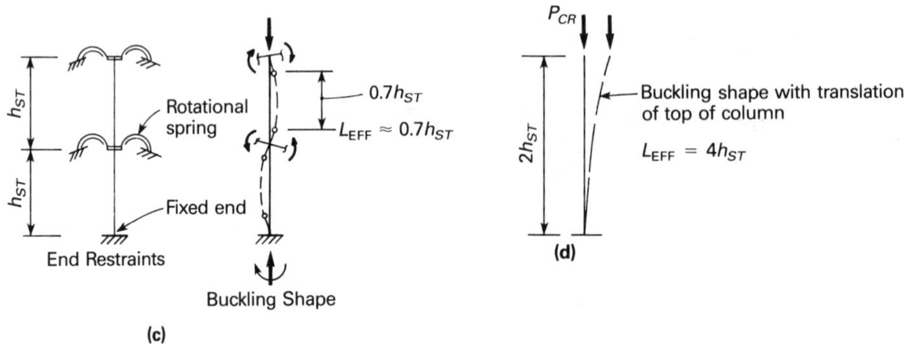

Figure 9.1 Column instability leads to construction failure: P_{CR} = load at which column can buckle. (a) Designed structure after completion; (b) condition during lifting; (c) column restraints, completed structure; (d) condition of column for lifting.

$$P_{cr} = \frac{\pi^2 EI}{L^2} \tag{9-1}$$

The columns did exactly what Euler's calculations and von Karman's later experiments indicated they would do—collapse.

In the second case, a boom was supplied by a manufacturer for a construction derrick. A boom is a compression member that must also resist some bending. The boom collapsed by buckling sideways the first time the user applied the full load for which it was "rated" by the manufacturer. A boom of the same cross section had served other users well when used for a boom length of 100 feet. The present user needed a length close to 200 feet, which had been supplied by simply adding identical pieces. The effective length was almost doubled. Our calculations, after the fact, indicated that the rated load caused compression almost exactly equal to P_{CR}. Once again, Euler's work was proven in a dramatic field test.

The primary difference between tension and compression action in steel members does not arise from differences in stress–strain relationships. The stress–strain curves of structural steels in compression are almost identical in the elastic and ductile regions to those for tension. The difference arises due to the proportions of the members and the phenomenon of buckling, which often takes place at stresses far below the yield stress. Euler demonstrated that the failure load for long, slender columns is governed by the parameter, EI/L_{eff}^2. E is a material property, the elastic modulus; I is a geometric property of the cross section; L_{eff} is the length of the column modified by a factor dependent on the nature of end restraints. The yield strength appears nowhere in this parameter. It only becomes significant when, for relatively short, stiff columns, the Euler equation leads to critical stress higher than yield, stress that the material cannot supply.

9.1.2 Eccentricity and Compression Failure

A second difference arises from the effect of bending moment when associated with axial load. We have already seen in Fig. 8.2 that axial tension tends to reduce member deflections caused by bending moment, thus tending to stabilize the member, while axial compression associated with end moment is destabilizing. Bending moment may arise in axially loaded members from a number of practical limitations on the manufacturing and fabrication processes (see Fig. 9.2). These moments are proportional to the applied axial loads and result in deflection of the member along its length. Such moments are destabilizing and reduce the available resistance to compressive stress.

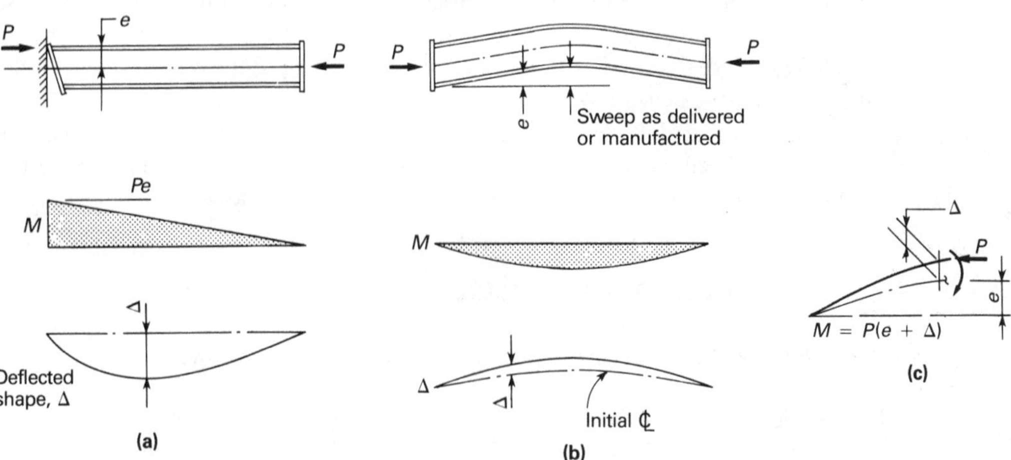

Figure 9.2 Eccentricity and $P\Delta$ moments: (a) axially applied load is resisted eccentrically; (b) initial member axis not straight; moment Pe causes added deflection; (c) partial free body.

9.2 SOURCES OF AXIAL COMPRESSION

The most easily recognizable compression member is a *column* in a building frame, which resists the gravity effects of dead and live loads on the various floors.* However, compression members are not limited to columns, as can be seen from Fig. 9.3. They may be oriented in any direction. They may resist loads that result from gravity or other forces acting in any direction.

In Fig. 9.3(a), the columns are carrying floor dead and live loads. Lateral forces are resisted by a bracing system that affects only those columns at the braced bay. There is little tendency by the system to bend the column. The columns are straight within usually accepted tolerances. The strut of Fig. 9.3(b) has been designed to prevent introduction of bending moment from end loads. In Fig. 9.3(c), the same is true, but gravity causes bending moment from the distributed weight of the member itself. Dead weight bending is often (not always) an insignificant effect compared to the axial loads. If the *strut* of Fig. 9.3(d) braces a wide excavation, the dead weight bending moment may be significant. The member in Fig. 9.3(e) occupies the entire space between two unyielding ends. Regardless of what its function may have been in the mind of the designer, increased temperature will cause it to act as a strut as the end supports inhibit the change in length that would otherwise occur from the change of temperature. In the *truss* and *trussed tower* of Figs. 9.3(f) and (g), members in compression are drawn with heavy lines to distinguish them from those in tension. Reversal of the forces would change the function of the members, interchanging compression and tension.

The *portal frame* in Fig. 9.3(h) has continuous joints and fixed base supports. The columns and the cross member are all beam columns. A *spandrel arch*, as shown in Fig. 9.3(j), can only be in pure compression if it has the shape of its moment diagram.† Since each load case has a different moment diagram, this is an impossible criterion for the usual steel arch, which must also resist bending moment. With tensile as well as compressive strength, steel can resist bending. But we will have to wait for later chapters to discuss how.

The methods here are sufficient for straight members resisting axial compression, where effects of bending are insignificantly small compared to axial loads. They also offer partial guidance in other, more complex cases.

9.3 FAILURE OF COMPRESSION MEMBERS

9.3.1 Failure Loads: Function of Cross Section and Length

In considering the failure of compression members, it is possible to define three different classes of columns, as in Fig. 9.4. They are defined in many texts on solid

*If its ends are continuously connected to other restraining members, it may also resist bending moment from dead and live loads. Combined effects are examined in Chapter 11.

†See Chapter 15.

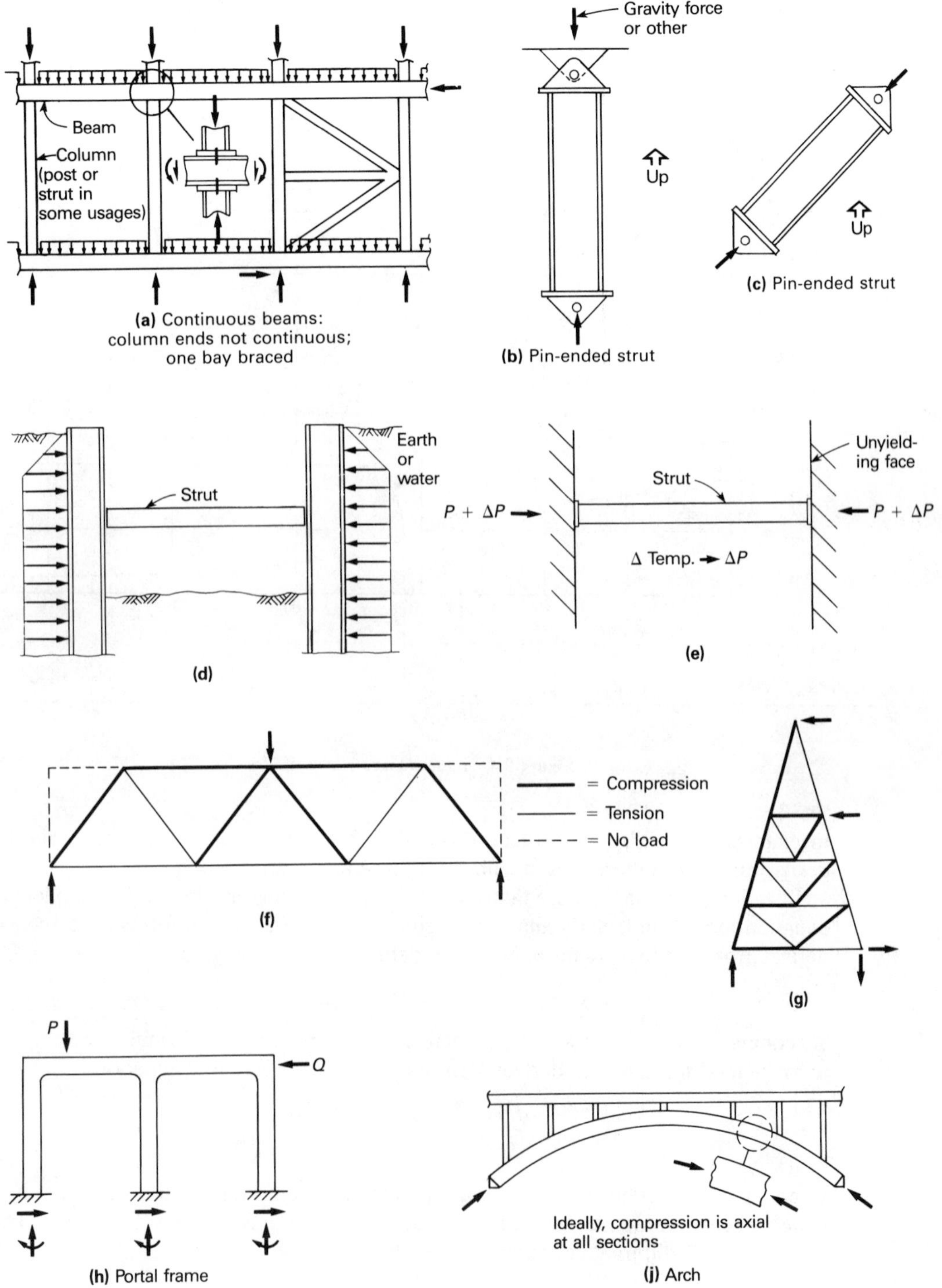

Figure 9.3 Sources of axial compression.

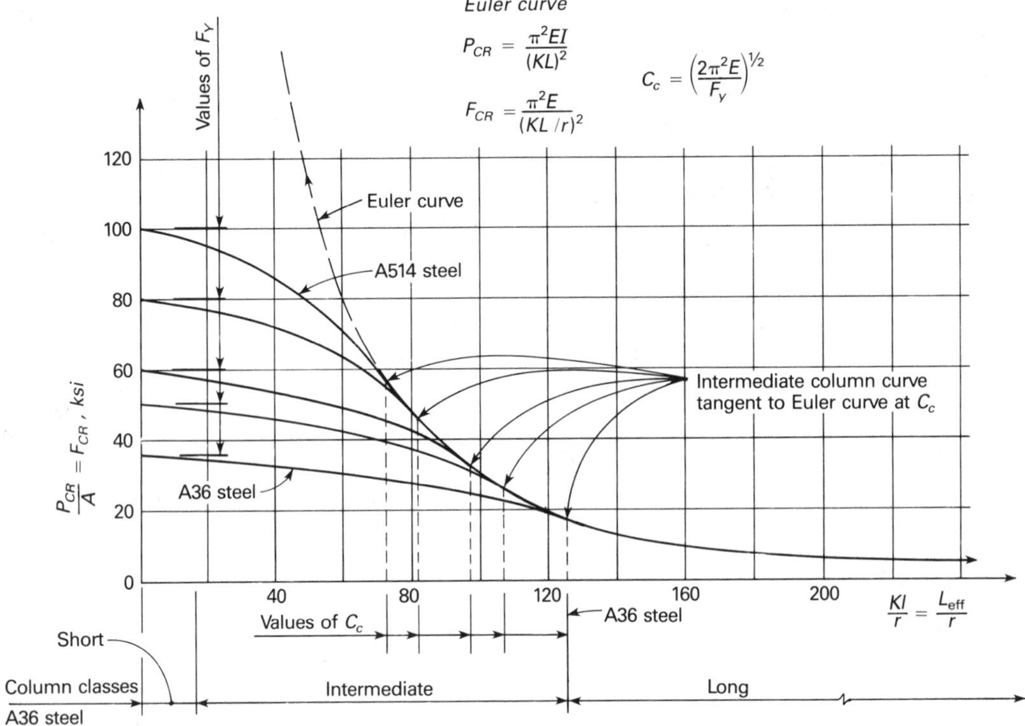

Figure 9.4 Failure loads versus KL/r.

mechanics[45,58] and by the Structural Stability Research Council[36], and underlie the design rules for compression members in the AISC Specification.

In long columns, failure takes place due to buckling instability. The applicable equation, based on Euler's analysis, is shown in Fig. 9.4 in two forms. The second derives from the first by introducing the definition of radius gyration,

$$r = \sqrt{I/A}$$

To compare L_{eff} to L, the length between ends, we first write the formula for P_{cr} in the more general form, which derives from its governing differential equation,

$$\frac{d^2\Delta}{dx^2} + \left(\frac{P_{cr}}{EI}\right)\Delta = 0$$

where Δ is the deflection of any point x along a column that is in equilibrium in a deflected condition under the action of the axial load, P_{cr}. See Popov[45] or others. The solution of the differential equation takes the form

$$P_{cr} = \frac{n^2\pi^2 EI}{L^2} \tag{9-2}$$

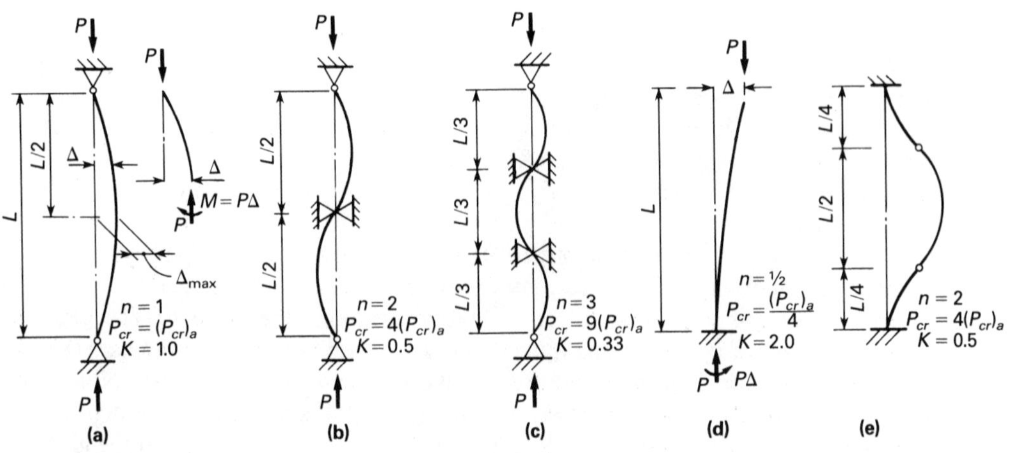

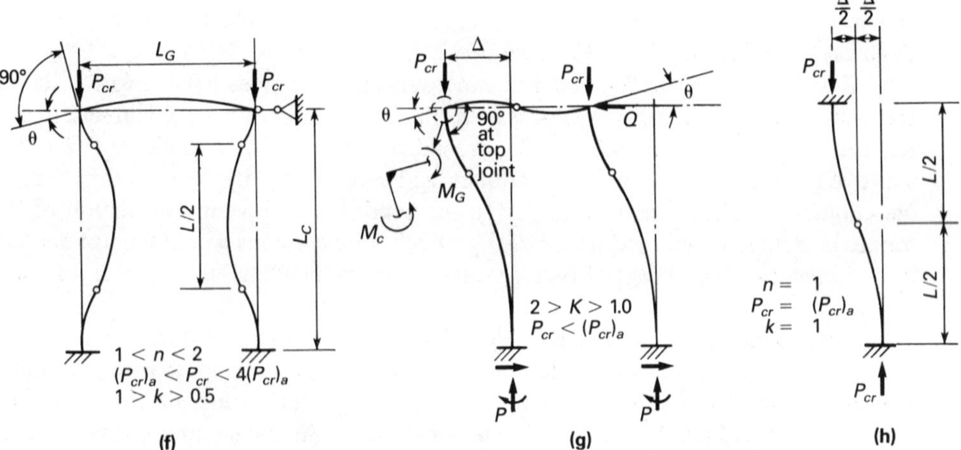

Figure 9.5 Buckling modes and lateral restraints: (a) first mode (pin–pin); (b) second mode; (c) third mode; (d) cantilever; (e) fixed ends; (f) simple portal, fixed bases; cross girder restrains rotation as elastic spring; sidesway inhibited; (g) simple portal, fixed bases; sidesway uninhibited except by elastic rotational restraint of cross girder; (h) translation at top uninhibited: fixed against rotation.

Figure 9.5 shows the deflection curves and P_{cr} corresponding to several values of n. Part (a) is the first, or *fundamental* buckling mode; parts (b) and (c) show the second and third modes and their greatly increased capacities. These columns are constrained against translation by intermediate lateral supports, forcing their buckling shapes into the higher modes. The fixed-base *flagpole* column in part (d), with $n = ½$, is equivalent to case (a) for a length of $2L$, while the fixed ended column in part (e) bends into two half-sine waves as does case (b). In all these ideal, regular cases, we can note that the value of n depends on the number of times the direction of curvature changes within the compressed length. At each change in direction of curvature, the internal moment equals zero; such a point is called a *point of contraflexure*. If we were to define *effective length*, KL, as the distance between two points of contraflexure, the equation for P_{cr} takes the form of Eq. (9-1) with $KL = L/n =$ *effective length*.

In the simple portal frame of Fig. 9.5(f), the restraint at the top prevents lateral translation (sidesway). The girder restrains rotation of the joint less perfectly than the fixed end in case (e), allowing elastic rotation through the angle θ. The effect is between a *pin end* condition and a *fixed* one. P_{cr} is also between that of cases (a) and (e), the exact value depending on the relative stiffnesses (EI/L) of the column and girder. The portal frame in part (g) is similar, but the sidesway restraint has been removed. The only resistance to translation is the frame stiffness itself. If the girder is infinitely stiff, the buckled shape of each column would be that of part (h). As the girder is made less stiff, the column action more and more resembles the cantilever in case (d), with corresponding loss in strength as the point of contraflexure shifts upward. As will be seen later in this chapter, columns in multistory frames without sidesway restraint may have K values $\gg 2.0$.

Recall Fig. 9.3(a). The columns not connected to braces have length L. If the end connections are not capable of transferring moment between column and beam, KL becomes L, as in Fig. 9.5(a). The braced columns would have $KL \approx L/2$ in the plane of the picture. With the ends of unbraced columns rigidly connected to beams, the columns would act as in Fig. 9.5(f); the braced bay prevents translation of the frame. In the unbraced rigid frame of Fig. 9.3(h), the columns would tend to act like that shown in Fig. 9.5(g). Load Q causes direct bending, the subject of later chapters.

Short columns are defined by very low values of the ratio KL/r. See Figs. 9.6(a) and 9.4. In this range, the Euler curve of critical load is approaching infinity. However, when the load becomes sufficient to cause yield stress, *failure* occurs by compression yielding. Actually, collapse is not likely. As the column shortens in the yield range, the elements of its cross section squash and thicken, increasing its real resistance. Short columns are not common in steel structures, but short special-purpose compression struts are.

Failure of *intermediate length* columns is affected by the tendency for buckling instability, the limit set by yield stress, and unmeasured effects such as eccentricity of end connections and lack of straightness [see Fig. 9.6(b)]. The curves in Fig. 9.4 show a smooth, gradual transition between the Euler curve and the yield condition. The two curves become tangent at a value of KL/r, dependent on F_y, that is called C_c. The value is somewhat arbitrarily chosen in the AISC Specification as

$$C_c = \sqrt{\frac{2\pi^2 E}{F_y}} \tag{9-3}$$

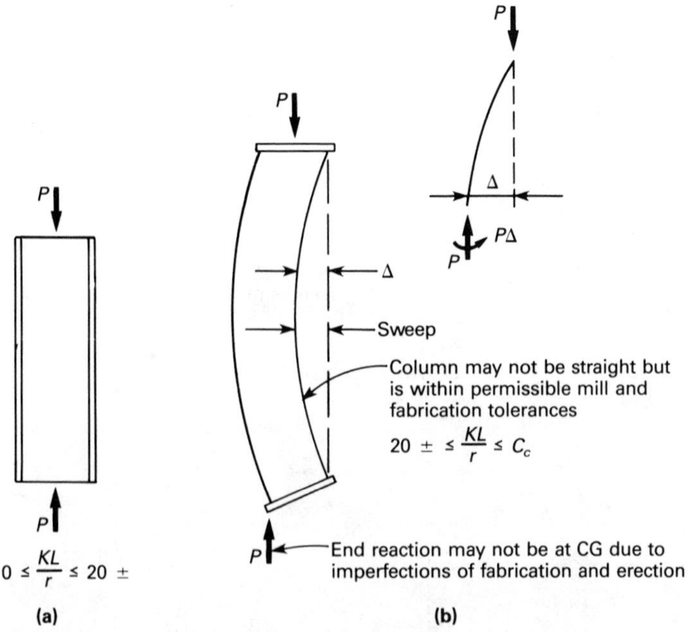

Figure 9.6 Short and intermediate columns: (a) short column; (b) intermediate column with initial imperfections.

As shown in Fig. 9.4, C_c varies from 76 to 126 for most common structural steels; the *intermediate* column range is, then, defined by

$$20 \pm \leq \frac{KL}{r} \leq C_c$$

The column sweep shown in Fig. 9.6(b) is permitted by usual mill practice (AISC Manual, pages 1–124 and 1–125) to be as much as 1 inch ± for long members. This is essentially not controllable by the designer. Moments caused by fabrication and erection error are somewhat controllable by the designer through strict specifications and inspection. But even there, perfection is costly and not usually achieved. The SSRC[36] identifies other common imperfections that increase effective compressive stresses, thus reducing the portion of F_y available for axial compression. When $KL > C_c$, the Euler critical stress becomes the failure stress.

9.3.2 Element Instability

Figure 9.7(a) shows an I-shaped compression member where a portion of one flange has buckled locally while the rest of the member remains straight. The reduction in compressive capacity should be obvious as the warped portion rejects load, transferring it to other elements. But why did the warping take place?

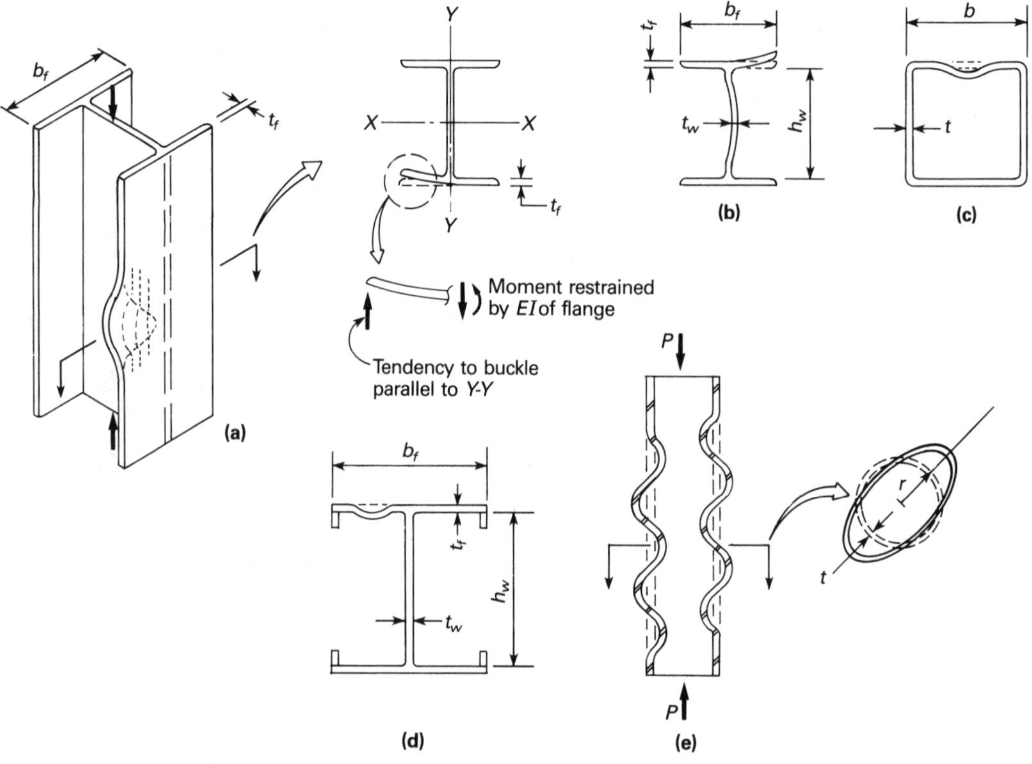

Figure 9.7 Local buckling of elements.

Consider a 1-inch-wide bar of thickness $t < 1$ inch. If we now consider the 1-inch-wide piece to be the outer edge of the flange of thickness t_f in Fig. 9.7(a), buckling parallel to the Y axis would be inhibited by the flange itself as a spring cantilevered from the web of the column. Figure 9.8 shows the cantilevered restraint and analogous springs. With high ratios of b_f/t_f, the spring restraint is insufficient to avoid buckling of the flange at low compressive stress, as shown in Fig. 9.7(a). Similar buckling potential exists in other elements of cross sections shown in Fig. 9.7, governed by the ratios of b/t or, in the case of round tubes, r/t.

Much of the attention of designers and design rules is directed to making individual elements of compression members stiff enough to eliminate local buckling as a control on failure load. If that is not done, the curves of Fig. 9.4 no longer describe the failure loads. Failure will be initiated at lower stress, being triggered by the instability of the vulnerable element. The controlling parameter is the ratio b/t, or its equivalent, r/t, in the case of thin-walled round tubes and pipes.

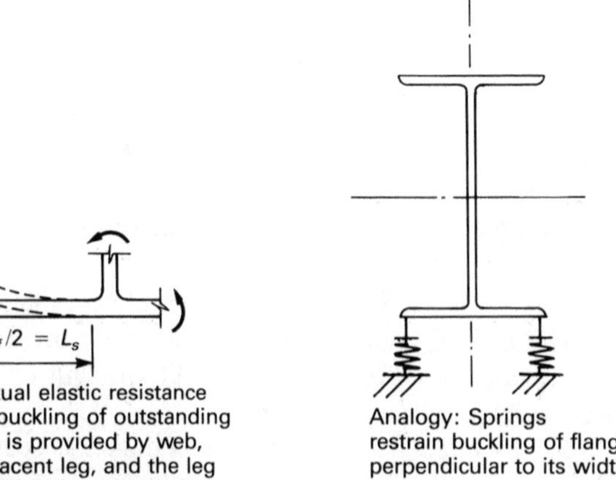

$b_f/2 = L_s$
Actual elastic resistance to buckling of outstanding leg is provided by web, adjacent leg, and the leg beyond area of potential buckling.
Analogy: Springs restrain buckling of flange perpendicular to its width.

Figure 9.8 Spring analogy for element stiffness.

9.4 SECTIONS USED FOR COMPRESSION MEMBERS

As with tension members, it is possible to use any type of steel cross section in compression. Compression members can be found with examples of all of the sections in Figs. 7.1 and 7.2(a) except cables and thin bars or rods. The designer is concerned not only with whether or not a section may be used, but also with the wisdom of the choice. Three considerations usually dominate the wisdom of choice: performance of the cross section, ease of connection, and least cost. However, there are often other considerations.

- The rectangular tube or pipe may offer distinct esthetic advantages.
- Round members have low wind and wave resistance.
- If least weight is more important than least cost, the choice may be tipped to a section built up from angles or small tubes at the corners made to act compositely through the use of *lacing*.
- A particular section may be advantageous because it is similar to sections chosen for other members in the structural system, thus simplifying overall problems of design, fabrication, and quality control.

We will look here at matters of structural performance and connectability. To examine performance, look once again at Fig. 9.4. For members of equal area, note two things. The yield strength of the material is very significant for short columns, of declining significance through the intermediate range, and of no significance in the performance of long columns. Increasing values of KL/r cause decline in the failure stress.

Decreasing values of f_{cr} must lead, for a given load, to the requirement of additional area and weight of steel (i.e., loss of efficiency in use of the material). A steep slope in the f_{cr} curve or low magnitude of f_{cr} indicates sensitivity to the inevitable errors in the calculation of design loads. On both counts, we are led away from the selection of *long columns*,* although we do not rule them out; there are instances where the compression is low enough to justify their use.

The *intermediate* column range usually offers the most desirable combinations of efficient use of the strength of steel and low sensitivity to calculation error. In that range, it is also possible to take advantage of the superior yield strength of high-strength steels and their favorable ratios of strength to cost.

Return now to Figs. 7.1 and 7.2(a) and consider the nature of the cross sections. To achieve large values of I and r for a given area A, it is necessary to distribute the area as far from its centroid as possible. From that point of view, tubes are superior to solid bars. I-shaped sections are also superior if I and r are calculated with respect to their X axes, but less so with respect to their Y axes. Double angles have different ratios of I_y/I_x and r_y/r_x, depending on which legs are placed back to back. Single channels and single angles have unfavorable ratios $r_{\min}/r_{\max}$. In addition, their shear centers do not coincide with their centroids of area, raising questions of torsional instability discussed later in this chapter. The unsymmetrical sections shown also suffer from this problem.

In a symmetrical section, buckling can take place parallel to either principal axis; restraint against buckling in one direction depends on the moment of inertia and radius of gyration with respect to the perpendicular axis. Figure 9.9 shows a W section with $r_y/r_x = 0.67$. In part (a), lateral restraint exists only at top and bottom. P_{cr} is based on r_y. Buckling takes place parallel to $x-x$. In part (b), a mid-height lateral restraint prevents movement in the $x-x$ direction at that level. For buckling to take place in that direction would require formation of the second-mode buckled shape shown in part (c) at $P_{cr} = 4(P_{cr})_a$. However, at $2.25(P_{cr})_a$, first-mode buckling takes place parallel to $Y-Y$. In part (c), the mid-height restraint inhibits both X and Y axis buckling, returning control to r_y. KL becomes $L/2$; $P_{cr} = 4(P_{cr})_a$.

The cases above assume that control of critical load lies with the KL/r ratio, not with the ratios b/t, which determine the local stability of elements of the cross section. The mathematical theory of buckling based on stability of section elements is beyond the scope of this book, but is explored in references on stability (e.g., references 21 and 59).

Returning to Figs. 7.1 and 7.2(a), we can draw some general conclusions. From the point of view of structural performance:

1. If there is no lateral restraint between the ends of a symmetrical column, the critical load is determined by the smaller of r_y and r_x. The most efficient sections

**Long* here implies value of $Kl/r > C_c$, not great length.

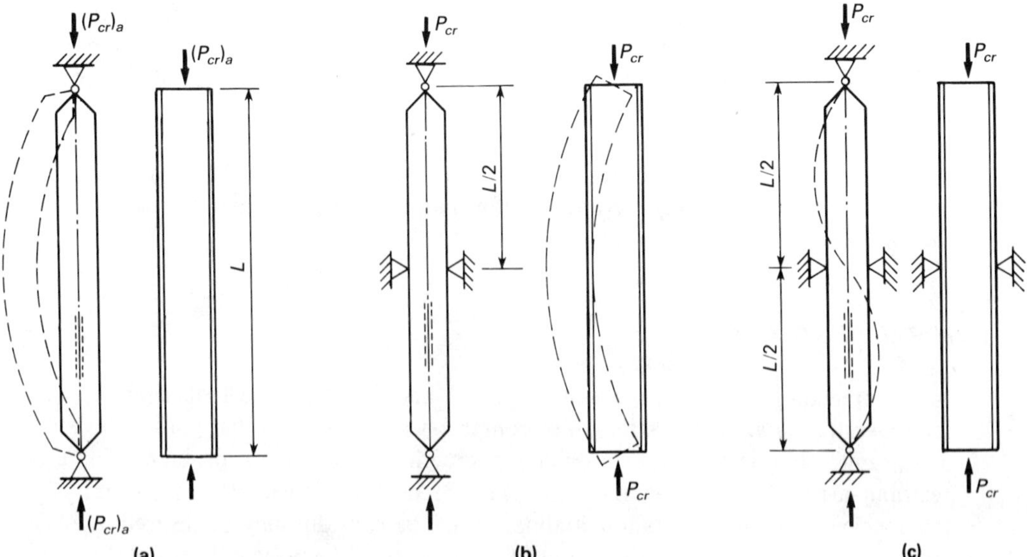

Figure 9.9 Effect of mid-height lateral restraints: (a) $r_x > r_y$; no lateral restraint; column buckles around y axis. (b) $r_y < r_x < 2r_y$; mid-height lateral restraint; column buckles around x axis; $P_{cr} = (r_x/r_y)^2\ (P_{cr})_a$. (c) $r_x > 2r_y$; mid-height lateral restraint; column buckles around y axis; $P_{cr} = 4(P_{cr})_a$.

are those with $r_y/r_x = 1$, the round or square tubes or composite sections in a square pattern. With other sections, efficiency increases as r_y/r_x approaches 1. In I-shaped sections, the larger the ratio of b/d, the more efficient. Of the available W sections, there are some W8, W10, W12, and W14's where $b/d \approx 1$, and others where $d \gg b$. The former are more desirable for compression, the latter usually for flexure. S sections, in almost all cases, have $d \gg b$, and r_y/r_x ratios smaller than W's of the same depth and weight.

2. The minor principal axis of a single angle is the Z axis shown in Fig. 7.1(a); the major axis is Z'; Z/Z' is usually $\ll 1$. Their small compression capacity on that account may be further reduced by torsional instability. However, if unequal leg angles are used in pairs, with their longer legs back to back, the r_y/r_x ratio becomes comparable to those of the "square" W sections.
3. If there are lateral restraints between column ends, the failure load is determined by the larger of the ratios $(K_xL)/r_x$ and $(K_yL)/r_y$. *Efficiency*, then, is sought by making $(K_xL)/r_x$ approach $(K_yL)/r_y$. This may lead to selection of W sections with $d \gg b$, rectangular tubes, or, less commonly, S sections. Double angles might be used with short legs back to back if Y movement is inhibited at mid-height and X movement is not.
4. These observations are only valid if ratios b/t for elements of cross section are low enough so that failure loads based on KL/r are lower than those based on

b/t. We will see later that the AISC Specification and usual practice encourage use of cross sections where that is true. However, the Specification also contains methods for evaluating the reduced capacity that results when b/t controls strength.

Tables of section properties in Part 1 and column tables in Part 3 of the AISC Manual have conveniently arranged information on r_x, r_y, their ratios, and b/t ratios.

9.5 LATERAL RESTRAINTS

In the preceding discussions and in Figs. 9.5 and 9.9, we have looked at the way lateral restraints affect the strength of compression members. What constitutes such a restraint? How is it usually provided in practice? The general problem of lateral restraint arises when it becomes necessary or desirable to inhibit movement of a member perpendicular to its longitudinal axis. The restraint may be desired in all or some directions, usually the directions of one or both of the principal axes of the cross section. The restraining system must have sufficient strength and stiffness to serve that purpose; that is, it must be capable of providing the necessary restraining force without itself buckling or translating excessively.

Examine Fig. 9.10. In part (a), the column is subjected to load P, which is known. The lateral strut, or the alternative brace, must be capable of resisting a force P_L (or P_B), which does not emerge from the analysis of applied loads on the structure. Without the restraining members, the member would buckle in its first mode. To force second-mode buckling and the attendant increase in capacity for load P requires a strut or brace that will hold the translation of point C at zero. To do this, the restraining member must be capable of resisting a force P_L (or $P_B = P_L/\sin\,\theta$), which may or may not exist. The magnitude of the force P_L is amazingly small, often evaluated conservatively* at $\pm 0.02P$. The usual control on such members is their own stiffness, KL/EI; they can be selected in the same manner as any other compression members.

Figures 9.10(b) and (c) illustrate the effect of longitudinal spring stiffness in lateral restraining systems. Depending on the stiffness of the spring, failure may take place in the first or second mode or a combination of the two; the critical load will vary with the mode of failure.

The condition of Fig. 9.10(a) assumes a spring of infinite rigidity, which is the condition usually sought. However, in some cases that is not feasible. Following the failure of some *pony truss* bridges (Fig. 9.11) in the late nineteenth century, F. Engesser and others sought a solution for the spring stiffness required to force buckling to take place between restraining springs rather than over the full length of the compression chord. The results and references are cited by the SSRC[36].

*SSRC[36].

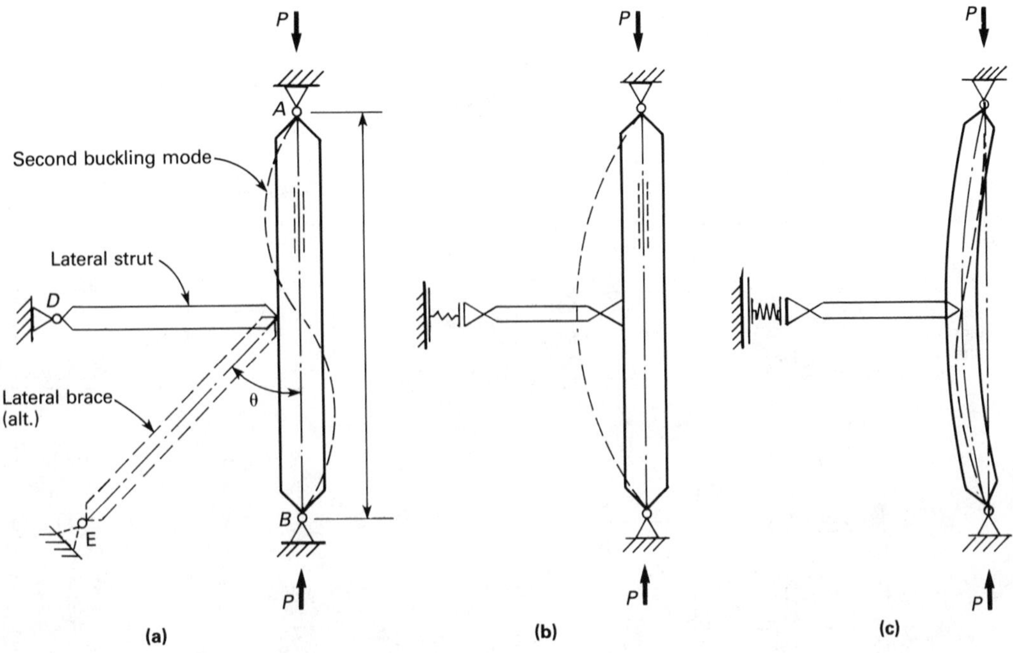

Figure 9.10 Mid-length lateral restraints: (a) strut prevents lateral translation; (b) weak spring, first-mode buckling; (c) stiff spring, mixed-mode buckling.

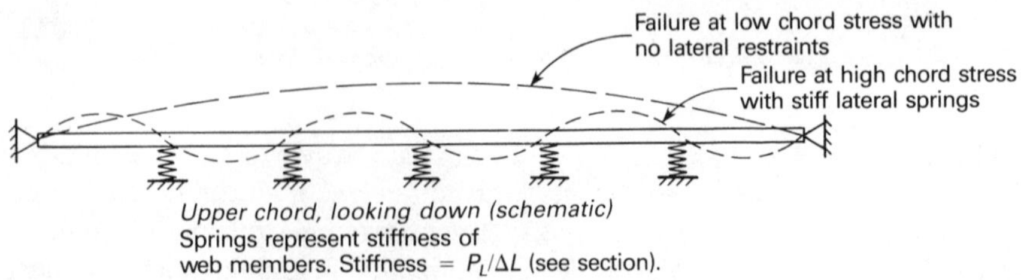

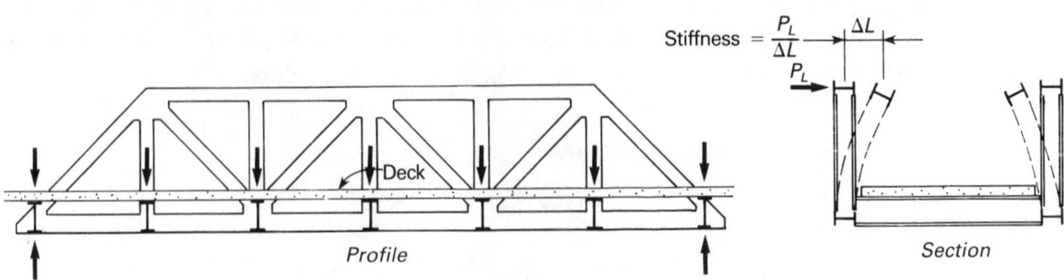

Figure 9.11 Pony truss; elastic lateral restraints.

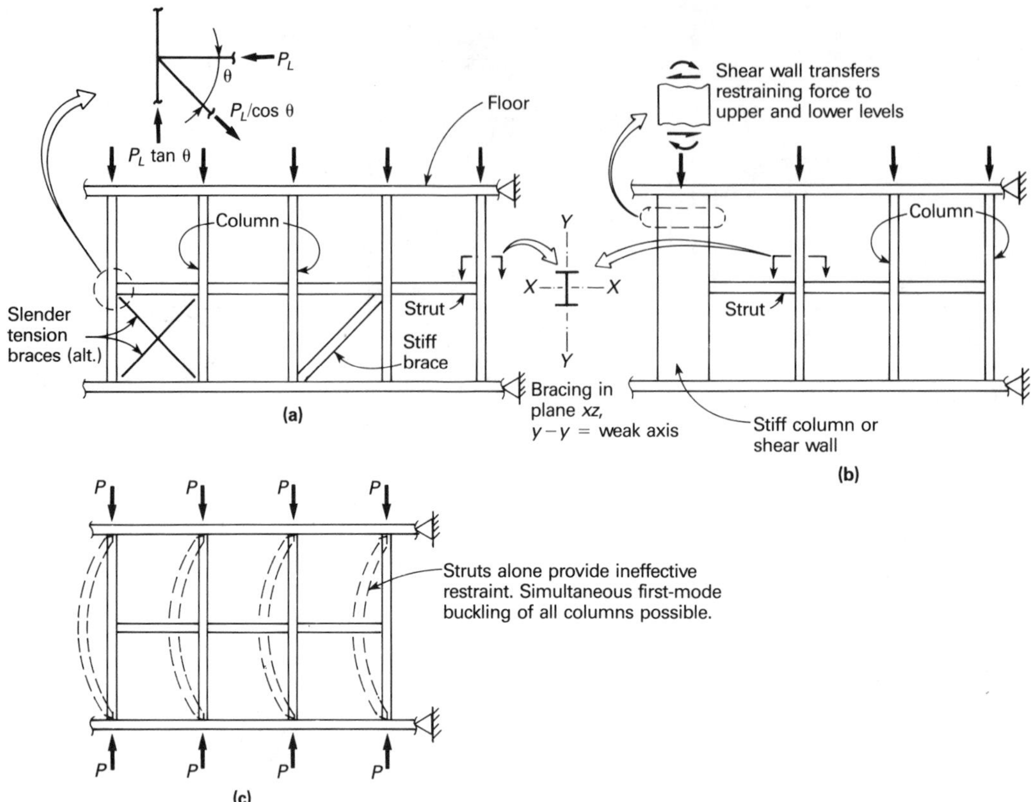

Figure 9.12 Mid-length restraints in structural systems.

The restraint systems in Fig. 9.12 illustrate ways to achieve mid-height restraints in columns. Lateral movement of the mid-height of all columns is prevented in the direction parallel to the *strong X* axis. The restraining struts must be ultimately connected to a member that can transfer the horizontal restraining force by shear to the next restrained level. The diagonals of Fig. 9.12(a) or shear walls of Fig. 9.12(b) are effective restraints. The system of Fig. 9.12(c) is not. Mid-height restraints such as shown are only permissible if the restraining members do not interfere with the required use of space. If that is permitted in one direction only, columns should be oriented with their strong axes in the plane of the restraints.

9.6 AISC RULES FOR COMPRESSION MEMBER DESIGN

9.6.1 Safety Factors and Allowable Stress

The AISC Specification (Part 1) establishes allowable stresses and other limitations intended to enforce required factors of safety against compression failures.

$$\text{FS} = \frac{\text{load at failure}}{\text{load permitted}} = \frac{\text{stress at failure}}{\text{stress permitted, } F_a} > 1.0$$

As will be seen, the required factor of safety (FS) is greater for long columns than for short and intermediate ones. Table 9.1 lists provisions of the Specification that apply to compression members. It can be used as a checklist to be sure that no requirements are overlooked. Sections 1.5.1.3.4, 1.5.1.3.5, 1.10.5.1, and 1.10.10 will be discussed in the chapter on plate girders. Appendix C is only invoked when the b/t limits in Section 1.9 are exceeded, a situation that designers usually try to avoid. In most of

TABLE 9.1 SECTIONS OF AISC SPECIFICATION, 1978, APPLYING TO COMPRESSION MEMBERS

		Allowable stresses					
		$Kl/r \leq C_c$	$Kl/r > C_c$				
AISC Specification section	Subject	F_a	F_a	F_{as}, secondary members	Stiffness and stability, members, (max Kl/r)	Stiffness and stability, elements, (b/t)	Details; stitching, lacing, battens
1.5.1.3.1	Allowable stress	●					
1.5.1.3.2	Allowable stress		●				
1.5.1.3.3	Allowable stress			●			
1.5.1.3.4	Plate girder stiffeners	●					
1.5.1.3.5	Web of rolled shapes at fillets	●					
1.5.6	Adjustment: Wind and seismic forces	●	●	●			
1.8	Stability, slenderness						
1.8.1	General				●		
1.8.2	Braced frames				●		
1.8.3	Unbraced frames				●		
1.8.4	Maximum ratios				●		
1.9	Width–thickness ratios						
1.9.1	Unstiffened elements					●	
1.9.2	Stiffened elements					●	
1.10.5.1	Plate girder bearing stiffeners				●		
1.10.10	Web crippling	●					
1.18.2	Built-up members						●
Appendix C	$b/t >$ limits in section 1.9		●			●	

our discussion, we will stay within the b/t limits of Section 1.9, avoiding potential element instability.

For most compression members, allowable stresses are set in paragraphs 1.5.1.3.1 and 1.5.1.3.2. They divide compression members into two classes by their values of KL/r, with the value C_c (see Fig. 9.4) dividing the two classes. For short and intermediate members, by AISC (1.5-1), if $KL/r \leq C_c$,

$$F_a = \frac{\left[1 - \frac{1}{2}\frac{KL/r^2}{C_c}\right]F_y}{\frac{5}{3} + \frac{3}{8}\left(\frac{KL/r}{C_c}\right) - \frac{1}{8}\left(\frac{KL/r}{C_c}\right)^3} \leq 0.60F_y \tag{9-4}$$

For $C_c \leq KL/r \leq 200$ (the upper limit of KL/r),

$$F_a = \frac{12}{23} \cdot \frac{\pi^2 E}{(KL/r)^2} \tag{9-5}$$

In Eq. (9-4), the numerator is a curve of failure stress, varying from F_y at $KL/r = 0$ to the Euler value at C_c. The denominator is a factor of safety, which varies from 1.67 to 1.92 (i.e., from $\frac{5}{3}$ to $\frac{12}{23}$). The reader will recognize in Eq. (9-5) the Euler formula of Fig. 9.4 divided by the constant factor of safety, $23/12 = 1.92$.

Figure 9.13 plots the values of F_a for A36 steel as well as the failure curves for that steel. Recall also Fig. 9.4. Except for minor differences due to variation of C_c, the

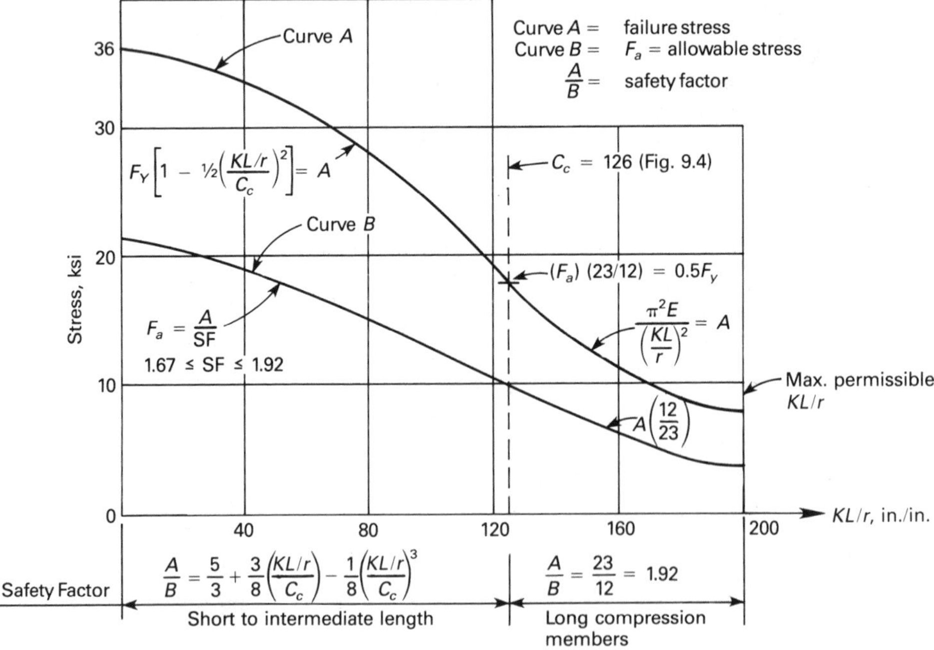

Figure 9.13 Failure stress and allowable stress versus KL/r; A36 steel.

long column region is identical for A36 to that of all other steels. The short and intermediate regions are different for other values of F_y. AISC Tables 3-36 and 3-50 give values of F_a.

Paragraph 1.5.1.3.3 permits an increase in allowable stress for "bracing and secondary" members with KL/r between 120 and 200:

$$F_{as} = \frac{F_a}{1.6 - (L/200r)} \tag{9-6}$$

The effect is to increase allowable stress by values varying from zero to a maximum of 67% at $L/r = 200$. (K is assumed to be 1.) The magnitude of this adjustment may be seen in AISC Tables 3-36 and 3-50, where separate columns tabulate F_{as} and F_a. The safety factor at $L/r = 200$ becomes $1.92/1.67 = 1.15$!

We have never found a universally accepted definition of a secondary member. Nor do we consider most bracing members as anything other than primary, particularly in the presence of high wind or seismic forces. We recommend that designers do not permit these higher stresses unless, in an unusual case, they are convinced that failure of such members will not endanger the basic structural system. The Commentary to the AISC Specification suggests other cautions on the use of Eq. (9-6).

Paragraph 1.5.6, AISC Specification, allows an increase of 33.3% in allowable stress for any combination of loads that includes the effect of wind or seismic loads. This applies to F_a as well as all other allowable stresses in the Specification, but not to fatigue loads covered in Appendix B. The increase may be justified by the transitory nature of wind and seismic effects. In the case of seismic loads, it would apply to levels of force prescribed for design* in codes prescribing static equivalents of forces due to earthquake accelerations.

9.6.2 Slenderness, Effective Length, K, KL/r

Formulas for F_a require determination of the effective length, KL. L is easily defined. K is more complex. Recall Fig. 9.5 and the preceding discussion (Sections 9.3 and 9.5). The effective length was seen to be the distance between points of contraflexure in the buckled shape. $K \leq 1.0$ if lateral translation is prevented at the ends of length L, but may be considerably greater than 1.0 if the frame exhibits sidesway.

The AISC Specification covers the slenderness issue in Section 1.8, which should be read with Section 1.8 of the Commentary. Refer to Table 9.2 and Fig. 9.14, both from the AISC Manual. In four of the six cases in Table 9.2, values of K are recommended somewhat higher than those resulting from idealized theory, a tacit, prudent acknowledgment that ideal end conditions are seldom realized. For example,

*The AISC Specification expects that *design-level* seismic forces will be resisted elastically. In an extreme seismic event, structural engineers anticipate that such design-level forces may be exceeded for a very short time, leading to yield stress and ductile strain. Rules for design of *ductile moment resisting frames* (e.g., in the *Uniform Building Code*) are intended to avoid collapse of structures in such an eventuality.

TABLE 9.2 *K* VALUES FOR COMPRESSION MEMBERS

	(a)	(b)	(c)	(d)	(e)	(f)
Buckled shape of column is shown by dashed line	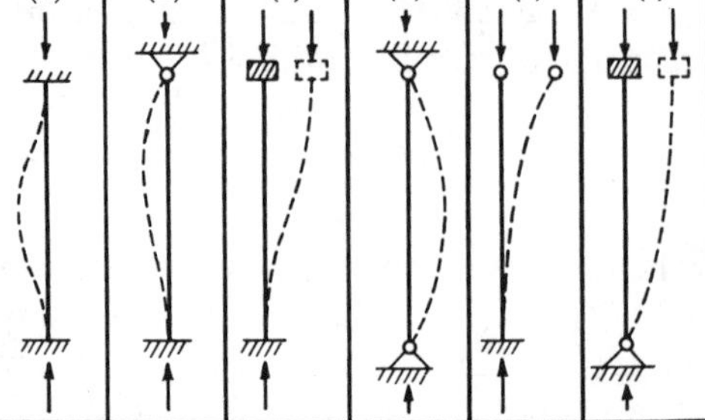					
Theoretical K value	0.5	0.7	1.0	1.0	2.0	2.0
Recommended design value when ideal conditions are approximated	0.65	0.80	1.2	1.0	2.10	2.0
End condition code		Rotation fixed and translation fixed Rotation free and translation fixed Rotation fixed and translation free Rotation free and translation free				

AISC Specification Table C1.8.1, reprinted with permission.

very small rotation of a nominally fixed end moves the point of contraflexure measurably.

The nomograms of Fig. 9.14 show two distinctly different cases. Frames with *sidesway inhibited* rely on some means other than the columns and girders of the steel frame to prevent horizontal translation of the joints of the frame (e.g., diagonal bracing or shear walls). Sidesway is (relatively) *uninhibited* when resistance to horizontal translation of the frame must be provided by the flexural strength and stiffness of a system of columns and girders with continuous joints. In Chapters 17 and 18, the reader will meet an instance of a braced system in a mill building and of a rigid frame with *sidesway uninhibited* in a multistory tier building. It should be obvious on reflection that the distinction is relative.*

The parameter G of Fig. 9.14 is the ratio of the sum of stiffnesses of all columns connected at a joint to the sum of stiffnesses of all girders or beams connected to the same joint.† If the girders are relatively very stiff, G approaches zero. Moments on

*By a curious semantic accident (?), the term *rigid frame* is applied to the less rigid of the two systems. The name is based on the rigidity of the joints, not of the framing system.

†Stiffness of a member is defined as the moment required to rotate one end of a member through a unit angle while the other end remains fixed. Stiffness is proportional to I/L.

$$G = \frac{\Sigma I_c/L_c}{\Sigma I_g/L_g}$$

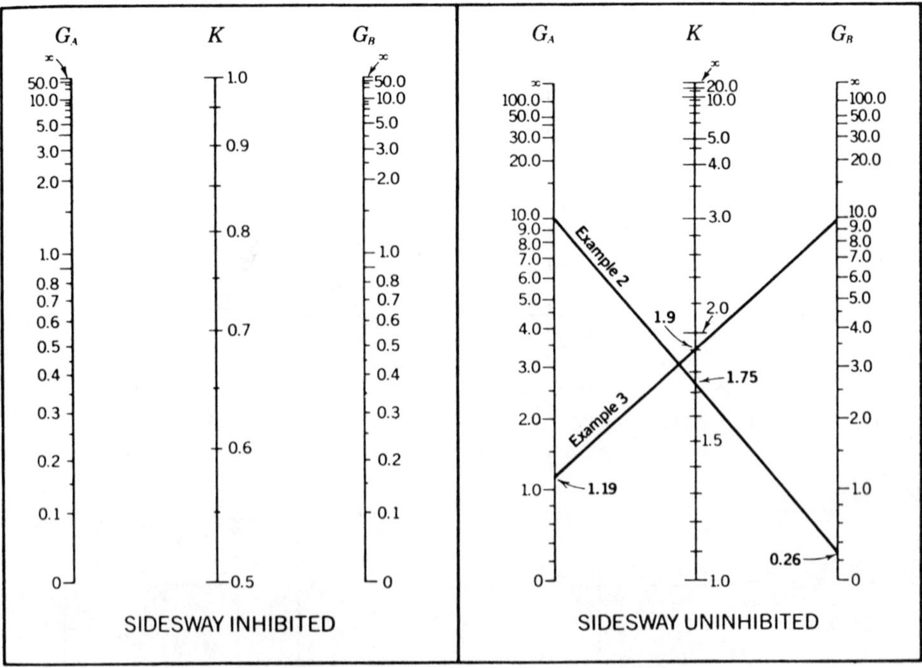

The subscripts A and B refer to the joints at the two ends of the column section being considered. G is defined as

$$G = \frac{\Sigma\,(I_c/L_c)}{\Sigma\,(I_g/L_g)}$$

in which Σ indicates a summation of all members rigidly connected to that joint and lying in the plane in which buckling of the column is being considered. I_c is the moment of inertia and L_c the unsupported length of a column section, and I_g is the moment of inertia and L_g the unsupported length of a girder or other restraining member. I_c and I_g are taken about axes perpendicular to the plane of buckling being considered.

For column ends supported by but not rigidly connected to a footing or foundation, G is theoretically infinity, but, unless actually designed as a true friction free pin, may be taken as "10" for practical designs. If the column end is rigidly attached to a properly designed footing, G may be taken as 1.0. Smaller values may be used if justified by analysis.

Figure 9.14 K versus ratio of column to girder stiffness (from AISC Manual, Part 3, Fig. 1, reprinted with permission). The examples noted in the nomograph are in the AISC Manual. To find K:

1. Select the applicable diagram.
2. Find G at bottom and top (G_A and G_B).
3. Join the value of G_A to that of G_B with a straight line.
4. Read K at the intersection of the line with the K scale.

The I applicable to each member is the one referred to the axis perpendicular to the plane of the bent.

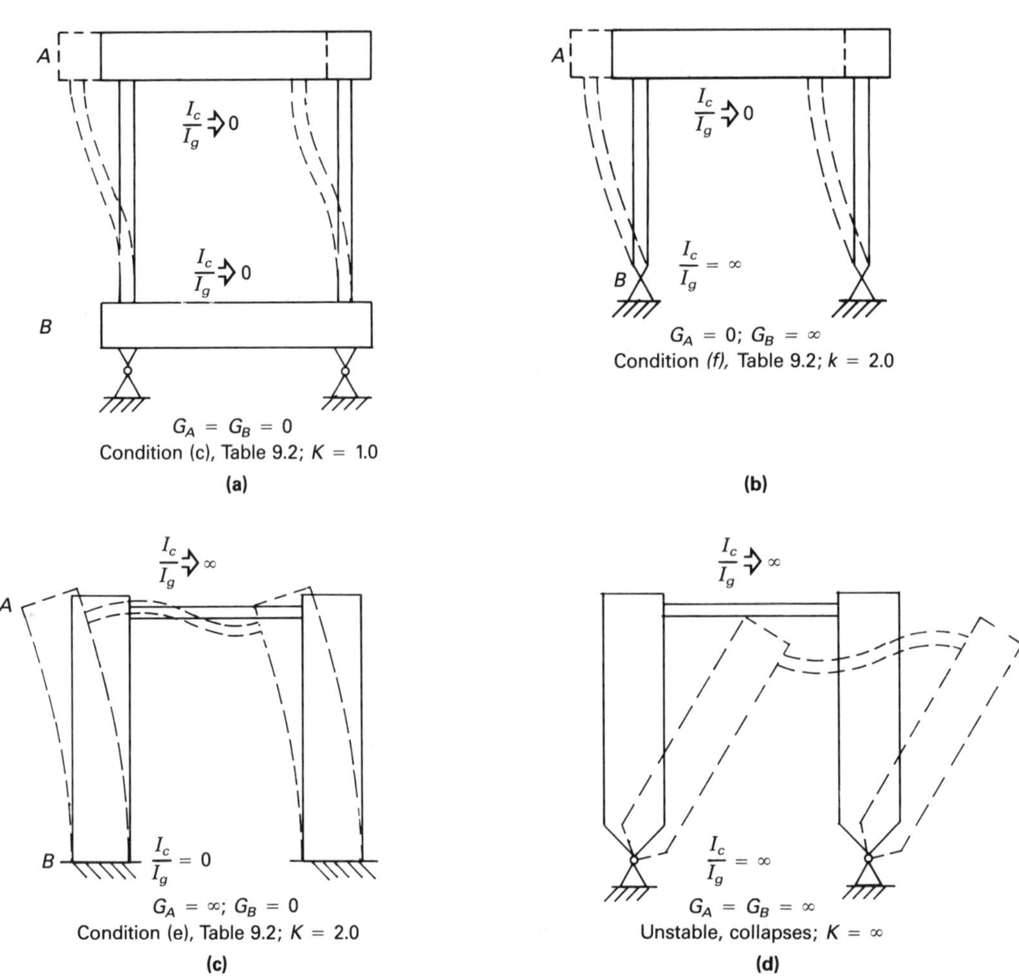

Figure 9.15 Values of K in extreme cases of simple one-story frames with sidesway.

the column are unable to rotate the joint, the condition for a *fixed* end shown in Table 9.2. When, as is usual, $G > 0$, rotational fixity is imperfect, and end rotational restraints fall between the fixed and pinned conditions. With sidesway inhibited, K falls by the left-hand nomogram of Fig. 9.14 between 1.0 and 0.5, the theoretical values of cases (a) and (d) of Table 9.2. With sidesway, case (c) of Table 9.2 sets the lower limit of values of K; there is no upper limit. The simple frames of Fig. 9.15 dramatize the extremes. In each of these cases, there is only one girder and one column at each joint. In the general case, where there may be a column above and below (or several columns other than vertical) and restraining girders on both sides,

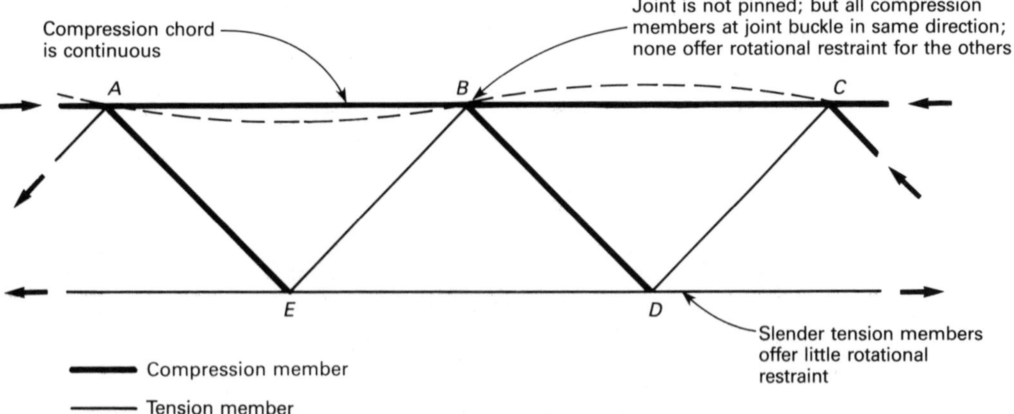

Figure 9.16 Buckling mode for compression members in a truss; $K = 1.0$ per AISC 1.8.2.

summations of $I_c L_c / I_g L_g$ are used. Examination of the right-hand nomogram of Fig. 9.14 will show how important stiff girders are in controlling KL/r of columns.

In the case of braced frames or trusses, Specification paragraph 1.8.2 requires $K = 1.0$. This seems to assume true pins at the ends of the compression members, although joints are seldom made with ideal pins. The justification may be seen in Fig. 9.16, which shows the shape of all the compression members buckling simultaneously.

9.6.3 Limits on b/t: Elements of Sections

Underlying AISC formulas for F_a [Eqs. (9-4) and (9-5)] is the assumption that individual elements of sections will not buckle at lower stress than that which causes general failure. Recall Fig. 9.7 and earlier discussion. Section 1.9, AISC, prescribes limits on b/t. Two main categories divide unstiffened elements from stiffened ones.

Table 9.3 may be helpful in interpreting Section 1.9. The limits for unstiffened elements are lower than those for stiffened elements, for reasons which should be clear from the buckled shapes in Fig. 9.7. Round tubes are self-stiffening within limits, but excessive values of b/t may invite accordion-type buckling. The Commentary to Section 1.9 warns against reliance on classical theory in evaluating the compressive strength of cylinders. The limit of $3300/F_y$ is based on test data, as is the absolute limit of $13{,}000/F_y$ found in Appendix C.

All limiting ratios in Table 9.3 are functions of F_y. The values for different values of F_y are tabulated in AISC Table 6. If these ratios are exceeded, the designer must choose between a change in member or the lower values of allowable stress covered in Appendix C. The former course is recommended for most cases. More on the latter in Chapter 15.

TABLE 9.3 LIMITING RATIOS OF WIDTH TO THICKNESS FOR ELEMENT STABILITY (AISC SPEC. SEC. 1.9)

Type of element in compression	Description	Examples	Limit of b/t*
Unstiffened	Single L Double L, separated	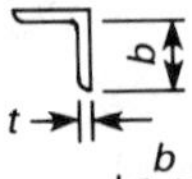	$\frac{76}{\sqrt{F_y}}$
	Double L, in contact Projecting L's or plates in compression	Stiffened element	$\frac{95}{\sqrt{F_y}}$
	Stems of tees	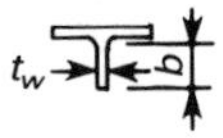	$\frac{127}{\sqrt{F_y}}$
Stiffened	Square or rectangular box sections		$\frac{238}{\sqrt{F_y}}$
	Unsupported width of cover plate perforated with access holes	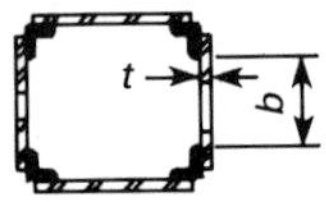	$\frac{317}{\sqrt{F_y}}$
	All other uniformly compressed stiffened elements	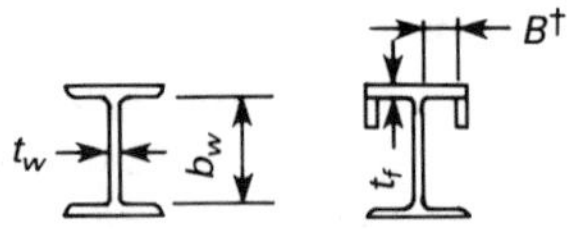	$\frac{253}{\sqrt{F_y}}$
	Round tubes	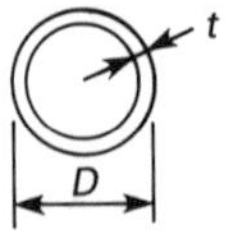	$\frac{D}{t} \leq 3300/F_y$

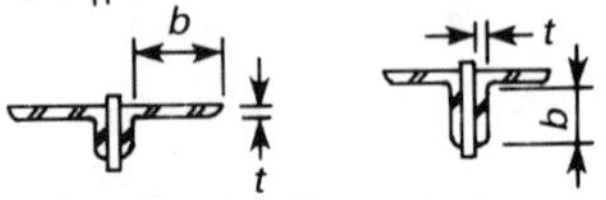

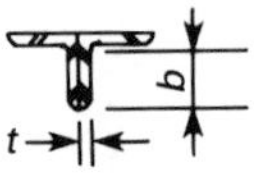

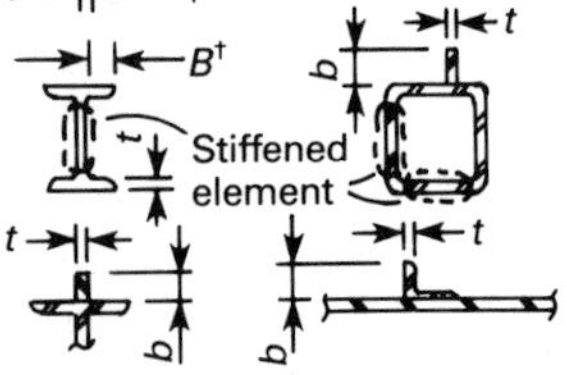

*Except for perforated cover plates, follow Appendix C, AISC Specification if limiting ratios here are exceeded.

†Ratio here is designated B/t, since b is usually used to refer to total flange width.

9.7 SELECTION OF COMPRESSION MEMBERS

9.7.1 General Comments

It should be no surprise to readers who have come this far that the problem of selecting sections for compression members is not simple and straightforward. The required area is

$$A_R = \frac{P}{F_a} \tag{9-7}$$

but F_a itself depends on other elements of cross section besides area (b/t, I_x, I_y, r_x, r_y), which are only known when an actual cross section is chosen. Similarly, it is governed by decisions about bracing and lengths between lateral supports (L_x, L_y), which may be made after the section is chosen. Procedures for selection are usually iterative; that is, they proceed from initial tentative selection through a series of refinements based on mathematical tests, to a final it is hoped optimum selection based on all considerations.

We will chart such procedures for several cases. The novice who is panicked by the multiplicity of seemingly independent, sometimes conflicting criteria for choice may be comforted to know that the field is not quite as open as it may seem. Before looking at individual members, decisions have often been made for the entire structural system. Type of steel, form of cross section, type of end connections, perhaps even depth of section—all may have been decided before the individual member is addressed. The system has been analyzed for all load cases and combinations. Some, usually small, adjustment in the dead loads may result from individual member selections. Small adjustments on the safe side may not require recalculation.

9.7.2 Procedures, Flow Charts, Examples

Flow Chart 9.1 shows procedures for selection for the common case where a member is always in compression. Part (a) is quite general for iterative selection processes and is not limited to the present case. By expanding the level of detail, part (b) becomes more specifically useful for axially loaded compression members. Part (c) is similarly limited, but is faster, relying on the tables of Part 3, AISC Manual, which list capacities for many common column sections calculated in accordance with the AISC Specification. It offers a special advantage by making possible rapid visual comparison of possible choices. All three variants are limited to cases where no load combination causes stress reversal to tension.

Often, a member functions primarily in compression, but some load combinations may cause net tension. Flow Chart 9.2, a variant of Flow Chart 9.1, requires tension checks before the final decision is recorded. The methods of the previous chapter apply. Flow Chart 9.3 may be used to find F_a or F_{as}.

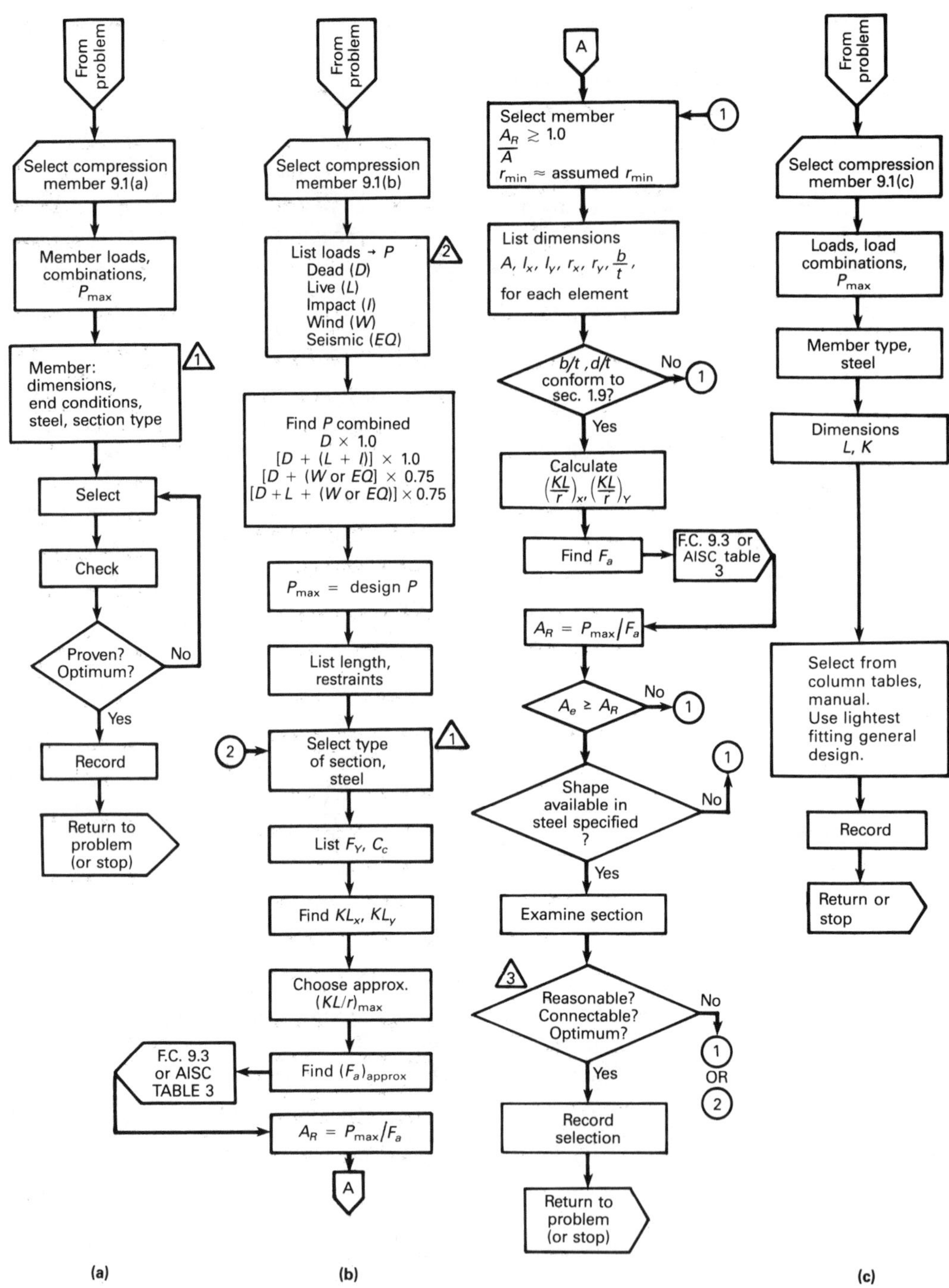

△1 From general design approach. If yields nonoptimum section, reconsider before return.

△2 Add others if apply.

△3 If academic exercise, may require more information than available. If so, skip this stage.

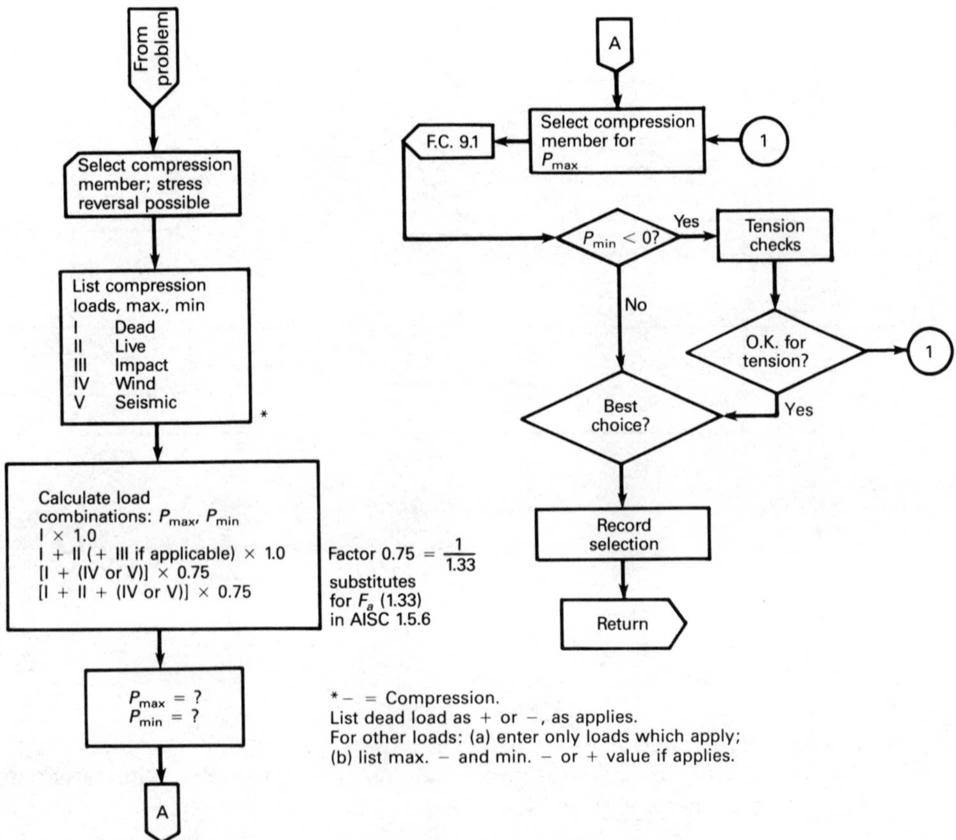

Flow Chart 9.2 Compression member selection, stress reversal possible.

Flow Chart 9.1 Select compression member (no reversal to tension; b/t conforms to AISC Section 1.9).

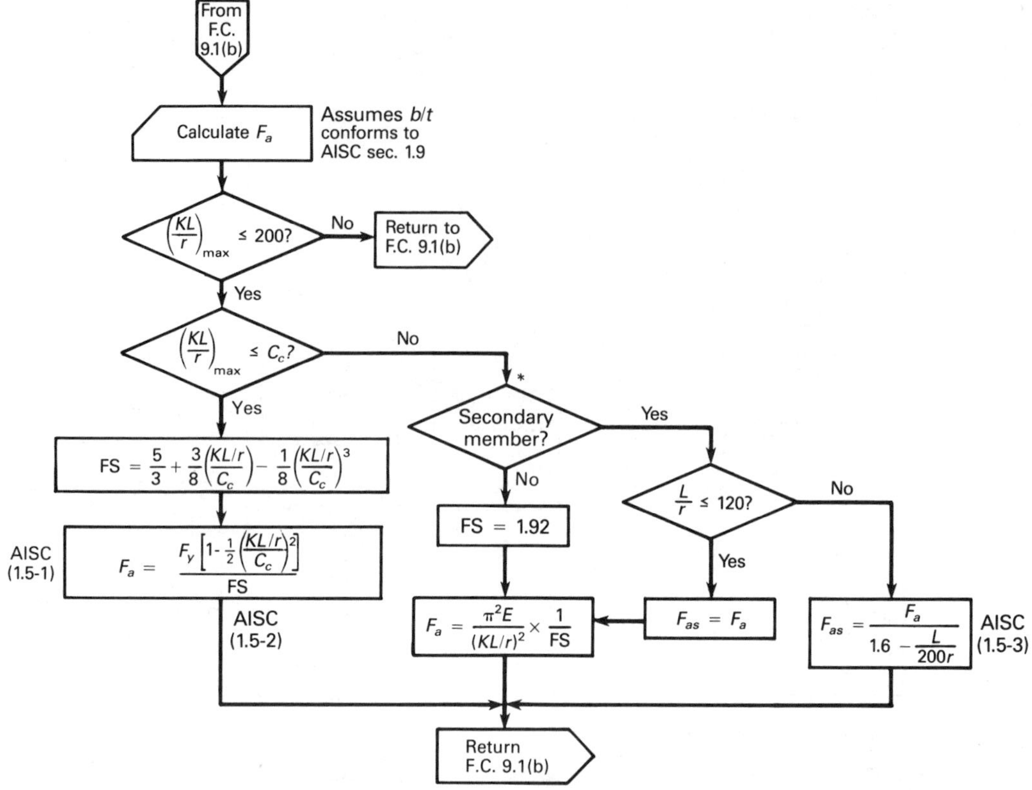

*See discussion of secondary members in text.

Flow Chart 9.3 Calculate F_a.

9.7.3 Example: Column Selection

A warehouse column is to be 16′ long between *pinned* end connections (i.e., end connections will be such that small rotations can take place with essentially no moment resistance). Design dead load is 200 kips, live load 285 kips. It has previously been decided that all columns in the building shall be of W shapes and no larger in nominal size than 10 in. Steel may be A36 or A572, Gr. 50. Select a column.

CALCULATIONS	*REMARKS*
Follow Flow Chart 9.1(b).	Since wind and seismic loads are not given, they may be considered to be not controlling. In an actual design, this should be verified.
(1) List loads, length, restraints:	

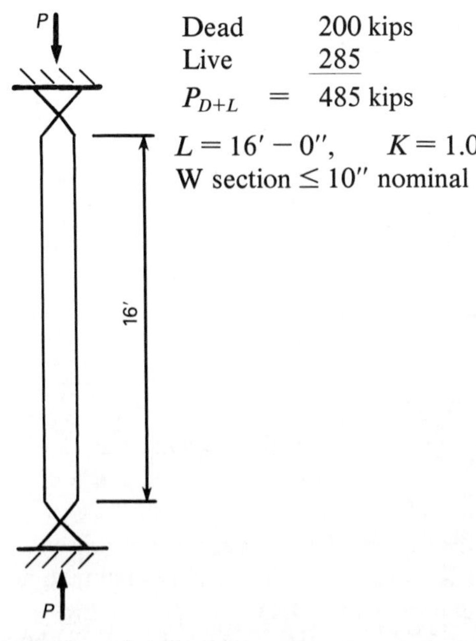

Dead 200 kips
Live 285
P_{D+L} = 485 kips

$L = 16' - 0''$, $K = 1.0$ — Conditions specified.
W section ≤ 10″ nominal

(2) Steel: Compare

ASTM A36: $F_y = 36$ ksi, $C_c = 126$
ASTM A572, Gr.50:
$F_y = 50$, $C_c = 107$

$$C_c = \left(\frac{2\pi^2 E}{F_y}\right)^{1/2}$$

(3) $(KL)_x = (KL)_y = 16.0'(12) = 192''$

(4) Approximate r_x, r_y, F_a, A_R.

Depth of W	W8		W10		
b_f, in., approx.	8	6	10	8	6
r_x	3.5	3.5	4.4	4.3	4.3
r_y	2.0	1.6	2.5	2.0	1.4
$(KL/r)_{max}$	96	120	77	96	137
$F_y = 50$ ksi					
F_a, ksi	15.6	10.4	19.6	15.6	8.0
A_R (in.^{2})	31.1	46.6	24.7	31.1	60.6
$F_y = 36$					
F_a, ksi	13.5	10.3	15.7	13.5	8.0
A_R (in.^{2})	35.9	47.1	30.9	35.9	60.6

r_x, r_y approximated by scanning Shape Tables for W8 and W10.

Table 3-50, Spec., App. A

Table 3-36, Spec., App. A

Try W10 with $b_f = 10''$, $F_y = 50$.

(5) Try W10 × 88, $F_y = 50$ ksi.

$A = 25.9$ in.^{2} $(KL/r_y) = 73$ — $A = 25.9 > 24.7 \approx A_R$, above
$r_x = 4.54$

$r_y = 2.63$	$F_a = 20.38$ ksi if satisfies AISC 1.9	But need check for b/t by AISC 1.9
$\dfrac{b_f}{2t_f} = 5.2 \ll \dfrac{95}{\sqrt{F_y}} = 13.44$	OK	Shape Tables for data; 1.9.2.1 and 1.9.2.2 for criteria
$\dfrac{d_w}{t_w} = 17.9 \ll \dfrac{253}{\sqrt{F_y}} = 35.8$	OK	
$P_{ALL} = 20.38(25.9)$ $= 528 \text{ k} < 485$ OK		
(6) Verify shape available in A572, Gr. 50 Available in all groups.		AISC Manual, Part 1
Use W10 × 88, ASTM A572, Gr. 50		

Flow Chart 9.1(b) led, fairly easily, to a selection based on minimum area and therefore minimum weight for the steel considered. Data from shape tables are easily accessible and easy to compare. Tables 3-36 and 3-50 were used for F_a, although the source equations diagrammed in Flow Chart 9.3 would give the same results. Try it.

The Column Tables of Part 3, AISC Manual, yield the same results. The procedure follows Flow Chart 9.1(c).

The column selected has $KL/r \ll C_c$, and therefore F_a is higher for $F_y = 50$ than for 36. Since weight per foot is proportional to area, the small unit price difference of the higher-yield steel will result in lower cost. This would not necessarily be true if $KL/r > C_c$.

9.7.4 Example: Column Adaptation

The column selected in Example 9.7.3 has been incorporated into the warehouse structure. Several years later, it is proposed to add a heavy permanent piece of ventilation equipment that will increase the total dead load on the column by 75 kips, without reducing the total design live load.

You are asked to check the column for the proposed new load condition. If the column is overloaded, you are to suggest possible ways to make it acceptable.

CALCULATIONS	*REMARKS*
(1) Check column: W10 × 88; $F_y = 50$ $L = 16'\text{-}0''$, ends pinned; $KL = 192''$ As designed: $P_{D+L} = 200 + 285 = 485$ k	

Add $\Delta P = 75 + 0$.

New $P_{D+L} = 275 + 285 = 560$ k

$P_{ALL} = 528$ kips < 560 NG — Example 9.7.3

Revision required.

(2) Revision: Alternative 1:
Reduce $(KL/r)_{max}$ by bracing against buckling in XZ plane.

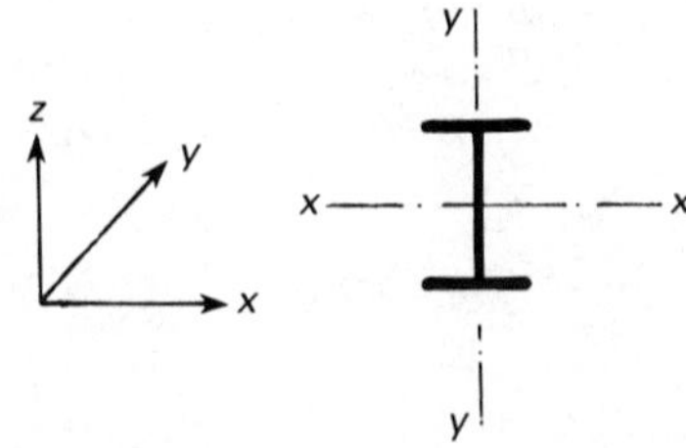

Column W10 × 88:

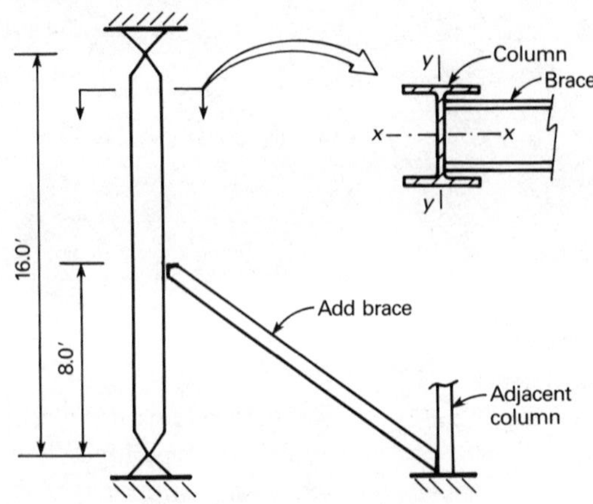

$r_x = 4.54$ in. — Shape Tables

$r_y = 2.63$ in.

$A = 25.9$ in.^{2}

$$\left(\frac{KL}{r}\right)_y = \frac{8(12)}{2.63} = 36.5$$

$(KL/r)_x > (KL/r)_y$; choose F_a based on $(KL/r)_x$.

$$\left(\frac{KL}{r}\right)_y = \frac{16(12)}{4.54} = 42.3$$

$F_a = 25.5$ ksi	Table 3-50, AISC
$P_{ALL} = 660.5 \text{ k} \gg 560$	
Use only if bracing acceptable without obstructing required space in XZ plane.	
(3) Revision: Alternative 2: Restrain either top or bottom of column against rotation by changing connection to floor beam in XZ plane.	Moment connections are discussed in Chapters 12 and 13. This alternative can only be considered if a relatively stiff beam exists.
$K_y = 0.80$ approximately	Table 9.2
$\left(\frac{KL}{r}\right)_y = \frac{0.8(16)(12)}{2.63} = 58.4$	
$\left(\frac{KL}{r}\right)_x = \frac{16(12)}{4.54} = 42.3$	
$F_a = 23.16$ ksi	Table 3-50
$P_{ALL} = 600 \text{ kips} > 560$ OK	
Use if floor beam is available, is stiff enough to restrain column rotation, and inexpensive connection is feasible. If welding is necessary to floor beam, provide temporary support.	Beam loses flexural strength when heated for welding. Requires temporary support.

Either alternative may be used to strengthen the column, making it acceptable for the added load. The choice depends on other conditions. Other approaches are also possible, limited only by the engineer's imagination and understanding of column theory.

9.7.5 Example: Column Selection, Compare Shapes

Compare the lightest adequate section available for a column in the following types of section [see Figs. 7.1 and 7.2(a)]:

W, S, Structural Tubing, Standard Pipes, Double Angles, Tees
$P_{D+L} = 200$ k
$L = 20'$
Ends fixed top and bottom
Out-to-out dimensions $\leq 12\frac{1}{2}'' \times 12\frac{1}{2}''$
$F_y = 36$ ksi

AISC tables may be used where helpful.

CALCULATIONS	*REMARKS*
(1) $KL = 0.65(20) = 13' = 156''$	Table 9.2
(2) W sections for $KL = 13'$; $F_y = 36$:	

	$b \times d$	P_{ALL}, kips
W10 × 45	8 × 10⅛	208 > 200
W12 × 45	8 × 12	202 > 200

AISC Column Tables

(3) S shapes:

	A	r_y	$(KL/r)_y$	F_a	P_{ALL}
S12 × 40.8	12.0	1.06	147	6.91	83 ≪ 200
S12 × 50	14.7	1.03	151	6.55	96 ≪ 200

Column Tables not available for S sections > 6″. F_a from Table 3-36.

No S section acceptable.

(4) Structural tubes:
Use square for $r_x = r_y$,
$KL = 13' = 156''$

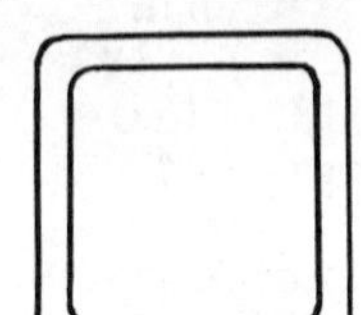

	A	Wt., lb/ft	P_{ALL}
7 × 7 × ½	12.4	42.05	207
8 × 8 × ⅜	11.1	37.6	196
10 × 10 × $\frac{5}{16}$	11.9	40.4	223

P_{ALL} values from Column Tables for $F_y = 46$, converted to $F_y = 36$ by factor 36/46 (approximate).

Try 8 × 8 × ⅜ by exact check:

$r_x = r_y = 3.09$

$$\frac{KL}{r} = 50.5$$

$F_a = 18.3$ ksi — Table 3-36, AISC

$P_{ALL} = 18.3(11.1) = 203 \text{ k} > 200$ — OK

Use Struct. Tube 8 × 8 × 3/8,
Wt. = 37.6 #/ft.

(5) Standard Pipe
10″ diameter (nominal)
Wt. = 40.48 lb/LF
$P = 226$ — AISC Column Table

(6) equal legs:

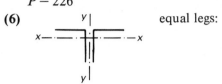

$r_y > r_x$ — Verify in ⅃L Shape Tables.

Use Column Tables for $KL = 13'$ and axis $X - X$. — $(KL/r)_x > (KL/r)_y$

	Wt., lb/ft	P_{ALL}
⅃L 6 × 6 × ⅝	48.4	210
⅃L 5 × 5 × ⅞	54.5	198

(7) ⅂Γ , long legs ⅜″ back to back:

	P_{ALL} $X-X$	P_{ALL} $Y-Y$	Wt., lb/ft
⅂Γ 8 × 6 × ½	216	213	46.0
⅂Γ 8 × 4 × ¾	292	217	57.4

Use smaller value of $X - X$ and $Y - Y$ for $KL = 13$, Column Tables.

(8) ¬⌐ , short legs ⅜″ back to back:

¬⌐ 8 × 6 × ¾	285	379	67.6
¬⌐ 8 × 6 × ⅝	235	271	57.8

Interpolated (availability of L 8 × 6 × ⅝ is uncertain).

(9) WT
WT 10.5 × 46.5 255 202 — Column Table

(10) Summary

Type	Lightest section, $P \geq 200$ k	Wt., lb/ft	P_{ALL}
W	10 × 45	45.0	208
or	12 × 45	45.0	202
S	None meet criteria		
Structural tube	8 × 8 × ⅜	37.6	203

The summary highlights some fairly general relationships related to efficiency of steel sections as compression members.

Pipe	10″ std.	40.5	226
⅃L (equal legs)	6 × 6 × ⅝	48.4	210
⅂Γ long legs b − b	8 × 6 × ½	46.0	213
¬⌐ short legs b − b	8 × 6 × ⅝	57.8	235
WT	10.5 × 46.5	46.5	202

In a less than perfect way, efficiency of steel usage in lb/ft correlates with efficiency in $/ft. The lightest section here is the structural tube.

9.8 SPECIAL CASES

9.8.1 Latticed Compression Members

The preponderant importance of the radius of gyration in determining allowable stress F_a may lead a designer to seek sections outside the limits of available single rolled sections. Several sections in Fig. 7.2(a) are made up of combinations of two or four rolled sections connected by lacing. Figure 9.17 shows such sections in more detail. Some of the special terms used in AISC 1.18.2 are defined graphically in that figure. Readers in most industrial areas will find many examples of such members in their daily travels (e.g., crane booms, members of bridge trusses, transmission towers).

The area resisting compression consists of the continuous chords only, as can be seen in Fig. 9.17(d). The lacing is necessary to act as lateral bracing for each of the individual chords. The members then act compositely, increasing r dramatically for the composite member. To avoid buckling of individual chord members, their L/r between lacing connections ($K = 1$) is prescribed to be less than or equal to the KL/r for the composite member. To brace the chords, the lacing must have the capacity to resist a nominal transverse shear arbitrarily set at 2% of the longitudinal compression. This is usually a more than sufficient criterion if there are no applied transverse loads causing shear and bending of the column. If such loads exist, a full analysis becomes necessary as for a truss. Such an instance arises in the boom design of the Stiffleg Derrick Case Study in Chapter 19.

Perforated plates are sometimes used in heavily loaded bridge members. The continuous net area of such plates is part of the compression area, while the portions between perforations act in a manner similar to lacing.

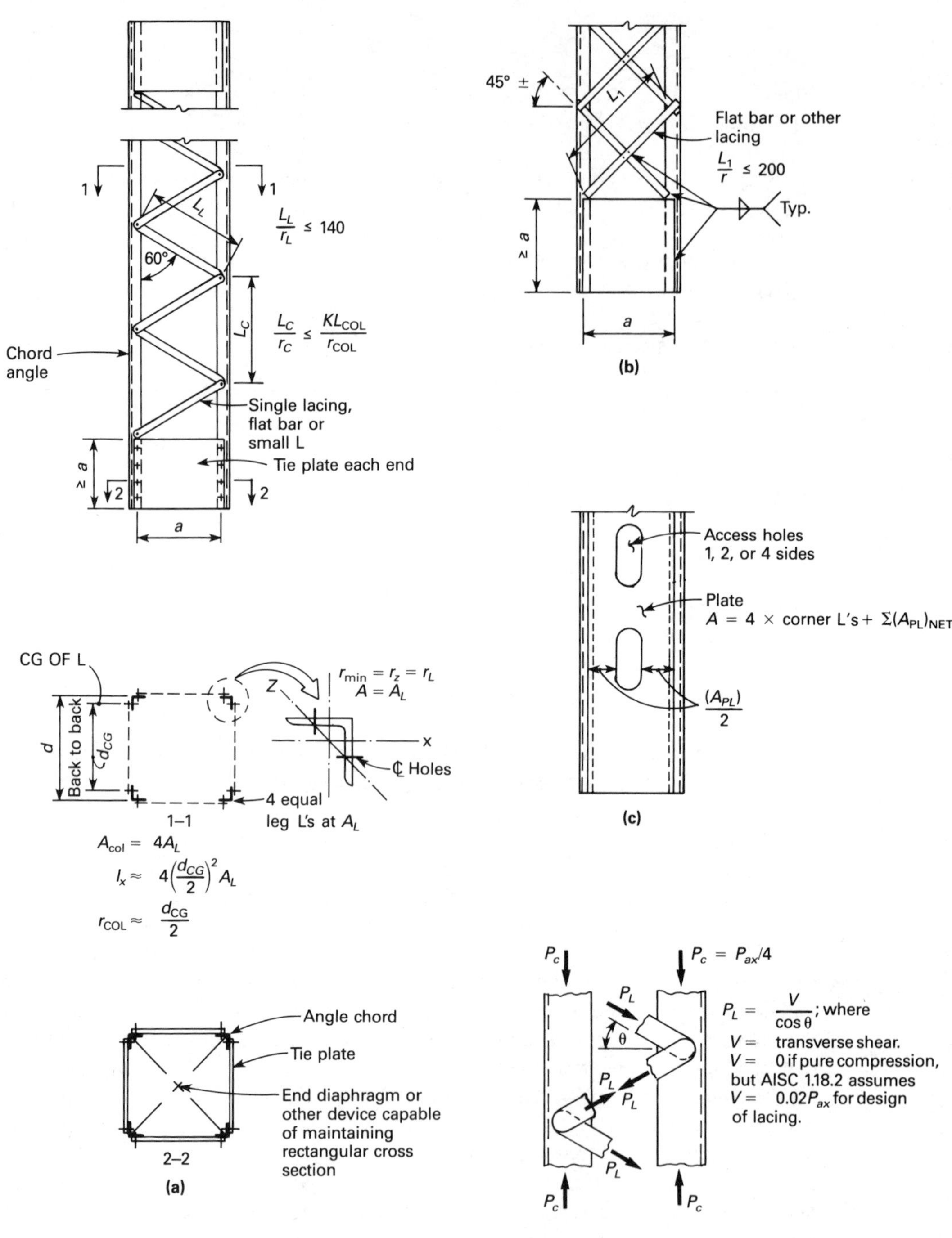

Figure 9.17 Built-up compression members: (a) riveted or bolted latticed compression member, single lacing; (b) welded latticed compression member, double lacing; (c) built-up compression member with perforated plates for access; (d) criterion, lacing design.

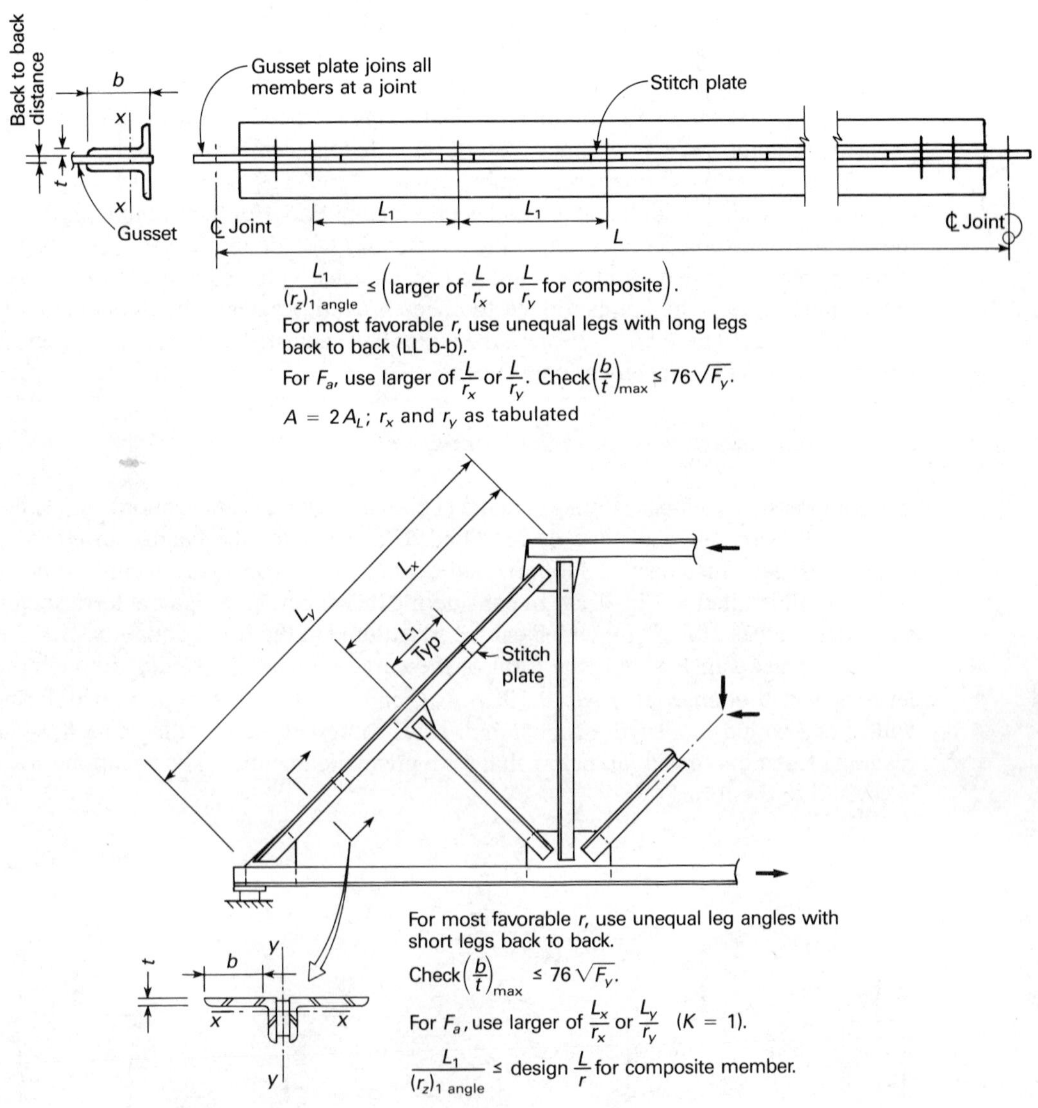

Figure 9.18 Double angles as compression members: (a) double angle, no intermediate lateral support; (b) compression diagonal with intermediate lateral support, one direction.

9.8.2 Double-Angle Compression Members

Double-angle compression members are commonly used in light trusses and columns with relatively light loads. They are symmetrical and easy to connect to other members with gusset plates. In Fig. 9.18, two common uses are illustrated. The decision about which legs to place back to back is based on the most favorable radius of gyration, which leads to long legs back to back if there is no intermediate lateral restraint and short legs back to back if restraint exists. Stitch plates are necessary to secure composite action, functioning like the lacing in latticed members. L/r for single angles is required to be no greater than $(L/r)_{max}$ for the composite member.

To find the most favorable combinations of area and radius of gyration, the designer can scan the double-angle tables provided in Part 1 of the Manual. Part 3 of the Manual lists compression capacities versus KL for double angles. Use of the tables is recommended to make comparative choices without unnecessary tedious calculations. Designers are cautioned to check b/t ratios against the limits in AISC Section 1.9.1.2. Where $b/t > 76/\sqrt{F_y}$, the tables have adjusted compression values by the criteria of AISC Specification, Appendix C.

9.8.3 Nonprismatic Compression Members

If a compression member is nonprismatic (i.e., has varying cross section), the Euler equation for critical load must be modified to account for the change in effective column stiffness that results. Nonprismatic columns exist in many forms, some of which are illustrated in Fig. 9.19. In considering this effect, investigators have sought equivalent values of I, (K_sI), which can be substituted in the Euler equation. If K_sI is used to replace I in Fig. 9.4, the form of the curve remains the same, although the level of it will change. In Figs. 9.19(b), (c), and (d), I_1 and I_2 represent differing values of I within the length of a column. If we represent the modified I as K_sI, the resulting Euler curve will fall below that for a prismatic member. The equation for P_{cr} is revised to the form

$$P_{cr} = \frac{\pi^2 E K_s I}{(KL)^2} \tag{9-8}$$

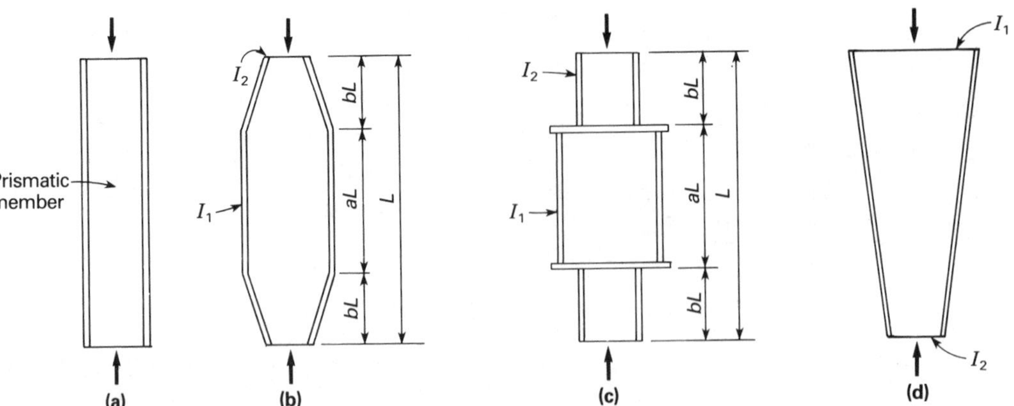

Figure 9.19 Nonprismatic members in compression.

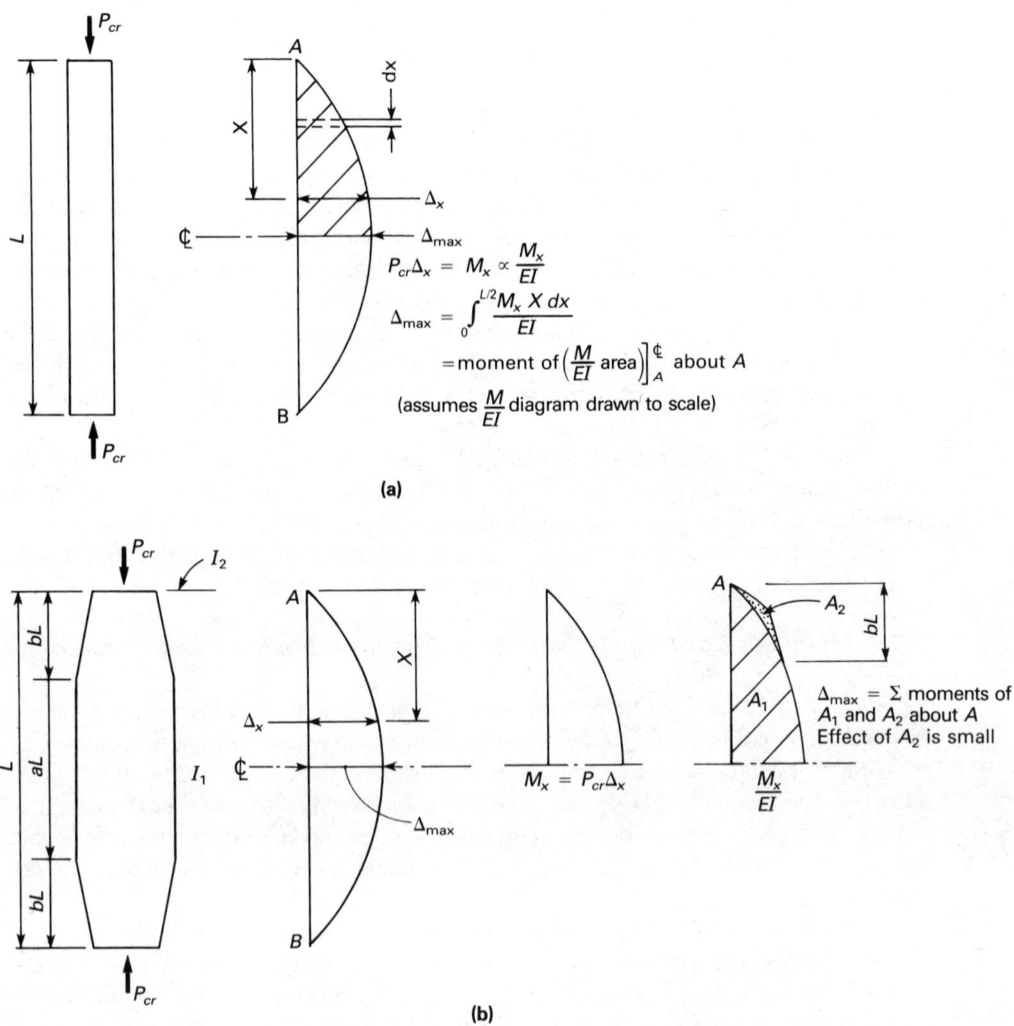

Figure 9.20 Effect of end taper on deflected shape: (a) prismatic case; (b) nonprismatic.

As in earlier discussion, K represents the nature of end constraints.

Values of the factor K_s are developed and tabulated for a number of cases in various references.[21,48,59] An important general observation deserves mention here. For columns like Figs. 9.19(b) and (c), the critical load is not very sensitive to even fairly large variations in I_2 compared to I_1.

Recall that the differential equation for P_{cr} was derived from the curve of deflection (Fig. 9.5). In Fig. 9.20, we examine qualitatively the effect of end tapering such as in Fig. 9.19(b). P_{cr} occurs when, for a deflected shape, the internal resisting moment at any point x is exactly enough to balance the external moment $P\Delta_x$. For the prismatic case, the familiar moment area theorems lead to a value of

$$\Delta_{max} = \text{moment of } \left(\frac{M}{EI} \text{ area} \right)_{A \text{ to } \mathbb{C}} \quad \text{about } A$$

Compare that case to the tapered strut shown in Fig. 9.20(b), where the reduced I in the end areas adds an additional area (A_2) to the area A_1 that would result if I_1 was constant. The value of Δ_{max} is increased by the moment of A_2 about point A. P_{cr} is reduced by that effect. The effect is small. Compared to a prismatic column with moment of inertia I, P_{cr} would be reduced only 10% (approximately) or less if $I_2/I_1 = 0.5$ or if $a = 0.5$. Similar results apply to Fig. 9.19(c).

We will have occasion to make use of tapered struts in the Stiffleg Derrick Case Study. The small reduction in capacity will be very helpful, making possible compact, efficient end details without requiring excessive material in the main portion of the members.

The AISC Specification, Appendix D, deals with regularly tapering compression members such as in Fig. 9.19(d) by establishing a method of calculating an equivalent K value, K_γ, for use with formulas (9-4) and (9-5), which are otherwise unchanged. For a number of common types of building frames, K_γ can be determined from curves in the Commentary, Appendix D. See also Chapter 15.

9.8.4 Members Not Doubly Symmetric: Torsional Buckling and Eccentricity

Another mode of buckling is possible due to instability of compression members in torsion. Torsional instability may result in failure at lower loads than is indicated by the Euler critical load curves. Such cases may occur in the case of open sections, such as angles and channels that do not have two axes of symmetry, so that the center of gravity and shear center do not coincide. Critical loads for such sections are developed theoretically by Bleich.[21] The methods covered in this book are not sufficient for safe design of such members except in simple cases.

The AISC Manual, in a preface to the double-angle column tables (Part 3), cautions against the use of single angles as struts for several reasons, including the (almost) inevitability of eccentric loading. The Manual does offer an interaction equation similar to those provided in Chapter 11 herein. Our suggestion:

1. Avoid the problem where possible by choosing doubly symmetric members for struts.
2. If others must be used, be sure to account for any bending moments due to eccentric loading or connections.
3. In such cases, tread carefully, and refer to other sources of authoritative advice where necessary.

9.8.5 Example: Double-Angle Compression Member

The compression diagonal in Example 8.9.1 has the same load and length as the tension diagonal. Select a double-angle member, using ASTM A36 steel, and without using the Column Tables.

CALCULATIONS	*REMARKS*
(1) Data:	
$P_{DES} = 70^K$	As in Example 8.9.1, wind load is not controlling.
Design for basic stress, F_a, per AISC 1.5.1.3:	
$K = 1.0$	AISC 1.8.2
$L = KL = 22' = 264''$	Given
Steel = A36 → $F_y = 36$ ksi	Given
$r_{min} \geq \frac{L}{200} = 1.32''$	AISC 1.8.4; r_{min} is smaller of r_x or r_y.
(2) Select double angle with long legs 3/8″ back to back.	ratio r_{min}/r_{max} closest to 1.0.
(3) $C_c = \sqrt{\frac{2\pi^2 E}{F_y}} = \left(\frac{2\pi^2(29{,}000)}{36}\right)^{\frac{1}{2}} = 126$	AISC 1.5.1.3.1
$F_a = \frac{12}{23}\,\frac{\pi^2 E}{(KL/r)^2} \quad \text{for } \frac{KL}{r} \geq C_c$	Assumes "long" strut, subject to verification. F_a may also be found in Table 3-36, Appendix A, AISC Spec. Truss diagonals are *main members*.

r_{min}	KL/r	F_a (ksi)	A_R (in.^{2})
1.32	200	3.73	18.77
1.50	176	4.82	14.52
1.75	151	6.55	10.69
2.00	132	8.57	8.17

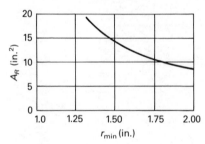

Use curve to check available area against A_R for r_{min} of tabulated ⌉⌈'s.

(4)

⌉⌈ (LLb-b)	A	r_{min}	Wt., lb/ft	Q_s	
8 × 4 × ½	11.5	1.51	39.2	—	NG
8 × 4 × ¾	16.9	1.55	57.4	—	
7 × 4 × ¾	15.4	1.62	52.4	—	
6 × 4 × ⅝	11.7	1.67	40.0	—	Use
5 × 3½ × ¾	11.6	1.53	39.6	—	NG

Data from Shape Tables, AISC Manual, Part 1. "—" in Q_s column indicates b/t conforms to AISC Sec. 1.9. If $Q_s < 1.0$, consult App. C, AISC Spec. Least weight ⌉⌈ is in long column range.

(5) Stitch plates:

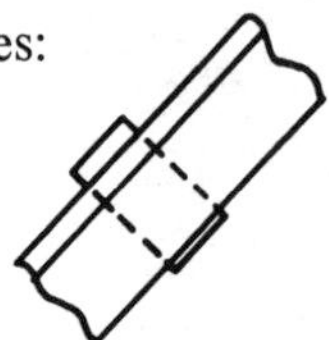

Provide 2 × ⅜″ PL's × 6½″, $l \leq 158r_z$, where r_z is r_{min} for single angle.

AISC 1.18.2.4; $KL/r = 158$ for composite section.

$$l_{max} = \frac{158}{12}(0.864) = 11.381$$

One stitch PL needed at mid-length between gussets.

(6) Choose ⅂⌈ 6 × 4 × ⅝, LL b-b. Use one 2″ × ⅜″ × 6 ½″ stitch plate at mid-length.

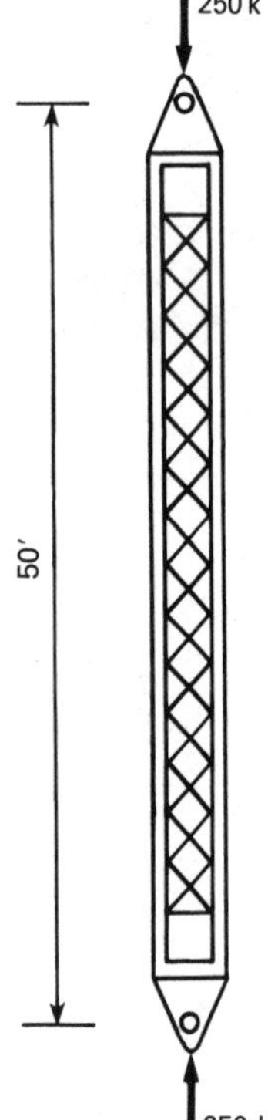

9.8.6 Example: Latticed Strut

An upright compression strut is to be designed for a total axial load (dead + live) of 250 kips, including weight of strut. Ends are to be pinned,* with connecting pins at 50 feet on centers. Design a latticed strut with welded details. (Design of end connections are not required.) Calculate weight per foot.

*Pinned connections are discussed in Chapter 13 and are encountered in the Stiffleg Derrick Case Study on several members.

CALCULATIONS	REMARKS
Follow Flow Chart 9.1(b), adapting as necessary. Follow AISC rules for built-up members (Section 1.18, Spec.).	See also Table 9.1
(1) Loads: $P_{D+L} = 250\text{ k} = P_{max}$ $K = 1.0$ $KL = 50' = 600''$	Given as control for design Pinned ends
(2) Open-latticed section:	See Fig. 9.17(a). L's are continuous.

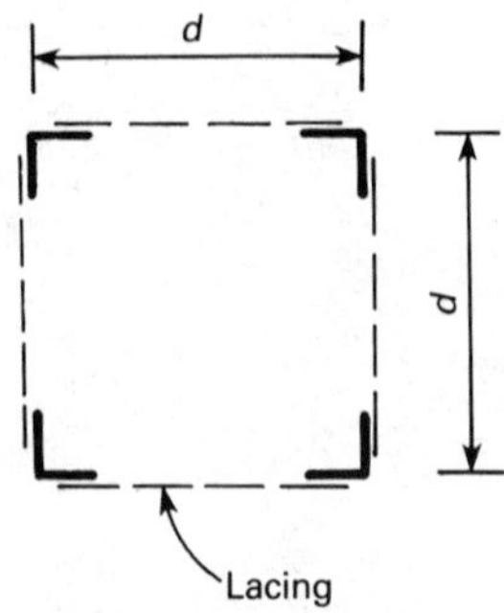

Type of steel:

Assuming $KL/r \approx 50 \ll C_c$

Since latticed section, designer can economically choose a relatively large r.

F_y	F_a	A_{CH}
36 Ksi	18.35	13.62
50	24.35	10.27
60	28.00	8.93

$A_{CH} = 4A_L$ = area of 4 L's. Tables 3-36, 3-50. For $F_y = 60$, use AISC Sec. 1.5.1.3.1. Extrapolation from Table 3-50 is not accurate.

High-strength steel is useful to reduce A_{CH}.
A588 is corrosion resistant; $F_y = 50$.
A572 available up to Gr. 65; $F_y = 65$, but doubtful for notch toughness for $F_y > 50$

See Table 4.1 and AISC Table 1, Part 1.

Use A588.

(3) *Select Chord L's*

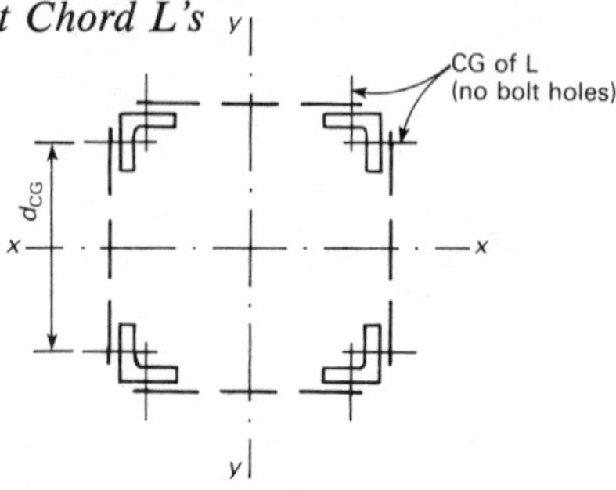

$(KL)_x = (KL)_y$ — Since square.

$$r = \frac{I}{A_{CH}}$$

$$\approx \frac{A_{CH}(d_{CG}^2)}{A_{CH}(4)}$$

$$r = \frac{d_{CG}}{2} \quad \text{Close enough}$$

Exact value of r is *slightly* higher than indicated since I includes a small term accounting for I of the L's around their own CG's.

Make $d_{CG} = 24''$; $KL/r = 50$.

$A_R = 10.27 = 2.57$ in.^{2} for each L — From step (2), above

Use L $3 \times 3 \times \frac{1}{2}''$ if conforms to AISC Sec. 1.9:

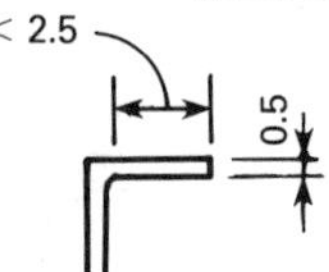

$$\frac{2.5}{0.5} = 5.0 < 10.75 = \frac{76}{\sqrt{F_y}}$$

AISC 1.9.1.2

Use L $3 \times 3 \times \frac{1}{2}$ each corner:

$$A_{CH} = 4(2.75) = 11.0 > 10.27$$

For single angle,

$$r_z = r_{min} = 0.584$$

L $4 \times 3\frac{1}{2} \times \frac{3}{8}$ is acceptable and somewhat lighter. Equal leg L is selected so that lacing connections to each leg will be identical; a debatable decision.

(4) Select lacing: Use 45° double lacing with flat bars, welded.

Follow AISC Sec. 1.18.2.6. Sketch, to scale, helps in decision to use double lacing. For weld symbols see Chap. 12.

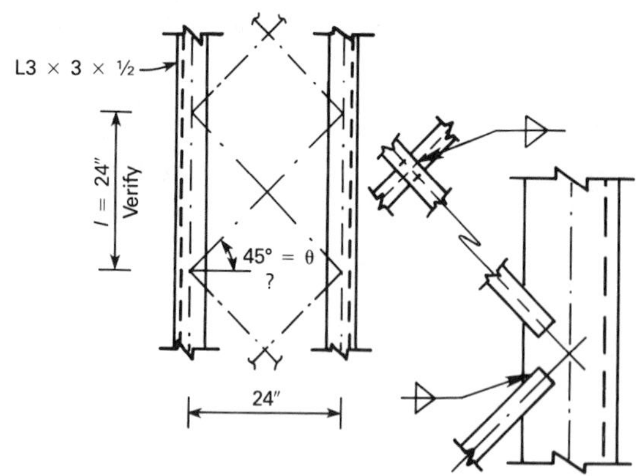

Check Kl/r for single chord angle:

$r_{min} = r_z = 0.584, \quad l = 24'', \quad K = 1$

$$\frac{KL}{r} = \frac{24}{0.584} = 41.10 < 50 \qquad \text{OK}$$

See Fig. 9.16 for K and AISC 1.18.2.6, first sentence.

Shear:

$$V = 0.02P_{AX} = 5 \text{ k} = 2.5 \text{ k/face}$$

AISC 1.18.2.6, sentence 2. Two parallel faces resist shear.

For bar, $l/r \leq 200$, where

$$l = \frac{0.7(24)}{\cos 45^\circ} = 23.76$$

1.18.2.6

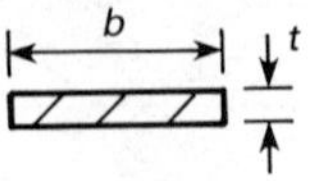

$r_{min} = 0.12$

$t_{min} = r_{min}\sqrt{12} = 0.42$

$$r = \sqrt{\frac{bt^3}{12bt}} = \frac{I}{A} = \frac{t}{\sqrt{12}}$$

Use $t \geq \frac{7}{16}''$, at $l/r = 200$, $F_{as} = 6.22$ ksi.

1.18.2.6 permits lacing to be treated as "secondary" members. Reasonable for this case.

Axial load on single bar:

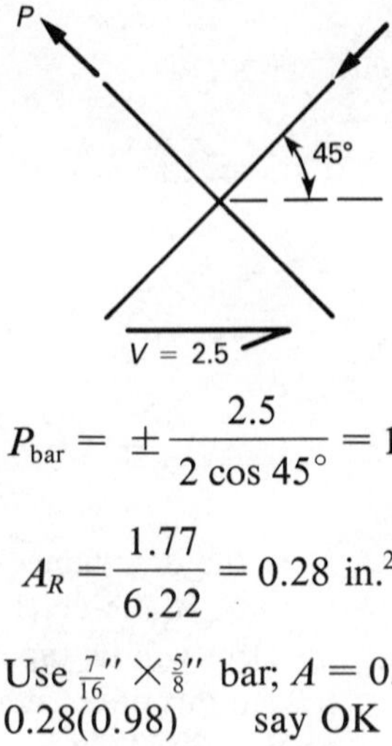

$$P_{bar} = \pm\frac{2.5}{2\cos 45^\circ} = 1.77 \text{ k}$$

$$A_R = \frac{1.77}{6.22} = 0.28 \text{ in.}^2$$

Use $\frac{7}{16}'' \times \frac{5}{8}''$ bar; $A = 0.27 = 0.28(0.98)$ say OK

Designer accepts $A < A_R$ (slightly), since assumed l was longer than distance between welds.

(5) End battens (tie plates):

AISC 1.18.2.5

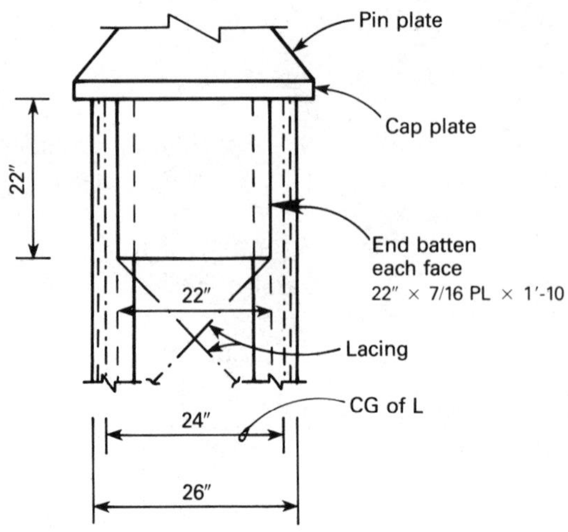

(6) Design end connection:

Not done here.

(7) Summary

This result may be compared with the design of the back leg in the Stiffleg Derrick Case Study.

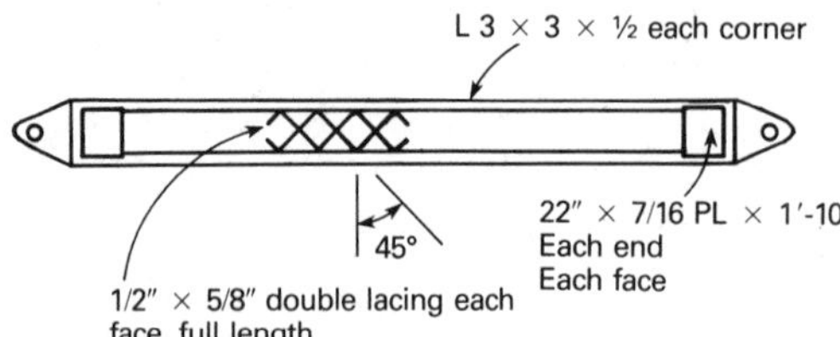

Weight, lb/LF:

$$4\text{L's } 4 \times 4 \times 3/8 = 4(9.8) = 39.2 \text{ lb/ft}$$

Lacing:

$$\frac{7}{16} \times \frac{5}{8} \times \frac{490}{144} = 0.93 \text{ lb/ft of bar}$$

$$4(2)(1.41)(0.93) = 10.5 \text{ lb/ft of strut}$$

Battens:

$$\frac{8}{50} \times \left(\frac{22}{12}\right)^2 \left(\frac{7}{16}\right)\frac{490}{12} = 9.6$$

$$\text{total} = 59.3 \text{ lb/ft}$$

Note the dominance of the end battens. In the derrick design, the strut will be tapered to reduce this effect. Density of steel is 490 lb/ft.[3]

Say 65 lb/LF with end connections.

The end detail is likely to be heavy; average weight/ft is increased significantly.

Readers may wish to compare the result here with the results of Problems 9.12 and 9.13.

PROBLEMS

In the following problems, the solution sought is the lightest acceptable member available with the type of cross section specified. Loads are dead + live unless otherwise noted. $F_y = 36$ unless $F_y = 50$ results in weight saving of at least 10%. Record the weight if not part of the section identification. Unless instructed otherwise by the problem statement or a class instructor, Column Tables in Manual, Part 3, are permitted where applicable.

9.1. $L = 40'$, $P = 150$ k:

(a) W section
(b) Structural tube
(c) Pipe
Do not use Column Tables for Problem 9.1.

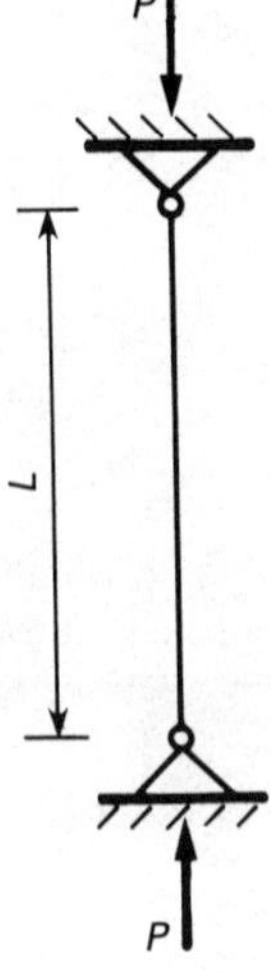

9.2. $L = 40'$, $P = 150$ k:

(a) W section
(b) Structural tube
(c) Pipe

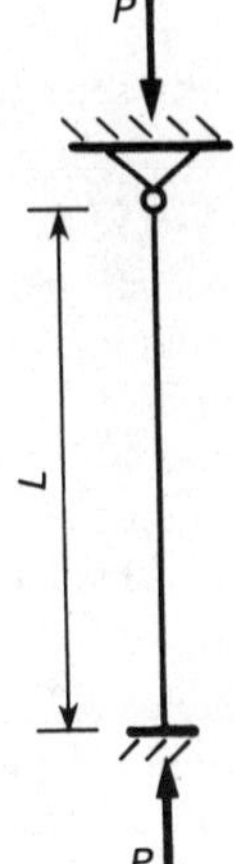

9.3. $L =$ story height $= 40'$
$P = 150$ k from dead + live loads
$Q = 0$

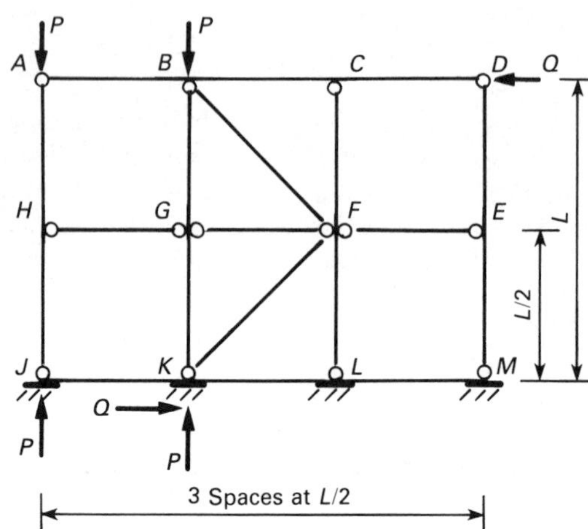

Mid heights of columns are braced in direction shown only. All restrained perpendicular to plane of bent at top and bottom.
Choose W sections for:
(a) *AJ* **(b)** *BK*
(c) *HG* **(d)** *FK*

9.4. Do Problem 9.1, parts (a) and (b), for all combinations of length and load as follows: $L = 15'$, $30'$, $45'$; $P = 50$ k, 100 k, 150 k.

9.5. Do Problem 9.2(a) for all combinations of length and load: $L = 15'$, $30'$, $45'$; $P = 50$ k, 100 k, 150 k.

9.6. Do Problem 9.3 with $Q = 75$ k from seismic effects. All other conditions are the same.

9.7. Do Example 9.8.6 for a length of 80 ft.

9.8. For the truss in Problem 8.7 and assuming the member forces given in the problem statement with corrections found in solution to Problem 8.7:

(a) Select sizes for U_0L_0, U_1L_1, and U_2L_2.

(b) Choose member sizes for U_0U_2 and U_2U_4, assuming lateral restraints at U_0, U_2, U_4, and U_6. The members are continuous over joints U_1 and U_3.

(c) If you introduced a member U_3L_3, what effect would that have on your selection for member U_2U_4? What size would you select for U_3L_3?

9.9. **(a)** Compare your solutions for Problems 8.7 and 9.8. Study the typical joint. If all your members are not of the same nominal depth, revise your selections as necessary to make all nominal depths the same.

(b) On a line diagram of the truss, enter all member forces and member sizes.

9.10. **(a)** If, in Problems 8.7 and 9.8, you were to decide to use steel conforming to ASTM A441, what would be your selections for member sizes? Show member forces and sizes on a line diagram of the truss.

(b) Assuming the cost of A441 steel, in \$/ton, delivered and fabricated, is 5% more than A36, which steel would you decide to use?

9.11. If you wish to use double angles for the truss of Problem 8.7 and connect by single welded gusset plates similar to Example 8.9.1, what sizes would you choose? Show on a line diagram of the truss. You should include a member U_3L_3. Lateral supports may be assumed at U_0, U_2, U_4, U_6, L_0, L_2, L_4, and L_6.

9.12.

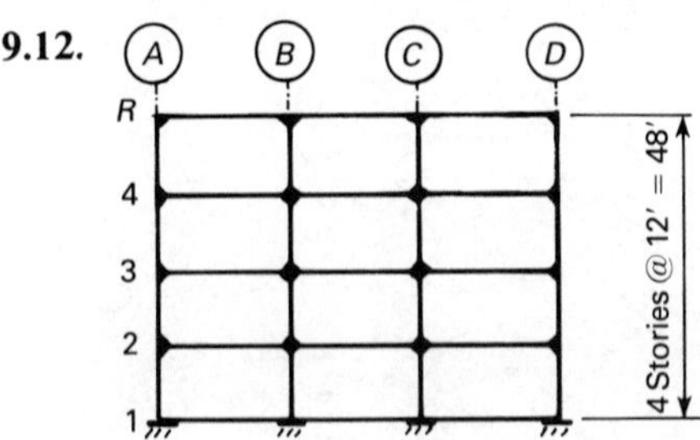

Elevation of typical rigid frame bent
(N–S bent shown; E–W bents similar).

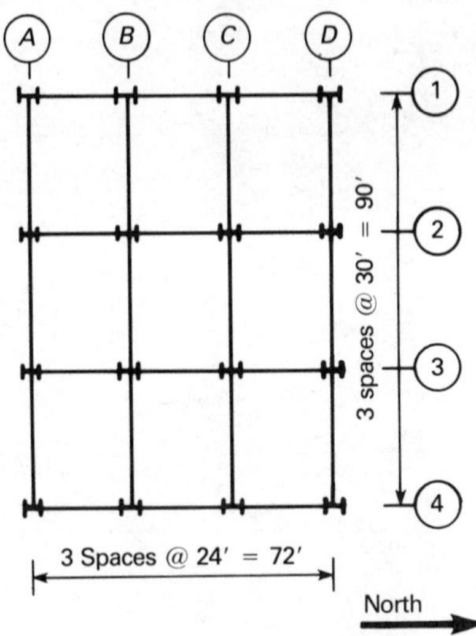

Typical plan: Floor framing

Symbols:

= rigid connection; small rotation of joint is possible, but angle between members is unchanged at joint

= fixed; rotation of column at joint = 0

Indicates orientation of W columns on floor framing plan; R = roof; 1, 2, etc. = floors

All E–W girders, W18 × 76
All N–S girders, W16 × 50
All columns W14 × 99
$F_y = 50$ ksi

(a) Using Fig. 9.14, what are the values of K for columns B1 and B2 in the first and second stories? Solve separately for north–south and east–west planes.

(b) For Column B2 between the second and third floors, what size girders would be required to make the values of $K = 1.2$? See Table 9.2, case (c).

The Tier Building Case Study, Chapter 18, involves a multistory rigid frame, similar to the one here.

9.13. Do Problem 9.12(a) assuming that all floors are connected by struts to an adjacent structure that prevents sidesway.

9.14. Do Example 9.8.6 using a structural tube. Find the size without column tables. The result may be checked with the tables.

9.15. Investigate the cost of steel and fabrication for the strut of Example 9.8.6 and Problem 9.14. On the basis of fabricated cost, which is less expensive and by what percent? (This is a longer problem than a student would do in a normal homework assignment. It addresses an important design issue that could be the subject of a design report. The problem would be more complete if the design of the end details was included, which would require methods such as those discussed in Chapter 12.)

10 Prismatic Members in Uniaxial Bending

10.1 SCOPE OF CHAPTER

Our consideration of flexural members opens with a narrowly defined but very common case: straight, prismatic beams subjected to uniaxial bending. A beam, by usual inexact definition, is a stiff member that supports transverse loads (i.e., loads perpendicular to its longitudinal axis). By mobilizing internal resistance to transverse shear and bending moment within its cross section, it transfers the effects of the transverse loads to remote supports. A more general term for bending resistance, with or without shear, is flexure. The term *beam* is general and covers a class that includes beams, girders, joists, purlins, lintels, and many other flexural members.

Figure 10.1 shows a number of beams, most of which, consistent with the common conception, are horizontal. The tree of Fig. 10.1(f), a cantilever beam in its response to wind pressure, demonstrates that beams need not be horizontal. The wood fibers of the tree can resist both tensile and compressive stresses, a characteristic shared with steel, which makes steel ideally suited to resist flexure.

Prismatic beams, having constant cross section, can usually be selected from rolled shapes. We will for the moment assume that there are no significant axial loads or torsion applied to the beams and that bending takes place about only one principal axis of the cross section. The next chapter will extend the discussion of flexure to cases where the members are subjected to bending about more than one axis, to both axial load and bending moment or to bending moment with torsion. When we get beyond the realm of utility of existing rolled shapes, it is often necessary to design

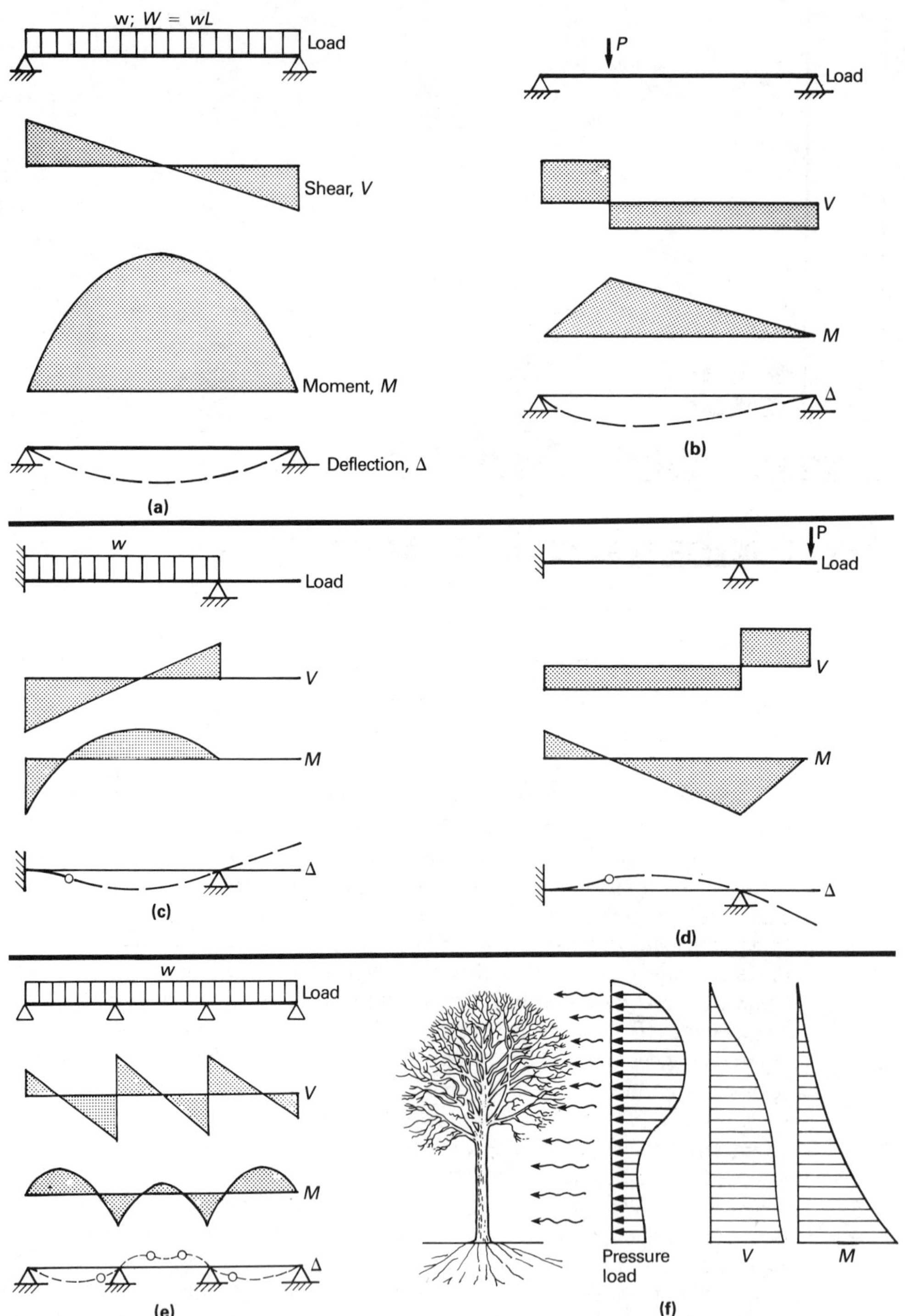

Figure 10.1 Load, shear, moment, deflection.

members of varying cross section. These may be made, as in Chapter 14, from plates or rolled shapes or from a combination of steel and concrete, using nonprismatic cross sections, varying with the changing requirements of shear and moment at different points along the member.

10.2 MECHANICS OF BENDING RESISTANCE: ELASTIC CONDITIONS

10.2.1 Background: Shear and Moment

Readers have no doubt met the mechanics of resistance to bending in studies of mechanics of materials and/or structural analysis. We will review some of the concepts here as background to further discussions of design selections. Many excellent standard texts offer more complete development.

In each case of Fig. 10.1, the characteristic shear and moment diagram is drawn. From those shear and moment diagrams, it is possible to derive a curve showing the deflection of any point along its length. The deflection curve is smooth except at pinned joints or locations where elastic stresses are exceeded. Many more cases are given in Part 2 of the AISC Manual under *Beam Diagrams*.

In both Fig. 10.1 and the AISC diagrams, the assumption is made that the beam cross sections are symmetrical and loads act in the plane of a principal symmetrical axis, thus avoiding torsional twisting. Bending takes place about the other principal axis.

Since structures usually must satisfy not one case of loading, but several different cases and combinations of cases, the designer must seek the worst possible combination and the shears and moment resulting from them.

10.2.2 Shear, Moment, M/EI Diagrams

The importance of shear and moment diagrams to a designer cannot be overemphasized. They are the key to intelligent selection of beams, and a powerful source of insight into the behavior of complex continuous flexural systems. Moment and shear diagrams show the designer in one graphic picture the values of shear and moment at all points along a beam simultaneously. By superposing (adding) diagrams drawn for different cases of loading, moment and shear *envelope* diagrams emerge, as in Fig. 10.2, which tell a designer the maximum values of design shear and moment at each point on a beam under all conditions governing the design.

Where conditions indicate the desirability of varying the cross sections these diagrams, and their derivative, the M/EI diagrams, are powerful tools available to the designer for understanding the effect of various design choices on structural behavior. M/EI diagrams and the areas bounded by them are used in *moment area* methods of analysis to derive the deflected shapes of flexural members, a central requirement in the analysis of indeterminate structures. By studying M/EI diagrams and making

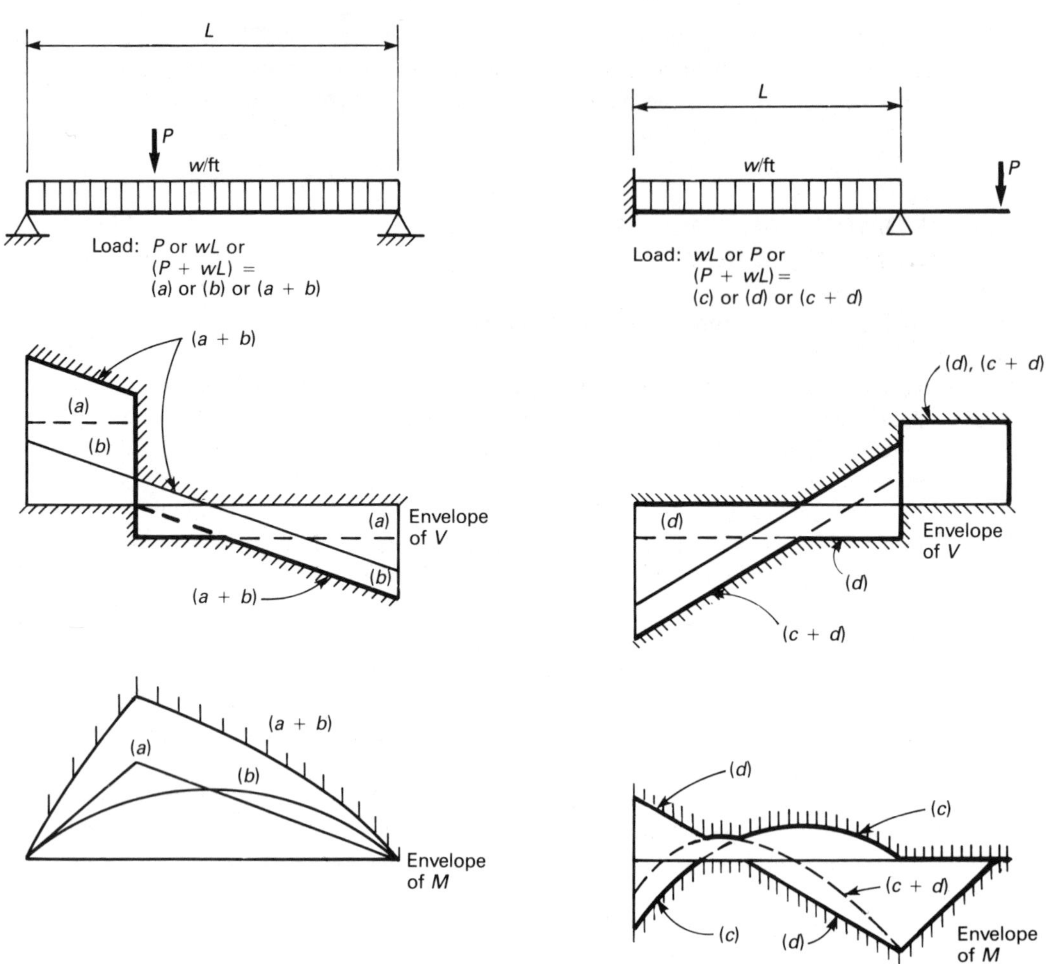

Figure 10.2 Shear and moment envelopes: cases (*a*) through (*d*) are as in Fig. 10.1. Design shear and/or moment is maximum of either case or combined case. Scale of moment or shear must be the same for all load cases in a single diagram.

appropriate design decisions, a designer can manipulate the deflected shape of a structure and in the process change the analysis itself, controlling such things as where maximum moments and shears will be and their magnitudes, the relative magnitude of support reactions, and the amount of deflection.

This chapter will only begin to reveal the uses of these diagrams for simple cases. They will appear again in many later chapters. Readers who are not accustomed to doing so are urged to use these diagrams regularly in the analysis and design of flexural systems. Drawn even in rough form, but preferably to scale, they will help guide the designer's thinking toward informed choices.

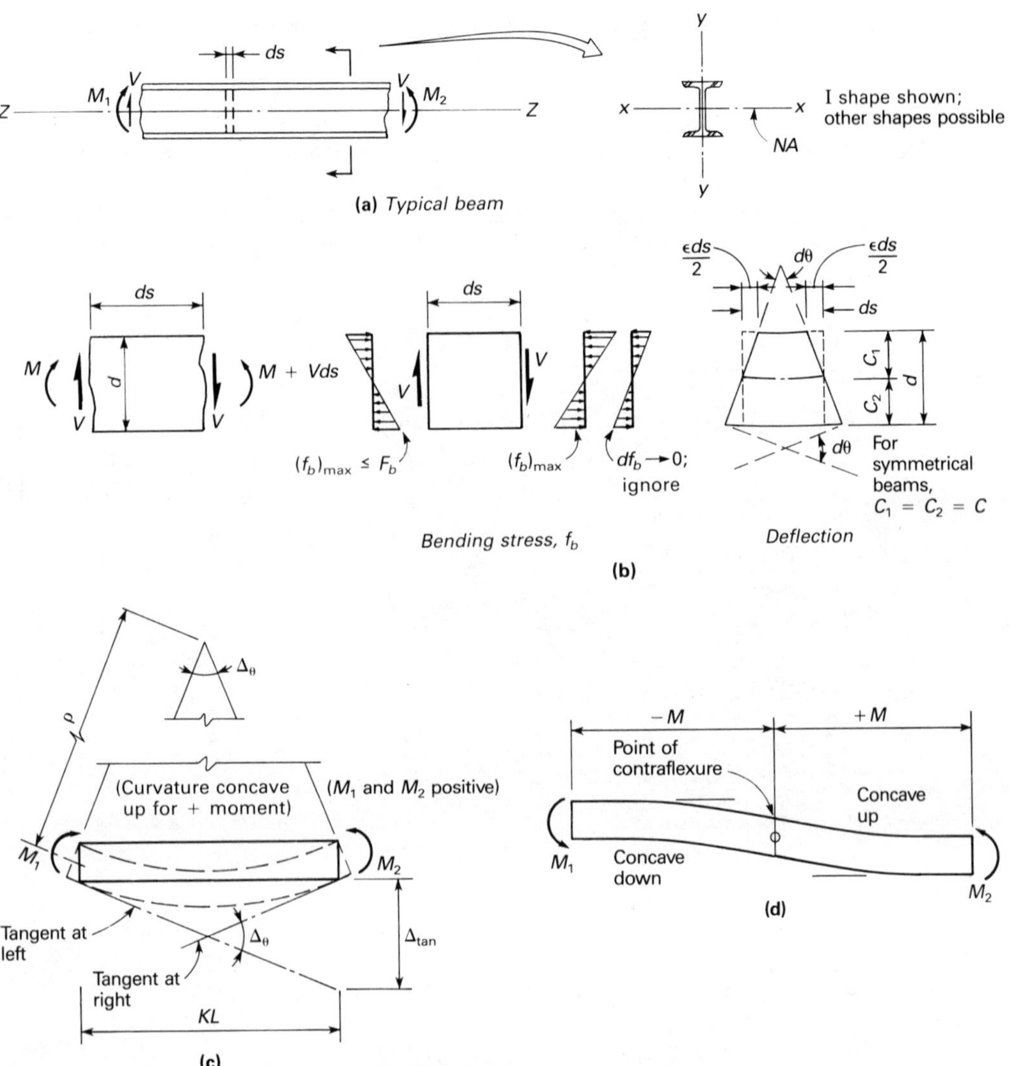

Figure 10.3 Bending moment and curvature.

10.2.3 Effect of Bending Moment

The effects of bending moment can be seen in Fig. 10.3(a), which could represent any piece of any of the beams shown in Fig. 10.1. In the case shown, bending moment causes compressive stress in the upper fibers and tensile stress in the fibers below the neutral axis. By common convention, this is arbitrarily defined as the case of positive moment, a condition that gives an elastic curve concave up (i.e., the center of curvature is above* the beam at the point under consideration). On a free body of a piece of the beam of differential length, *ds*, we have drawn both shear and moment on the *cut* faces. The right-hand moment differs from the left-hand moment by the amount *Vds*. This reveals the mechanism of growth of moment due to shear. However, the change of moment within the infinitesimal length *ds* is also infinitesimal. We will ignore its effect on longitudinal stress in what follows.

For linearly elastic materials such as structural steels in the range below yield stress F_y, stresses and strains both vary linearly with distance from the neutral axis. Bending stress,

$$f_b = \frac{My}{I} \tag{10-1}$$

and is maximum at the outer fibers, where

$$(f_b)_{\max} = \frac{Mc}{I} \tag{10-2}$$

Defining *section modulus* as,

$$S = \frac{I}{y_{\max}} = \frac{I}{c}$$

formula (10-2) can be rewritten as

$$(f_b)_{\max} = \frac{M}{S} \tag{10-3}$$

Noting that strain

$$\varepsilon = \frac{f_b}{E}$$

$$(f_b)_{\max} = \frac{Mc}{I} = \varepsilon E$$

$$\frac{\varepsilon}{c} = \frac{M}{EI} \tag{10-4}$$

From Fig. 10.3 it can be seen that $\varepsilon/c = d\phi$, the change in angle between tangents to the elastic curve at the ends of the differential length. Over a longer length k, the change in angle between tangents to the elastic curve is

*This common convention is easy to follow when the member is horizontal, loads are vertical, and applied moments are around a horizontal axis. It breaks down when these conditions are not fulfilled. In general, whenever the sign of bending moment changes, the direction of curvature changes as well. The transition point between positive and negative moment is a position of zero moment, called a point of contraflexure since curvature is of opposite sense on the two sides of such a point. It should be easy to identify points of contraflexure in Fig. 10.1.

$$\Delta\phi = \int_{KL} \frac{Mds}{EI} \tag{10-5}$$

The shape of the elastic curve is thus established by the progressive angle changes that take place in response to the relationship M/EI. *Curvature* may be defined as $1/\rho$, where ρ is the radius of a circular curve.* The curvature of an initially straight beam at any point is almost exactly

$$\frac{1}{\rho} = \frac{M}{EI} \tag{10-6}$$

Thus, either high values of moment or low values of moment of inertia lead to increased curvature and, therefore, increased deflection of the elastically curved beam from the initially straight line joining its ends. Elastic deflections can be calculated by, among others, the classical slope deflection methods or the moment-area relationships derived from them.

10.2.4 From Load to Deflection by Integration

A four-stage process of integration leads from a load q at any point on a beam to its effects on shear, moment, slope, and deflection of the beam. The process, expounded in detail in many standard references (e.g., Popov[46]) is set down here using the symbols of this chapter. Z is the longitudinal axis of a beam; Δ is deflection of the elastic curve from the original straight line joining its ends.

$$\text{Load} \quad EI\frac{d^4\Delta}{dz^4} = q(z)$$

$$\text{Shear} \quad EI\frac{d^3\Delta}{dz^3} = \int_0^z q\,dz + C_1$$

$$\text{Moment} \quad EI\frac{d^2\Delta}{dz^2} = \int_0^z \int_0^z q\,dz\,dz + C_1 z + C_2 \tag{10-7}$$

$$\text{Slope} \quad EI\frac{d\Delta}{dz} = \int_0^z \int_0^z \int_0^z q\,dz\,dz\,dz + \frac{C_1 z^2}{2} + C_2 z + C_3$$

$$\text{Deflection:} \quad EI\Delta = \int_0^z \int_0^z \int_0^z \int_0^z q\,dz\,dz\,dz\,dz + \frac{C_1 z^3}{3\cdot 2} + \frac{C_2 z^2}{2} + C_3 z + C_4$$

C_1 through C_4 are the values of shear, moment, slope, and deflection at the lower limit of integration and are derived from known conditions at the boundaries.

*The elastic curve is not circular, but a differential portion of the curve is approximately circular. The radii of the differential circles change as the moment changes.

The same relationships can be linked in reverse order by differential equations.

$$\Delta = \text{deflection of the elastic curve}$$

$$\theta = \frac{d\Delta}{dz} = \text{slope of the elastic curve}$$

$$M = EI\frac{d^2\Delta}{dz^2}$$

$$V = \frac{dM}{dz} = \frac{d}{dz}\left(EI\frac{d^2\Delta}{dz^2}\right)$$

$$q = \frac{dV}{dz} = \frac{d^2}{dz^2}\left(EI\frac{d^2\Delta}{dz^2}\right) \tag{10-8}$$

Figures 10.4 and 10.5 illustrate the stages of successive graphical integration of areas that lead from a diagram of load, to a diagram of shear, then to a diagram of moment. Applying $1/EI$ to ordinates of the moment diagram gives the M/EI diagram, which in the next two stages of area integration leads to slope, and then to deflection. The beam in Fig. 10.4 is prismatic; all M/EI ordinates are proportional to moments. Figure 10.5 has a beam with different I's applicable to different parts of the beam. In each stage except conversion from M to M/EI, the ordinate of the prior diagram is the

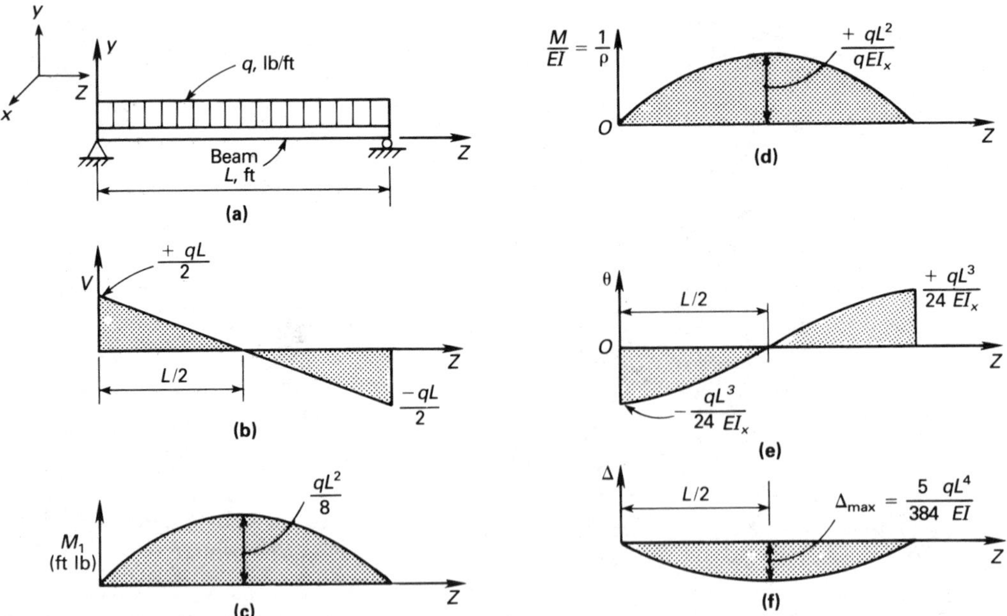

Figure 10.4 Load to deflection, prismatic beam.

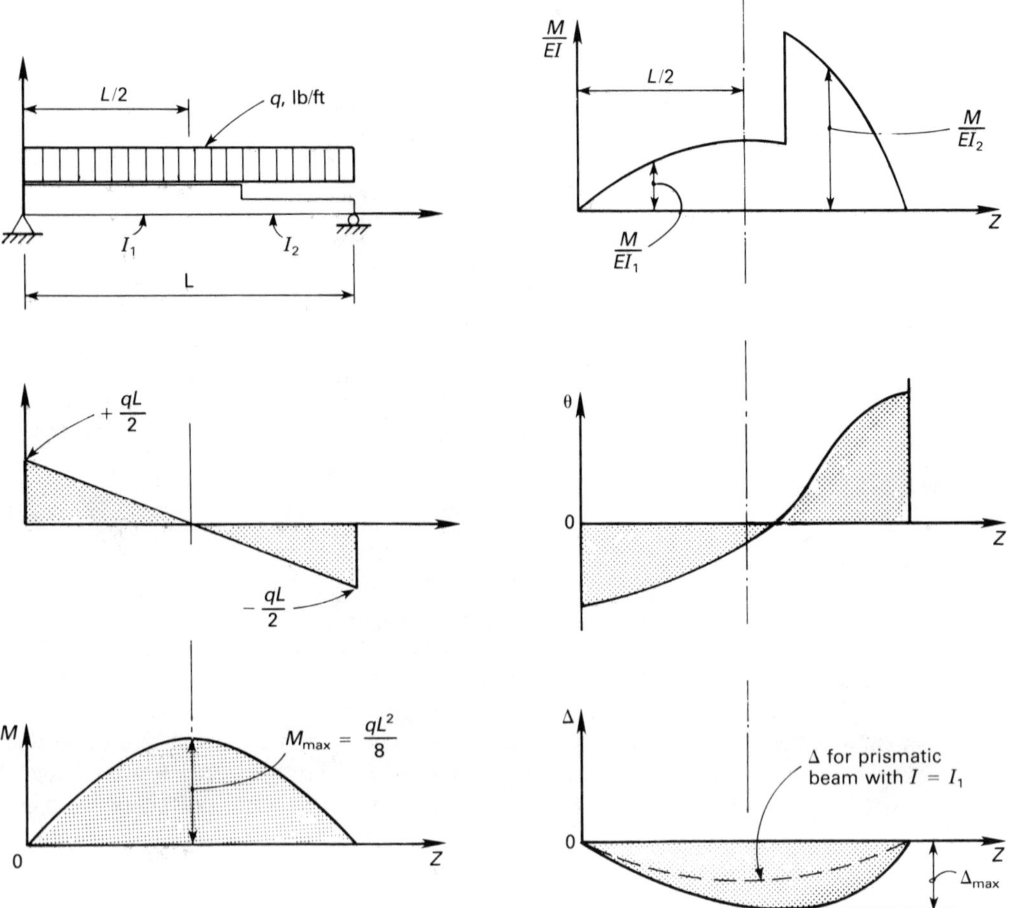

Figure 10.5 Load to deflection, nonprismatic beam.

slope of the diagram being drawn. These relationships and procedures are general and can be applied to any portion of a beam if the conditions at its end boundaries are known.

10.2.5 Effect of Shear

In addition to changes in bending moment, shearing forces also cause shear stresses and the additional beam deflection attendant on them. These are shown qualitatively in Fig. 10.6. A cross section subjected to a shear force develops resisting shear stresses that vary continuously from zero at the free top and bottom surfaces to a maximum at the neutral axis of the cross section. A square element of the cross section is subjected to equal and opposite couples, causing angle change, γ,

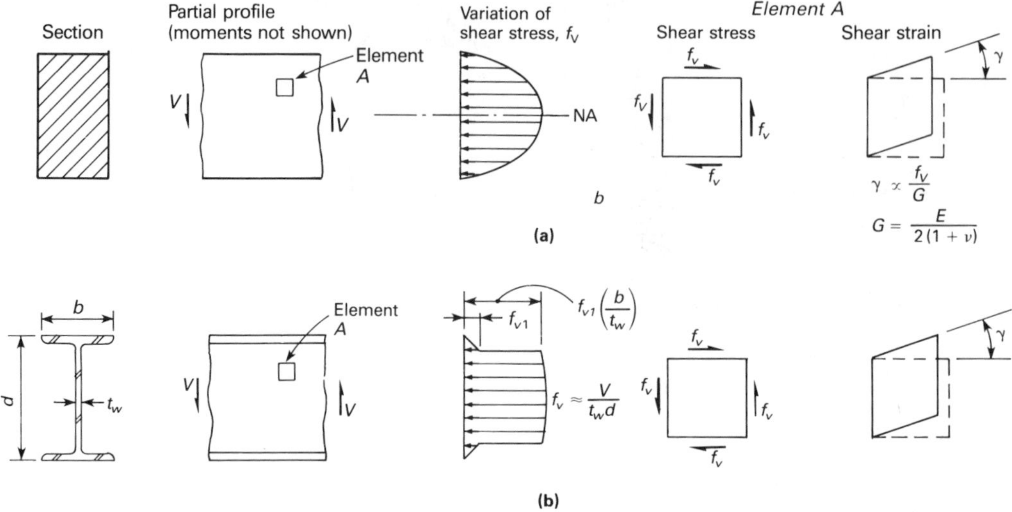

Figure 10.6 Variation of shear stress and shear strain.

proportional to f_v/G. G is the shear modulus, equal, for steel, to $0.38E$. The maximum value of shear stress is always a significant factor in considering the strength of a beam. The beam deflection caused by shear is usually very small compared to the flexural deflection. Except in the case of very deep, short beams subjected to high shears, it can usually be ignored in beam design. It is not ignored in more advanced elastic theory, which must be invoked to understand the localized action of many details.

10.2.6 Design Considerations: Bending and Shear Effects

Several criteria for judging the adequacy of a prismatic elastic beam emerge from the preceding discussion:

1. To satisfy requirements of strength, the maximum bending stress, f_b, must not exceed an allowable bending stress, F_b. From Eq. (10-3), it can be seen that the governing parameter controlling $(f_b)_{\max}$ is the section modulus (S) of the beam, a geometric property related to both the moment of inertia and the depth. $S = 2I/d$ if the section is symmetrical around the axis of bending (X axis in Fig. 10.3).
2. As an additional requirement of strength, the maximum shear stress, f_v, must not exceed an allowable shear stress, F_v. The value of maximum shear stress depends on the nature of the cross section, particularly of the area resisting shear (A_v).

3. Often, considerations of use of the structure set limits on permissible deflections. From Eq. (10-4) and (10-5), it can be seen that the geometric property of cross-sectional area that limits deflection is the moment of inertia, I. Deflection is inversely proportional to I.
4. Maximum shear and maximum moment do not usually exist at the same point along the length of a beam. Beam selection must consider both.
5. As a consequence of Criterion 4, it is not usually possible to select a beam that fully utilizes both its shear capacity and its moment capacity. Usually, a beam selected for moment is underutilized with respect to its shear strength. Occasionally the converse is true. The common practice of selecting beams for bending strength alone sometimes leads to shear failure.
6. All other things being equal, to seek economy in the selection of beam cross sections, we usually look for the lightest available section (smallest area of cross section) that satisfies simultaneously requirements of bending strength (S), shear strength (A_v), and stiffness (I).

The first three criteria can be summarized in the following equations, which can be used to help guide beam selection.

$$F_b \geq (f_b)_{\max} = \frac{M}{S}$$

$$\text{Provide} \quad S \geq S_R = \frac{M}{F_b} \tag{10-9}$$

$$F_v \geq (f_v)_{\max} = \frac{V}{K_v A_v}$$

$$\text{Provide} \quad A_v \geq (A_v)_R = \frac{V}{K_v A_v} \tag{10-10}$$

$$\text{Provide } I \geq I_R \tag{10-11}$$

where the subscript R = *required*. In Eq. (10-10), A_v is the part of the cross-sectional area that is effective in resisting shear (e.g., the web of a W section that is loaded in the plane of the web):

$$K_v \approx \frac{(f_v)_{\text{ave}}}{(f_v)_{\max}}$$

I_R is the value of I that will limit deflection to the maximum amount permitted.

10.3 MECHANICS OF BENDING RESISTANCE: STABILITY AND PLASTIC BEHAVIOR

10.3.1 Limitations of Elastic Behavior

Two basic assumptions underlie the discussions of the previous section on elastic behavior of beams:

1. There would be no problems of buckling stability that would change the linear response of the beams to loads.
2. The limiting or failure moment is the yield moment, the moment that causes the outer fibers only to be stressed to F_y.

Readers who have come this far may already be concerned about the buckling problem. The compression portions of beams have buckling problems similar to compression members. Allowable stresses require similar adjustments. Assumption 2 raises an issue that was not pertinent to axially loaded members—the definition of the moment that represents failure.

10.3.2 Buckling of Beams: Lateral Instability

Figure 10.7 shows a uniformly loaded continuous prismatic beam and its moment diagram. We will focus on the portion between points of contraflexure and consider first the I-shaped beam. The portion above the neutral axis is in compression throughout this region, acting in a manner similar to a column. Unless there are closely spaced lateral restraints, we can expect that there is some value of critical moment, M_{cr}, analogous to the critical column load, P_{cr}, at which the compression portion will buckle laterally. Recall the Euler critical load curve shown in Fig. 9.4. Figure 10.7(d) shows the characteristic mode of buckling of an I-shaped section, which consists in part of twisting (torsion) and warping (opposite lateral bending of the two flanges).

The formulas for M_{cr} in what follows are adapted from a more complete treatment in Gaylord.[29]

$$M_{x,cr}^2 = C_b^2 \left[\frac{\pi^2 EI_y}{(KL)^2} \left(GJ + \frac{\pi^2}{(KL)^2} EC_w \right) \right] \tag{10-12}$$

where C_w is a *warping coefficient* related to the lateral bending stiffness of the flanges, J is the *polar moment of inertia* of the cross section, and C_b is a modifier based on the *moment gradient* within the length between lateral supports of the compression flange. Figure 10.8 gives the formula for C_b, which can be seen to be a function of the ratio of moments at the ends of the laterally unsupported length and the shape of the moment diagram between. It can be seen that reverse-curvature bending is stabilizing, while single-curvature bending destabilizes.

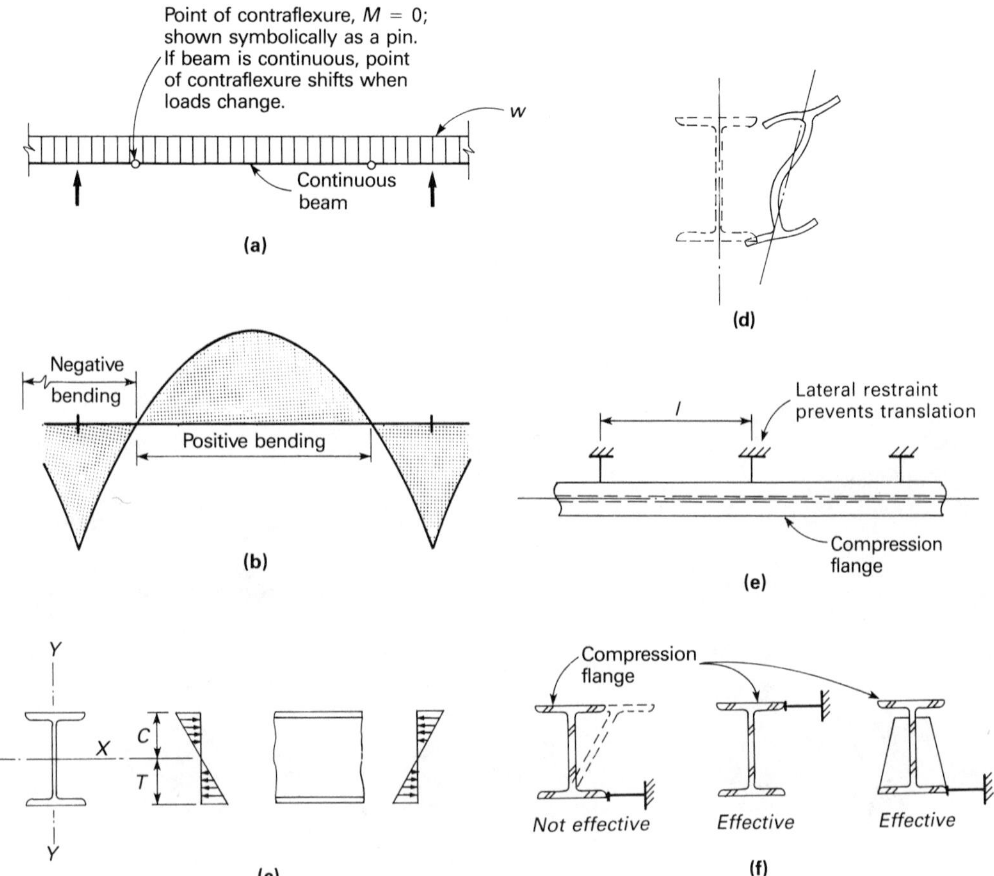

Figure 10.7 Lateral buckling of beams: (a) continuous beam, uniformly distributed load; (b) bending moment diagram; (c) portion of positive bending region, with section; (d) lateral–torsional buckling of I beam; (e) lateral restraints limit effective length of compression flange; (f) effective lateral restraint must prevent translation of compression flange.

For I-shaped beams,

$$C_w \approx \frac{I_y d^2}{4}$$

and has very large values for deep beams.

$$J = \Sigma \frac{bt^3}{3}$$

and has very small values. Then, after substituting F_{cr} for M_{cr}/S_x and ignoring the first term of Eq. (10-12),

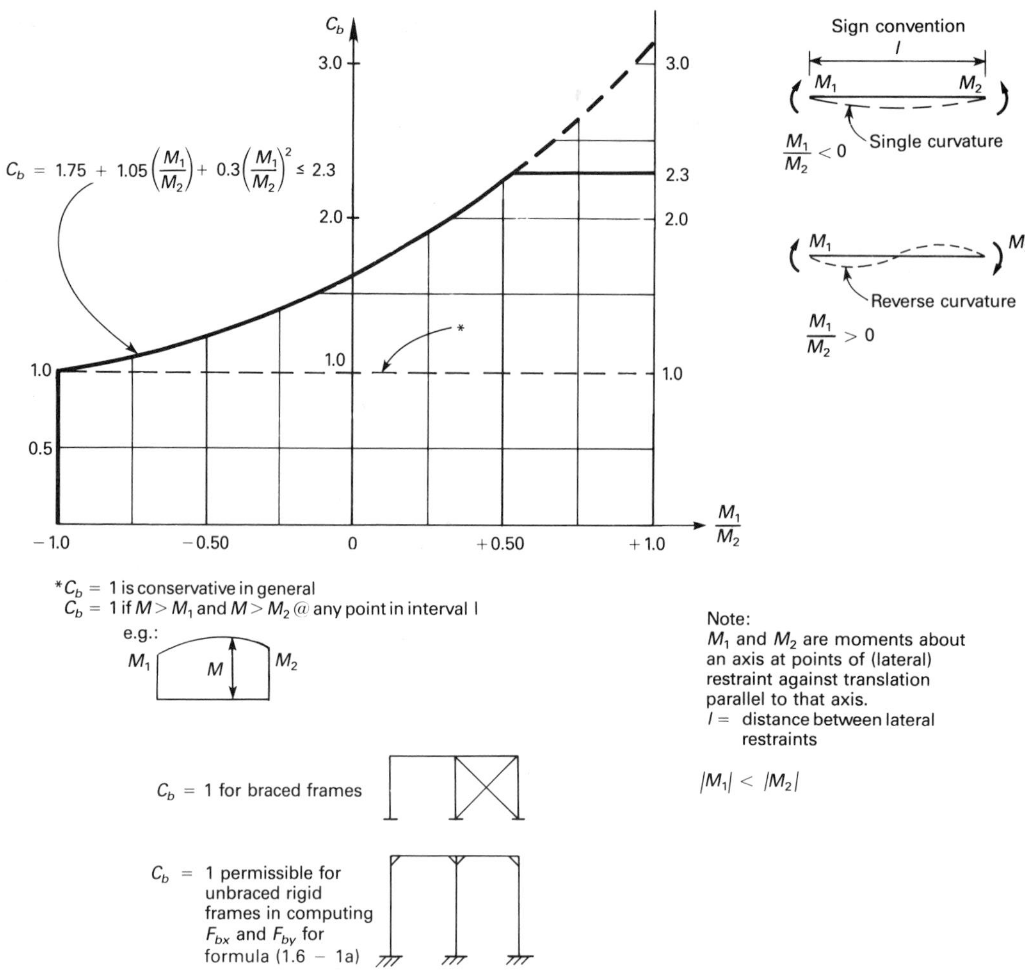

Figure 10.8 M_1/M_2 versus C_b per AISC Specification 1.5.1.4.5, paragraph 2.

$$F_{cr} \gtrsim \frac{C_b}{S_x} \frac{\pi^2 E I_y}{(KL)^2} \cdot \frac{d}{2} \tag{10-13}$$

Noting for deep, thin-webbed I shapes, that

$$S_x = \frac{2I_x}{d} \approx \frac{2}{d}\left[2A_f\left(\frac{d}{2}\right)^2 + A_w \frac{d^3}{12}\right]$$

$$= \frac{d}{2}\left(2A_f + \frac{A_w}{3}\right) \approx A\frac{d}{2}$$

$$r_y^2 = I_y/A$$

and

$$E = 29000 \text{ ksi}$$

$$F_{cr} \approx \frac{296{,}000C_b}{(KL/r_y)^2} \tag{10-13a}$$

Similar values apply for C_w, J, and F_{cr} in other open, thin-walled sections with two flanges parallel to the X axis.

Shallow I-shaped beams may be dominated by the twisting term (*St. Venant torsion*) of Eq. (10-12). If the warping term of Eq. (10-12) is neglected, substituting E for $G/2.6$, and rearranging,

$$F_{cr} \approx \frac{0.62C_b}{S_x} \frac{\pi E}{KL} \sqrt{JI_y} \tag{10-14}$$

Furthermore, ignoring the relatively thin web in I_y and J, and for $F_{cr} < F_y$ after some algebraic manipulation Eq. (10-14) leads to

$$F_{cr} \gtrsim \frac{20{,}000C_b}{KLd/A_f} \tag{10-14a}$$

where A_f is the area of the compression flange.

It can be seen that either Eq. (10-13) or (10-14) gives conservative values of F_{cr}. When used in specifications, if F_b is based with the desired factor of safety on the larger value of F_{cr} resulting from the two separate equations, F_b will be on the safe side. However, as with the Euler curve for P_{cr}, small values of L lead to values of F_{cr} above F_y. Both equations lead to curves similar in form to those of Fig. 9.4, with a *short* range dominated by F_y, a *long* range dominated by either Eq. (10-13) or (10-14) and independent of F_y, and an *intermediate* range between. AISC equations for F_b will be found to be of these forms.*

Closed thin-walled rectangular tubes are very stiff torsionally. SSRC[36] suggests

$$F_{cr} = \frac{\pi}{LS_x} \sqrt{JGEI_y} \tag{10-15}$$

where

$$J = \frac{2b^2d^2}{(b/t) + (d/t_w)}$$

For a tube of constant thickness,

$$J = \frac{2tb^2d^2}{b + d}$$

*The reader may look ahead to Fig. 10.10.

The beam strength of rectangular tubes is usually limited by F_y, as is true with round pipes.

Parts (e) and (f) of Fig. 10.7 show arrangements of lateral restraints. Limiting effective length l between lateral restraints is the key to the compression stability of beams.

10.3.3 Buckling of Beams: Element Instability

Recall now the discussion of element instability in Section 9.3.2 and the empirical rules discussed in Section 9.6.3 to avoid that problem. The problem is similar when compressive stress results from bending. The principal control is a limiting value of the ratio of width to thickness (b/t) of an element of cross section. Below that ratio, element instability is not a problem. Above that ratio, allowable stresses must be reduced to avoid element buckling. Section 1.9 and Appendix C of the AISC Specification treat element buckling in the same way for axial compression and compression due to bending.

10.3.4 Plastic Action in Beams

If we define failure as that condition where the outer fibers only of a beam cross section are stressed to $f_b = F_y$, the failure condition for a beam is M_y in Fig. 10.9(a).

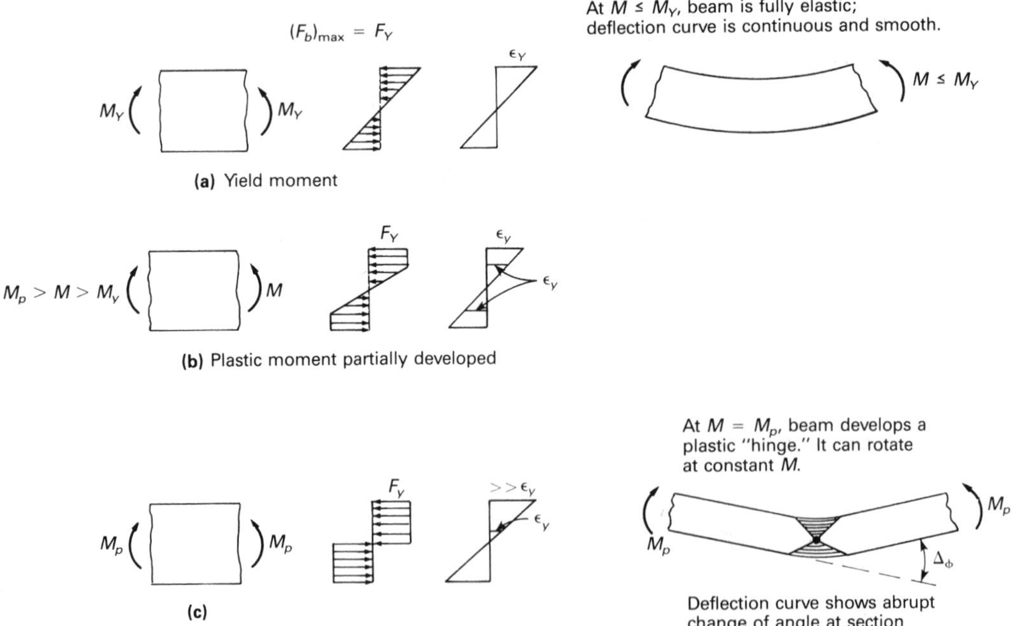

Figure 10.9 Yield moment and plastic moment.

The outer fibers have reached their yield level for both stress and strain. In a tension or compression member, yield stress is a clear limit on resistance. For a beam, the case is not so clear. Only the fibers at the outer surfaces are yielding. All others have reserve strenth ($F_y - f_b$) still available to resist additional moment. In Fig. 10.9(b), additional moment has been added, which has now brought more of the cross section to yield stress, but still does not threaten collapse. A new limiting case can be seen in Fig. 10.9(c), where all fibers are yielding. Moment greater than M_p is only possible with very large strains in the strain hardening range of the steel.

For a solid rectangular bar or an I-shaped section bent about its *minor*, *Y*, axis,

$$\frac{M_p}{M_y} = 1.5$$

In the more usual case, bending of an I shape is around the *major, X*, axis. Since much of the area is concentrated in the flanges, that part is all at or close to yield at M_y. The plastic reserve is much less, but still significant.

$$1.10 \lesssim \frac{M_p}{M_y} \lesssim 1.2$$

the value depending on the actual section.

If beam selection is to be based on the plastic moment, the section modulus S in Eq. (10-9) must be replaced by a new plastic modulus Z, based on the relationship

$$Z = \frac{M_p}{F_y} \tag{10-16}$$

Formula (10-16) is a partial basis for design in accordance with Part 2 of the AISC Specification or other load factor approaches.

Although Part 1 of the AISC Specification is based on *elastic* design methods, it will be seen that the plastic capacity of beams is recognized in significant ways even in Part 1. It must be noted, however, that the plastic moment, M_p, cannot be reached if buckling, whether due to beam or element instability, takes place at a lower moment. All sections chosen under Part 2 and *compact sections* chosen under Part 1 of the AISC Specification must be proportioned to avoid such premature buckling.

10.4 FATIGUE IN BEAMS

The fatigue phenomenon in beams arises, in very much the same way as was discussed in connection with tension members in Chapter 8, when a large number of stress cycles can be expected during the life of a structure, with at least one limit of cyclical stress being tensile. Recall Fig. 8.4. The problem is aggravated by any abrupt change in cross section that acts as a notch. In addition to considering allowable

TABLE 10.1 GUIDE TO SELECTION OF ROLLED SHAPES AS BEAMS

Type of section	$\frac{d}{b}$	$\frac{I_x}{A}$, $\frac{r_x}{A}$, $\frac{S_x}{A}$	$\frac{I_y}{A}$, $\frac{r_y}{A}$, $\frac{S_y}{A}$	Compact sections $\left(\frac{M_p}{M_y}\right)_x$	Compact sections $\left(\frac{M_p}{M_y}\right)_y$	Usual uses: Loads in YZ plane M_x	Usual uses: Loads in XZ plane M_y	M_z	Lateral and torsional stability	Remarks
W, HP, M	≫1	■	□	1.15±	1.50	●	○	x	x	Most common beams for M_x
	≈1	⊠	⊠	1.20±	1.50	⊘	⊘	⊘	⊘	Allows increased l, some biaxial bending and torsion
S	≫1 for large d	■	□	1.20±	1.50	●	○	x	x	No longer common; W sections usu.illy superior and offer more variety of sizes in each depth
	>1 for small d	⊠	□	1.20±	1.50	⊘	○	x	x	
(channels back to back)	Similar to W sections									Used in place of W if particular connections desired
Box, $d/b < 6$	≫1	■	⊠	1.15±	1.20±	●	⊘	●	●	Maximum mill-supplied depth = 20″; selection of steels may be limited; end connections may be difficult
	≈1	⊠	⊠	1.20±	1.20±	⊘	⊘	●	●	

 $\phi/t \le \frac{13{,}000}{F_Y}$	1	⊠	⊠	Not compact		⊘	⊘	●	●	Unusual for beams; hydraulic resistance favorable
$\phi/t \le \frac{3300}{F_Y}$	1	⊠	⊠	1.7(−)	1.7(−)	●	⊘			
		■	■	1.7(−)	1.7(−)	●	⊘	x	Lat. ● Tors. x	Common crane rail support; good for biaxial bending
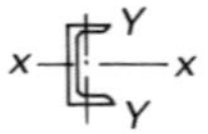	≫1	■	□	Not qualified as compact		●	x	x		Shear center not at CG, invites torsion. Use as beam only if laterally stabilized
	No general range Check each selection					Subject to check	x	x	Lat. ⊘ x Tors. x	Unusual for beam, but may be used for combined axial load and flexure
										Not recommended for flexure; if used, special analysis recommended

Notes

1. Symbols:
   - ■ High
   - ⊠ Intermediate
   - □ Low
   - ● Common
   - ⊘ Less common
   - ○ Occasional use, particularly in biaxial bending with $M_x/M_y \gg 1$
   - x Not suitable
2. This is a general guide only based on opinions of the authors. Suitability of a particular section for particular uses must be determined based on calculation and engineering judgment.

stresses, F_b, it becomes necessary to consider allowable ranges of cycling stress, F_{sr}. The special rules of AISC Specification, Appendix B, apply.

10.5 SECTIONS USED AS BEAMS

Any of the cross sections shown in Figs. 7.1 and 7.2 has at least some flexural strength and therefore can be used for a beam, except nonstiff sections such as wire rope and small bars. The problem in design is how to get sufficient bending strength and stiffness with efficient use of the materials so as to minimize the weight of the steel used. Weight per linear foot is proportional to cross-sectional area. So the problem of efficient use of steel translates into one of effective distribution of the cross-sectional area in order to maximize bending strength and stiffness.

Although this chapter set out to discuss bending without torsion, we have discovered that we cannot avoid consideration of torsional stability. However, this is a separate question from applied torsional moments (i.e., moments around the longitudinal Z axis of the beam). To avoid torsion requires that loads be applied through the shear center. This is relatively easy to do if we use doubly symmetric sections, where the shear center is at the intersection of the two principal axes, which is also the centroid of area. Loads applied in the XZ and YZ planes* do not cause torsion.

In Table 10.1, we have assembled a number of rolled shapes that are used as beams and indicated some of their general characteristics as a guide to selection. For the most part, the table should be self-explanatory. However, a few comments may help.

- For uniaxial moment, both strength and stiffness are maximized by loading through the minor axis (bending about the major axis) and locating as much of the area as possible in the flanges, which should be as far as possible from the neutral axis (e.g., deep sections of I shape).
- When I-type sections or rectangular tubes are used as beams bent around the X axis, the web area resists transverse shear, acting as A_v in Eq. (10-10). This sets a limit on the feasible depth of a beam. Since an element in shear experiences compression in one diagonal direction, the ratio of depth to web thickness (d/t_w) must be high enough to avoid compression buckling due to shear.
- If d/b is high and a beam is bent about its major axis, it tends to become unstable for lateral buckling around the minor axis. This may require closely spaced lateral restraints to avoid buckling.
- If closely spaced lateral restraints are not desirable because of design conditions, use of closed, box-type beam sections may make them unnecessary.

*X and Y axes are axes of cross section, as used in Part 1, AISC Manual. Z is the longitudinal axis of the member through its centroid.

- Channels and single angles should be used with caution for reasons noted in Table 10.1.

The proportions of rolled sections are chosen by the steel industry to satisfy, as closely as possible, common design requirements. The most common sections provided for beams in uniaxial bending are W sections of relatively large d/b ratio. In most cases, thicknesses of beam elements are selected to avoid element buckling. W shapes have wide enough flanges (they were formerly called *wide-flange* shapes) to be relatively stable laterally when compared to S shapes. Closed sections are available in a relatively limited selection of sizes. By creating their own sections from plates, designers can often serve their strength and stiffness needs with less weight (not necessarily less cost) than they could by selecting from available rolled shapes. In such instances the proportioning of elements becomes a major consideration. Chapter 14 addresses plate girder design.

10.6 PROVISIONS OF AISC SPECIFICATION APPLICABLE TO BEAMS

The primary equations used in the selection of shapes for beams have already been stated in Eqs. (10-9) through (10-11). Although all three require checking, the most important may be Eq. (10-9);

$$S \geq S_R = \frac{M}{F_b} \tag{10-9}$$

This is usually the starting point of the selection process.

Experienced designers often recall fondly although somewhat imprecisely the "good old days" when F_b was "always" 20 ksi and, presumably, an engineer needed only that one equation to make proper selections. They tend to forget that in those days their experience-based judgment and the heavier sections that were common when steel was much less expensive helped to avoid failures. In addition, they could usually select from a grand total of one type of structural steel, ASTM-A7.

However, their impatience with the current specification is understandable when we examine Table 10.2 and see that there are six sections of the Specification that deal with establishment of F_b and two dealing with F_{sr} as bending stress. Table 10.2 lists the various sections of the Specification related to bending, and notes which aspect of bending is covered, whether stress, deflection, stability, or other. The list is truly long. Without the table, some of the pertinent provisions might escape the designer. The complexity arises from attempts by the Specification writers to take the fullest possible advantage of the high strength of the material, while avoiding the problems of potential instability. Further discussion will not follow the provisions in numerical order. Nor will we attempt to deal with all that are listed. Some will arise in later chapters. Readers are referred to the Specification itself for complete provisions.

TABLE 10.2 FLEXURAL ISSUES ADDRESSED IN PROVISIONS OF AISC SPECIFICATIONS, 1978

Para. AISC Spec.	Subject	Allow. stress								
			F_b							
		F_v	T and C	T	C	F_{sr}	Deflection	Unbraced length and stability	Adjustments to analysis	Details to maintain integrity	Other
1.5.1.2	Shear	●									
1.5.1.4	Bending										
1.5.1.4.1	Bending, compact section (M_x)		●					●[a]	●	●	
.2	Bending, compact section		●							●	
.3	Bending, compact section (M_y)		●								
.4	Bending, box-type member		●					●[a]		●	
.5	Bending, noncompact			●	●			●[a]		●	
1.5.6	Wind and seismic loads	●	●	●	●						
1.6	Combined stresses, axial load and bending										●
1.7.1 and Appendix B	Fatigue					●					

1.9 and Appendix C	Width–thickness ratios	•	•	• / •	•				•	
1.10.1[b]	Plate girders and rolled beams, proportions								•	
1.10.11[b]	Plate girders and rolled beams, support restraints								•	
1.11[c]	Composite construction	•	•			•			•	•
1.12	Simple and continuous beams									
1.13	Deflections, variations, ponding					•	•[d]	•[d]		
1.18.1	Built-up members, open box-type beams and grillages								•	
1.19	Camber					•				

[a]Stability (and instability) as used here refers to the phenomenon of compression buckling at low stress ($<F_y$). The compression flange may buckle between lateral supports if l/r_t is too large; individual elements in compression may buckle if b/t is excessive. In either case, the member cannot mobilize its full yield strength.

[b]Provisions of 1.10 apply to plate girders and built-up flexural sections as well as rolled beams. Plate girders and built-up members are covered in a later chapter.

[c]Composite construction as covered here applies to composites of reinforced concrete compression flanges with webs and tension flanges of structural steel. Composite beams and girders are discussed in a later chapter.

[d]Instability may result from ponding, the phenomenon of progressively increasing live load from water trapped within a sagging surface (similar to water on a stretched membrane). The increased water load causes additional deflection and traps more water. If flexural members have insufficient stiffness, EI/L, the deflection is not self-limiting. Failure may be at yield moment, unless member proportions invite earlier unstable buckling as in note a.

10.7 FINDING F_b: THE KEYS TO THE MAZE

10.7.1 Compact and Noncompact Sections

The provisions of Sections 1.5.1.4.1 through 1.5.1.4.5 may initially appear to the user as a maze of legalistic language to be painfully unraveled in order to arrive at an allowable bending stress, F_b. There is a logic based on the search for a consistent factor of safety accounting for the strength of the steel, yield and plastic moment capacities, and potential for instability. Figure 10.10, Tables 10.3 and 10.4, and Flow Charts 10.1 through 10.3, each in its own way, are intended to help chart the maze. The reader should, however, ultimately rely on the complete wording of the Specification, which is not repeated here.

There are two main categories of beams, *compact* and *noncompact*. Compact beams, by virtue of special controls on their geometry, are particularly stable. *Semicompact* beams pass all tests for compactness except one, and are close on that test. All other beams are noncompact.

10.7.2 Compact Sections

Recall from the discussion of plastic action, Section 10.3.4, that beam sections can call on their plastic reserve to resist moments higher than their yield moments, M_y. For rolled I-shaped sections bent about their major axis, the range of augmented strength is approximately

$$1.10 < \frac{M_p}{M_y} < 1.20$$

This is only true if the plastic moment, M_p, can be mobilized without earlier failure due to instability of elements or of the member. Members proportioned so that the plastic moment can be mobilized are called *compact*. Criteria for deciding whether or not a member is compact are given in AISC Specification 1.5.1.4.1 and summarized here in Table 10.3 and Flow Chart 10.1. *Compact section criteria* are based on the yield strength of the steel, the type of cross section, the ratios of width to thickness of the elements of cross section, and the ratio (l/r_T)* of length between lateral supports of the compression flange to the *weak way* radius of gyration of that flange. Readers will recognize that these are the controls on member and element buckling.

Although Table 10.3 and Flow Chart 10.1 look forbidding, determination of compactness for most rolled sections is usually fairly simple, particularly with the help of the Compact Section Criteria columns in the AISC Shape Tables.

*The lowercase l is used to denote distance between lateral supports of the compression flange. Uppercase L is span length between supports in the direction of the load. L_c and L_u are specific values of l. The symbols of Chapter 9 differ slightly. r_T is defined in Fig. 10.10. For rolled I-shaped beams and other I shapes with identical top and bottom flanges, r_T is slightly higher than r_y. Values of r_T are tabulated for I-shaped sections in the Shape Tables, Manual, Part 1.

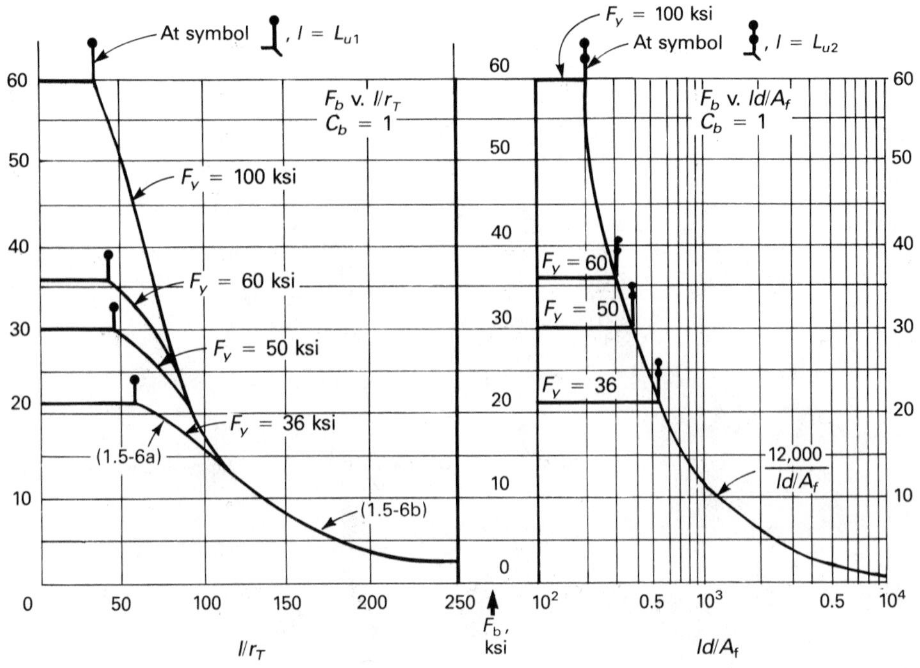

Use larger value of F_b based on l/r_T or ld/A_f

Bending is about major axis (x − x)

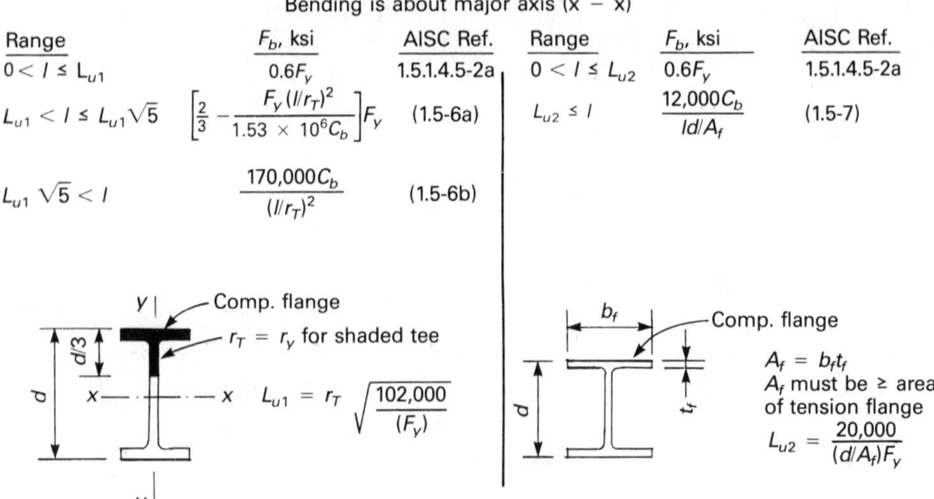

Range	F_b, ksi	AISC Ref.
$0 < l \le L_{u1}$	$0.6F_y$	1.5.1.4.5-2a
$L_{u1} < l \le L_{u1}\sqrt{5}$	$\left[\frac{2}{3} - \frac{F_y(l/r_T)^2}{1.53 \times 10^6 C_b}\right]F_y$	(1.5-6a)
$L_{u1}\sqrt{5} < l$	$\frac{170{,}000C_b}{(l/r_T)^2}$	(1.5-6b)

Range	F_b, ksi	AISC Ref.
$0 < l \le L_{u2}$	$0.6F_y$	1.5.1.4.5-2a
$L_{u2} \le l$	$\frac{12{,}000C_b}{ld/A_f}$	(1.5-7)

Figure 10.10 Values of F_b (compression) for noncompact I-shaped sections, by AISC Specification, 1978. For tension, $F_b = 0.60F_y$.

TABLE 10.3 COMPACT SECTION CRITERIA (PER AISC SPEC. 1.5.1.4.1)[a]

$F_{bx} = 0.66F_y$

Steel: Entire section of one type (nonhybrid); exclude A514 steel

Direction of load: In plane of minor (*Y-Y*) axis, which must be an axis of symmetry

I-type sections	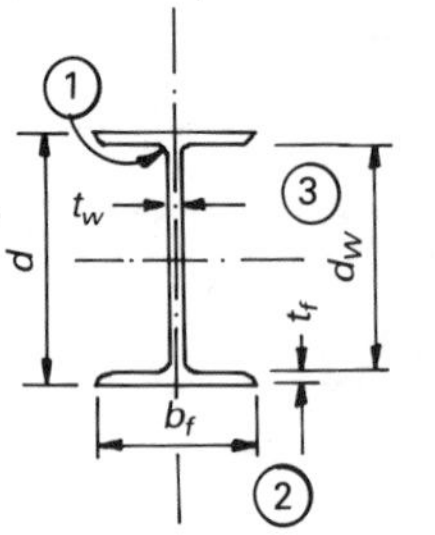 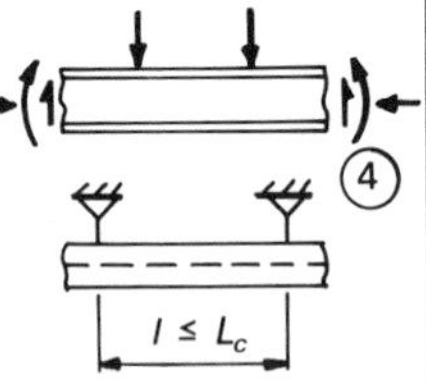	① Continuous connection between flange and web ② $\dfrac{b_f}{2t_f} \le \dfrac{65}{\sqrt{F_y}}$ for compression flange (see F'_y in shape tables) ③ $\dfrac{d_w}{t_w} \le \dfrac{640}{\sqrt{F_y}}\left(1 - 3.74\dfrac{f_a}{F_a}\right)$ if axial compression is minor, i.e., $\dfrac{f_a}{F_y} \le 0.16$ or $\dfrac{d_w}{t_w} \le \dfrac{257}{F_y}$, if $\dfrac{f_a}{F_y} > 0.16$ (see F'''_y in shape tables) ④ $L_c \le \dfrac{76b_f}{F_y}$ and $l \le \dfrac{20{,}000}{(d/A_f)F_y}$ where l is the length between lateral restraints of the compression flange
Box-type sections	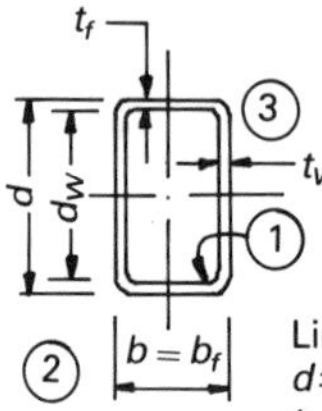 Limits: $d \le 6b$, $t_f \le 2t_w$	① Same as 1 for I sections ② $\dfrac{b_f}{t_f} \le \dfrac{190}{\sqrt{F_y}}$ ③ Same as 3 for I sections ④ $\dfrac{L}{b} \le \dfrac{1950 + 1200(M_1/M_2)}{F_y} \le \dfrac{1200}{F_y}$ (for M_1/M_2 see Fig. 10.8)
Pipes, round tubes	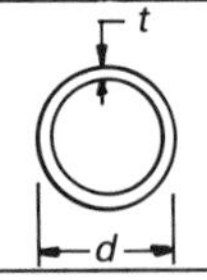	$\dfrac{d}{t} \le \dfrac{3300}{F_y}$

[a]If all applicable criteria are met, section may be considered compact.

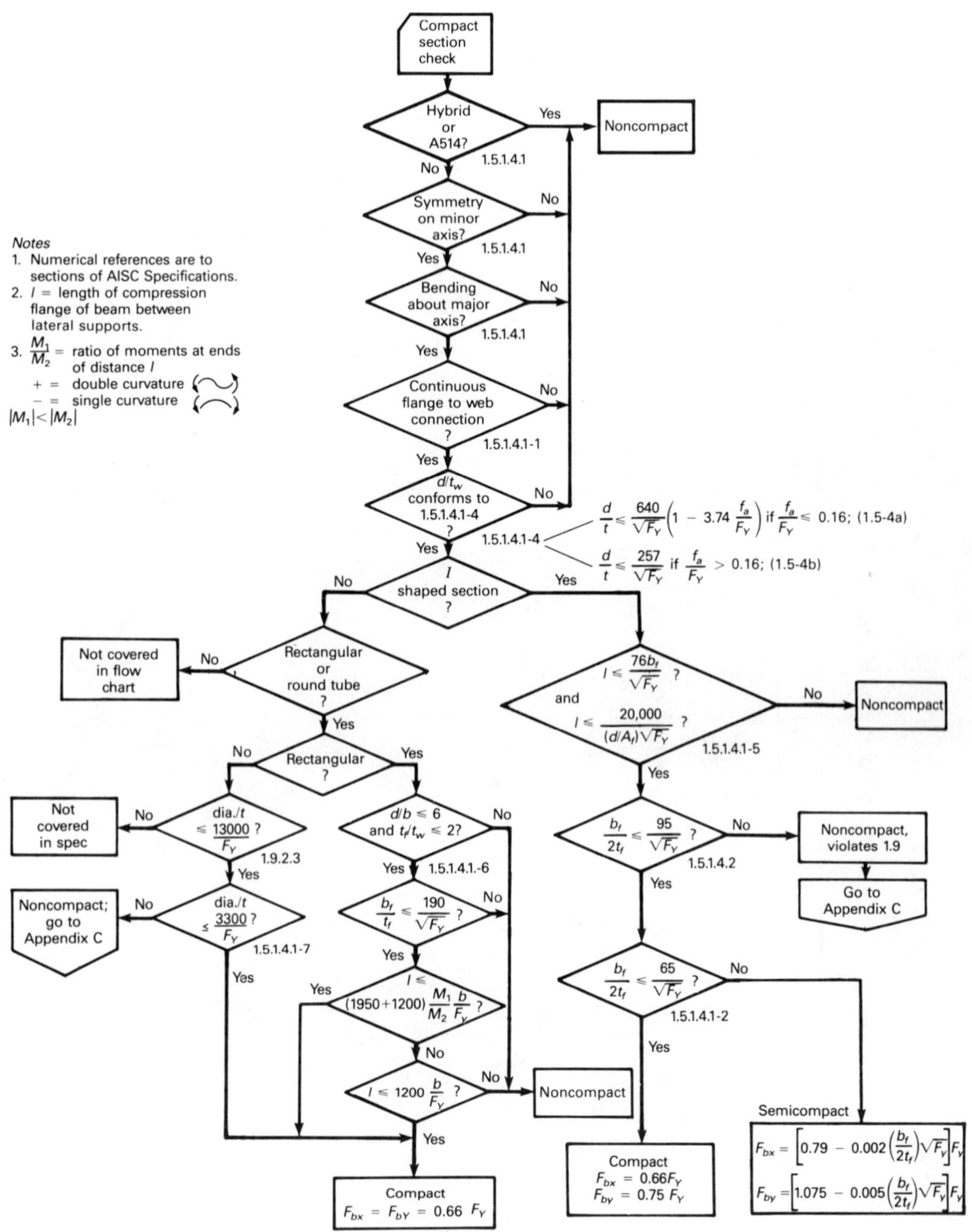

Flow Chart 10.1 Test for compact section; AISC Specification (1978), Section 1.5.1.4.1.

TABLE 10.4 ALLOWABLE BENDING STRESS, F_b, BY AISC SPECIFICATION 1978: VARIATION WITH UNSUPPORTED LENGTH, COMPACT OR NONCOMPACT*

Unsupported length, l†	Compact or noncompact	Stress Type, T or C	Bending axis	AISC Limits: Reference	AISC Limits: Formula	AISC Limits: F_b/F_y or range of F_b/F_y	Remarks
$0 \le l \le L_c$	Compact	T or C	Major	1.5.1.4.1	$F_b = 0.66F_y$	0.66	See also Table 10.4
$0 \le l \le L_c$	Compact except b/t ratios qualified (semi-compact)	↓	Major	1.5.1.4.2 eq. (1.5–5a)	$\frac{65}{\sqrt{F_y}} \le \frac{b_f}{2t_f} \le \frac{95}{\sqrt{F_y}}$; $F_b = F_y\left[0.79 - 0.002\frac{b_f}{2t_f}\sqrt{F_y}\right]$	$0.66 \le \frac{F_b}{F_y} \le 0.60$	↓
Not specified by AISC	Compact I ● ■	↓	Minor	1.5.1.4.3	$F_b = 0.75F_y$	0.75	↓
	Compact except b/t ratios qualified (semi-compact)	↓	Minor	1.5.1.4.3 eq. (1.5–5b)	$\frac{65}{\sqrt{F_y}} \le \frac{b_f}{2t_f} \le \frac{95}{\sqrt{F_y}}$ $F_b = F_y\left[1.075 - 0.005\frac{b_f}{2t_f}\sqrt{F_y}\right]$	$0.75 \le \frac{F_b}{F_y} \le 0.60$	↓
$0 \le l \le L_{c1}$	Compact □	↓	Major	1.5.1.4.1	$F_b = 0.66F_y$	0.66	↓
Not specified	Compact except limit on l not applicable	↓	Minor	1.5.1.4.3	$F_b = 0.66F_y$	0.66	↓

	$l > L_c$	All	T	All	1.5.1.4.5–1	$F_b = 0.60F_y$	0.60	See Figs. 10.10 and 10.8
Control by l/r_T‡	$L_c \leq l \leq L_u$	Compact except for l	C	Major	1.5.1.4.1 and 1.5.1.4.5–2a	$F_b = 0.60F_y$	0.60	↓
	$0 \leq l \leq L_u$	Noncompact	C	All	1.5.1.4.5–2a	$F_b = 0.60F_y$	0.60	↓
	$L_{u1} \leq l \leq L_{u1}\sqrt{5}$	All	↓	Major	1.5.1.4.5–2a Eq. (1.5–6a)	$F_b = \left[0.67 - \dfrac{F_y(l/r_T)^2}{1.53 \times 10^6 C_b}\right] F_y$	$0.60 \leq \dfrac{F_b}{F_y} \leq 0.33$	↓
	$l > L_u\sqrt{5}$	All	↓	Major	1.5.1.4.5–2a Eq. (1.5–6b)	$F_b = \dfrac{170{,}000 C_b}{(l/r_T)^2}$	F_b independent of F_y	↓
Control by l_d/A_f‡	$l \leq L_{u2}$	Noncompact	C	All	1.5.1.4.5–2a	$F_b = 0.60F_y$	0.60	See Fig. 10.10
	$l > L_{u2}$	Noncompact	↓	Major	Eq. (1.5–7)	$F_b = \dfrac{12{,}000 C_b}{ld/A_f}$	F_b independent of F_y	↓
	No limit specified	Noncompact for b/t	T or C	All	1.5.1.4.4	$F_b = 0.60F_y$	0.60	
	$l < \dfrac{76 b_f}{\sqrt{F_y}}$	Noncompact	T or C	Major	1.5.1.4.5–2b	$F_b = 0.60F_y$	0.60	Sections not included in 1.5.1.4.5–2a
	No limit specified	Noncompact	T or C	Minor	↓	↓	↓	

* Excludes sections not conforming to AISC 1.9.

† l = length of compression flange between supports.

l_c = smaller of $76\, b_f/\sqrt{F_y}$ or $20{,}000/(d/A_f)F_y$.

$$L_{c1} = \frac{b}{F_y}\left(1950 + 1200\,\frac{M_1}{M_2}\right) \leq 1200\,\frac{b}{F_y}$$

$$L_{u1} = r_T\sqrt{\frac{102{,}000 C_b}{F_y}}\,;\; L_{u2} = \frac{20{,}000}{(d/A_f)F_y}$$

See Fig. 10.8 for M_1/M_2 and C_b.

For r_T, see Fig. 10.10 or AISC 1.5.1.4.5, para. 2a.

‡ Use larger of the values of F_b as controlled by l/r_T or l_d/A_f.

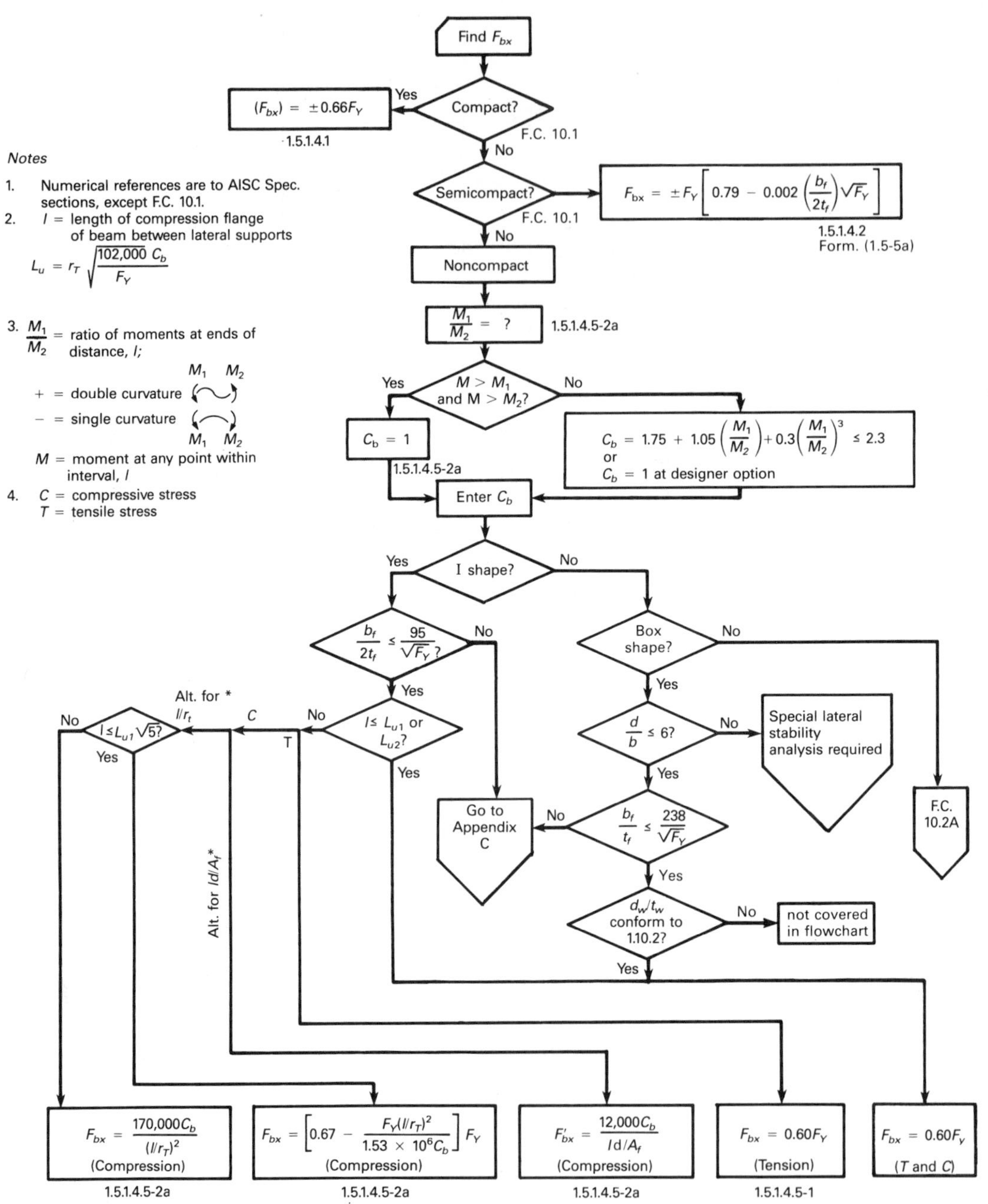

*Use larger value of F_{bx} or F'_{bx}.

$$\dagger L_{u1} = r_T\sqrt{\frac{102{,}000C_b}{F_y}} \cdot L_{u2} = \frac{20{,}000C_b}{ld/A_f}$$

Flow Chart 10.2 Bending about major (X) axis; find F_{bx}.

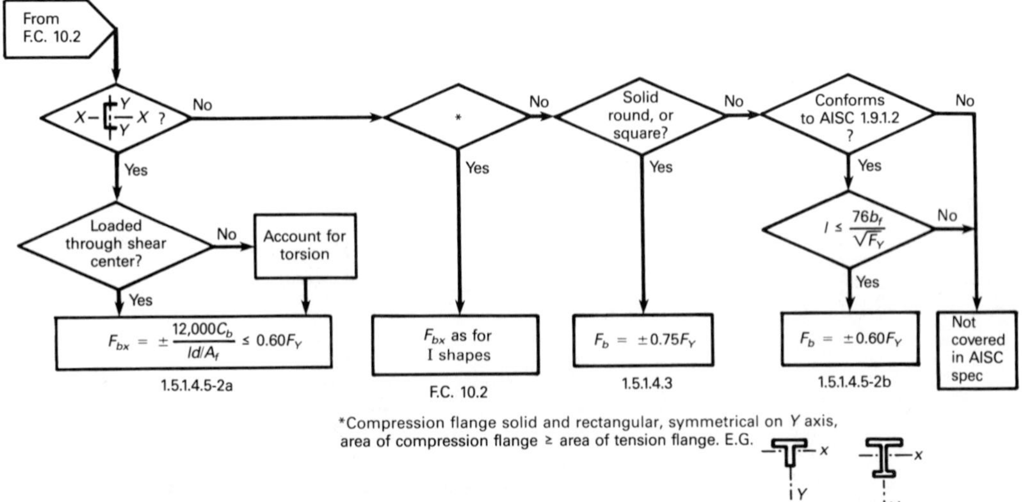

Flow Chart 10.2A Continuation of Flow Chart 10.2.

Members fulfilling the criteria are compact whether they are rolled shapes or shapes made from plates. To fulfill all criteria, the member must be symmetrical about its minor axis, be bent about its major axis, meet limits of b/t and d/t stricter than those in AISC Specification, Section 1.9, and have unsupported compression lengths less than L_c. Unless all criteria are fulfilled, the member is noncompact.

The limit of unsupported length for otherwise compact box sections is very much higher than for I-shaped sections, due to the immensely superior torsional stability of box sections. A round tube is also stable torsionally provided its wall does not buckle. The diameter/thickness limit of $3300/F_y$ in AISC 1.5.1.4.1 is intended to avoid such wall buckling.

If all criteria are fulfilled and if bending is about the major axis, $X - X$,

$$F_{bx} = 0.66F_y$$

which is 10% higher than for the most favorably proportioned noncompact sections. Recall from the previous two chapters that $F_t = 0.60F_y$ and, for $KL/r = 0$, $F_a = 0.60F_y$. A compact I shape bent about its minor axis consists essentially of two rectangles, since the web contributes little to I_y. In that case, $M_p/M_y = 1.5$, and

$$F_b = 0.75F_y \qquad \text{(AISC 1.5.1.4.3)}$$

The criterion for unsupported length is dropped since the major-axis stiffness provides lateral supported in minor-axis bending. For compact rectangular tubes, in minor axis bending,

$$F_b = 0.66F_y \qquad \text{(AISC 1.5.1.4.3)}$$

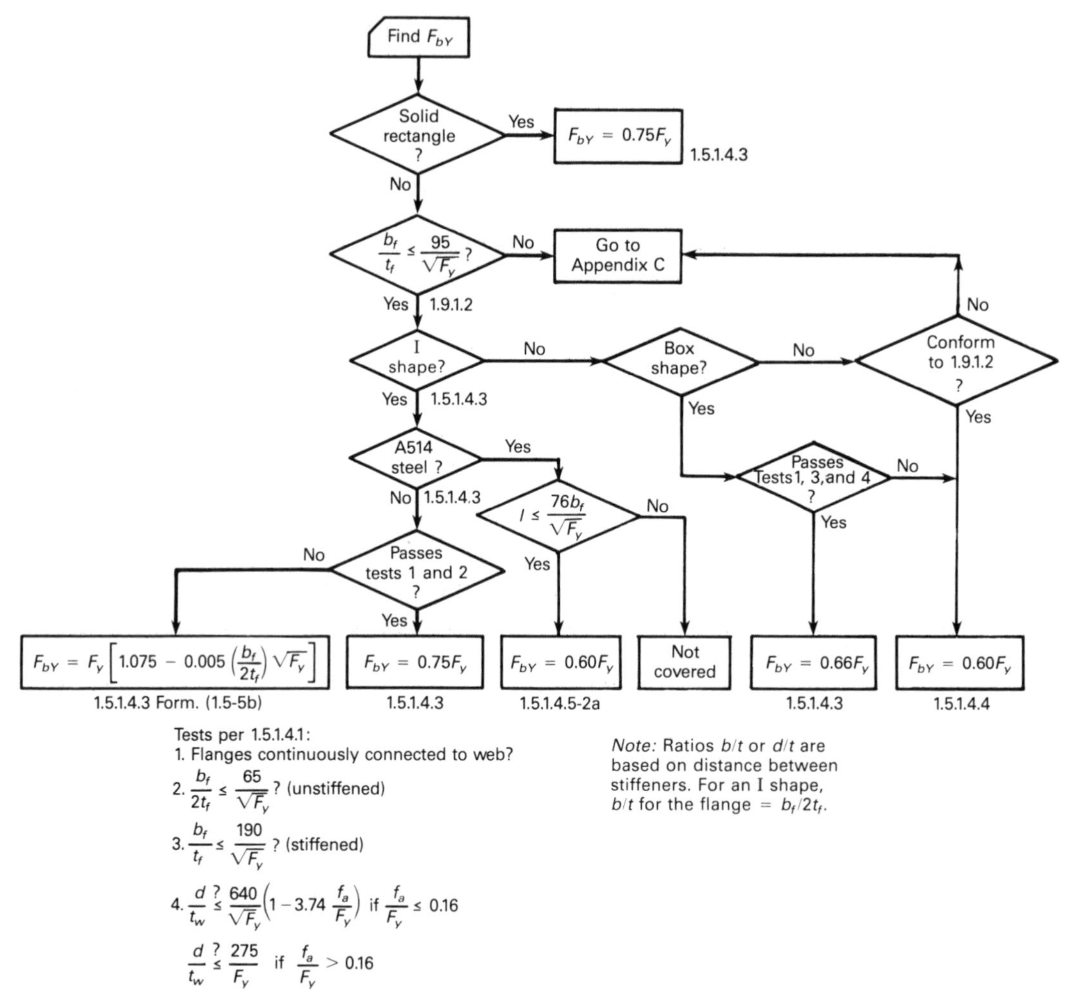

Flow Chart 10.3 Bending about minor (Y) axis; find F_{bY}.

10.7.3 Example: Test for Compact Section

The beam shown is simply supported at its ends. The end connection is effective in preventing rolling. There is no other lateral support. (1) Is the member compact? (2) If not, can it be made compact? If so, how? The solution follows Flow Chart 10.1.

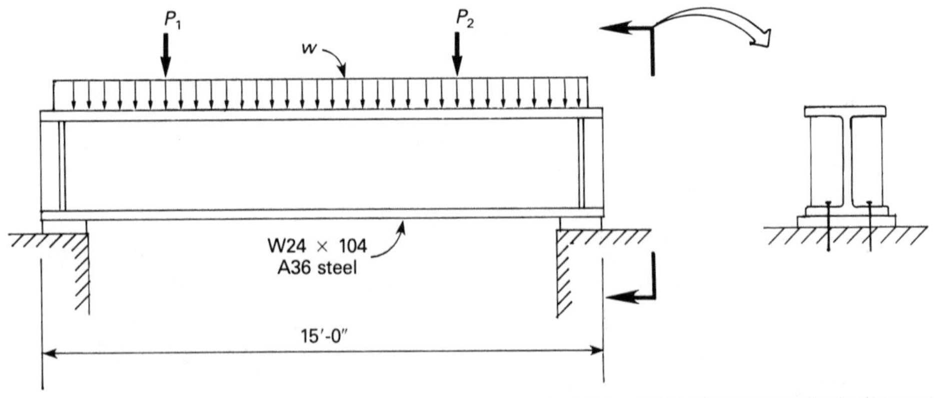

CALCULATIONS		*REMARKS*
(1) *List properties*:		Source: Shape Tables, AISC Manual. Only properties pertinent to tests for compactness are listed.
$d = 24.06''$ $\quad r_T = 3.35''$		
$t_w = 0.50''$ $\quad \dfrac{d}{A_f} = 2.52\text{ in.}^{-1}$		
$b_f = 12.750''$		
$t_f = 0.750''$		
Follow Flow Chart 10.1:		
Hybrid or A514?	No	Rolled sections not hybrid; steel is A36.
Symmetry on minor axis?	Yes	All W's symmetrical on Y axis.
Bending around major axis?	Yes	Loads parallel to Y; bending around X.
Continuous flange to web connection?	Yes	Rolled sections all conform.
d/t_w conforms to Sections 1.5.1.4.1–4?		
$\dfrac{f_a}{F_y} = 0$		No axial load given.
$\dfrac{d}{t_w} = 48.12 < \dfrac{640}{\sqrt{F_y}}(1 - 0) = 106.7$		AISC (1.5-4a)

$\therefore d/t_w$ conforms.

I shape? Yes All W's are I shaped.

$$l \leq \frac{76b_f}{\sqrt{F_y}} ?$$

$$15(12) = 180 > \frac{76(12.75)}{6} = 161.5$$

$\therefore$ Noncompact.

(2) To make compact, provide lateral support for top flange. Try $l = L/2$. *Continue tests*:

Assumes no other design requirement is violated.

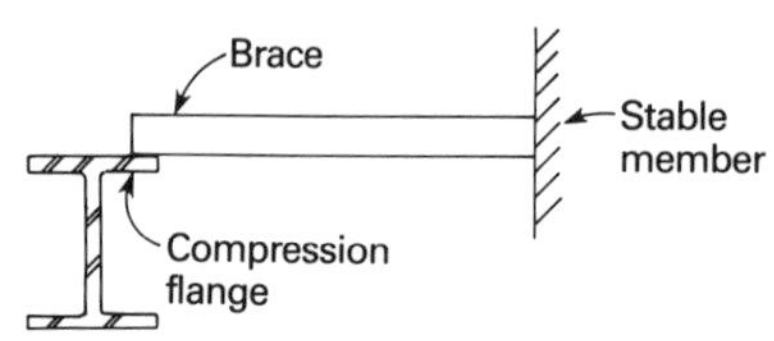

$$l = 7.5(12) = 90 \ll \frac{76(12.75)}{6} \quad \text{OK}$$

$$l \leq \frac{20{,}000}{(d/A_f)F_y} ?$$

$$7.5(12) = 90 \ll \frac{20{,}000}{2.52(36)} = 166 \quad \text{OK}$$

$$\frac{b_f}{2t_f} \leq \frac{65}{\sqrt{F_y}} ?$$

$$\frac{12.75}{2(0.75)} = 8.50 < \frac{65}{6} = 10.83 \quad \text{OK}$$

$\therefore$ *Compact*, if lateral support at midspan.

All criteria fulfilled.

Values of $b_f/2t_f$ and d_w/t_w are listed in the AISC Shape Tables among *Compact Section Criteria*. The column labeled F_y' gives maximum values of F_y for which the section conforms to the $b_f/2t_f$ criterion for compactness. The column labeled F_y''' gives maximum values of F_y for which $d/t_w \leq 257\sqrt{F_y}$. That limit applies only if the beam has axial load as well as bending moment and $f_a/F_y > 0.16$. [AISC (1.5-4b)] In the preceding problem, $f_a = 0$. The less restrictive criterion of AISC (1.5-4a) was applied. When no values are given for F_y' and F_y''', all structural steels except A514 conform to the criterion.

10.7.4 Semicompact Sections

When $b_f/2t_f$ is in the range between the special *compact* limit of $65/\sqrt{F_y}$ and the limit of $95/\sqrt{F_y}$ set in AISC Section 1.9, the beam is *semicompact*. Linear adjustments are made in F_b for both major and minor axis bending. The equations [AISC (1.5-6a) and (1.5-6b)] are given in Table 10.3.

10.7.5 Moment Adjustment in Compact Sections

The plastic reserve moment in compact sections is also recognized in another way in AISC 1.5.1.4.1. This applies to continuous beams and girders except cantilevers subjected to gravity loads, or to beams or girders rigidly connected to columns at their ends. A multispan beam case is shown in Fig. 10.11. The moment diagram shown in solid lines results from elastic analysis. Very often in such cases the midspan positive moment is much less than the negative moment at the supports. Actual "failure" does not occur until multiple plastic hinges form in one of more spans as shown. When $M = M_p$ as in Fig. 10.9(c), all fibers are stressed to F_y, leaving no resistance to rotation around the transverse axis of bending—the condition for a plastic hinge. If $M_4 \ll |M_1|$ and $|M_2|$, the moment reserve $(M_p - M_4)$ is much greater than $(M_p - M_1)$ or $(M_p - M_2)$. The conditions for the end span are similar. In such a case, it is permitted to decrease the design negative moment by 10% provided the design positive moment is increased by 10% of the average of the two end negative moments. In the Tier Building Case Study in Chapter 18, this provision results in considerable saving of steel in both beams and girders without sacrificing safety. Definition of

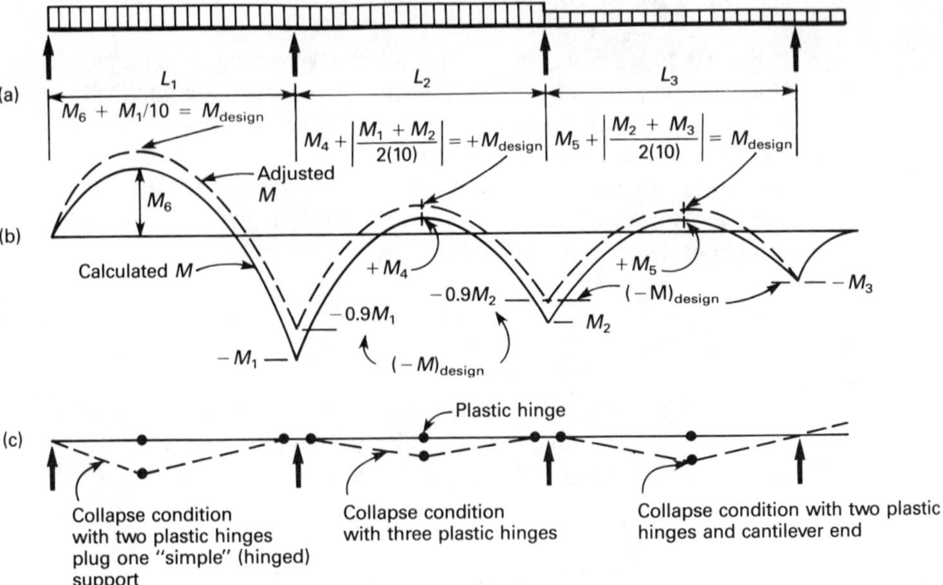

Figure 10.11 Adjustment of design moments for continuity based on plastic hinges as mechanism of collapse: (a) loading; (b) moments; (c) collapse mechanisms.

failure in this way underlies much of the thinking that led to adoption of *plastic design* methods in Part 2 of the AISC Specification.

10.7.6 Noncompact Sections

If a member does not qualify as compact, it becomes unsafe to take advantage of the plastic reserve, $(M_p - M_y)$. F_b will vary with l/r_T and ld/A_f. The governing equations are shown in Table 10.4 and, for major axis bending, Fig. 10.10. They assume width/thickness ratios within the limits of AISC Section 1.9. If b/t ratios exceed the limits of Section 1.9, F_b is reduced in accordance with AISC, Appendix C.

Figure 10.10 reveals the nature of the curves of F_b for noncompact I-shaped sections bending about their major axes. The resemblance to the curve of F_a in Fig. 9.13 should be clear and not surprising. There are two separate curves. That for F_b versus l/r_T is based in its outer region ($l > L_{u1}\sqrt{5}$) on a simplification of Eq. (10-13). The sloping portion of the ld/A_f curves ($l > L_{u2}$) is derived from Eq. (10-14). For reasons previously discussed, it is reasonable to use, in each case, the larger of the values derived from either the l/r_T curve [AISC equations (1.5-6a) or (1.5-6b)] or the ld/A_f curve (AISC 1.5-7).

All equations contain the modifier, C_b (see Fig. 10.8). Values of $C_b > 1.0$ can increase F_b up to the limit of

$$F_b = 0.60F_y$$

10.7.7 More Keys: Table 10.4, Flow Charts 10.2 and 10.3

In Table 10.4, we have assembled all the governing equations for F_b and the conditions under which each applies. It covers *compact* and *noncompact*, *symmetrical*, and *nonsymmetrical* sections. It omits sections that do not conform to Specification 1.9. Flow Charts 10.2 and 10.3 suggest a procedure for finding F_b, the former for *major-axis bending*, the latter for *minor-axis bending*. With a modicum of patience, the flow charts should lead the reader through the various branches of the F_b maze without neglecting any of the necessary controls.

10.8 CONTROL OF ld/A$_f$ *AND* l/r$_T$

It is comforting to realize that a very large number of beam problems fall in the short range of l/r_T and ld/A_f, and most rolled beams have stable width/thickness ratios. For these, there are only two values of F_b for bending about the major axis.

$$F_{bx} = 0.66F_y \qquad \text{if compact } (l \leq L_c)$$

$$F_{bx} = 0.60F_y \qquad \text{if noncompact}$$

but

$$l \leq L_u \text{ or } l \leq \frac{20{,}000}{d/A_f}$$

It is usually within the power of the designer to limit l/r_T, ld/A_f, and b/t in order to take advantage of these high values of F_b. Sometimes, however, it becomes uneconomical to do so.

Many beams are used to support concrete floor slabs in buildings or bridges. Figure 10.12 shows how the slab itself can be made to provide essentially continuous lateral support for a beam with simple details. In such cases, l approaches 0. Where beams are not connected to such floor slabs, it is very often possible for the designer to provide support at spacings less than L_c or L_u. The methods are similar to those discussed for compression members in Chapter 9. It is necessary, as shown in Fig. 10.13, to be sure that the lateral restraint prevents translation of the compression flange.

r_T and d/A_f are among properties tabulated for I-shaped sections in the Shape Tables of the Manual, Part 1. d/A_f is listed for channels; r_T does not apply. L_c and L_u are tabulated in the Beam Tables of the Manual, Part 2. Their effects can be seen in the Allowable Moment Curves in the same part of the Manual. These curves

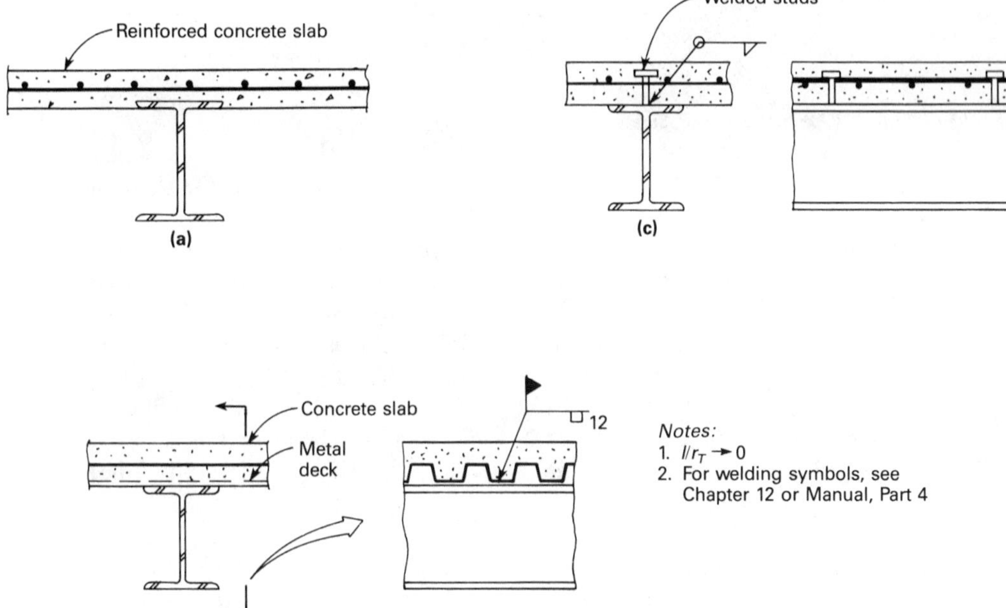

Figure 10.12 Details providing continuous lateral support of beam by slab.

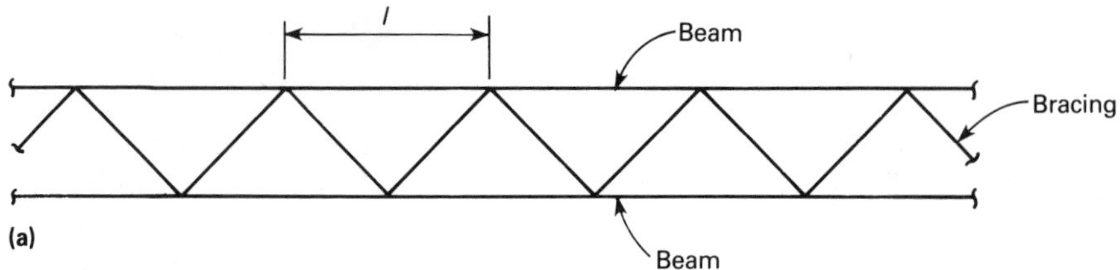

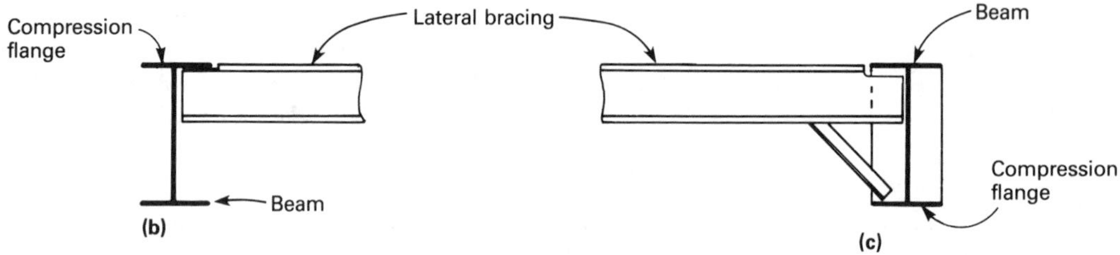

Figure 10.13 Lateral restraint of compression flange in braced system.

incorporate, for each section, the controlling values of F_b from the two choices in Fig. 10.10, plus, for its range, the compact section values.

10.9 SHEAR STRESS IN BEAMS

Referring to Table 10.2, we note that the value of F_v, the allowable shear stress, is covered in Section 1.5.1.2 of the AISC Specification. Sections 1.5.6, 1.9, and Appendix C modify F_v. The reader no doubt already knows that Section 1.5.6 provides for a one-third increase in F_v, as well as other allowable stresses, for wind and seismic loads. He can also infer from previous discussion that Section 1.9 sets limits on b/t or d/t for usual values of allowable stress, and Appendix C contains rules for the reduction of F_v when those limits are exceeded.

Let us examine the provisions of Section 1.5.1.2. We will delay until Chapter 12 discussion of Subsection 1.5.1.2.2, which relates to local stresses near fasteners. Section 1.5.1.2.1 gives the basic equation for allowable shear stress in a beam:

$$F_v = 0.40F_y$$

two-thirds of the basic value for tension. The stress is applicable on "the cross-sectional area resisting shear." As discussed earlier, this leads to the equation

$$A_v \geq (A_v)_R = \frac{V}{K_v F_v} \tag{10-10}$$

where A_v is the area resisting shear. We introduced the modifier K_v to account for the ratio $(f_v)_{ave}/(f_v)_{max}$.

In the usual case of an I section bending about its major axis or a rectangular box section bending about either principal axis of cross section, the distribution of shear stress is as shown in Fig. 10.6(b), and the maximum shear stress is only slightly higher than the average. The difference is ignored in usual practice, setting K_v equal to 1. For such cases,

$$(A_v)_R = \frac{V}{0.4F_y} \tag{10-17}$$

$$A_v = d(t_w) \tag{10-18}$$

This procedure would be unconservative for minor axis bending. However, shear stresses are seldom high for that case.

The AISC Commentary, Section 1.5.1.2, notes that rolled shapes are usually proportioned with webs sufficiently strong to avoid shear failure. However, we suggest that designers be alert for the unusual case where shear failure is possible.

10.10 DEFLECTION CRITERIA FOR BEAMS

The subject of deflection has arisen in a number of ways in earlier parts of this chapter. For the most part we have looked at the deflected shape of beams and the effect on that shape of the moment diagrams resulting from various cases of loading. We have also been concerned with the large, uncontrolled deflections that result from instability. Deflection of beams may also affect the usefulness of a structure even if there is no resulting lack of safety.

Sagging of a floor between supports may make the floor hard to maintain or hard to use. An obvious example would be the lanes of a bowling alley. A plaster ceiling may crack if the structural supporting beams deflect excessively. Flat roofs supported by slender beams may *pond* water, leading to progressively increasing deflection as additional water is attracted to the deflected area, possibly causing eventual collapse. (This is a safety problem.) Sagging of floors may cause doors to stick. Lateral flexural deflection (*drift*) may break windows or other nonstructural elements of structures.

The AISC Specification 1.13.1 limits *live load deflection* of some floors to 1/360 of the span. In Section 1.13.3 it sets stiffness criteria intended to guard against *ponding* on roofs. Section 1.13.2 alerts the designer to the effects of vibration. It is otherwise silent on the matter of deflection except for recommendations in the Commentary, Section 1.13. However, designers must deal with deflections whenever they can have adverse effects, whether the Specification covers it or not. They may be guided by provisions of design codes for common situations, or they may need to research the implications of deflection on function for the particular case.

If large *dead load deflections* are calculated, the designer can require that beams be *cambered*; that is, deflections opposite to dead load deflections are to be built in to the fabricated beams, which will then become straight when the dead load is applied. For deflections caused by live load with or without impact or lateral deflection caused by wind or earthquake, the designer must provide enough stiffness to keep deflections within acceptable limits. The chief tools available are moment of inertia and span length. If Δ is the deflection of a beam from a chord joining its ends, for any case of loading with total load W and length L between supports

$$\Delta_{\max} = \frac{kWL^3}{EI} \tag{10-19}$$

where the value of k depends on the nature and distribution of W. Negative end moments reduce deflection; positive increase it.

Noting that for most steel sections

$$I = d^n$$

where $2 \leq n \leq 3$, then

$$\frac{\Delta_{\max}}{L} \approx k_1 \frac{W}{E} \left(\frac{L}{d}\right)^2 \tag{10-20}$$

The ratio L/d is commonly used as an indicator of acceptable beam proportions for deflections, and Δ/L is the usual parameter used to set service limits on deflection. Common values of Δ/L used in various codes for live load (plus impact if applicable) are in the range

$$\frac{1}{180} \leq \frac{\Delta}{L} \leq \frac{1}{1000}$$

Span/depth ratios are commonly in the range

$$20 \leq \frac{L}{d} \leq 30$$

for simple spans, sometimes greater than 30 for continuous beams, and less than 20 for tighter limits of Δ/L.

The actual deflection relates not so much to F_y as to the actual values of f_b that result from the load that causes deflection. If a simple span beam is subjected to uniform load w in pounds per foot, leading to total load W,

$$M_{\max} = \frac{WL}{8}$$

$$\Delta_{\max} = \frac{5}{384} \frac{WL^3}{EI} = \frac{5}{48} \left(\frac{M_{\max} L^2}{EI}\right)$$

Since

$$(f_b)_{max} = \frac{M_{max}}{S} = \frac{M_{max}}{I}\left(\frac{d}{2}\right) \qquad \text{and} \qquad E = 29{,}000 \text{ ksi}$$

$$\frac{\Delta_{max}}{L} = \frac{5}{48}\left[\frac{2(f_b)_{max}I}{(I)E}\right]\left(\frac{L}{d}\right) = \frac{(f_b)_{max}}{139{,}200}\left(\frac{L}{d}\right)$$

If $L/d = 24$ and $(f_b)_{max} = 24$ ksi, this yields

$$\frac{\Delta_{max}}{L} = \frac{1}{242}$$

Table 10.5 compares the effects of a few simple load cases and end conditions on the relationship between Δ/L and $(f_bL)/d$.

TABLE 10.5 Δ_{max} AND Δ/L; A FEW COMMON CASES OF LOAD AND SUPPORT (PRISMATIC BEAMS)

	Loading	Moments	M_{max}	Δ_{max}	Δ/L	Compare to case 1; relative value, Δ_{max} or $(\Delta_1)_{max}$
1	$wL = W$; L	M	$\frac{WL}{8}\ \frac{5}{48}$	$\frac{M_{max}}{EI}$	$\frac{1}{139{,}200}\ \frac{f_bL}{d}$	1.0
2	W; $L/2$, $L/2$		$\frac{WL}{4}\ \frac{1}{12}$	$\frac{M_{max}}{EI}$	$\frac{1}{174{,}000}\ \frac{f_bL}{d}$	0.80
3	W; $L/2$, $L/2$	M_{max}; $-M_{max}$	$\pm\frac{WL}{8}\ \frac{1}{24}$	$\frac{M_{max}}{EI}$	$\frac{1}{348{,}000}\ \frac{f_bL}{d}$	0.40
4	$wL = W$	$+\frac{M_{max}}{2}$; $-M_{max}$	$-\frac{WL}{12}\ \frac{1}{32}$	$\frac{M_{max}}{EI}$	$\frac{1}{464{,}000}\ \frac{f_bL}{d}$	0.30

10.11 PROCEDURES FOR BEAM SELECTION

10.11.1 General Procedure

The process of beam selection may be very simple or very complex, depending on what is sought in the design. In complex cases, the procedure may require comparison among a number of choices based on the various design constraints. It may involve several iterations with different spacings and variations of other design controls.

A procedure for selection is presented in Flow Chart 10.4, with comments regarding each stage. The procedure is very general, and it is left to the designer to set

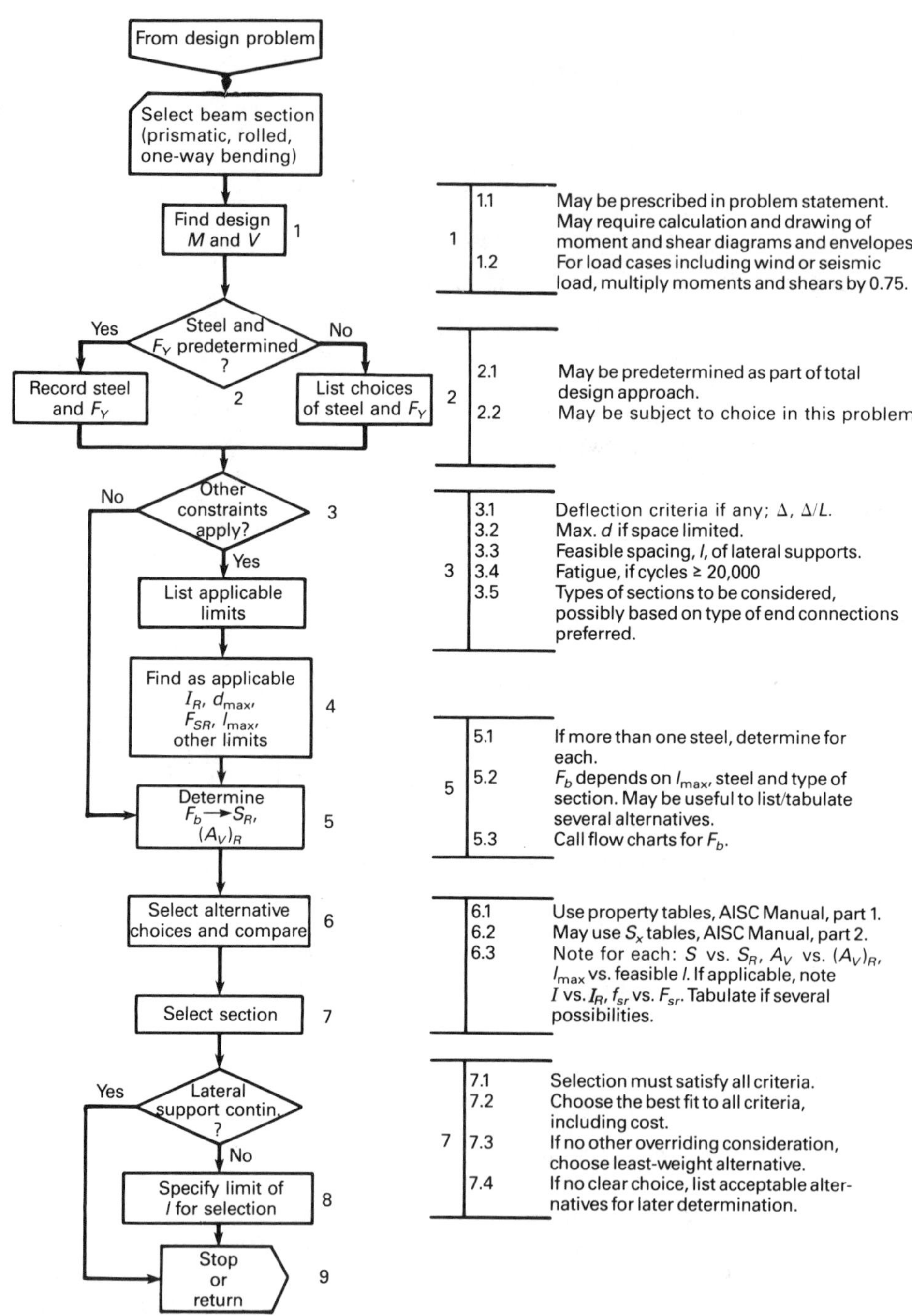
From design problem
Select beam section (prismatic, rolled, one-way bending)
Find design M and V
1
Steel and F_Y predetermined ?
Yes
No
2
Record steel and F_Y
List choices of steel and F_Y
Other constraints apply?
No
Yes
3
List applicable limits
Find as applicable I_R, d_{max}, F_{SR}, l_{max}, other limits
4
Determine $F_b \rightarrow S_R$, $(A_V)_R$
5
Select alternative choices and compare
6
Select section
7
Lateral support contin. ?
Yes
No
Specify limit of l for selection
8
Stop or return
9
1
1.1 May be prescribed in problem statement. May require calculation and drawing of moment and shear diagrams and envelopes.
1.2 For load cases including wind or seismic load, multiply moments and shears by 0.75.
2
2.1 May be predetermined as part of total design approach.
2.2 May be subject to choice in this problem.
3
3.1 Deflection criteria if any; Δ, Δ/L.
3.2 Max. d if space limited.
3.3 Feasible spacing, l, of lateral supports.
3.4 Fatigue, if cycles ≥ 20,000
3.5 Types of sections to be considered, possibly based on type of end connections preferred.
5
5.1 If more than one steel, determine for each.
5.2 F_b depends on l_{max}, steel and type of section. May be useful to list/tabulate several alternatives.
5.3 Call flow charts for F_b.
6.1 Use property tables, AISC Manual, part 1.
6.2 May use S_x tables, AISC Manual, part 2.
6.3 Note for each: S vs. S_R, A_V vs. $(A_V)_R$, l_{max} vs. feasible l. If applicable, note I vs. I_R, f_{sr} vs. F_{sr}. Tabulate if several possibilities.
7
7.1 Selection must satisfy all criteria.
7.2 Choose the best fit to all criteria, including cost.
7.3 If no other overriding consideration, choose least-weight alternative.
7.4 If no clear choice, list acceptable alternatives for later determination.

Flow Chart 10.4 Procedure for beam selection.

detailed steps for implementation of each stage. We will present several examples next. If there are no applicable constraints other than shear and moment and/or if the steel is predetermined, some parts of the process may be bypassed. If, as in a building floor, continuous lateral support of the compression flange is easy to guarantee, further consideration of l_{max} is unnecessary. If design conditions permit easy and inexpensive placement of lateral supports wherever desired, F_b can be set simply as $0.66F_y$ if compact or $0.60F_y$ if noncompact. Flow Chart 10.1 can help separate compact and noncompact sections. If F_b cannot be determined easily, it becomes helpful to use Flow Charts 10.2 and 10.3.

The process usually leads to several possible choices, among which the designer should try to select one that best fulfills all criteria for choice. If no other criteria control, the least-weight selection is often the least cost choice and should be used. If a single best choice is not clear among two or three contenders, it is useful to list the contenders and make the final selection later, based possibly on considerations not yet clear at this stage of the design.

10.11.2 Example: Beam Selection, Laterally Supported

Select a beam to span 24 ft between simple supports to carry the dead plus live load of a concrete floor slab plus its own weight. Dead and live load from the floor are each 100 lb/ft². Beams are spaced 8 ft on centers. The beam is part of a building design that uses ASTM-A36 steel throughout. No other constraints control.

CALCULATIONS

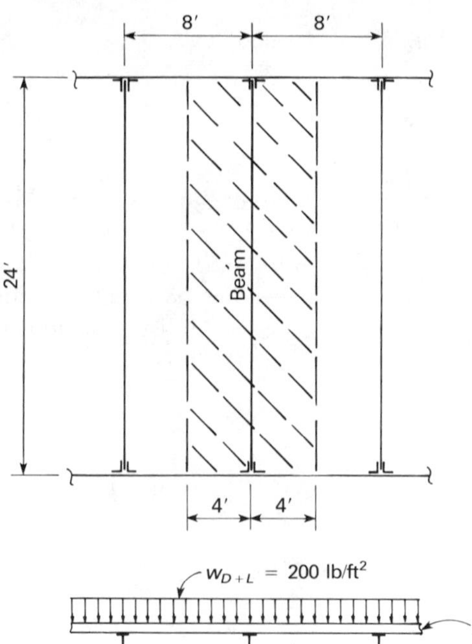

REMARKS

Sketch helps to clarify problem. Load per lineal foot of the beam will be contributed by 8 ft of the width of the slab.

From slab, $w_{D+L} = \dfrac{8'(200)}{1000} = 1.60$ k/LF	Data given.
Add for beam (assume $\Delta w_D = 0.10$): $w_{D+L} = 1.70$ k/LF; $W_{D+L} = wL = 40.8$ k	A guess, probably on the high side. Note that $0.1/1.7 = 0.06$. Large error in assumed beam weight contributes small error in total w.
M_{max} at $\dfrac{WL}{8} = 122.4$ ft k at midspan	
$V_{max} = \dfrac{W}{2} = 20.4$ k at ends	
$F_y = 36$ ksi	A36 steel specified.
$F_b = \begin{cases} 0.66F_y = 24 \text{ ksi} \\ 0.60F_y = 21.6 \text{ ksi} \end{cases}$	If compact / Noncompact } Lateral support is continuous.
$F_v = 0.40F_y = 14.4$ ksi	
$S_R = \dfrac{M}{F_b} = \dfrac{122.4(12)}{24} = 61.2 \text{ in.}^3$	Compact
$S_R = \dfrac{M}{F_b} = \dfrac{122.4(12)}{21.6} = 68.0 \text{ in.}^3$	Noncompact
$(A_v)_R = \dfrac{V}{F_v} = 1.42 \text{ in.}^2$	

Consider:

	Section	S_x
→	W18 × 40	$S_x = 68.4$
	W10 × 60	66.7
→	W16 × 40	64.7
	W12 × 50	64.7
	W14 × 43	62.7

Source: S tables, AISC Manual, Part 2. Since $l = 0$, F_b will be either $0.6F_y$ or $0.66F_y$, depending on compact section criteria. All listed sections lighter than 100 lb/ft.

For lightest sections:

Check if f_v is less than F_v?

W18 × 40: $A_v = d(t_w) = 17.9(0.315) = 5.64 \text{ in.}^2 \gg 1.42$ — AISC Manual,—Part 1, Shape Tables

W16 × 40: $A_v = 16.01(0.305) = 4.88 \gg 1.42$

Check elements for "Compact": — AISC Specification 1.5.1.4.1; Limits tabulated, App. A, Table 6.

	Compact limit	W18 × 40	W16 × 40
$\frac{b_f}{2t_f}$	10.8	5.7	6.9
$\frac{d_w}{t_w}$	106.7	56.8	52.5
Conforms?		OK	OK

$b_f/2t_f$ and d/t_w from Shape Properties, Part 1, AISC Manual.

Choose W16 × 40. — Either W18 × 40 or W16 × 40 is OK. W16 takes less space.

The procedure followed 1 → 2 → 4 → 5 → 6 → 7 → STOP, on Flow Chart 10.4. The sections considered were all compact. Since the S_x table offered no lighter noncompact section, there was no need to make selections based on $F_b = 0.60F_y$. Shear stress is very low compared to the allowable.

The reader might try this same problem with a steel of $F_y = 60$ ksi. Try also a solution of the same problem using the *Beam Tables*, Part 2, AISC Manual. These tables are designed specifically for *simple spans* with uniformly distributed loads and lateral support at $l \leq L_u$. If deflection were a criterion, it could be easily found from the coefficients and formulas in the table.

10.11.3 Example: Beam Selection, Laterally Unsupported

The example that follows is much more tedious than the first example. The beam is not permitted to have lateral support except at its ends. The long unsupported length leads to unfeasibility of compact sections or even noncompact sections with $F_{bx} = 0.6F_y$. The beam must be found by "cut and try" or iterative methods. Three steels are considered, but it turns out that the size of the beam is the same whichever steel is used.

The procedure follows Flow Chart 10.4: 1 → 2 → 3 → 4 → 5 → 6 → STOP. Step 5 requires enlistment of Flow Charts 10.1 and 10.2. Shear check is left to the end since the solution for moment is so much more complex.

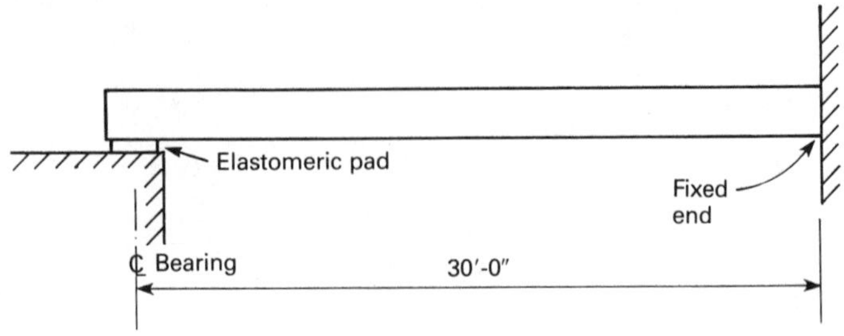

The beam in the figure is part of a structure the design of which does not permit lateral support except at the end vertical supports. It supports a continuously distributed load (dead + live) of 500 lb/ft including its own weight. It must occasionally (about 20 times/year) support concentrated loads at the 1/3 points of its length, each equal to 30 kips. Select a suitable beam of I shape, considering $F_y = 36$ or 50 or 60 ksi. The solution will follow Flow Charts 10.4 and 10.2.

CALCULATIONS

(1) Find design shear and moment:

***Load case 1, distributed* D + L:**

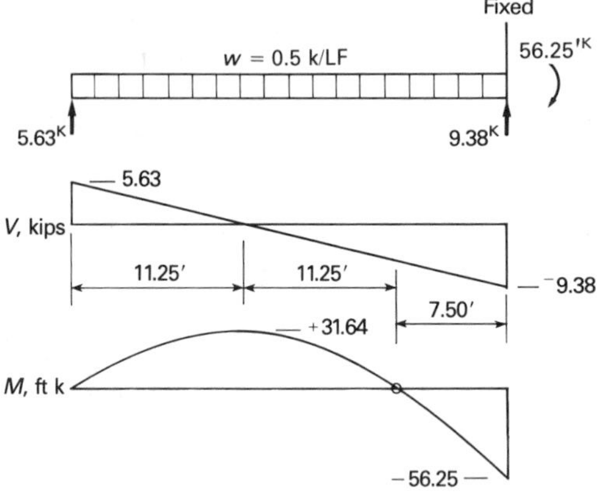

Load case 2, add concentrated loads:

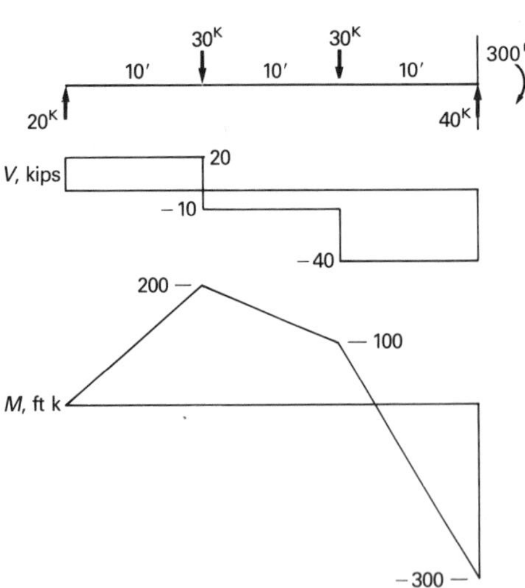

REMARKS

Calculations of shear and moment were done using AISC Beam Diagrams 12 and 14. Numerical work is omitted here in the interest of space. (We do not recommend omitting numerical work in record calculations.)

Diagrams are drawn to clarify the problem. Maximum shear and maximum moments must be found.

Envelope diagrams: Case 1 + Case 2

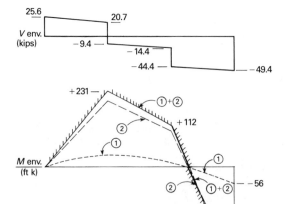

The envelope diagrams show the worst combined effects of the two load cases. Notice the short stretch where there can be either + or − moment.

In drawing envelopes, each load case should be drawn to the same scale. Ordinates can then be added by scaling.

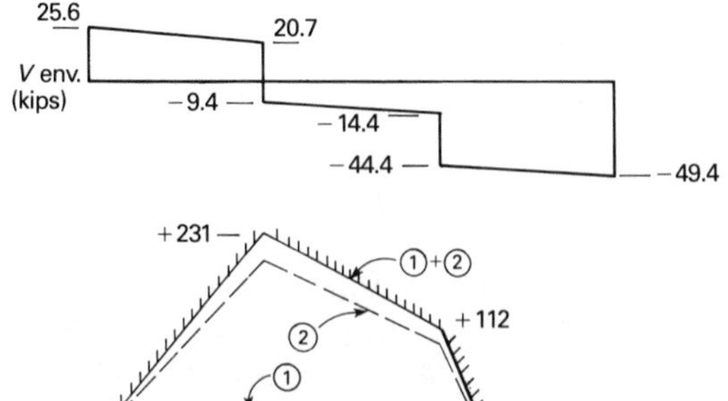

Design shear and moment:

From envelopes

$$V_{max} = 49.4 \text{ k}$$

$$M_{max} = -356 \text{ ft k}$$

$$+230 \text{ ft k}$$

Adjust design moment by plastic action?

AISC 1.5.1.4.1
Use adjusted moments if compact section is chosen.

$$\text{Design } M_{max} = -356(0.9) = -320 \text{ ft k}$$

$$\text{or} \quad = +230 + \frac{356}{2}(0.1)$$

$$= +248 \text{ ft k}$$

Use −320 ft k.

(2) Choice of Steel
Consider:

A36	$F_y = 36$ ksi
A572, Gr. 50	$= 50$
A572, Gr. 60	$= 60$

(3) Other constraints?

Deflection limit of I_R	No	Not stipulated.
d_{max}	No	Not stipulated but l/d should be < 30.

F_{sr}	No	20 cycles/year × any reasonable
l_{max}	Given	life ≪ 20,000.

(4) $l_{max} = l_{min} = L = 30(12) = 360''$ — Problem statement

(5) Find $F_b \rightarrow S_R$:

(a) If *compact*, $F_b = 0.66F_y$. Then — Flow Chart 10.1

$$(S_x)_R = \frac{M}{F_{bx}} = \frac{320(12)}{0.66F_y} = \frac{5818}{F_y}$$

With $30' = l$, the most likely test to fail seems to be the unsupported length test.

and requires $l \leq 76b_f/\sqrt{F_y}$ plus all other tests. Check if $L_c < 30'$ for section chosen for $F_b = 0.66F_y$.

F_y	F_b	$(S_x)_R$	Section	S_x	L_c (ft)	
36	24.0	160	W24 × 76	176	9.5	NG
			W18 × 86	166	11.7	NG
50	33.0	116	W21 × 62	127	7.4	NG
			W16 × 67	117	9.2	NG
60	39.6	97	W18 × 55	98	6.1	NG
			W12 × 72	97	9.9	NG

Refer to S_x table. L_c is listed in S_x table, Manual, Part 2. For $F_y = 60$, calculate

$$L_c = \frac{76b_f}{\sqrt{F_y}}$$

and

$$L_c = \frac{20{,}000}{(d/A_f)F_y}$$

and use smaller.

∴ *Compact section not feasible.*

(b) Consider noncompact with

$$l < L_u, \qquad C_b = 1$$

$$M = -356^{\text{ft k}}, \qquad F_b = 0.6F_y$$

Adjusted moments not permitted if noncompact.

$$(S_x)_R = \frac{356(12)}{0.6F_y} = \frac{7120}{F_y}$$

F_y	F_b	$(S_x)_R$	Section	S_x	L_u (ft)	
36	21.6	198	W27 × 84	213	11.0	NG
			W18 × 106	204	26.0	NG
50	30.0	142	W24 × 68	154	8.5	NG
			W14 × 90	143	24.5	NG
60	36.0	119	W21 × 62	127	7.2	NG
			W14 × 82	123	16.8	NG

S_x table lists L_u for $F_y = 36$ and 50. For $F_y = 60$, $C_b = 1$ from moment envelope. Find limits of l/r_T and ld/A_f from Fig. 10.10. Search for possible sections in Shape Tables. L_u is larger of

$$\frac{r_T}{12}\sqrt{\frac{102{,}000}{F_y}} \quad \text{or} \quad \frac{20{,}000}{12(d/A_f)F_y}$$

∴ $l < L_u$ *not feasible.*

(c) Consider noncompact with

$$L_{u1} < l \leq L_{u1}\sqrt{5}$$

Note that lighter section in each case is deeper, has greater S, but

$C_b = 1$ since midspan $M >$ both end moments.

is less stable laterally (i.e., L_u is smaller).
The l/r_T branch of Flow Chart 10.2 is tried first.

F_y	L_{u1}/r_T	$\frac{L_{u1}\sqrt{5}}{r_T}$	Limit for $b_f/2t_f$	Range of F_b
36	53	120	15.8	$21.6 \geq F_b \geq 12$
50	45	101	13.4	$30 \geq F_b \geq 16.5$
60	41	92	12.3	$36 \geq F_b \geq 19.8$

Refer to Figure 10.10 for L_{u1}/r_T. Formulas shown on figure. Limit of

$$\frac{b_f}{2t_f} = \frac{95}{\sqrt{F_y}}$$

Range of F_b is $0.6F_y < F_b \leq 0.33F_y$.

(d) Tentative selection, *A36 steel*; $F_y = 36$.

$$(S_x)_R = \frac{356(12)}{F_b} = \frac{4272}{F_b}, \qquad f_b = \frac{4272}{S_x}$$

$$(r_T)_{\min} = \frac{30(12)}{120} = 3.00$$

Trial F_b	S_R	Section	Adjust S_x	r_T	l/r_T	F_b	f_b	
20	214	W24 × 94	222	2.33	NA	$r_T < 3.00$		
18	237	W27 × 94	243	2.56	NA			
		W24 × 104	258	3.35	107	14	16.6	NG
16	267	W24 × 117	291	3.37	107	14	14.7	NG
		W24 × 131	329	3.40	106	14	13.0	OK

Successive trials are necessary. S_x tables list possible sections. r_T is found in Shape Tables; F_b from Fig. 10.10 or formula.

Check W24 × 131 and W24 × 117 for $b_f/2t_f$:

$$\frac{b_f}{2t_f}\sqrt{F_y} = 7.5(6) \ll 95 \qquad \text{OK}$$

W24 × 117 may still qualify if $F'_{bx} > F_{bx}$(see Flow Chart 10.2).

Check same sections for F'_{bx}:

$$\text{W24} \times 117\text{:} \quad \frac{ld}{A_f} = 360(2.23)$$

$$\frac{12{,}000}{ld/A_f} = F'_{bx}$$

$$= 14.95 > 14.7 \qquad \text{OK}$$

Use W24 × 117 *if* A36 steel and *if* checks for shear.

Check shear:

$$F_v = 0.4F_y = 14.4 \text{ ksi}$$

$$\max f_v = \frac{V}{d(t_w)} = \frac{49.4}{24.26(0.550)}$$

$$= 3.7 \text{ ksi} \ll 14.4$$

Use W24 × 117 if A36.

(e) A572, Gr. 50; $F_y = 50$:

$$S_R = \frac{4272}{F'_{bx}}, \qquad f_b = \frac{4272}{S_x}$$

Use W24 × 117 if $F_y = 50$.

(f) *Use W24 × 117 for any steel used.*

Since $F'_{bx} > F_{bx}$, F_b is on the sloping portion of the ld/A_f curve in Fig. 10.10 and $F_b < 0.6(36)$. The selection is the same for all steels.

The selection is the same for all steels. L/d is 15, surprisingly low, but not objectionable since that reduces deflection. Low values of L/d are consistent with control by ld/A_f criteria. The low stress utilized makes us question the solution. This seems to be the best we can do with an I section for the problem as stated. However, could we do better if we were not limited to an I? This is food for thought.

The large values of l/r_T and ld/A_f also make us concerned about the ability of this beam to resist even small lateral forces combined with the vertical forces specified here. We will look at that problem in the next chapter.

10.12 SPECIAL ISSUES

10.12.1 Web Crippling

A large concentrated load applied over a short distance on a beam flange can cause high compressive stresses in the thin web. The condition is shown in Fig. 10.14. As shown in Figs. 10.14(b) and (f), the area resisting the bearing stress, F_a, is the contact length plus the effective spread (approximate at 45°) of that length in the distance, k, between the surface and the toe of the fillet; this is the level at which the web thickness becomes uniform. The allowable limit of such bearing stress, f_a, is

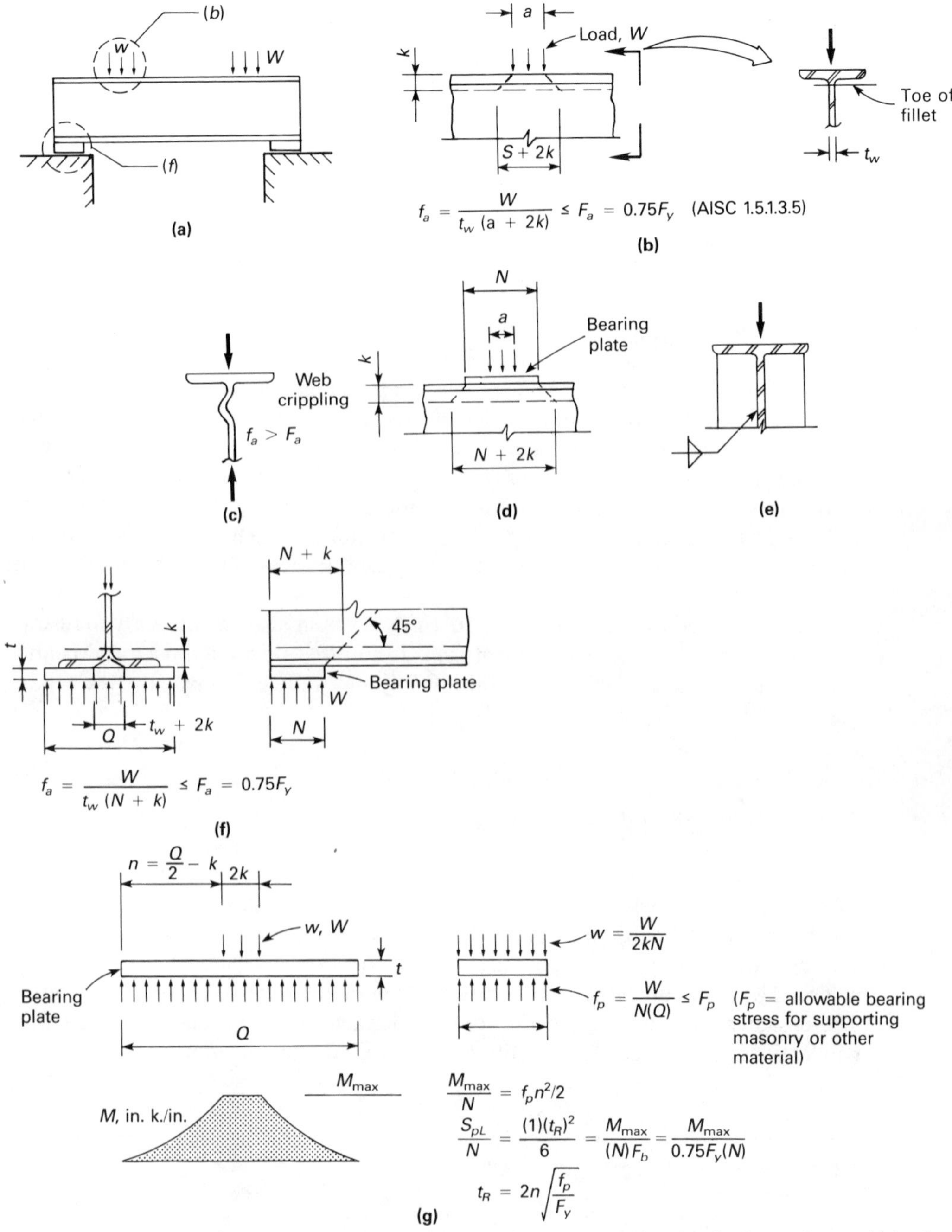

Figure 10.14 Web compression and crippling; bearing plates: (a) simple span beam with load distributed over short distance; (b) compressive stress at toe of fillet; (c) web crippling from excessive f_a; (d) bearing plate under load reduces f_a; (e) stiffeners on web prevent web crippling (symbol indicates weld); (f) compressive stress at toe of fillet at beam end with bearing plate; (g) plan dimensions of bearing plate are selected to reduce bearing pressure to allowable pressure for supporting material; thickness of bearing plate is selected to satisfy bending moment.

$$F_a = 0.75F_y \qquad \text{(AISC 1.5.1.3.5)}$$

The consequence of excessive stress can be to buckle the web as in Fig. 10.14(c). f_a can be reduced in several ways. Introduce a thick plate under the load of length greater than a [Fig. 10.14(d)]. Alternatively, attach bearing stiffeners to the web tightly fitted under or welded to the flange [Fig. 10.14(e)]. The latter solution will appear again in Chapter 14 and the Highway Bridge Case Study (Chapter 20).

10.12.2 End Bearing

Figure 10.14(f) addresses the same problem of potential crippling as in Fig. 10.14(b). The context is somewhat different since the load must be resisted not only by the web of the beam, but also by the material of the support. If the support is a masonrylike material, permissible bearing pressures, F_p, will be much less than for steel. See AISC Specification 1.5.5. The load must be spread, as in Fig. 10.14(g), over a large enough area to reduce the pressure below F_p. Bending moment is introduced into the bearing plate. The equation shown determines the thickness of plate required to resist that moment.

For the condition shown in Fig. 10.14(f), it would also be necessary to restrain the upper flange against lateral translation. If no other restraint exists, end bearing stiffeners similar to the stiffeners of Fig. 10.7(f) may suffice. In that case, the bearing plate could be made thinner or possibly eliminated.

10.12.3 Open Web Joists

For lightly loaded spans, with beams closely spaced, it often happens that available rolled sections will have very low stress (inefficient) or unacceptable deflection. A special type of standard manufactured beam is available in the form of a light truss, such as shown in Fig. 10.15. Chords and webs are very light. Strength and stiffness is achieved by relatively large values of depth. Continuous lateral support is imperative. The designer may find information on the characteristics of open web joists from publications of the Steel Joist Institute or individual manufacturers.

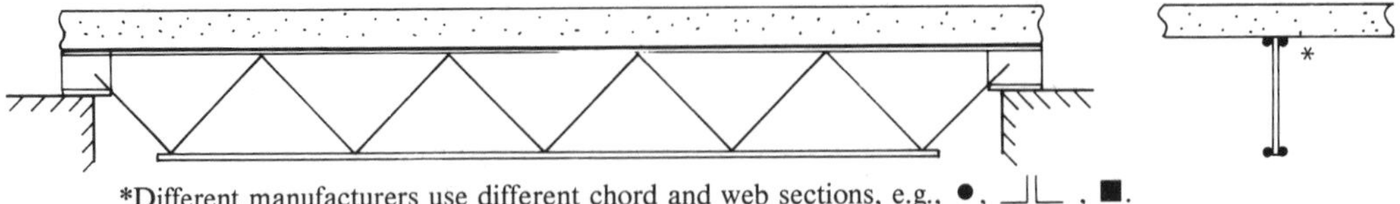

*Different manufacturers use different chord and web sections, e.g., ●, ⅃L, ■.

Figure 10.15 Open web joists.

10.13 AISC MANUAL: BEAM TABLES AND DESIGN AIDS

We have had a number of occasions in this chapter to refer to sections of the AISC Manual. The reader would do well to become familiar with the aids to beam design and selection found particularly in Parts 1 and 2 of the Manual. Some have been mentioned previously.

Shape tables for W, S, and other I-shaped sections list pertinent beam properties for each section. In columns under "Compact Section Criteria," $b_f/2t_f$ and d/t_w are given; alongside each is the maximum value of F_y (F'_y and F'''_y) for which the section conforms to the criteria. Where no value is given, all structural steels except A514 conform. Values shown for F'''_y are based on AISC formula (1.5-4b), which applies when there is considerable axial load in addition to bending. In the case of beams with small or no axial load, if $F'''_y < F_y$, it is worth checking formula (1.5-4a). Values of r_T and d/A_f are also listed. Readers will easily recognize the purpose of the other properties listed.

Part 2 of the Manual is specifically intended to ease the pains of beam analysis, design, and selection. Tables list values of S_x, Z_x, and I_x for beam shapes in descending numerical order, a considerable help in selecting from precalculated required values. The S_x table also offers, for $F_y = 36$ and 50 ksi, values of L_c, L_u, and the resisting moment M_r for *compact* and *semicompact* sections. The Beam Tables can offer shortcuts to beam selection for $F_y = 36$ and 50 ksi. These apply to laterally supported beams. For beams where unbraced length l needs investigation, curves of allowable moment versus l are provided for most beam sections. These are drawn separately for $F_y = 36$ and 50 ksi. Due to the large vertical scale, the curve for each beam is on several pages. This may obscure the fact that the shape of these curves is the same as our Fig. 10.10 with special allowance at the short end for compact and semicompact sections. Beam diagrams give the shape of shear and moment curves for 33 combinations of load and support conditions in single spans. They also provide formulas for calculation of shear, moment, and deflections. Six additional diagrams cover multiple spans and four more deal with moving loads.

In Part 5, Tables 1, 6, and 7 in Appendix A of the Specification provide values of various variables in the formulas governing beam design.

PROBLEMS

10.1. Classify the following sections as *compact, semicompact,* or *noncompact* by AISC criteria. Make your determinations from the requirements as given in the Specification. Verify your results with the Compact Section Criteria in AISC tables where available.

	Shape	Steel	M_x and lateral support
			Positions of lateral support of compression flange are designated thus: ⊗ Positions of vertical support, thus: ↑
(a)	W14 × 30	A36	M_x 3 Spaces @ 6.7′ = 20′
(b)	W14 × 30	A572, Gr. 50	Same
(c)	C15 × 50	A588	Same
(d)	W36 × 300	A36	M_x 12′
(e)	W14 × 90	A588	M_x 2 Spaces @ 10′ = 20′

10.2. Find F_{bx} and F_{by} for the following beams, assuming $C_b = 1$.

	Shape	F_y, ksi	Laterally unsupported length, ft
(a)	W14 × 30	50	30
(b)	W36 × 300	50	12
(c)	W36 × 300	50	30
(d)	C15 × 50	36	12
(e)	WT18 × 150*	36	12

*(Flange in compression for M_x)

10.3.

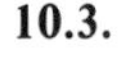

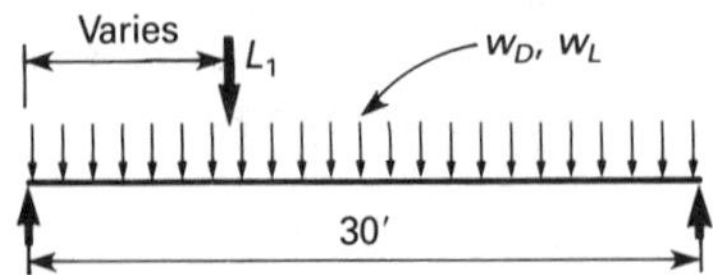

Select a beam, checking all pertinent criteria: $w_D = 1$ k/ft (beam not included), $w_L = 1$ k/ft, $L_1 = 20$ k, continuous lateral support.

(a) A36 steel

(b) A588 steel

10.4. Same as Problem 10.3, but lateral support provided at ends and midspan only. For deflection limits, use Table 23-D, Appendix A.

10.5. Same as Problem 10.3, but lateral support provided at ends only.

10.6. Given the following moment diagrams and support locations, select a suitable beam of I shape.

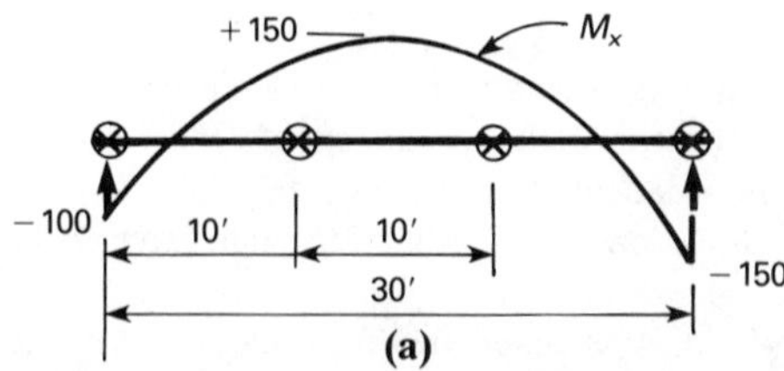

(a)

(a) This diagram results from uniform load on the span. Can you calculate the magnitude of the load? Can you calculate the deflection for the beam you chose?

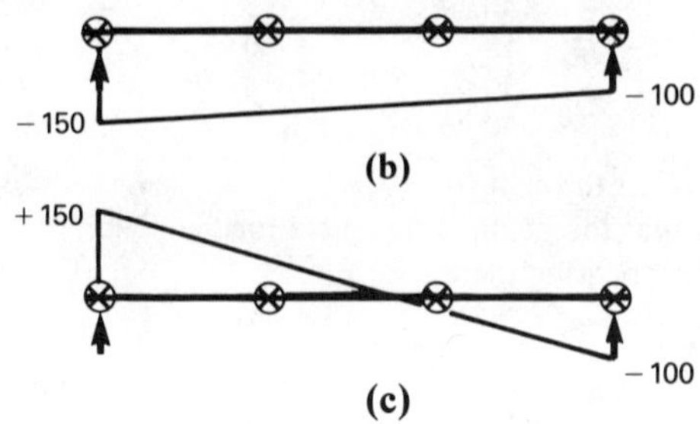

(b)

(c)

(b and c) Can you reconstruct the loading on these spans?

10.7. The beam is to be supported laterally at the two vertical supports and at midspan.

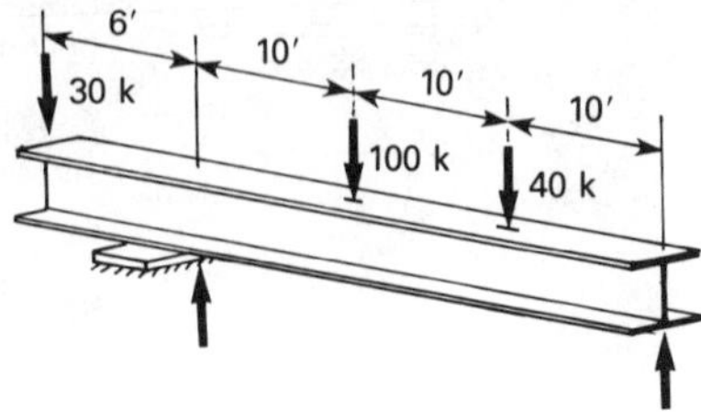

Using A36 steel:

(a) Select an I-shaped beam suitable for all three loads applied simultaneously.

(b) Select an I-shaped beam, assuming that any combination of one, two, or all three loads is possible.

(c) Which flange needs to be laterally restrained?

10.8. Redo Example 10.11.3 using structural tubes. (*Note*: Table 5.1 does not list any steels with $F_y > 50$ ksi for rectangular tubes.)

10.9. The beam of Example 10.11.2 is to be made continuous over its supports for four spans as shown. Select a beam size. (Unless instructed otherwise, the multispan beam diagrams in the Manual, Part 2, may be used. The maximum moments and shears occur with dead loads on all spans and live loads on some spans only.)

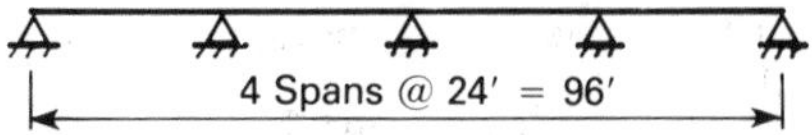

10.10. Conditions are as shown in Fig. 10.14(b). Assume the beams listed are adequate for shear and moment, including the effects of *W*. For each of cases (a) through (d):

(1) Is there an adequate factor of safety against crippling?

(2) If no, how long does a bearing plate have to be to provide the required safety factor? How thick?

(3) If the load *W* is suspended from the lower flange and the detail creates no other problems, what is the effect on the potential for crippling?

	W (kips)	*a* (in.)	Section	Steel
(a)	140	3	W27 × 146	A588
(b)	15	0	W18 × 50	A36
(c)	70	5	W10 × 33	A572, Gr. 60
(d)	70	5	W10 × 33	A36

10.11. For each of (a) through (d):

(1) Verify that the beam is adequate for the shear.

(2) Design a bearing plate.

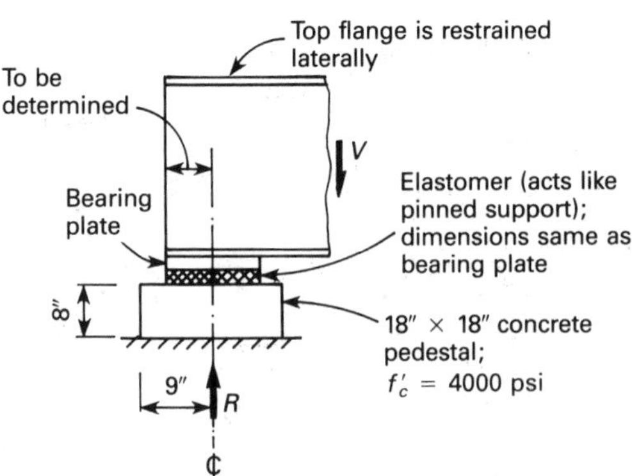

	V (kips)	Section	Steel
(a)	300	W27 × 146	A588
(b)	40	W18 × 50	A36
(c)	50	W10 × 33	A572, Gr. 60
(d)	40	W10 × 33	A36

11 Members under Combined Loading

11.1 INTRODUCTION: INTERACTION EFFECTS

The previous three chapters have each dealt with one of the principal kinds of action to which structural members are subjected: axial tension, axial compression, and flexure. In each case, we were able to see the effects of the single action on the members and develop procedures, based on the AISC Specification, for selecting member cross sections to resist the effect. As it happens, few structural members are single-effect members in their total function. The aching backs that are the plague of so many humans past 40 often result from the difficulty improperly toned muscles and worn vertebral joints have in resisting simultaneously the axial loads carried by our vertebrae, the flexure about our two principal transverse axes, and the twisting (torsion) that our muscles must control. Fortunately for us, the design of the human structure is more efficient for some other human tasks than it is for lifting. In a less painful example from nature, a tree trunk in the wind is at once a compression member, a beam bending about any one of an infinite number of transverse axes, and a torsional shaft resisting the twisting effect of unbalanced wind drag on the unsymmetrically arranged leaf surfaces in the wind stream.

Even in our man-made structural members in the single-effect cases of Chapters 8 through 10, we had to recognize the many ways in which each such member was exposed to other incidental effects. A glance back at the figures in these chapters will remind the reader of these incidental, *secondary* effects. However, combined effects are not always secondary. The moment $P\Delta$ in situations such as shown in Figs. 9.2

and 9.5(g) may be unwanted but real effects large enough to require attention in calculation and member selection. The force Q in Fig. 9.5(g) causes moments in the columns that are intended as part of the primary resisting mode of the portal frame.

The beam selected in Example 10.11.3 was intended to resist loads in the YZ plan. The W24 beam selected was stressed to 98% of its allowable stress, F_{bx}. If the 30 k loads were directed just 5° away from vertical, they would cause additional moments,

$$M_y = 30 \tan 5^\circ \, (10') = 26.25 \text{ ft k}$$

and

$$f_{by} = \frac{M_y}{S_y} = \frac{26.25(12)}{46.25} = 6.81 \text{ ksi}$$

by itself 25% of F_{by}. If, furthermore, the lateral components of force of 30 tan 5° were actually applied at the top flange level, there would be a further torsional effect of 30 tan 5° (12.13″) = 31.84 inch kips, which would seek a resisting mode from a member with little torsional strength or stiffness.

A designer can often avoid combined effects by creating systems in which they do not exist. The slab of Fig. 10.11 resists any lateral forces, relieving the beam of that responsibility. However, it is often preferable, as in the frame of Fig. 9.5(g), to select members suitable for combined effects. It becomes necessary to establish a systematic approach to do so without sacrificing safety. The level of safety should be at least not less than intended for axial load by itself or uniaxial bending.

If there was one level of allowable stress common to all superposable effects, it would be easy. We would write a simple equation for the most highly stressed point on the cross section,

$$\sum_{i=1}^{I} f_i \leq F \tag{11-1}$$

where each f_i is the stress caused by the individual ith effect. But we have a multiplicity of different values of allowable stress F, each based on different considerations—whether yield, rupture, stability, or other. The most common approach is to seek some form of interaction equation that accounts for the differences in allowable stresses. The approach taken by the AISC Specification is to assign a *utilization* term to the proportion of capacity used for each separate effect, f_i/F_i. The sum of such utilization terms is limited to 1.0:

$$\sum_{i=1}^{I} \frac{f_i}{F_i} \leq 1.0 \tag{11-2}$$

limiting the total utilization of allowable capacity to 100%. For load cases involving wind and seismic effects, this is modified to allow for the generally permitted 33% increase in allowable stresses (AISC 1.5.6):

$$\sum_{i=1}^{I} \frac{f_i}{F_i} \leq 1.33 \tag{11-3}$$

As long as the effects of one action do not affect the capacity for another, this is a relatively straightforward approach, which at least does not reduce the general factor of safety. We will see how equations and provisions of the AISC Specification account for the interactive effect of each action on the capacity for the other by the use of modified utilization terms.

Provisions and equations relating to the interaction of axial load and bending about the two principal axes are in Section 1.6, AISC Specification.

11.2 BIAXIAL BENDING

11.2.1 Interaction of Bending Stress

The nature of interaction in biaxial bending can be seen in the beam of Fig. 11.1, which is doubly symmetric in cross section, is being subjected to simultaneous loads in the direction and planes of both principal axes, and offers reactions at its supports in the same plane as the loads. The result is biaxial bending without torsion. Bending stress at any point on the cross section is, for stresses below yield,

$$f_b = f_{bx} + f_{by} = \pm \frac{M_x y}{I_x} \pm \frac{M_y x}{I_y}$$

and is maximum at diagonally opposite corners,

$$(f_b)_{\max} = \left(\pm \frac{M_x c_y}{I_x} \pm \frac{M_y c_x}{I_y} \right)$$

which is physically true but does not reveal the factor of safety.

Figure 11.2 is an interaction diagram that takes into account the fact that F_{bx} and F_{by} are not usually the same. At any point along the diagonal line, the sum of the percentages of capacity utilized in X and Y bending equals 100. For any point within the right triangle, the sum is less. The equation of the diagonal, shown in Fig. 11.2, is

$$\frac{f_{bx}}{F_{bx}} + \frac{f_{by}}{F_{by}} = 1.00 \tag{11-4}$$

the same as Formula (1.6-2) in the AISC Specification with the axial load term equal to zero. Values of F_{bx} and F_{by} are those found in the previous chapter. Each term may be considered a *utilization factor* applied to capacity in either X or Y bending.

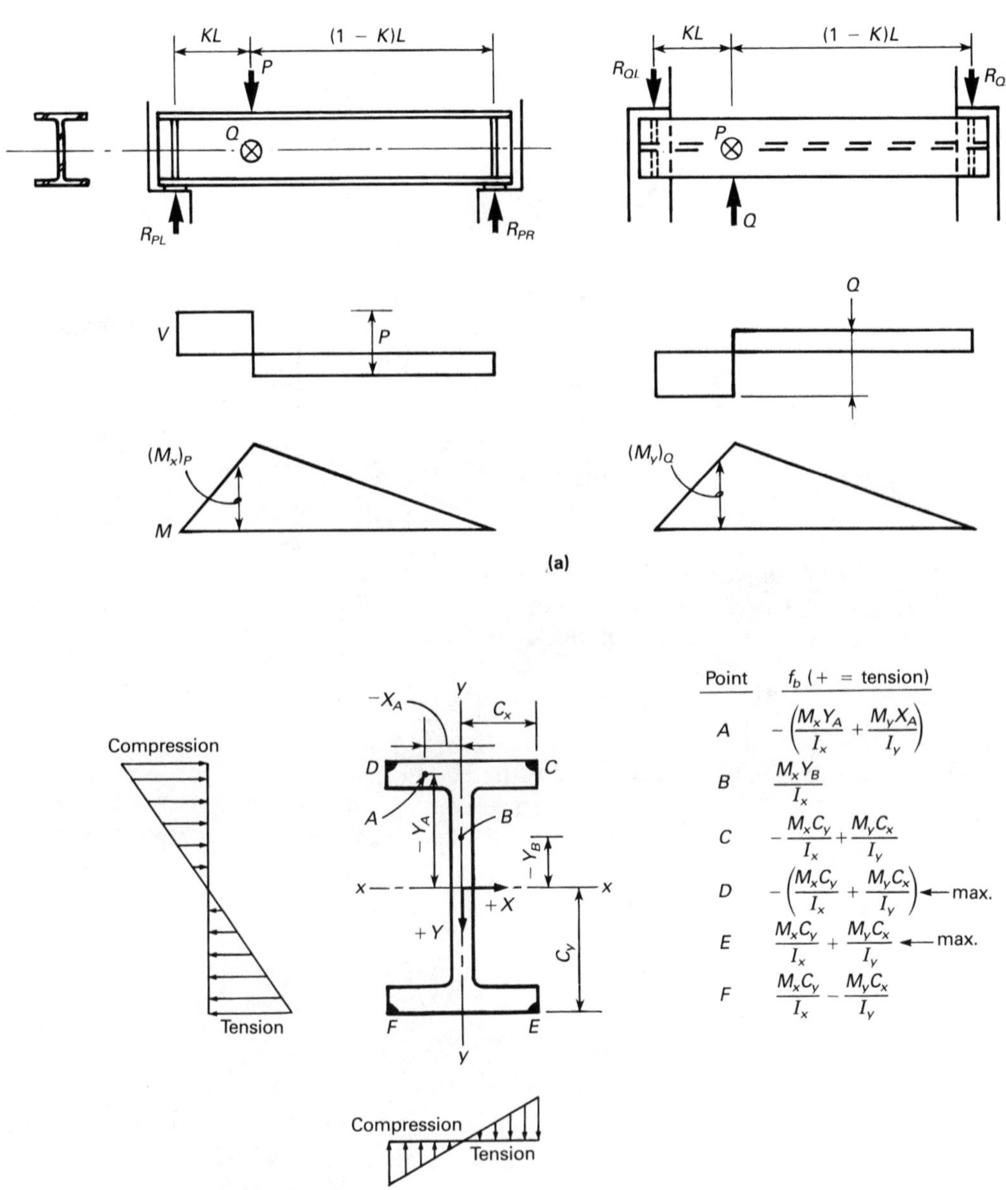

Figure 11.1 Combined stresses, biaxial bending: (a) load, shear, moment; (b) bending stress.

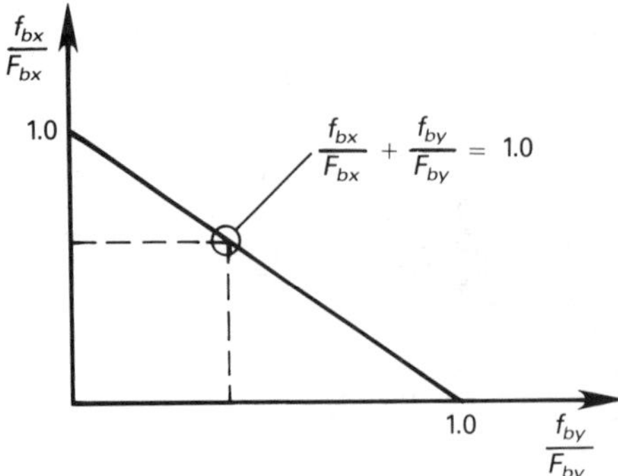

Figure 11.2 Interaction diagram, stresses in biaxial bending.

11.2.2 Shear Stress

The limiting value of f_v in the AISC Specification is set at

$$F_v = 0.40F_y \qquad \text{(AISC 1.5.1.2.1)}$$

and for shear parallel to the Y axis is to be based on "the cross-sectional area resisting shear." "The effective area . . . of rolled and fabricated shapes may be taken as the overall depth times the web thickness," giving

$$f_v = \frac{V}{d(t_w)}$$

a reasonable approximation for V_y in Fig. 11.1.

The effect of Q is to cause shear stress in the beam flanges, now acting as two parallel *webs*. It is not necessary to add the values of F_v resulting from P and Q, since they mostly affect different elements of the cross section. The average calculated value of f_v that results in the flanges is

$$(f_v)_{\text{ave}} = \frac{V_x}{2b(t_f)}$$

$$(f_v)_{\text{max}} = \frac{1.5\ V_x}{2b(t_f)} \leq 0.4F_y = F_v$$

Figure 11.3 illustrates the application of AISC Section 1.5.1.2 to the biaxial bending case of Fig. 11.1. It also indicates that both loads P and Q actually cause shear stress in the flanges. (See reference 45 or others on shear flow.) Load P does in fact cause shear flow and, therefore, shear stress in the flanges. This is ignored in the approximation of AISC 1.5.1.2, but should be kept in mind when shear stress is high and bending is biaxial.

A selection process for biaxial bending is illustrated in Flow Chart 11.1 and followed in general in Example 11.2.3.

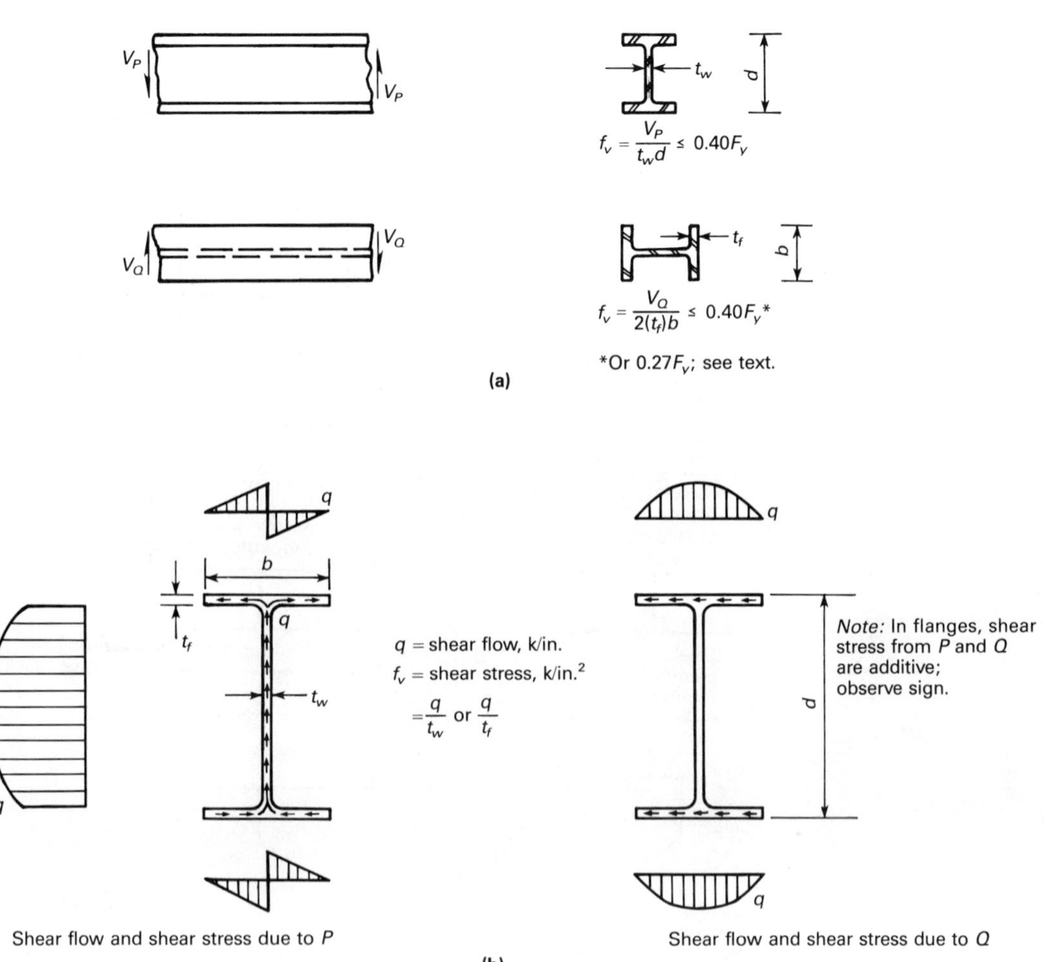

Figure 11.3 Shear in biaxial bending: (a) shear stress, beam of Figure 11.1, following AISC 1.5.1.2; (b) shear flow.

11.2.3 Example: Roof Purlin

Select a rolled shape for the roof purlin shown in Fig. 11.4. (The purlin selected will be similar to the one used in the Mill Building Case Study, Chapter 17.) Use A36 steel. The roof covering is not designed as a diaphragm that could provide continuous lateral restraint of the compression flange.

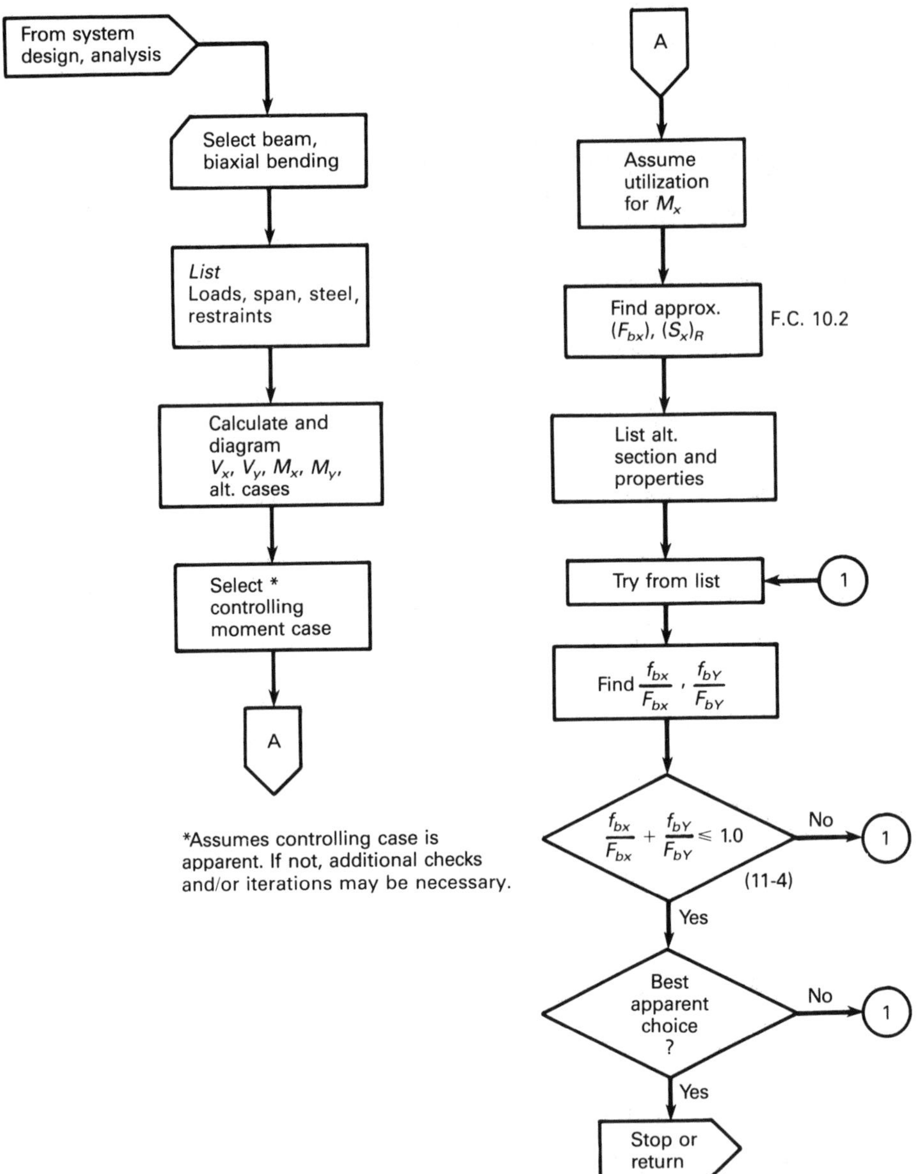

Flow Chart 11.1 Select beam, biaxial bending.

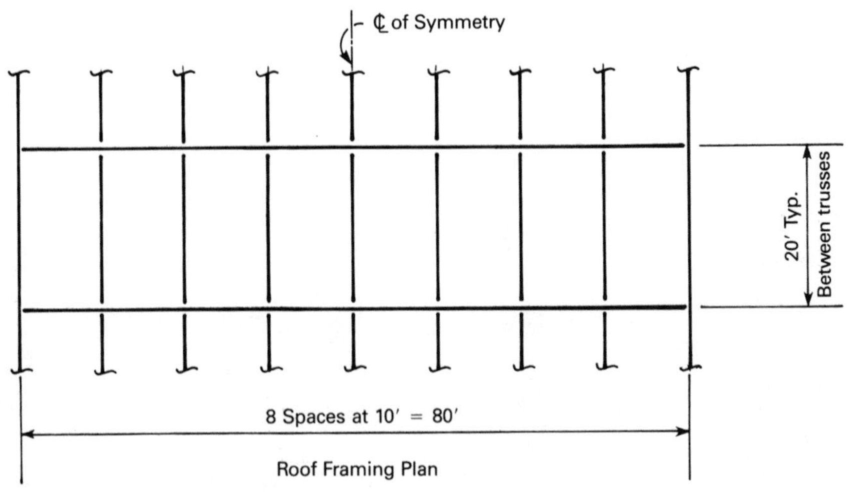

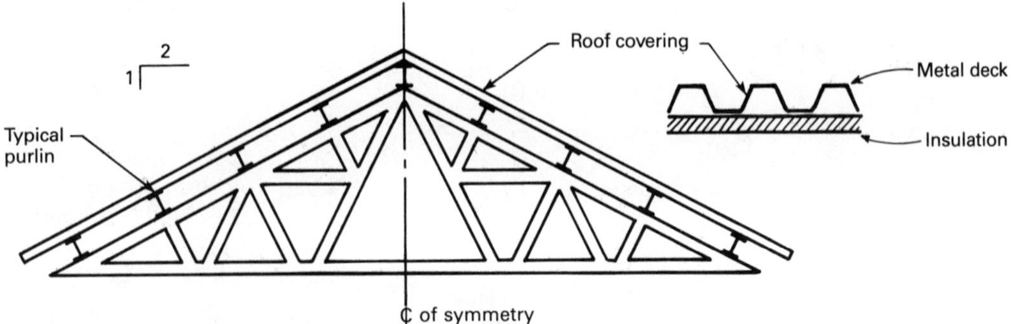

Loads to Purlin

Dead load: Insulation: 2 lb/ft^2

Metal Deck: 3 lb/ft^2

5 lb/ft^2 of roof surface plus weight of purlin

Live load: Snow at 30 lb/ft^2 of horizontal area

Figure 11.4 Roof framing for industrial building (Example 11.2.3).

CALCULATIONS

REMARKS

(1) Evaluate dead load w_D and live load w_L.

Convert dead load of roof from pounds per square foot of surface to pounds per square foot of horizontal projection.

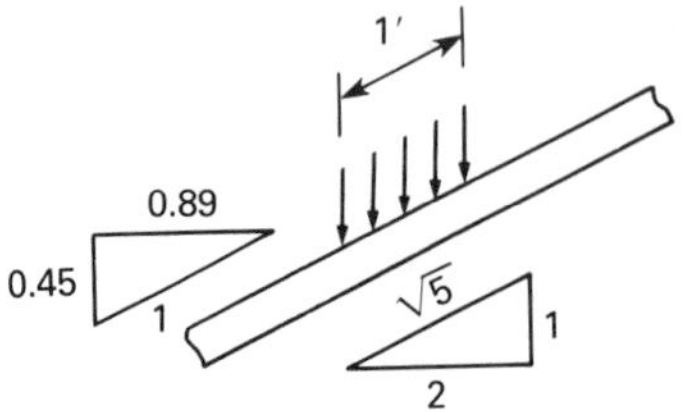

Roof covering:

$$w_{D1} = \frac{5}{0.89} = 5.62 \text{ lb/ft}^2 \text{ of horizontal projection}$$

To purlin:

$$5.62(10') = 56.2 \text{ lb/LF of beam}$$

Purlin will support, in lb/LF, the load from 10 ft^2/LF.

Assume beam weight = 19.8:

$$w_D = 56.2 + 19.8 = 76$$

Assumed beam weight in lb/LF is subject to later revision.

Live load:

$$w_L = 30(10) = 300$$

$$w_{D+L} = 376 \text{ lb/LF}$$

(2) Convert w to components:

Load needed in principal directions of bending, parallel to principal axes.

Note that the lateral component (parallel to x) acts at the top flange. I sections are not stiff enough torsionally to transfer the effective lateral force to the x axis.

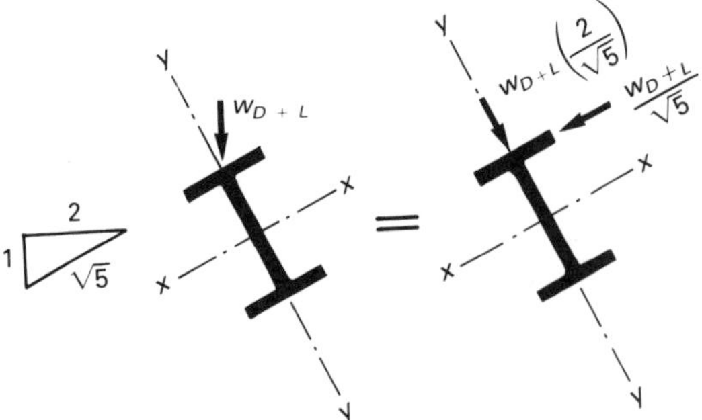

(3) Formulas for f_{bx} and f_{by}:

$$w_{D+L} = 376\left(\frac{1}{\sqrt{5}}i + \frac{2}{\sqrt{5}}j\right)$$

$S_y/2 = S_y$ for one flange; lateral load is applied and resisted at upper flange level. See further discussion after this solution.

$$f_{bx} = \frac{M_x}{S_x}$$

$$f_{by} = \frac{M_y}{S_y/2} = \frac{2M_y}{S_y}$$

(4) Load YZ plane; V_y, M_x:

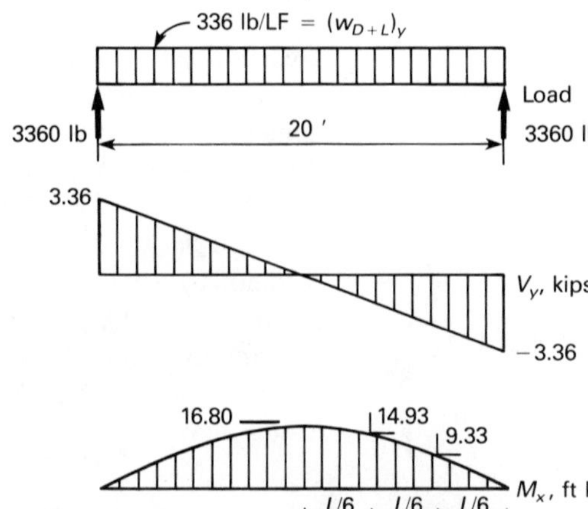

$$V_{max} = \frac{wL}{2} = \frac{W}{2}$$

$$M_{max} = \frac{WL}{8}$$

At 1/6 points of span, reduce by

$$\frac{M_{max}}{9}, \frac{4}{9}M_{max}$$

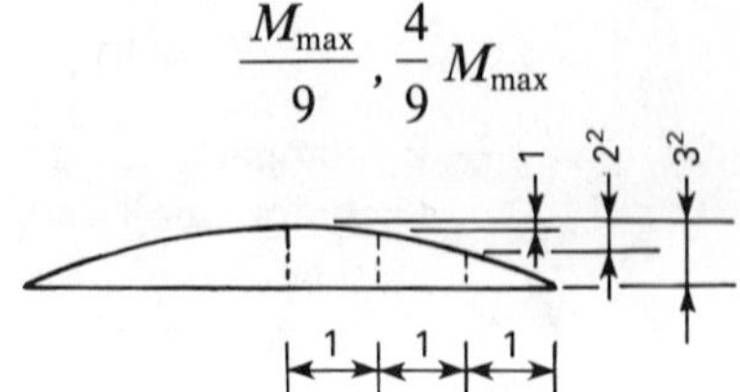

(5) Load, XZ plane; V_x, M_y

Try lateral restraint at 1/3 points:

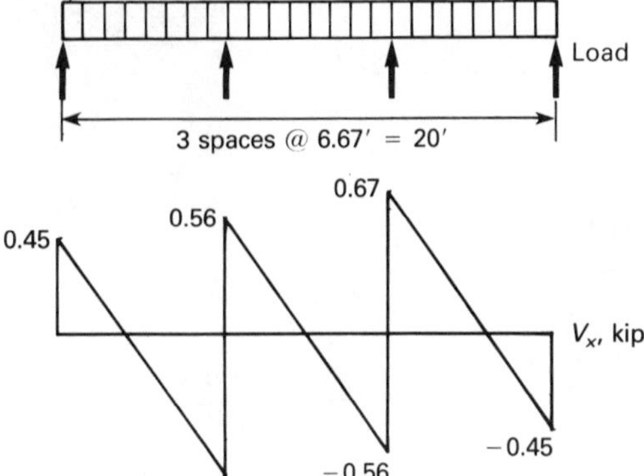

It will be necessary to restrain the compression flange to reduce l/r_T, to increase F_{by}, *and* to reduce bending moment and deflection for $y-y$ bending. l (laterally unsupported length) = L (span).

Shear and moment calculated, using AISC Part 2, beam diagrams, Case 36.

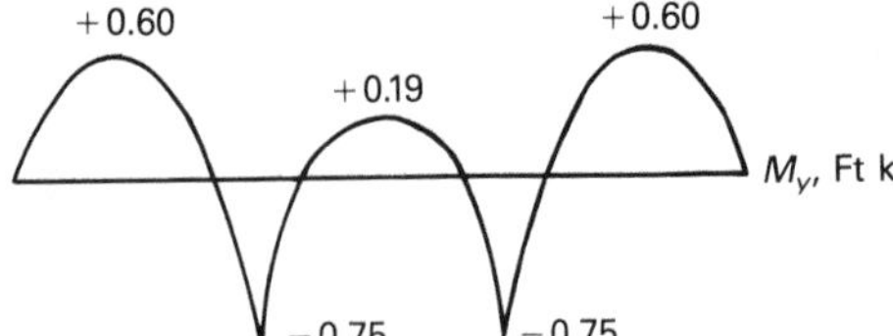

(6) M_x and M_y for design:

(**a**) At midspan:

$$M_x = 16.8 \text{ k ft}$$

$$M_y = 0.19$$

(**b**) At 1/3 point of span:

$$\left.\begin{aligned} M_x &= 14.93 \text{ k ft} \\ M_y &= -0.75 \text{ k ft} \end{aligned}\right\} \xleftarrow{\text{use for}\ \ \text{design}}$$

Design combination of moments is set at point where $(M_y)_{max}$ is coincident with a value of M_x that is almost max. M_y is varying rapidly, M_x slowly. As a general rule, both combinations, (a) and (b), should be checked. Readers should check the interaction equation for combination (a) after the beam is selected.

(7) Find tentative $(S_x)_R$:

$$L = 20', \qquad l = 6.67'$$

If compact, L_c must be $\geq 6.67'$, not likely for small and light W sections; L_u may be $> 6.67'$ for some shallow sections with wide flanges.

See Flow Chart 10.2. S_x table, AISC Manual, will confirm these judgments. See members with $l/d \approx 20$; $d \approx 20(12)/20 = 12$; say $8 \leq d \leq 12$.

If noncompact, $l > L_u$, and *utilization* in $x - x$ bending $\approx 75\%$, and $F_{bx} \approx 0.5F_y = 18$ ksi; then

Assumptions necessary to start search for appropriate section.

$$(S_x)_R = \frac{M}{F_{bx}(0.75)}$$

$$= \frac{14.93(12)}{(18)(0.75)} = 13.27 \text{ in.}^3$$

Try S_x in range, say, 11 to 17.

(8) List possible choices; tabulate properties:

Section	S_x	S_y	L_c	L_u	r_T	d/A_f	l/r_T	ld/A_f
M12 × 11.8	12.0	0.64	2.7	3.0	0.68	17.4	118	1393
W10 × 12	10.9	1.10	3.9	4.3	0.96	11.9	83	952
W8 × 15	11.8	1.70	4.2	7.2	1.03	6.41	78	513
W10 × 15	13.8	1.45	4.2	5.0	0.99	9.25	81	740
W12 × 16	17.1	1.41	4.1	4.3	0.96	11.3	83	904
W10 × 17	16.2	1.78	4.2	6.1	1.01	7.64	79	612
W8 × 18	15.2	3.04	5.5	9.9	1.39	4.70	58	376

Choices are found with help of S_x tables in Manual, Part 2. Properties come from S_x table and section properties table.

(9) Try sections from list.

(a) Try W10 × 12:

M12 × 11.8 is slightly lighter, but looks less stable laterally. Compare L_u.

$$F_{bx} = \left[0.67 - \frac{F_y(l/r_T)^2}{1{,}530{,}000}\right]F_y = 18.28 \text{ ksi} \longleftarrow$$

AISC Formula (1.5-6a); $C_b = 1$

or

$$F'_{bx} = \frac{12{,}000}{ld/A_f} = 12.61 \text{ ksi} \ll 18.28$$

AISC Formula (1.5-7)

Check: $b_f/2t_f = 3.96/2(0.210) = 9.4.$

Properties Table

$$9.4 < \frac{65}{\sqrt{F_y}} = \frac{65}{6} = 10.8 \qquad \text{OK}$$

Satisfies Section 1.5.1.4.1, para. 2

$$F_{by} = 0.75F_y = 27.0 \text{ ksi}$$

AISC 1.5.1.4.3

Is $$\frac{f_{bx}}{F_{bx}} + \frac{f_{by}}{F_{by}} \leq 1.00$$

$$f_{bx} = \frac{M_x}{S_x} = \frac{14.93(12)}{10.9} = 16.44 \text{ ksi}$$

$$f_{by} = \frac{2M_y}{S_y} = \frac{2(0.75)(12)}{1.1} = 16.36 \text{ ksi}$$

$$\frac{f_{bx}}{F_{bx}} + \frac{f_{by}}{F_{by}} = 0.90 + 0.61 = 1.51 \gg 1.0 \qquad \text{NG}$$

New try required.
Utilization for $M_x >$ assumed.
Utilization for $M_y \gg$ assumed.

(b) Try W8 × 15:

$$F_{bx} = 0.6F_y = 21.6 \text{ ksi}$$

Since $L_c < l < L_u$,

$$F_{by} = 0.75F_y = 27.0 \text{ ksi} \quad \text{as above}$$

W10 × 15 has same weight but, for W8 × 15, $l < L_u \rightarrow F_{bx} = 0.6F_y$ and S_y is larger than for W10.

$$\frac{f_{bx}}{F_{bx}} + \frac{f_{by}}{F_{by}} = \left[\frac{14.93(12)}{11.8(21.6)} + \frac{0.75(12)}{(1.7/2)27}\right]$$

$$= 0.70 + 0.39 = 1.09 > 1.00$$

NG — Closer; try again.

(c) Try W8 × 18:

$$F_{bx} = F_{by} = 21.6 \text{ ksi}$$

W12 × 16, and W10 × 17 look less desirable than W8 × 18 because of their narrow flanges. W8 × 18 has wider flange than W8 × 15; ∴ larger S_y.

$$\frac{f_{bx}}{F_{bx}} + \frac{f_{by}}{F_{by}} = \left[\frac{14.93(12)}{15.2(21.6)} + \frac{0.75(12)}{(3.04/2)27}\right]$$

$$= 0.55 + 0.22 = 0.77 < 1.0$$

OK

(10) Select W8 × 18. *Note*: 18.0 < 19.8 = assumed weight of beam.

Reader may wish to verify that W10 × 17 is also acceptable.

The procedure took us through several iterations before we could find the best combination of properties for selection in this biaxial bending problem. The $y-y$ bending portion contributed a large part of the total utilization, although M_y is very small. The S_y modulus is very small, and only half of it is effective because of lack of torsional stiffness and strength. It would be of interest to see whether a section more effective torsionally can help. To try that, we must first establish procedures for combining bending and torsion. A problem at the end of the chapter invites the reader to try this.

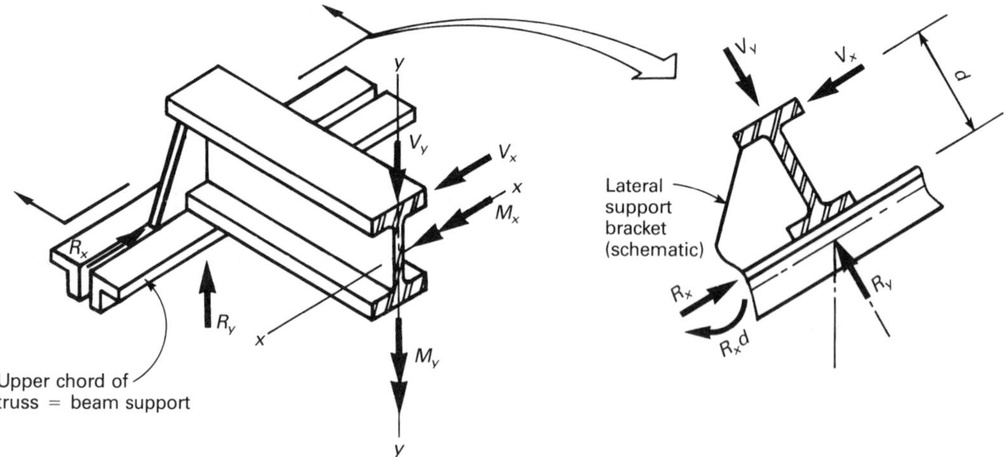

Figure 11.5 Purlin end support.

For the selection as made in this problem, we made the assumption that the lateral force was resisted by the flexure of the upper flange without torsion. The solution can only be correct if the end detail makes that possible (i.e., if a horizontal end support acts at the upper flange level). In the detail of Fig. 11.5, a special bracket at the end of the purlin intercepts the horizontal reaction at the top flange and transfers it and its overturning effect to the top chord level on the truss.

Similarly, the intermediate lateral supports should be at or very close to the upper flange level. In the Mill Building Case Study, *sag rods* are used for these lateral supports.

11.3 AXIAL LOAD WITH BENDING

11.3.1 General Comments

Combinations of axial load with bending are the rule rather than the exception. The members shown in Figs. 8.1 and 9.3 are axially loaded. All except the vertical members experience bending from their own weight. When joints are not pinned, even the vertical members have bending moment. In the case of the portal frame of Figs. 9.3(h) and 9.5(g), the principal mechanism for resisting lateral forces utilizes the bending strength of both the columns and the girders and the ability of continuous joints to transfer moments from member to member. All members have both axial and bending effects.

The relative significance of bending and axial load varies greatly. Dead weight bending stresses in axially loaded truss members are often small enough in comparison to axial stresses so that, after an initial evaluation shows that fact, they can thereafter be ignored. Lateral forces in portal frames, such as in Fig. 9.5(g), are usually much smaller than the vertical forces. So it is common that axial forces in the girders have little effect when compared to bending. Columns, on the other hand, always have significant axial stresses, usually combined with significant bending stress. This will be found to be the case in Part 3 in both the mill building and tier building case studies, as well as the mast and boom of the stiffleg derrick.

As with biaxial bending, the problem of combined stresses cannot be solved simply by adding the stresses from each effect. Allowable stresses F_a, F_{bx}, and F_{by} are usually different and are controlled by different member properties. In addition, new complications arise from the cross effects of axial load and bending. However, once again interaction equations establish the limit of combined stress effect in the form

$$\sum_{i=1}^{I} \frac{f_i}{F_i} \leq 1.00$$

with each utilization term representing the percentage utilized of the separate allowable capacity for a single effect. The usual maximum permitted utilization is 100%; for load cases including wind or earthquake 133% is permitted.

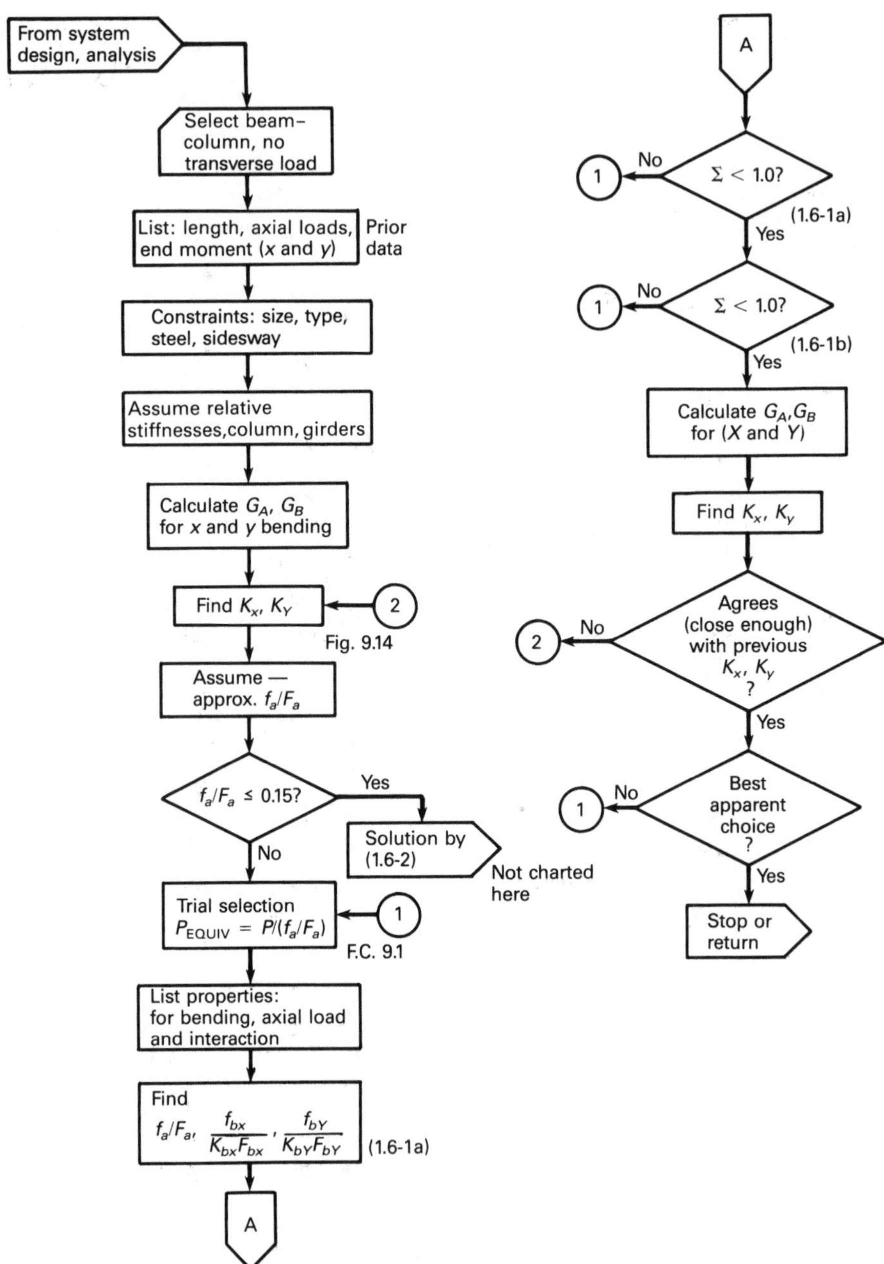

Flow Chart 11.2 Select beam column, no transverse loads.

11.3.2 Axial Compression Plus Bending

In both separate cases, axial compression and axial bending, earlier chapters showed that allowable stresses (F_a and F_b) were influenced by a combination of the yield strength of the steel, F_y, and a factor representing the potential for instability (or buckling). The combined effects are influenced by the same factors, plus the interaction of the different destabilizing factors on each other. In this light we can view the various forms of interaction equations in Specification Section 1.6.1:

$$\frac{f_a}{F_a} + \frac{C_{mx} f_{bx}}{\left(1 - \frac{f_a}{F'_{ex}}\right) F_{bx}} + \frac{C_{my} f_{by}}{\left(1 - \frac{f_a}{F'_{ey}}\right) F_{by}} \leq 1.0 \qquad \text{(AISC 1.6-1a)}$$

$$\frac{f_a}{0.6F_y} + \frac{f_{bx}}{F_{bx}} + \frac{f_{by}}{F_{by}} \leq 1.0 \qquad \text{(AISC 1.6-1b)}$$

$$\frac{f_a}{F_a} + \frac{f_{bx}}{F_{bx}} + \frac{f_{by}}{F_{by}} \leq 1.0 \qquad \text{(AISC 1.6-2)}$$

Formulas (1.6-1a) and (1.6-1b) may be used for all combinations of axial compression and bending, in which case both must be satisfied. In special cases, where $f_a/F_a \leq 0.15$, a relatively minor utilization of axial load capacity, the single simple equation of (1.6-2) may be used instead of the two forms of Eq. (1.6-1). The reasonable implication is that minor axial compression does not materially affect the member's capacity in bending. Equation (1.6-2) is the extended version of our Eq. (11-4), with the axial term now included.

Equation (1.6-1b) is useful primarily in investigating stress conditions near a support, where continuous members without transverse loading within a span have maximum moments and f_b. Equation (1.6-1a) is intended to account for the possibility of buckling, which may be increased by bending deflection between supports.

The selection process for beams with minor axial load ($f_a/F_a \leq 0.15$) could be a simple extension of that in Flow Chart 10.2. There would be an axial load term, plus one or two bending terms depending on whether bending is uniaxial or biaxial. Flow Chart 11.2 would be applicable to many problems of selection for major axial load and bending.

11.3.3 Example: Use of AISC Equation (1.6-2)

The purlin selected in Example 11.2.3 utilized a total of 82% of its capacity to resist biaxial bending. Of the 18% remaining for full utilization, up to 15% may be used for axial compression, if Eq. (1.6-2) is to be used. How much axial load is permitted using this equation?

CALCULATIONS	*REMARKS*
W8 × *18* $A = 5.26$ in.^{2}, $r_y = r_{\min} = 1.23$ in. $L = 20' = 240''$, $K = 1.0$ $\dfrac{KL}{r} = 195 < 200$	Final selection Example 11.2.3. The upper flange is laterally restrained at 6.67′ intervals, the lower flange only at the ends, 20′ apart. We will consider the unbraced length l to be 20′.
A36 steel.	
$F_a = 3.93$ ksi Axial capacity in kips: $AF_a = 5.26(3.93) = 20.67$ k $0.18(20.67) = 3.72$ k	Table 3-36, AISC. We treat this as a primary member for reasons discussed in Chapter 10.
But axial load permitted by AISC (1.6-2): $0.15(20.67) = 3.1$ k	Assumes the utilization of bending effects in the example is concurrent with the axial load.

Noting the large unused capacity, we could, momentarily, be tempted to reduce the member size. However, the member we chose in Example 11.2.3 was almost the lightest acceptable for the dead + live load case. The permitted stress increase for wind load does not supersede the requirements of the basic dead + live load case.

11.3.4 Equations (1.6-1): Axial Compression Plus Bending

Equation (1.6-1a). One look at AISC equation (1.6-1a) is often enough to make the novitiate designer want to run for cover and the veteran designer yearn for the days of the 1947 or 1956 versions of the AISC Specification when the whole interaction problem was settled by reference to the formula

$$\frac{f_a}{F_a} + \frac{f_b}{F_b} \le 1.0$$

Since 1963, designers using the AISC Specification* have had to deal with the factors C_m and f_a/F'_e as they appear in Eq. (1.6-1a), with good reason.

Consider the factor $(1 - f_a/F'_e)/C_m$ as a modifier to be applied to the allowable bending stress, F_b. There is a separate such factor for F_{bx} and F_{by}. So AISC (1.6-1a) can be rewritten in the form

*Readers will find it useful to refer to Section 1.6.1 of the AISC Commentary.

$$\frac{f_a}{F_a} + \frac{f_{bx}}{K_{bx}F_{bx}} + \frac{f_{by}}{K_{by}F_{by}} \leq 1.0 \tag{11-5}$$

where

$$K_{bx} = \frac{1 - f_a/F'_{ex}}{C_{mx}}$$

and (11-6)

$$K_{by} = \frac{1 - f_a/F'_{ey}}{C_{my}}$$

The subscripts x and y are used for different values of C_m and F'_e applicable to the axis around which bending is being considered. f_a/F'_e is always positive, so $(1 - f_a/F'_e)$ always has the effect of reducing F_b in the presence of axial load. C_m varies between 1.0 and 0. Values of C_m less than 1.0 have the effect of increasing F_b. The total effect of K_b may be to increase F_b dramatically, a possibly unsafe result against which Eq. (1.6-1b) offers protection. Both equations must be satisfied.

The factor $(1 - f_a/F'_e)$ is a reduction in bending capacity that results from axial load. We have met the *$P\Delta$ effect* in several ways earlier in this book. $P\Delta$ is the moment that results from application of axial load to a member whose longitudinal axis has been bent by other effects such as bending moments around a transverse axis. It is illustrated in different contexts in Figs. 9.2, 9.5, 9.6, and elsewhere. The usual calculation of f_b does not take into account the stress resulting from the moment $P\Delta$. The factor $(1 - f_a/F'_e)$ is a device used to reduce the value of F_b to account for the $P\Delta$ effect. f_a is the uniformly distributed longitudinal stress from axial load. F'_e is the by now familiar critical column stress determined by Euler divided by the same factor of safety, 23/12, that governs long columns in AISC formula (1.5-2). However, two separate values of F'_e are calculated according to Section 1.6.1, AISC Specification:

$$F'_{ex} = \frac{12}{23}\frac{\pi^2 E}{(Kl_x/r_x)^2} \quad \text{and} \quad F'_{ey} = \frac{12}{23}\frac{\pi^2 E}{(Kl_y/r_y)^2} \tag{11-7}$$

The subscripts x and y are axes both of bending and buckling (e.g., buckling around the x axis is considered in the term for x-axis bending).

Thus, the beam in Fig. 11.6 spans the distance L in the YZ plane and $L/3$ in the XZ plane.The force P causes displacements Δ_y, and bending moments M_x which can usually be safely ignored if P_{ax} is small. However, when the $P\Delta$ moment is large, Δ itself is increased, potentially without limit. The factor $(1\text{-}1/F'_e)$ reduces the F_{bx} to account for the decreased bending resistance in the presence of large axial load. Bending around the Y axis is treated similarly.

F_{bx} and F_{by} are evaluated by the methods of Chapter 10. In Fig. 11.6, both flanges are restrained against translation parallel to x, making $l_x = L/3$ for evaluating F_{bx}. To evaluate F_a, as in Chapter 9, we use the larger value of KL/r.

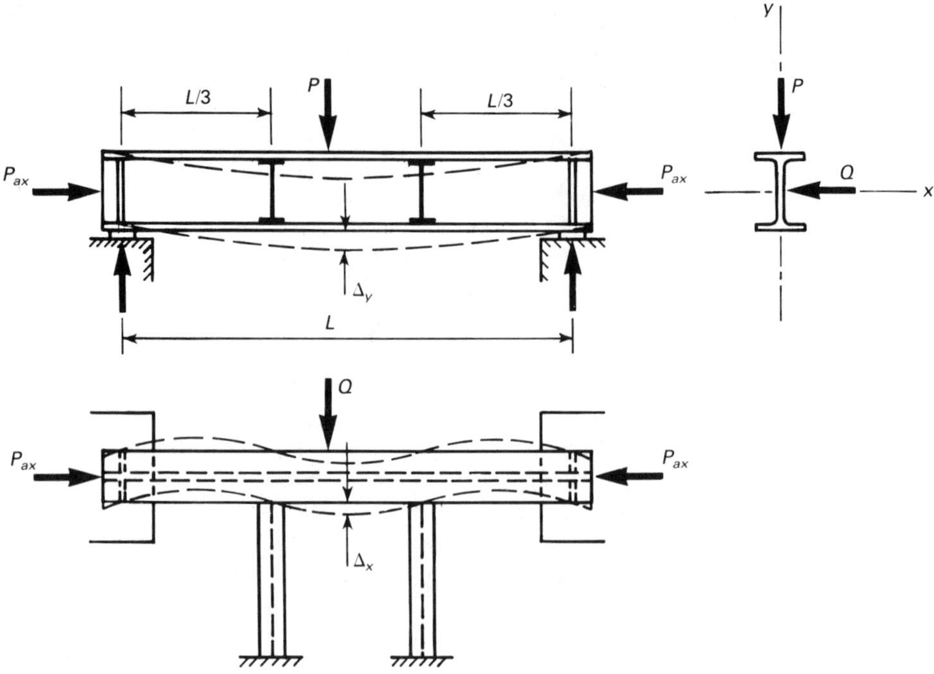

Figure 11.6 Biaxially bent purlins with axial compression.

The effect of $1/F'_e$ in reducing the values of F_b can be seen dramatically from the form of the curve in Fig. 11.7, which is derived from the Euler curve. Like the Euler curve itself, those of Fig. 11.7 are independent of F_y. In the lower-story

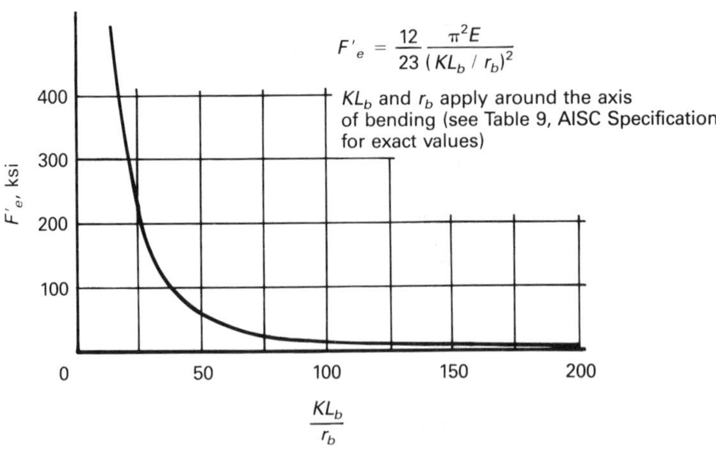

Figure 11.7 F'_e versus KL_b/r_b for use in AISC Eq. (1.6-1a).

columns of our Tier Building Case Study (Chapter 18, Sheet 16 of Calculations), F_{by} is reduced dramatically by the term f_a/F'_{ey}, forcing the designer to greatly increase the weight of the columns. Although M_x is larger than M_y, f_a/F'_{ex} causes a much smaller reduction in F_{bx}, which does not control the column selection.

The bending terms in Eq. (1.6-1a) may be further modified by the factors C_{mx} or C_{my}, applied to the term for X-axis or Y-axis bending. AISC 1.6.1 established three categories for C_m, represented in Table 11.1. The range of C_m is

$$1.0 \geq C_m \geq 0.4$$

TABLE 11.1 VALUES OF C_m IN AISC 1.6.1

Category	Characteristics of frame	C_m
1	Joint translation not inhibited (as by bracing)	0.85
2	Joint translation inhibited by bracing, no transverse loads between supports	$\left(0.6 - 0.4\dfrac{M_1}{M_2}\right)$, but ≥ 0.4*
3	Joint translation inhibited by bracing, transverse loads between supports	
	(a) Members restrained (against rotation) at supports	0.85
	(b) Members unrestrained at supports	1.00

*M_1 and M_2 are end moments.

$|M_1| < |M_2|$

$\dfrac{M_1}{M_2}$ is positive for members bent in double curvature.

$\dfrac{M_1}{M_2}$ is negative for members bent in single curvature.

Recall the definition of K_{bx} and K_{by} in Eqs. (11-5) and (11-6). Values of C_m less than 1.0 increase F_b, offsetting the effects of f_a/F'_e when the shape of the elastic curve enhances stability. This is seen in reduction to 0.85 in categories 1 and 3a, due to rotational restraint at ends of the member. In category 2, where the shape of the curve is not affected by transverse loading, reverse curvature bending may reduce C_m to as little as 0.4. Considerations were similar and the definition of M_1/M_2 the same when C_b was evaluated in Chapter 10 (see Fig. 10.8).

The values of C_m in category 3 are arbitrary and may be adjusted by *rational analysis*. Table 11.2, from the AISC Commentary, gives the results of such analysis for several standard cases. Values from Table 11.2 are in the range from 0.6 to 1.0.

TABLE 11.2 VALUES OF C_m FOR VARIOUS CONDITIONS OF TRANSVERSE LOADING

Case	ψ	C_m
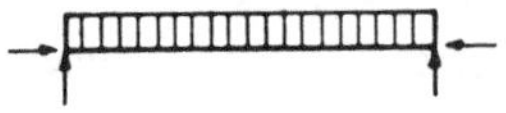	0	1.0
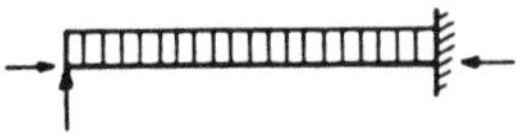	−0.4	$1 - 0.4\frac{f_a}{F'_e}$
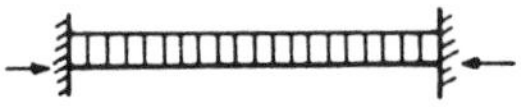	−0.4	$1 - 0.4\frac{f_a}{F'_e}$
	−0.2	$1 - 0.2\frac{f_a}{F'_e}$
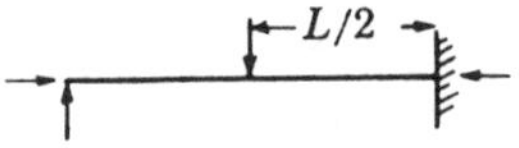 $L/2$	−0.3	$1 - 0.3\frac{f_a}{F'_e}$
	−0.2	$1 - 0.2\frac{f_a}{F'_e}$

Reproduced with permission from AISC Table C1.6.1.

When in doubt, C_m may be conservatively taken as 1.0. On the other hand, with small axial load, note that K_{bx} or K_{by} may approach 2.5, giving very large values of KF_b. The requirement that both Eqs. (1.6-1a) and (1.6-1b) be satisfied avoids any danger on that account.

Equation (1.6-1b). We have already seen one of the functions of Eq. (1.6-1b). Its primary purpose is for use in examining the interaction of axial and bending stresses in the regions of the column ends. The axial load utilization term in Eq. (1.6-1b) is

$$\frac{f_a}{0.6F_y}$$

AISC Eq. (1.5-1), the formula applicable to short and intermediate columns, yields its maximum value,

$$F_a = 0.60F_y$$

for $Kl/r = 0$. The lower values of F_a that result from increasing values of Kl/r reflect the increased potential for lateral translation between the ends of the column. For local stresses at the ends, the safety factor of 1/0.60 seems adequate.

In considering the interaction of axial and bending effects in the end regions, the term $f_a/0.6F_y$ gives a lower utilization factor than the corresponding term, f_a/F_a, in Eq. (1.6-1a), allowing correspondingly increased utilization terms for bending moments, which are likely to be maximum at the ends of columns with restrained ends. In this case, the terms f_b/F_b are also used in unmodified form.

11.3.5 Example: Use of Equations (1.6-1) with Transverse Loads

The W8 × 18 purlin selected in Example 11.2.3 is part of a wind-bracing system. An axial load of 7 kips due to wind must be resisted in addition to full dead plus live load. Is the member acceptable for this load combination? We note that 7 kips is more than the axial load permitted under formula (1.6-2). Equations (1.6-1a) and (1.6-1b) may permit more.

CALCULATIONS	*REMARKS*
(1) Previous results: $\dfrac{f_{bx}}{F_{bx}} + \dfrac{f_{by}}{F_{by}} = 0.55 + 0.27 = 0.82$	Solution will rely on information in the earlier example; not all repeated here.
(2) State interaction equations: $\dfrac{f_a}{F_a} + \dfrac{C_{mx}}{(1 - f_a/F'_{ex})} \cdot \dfrac{f_{bx}}{F_{bx}} + \dfrac{C_{my}}{(1 - f_a/F'_{ey})} \cdot \dfrac{f_{by}}{F_{by}} \le 1.33$	AISC (1.6-1a) and 1.5.6
$\dfrac{f_a}{0.6F_y} + \dfrac{f_{bx}}{F_{bx}} + \dfrac{f_{by}}{F_{by}} \le 1.33$	AISC (1.6-1b) and 1.5.6
(3) Check by Eq. (1.6-1b): $\dfrac{f_a}{0.6F_y} = \dfrac{P}{A(0.6F_y)} = \dfrac{7}{5.26(21.6)} = 0.06$	

$$\frac{f_a}{0.6F_y} + \frac{f_{bx}}{F_{bx}} + \frac{f_{by}}{F_{by}} = 0.06 + 0.55 + 0.27$$

$$= 0.88$$

$$< 1.33 \qquad \text{OK}$$

(4) Check by Eq. (1.6-1a):

$l = 20'(12) = 240''$, $r_y = 1.23''$ — Lower flange is not laterally restrained. Use $l = L$.

$A = 5.26 \text{ in.}^2$, $\dfrac{KL}{r_y} = 195 < 200$

$$\frac{KL}{r_x} = \frac{240}{3.43} = 70$$

$F_a = 3.93$ ksi — AISC Table 3-36, Main Members

$F'_{ex} = 30.5$ — AISC Table 9

$F'_{ey} = 3.93$

$C_{mx} = 1.0 = C_{my}$ — Table 11.1, Category 3b

$$f_a = \frac{7}{5.26} = 1.33 \text{ ksi}$$

Substituting in (1.5-1a)

$$\frac{1.33}{3.93} + \frac{1}{1 - \dfrac{1.33}{30.5}}(0.55) + \frac{1}{1 - \dfrac{1.33}{3.93}}(0.27)$$

$$= 0.34 + 0.58 + 0.41 = 1.33 \qquad \text{OK}$$

(5) *W8 × 18 may be used.*

The very small axial stress had a drastic effect because of the instability of the lower flange. If the axial load were larger, lateral restraints as in Fig. 11.6 would increase F_a and F'_{ey} dramatically. However, that detail is more expensive than the commonly used sag rods, which will be found in the Mill Building Case Study. When the wind problem arises in that study, a different approach, tied to the connection details, validates the purlin without abandoning the use of sag rods.

11.3.6 Example: Beam Column without Transverse Loads.

We are asked to select a column for a building frame such as is shown in Fig. 11.8. In the *weak* direction of the columns (*xy* plane), the building has lateral bracing and rigid

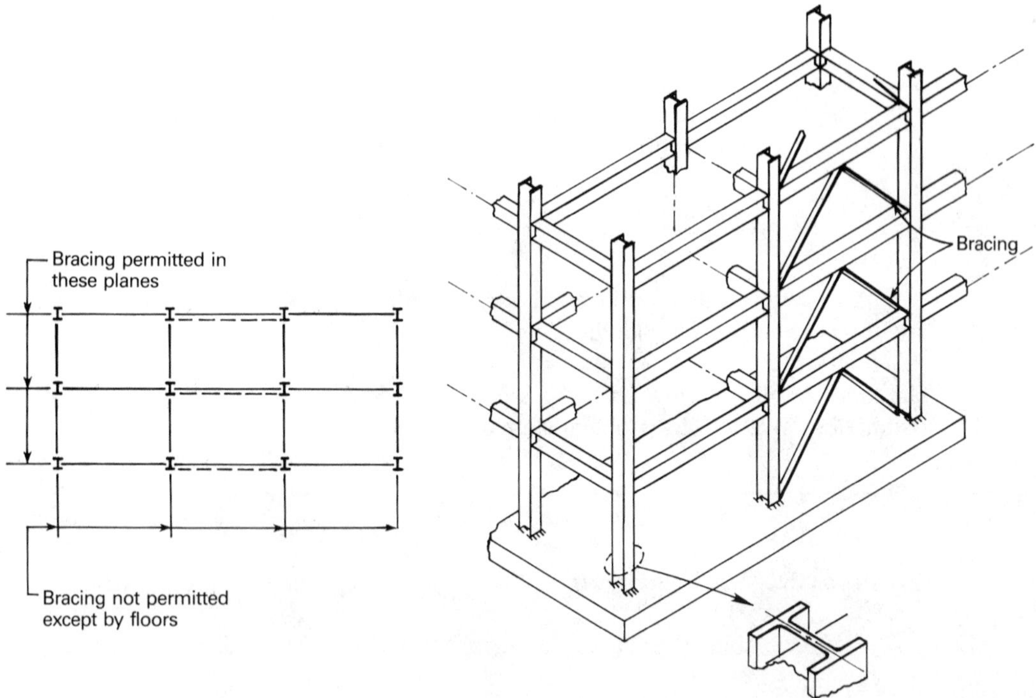

Figure 11.8 Building frame, braced one way.

beam-to-column joints. In the *strong* direction of the columns, the frame is not braced; columns and girders are rigidly connected, forming moment-resisting frames. Conditions to be used for column selection are as follows:

Story height = 10'-0"
Axial compression = 500 kips
$M_x = \pm 100$ ft k at both top and bottom of each story
$M_y = \pm 30$ ft k at both top and bottom of each story

All the forces and moments cited arise from dead plus live load conditions and may occur simultaneously. The joints between the columns, beams, and girders are continuous, and M_x and M_y arise from analysis of the condition where the beam span on only one side of the column experiences live load.

The senior engineer has previously decided that all columns in the building shall be:

1. Of I shape for ease of connections.
2. Of 12 or 14 in. nominal depth because of the requirements of lateral frame action and available space.
3. Of steel conforming to ASTM A572, Gr. 50.

Based on a prior preliminary analysis of the frame, we can reasonably assume that:

4. The relative stiffnesses of beams (and girders) compared to columns at each joint, in both directions, is

$$\frac{\Sigma(I_c/L_c)}{\Sigma(I_g/L_g)} = 2.0^*$$

5. A column selected for the conditions specified will also satisfy the requirements of wind and seismic loading.

The solution follows Flow Chart 11.2 in general. Some steps are omitted because of assumptions given in the problem statement.

CALCULATIONS	*REMARKS*
(1) Steel: ASTM A572, Gr. 50	Prescribed
$F_y = 50$ ksi	
Available all groups, 1 through 5	AISC Manual, Part 1, Table 1
(2) Assume $f_a/F_a \approx 0.80$.	Assumes 80% utilization for axial load; bending assumed relatively minor.
Then find W12 and W14 in AISC Column Tables with capacity $500/0.8 = 625$ k	Column Tables in Part 3, AISC Manual. Use $F_y = 50$.
To determine $(KL)_x$, $(KL)_y$, need K.	
Braced direction: $G_A = G_B = 2.00$	Given
$K_y = 0.86, \quad (KL)_y = 10(0.86)' = 8.6'$	Fig. 9.14 (AISC Manual, Part 3, Fig. 1). No sidesway.
Unbraced direction: $G_A = G_B = 2.0$	
$K_x = 1.60, \quad (KL)_x = 10(1.60) = 16.0'$	Fig. 9.14, with sidesway.
$\dfrac{(KL)_x}{(KL)_y} = 1.86$	
(3) First trials from Column Table:	Note that column tables assume $(KL/r)_x < (KL/r)_y$.
$\left(\dfrac{KL}{r}\right)_y = 9.0'$	

*At some time before a final irrevocable decision, this assumption must be checked. F_a, F'_{ex}, and F'_{ey} are sensitive to these ratios. Since we do not know the girder characteristics, we will use the assumption in our calculation.

	W14 × 90	W12 × 87	
Allow P_{ax}	722 k	678 k	
r_y	3.70	3.07	
r_x/r_y	1.66 < 1.86	1.75 < 1.86	See bottom of table for r's.
r_x	6.14	5.37	See $(KL/r)_x > (KL/r)_y$.
$\frac{(KL/r)_x}{(KL/r)_y}$	$\frac{1.86}{1.66} = 1.12$	$\frac{1.86}{1.75} = 1.06$	Column strength is controlled by $(KL/r)_x$.
Capacity for $(KL/r)_x$:			
$(KL/r)_x$	$\frac{16.0(12)}{6.14} = 31.27$	$\frac{16(12)}{5.37} = 35.75$	
F_a, ksi	26.93	26.4	AISC Spec., Table 3-50
A	26.5	25.6	
f_a, ksi	$\frac{500}{26.5} = 18.87$	$\frac{500}{25.6} = 19.53$	
$\frac{f_a}{F_a}$	0.70	0.74	

(4) Check Bending utilization by AISC Eq. (1.6-1a) — $f_a/F_a > 0.15$

Compact?	W14 × 90	W12 × 87	
Cross section	No	Yes	
L_c, ft	NA	10.9 > 10.0 OK	Noncompact sections are noted in Column Table, see footnote.
L_u, ft	24.5 ≫ 10	NA	L_c and L_u are in Column Tables.
F_{bx}, ksi	0.60(50) = 30	0.66(50) = 33	
F_{by}, ksi	0.60(50) = 30	0.75(50) = 37.5	
S_x, in.^{3}	143	118	
S_y, in.^{3}	49.9	39.7	
r_x, in.	6.14	5.38	
r_y, in.	3.70	3.07	
$(KL/r)_x$	31.3	35.8	As before
$(KL/r)_y$	27.9	33.6	
F'_{ex}, ksi	152	117	Table 9, AISC Spec. See also Fig. 11.7.
F'_{ey}, ksi	192	132	

(Table continues)

Compact?	W14 × 90	W12 × 87
C_{mx}	0.6 + 0.4(1) = 1.0	1.0
C_{my}	1.0	1.0
$K_{bx} = \dfrac{1 - f_a/F'_{ex}}{C_{mx}}$	0.88	0.83
$K_{by} = \dfrac{1 - f_a/F'_{ey}}{C_{my}}$	0.90	0.85
$f_{bx} = \dfrac{M_x}{S_x}$, ksi	8.39	10.17
$f_{by} = \dfrac{M_y}{S_y}$, ksi	7.21	9.07
$\dfrac{f_{bx}}{K_{bx}F_{bx}} + \dfrac{f_{by}}{K_{by}F_{by}}$	0.32 ± 0.27	0.37 ± 0.28

Since M_x and M_y may both be ±, the least stable combination is $\left(\dfrac{M_1}{M_{2_x}}\right) = \left(\dfrac{M_1}{M_{2_y}}\right) = +1.0$

(5) Interaction: Axial load + bending

$$\Sigma\frac{f}{F}:\quad 0.70 + 0.32 + 0.27 = 1.29 > 1.0 \quad \text{NG}$$

$$0.74 + 0.37 + 0.28 = 1.39 > 1.0 \quad \text{NG}$$

(6) Trial 2: Try W14 × 109.

$\dfrac{109}{90} = 1.21$ $\qquad L_c = 13.1' > 10.0$

$A = 32.0$

$K_x = 1.60$ $\qquad K_y = 0.86$

$S_x = 173$ $\qquad S_y = 61.2$

$r_x = 6.22$ $\qquad r_y = 3.73$

$\left(\dfrac{KL}{r}\right)_x = 30.9$ $\qquad \left(\dfrac{KL}{r}\right)_y = 27.7$

$F_a = 27.0$

$f_a = 15.63$

$F'_{ex} = 156$ $\qquad F'_{ey} = 195$

Since bending utilization is more important than assumed, the W12 is dropped from consideration. Although bending utilization looks almost the same for W12 × 87 and W14 × 90, we expect this to shift in favor of the W14, since W14 × 109 is compact, according to column table; W14 × 90 is noncompact.

$$\text{ratio}\,\frac{109}{90} = 1.21 < 1.33,$$

But we expect S to increase with weight faster than A.

$C_{mx} = 1.0 \qquad C_{my} = 1.0$

$F_{bx} = 0.66(50) = 33,$

$F_{by} = 0.75(50) = 37.5$

Cross section compact, $L_c = 13.1\text{ ft} > 10.0$.

$$f_{bx} = \frac{100(12)}{173} = 6.94,$$

$$f_{by} = \frac{30(12)}{61.2} = 5.88$$

$K_{bx} = 0.90, \qquad K_{by} = 0.92$

$$\frac{f_a}{F_a} + \frac{f_{bx}}{K_{bx}F_{bx}} + \frac{f_{by}}{K_{by}F_{by}}$$ AISC Eq. (1.6-1a)

$$= \frac{15.63}{27.0} + \frac{6.94}{0.9(33)} + \frac{5.88}{0.92(37.5)}$$

$$= 0.58 + 0.23 + 0.17 = 0.98 < 1 \qquad \text{OK}$$

$$\frac{f_a}{0.6F_y} + \frac{f_{bx}}{F_{bx}} + \frac{f_{by}}{F_{by}}$$ AISC Eq. (1.6-1b)

$$= \frac{15.63}{30} + \frac{6.94}{33} + \frac{5.88}{37.5}$$

$$= 0.52 + 0.21 + 0.16 = 0.89 < 1.0$$

(7) *Use W14 × 109.*

The trial and error procedure was tedious. However, the selection came after only two cycles of iteration. It sometimes takes more. However, the number of cycles was reduced, and usually can be, by judicious use of the available tools.

Although the AISC column tables did not give a direct solution, they did put us in range the first time. Although our first guess, which assumed axial utilization of 80%, was far off the mark, it was clearly better than a guess of 100% and served as a useful starting point. We kept the three utilization terms separate, rather than, as is often done, recording only the sum. Examination of the separate terms helps guide us in deciding which way to go in the next cycle. After the first trial, we could have kept trying W12's as well as W14's. We could, in the W14 series, have increased first to 99 lb/ft, rather than jumping from 90 to 109. By examining the magnitude of change needed (from 1.33 to 1.00) and the relative rate of change of axial and bending strength with increase of weight, we arrived at our final choice, happily, with the second trial.

A tedious, iterative process like this seems to cry out for a computer program. The engineer with many such problems can soon find or write one. We suggest that he or she first use procedures like the one here in order to develop judgment about the effects of various parameters on the interaction. If this type of problem arises only occasionally in the individual's practice, it may take longer to dig up the automatic method than to do the problem without it.

The selection of columns in the Tier Building Case Study, Chapter 18, is based on the procedure of this example.

11.3.7 Alternative AISC Method: Selection of Beam Columns

The previous example started out with an assumption as to the magnitude of the axial utilization term in the interaction equations. This led to an initial trial selection based on an equivalent axial load larger than the real load, the difference being the estimated effect of the bending terms on axial capacity. It is possible to refine the assumption of bending effects by use of the coefficients B_x, B_y, a_x, and a_y, which are listed in the AISC Column Tables for each section tabulated. The procedure is explained in the AISC Manual immediately preceding the Column Tables. It is also based on assumption of an equivalent axial load; it is also iterative. Those readers who might prefer that method should refer to the explanation in the Manual, which needs no elaboration here.

11.3.8 Axial Tension Plus Bending

We have seen how axial compression added secondary ($P\Delta$) moment to a beam column, contributing increased actual moment and sometimes reduction of strength through potential instability. The effect of axial tension on a member subjected to bending is exactly the opposite. The $T\Delta$ moment reduces bending moment and increases stability.

The AISC Specification deals with this relatively simple interaction problem very simply. It enlists interaction equation (1.6-1b), applying it to bending stress. The designer is also cautioned to check the maximum compressive bending stress, taken alone, an important consideration. Since axial tension reduces the net compressive stress, it would be unsafe to rely on it to do so unless axial tension and bending exist always in the same proportions.

In a similar vein, if axial tension dominates the interaction equation, it would be prudent to check both criteria for F_t used in AISC 1.5.1.1.

11.4 BENDING WITH TORSION

11.4.1 General Remarks on Torsion

Many members that are efficient in bending are poorly suited to resist torsion. To a somewhat lesser degree, circular sections, ideal for machinery shafts because of their excellent torsional properties, are less efficient in bending and require closely spaced transverse supports.

The sections in most common use as beams are of I shape with $b \ll d$, chosen usually because bending is uniaxial or almost so. Under these conditions, they are both efficient in use of steel and easy to connect to other members in the system. System design is usually directed to avoiding torsional problems rather than forcing the member to do what it is ill suited to do. In those cases where torsion is unavoidable, we are often forced to seek other cross sections more suited for torsional resistance, but often at considerable expense in weight of the member or ease of assembly.

Adequate discussion of torsional effects on steel structural members is beyond the scope of this book. Many advanced texts in strength of materials, (e.g., Timoshenko[58]) are available. The AISC Code does not attempt to establish rules for torsional design or its interaction with flexure. It does have some discussion in the Commentary to Section 1.5.1.4, Specification. We will examine some of the simpler aspects of the problem, which should be sufficient for the reader to handle many designs. However, when in doubt, we suggest use of the more advanced material.

11.4.2 Bending with Torsion: Open Sections

The beam in Fig. 11.9(a) cantilevers from a support that fixes it against rotation about all three of its principal axes. It is subjected to biaxial bending plus the twisting effect of the torque M_z. Figures 11.9(b) and (c) illustrate two ways of applying bending loads that give rise to twisting moments. In these cases the beams themselves are doubly symmetrical. M_z arises from eccentricity of the load with respect to the centroid.

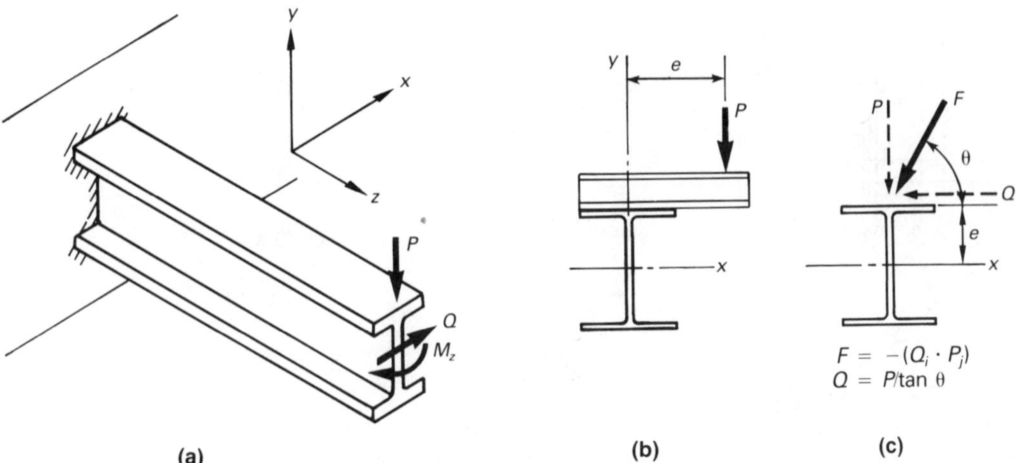

Figure 11.9 Bending with torsion: (a) loads on transverse axes plus applied torque; (b) eccentric load, $M_z = Pe$; (c) skewed load, $M_z = Qe$.

The channel in Fig. 11.10(a) is symmetrical about its major axis only. Vertical load, P, causes shear. Longitudinal stresses, f_b, and longitudinal shear flow results as shown in Figs. 11.10(c) through (e). The flow of shears in the cross section is illustrated in Fig. 11.10(f). If there is to be no torsion on the section, equilibrium requires that P act through a particular point, S, called the shear center, which is not, in general, at the centroid of area. In the doubly symmetric section of Fig. 11.10(g), the symmetry of the flanges causes the shear center to coincide with the centroid.

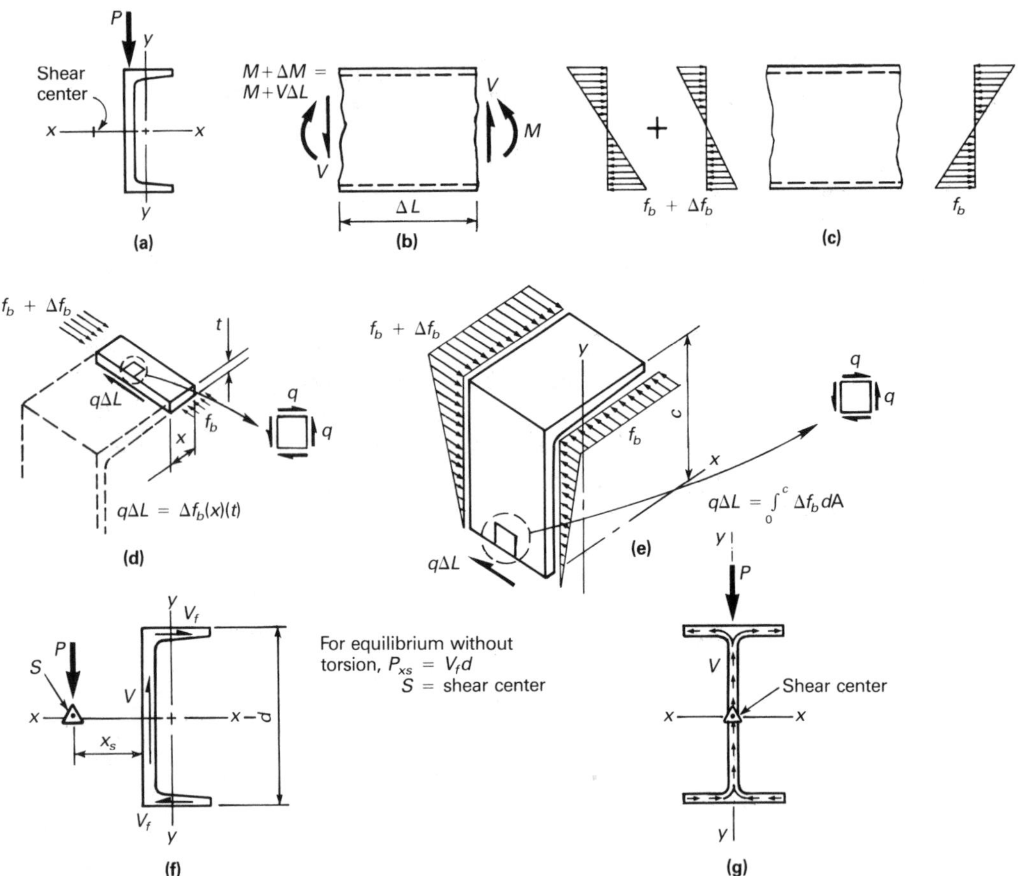

Figure 11.10 Shear flow in C and I sections

C: (a) Shearing load applied

(b) (c) Moment and shear; shear causes ΔM and Δf

(d) Longitudinal stress change requires longitudinal shear flow

(e) Shear flow is maximum at neutral axis

(f) Load must go through shear center to avoid net torque

I: (g) Shear center of symmetrical section is on axis of symmetry

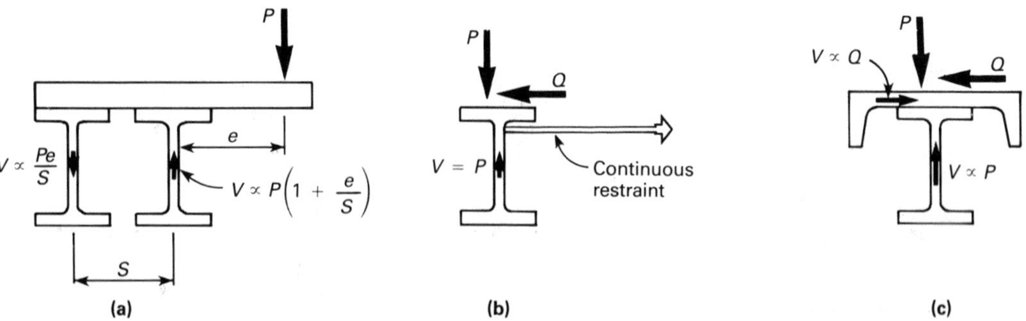

Figure 11.11 Avoiding torsion with eccentric loads.

In the usual design situation, the designer will try to avoid torsional effects on open sections such as I or C shapes. Thus, for the situation of Fig. 11.9(b), one could provide two beams to support the outrigger as in Fig. 11.11(a). For the load case of Fig. 11.9(c), an independent lateral restraint may be used to relieve the I beam of y bending. This is true when a beam is joined to a continuous slab as in Fig. 10.12. Where no such slab exists, as in the common problem of crane rail support girders in industrial buildings, a horizontal channel may be used as in Fig. 11.11(c) to resist Q, while the combined section is effective in resisting P.

Left on their own to resist twisting moments, I and C sections (or open sections in general) are adequate for small moments only. However, it is important to examine briefly the nature of such resistance when it is called on.

From texts on strength of materials, the reader will recall that torsion in circular shafts occurs without warping. Resistance at any plane cross section perpendicular to the shaft central axis results from mobilization of shear stresses, as shown in Fig. 11.12(a).

$$f_v = \frac{M_z r}{J}$$

$$(f_v)_{\max} = \frac{M_z c}{J} \qquad (11\text{-}8)$$

Along any length ΔZ, the shaft is twisted through an angle $\Delta\theta$. When torque and shaft are constant,

$$\Delta\theta = \frac{M_z}{JG}\,\Delta z \qquad (11\text{-}9)$$

$$G = \frac{E}{2(1+\nu)} = 11{,}000 \text{ ksi for structural steels}$$

Rectangular sections also resist torque by mobilization of shear resistance. However, they must warp to do so, as shown in Fig. 11.12(b). This is often called *St. Venant*

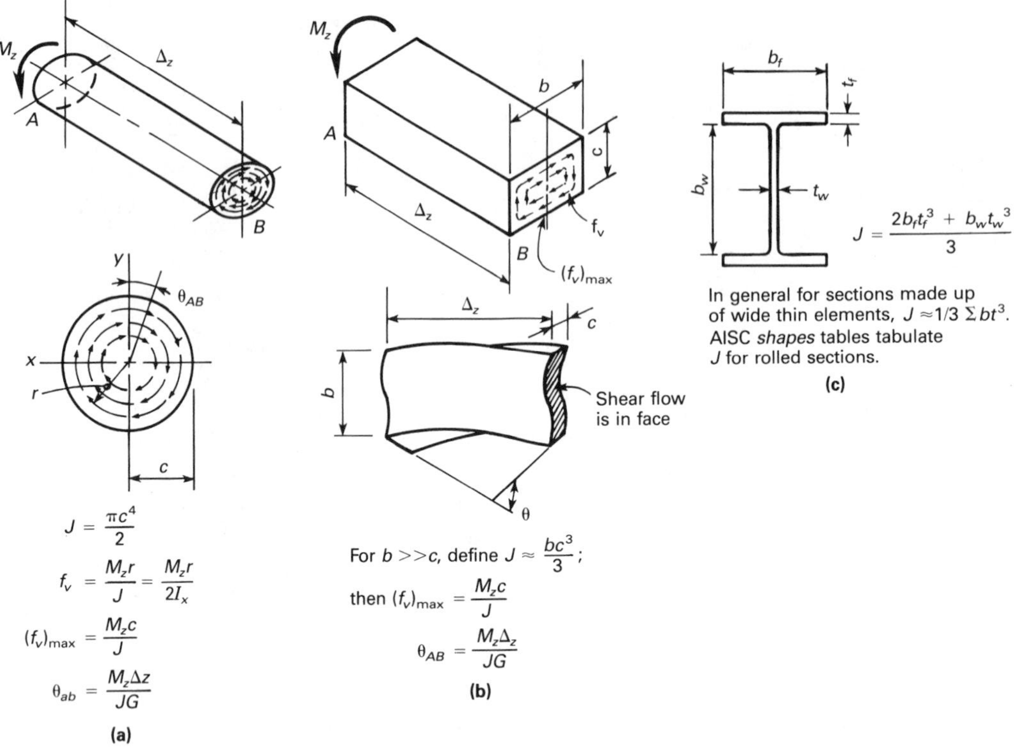

Figure 11.12 Shear flow in torsion: (a) torsion without warping, circular shaft; (b) warping (St. Venant) torsion on rectangular section; (c) polar moment of inertia for I shaped section (C similar).

torsion, after the French engineer credited with the first elaboration of the general theory. If $b \gg t$, the corresponding stress and rotation equations can be written in the same form:

$$(f_v)_{max} = \frac{M_z t}{J^*} \tag{11-10}$$

$$\Delta\theta = \frac{M_z \Delta z}{JG}$$

Where, as in elements of most steel sections, $b \gg t$,

$$J \approx \frac{bt^3}{3} \tag{11-11}$$

*J is defined here as kbc^3, where k depends on b/c. It is not $I_x + I_y$, which would result from the equation for polar moment of inertia, $J = \int(x^2 + y^2)dA$. AISC uses the definition of J used here, which is convenient for the twisting problem of thin, wide elements.

In a section such as an I or C made up of a number of wide thin rectangles,

$$J \approx \frac{1}{3} \sum bt^3 \tag{11-12}$$

If the reader examines in the AISC Shape Tables the values of J applicable to rolled sections, he or she will be struck by how small they are compared to the moments of inertia, an indication of the very large shear stresses and rotations resulting from small twisting moments, usually unacceptable. Fortunately, the lack of rotational stiffness usually makes it possible to avoid the shear stress itself. A designer may introduce other, stiffer torsion-resisting elements that shield the beam against large angular rotation, setting a limit on torsional shear stress.

If warping of the cross section is prevented, as, for example, by the fixed support in Fig. 11.9(a), St. Venant torsion is prevented. However, another mechanism exists to resist torque, the flexural resistance of each of the flanges. In Fig. 11.13, the torque M_z has been converted to a couple of forces, M_z/d, each of which bends a flange; both the direction of the forces and the direction of bending are opposite. Each of the flanges has a section modulus equal to $S_y/2$ of the beam, usually a small value, indicating that this mode of resistance is also limited to small torques.

At any point along a beam, the actual resistance is ordinarily a mixture of St. Venant torsion and bending resistance to torsion, with the latter applying primarily close to fixed supports and for relatively stiff beams. In most cases, if St. Venant torsion is ignored, conservative values of f_{by} will appear in the calculation, but torsional shear stresses are underestimated. The net effect is usually on the safe side. The applicable interaction equations are those used for biaxial bending.

Examine once again the conditions of Example 11.2.3. If the lateral load is applied at the top flange as in that problem, the net effect is to apply the load wz as in

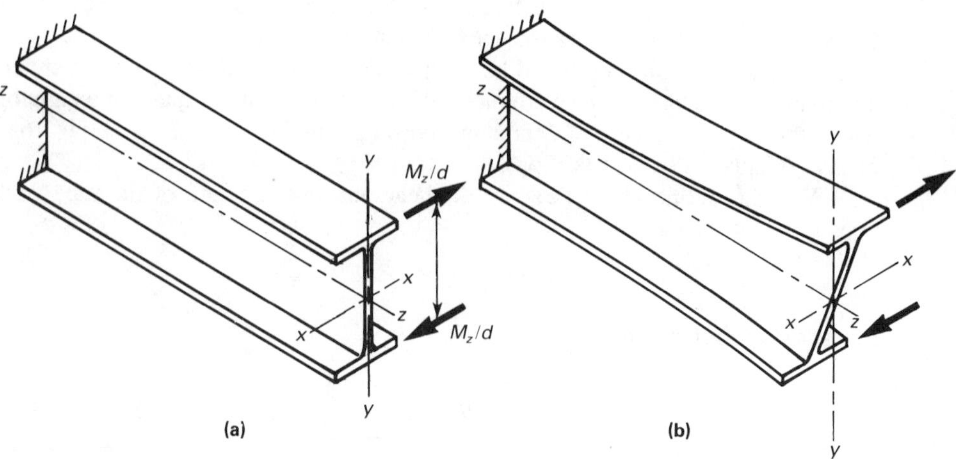

Figure 11.13 Resisting torsion by flange bending.

TABLE 11.3 *w* PER FLANGE, BENDING WITH TORSION

	Lateral load	m_z/d	Net w_{fl}
Top flange	$\frac{w}{2\sqrt{5}}$	$\frac{w}{2\sqrt{5}}$	$\frac{w}{\sqrt{5}}$
Bottom flange	$\frac{w}{2\sqrt{5}}$	$\frac{-w}{2\sqrt{5}}$	0

Table 11.3. The distributed load of $w/\sqrt{5}$ per foot is shifted to the axis, but the shift must be compensated by a moment of $(w/\sqrt{5})(d/2)$. The result justifies the intuitive way lateral force was applied in the calculation of Example 11.2.3. Recall in that calculation that even after we reduced the span for y bending to one-third of that for x bending, we were forced to choose a relatively squat beam with wide flanges. Even then, the utilization term for y bending is half as much as for x. Had we not introduced the intermediate lateral supports, the calculated bending moment around the Y axis would be nine times as big. This would be a completely unacceptable result, made even more so by the fact that St. Venant torsion would actually dominate for most of the beam length, and rotation of the beam around its Z axis would be very large.

11.4.3 Bending with Torsion: Closed Sections

In some structures, the conditions of design make it both necessary and/or desirable to seek torsionally resistant beams. The solutions in Fig. 11.11 are not necessarily the most desirable. Figure 11.14 shows three instances where the solutions of choice are torsionally resistant, thin-walled closed sections.

Examine the flow of shear that resists torque M_z in the thin-walled rectangular tubes of Fig. 11.15. Compare the shear flow to that in the sections of Figs. 11.12(b) and (c). If the sections compared have equal area of steel cross section, it should be apparent that the closed tubes are more efficient in resisting torque. (In an analogous way, efficiency in resisting M_x is achieved by placing most of the area far from the neutral X axis, as in an I section).

We can define *shear flow*, q, as shear per unit of width of the tube wall, so that

$$q = tf_v \tag{11-13}$$

It is shown in reference 58* that

$$q = \frac{M_z}{2A_0} \tag{11-14}$$

*The notations used in the reference are not the same as ours. The results are.

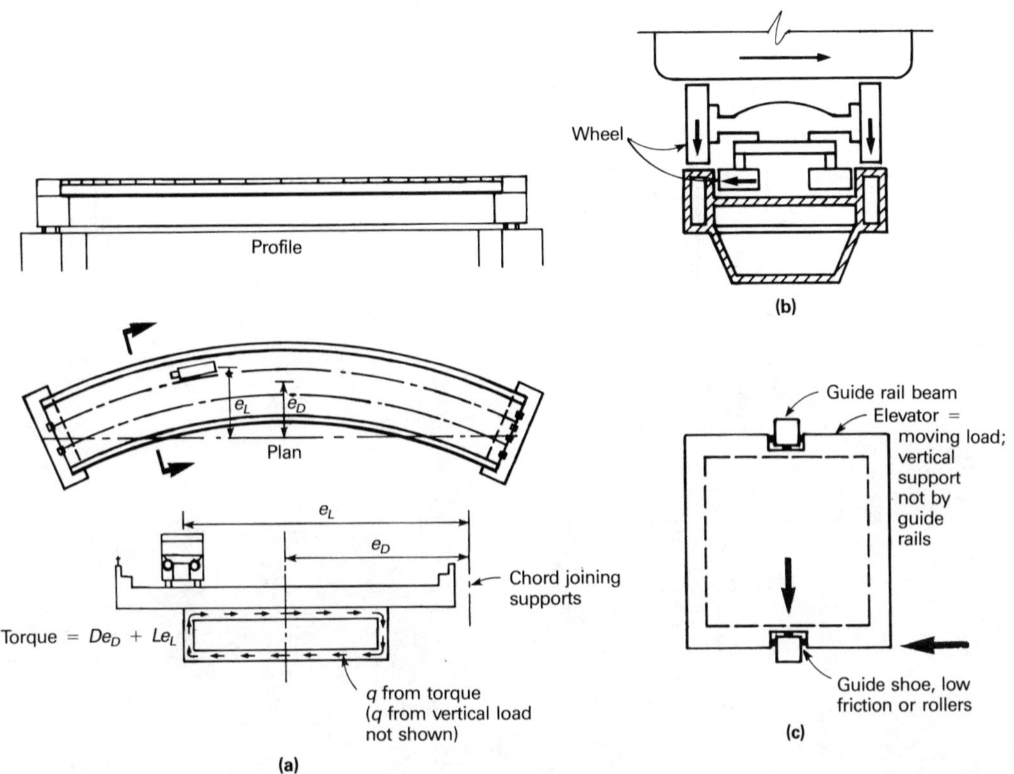

Figure 11.14 Resisting torsion with box sections: (a) highway bridge, curved in plan; (b) vehicular guideway, straight or curved; (c) elevator guide rails.

which is independent of thickness, or

$$f_v = \frac{M_z}{2A_0 t} \tag{11-15}$$

where A_0 is the area bounded by the tube [e.g., bxd in Fig. 11.15(a)]. The relationships apply to any closed thin-walled tube. The formula for angular rotation due to a torque M_z applied over a length ΔZ of a closed tube of constant thickness t is

$$\Delta\theta = \frac{M_z \Delta Z}{4(A_0)^2 G} \left(\frac{\text{perimeter}}{t}\right) \tag{11-16}$$

Formulas (11-15) and (11-16) may be compared to (11-9) and (11-10). For the same area of steel cross section, the closed tube will yield much lower stress f_v and much less torsional rotation.

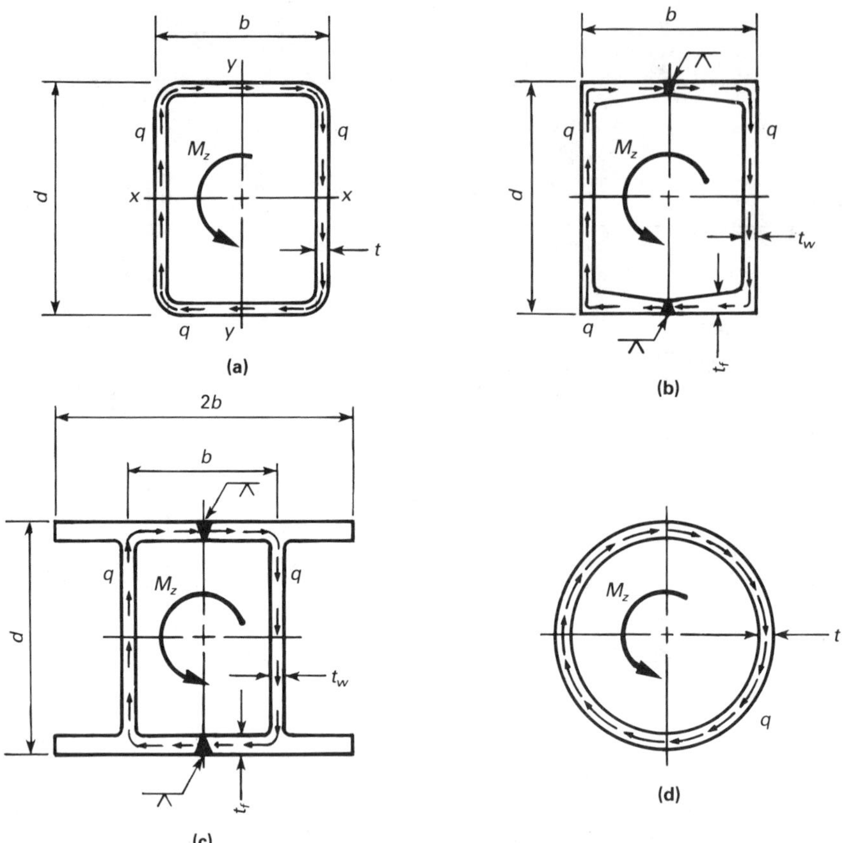

Figure 11.15 Shear flow from torsion in closed sections; q = shear flow = tf_v. Units of q are force/unit length.

Interaction of torsion with bending is not covered in any formulas in the AISC Specification. However, note that the resistance to torsion in closed sections is supplied by shearing stress. In the usual case, as in a beam subjected to transverse shear and bending, longitudinal stresses f_b can be calculated from the bending moment and compared to F_b, determined in accordance with AISC Section 1.5.1.4 and/or the flow charts of Chapter 10. If bending is around two axes, the interaction formulas apply. The maximum shear stress can be found by superposing the effects of the transverse and torsional shears. Thus, if a rectangular tube, such as in Fig. 11.15(a) or (b) is subjected to a transverse shear V_y as well as the torque M_z,

$$f_v = (f_v)_V + (f_v)_M$$

$$= \frac{V_y}{2(d)t} \pm \frac{M_z}{2A_0 t}, \quad \text{Eq. (11-15) and Fig. 10.6(b); rectangular tube similar to I}$$

$$(f_v)_{max} = \frac{1}{2t}\left(\frac{V_y}{d} + \frac{M_z}{A_0}\right)$$

the allowable limit being

$$F_v = 0.40F_y \qquad \text{(AISC Sect. 1.5.1.2.1)}$$

Flow Chart 11.3 is applicable to problems of selection of closed sections for bending with torsion. It guides the solution of Example 11.4.4.

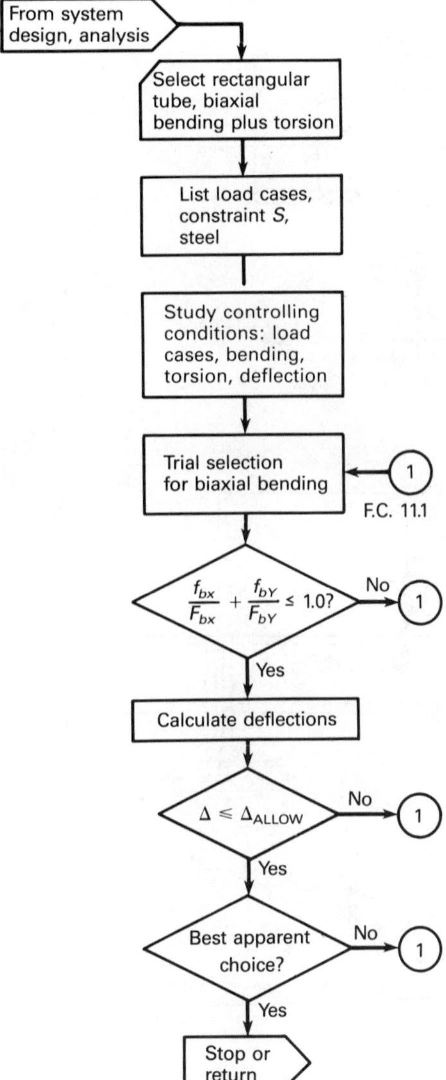

Flow Chart 11.3 Select rectangular tube, biaxial bending plus torsion.

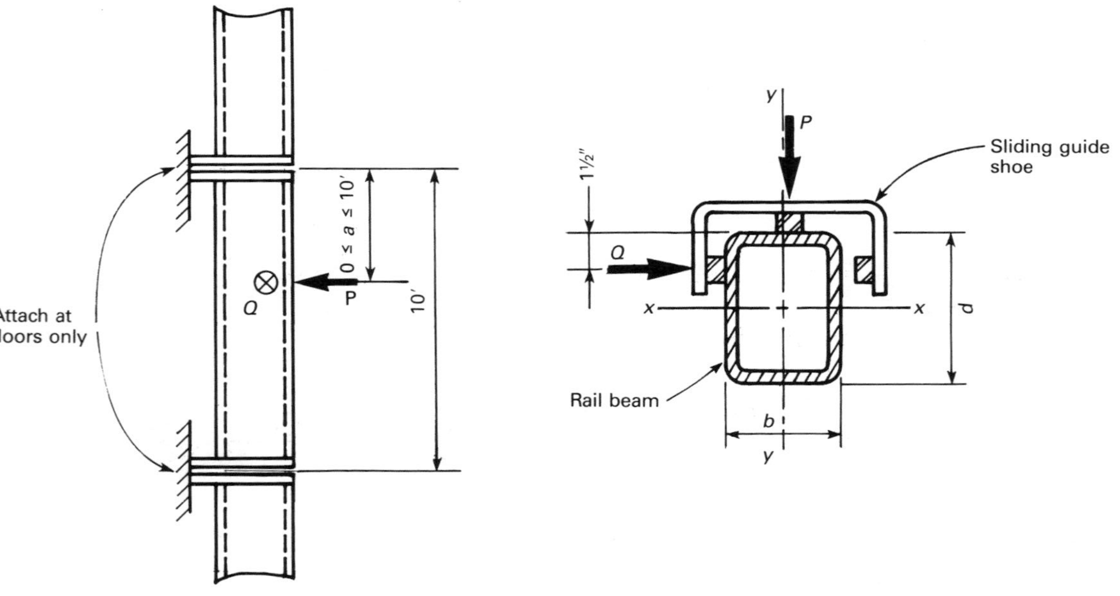

Figure 11.16 Elevator guide rail (Example 11.3.4).

11.4.4 Example: Bending with Torsion, Closed Section

A beam is to be selected for a guide system for an elevator [see Fig. 11.16]. The controlling values of horizontal forces P and Q are expected to occur in an earthquake and have been evaluated for three alternates at

Case A: $P = 7$ k, $Q = 0$

Case B: $P = 0$, $Q = 3.5$

Case C: $P = 5$ k, $Q = 2.5$

The weight of the elevator is not supported by the guide beam. The beam spans between floors 10 ft apart. Supports will be capable of resisting shear in both of the x and y directions, as well as torque. No point on the beam shall deflect more than ½ in. in either the x or y direction from application of P and/or Q.

Select an appropriate beam using a rectangular tube of ASTM A501 steel. (ASTM A501 is a steel used in the production of thin-walled tubes. $F_y = 36$ ksi; see Table 5.1.)

CALCULATIONS

(1) Find the conditions controlling moments, shears, and deflections.

(a) Maximum bending moments and deflections:

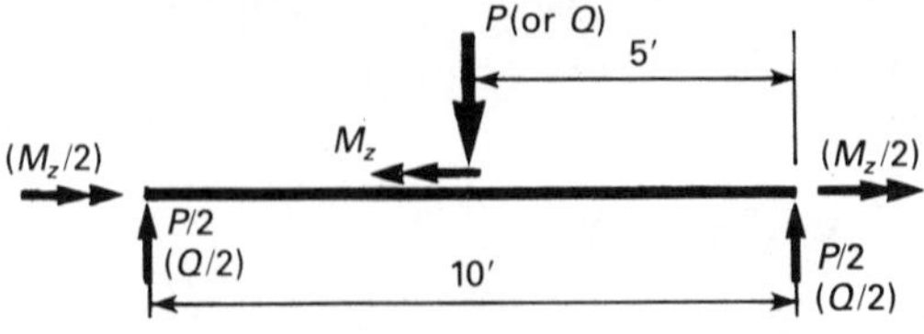

Effect of loads* P *and* Q *without Torque:

$$M = \frac{(P \text{ or } Q)L}{4}, \qquad \Delta = \frac{(P \text{ or } Q)L}{48EI}$$

$L = 120''$, $E = 29{,}000$ ksi

Case	Kips		Ft kips		Inches	
	P	Q	M_x	M_y	y	x
A	7	0	17.5	0	$8.69/I_x$	0
B	0	3.5	0	8.75	0	$4.39/I_y$
C	5	2.5	12.5	6.25	$6.21/I_x$	$3.14/I_y$

REMARKS

(arrow with double head) represents moment about axis in direction of arrow. Arrow heads define direction of rotation by usual right-hand rule. For the effects of P and Q, this is a standard case, covered in AISC Beam Diagrams 7.

M_z is the applied torque. If the rotational restraints at the ends are identical, the reacting moments and internal M_z will each equal $M_z/2$ when P and Q are at midspan.

M_z and M_y can only occur simultaneously in this problem.

Effect of torque:

$$Q\left(\frac{d}{2}-1\frac{1}{2}\right)=M_z$$

$$f_v=\frac{M_z}{4A_0t} \qquad \text{Eq. (11-15)}$$

$$\theta_z=\frac{M_z\Delta z}{2A_0G}\cdot\frac{(2)\,(b+d)}{t} \qquad \text{Eq. (11-16)}$$

$$=\frac{M_z(60)(b+d)}{2bd(11{,}000)t}$$

$$=0.00273Q\left(\frac{b+d}{bdt}\right)\left(\frac{d}{2}-1.5\right)$$

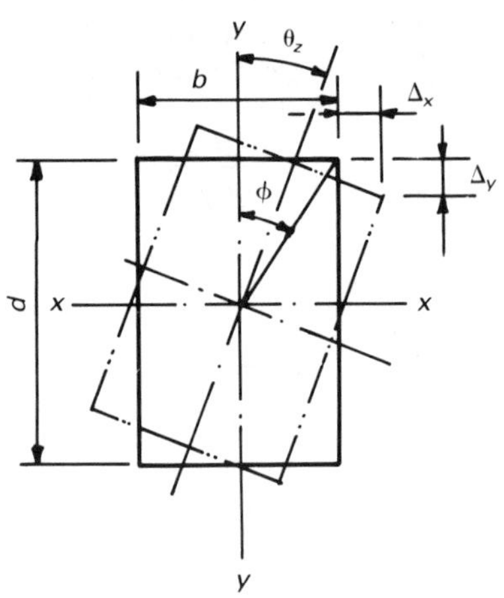

Case	P	Q	θ_z
A	7	0	0
B	0	3.5	$0.0096\left(\frac{b+d}{bdt}\right)\left(\frac{d}{2}-1.5\right)$
C	5	2.5	$0.0068\left(\frac{b+d}{bdt}\right)\left(\frac{d}{2}-1.5\right)$

$$\phi=\tan^{-1}\left(\frac{b}{d}\right)$$

Assuming small rotation θ_z,

$$\Delta_x=r\sin\theta_z\cos\phi$$

$$\Delta_y=r\sin\theta_z\sin\phi,\quad \text{where}$$

$$r=\sqrt{b^2+d^2}/2$$

(b) Maximum shear and torsion:

$$\theta_z=0,\qquad M_x=M_y=0$$

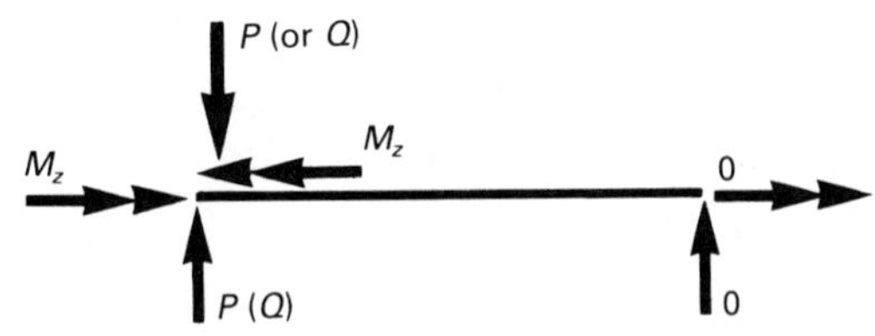

Maximum shear in either direction and maximum internal torque, all arise from load position shown.

Case	V_y	V_x	M_z
A	7.0	0	0
B	0	3.5	$3.5\left(\frac{d}{2} - 1.5\right)$
C	5.0	2.5	$2.5\left(\frac{d}{2} - 1.5\right)$

(2) Tentative selection for M_x and M_y: Flow Chart 10.1

$$F_{bx} = F_{by} = 0.66F_y$$

$$\frac{f_{bx}}{F_{bx}} + \frac{f_{by}}{F_{by}} = \frac{f_{bx} + f_{by}}{23.8}$$ AISC (1.6-2)

$$= \frac{1}{23.8}\left(\frac{M_x}{S_x} + \frac{M_y}{S_y}\right) \leq 1.0$$

Case A:

$$(S_x)_R = \frac{M_x}{23.8} = \frac{17.5(12)}{23.8} = 8.82$$

Case B:

$$(S_y)_R = \frac{M_y}{23.8} = \frac{8.75(12)}{23.8} = 4.41$$

Case C:

If $S_x = 8.82$ and $S_y = 4.41$,

$$\frac{12}{23.8}\left(\frac{12.5}{8.82} + \frac{6.25}{4.41}\right) = 1.43 \quad \text{NG}$$

Look for $S_x \approx 1.43(8.82) = 12.61$

$S_y \approx 1.43(4.41) = 6.31$

Structural tube	Wt/ft	S_x	S_y	Utilization, Case C	
$7 \times 4 \times \frac{5}{16}$	21.21	11.0	7.98	$0.57 + 0.39 = 0.96$	OK
$6 \times 6 \times \frac{1}{4}$	19.02	10.1	10.1	$0.62 + 0.31 = 0.93$	OK

(3) Check deflections:

$$\Delta_x \leq 0.50'' \text{ and } \Delta_y \leq 0.50''$$

Problem statement

Check deflection of 6 × 6 × ¼ tube:

$$I_x = I_y = 30.3, \qquad S_x = 10.1 = S_y$$

$$t = 0.25'', \qquad d = b = 6'',$$

$$A_0 = 36 \text{ in.}^2$$

$$r = \sqrt{2(9)} = 4.24''$$

Add translations from bending and torsion:

Case	P	Q	Δ_y (in.)	Δ_x (in.)
A	7	0	$0.29 + 0 = 0.29$	0
B	0	3.5	$0 + 0.06 = 0.06$	$0.15 + 0.06 = 0.21$
C	5	2.5	$0.21 + 0.04 = 0.25$	$0.11 + 0.04 = 0.15$

Recall positions 1a and lb earlier.

$$\Delta_{max} = 0.25'' < 0.50''$$

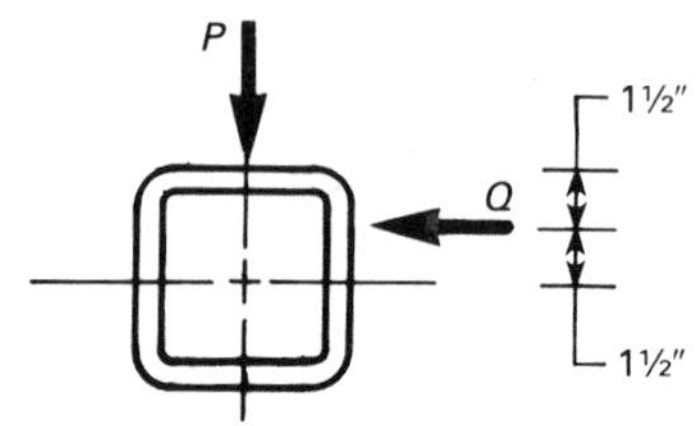

(4) Check 6 × 6 × ¼ tube for $f_v \leq F_v = 0.40F_y$:

Case	Load Pos.	V_x	f_{v1}	V_y	f_{v2}	M_z	f_{v3}	f_v
A	2a	3.5	1.17	0	0	0	0	1.17
	2b	7.0	2.33	0	0	0	0	2.33
B	2a	0	0	1.75	0.58	1.75	0.15	0.68
	2b	0	0	3.50	1.17	3.50	0.28	1.36
C	2a	2.5	0.58	1.25	0.42	1.25	0.10	1.10
	2b	5.0	1.17	2.5	0.84	2.50	0.20	2.21

$$(f_v)_{max} = 2.21 \text{ ksi} \ll 0.4F_y = 14.40$$

(5) Use 6 × 6 × ¼ tube.

The section was tentatively selected to satisfy the bending effects of case C, which involved biaxial bending. The effect of torsion was so small in this case that the exercise may seem trivial. The reader might find it edifying to seek the lightest I shape that could do this job. M_z did turn out to be small with the section selected. However, the power of this type of closed section in limiting the deflection and stress effects of biaxial bending with torque should be clear from examination of the detailed calculations. These virtues are even more useful in larger problems.

At this time, the reader might recall Example 11.2.3. In that case, with an I section, we could only count on half of S_y to be effective. In Problem 11.4, the reader is invited to redo Example 11.2.3 using a closed tube.

PROBLEMS

11.1. Find appropriate I-shaped sections. w_x and w_y are uniformly distributed loads acting simultaneously over the entire span. The beam is to be simply supported for the span given in the *YZ* plane. Intermediate supports are to be provided at the spacing indicated in the *XZ* plane, similar to Fig. 11.6. Assume the given loads include the effect of the weight of the beam. $F_y = 36$ ksi. *Hint:* AISC Manual, Part 2, Beam Diagram 36, provides *V* and *M* diagrams for w_x.

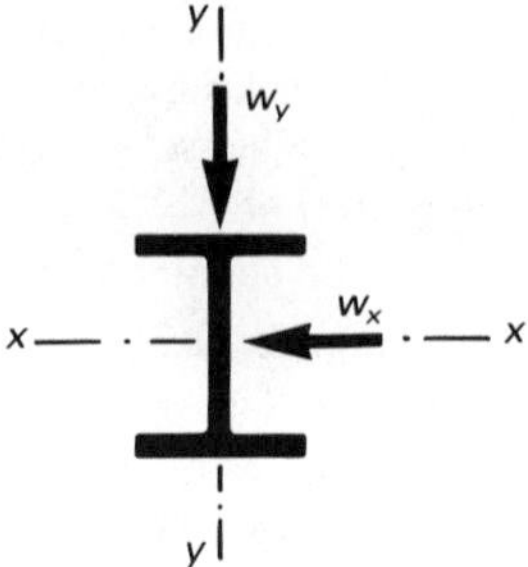

	Span, *YZ* plane, ft	Spans, *XZ* plane, ft	w_y, lb/ft	w_x, lb/ft
a	24	8	500	100
b	24	8	500	200
c	24	8	500	500
d	30	10	3000	1000
e	30	10	3000	2000
f	30	10	3000	3000

11.2. Conditions similar to Problem 11.1, but you are to choose spacing of supports in *XZ* plane to help select an efficient I section. Use $F_y = 50$ ksi.

	Span, YZ, ft	w_y, lb/ft	w_x, lb/ft
a	30	4000	1000
b	30	4000	2500
c	30	4000	4000
d	24	4000	1000
e	24	4000	2500
f	24	4000	4000

11.3. Do Problem 11.2, parts *d* through *f*, using structural tubes.

11.4. Select a structural tube for the purlin of Example 11.2.3. You may change the spacing of sag rods or eliminate them. Explain your decision.

11.5. Redo Problem 11.1 with w_x applied at the top flange level.

11.6. Redo Problem 11.2 with w_x applied at the top flange level.

11.7. Redo Problem 11.3 with w_x applied at the top flange level.

11.8. Redo Example 10.11.3, with the 30 k forces acting at the midpoint of the top flange at 5° to the vertical in xy planes.

11.9. For the purlin of Example 11.2.3, check the acceptability of W10 × 17 and W12 × 16.

11.10. Design a guide beam for the loads of Example 11.4.4 if structural tubes are not available. Provide at least two acceptable solutions. You may adapt the form of the guide shoe to suit the type of section you choose for the guide beam.

11.11. The member of Example 11.3.6 is suspended from a rigid overhead system so that it supports axial tension of 500 k instead of axial compression. End moments are the same. End connections are to be welded. Nominal sizes are to be no larger than W14. Select a member, using steel of ASTM A572, Gr. 50.

11.12. Redo Problem 11.2 with the addition of a 50-k axial force in each part.

11.13. For the conditions of Example 11.3.6, select the lightest satisfactory W12.

11.14. A W16 × 77 of ASTM A36 is used as a continuous beam in a frame restrained against sidesway. The member itself has no lateral restraint between its vertical supports. The moment envelope for the critical span is given. How much axial load may be applied to the member concurrently with the bending moment? Use the basic allowable stresses permitted by the AISC Specification for the combined loading (i.e., do not increase allowable stresses as is sometimes permitted for particular load combinations). Steel is A36.

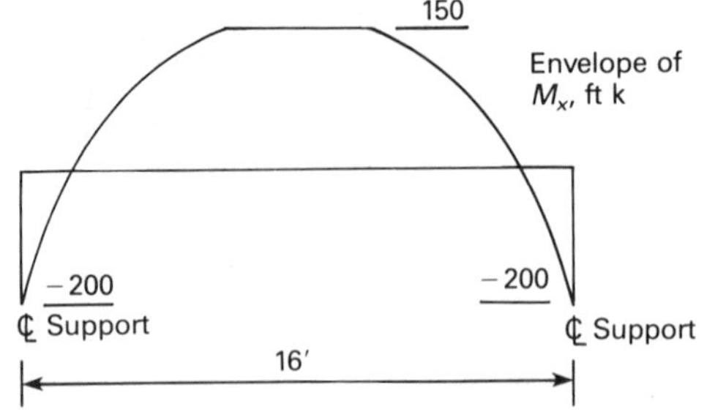

12 Connections: Part 1

12.1 JOINT DESIGN: ART AND SCIENCE *

A structural system is an assemblage of parts that by virtue of their arrangement, displays composite action different from that of the individual members. We have seen several such systems in the previous chapters and have separated the individual members for scrutiny and selection. To reconstitute the system from its parts, it is necessary to connect members to each other. The connection must transmit from one member to others the forces and moments that the members must resist at the point of connection. It must do so in a manner consistent with the assumptions of the system analysis regarding relative movements at the joint.

There is little that tests the designer's mastery of both the science and the art of structural design more than the problem of joint design. The starting point is to specify the nature of the problem: the forces and moments in the members being joined at the location of the joint, and the constraints set on the design. The former come from the mathematical analysis. Constraints on the design also involve, among other things, the mathematical analysis. Any movements within the joint that are not consistent with the mathematical model invalidate the numerical results, sometimes in trivial ways, sometimes in very significant ways, making the numerical results useless. Among other constraints are the following:

*Reference 67, by AISC, which appeared while this book was in the process of publication, is devoted almost exclusively to the design of connections. It is recommended to supplement the material in Chapters 12 and 13.

CONSTRAINT	*COMMENT*
1. The space available for the materials of the joint.	Joints usually occupy a larger cross-sectional area than the members being joined.
2. The location where the joint will be made up and, consequent of that:	
a. Its effect on noise limitations.	Riveting is usually unacceptable in congested areas today. Even the somewhat less noisy high-strength bolting might be ruled out by ordinance.
b. Its effect on degree of construction hazard.	Probably weighs against riveting; could weigh against either bolting or welding depending on circumstances.
c. Quality control and inspection.	Shop conditions as close to ideal as possible. If field conditions do not permit proper inspection, design a different type of joint.
d. Cost.	Relative cost of joint types is very much dependent on location. Field location may speak for bolts, shop assembly for welds. But will vary with plant, structure, and region.
3. Temperature	Low service temperature demands tough steels and notch-free joints.
4. Availability of equipment, trained personnel, and qualified quality-control personnel.	Particularly with the more sophisticated systems, it becomes unsafe to ascribe capacity to a joint that cannot confidently be expected to result in practice. Even U.S. conditions are not uniform. Conditions elsewhere may, but often do not, approximate U.S. conditions. A large project may warrant importation of all necessary parts of the system; a small project seldom would.

5. Repetitive loading	Requires attention to possibility of fatigue failure. The designer must choose between designing notch-free joints (to the extent possible) or reducing the permissible range of cycling stresses.
6. Quality control	Varies with location as in item 4. Ability to achieve reliable quality control may also vary with the form of the joint (e.g., tight, enclosed spaces are hard to work in and to inspect).
7. Ease of maintenance (*housekeeping*)	Avoid pockets that trap dirt or water, causing problems of sanitation and/or corrosion.
8. Cost	Varies with location as noted in item 4. Varies even more with form of joint. Joints difficult to assemble tend to be time consuming and costly. However, a difficult joint becomes less costly if repeated many times, sufficient to justify special tooling.

The connection itself, isolated as a free body, or parts of the connection so isolated, must fulfill the same requirements of equilibrium and compatibility as the total structure or any of its members. The actual stress conditions are often too complex for complete analysis by simple elastic methods, forcing designers to rely much more on test results than they do in member selection. However, simple free bodies are useful in planning and understanding the connections.

Connections or *joints* refer to the entire system of parts through which forces must be transmitted in joining members. They consist of special forms of connecting materials called *connectors* or *fasteners* (bolts, rivets, pins) or *weld metal*, plus other plates or shapes intervening between the members connected (*gussets, splice plates*, etc.). Although there are standard joints that have been analyzed and validated by tests for assigned loads, joints are not mill supplied, but fabricated either from standard details or special details devised by the designer. Many of the special requirements of both joints and connecting materials are regulated by the AISC Specification in various of its sections. Table 12.1 lists applicable sections and their subjects.

TABLE 12.1 PROVISIONS OF AISC SPECIFICATION, 1978, SPECIFICALLY APPLICABLE TO CONNECTIONS

Section	Subject	Remarks
1.5.1	Allowable stress, structural steel	
1.5.1.1	Tension	Splice material
1.5.1.2	Shear	Connectors
1.5.1.3	Compression	Splice material
1.5.1.5	Bearing	Splice material, bearing splices, bolts, and rivets
1.5.2	Rivets, bolts, threaded parts (allowable stresses)	Includes Table 1.5.2.1
1.5.3	Welds (allowable stresses)	Includes Table 1.5.3
1.14	Gross and net areas	All applies
1.14.2.3	Areas, splice plates	
1.14.5	Pin-connected members	
1.14.6	Effective area, welds	Includes Tables 1.14.6.1.2 and 1.14.6.1.3
1.15 All	Connections	*Applies to all connections.* Distinguishes flexible (unrestrained) from restrained connections (restraint is rotational restraint, causing moment)
1.16 All	Rivets and bolts	For A307 bolts and A502 rivets, apply *directly*. For high-strength bolts, refers to specifications for structural joints using ASTM *A325* or *A490* Bolts. Includes tables for spacing, end, and edge distances.
1.17 All	Welds	Includes tables for min. fillet weld sizes and effective thickness of partial penetration welds; other limits
1.23	Fabrication	
1.23.2	Thermal cutting	Preparation of joints
1.23.3	Planing of edges	Preparation of joints
1.23.4	Riveted and bolted construction	Holes
1.23.5	Riveted and high-strength bolted connections	Assembly
1.23.6	Welding	Refers to AWS Spec. D1.1
1.23.7	Compression joints	Preparation

12.2 CONNECTING MATERIALS: CLASSIFICATION

Figure 12.1 shows four different ways of transferring a compressive force from one to the other of the two members shown.

1. *Multiple connectors* in prepared holes (predrilled or prepunched). Connectors may be bolts or rivets and occur in groups of two or more [Fig. 12.1(a)].

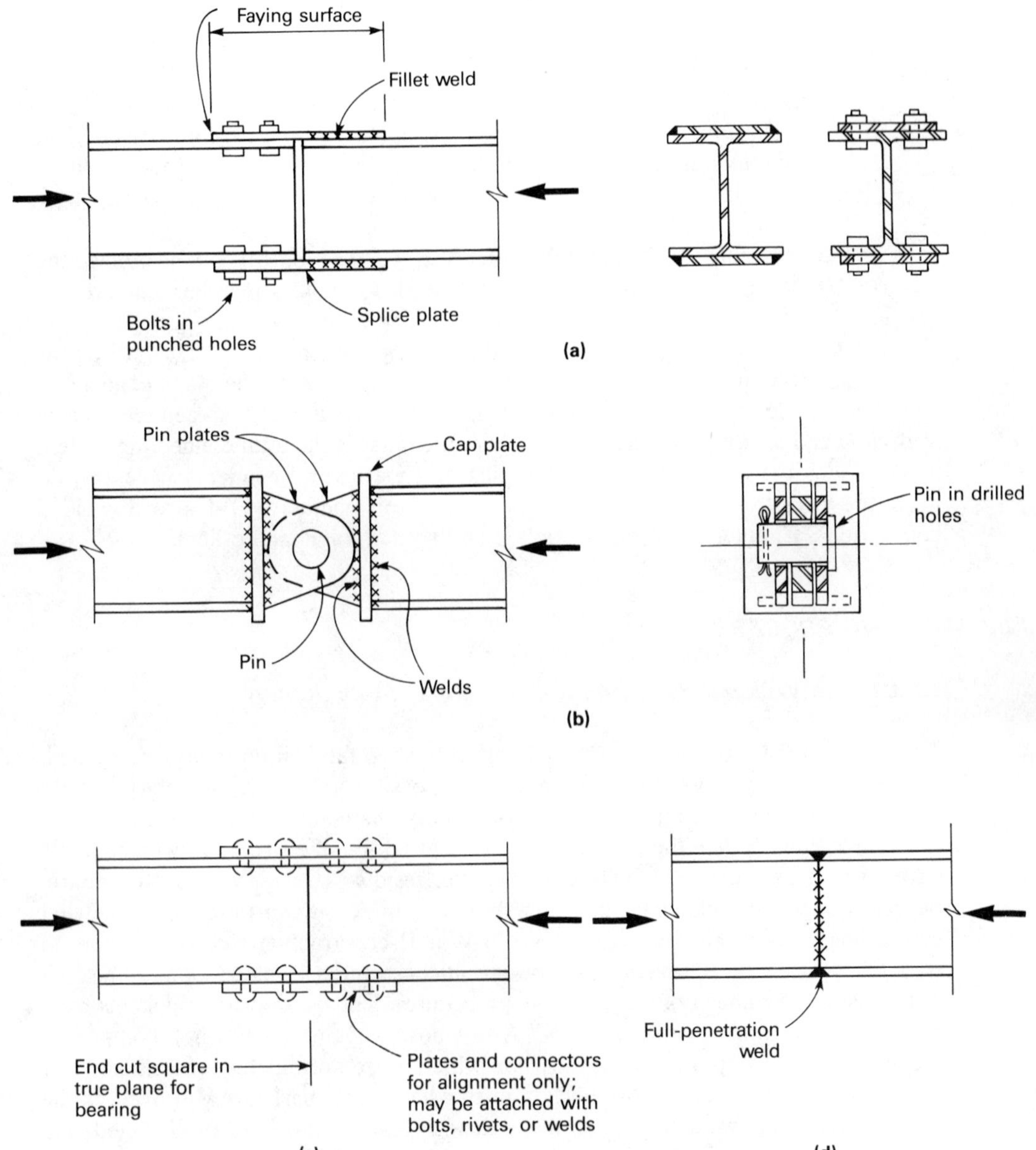

Figure 12.1 Various mechanisms for transferring axial compression.

2. *Welds*: metal fused to members being joined at melting temperature of the steel. Weld may be of *fillet* type, as in Fig. 12.1(a), or *penetration* type, as in Fig. 12.1(d).
3. *Pin*: a single large, solid cylinder introduced into a predrilled hole [Fig. 12.1(b)].
4. *Bearing*: carefully prepared ends of the members transmit the force by bearing. Auxiliary materials are used only to keep the members in their proper position [Fig. 12.1(c)].

In the preceding list, it should be noted that we have focused on the connecting materials (bolts, rivets, welds, pins). There are also auxiliary members intervening between the principal members being connected. The splice plates in Fig. 12.1(a), the cap and pin plates in Fig. 12.1(b), the alignment plates in Fig. 12.1(c), all act as intervening structural elements functioning to make possible the use of the connectors or welds themselves. We will distinguish, in further discussion, between *connecting materials* and *joints* (or *connections*), where the latter is the total connecting system and includes the connecting materials plus other structural elements necessary to make the force transfer possible. The balance of this chapter will deal with welded connections. Chapter 13 discusses bolted and riveted joints and others.

12.3 WELDING PROCESSES

12.3.1 The Welding Revolution in Structural Steel Joinery

The most common types of connections used in shop fabrication today are welded. Welded connections are also very common (possibly the most common) in field assembly practice. This represents a revolutionary change from the time, as recently as the late 1930s, when the primary joining material was the rivet, and welded joints were suspect as a cause of high, uncalculable, residual stresses and brittle failure. Several things have changed in the years since; in American practice (somewhat earlier in Europe), the demands of World War II construction forced designers to take advantage of the simplicity and speed that characterizes welded joinery, in some instances moving ahead of fully reliable procedures. The basic structural steels have been changed from a steel (ASTM A7) whose chemistry was not completely specified to a group of steels most of which were formulated specifically for weldability (see Chapter 5). Methods of testing have been markedly improved, to the point where most weld and base metal defects can be avoided or detected and corrected. The American Welding Society (AWS) has detailed specifications directed to quality control and reliable welding joinery. Highly skilled engineering specialists are available with detailed understanding of the total welding process, who develop welding procedures intended to minimize cost while avoiding the problems that worried earlier designers.

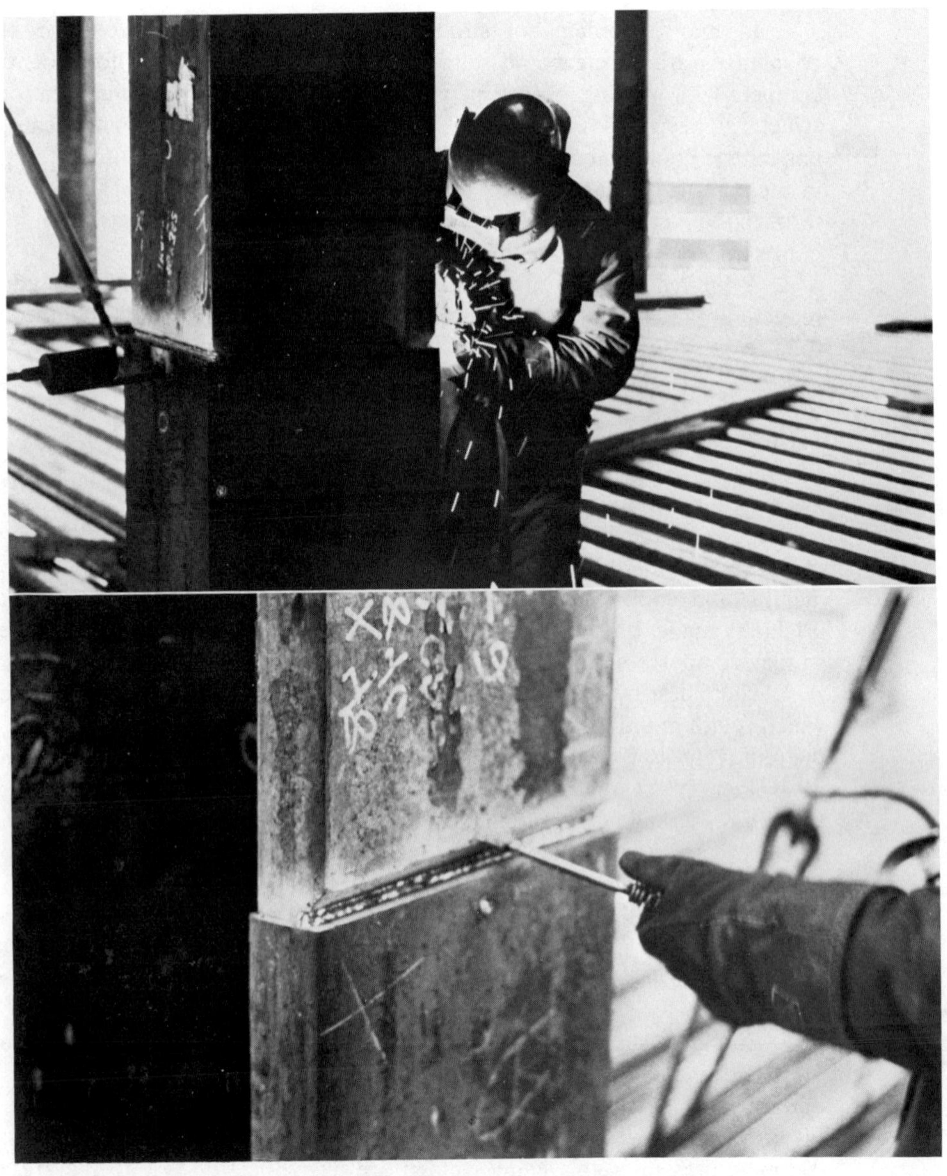

A welder is shown welding a pass of a partial penetration butt joint of a building column and then removing the deposited slag. The process is *shielded metal arc welding* (SMAW). (Courtesy of American Bridge Division, United States Steel Corporation)

In many problems of structural joinery today, the economic and design advantages are so clear and reliability so high that no other form of joinery can compete with welding. However, welding is not without problems even today. Most structural designers cannot hope to be welding specialists. They can and must understand enough about the technology of welding to use it effectively and prudently in their designs. What follows may be considered a start. Much more is available through the American Welding Society (AWS), the Lincoln Electric Co.,[22] and many books and specialists in welding technology.

A limited glossary of welding terms is offered at the end of the chapter. The reader unfamiliar with welding terminology will need to refer to it to understand what follows.

12.3.2 Processes: General

Welding is accomplished by fusing two pieces of steel together so that they become continuous. In the most common processes, fusion occurs at the melting temperature of the steel, with solidification occurring as the temperature drops. The process of melting and cooling resembles the conditions under which the steel parts were originally made; the characteristics of the resulting weld are very much like steels of similar composition cooled under similar conditions.

The oldest welding method (several thousand years) is probably forge welding, which is still practiced in specialized uses. Light-gage steels are sometimes joined by the relatively slow process of oxyacetylene gas welding.* An AWS chart shows over 40 welding processes. However, the methods that brought about the revolution in structural welding joinery are a more limited number, which derive their energy from an electric arc, operate much faster than any of the older methods, and are specifically adapted to the needs of structural practice.

In most structural welding, additional steel called weld metal or filler metal is added at a linear juncture between two pieces of *base metal.* Figure 12.2 shows the two basic types of joints: *fillet* and *butt-welded* made by the *shielded metal arc process*.

Energy is supplied when an electric arc jumps a gap between the metal of the electrode and the base metal, which are at different electrical potentials. Enough current must be supplied by the welding machine to cause the electrode and both pieces of base metal to melt as a filler metal is carried in a stream across the gap. Temperatures in metals being welded are over 3000°F. Within the arc, they may reach as high as 10,000°F. The heated electrode coating forms a gaseous barrier to the introduction of oxygen in the molten pool and flux materials that float on the molten pool. When the electrode moves away from the molten metal, the metal

*Soldering is a process similar to welding. However, temperatures are lower; the base metal is not melted; the filler metal adheres to, but does not fuse with the base metal.

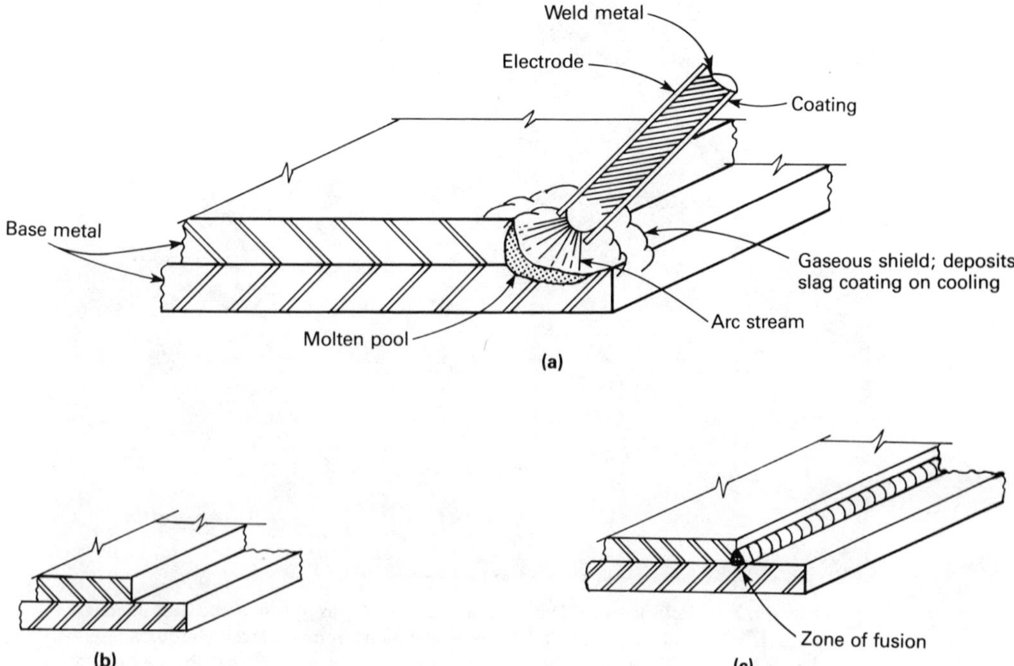

Figure 12.2 Shielded metal arc welding (SMAW) showing fillet welded joints; process similar for butt welds: (a) weld in progress; (b) pieces prepared for welding; (c) after welding and slag removal.

solidifies while adjacent metal melts. The flux forms a low-density slag, which helps to protect the cooling steel. Slag must be removed after cooling.

Welding proceeds in a continuous (or sometimes discontinuous) line as shown in the "after" sketches of Fig. 12.2. The line may be straight or curved, but usually has considerable length. Many figures in this chapter and other books show only the cross section through a weld. It is common also to find figures showing the weld filling only the space outside the shape of the originally prepared surface, ignoring the zone of fusion and the heat-affected zone (HAZ). See Fig. 12.3. The reader should recognize that these are drafting devices only. Unfused welds are ineffective. The HAZ is a zone of altered microstructure in the base metal adjacent to the weld, and is potentially dangerous if the steels or the welding procudures are improperly controlled.

Recall from Chapter 5 that one of the reasons for the selection of specific steels as "structural" steels is because of relative immunity to alteration of their physical properties through temperature changes such as with welding. This is only relatively true and, as noted earlier, not true of the high-strength, quenched and tempered steel governed by ASTM A514.

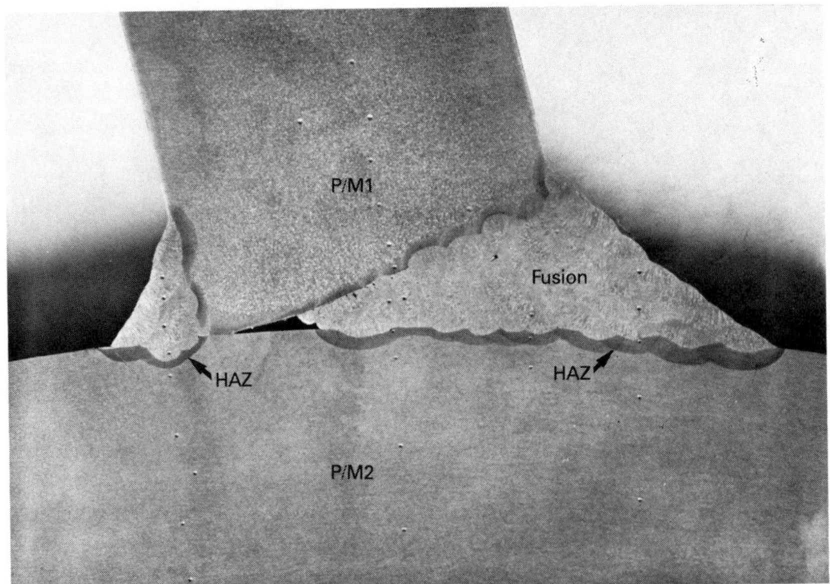

Figure 12.3 Polished sample cut through a weld. Areas of fusion and heat-affected zone (HAZ) can be seen. Dark dots are indentations from Rockwell Hardness tests. The HAZ has altered microstructure. Adapted with permission from Preece (47).

12.3.3 Principal Welding Processes in Structural Practice

The AWS *Structural Welding Code—Steel* (AWS D1.1)[17] is the most widely recognized governing standard for structural welding in U.S. practice. It is incorporated by reference almost without exception into the AISC Specification. It accepts as *prequalified* (i.e., acceptable without further proof of suitability if applied under specified conditions) four welding processes, all using electric arcs. These are shielded metal arc welding (SMAW), submerged arc welding (SAW), gas metal arc welding (GMAW), and flux-cored arc welding (FCAW). Of these, SMAW is the principal method used for *hand welding*; the others are *automatic* or *semiautomatic* processes. Shop practice on major weldments is mostly automatic, offering the advantages of much higher speed and greater reliability. Hand welding today is mostly limited to relatively short, small production welds, *tack welds* used in fitting up (holding pieces in position for production welding), and field applications where automatic machinery is too difficult or costly to set up. Both fillet and butt welds are made with all the methods named.

Acceptable procedures using these processes or others require testing of welding operators and of welds, use of metal *matching* the base metal, preparation of joints in qualified form, controlled *preheat* and *interpass temperatures* based on the steel, the thickness of parts, and the welding process.

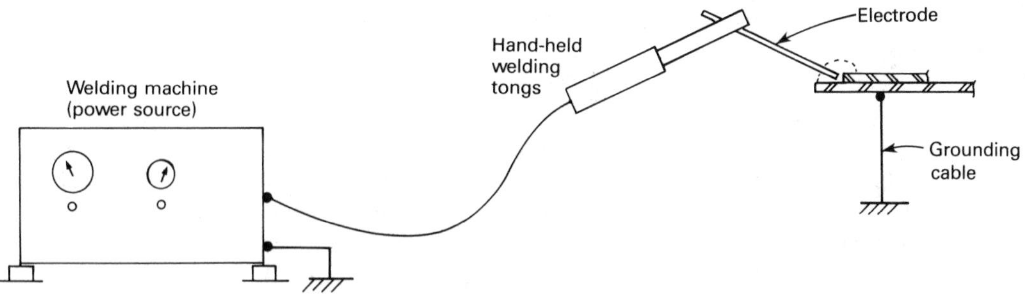

Figure 12.4 Circuit for SMAW.

Shielded metal arc welding (SMAW). Refer to Fig. 12.4. An operator holds the end of a short *electrode* in the jaws of a set of welding *tongs.* The circuit is completed when the electrode is brought close enough to the joint so that an arc jumps the gap, carrying weld metal with it. Electrodes consist of weld metal coated by materials that form shielding gases and flux, both intended to protect the molten metal from the air. The usual electrode is consumed in a few minutes, necessitating a short interruption to clamp a new electrode into the tongs. Continuous feed from a reel is possible in *semiautomatic* welding. The solidified flux, or *slag*, must be removed before additional passes are made over a prior pass.

Electrode diameters are in the range from $\frac{5}{32}$ to $\frac{5}{16}$ in. Welds may be made in any position, flat, vertical, horizontal, or overhead (see Fig. 12.8). Sizes of single *passes** are limited depending on the type of joint and position. If welds are to be larger than permitted for a single pass, they are made with multiple passes.

Submerged arc welding (SAW). Figure 12.5 helps to show the process. The electrode is fed continuously from a reel. Flux is deposited from a separate

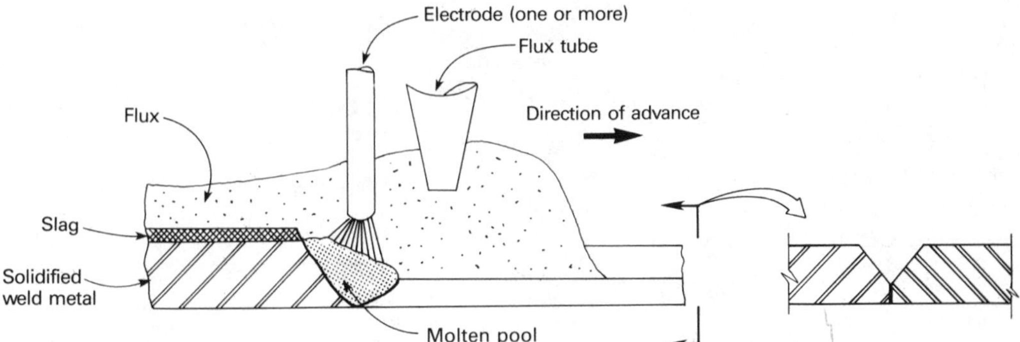

Figure 12.5 Submerged arc welding (SAW); schematic longitudinal section through joint.

*See Glossary at the end of the chapter.

source and submerges the end of the electrode and the arc, shielding it from the outside air. The electrode, flux hopper, and power source are attached to a frame on rollers, which advances the weld at a controlled rate. The process is continuous until stopped at the end of a pass.

The size of a single pass is limited if a single electrode is used. Multiple electrodes are permitted with no limit on weld size. Positions for welding must be flat or horizontal, so work must often be turned on jigs into the proper position for welding.

Gas metal arc welding (GMAW). Also known as *metal inert gas welding (MIG)*, GMAW welding is a semiautomatic process that uses welding wire continuously fed from a reel and a separate supply of shielding gas (carbon dioxide, helium, argon, or other type permitted under the AWS Specification). It is similar to the FCAW process discussed next, except that the freely discharged gas is sensitive to wind currents. AWS requires wind shelters to keep wind speeds at the work below 5 mph.

Flux-cored arc welding (FCAW). As shown in Fig. 12.6, this process uses a steel tube electrode that is filled with flux, continuously fed from a reel. Gas shielding and protective slag are formed from the flux. This is a semiautomatic process that can be used in all welding positions. AWS limits the size of electrode, the size of welds made in a single pass, and the nature of the shielding gas.

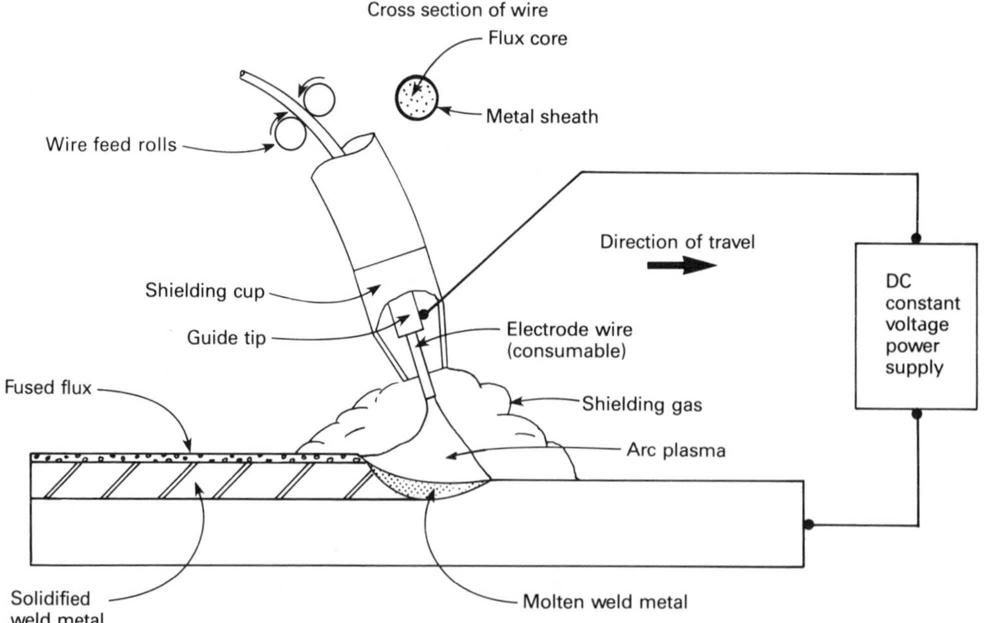

Figure 12.6 Schematic illustration of flux-cored arc welding.

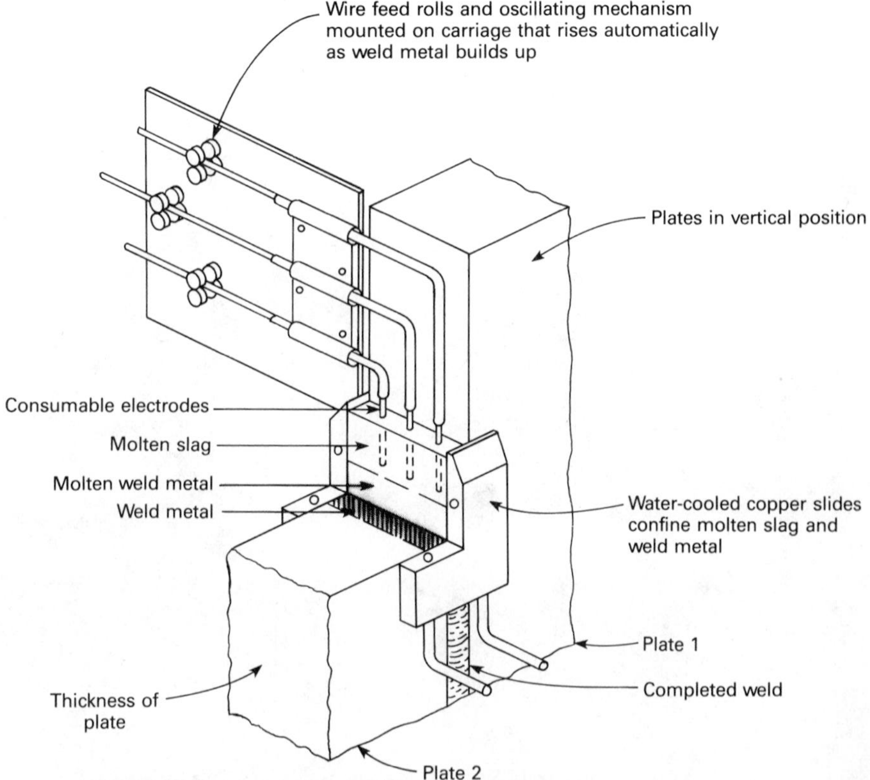

Figure 12.7 Electro-slag welding.

Additional processes. *Electro slag welding* uses a number of flux-cored wire electrodes and gas shielding to fill a square butt gap between two thick, wide plates in a single operation, fed from above as in Fig. 12.7. A similar process, *electro gas welding* uses one flux-cored wire to weld thinner plates. These processes must be qualified by a contractor who proposes to use them in accordance with AWS procedures for qualification.

12.4 FORMS OF WELD CROSS SECTIONS

Figure 12.8 shows samples of fillet and butt welds, as well as the positions they were in before welding. These are the primary types of welded joints. Butt welds are the stronger, but fillet welds are probably the most common. Typical cross sections of fillet welds are shown in Fig. 12.9. To achieve fusion through the full thickness or to desired depth of a butt joint, it is necessary to prepare the pieces with sufficient gap so that the arc is not formed when the electrode is too close to the surface. Figure 12.10

Butt (Groove)

Fillet, tee joint

Fillet, lap joint

Flat

Horizontal

Vertical

Overhead

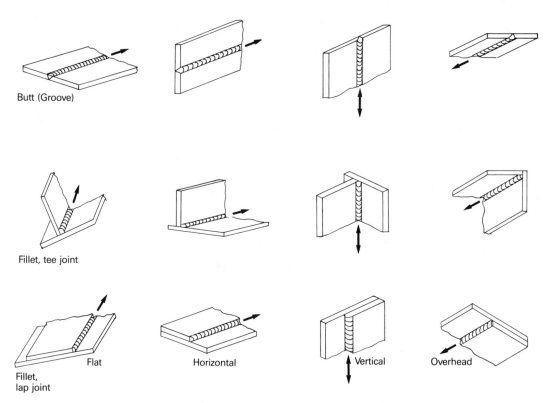

Figure 12.8 Fillet and butt welds: positions for welding.

shows a number of forms of joint preparation. Some allow full penetration (i.e., fusion through the full thickness). Some achieve only partial penetration of the welds. The welding symbols indicate the type of preparation. The lettering in the tail of the bent arrow designates a form of preparation fully dimensioned in an AWS table, which is reproduced in the AISC Manual, Part 4. If the required dimensions are used with any

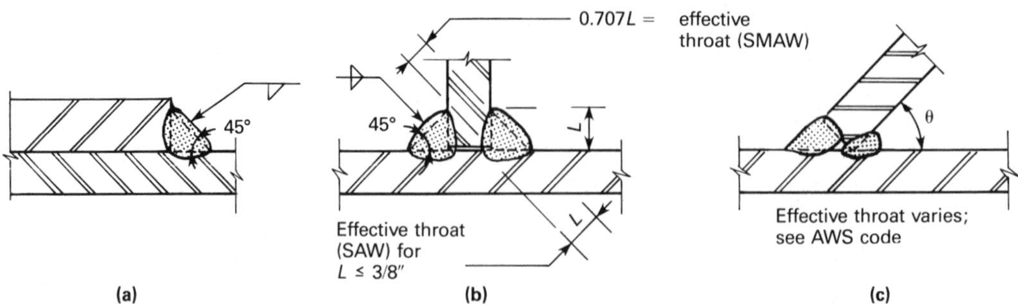

Figure 12.9 Fillet welds in sections: (a) lapped fillet; (b) double fillet; (c) skewed fillets.

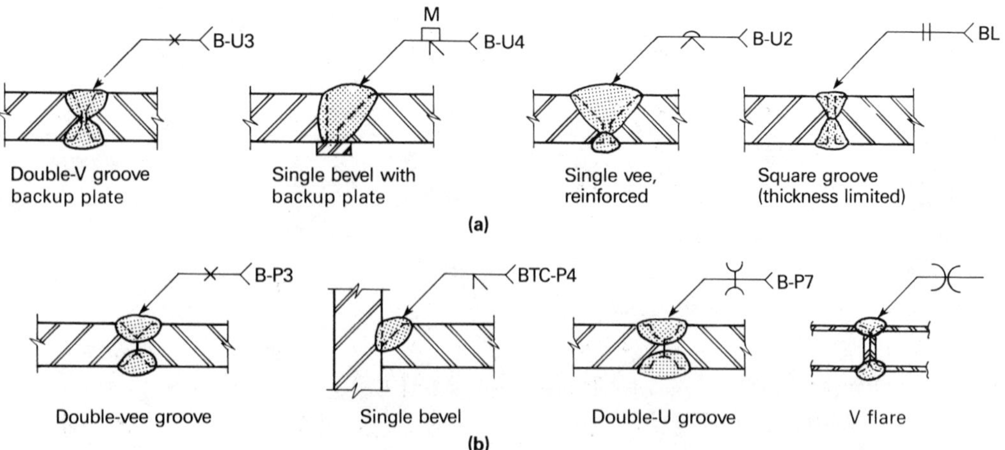

Figure 12.10 Groove welds in section: (a) complete penetration; (b) partial penetration. Symbols in tails of arrows are joint forms prequalified by AWS. See AISC Manual, Part 4.

of the prequalified welding processes properly used, the weld itself is prequalified as to the degree of penetration.

Most of these welds must be made partially from each side. Weld B-U4 in Fig. 12.10(a) achieves full penetration from one side, with the help of a backup plate that prevents the first pass from blowing through at the bottom. Figure 12.11 shows examples of plug and slot welds used on overlapping plates. A round or oval hole is cut in one plate, large enough so the arc is formed at the surface of the other. Welding fuses the two plates in the limited area of the hole, which can be filled completely or partially, as desired.

Some instances occur where it is necessary or desirable to combine fillets and groove welds. Two such are illustrated in Fig. 12.12.

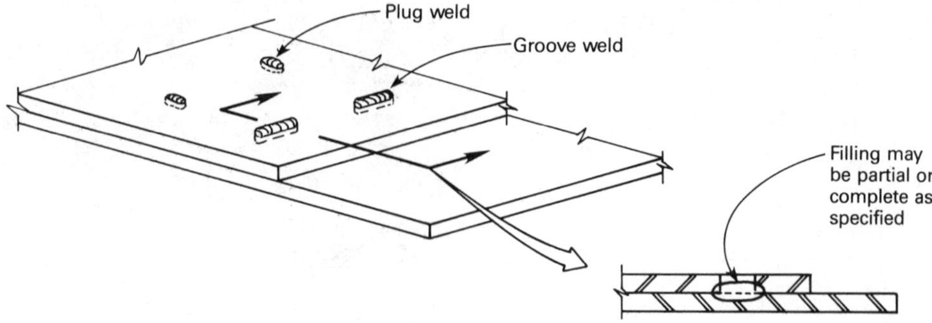

Figure 12.11 Slot and plug welds.

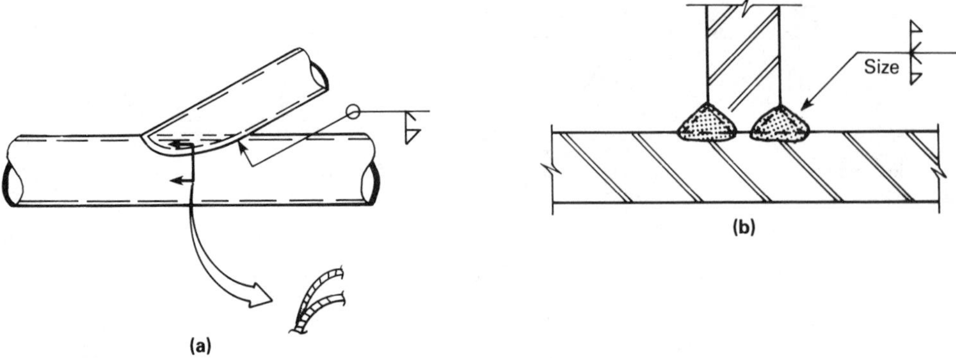

Figure 12.12 Combined fillets and grooves in tee joints: (a) tee joint of pipes; (b) tee joint of plates; partial bevel with fillets.

Cutting of holes and edges to required contours can be done with planing machines or more commonly today with oxyacetylene torches. Guided or automatically controlled torches can cut very regular, clean contours in straight or curved lines. Hand cutting by oxyacetylene torch is sometimes required, but it is slower and requires great skill to avoid unsatisfactory, irregular surfaces. Electric arc cutters are also used.

12.5 WELDING SYMBOLS

It is customary in structural drafting to draw the pieces of steel to be joined by welding without attempting to show the actual deposited weld metal or grooves, as we have shown them in the figures up to this point. Welding symbols are used as a shorthand way to give instructions about the form, size, and other characteristics of the welds to be made. The forms of the symbols are precisely prescribed by the AWS; when properly used, the instruction is clear and unambiguous. If not used exactly as prescribed, the meaning may be ambiguous, leading to a multiplicity of problems for everyone concerned.

The basis of the system of symbols is the diagram shown in Fig. 12.13. The

Figure 12.13 Welding symbols. Reproduced with permission from AISC Manual.

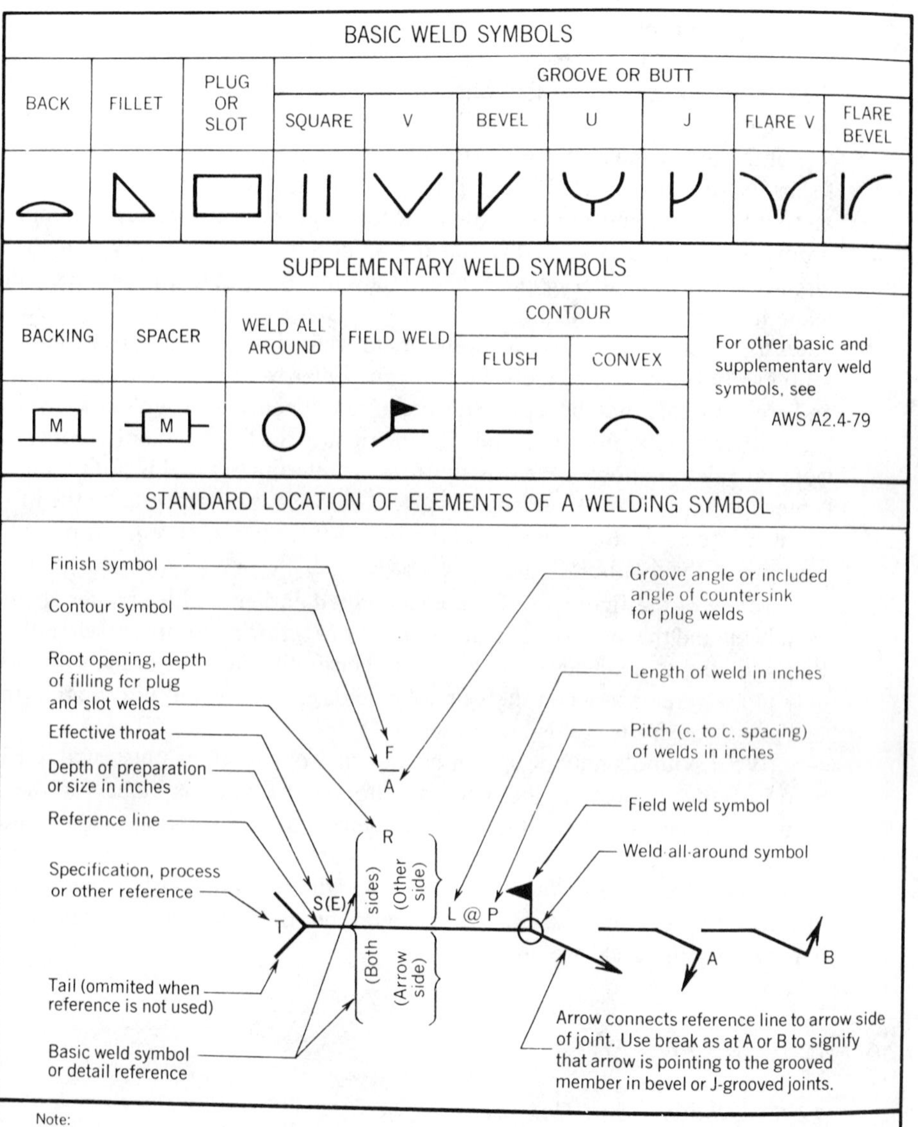

Note:

Size, weld symbol, length of weld and spacing must read in that order from left to right along the reference line. Neither orientation of reference line nor location of the arrow alter this rule.

The perpendicular leg of ◺, V, ⊬, |⌐ weld symbols must be at left.

Arrow and Other Side welds are of the same size unless otherwise shown. Dimensions of fillet welds must be shown on both the Arrow Side and the Other Side Symbol.

The point of the field weld symbol must point toward the tail.

Symbols apply between abrupt changes in direction of welding unless governed by the "all around" symbol or otherwise dimensioned.

These symbols do not explicitly provide for the case that frequently occurs in structural work, where duplicate material (such as stiffeners) occurs on the far side of a web or gusset plate. The fabricating industry has adopted this convention: that when the billing of the detail material discloses the existence of a member on the far side as well as on the near side, the welding shown for the near side shall be duplicated on the far side.

symbol consists of an arrow, always bent at least once, often several times before the tip that points to the joint being welded. On each side of the arrow shaft, one representing the *arrow* or *near* side of the joint, the other the *far* side, are placed symbols from among those at the top of the sheet, which indicate the form of the weld desired. The *tail* of the arrow may be used to refer to one of the standard AWS weld preparation forms or for other notes. If none is desired, the tail may be omitted. An *open circle* around the first bend in the arrow indicates that the weld makes a complete circuit around the joint, ending at the starting point. (It does not imply that the circuit is circular.) The triangular *pennant* above the same bend indicates that the weld is to be made in the *field** as distinguished from the shop. If the pennant is omitted and no other instructions are given, a shop or field weld is permitted. On the left side of the form symbol, the size of the weld is given. Fillet weld sizes must be specified. Butt and groove welds may be assumed to require full penetration unless a size is given on the shaft or by reference in the tail. To the right of the form symbol is either a length of weld or a designation of the length of short, intermittent welds with their center-to-center spacing (pitch). If no length is given, the weld is continuous for the full length of the line. The notes at the bottom of the sheet set further requirements on the form and meaning of the symbols.

Many of the figures we have encountered earlier in this chapter show both the weld itself and the symbol. The diagrams of *prequalified joints* shown in the Manual, Part 4, do the same. These should help to clarify the meaning of the symbols, but it is only in the rare case when the symbol cannot give sufficient information that actual sections through the welds are drawn.

Weld symbols may be shown on any view of a joint, as illustrated in Fig. 12.14. Study of this figure will help in interpreting the various rules for the symbols. Remaining figures in this and later chapters will use symbols without necessarily showing the welds themselves. Where welds are shown in cross section, we will show the original prepared surfaces and the weld without necessarily showing the zone of fusion. This is common drafting convention, but the reader will recognize that fusion is necessary in actual welds.

12.6 PROBLEMS AND PITFALLS

12.6.1 Knowns and Unknowns

Most welded joints operate successfully and merit the confidence that designers place in welding technology. However, every year the engineering and construction journals report significant instances of cracking of welded joints, sometimes leading to collapse of structures. Often the distressed structures were designed in strict accordance with the applicable specifications and built in strict conformance with the designer's drawings and construction specifications. Clearly, there is still much about welding

*In older drawings, a filled in circle at the bend of the arrow had the same meaning.

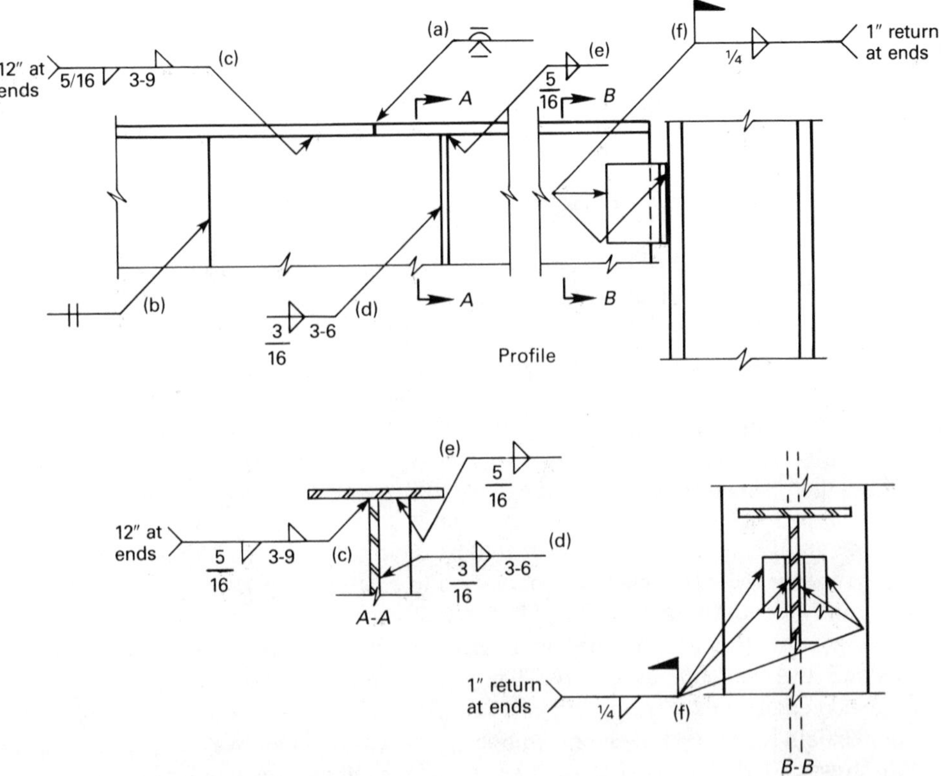

Figure 12.14 Examples of use of welding symbols: (a) Full penetration V groove on top (arrow) face with single reinforcing pass on lower face. Reinforcing crown top and bottom ground flush. (b) Square butt joint welded from both sides. Provides full penetration only for limited thickness, which varies with welding process. (c) Intermittent 5/16" fillet welds 3" long, spaced 9" on centers. Same welds on far side but staggered with near side; i.e., far side welds are located between near side welds. (Staggering helps to prevent compression buckling in the plate.) The intermittent welds continue for the full length of the joint, except for the ends where a 12" long weld is required both sides. (d) 3/16" fillet both sides, intermittent, 3" at 6" centers; not staggered. (e) 5/16" fillet both sides for length of contact. (f) 1/4" fillets at both toes of two angle clips. The welds are vertical, but continue for one horizontal inch at top and bottom. Black pennant indicates these welds are to be made in the field.

technology that is known, but not widely enough, and there is much still not known.

There is little to be said here about problems due to defective workmanship, except to emphasize the great importance of quality control in this technique of joining materials, which seems so easy to learn, and succeeds so often, but is very vulnerable to neglect of proper procedure. Selection of welding materials; clean, well-prepared surfaces; avoidance of trapped foreign matter, moisture, or air; shielding; control of the temperature of the steel—these are among the requirements that are carefully specified in the AWS Code and with proper enforcement, serve to avoid problems due to defective workmanship.

However, there are other pitfalls of which the wary designer should be aware. Many of these relate to the nature of steel and of welding. Some others come from the design of joints. We will look at a few of both.

12.6.2 Shrinkage and Restraint: Locked-in Stress

Most of the problems inherent in welding arise from the necessity to raise the local temperature very high very rapidly and then cool it at an only partially controlled rapid rate. Cooling causes contraction. If the steel is to be stress free while shrinking in size, it must be free of any restraints against movement. If it is not free and cannot shrink, the negative of the shrinkage becomes elongation associated with tensile stress. Beam A of Figure 12.15 is held temporarily in place while its ends are to be welded to the columns. If the left end is welded first, it will move toward the left column, opening a gap at the right end, possibly between $\frac{1}{16}$ and ⅛ in., with no apparent distress. Welding the right end is another matter. The equivalent movement toward the right column is prevented by the rigid structure, forcing Beam A to stretch as the weld cools. The reader may wish to calculate the stress that is *locked-in* to the joint. Recall that $\Delta L = PL/AE = f_t L/E$.

Small assemblies may be stress relieved by heating in a furnace after welding and then cooling uniformly. The result, relief of the locked-in stress, is often accompanied by warping and twisting of the assembly. Furnace stress relieving is not feasible for very large subassemblies or structures. One way to partially relieve the stresses in the case of Fig. 12.15 is to peen each pass. A much more effective way is

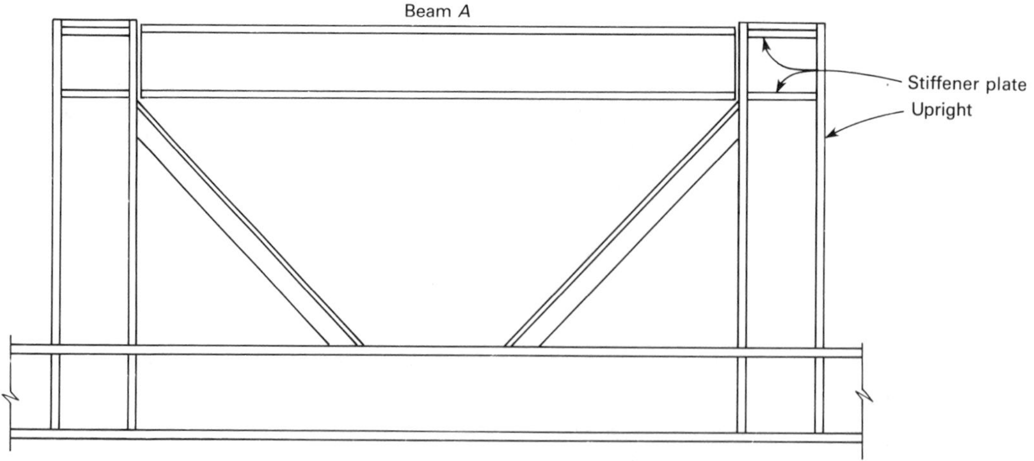

Figure 12.15 Restraint in welding. Both ends of beam *A* are to be welded to the very rigid frame. How does the designer cope with cooling shrinkage?

to change the sequence of assembly so that the locked in stresses either do not arise or are minimized.

12.6.3 Warping and Inhibited Warping: Lamellar Tearing

Figure 12.16 illustrates a problem of shrinkage that is easy to control if recognized. Weld shrinkage bends the relatively thin flange of the tee joint in part (a). This can be easily prevented by prebending the plate or clamping it while welding.

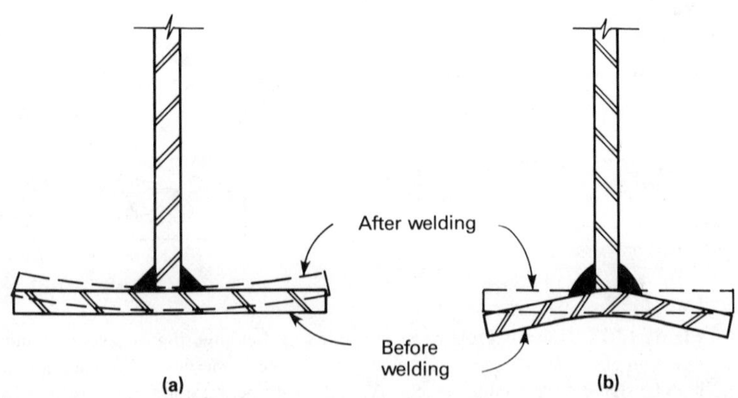

Figure 12.16 Warping in tee joints: (a) straight plate bends from welding shrinkage (rotation of several degrees is not uncommon); (b) prebent plate becomes straight when welded.

A similar shrinkage problem in Fig. 12.17 can cause much more serious consequences. The full penetration bevel and fillet welds must shrink. If the plates are thick, two problems arise. The flange plate is too stiff to warp, as it does in Fig. 12.16, and acts as a restraint against shrinkage. After the first inner pass has solidified, it becomes a fulcrum around which contraction strains try to take place, but are, in this case, inhibited. High tensile stress results perpendicular to the flange.

A third problem may arise, which results from the rolling process used in making the steel. The stress–strain curves seen in this and most books, which are the usual basis for strength and ductility requirements of the steel, are derived from specimens pulled in the direction of rolling, the X direction of Fig. 12.17(b). Tests taken in the transverse (Y) and the through-thickness (Z) directions exhibit lower ductility and impact toughness. Discontinuities such as nonmetallic inclusions and grain nonuniformities may serve as crack starters under high tensile stresses. The effect is most critical under Z direction restraint where the ductility may be quite low and cracks can join together in step fashion, producing *lamellar tearing*, as in Fig. 12.17(c). The *reduction of area* value from the tension test is the most sensitive to the propensity for lamellar tearing in any given steel, with low-digit readings that may

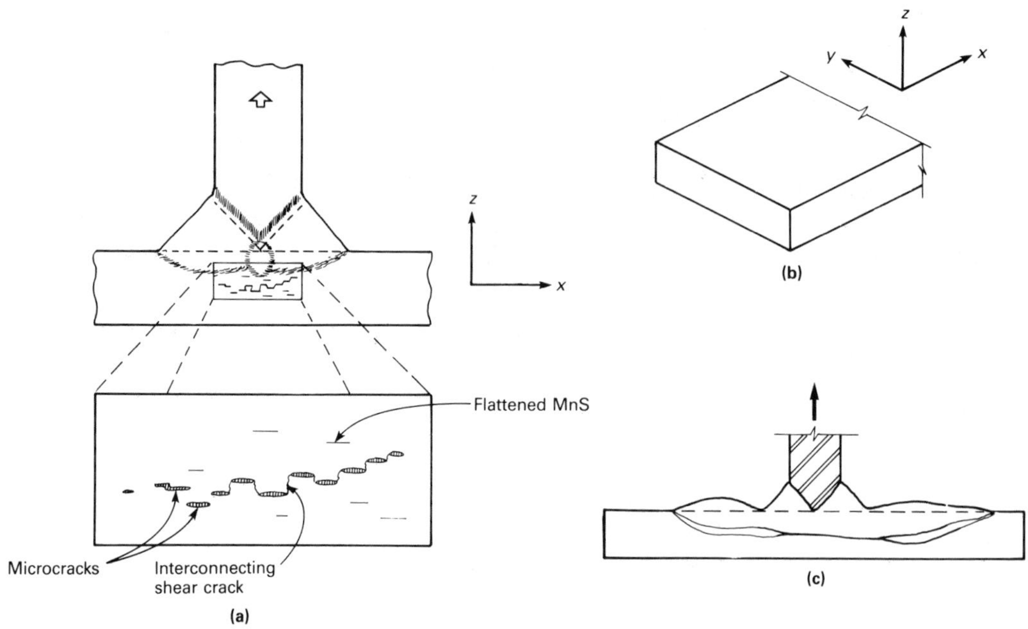

Figure 12.17 Lamellar tearing in a tee joint: (a) flat impurities in steel form microcracks in the direction of rolling; (b) axes for discussion of plate properties; (c) linking of microcracks under high tension causes a lamellar tear. Adapted with permission from Preece (47).

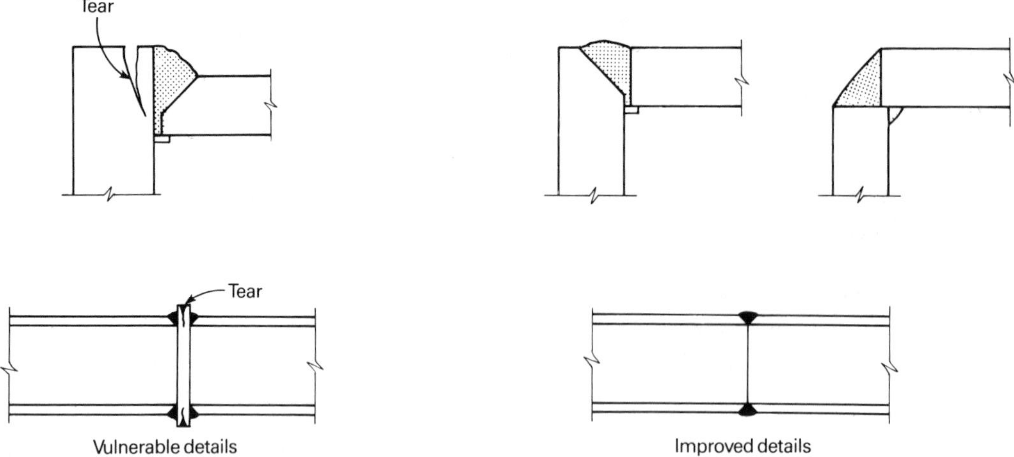

Figure 12.18 Improving details vulnerable to lamellar tearing.

approach zero. Reference 4 is recommended for its more complete discussion and bibliography.

The phenomenon of lamellar tearing is not limited to thick plates, but it is much less likely if the plates are thin. It is characteristic of rigid, highly restrained joints heavily welded. It may be aggravated by external tension applied after welding. It may show up as fatigue cracking after a number of cycles of varying loads. It may be detected by ultrasonic testing of the plates, which is designed to reveal impurities. Much more reliable are details that avoid the cross-thickness tension, such as those suggested in Fig. 12.18. The alert designer will find many other vulnerable details that can be made more reliable if the problems of *through-thickness* weakness and *lamellar tearing* are recognized.

12.7 INSPECTION AND TESTING

Indispensable to the reliable use of welding joinery is a systematic program of inspection and testing. Inspection is done at the shop or field site. The function is to guarantee that specified materials and procedures are used under conditions where proper welding is possible. If the sequence of welding has been specified, the inspector should be able to certify conformance. Inspectors often observe the actual welding as it is going on.

In spite of careful inspection, weld defects may escape detection unless all or part of the work is subjected to tests. Samples cut out from the work may be tested for strength, ductility, and notch toughness. Highly polished specimens through the weld will reveal the depth of fusion and the heat-affected zone. Hardness tests of the weld and HAZ may be revealing. However, such *destructive* tests require repair of the work and are used sparingly. A number of *nondestructive tests* are in common use.

Magnetic particle testing (MT) is done by covering the surface of the weld with a suspension of ferromagnetic particles and then applying a strong magnetic field. Cracks in the weld interrupt the magnetic force lines, causing the particles to concentrate in the vicinity in patterns easy for the inspector to interpret.

Dye penetrant testing (PT) uses a dye in liquid form to search out cracks. Capillary tension in the liquid causes the dye to penetrate into the crack, remaining behind after the surface is cleaned. A developer can cause the dye in the crack to become visible.

Both MT and PT interrupt the work if, as is often desirable, they are applied after the first or early passes. *Radiography* (RT) can find cracks and inclusions after the weld is completed, provided a sensitized plate can be placed on the side opposite a source of x rays or gamma rays. Shadows on the exposed film indicate cracks or inclusions in the weld or adjacent areas. Radiography is most effective on full penetration butt joints with ready access to both sides.

Ultrasonic testing (UT) is the latest (circa 1960) method in widespread use. It relies on the reflection patterns of high-frequency sound waves, which are passed at an angle through the work. Cracks and defects interrupt the sound transmission, altering the patterns that are displayed on an oscilloscope. A skilled operator can interpret the patterns by reference to standard patterns of similar sound joints. The method can reveal many defects that other methods do not. But it has limits, not the least of which is that it relies very heavily on the interpretive skill and integrity of the operator. Partial penetration welds have internal "cracks" outside the weld cross section, causing difficulty in interpreting UT tests.

12.8 AISC STRENGTH CRITERIA FOR WELDS

12.8.1 Type of Stress Transferred versus Stress in Weld

Refer once again to the welded details of Fig. 12.1. The type of stress to be transferred is not necessarily of the same nature as the stress in the weld. The latter results from both the force to be transferred and the form of the joint. Figure 12.1 shows compression joints, but, except for Fig. 12.1(c), could also apply to tensile joints.

In Fig. 12.1(a), the welded side of the joint transfers compression from member to splice plate by shear in the fillet welds. In Fig. 12.1(c), if the faces of the two members are in full contact prior to welding, and if there is no possibility of stress reversal or moment, we could eliminate the overlapping, riveted plates shown and apply enough weld to maintain the position of the members and, possibly, to seal against corrosion. A seal weld applied at the tight butt would be sufficient. The full penetration butt (or *groove*) welds of Fig. 12.1(d), properly made and with weld metal at least equal in strength to the steel in the members, create joints capable of resisting all the stresses that the members themselves can resist. They are said to *fully develop* the members. The welds in Fig. 12.1(b) are not described sufficiently to judge their strength. If the pin plates and W shape are in full contact with the cap plate before welding, compressive loads transfer from one to the other through the cap plate in bearing. (The assumption is made that the cap plate is heavy enough to make the transfer without itself being overstressed in bending.) The weld only maintains alignment. If there is a gap between the cap plate and the other parts before welding, a fillet weld would be in shear at the side contact surface and compression at the other

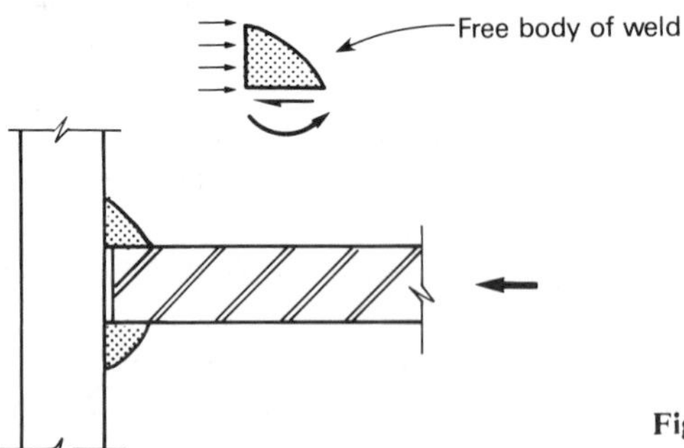

Figure 12.19 Fillet weld in shear to transmit compression.

leg, as can be seen in Fig. 12.19. For reasons discussed earlier, this detail should be approached carefully if the member is in tension.

Further discussion will set forth the methods of calculating the strength of welds in accordance with AISC criteria. Joints will be examined using both fillet and groove welding, each of which has advantages under different circumstances.

12.8.2 Provisions of AISC Specifications Related to Welding

General provisions. Note in Table 12.1, that, among provisions of the AISC Specification related to connections, several apply to welded connections only. AISC Table 1.5.3 gives allowable stresses for welds: they may be used directly for nonfatiguing conditions or modified in accordance with Appendix B where fatigue might otherwise threaten early failure. Table 12.2, herein, is a reproduction of AISC Table 1.5.3. Some things should be noted.

1. Weld metal must be *matched* by strength to the base metal, in accordance with Table 4.1 of the AWS *Structural Welding Code D1.1.* Table 12.3 is derived from that much longer table. Strength of weld metals is *not permitted* to be more than "one strength level stronger than 'matching' weld metal . . . ." In some instances, strengths less than "matching" are permitted.
2. Control of preheat and interpass temperatures in accordance with AWS D1.1 is assumed in setting the allowable stresses. Table 12.4 is based on the longer Table 4.1 of AWS D1.1.
3. Allowable stresses are given in the usual dimensions of F/L^2, (ksi) as is customary in the AISC Specification. Design interest is usually focused on *allowable force/unit length*, requiring, for conversion to stress, determination of an effective transverse dimension.
4. The effective transverse dimension depends on the type of weld. In full-penetration butt welds, it is the thickness of the thinner part joined. For partial-penetration welds, it is a transverse dimension based on the depth of the zone of fusion. For fillet welds it is based on the smallest transverse dimension across the roughly triangular fusion zone. Weld strength is often calculated by the dimensions of shear flow, kips/inch, and is based on

$$\text{allowable } q = \text{allowable stress} \times \text{effective transverse dimension}$$

5. Two different allowable stresses much be considered, one based on the strength of the weld metal, the other on that of the base metal, each applied to appropriate effective widths or thicknesses. Stresses on the base metal are based on F_y; those on the weld metal on its F_u.
6. A footnote, "e", to Table 12.2 addresses an issue not immediately apparent, which is explained immediately below.

TABLE 12.2 ALLOWABLE STRESSES ON WELDS

Type of Weld and Stress[a]	Allowable Stress	Required Weld Strength Level[b,c]
Complete-Penetration Groove Welds		
Tension normal to effective area	Same as base metal	"Matching" weld metal must be used.
Compression normal to effective area	Same as base metal	Weld metal with a strength level equal to or less than "matching" weld metal may be used.
Tension or compression parallel to axis of weld	Same as base metal	
Shear on effective area	0.30 × nominal tensile strength of weld metal (ksi), except shear stress on base metal shall not exceed 0.40 × yield stress of base metal	
Partial-Penetration Groove Welds[d]		
Compression normal to effective area	Same as base metal	Weld metal with a strength level equal to or less than "matching" weld metal may be used.
Tension or compression parallel to axis of weld[e]	Same as base metal	
Shear parallel to axis of weld	0.30 × nominal tensile strength of weld metal (ksi), except shear stress on base metal shall not exceed 0.40 × yield stress of base metal	
Tension normal to effective area	0.30 × nominal tensile strength of weld metal (ksi), except tensile stress on base metal shall not exceed 0.60 × yield stress of base metal	
Fillet Welds		
Shear on effective area	0.30 × nominal tensile strength of weld metal (ksi), except shear stress on base metal shall not exceed 0.40 × yield stress of base metal	Weld metal with a strength level equal to or less than "matching" weld metal may be used.
Tension or compression parallel to axis of weld[e]	Same as base metal	
Plug and Slot Welds		
Shear parallel to faying surfaces (on effective area)	0.30 × nominal tensile strength of weld metal (ksi), except shear stress on base metal shall not exceed 0.40 × yield stress of base metal	Weld metal with a strength level equal to or less than "matching" weld metal may be used.

[a] For definition of effective area, see Sect. 1.14.6.
[b] for "matching" weld metal, see Table 4.1.1, AWS D1.1-77.
[c] Weld metal one strength level stronger than "matching" weld metal will be permitted.
[d] See Sect. 1.10.8 for a limitation on use of partial-penetration groove welded joints.
[e] Fillet welds and partial-penetration groove welds joining the component elements of built-up members, such as flange-to-web connections, may be designed without regard to the tensile or compressive stress in these elements parallel to the axis of the welds.

Reproduced with permission from AISC Specification, 1978.

TABLE 12.3 REQUIREMENTS FOR MATCHING FILLER METAL, BASED ON AWS D1.1–80

Steel specification requirements			Filler metal requirements		
Steel specification[a]	Minimum yield point, ksi	Tensile strength range, ksi	Electrode specification	Minimum yield point, ksi	Tensile strength range, ksi
			SMAW		
ASTM A36	36	58–80	AWS A5.1 or 5.5		
			E60XX or	50	67 min.
			E70XX	57	77 min.
			SAW		
ASTM A500 Grade A	33–39	45 min.	AWS A5.17 or A5.23		
Grade B	42–46	58 min.	F6X-EXXX or	50	62–80
ASTM A501	36	58 min.	F7X-EXXX	60	70–95
			FCAW		
			AWS A5.20		
			E60T-X	50	62 min.
			E70T-X	60	72 min.
ASTM A441	42–50	63–70 min.	SMAW		
			AWS A5.1 or A5.5		
			E70XX[b]	57	77 min.
ASTM A572 Grade 42	42	60 min.	SAW		
Grade 45	45	60 min.	AWS A5.17 or A5.23		
Grade 50	50	65 min.	F7X-EXXX	60	70–95
ASTM A588 (4 in. and under)	50	70 min.	FCAW		
			AWS A5.20		
			E70T-X	60	72 min.
			SMAW		
ASTM A572 Grade 60	60	75 min.	AWS A5.5		
Grade 65	65	80 min.	E80XX[b]	67	72 min.
			SAW		
			AWS A5.23		
			F8X-EXXX	68	80–100
			FCAW		
			Grade E80T	68	80–95
			SMAW		
ASTM 514 (over 2½ in.)	90	105–135	AWS A5.5		
			E100XX[b]	87	100 min.
			SAW		
			AWS A5.23		
			F10X-EXXX	88	100–130
			FCAW		
			Grade E100T	88	100–115

TABLE 12.3 *(continued)*

Steel specification requirements			Filler metal requirements		
Steel specification[a]	Minimum yield point, ksi	Tensile strength range, ksi	Electrode specification	Minimum yield point, ksi	Tensile strength range, ksi
ASTM A514 (2½ in. and under)	100	115–135	SMAW ASW A5.5 E110XX[b]	97	110 min.
			SAW AWS A5.23 F11X-EXXX	98	110–130
			FCAW Grade E110T	98	110–125

[a]In joints involving base metals of two different yield points or strengths, filler metal electrodes applicable to the lower-strength base metal may be used, except that, if the higher-strength base metal requires low hydrogen electrodes, they shall be used.

[b]Low hydrogen classifications only.

Reproduced with permission from reference 47.

TABLE 12.4 MINIMUM PREHEAT AND INTERPASS TEMPERATURES,[a,b] BASED ON AWS D1.1-80

Steel specification	Welding process	Thickness of thickest part at point of welding, in.	Minimum temperature, °F
ASTM A36 ASTM A500 Grade A Grade B ASTM A501	Shielded metal arc welding with other than low hydrogen electrodes	Up to ¾	None[c]
		Over ¾ through 1½	150
		Over 1½ through 2½	225
		Over 2½	300
ASTM A36 ASTM A572 Grades 42, 45, 50, 55 ASTM A441 ASTM A500 Grade A Grade B ASTM A501	Shielded metal arc welding with low hydrogen electrodes, submerged arc welding, gas metal arc welding, flux-cored arc welding	Up to ¾	None[c]
		Over ¾ through 1½	50
		Over 1½ through 2½	150
		Over 2½	225

TABLE 12.4 *(continued)*

Steel specification	Welding process	Thickness of thickest part at point of welding, in.	Minimum temperature, °F
ASTM A572 Grades 55, 60, 65	Shielded metal arc welding with low hydrogen electrodes, submerged arc welding, gas metal arc welding, flux-cored arc welding	Up to ¾	50
		Over ¾ through 1½	150
		Over 1½ through 2½	225
		Over 2½	300
ASTM A514 over 2½ in.	Shielded metal arc welding with low hydrogen electrodes, submerged arc welding with carbon or alloy steel wire, neutral flux, gas metal arc welding or flux-cored arc welding	Up to ¾	50
		Over ¾ through 1½	125
		Over 1½ through 2½	175
		Over 2½	225
ASTM A514 2½ and under	Submerged arc welding with carbon steel wire, alloy flux	Up to ¾	50
		Over ¾ through 1½	200
		Over 1½ through 2½	300
		Over 2½	400

[a]Welding shall not be done when the ambient temperature is lower than 0°F. When the base metal is below the temperature listed for the welding process being used and the thickness of material being welded, it shall be preheated (except as otherwise provided) in such manner that the surfaces of the parts on which weld metal is being deposited are at or above the specified minimum temperature for a distance equal to the thickness of the part being welded, but not less than 3 in. both laterally and in advance of the welding. Preheat and interpass temperatures must be sufficient to prevent crack formation. Temperature above the minimum shown may be required for highly restrained welds. For quenched and tempered steel the maximum preheat and interpass temperature shall not exceed 400°F for thicknesses up to 1½ in. inclusive and 450°F for greater thicknesses. Heat input when welding quenched and tempered steel shall not exceed the steel producers recommendation.

[b]In joints involving combinations of basic metals, preheat shall be as specified for the higher-strength steel being welded.

[c]When the base metal temperature is below 32°F, the base metal shall be preheated to at least 70°F and this minimum temperature maintained during welding.

Reprinted with permission from reference 47.

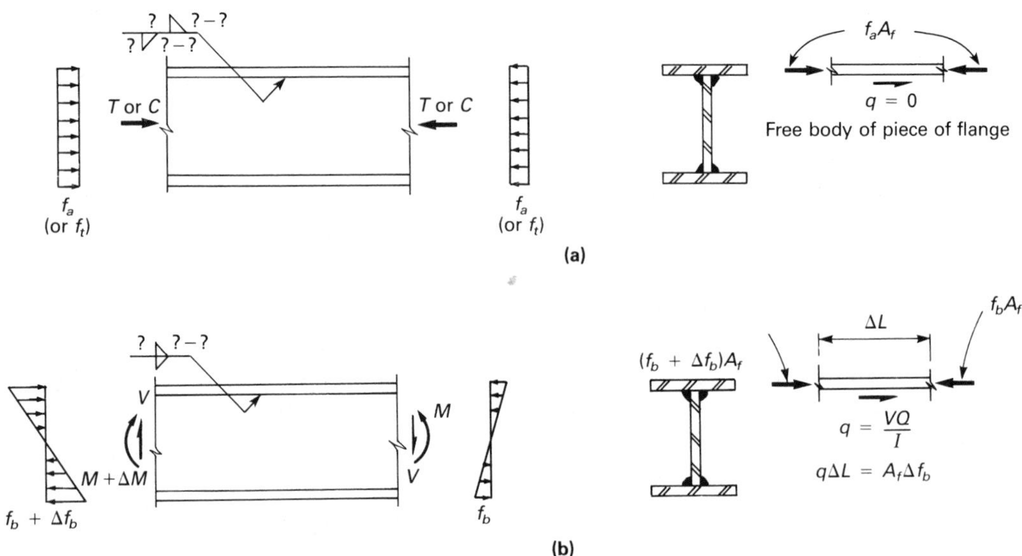

Figure 12.20 Shear on welds parallel to longitudinal stress: (a) axial load; (b) flexure.

Footnote e of Table 12.2 says that the weld required to join the separate elements of a cross section is independent of the longitudinal (axial) stress on the elements. Figure 12.20 may clarify the reasons, if not immediately apparent. Figure 12.20(a) shows a uniformly compressed member made from three separate plates. The question to be answered is how much weld is required from the flanges to the web. A free body of a piece of flange reveals that there is no stress to be transferred from the flange to the web. The functions of a weld in this case are:

To hold the relative positions of the flanges and the web.

To prevent buckling of the separate plate elements by using the adjacent plate as a stiffener.

To prevent corrosion if the environment is corrosive.

The first two requirements can be satisfied by intermittent welding, with short fillet welds widely spaced, the maximum spacing depending on the thickness of the plate elements. Corrosion protection requires seal welding (i.e., continuous welds that provide a barrier to moisture and oxygen). Seals can be achieved in other ways if the exposure is not too severe. In all these cases, the size of the weld can be the minimum permitted by considerations other than stress.

A similar question arises in Fig. 12.20(b), where an I shape made from three plate elements is to be used as a beam subjected to moments, varying along its length due to shear. A free body of the flange in this case reveals the shear flow that must be satisfied at the flange to web connection. The equation for shear flow, q, given in Fig. 12.20(b) should be familiar from study of *strength of materials*. Its dimensions are force per unit length. It is independent of the magnitude of the longitudinal stress, but very directly related to the rate of change in that stress. In most beams the shear flow demand is very low and can be satisfied by intermittent welding with small fillet welds, often less than the minimum weld required in cases like Fig. 12.20(a).

If the beam is to be treated as compact, continuous welds are necessary. Allowable stresses in welds are governed by Table 12.2.

Complete-penetration groove welds. The provisions for allowable stress on *complete (full) penetration* welds should require little detailed explanation. Steel is continuous through the joint. The thickness is not decreased. The smaller strength based on either the base metal or weld metal determines the allowable stress. Full penetration groove welds are the most satisfactory to avoid fatigue, and in some specifications are the only ones permitted where fatigue is a possibility. Values of F_{sr} are increased in AISC Specification, Appendix B, if the crowns of the groove weld are ground flush with the base metal, reducing the notch effect. On this account, details with attached backup plates are objectionable.

If plates of different thicknesses are joined, the strength of the weld is based on the thinner one. If plates of different strengths are joined, the strength of the weld is based on the lower strength. Groove dimensions for prequalified full-penetration groove welds are fully specified in Part 4 of the Manual. Dimensions vary with the welding process. A contractor wishing to qualify other forms of groove must submit evidence of their effectiveness according to AWS procedures.

Partial-penetration groove welds. Some examples of partial-penetration groove welds have already been met in Fig. 12.10(b). The *effective* thickness is the depth of actual penetration. It can be determined by cutting out representative samples of the weld and adjoining material and polishing them to reveal the fused and unfused portions (recall Fig. 12.3). When partial-penetration joint forms are prequalified by AWS, all dimensions of the groove and gaps are specified as well as the welding process. Effective thicknesses are stipulated by AWS based on the results of extensive programs of inspection and testing of sample welds. Table 12.5, reproduced from the AISC Specification, gives effective thicknesses for a few forms of cut and flare grooves. See also Fig. 12.10. Part 4 of the Manual contains a much longer table of prequalified joint forms. Welds can be specially qualified with acceptable test evidence.

Fillet welds. To consider fillet welds, return to Fig. 12.9. In the 45° fillets of Figs. 12.9(a) and (b), there are two critical dimensions of the cross section involved in transfer of shear from member to member. The nominal weld size is the dimension L, a leg of the 45° right triangle. The effective throat for hand (SMAW) welds is $0.707L$. Automatic processes such as SAW have larger effective throats due to greater depth of the fusion zone. See AISC 1.14.6.2. Cases such as Fig. 12.9(c) require special

TABLE 12.5 EFFECTIVE THROAT OF PARTIAL-PENETRATION GROOVE WELDS AND FLARE GROOVE WELDS

Welding Process	Welding Position	Included Angle at Root of Groove	Effective Throat Thickness
Shielded metal arc or submerged arc	All	<60° but ≥45°	Depth of chamfer minus ⅛-inch
		≥60°	Depth of chamfer
Gas metal arc or flux cored arc	All	≥60°	Depth of chamfer
	Horizontal or flat	<60° but ≥45°	Depth of chamfer
	Vertical or overhead	<60° but ≥45°	Depth of chamfer minus ⅛-inch
Electrogas	All	≥60°	Depth of chamfer

EFFECTIVE THROAT THICKNESS OF FLARE GROOVE WELDS

Type of Weld	Radius (R) of Bar or Bend	Effective Throat Thickness
Flare-bevel-groove	All	$\frac{5}{16}R$
Flare-V-groove	All	$\frac{1}{2}R$[a]

[a] Use $\frac{3}{8}R$ for Gas Metal Arc Welding (except short circuiting transfer process) when $R \geq 1$ inch.

Reproduced with permission from AISC Specification, 1978.

geometric study. The allowable stress on the plane through the throat is, according to Table 12.2, 0.3 times F_u of the weld metal. The other critical plane is where the weld metal joins the base metal at the leg L. On that plane, the smaller of $0.3F_u$ for the weld metal or $0.4F_y$ for the base metal is the strength control.

In Fig. 12.12(a), a weld is specified all around the juncture of one pipe to another. It will be recognized that the angle between the two pipe walls and, consequently, the form of the fillet are continuously variable around the joint. In Fig. 12.12(b), a *tee* joint is made with a combination groove and fillet weld. The critical plane must be determined after geometric analysis of the joint and consideration of both the weld and base metals.

Shear strengths of fillet welds are established as noted previously for the case where the longitudinal axis of the weld is in the direction of the shear. Fillet welds perpendicular to the shear direction test at least as strong as (usually stronger than) parallel welds. Assigned values for such welds are the same as for parallel ones.

For steels of commonly used yield strengths with matching weld metals, Table 12.6 shows the strength parameters used in the design of fillet-welded joints.

TABLE 12.6 ALLOWABLE SHEAR IN FILLET WELDS USING MATCHING WELD METAL

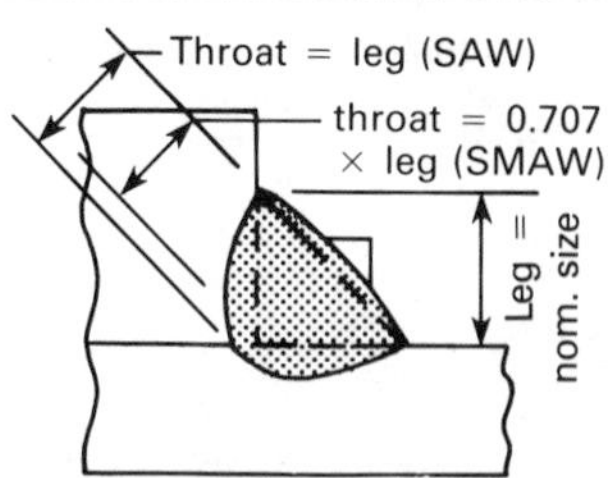

45° fillet
Weld size in fractions of 1 in.
Values in kips per inch per $\frac{1}{16}$ in. of nominal size
Values for shear parallel to and perpendicular to weld are identical
SMAW = shielded metal arc welding
SAW = submerged arc welding
Use smaller of values on base or weld metal

Base metal, F_y, ksi	Matching weld metal		F_v, ksi		Allowable q, kips/in./$\frac{1}{16}$ in. SMAW		Allowable q, kips/in./$\frac{1}{16}$ in. SAW	
	Type	F_u, ksi	Base metal	Weld metal	Base metal	Weld metal	Base metal	Weld metal
36	E60XX	67 min.	14.4	20.1	0.90	0.89	0.90	1.26
	E70XX	77 min.	↓	23.1	0.90	1.02	0.90	1.44
50	E70XX	77 min.	20.0	23.1	1.25	1.02	1.25	1.44
	E7X-EXXX	70–95	↓	21.0		0.93	1.25	1.31
	E70T-X	72 min.	↓	21.6	↓	0.95	1.25	1.34
60	E80XX	75 min.	24.0	22.5	1.50	0.99	1.50	1.40
	F8X-EXXX	80–100	↓	24.0		1.06	1.50	1.50
	E80T	80–95	↓	24.0	↓	1.06	1.50	1.50

Based on AISC Specification, 1978.

Strengths are given per $\frac{1}{16}$ inch of nominal weld size and are applied as force per inch of weld length. It will be seen that the controlling strength is sometimes based on the base metal strength, sometimes on that of the weld metal. The smaller value would control in each case. The AWS Code gives more complete information on matching weld metals.

Minimum weld sizes. Calculations of strength demand of welds often lead to very small required size for butt welds, small enough so that the energy input in welding could be rapidly dissipated through heavy base metals, making for incomplete fusion to the heavy member. One traditional way to avoid poor fusion is to specify minimum weld sizes as a function of the thickness of the plates joined. A weld of large size, if made in a single welding pass, requires a high rate of energy input capable of melting both the thick as well as the thin plate at the weld simultaneously, and achieving fusion to both on cooling. The AWS Code D1.1, which is adopted with some limited exceptions (see AISC 1.17.1) as part of the AISC Specification, also requires control of preheat and interpass temperatures as a function of the type of

TABLE 12.7 MINIMUM SIZES OF WELDS

Material Thickness of Thicker Part Joined (Inches)	Minimum[a] Size of Fillet Weld (Inches)
To 1/4 inclusive	1/8
Over 1/4 to 1/2	3/16
Over 1/2 to 3/4	1/4
Over 3/4	5/16

[a] Leg dimension of fillet welds.

MINIMUM EFFECTIVE THROAT THICKNESS OF PARTIAL-PENETRATION GROOVE WELD

Material Thickness of Thicker Part Joined (Inches)	Minimum Effective[a] Throat Thickness (Inches)
To 1/4 inclusive	1/8
Over 1/4 to 1/2	3/16
Over 1/2 to 3/4	1/4
Over 3/4 to 1 1/2	5/16
Over 1 1/2 to 2 1/4	3/8
Over 2 1/4 to 6	1/2
Over 6	5/8

[a]See Sect. 1.14.6.

Reproduced with permission from tables of AISC Specification, 1978.

steel, the thickness of parts, and the welding process. Table 12.7, which reproduces two tables of the AISC Specification, gives minimum sizes of fillet and partial-penetration groove welds as a function of thickness. The fillet welds so indicated are single-pass sizes.

12.9 WELDED JOINTS

12.9.1 General Remarks

The variety of joints that can be and are made with welds is very large. Most types that can be made with bolts or rivets can be made with welds simply by substituting welding for mechanical fasteners. A number of additional joint forms are possible using direct welding of member to member without intervening splice materials. Thus, welded joints can be less bulky than bolted joints. Under shop conditions and sometimes under field conditions, they are usually less expensive than equivalent bolted connections. The direct, nonslip action of stress transfer through welds is an advantage where slip* of bolted joints is objectionable. It may be a disadvantage or a dangerous procedure where ductile relief of highly restrained points is necessary.

*Slip of bolted joints is discussed in Chapter 13.

12.9.2 Types of Welded Connections: Table 12.8

In Table 12.8 we have shown a number of types of joints made with welds. Many have very similar bolted counterparts, which will be seen in the next chapter. It is necessary for each joint to note separately the type of stress in the member and in the welds.

12.9.3 Axial Member Loads, Types A and B with Example

To transfer a force by fillet welds in shear, we need only determine the *value* of the desired weld in force per inch and, by dividing that into the force, calculate the required length. Thus, if in joint A(1) of Table 12.8, we want to transfer a tension or compression of 100 kips from a 6 × 3/4 in. plate to one 7 × 5/8 in., both of A36 steel, the following calculation would be used:

Minimum fillet weld size = ¼ in. (Table 12.5)
Using E60XX weld metal, and SMAW
Value of weld = 0.9 kips/in./$\frac{1}{16}$ in. (Table 12.2)
For ¼ in. weld, value = 0.9(4) = 3.6 kips/in.
Length of weld required = 27.78; use 28
Alternative 1: ¼-in. fillet × 14 in. each edge
Alternative 2: ¼-in. fillet all around

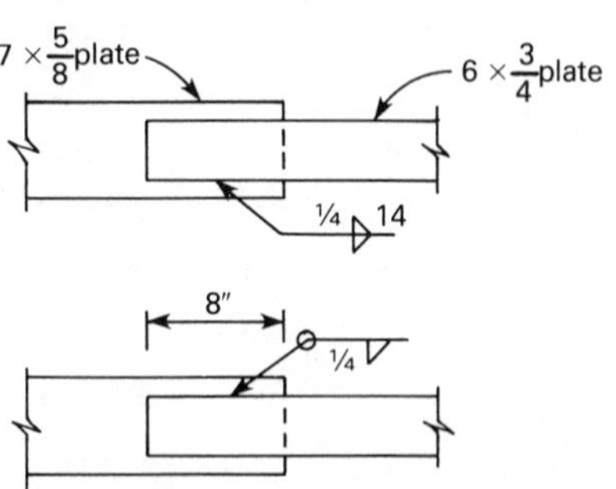

The weld size selected can be made with one pass. Effective costs go up dramatically with multiple passes. Of alternatives 1 and 2, number 2 requires less overlap. However, welding all around a small joint may cause high locked-in (residual) stresses from shrinkage. In the absence of corrosion problems, the author would prefer alternative 1. Alternative 2 can be used successfully with careful attention to preheating and cooling procedures.

Joint A(2) in Table 12.8 is similar to A(1), but has two *faying* (overlapping) surfaces. Its symmetry on the line of action of the force makes it superior, particularly for compressive forces. A(4) employs two splice plates at the flanges to transfer load between two rolled sections. Each end of each splice plate functions like joint A(1). If the force to be transferred is large or the flange splice plates short, it is desirable to splice the web separately. Beyond the connection, the member is stressed in both its web and flanges. If only the flanges are spliced, stress in the web must flow into the flanges before transfer. Variants of A(4) may occur when the members being joined are of different sizes. Two such are shown in A(5) and A(6). A(5) would occur if two W sections of the same nominal size but different overall depths are joined. The fill plates, separately welded to the shallower member, compensate for the difference. Differences of depth up to $\frac{3}{16}$ inch (approximately) do not require shims. Since the inside surfaces usually match, the designer may be tempted to avoid the shims by

TABLE 12.8 ACTIONS OF SOME TYPICAL WELDED CONNECTIONS

Type	Form of connection	Action on members	Action on: Weld	Action on: Splice material	Comments
A	+ OR −				
(1)	ⓐ; Alt. 1; Alt. 2	Tension or compression	Shear	NA	ⓐ Note that one member must be wider than the other for fillet welding.
(2)	Optional; ⓐ; + OR −	Tension or compression	Shear	NA	ⓑ Groove welds, full penetration unless noted. Symbol should reflect form of preparation.
(3)	+ OR −; ⓒ; ⓑ	Tension or compression	Tension or compression	NA	
(4)	Each plate; ⓐ; Optional; Web splice optional	Tension or compression	Shear	T or C	ⓒ If $t_1 \neq t_2$, capacity set by thinner member.
					ⓓ If stress reversal possible, weld must be sufficient for max. tension.
(5)	Each flange; Fill (shim) plate; Splice plate	Tension or compression	Shear	T or C	ⓔ Web connection may be unnecessary.
	ⓐ; Each Flange; 2 Plates; 2 Plates; ⓑⓔ; Each side of web				ⓕ Compression is transferred at fitted surfaces in bearing. See also AISC 1.15.8.
(6)	ⓑⓒ; ⓙ	Tension or compression	Shear and T or C	T or C	
(7)		Tension or compression	T or C	NA	

TABLE 12.8 *(continued)*

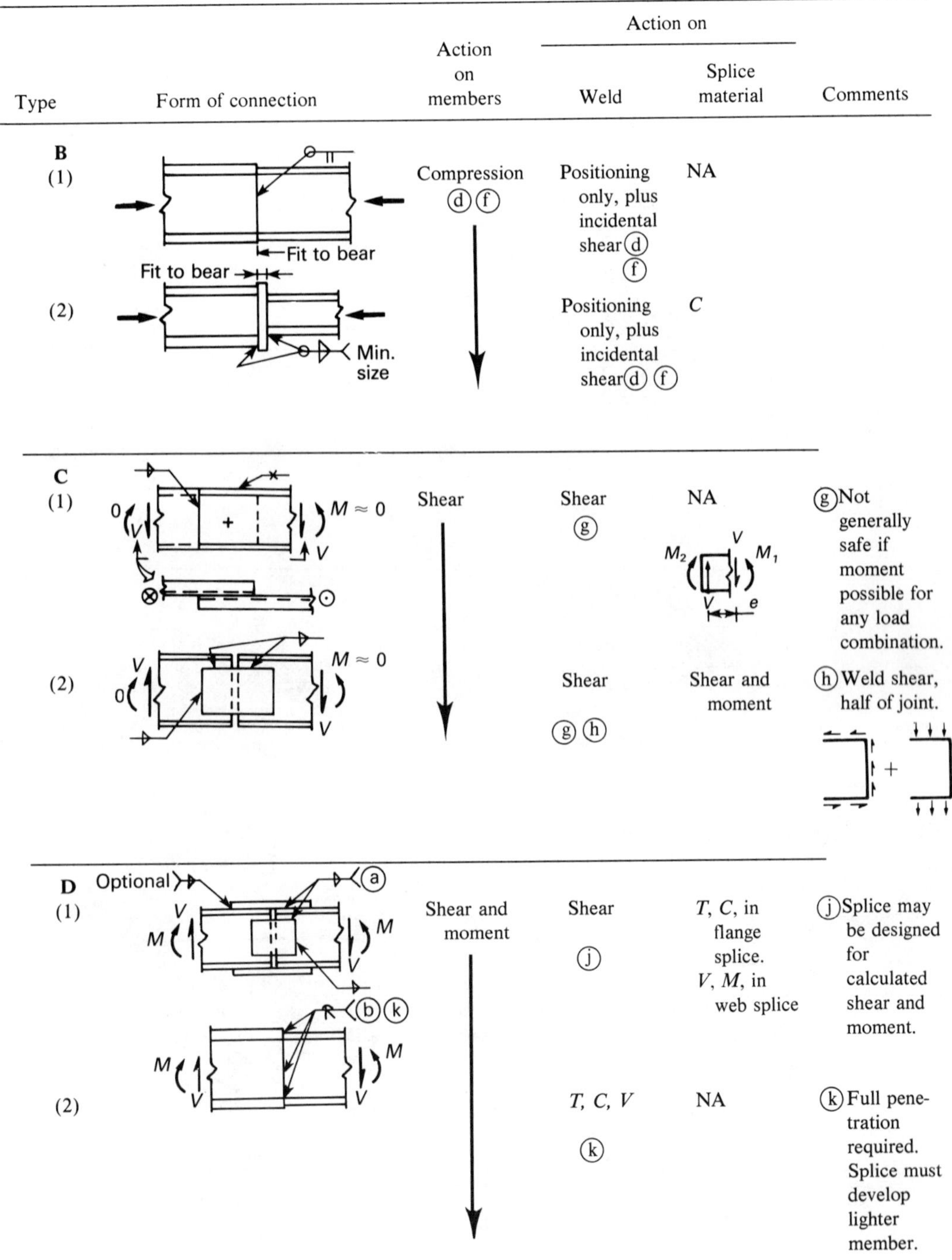

Type	Form of connection	Action on members	Action on: Weld	Action on: Splice material	Comments
B (1)		Compression (d) (f)	Positioning only, plus incidental shear (d) (f)	NA	
(2)			Positioning only, plus incidental shear (d) (f)	C	
C (1)		Shear	Shear (g)	NA	(g) Not generally safe if moment possible for any load combination.
(2)			Shear (g) (h)	Shear and moment	(h) Weld shear, half of joint.
D (1)		Shear and moment	Shear (j)	T, C, in flange splice. V, M, in web splice	(j) Splice may be designed for calculated shear and moment.
(2)			T, C, V (k)	NA	(k) Full penetration required. Splice must develop lighter member.

TABLE 12.8 *(continued)*

Type	Form of connection	Action on members	Action on: Weld	Action on: Splice material	Comments
E (1)	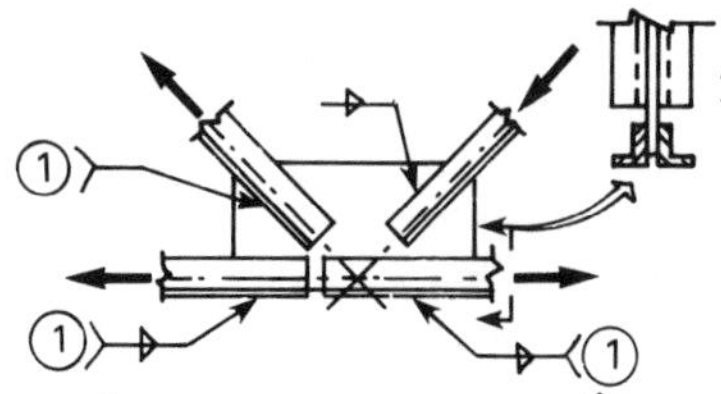 (1)	T or C	Shear (1)	Shear, moment	(1) If fatigue conditions, balance welds for no eccentricity.
(2)	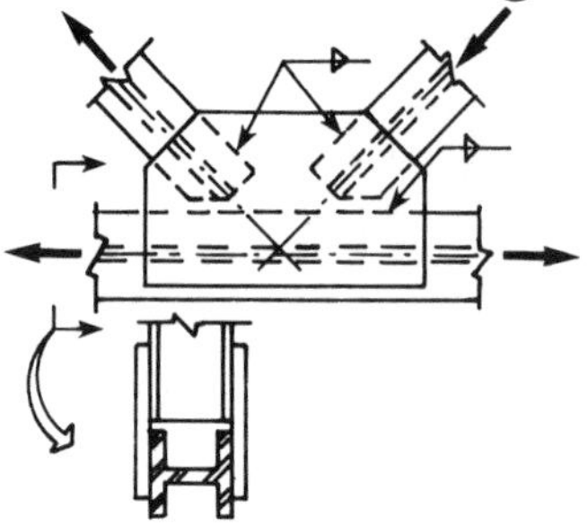				(m) If members in tension, consider lamellar tearing. (n) Stiffener may be optional.
F (1)	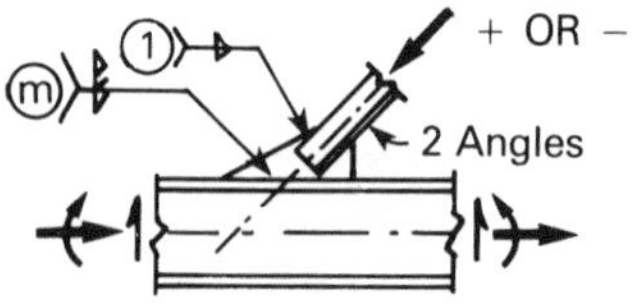 (m) (1) + OR − 2 Angles	T or C on diagonal	Shear to gusset; shear and axial to W	Axial and bending	
(2)	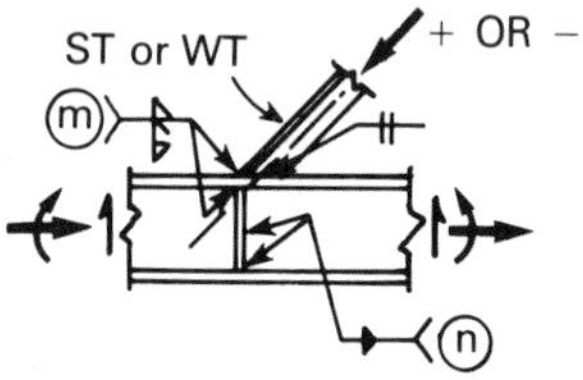 ST or WT + OR − (m) (n)		Shear + axial	(n) Stiffeners 	
G (1)	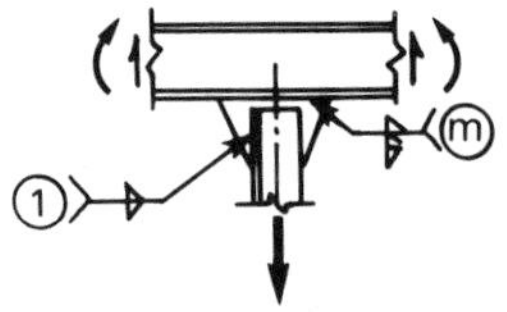 (m) (1)	Tension on hanger (or compression on strut); shear and moment on beam	Shear to gusset; axial to beam	T (or C) if symmetrical on line of load	
(2)	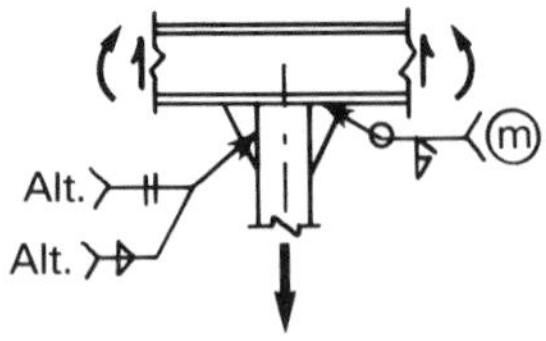 (m) Alt. Alt.		Axial to beam; shear to gussets	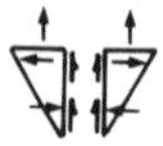	

TABLE 12.8 *(continued)*

Type	Form of connection	Action on members	Action on: Weld	Action on: Splice material	Comments
H (1)	Brackets	Shear and bending + axial	(p) Shear	NA	(p) vert. + hor. + torque
(2)			(q) Shear, axial, as for beam splice	NA	Shear in weld resists all forces and torque. (q) $q_{max} = 1.5q_{ave}$ shear; long'l stress
J	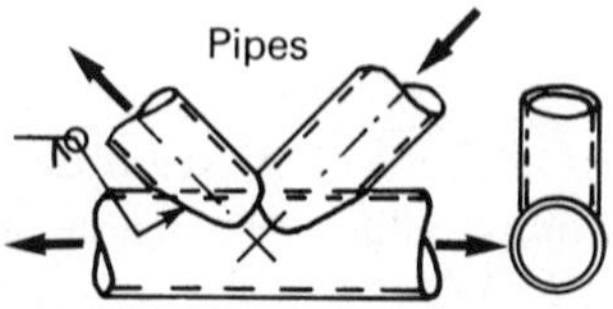 Pipes	Axial, *T* or *C* (r)	Shear and axial	NA	(r) If pipe or tube sizes differ, consider punching shear.
K	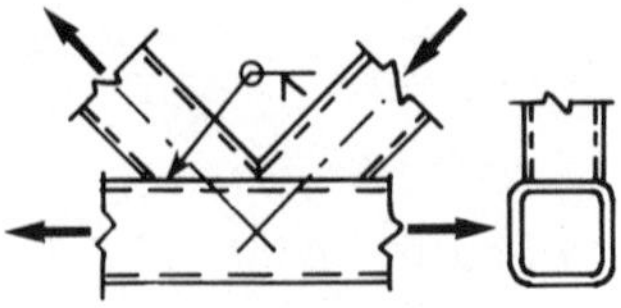	(r) Axial, *T* or *C* (r)	Shear and axial	NA	

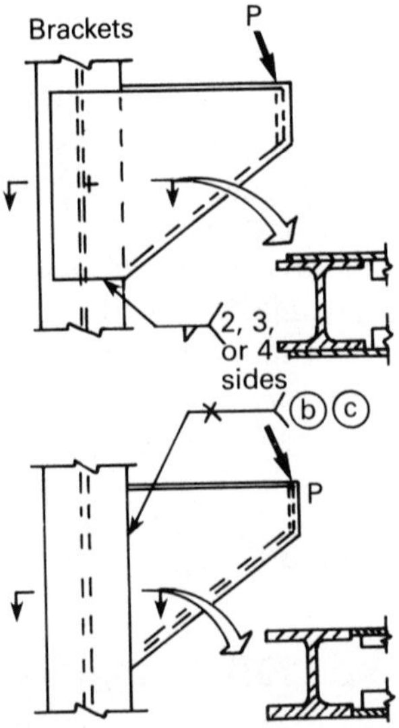

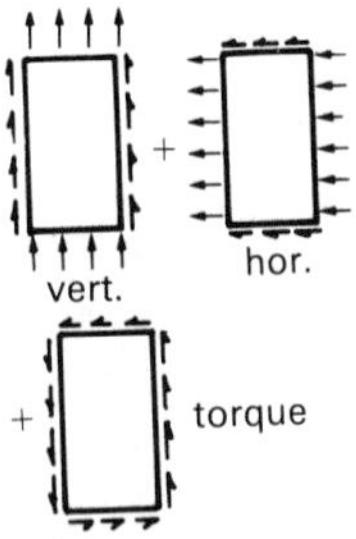

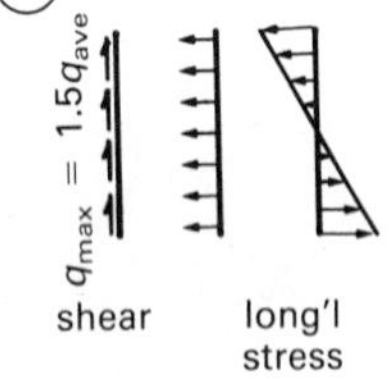

putting the splice plates below the flange. Welding the edge of a splice plate to the flange near the web is difficult, costly, and often ineffective.

Joints A(7) and B(1) have no counterpart in bolted joinery. A bolted joint similar to B(2) is possible with the addition of connecting (clip) angles. In these cases, both member stress and weld stress are axial.

Joints B require only enough weld to maintain alignment. However, AISC 1.15.8 requires, for compression members other than columns, that the welds be capable of transmitting 50% of the computed member load. The possibility of stress reversal must be checked, and the welds must be able to resist any net tension. Net tension may arise if the live load causing compression is removed and another possible load case causes tension sufficient to overcome dead load compression. For this check, to maintain a safety factor, only 75% of the dead load and no live load may be used to offset the effects of tension.

12.9.4 Example: Welded Butt Splice

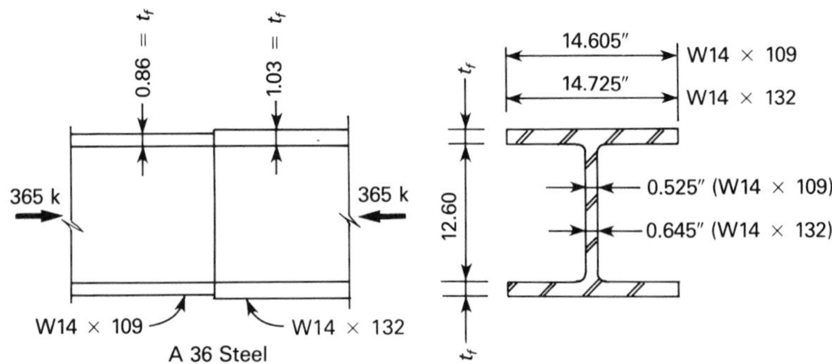

Capacity of W14 × 109 = 600 kips based on its KL/r and AISC Specification. Capacity of W14 × 132 > 600. No load reversal is possible. Dimensions are from Shape Tables. Design a welded butt splice using SMAW.

CALCULATIONS	*REMARKS*
(1) Base splice on 365-k load.	Given
(2) $365/600 = 0.61 > 0.50$.	Satisfies AISC 1.15.7.
(3) $2A_f = 14.605(0.86)(2) = 25.12 \approx 0.79A$	This is lighter member.
$A_w = 12.60(0.525) = \dfrac{6.62}{31.74} \approx 0.21A$	
Total area with fillets = 32.00	Checks sum by Shape Table.

(4) If E60XX electrodes: use $F_a = 0.6F_y$ through weld:

Matching Metal, Table 12.3
$F_y = 36$ ksi for base metal

$$\text{Req'd } A_{\text{EFF}} = \frac{365}{21.6} = 16.90 \text{ in.}^2$$

$$\frac{(A_{\text{EFF}})_R}{A} = \frac{16.9}{32} = 0.53$$

See Table 12.2.

Develop:

Alt. 1: 53% of each of web and flanges

Alt 2: 53% of area $= \dfrac{0.53A}{0.79} = 0.67A_f$

Alt. 2 assumes no web splice.

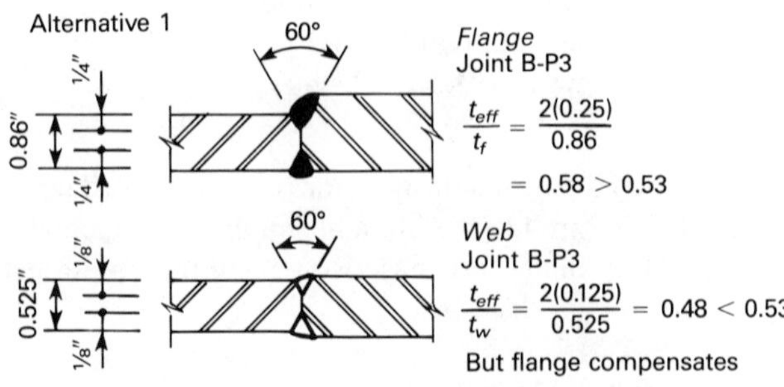

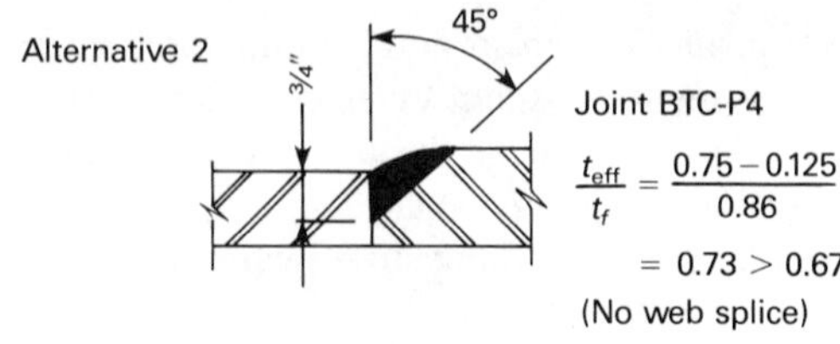

See AISC Manual, Part 4, for Joint Forms and Values of t_{EFF}. Prequalified, partial penetration, SMAW.

Either alternative is acceptable. Alternative 2 requires more welding, but also more preparation work. Full penetration welding would be overkill if 365 kips is the maximum feasible load on the member and there is no significant bending moment. This would not be true if loads were tensile and cycling.

12.9.5 Splices for Member Shear, No Moment

Joints C(1) and C(2) of Table 12.8 are intended to resist shear forces only, with no bending moment. A caution is necessary. The welded joint is rigid, allowing no relative rotation around the axis perpendicular to the plane of the joint. (Angle θ between the members is 0 at the joint.) This is a condition of full continuity. If a stiffness analysis for any load combination shows significant bending moment, a shear connection is likely to fail. (In the corresponding bolted case, slip and ductility at the bolts may permit a small angle of relative rotation, preventing large bending moment.) A prudent rule would be, don't design a rigid welded connection for shear alone if there is a chance that bending moment may occur under any circumstances.

In joint C(2), even if the members have no bending moment, a small moment exists in the splice plate. The weld pattern also must resist torque as well as transverse shear. The weld pattern on each half of the joint is three sided. Unlike the rolled channel, which it resembles, all three legs of the weld resist both the transverse shear and the torque. The calculation requires the vector addition, at each point along the weld, of the weld shear from transverse shear and that from twisting moment.

$$q = \frac{V}{L} \pm \frac{Ver}{J} \qquad \text{(added vectorially)}$$

where q = shear flow, in force/inch, L is the total length of weld
r = distance from the centroid of the weld pattern to a point on the weld
J = polar moment of inertia of the weld pattern $= I_x + I_y$
q_{max} occurs at the position of r_{max}

The calculation that follows for joint H(1) will illustrate an elastic procedure for determining the required weld. Tables in Part 4 of the Manual simplify the calculation for this and other standard cases. The tables are based on an ultimate strength procedure.

12.9.6 Splices for Shear and Bending Moment

Splices D(1) and (2) of Table 12.8 transfer both shear and bending moment between members. They work essentially as the members themselves do, with most moment being transferred through the flange splice and shear through the web splice. D(1) must be designed for at least the maximum calculated shear and moment. For simple cases it is sufficient to design a splice plate for each flange to transfer, as in joint A(1), a tensile or compressive force of

$$T \quad (\text{or } C) = \frac{M}{d}$$

The web splice is designed as for joint C(2), but in this case with no special concern for the possibility of moment overstress.

Butt splices, such as D(2), for beams and girders must, according to AISC 1.10.8, develop the full strength of the lighter of the two sections joined. This requires full-penetration welds with properly matched weld metal.

12.9.7 Example: Beam Splice, Shear and Bending

We are to splice a W24 × 104 to a W24 × 68. Steel is ASTM A 588. At the location of the splice,

$$V_{max} = 110 \text{ k}$$

$$M_{max} = 350 \text{ ft k}$$

The desired splice is to use lapped splice plates as in Table 12.8, D(1). The W24 × 68 has been checked previously for V and M and is OK.

CALCULATIONS	*REMARKS*

(1) Sketch joint:

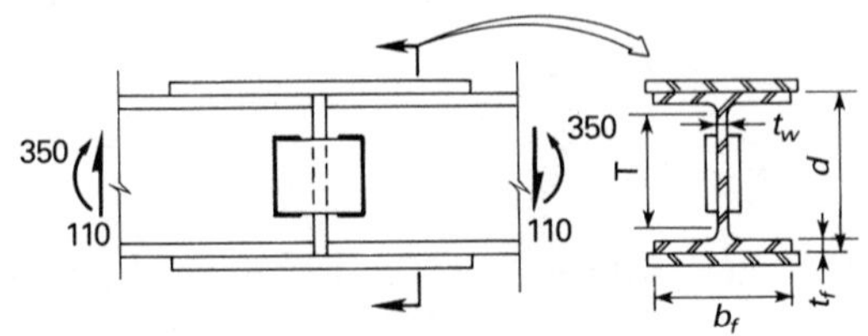

(2)

Dimensions	W24 × 104	W24 × 68
b_f	12.75	8.97
t_f	0.75	0.59
d	24.06	23.73
t_w	0.50	0.415
T	21	21

Shape Tables. Note d dimensions are close enough so that fills can be avoided.

(3) Select flange splice plates:

$$T = C = \frac{M}{d} = \frac{350(12)}{24} = 175 \text{ k}$$

Assumes flange splices transfer all of M. Assumes no reduction necessary for unbraced compression length. Note $A_R > A_f b_f$. Why is this not objectionable?

Size flange plates for $f \leq 0.60F_y$:

$$A_R = \frac{175}{0.6(50)} = 5.83 \text{ in.}^2$$

Use 10 × ⅝ plate; $A = 6.25$.

Must be wider than flange for fillet welding; also narrower than larger flange.

(4) Weld required:

Use $\frac{5}{16}''$ weld, E70XX, SMAW:

Table 12.6. Conditions do not warrant SAW.

$$\text{value} = 5(1.02) = 5.10 \text{ k/in.}$$

$$\text{length required} = \frac{175}{5.10} = 34.3 \text{ in.}$$

Use 36.

(5) Select splice plates for web:

$$\frac{V}{F_v} = \frac{110 \text{ k}}{0.4(50)} = 5 \text{ in.}^2$$

Weld required, each side of web:

$$\frac{50}{\text{value}} = L$$

If $\frac{3}{16}''$ fillets with ¼″ plates, — See AISC 1.17.3, Table 12.6

value = 3(1.02) k/in.
required $L = 16.50''$
Use 2 plates 3 × 16 × ¼″; $A_v = 8.0 > 5$.

$$L = 16 + 2(1) = 18'' \text{ each plate}$$

$$16'' < T = 21''$$

Size determined by required weld. Narrow plate used to minimize torque on web splice.

(6) Draw joint:

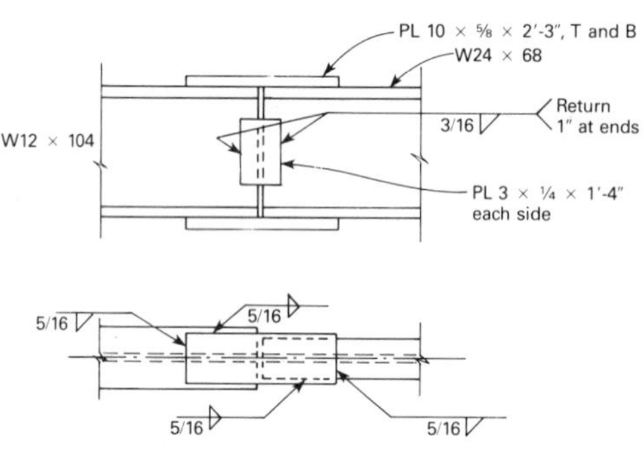

12.9.8 Gusset Plates and Tee Joints

Joint E(1) in Table 12.8 is typical of trusses with double-angle members, which were met earlier in Chapters 8 and 9. A *gusset plate* inserted between the parallel legs is used to splice all the converging members together. The double fillet weld symbols in E(1) indicate the toe and heel weld on the near angle. By industry convention (see Fig. 12.13, footnotes), the far side angle takes the same weld. Welds must be sufficient to transfer the maximum calculated force in each member or (AISC 1.15.7), if that is greater, at least 50% of "the effective strength of the member," which would be different if the member is evaluated for compression or for tension. Note that all member centroidal axes meet. If they do not, the eccentricity of forces within the joint causes bending moment in the members.

A similar gusset plated joint, E(2), joins I-shaped members of the same depth. In this case, the chord member is shown to be continuous, a common situation. The welds between chord and gusset must transfer only the difference in tension in the chords on each side, which is the sum of the horizontal components of the forces in the two diagonal web members. ΔT may be very small. The symbol indicates intermittent welding, which must satisfy at least that force, but also be spaced close enough to maintain close contact on the faying surfaces.

The shape and size of gusset plates are determined by the length of welds necessary to make all force transfers and by the demands of simplicity for shop operation. Each joint is studied separately. If a reasonable minimum thickness is established for all gussets, little further study is needed. However, the designer should realize that gusset plates, as well as other members, are subject to overstress and buckling. A free-body analysis of the gusset plate itself can be used to determine internal stresses.

Joints F and G in Table 12.8 are variants of joints E for connecting two members in a *tee** configuration. F(2) needs no gusset but may need stiffeners if the portion of load in the outstanding leg of the T-shaped diagonal is to be transferred directly to the beam flange. A study of transverse bending of the flange will reveal the problem. The gussets in F(1), G(1), and G(2) are shown symmetrical and flaring to ease stress flow.

12.9.9 Examples: Tee Configured Joints

Two examples are shown, illustrating types F(1) and F(2). Type E joints are similar to F(1), but collect more members.

(a) *Double angle knee with gusset to column*: Design a gusset plate and weld to ⅂⌈ and column. The solution will balance the welds so that the resultant resistance is

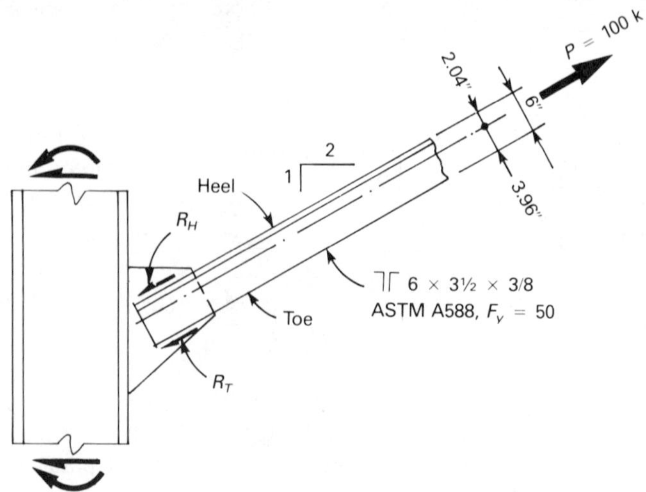

*This is a slightly different use of the word *tee* from that used earlier to describe two plates welded into the form of a T. The word is also used to describe ST and WT rolled sections.

on the axis of the ⊓⊏. The CG of the gusset will be on the member axis at its juncture with the column.

CALCULATIONS	*REMARKS*
(1) Size of fillet welds and value, k/in.:	
Min = 3/16″	Table 12.7
Max = 5/16″	AISC 1.17.3
Use ¼″ for single-pass weld	
Welding rods, E70-XX, SMAW	
Value = 4(1.02) k/in.	Table 12.6
(2) Length of weld, each L:	
$$\frac{50}{4.02} = 12.44$$	
(3) Percent of required weld to heel and toe:	
0.66(12.44) = 8.21; use 9″	$\Sigma M_0 = 0$
0.34(12.44) = 4.73; use 4½″	
(4) Detail:	The welds to ⊓⊏ have been calculated. The gusset-to-column weld shears on the gusset and pulls on the column flange. Symmetry keeps weld stresses uniform. The full-penetration weld form (Manual Part 4) with reinforcing fillet improves stress flow. The connection is in the Z direction of the column flange (see Fig. 12.17). Ultrasonic testing would be useful.

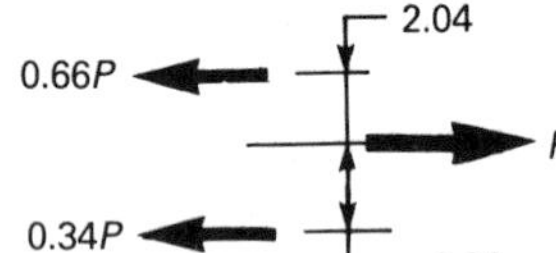

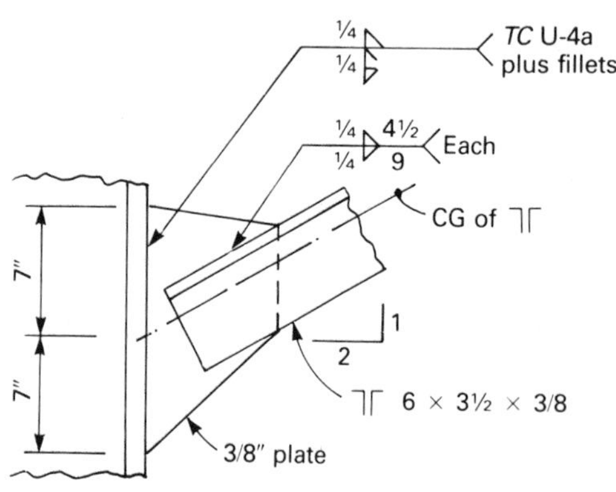

(b) *WT knee to column*: Figure 12.21 shows a connection similar to the previous example where the knee is a WT. Gusset plates do not work easily with WT's. Numerical work is not included. The shaded area in Fig. 12.21(c) can transmit force directly without cross-bending of the legs of the column flange. If that is not adequate, stiffeners are necessary as indicated.

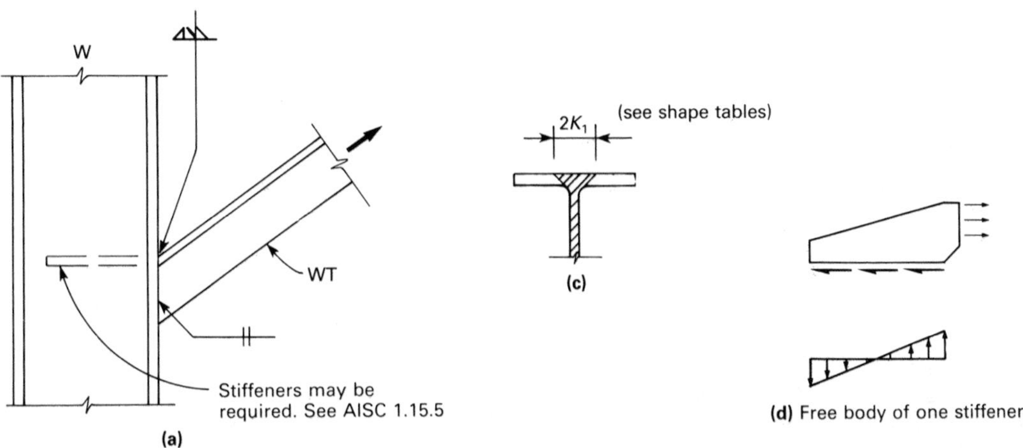

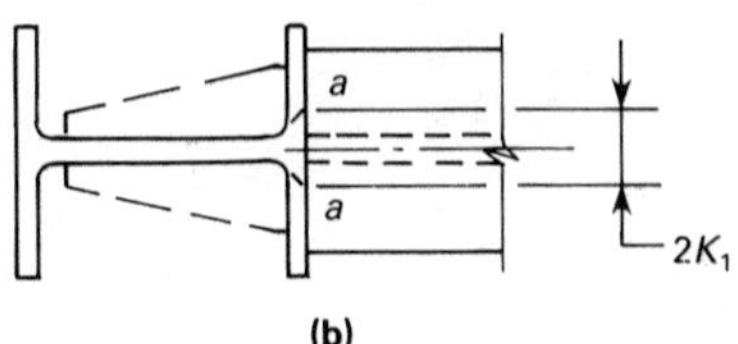

Figure 12.21 Connection, structural tee to column.

12.9.10 Eccentric Load on Weld Pattern

Before attempting an example of a bracket such as in joint type H(1), we should examine the mechanics of elastic resistance of a weld pattern in shear that is eccentrically loaded. A simple, four-sided, doubly symmetric instance is illustrated in Fig. 12.22. The load P acts [Figs. 12.22(a) and (b)] in a plane parallel to the weld pattern, but eccentric to the center of resistance. The reaction forces and torque are noted on Fig. 12.22(a). A pattern of welding is shown in Fig. 12.22(c) of length $L = 2(b + d)$. The equations for I_x and I_y are given in Fig. 12.22(c). They should be familiar except, possibly, for one thing. Since the weld pattern is a line without width, it has the dimensions of length, and the moments of inertia have the dimensions, $(\text{length})^3$.* Polar moment of inertia, J, is also given.

The shear flow resisting the loads is given in Fig. 12.22(d). Note that all four sides resist each component of force. q is uniform and consists of the vector sum of

*If a unit width were used for the weld, the equation would be multiplied by 1 and I would have dimensions, $(\text{length})^4$ as usual.

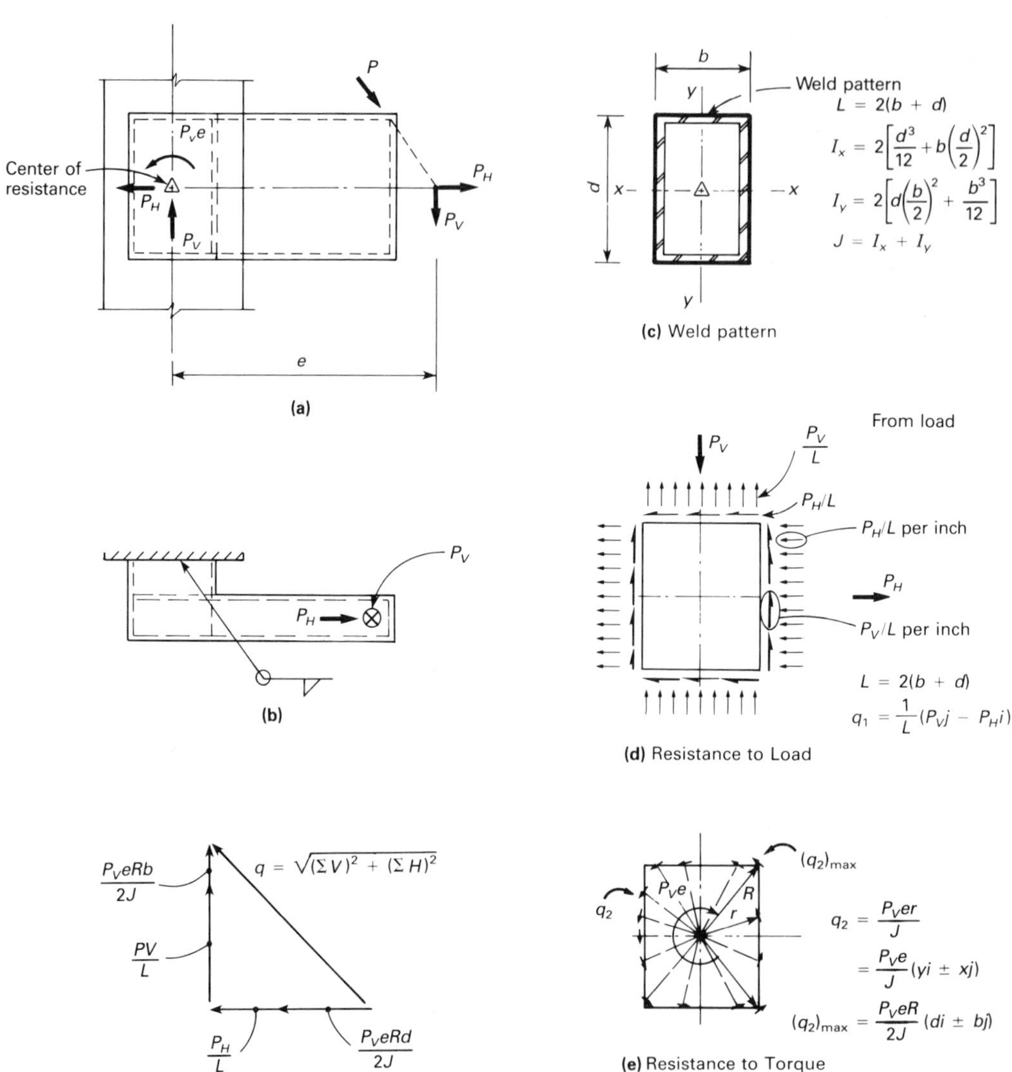

Figure 12.22 Eccentrically loaded fillet welds.

horizontal and vertical components. The shear flow that resists torque, as shown in Fig. 12.22(e), is designated as q_2.

Recall from the study of *strength of materials* the equations for shear stress on a circular shaft, subjected to a torque, T,

$$f_v = \frac{T\rho}{J} \quad \text{and} \quad (f_v)_{\max} = \frac{TR}{J}$$

where ρ is any radius from the central axis and R is the outer radius. The shear stress is perpendicular to a radius and proportional to the distance from the central axis. J is the polar moment of inertia, $2I_x$. The equivalent equations in the case of a weld pattern are the equations in Fig. 12.22(e), where J, as in Fig. 12.22(c), is $I_x + I_y$, with the dimensions of $(\text{length})^3$. Shear flow, q_2, is perpendicular to the radius from the centroid to any point, is proportional to the radius and maximum at the corners, and has the dimensions of force per unit of length.

Adding vectorially as in Fig. 12.22(f),

$$q = q_1 + q_2$$

$$q_{\max} = q_1 + (q_2)_{\max}$$

$$= \left[\left(\frac{P_H}{L} + \frac{P_v eRd}{2J}\right)^2 + \left(\frac{P_v}{L} + \frac{P_v eRb}{2J}\right)^2\right]^{1/2}$$

If b and d are known (or the dimensions and centroid of any other pattern of welding), the required weld size may be calculated to satisfy $q_{\max}$.

The method described is presented in somewhat different form in the AISC Manual, Part 4, in discussion preceding AISC Tables XIX–XXVI. The tables themselves are based on an ultimate strength method, also presented, which is somewhat less conservative. The tables give, for several standard weld patterns, coefficients to be used in determining the required lengths of weld or weld size for a given eccentric load.

12.9.11 Example: Weld Bracket to Column, Joint Type H(1)

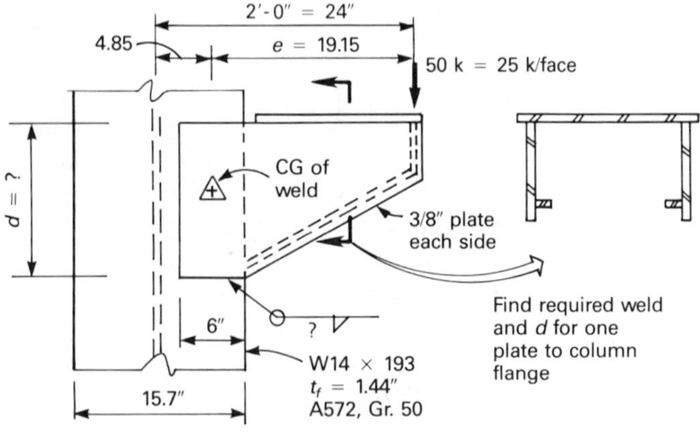

CALCULATIONS		REMARKS
Min. weld to column $= \frac{5}{16}''$	Use $\frac{5}{16}$	Table 12.7
Max. weld to bracket PL $= \frac{5}{16}''$		AISC 1.17.3

$$V = 25 \text{ k}; \frac{Pe}{2} = M_T = 25(19.15) = 479 \text{ in. k}$$

Use E70XX electrodes for SMAW

Weld shear value = 1.02 k/in./$\frac{1}{16}$″ — Table 12.6

$$q_w = 5(1.02) = 5.10 \text{ k/in.}$$

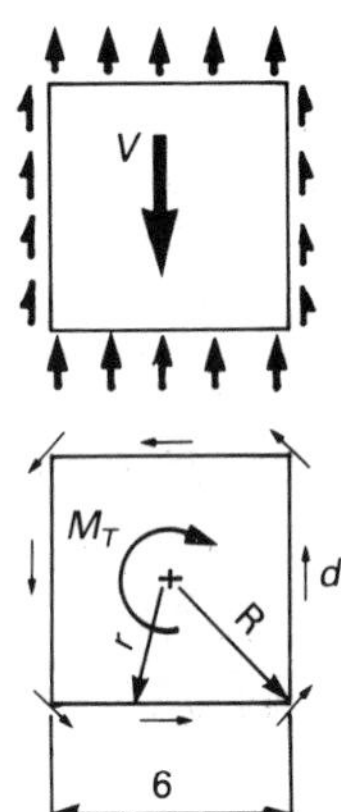

$$q = (q_H^2 + q_v^2)^{1/2}$$

$$V = \frac{P}{2} = 25$$ Effect of shear

$$q_v = \frac{25}{2(6+d)} \text{ k/in.} = q_{v1}\uparrow$$

$$R = r_{\max} = \sqrt{3^2 + \left(\frac{d}{2}\right)^2}$$ Effect of torque

$$M_T = 479 \text{ in. k}$$

$$q_M = \frac{M_T}{J}(x\uparrow + y\rightarrow)$$

$$(q_M)_{\max} = \frac{M_T}{J}\left(3\uparrow + \frac{d}{2}\leftarrow\right)$$

$$= \frac{M_T}{J}(q_{v2}\uparrow + q_H\leftarrow)$$

$$J = I_x + I_y = 2\left[6\left(\frac{d}{2}\right)^2 + \frac{d^3}{12} + d(3)^2 + \frac{(6)^3}{12}\right]$$

To solve for q in k/in., J must have dimensions of in.^{3}, not in.^{4}. Thickness of weld dimensionless.

$$J = \frac{d^3}{6} + 3d^2 + 18d + 36$$

d	J	q_{v1}	$(q_M)_{\max}$ $q_{v2}\uparrow$	$(q_M)_{\max}$ $q_H\leftarrow$	$q_{\max}$
7.5	410	0.93	3.50	4.38	6.23
10.0	683	0.78	2.10	3.51	4.54
12.5	1055	0.68	1.36	2.84	3.50

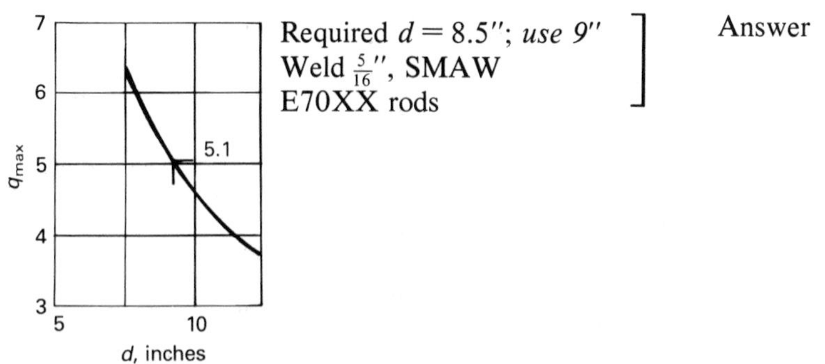

Required $d = 8.5''$; *use 9''*
Weld $\frac{5}{16}''$, SMAW
E70XX rods

Answer

The solution was done with a four-sided welding pattern. To weld the right-hand vertical leg may be awkward and costly. The same procedure could be used with a three-sided pattern (top, bottom, and left leg). The x position of the CG of the weld would vary slightly with d and, of course, d would become longer. One might worry about the long unwelded contact surface at the toe of the column flange. Two or three small plug welds made from the outside could put that worry to rest.

12.9.12 Example: Butt Welded Bracket

In bracket type H(2), the side plates are groove welded to the toe of the column flange. The analysis is much simplier than the previous one. Note q in Table 12.8 shows the pattern of stress, which is the same as that for shear, moment, and axial load on a rectangular beam. Figure 12.23 gives the simple calculation necessary for conditions similar to the previous example.

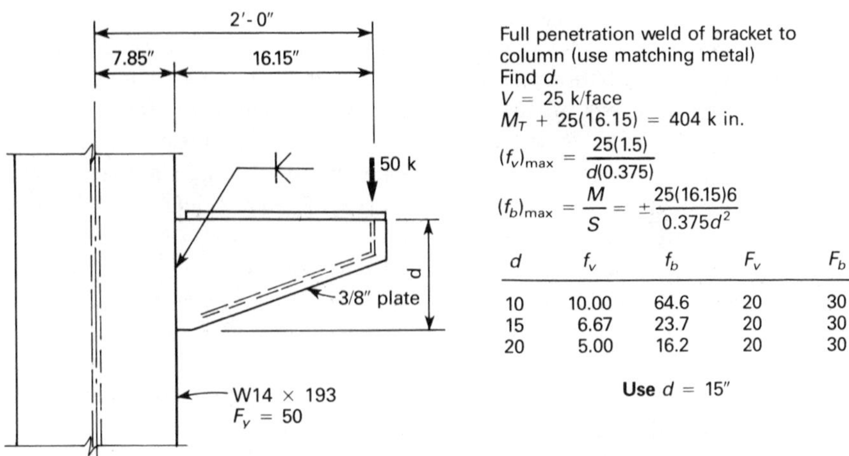

d	f_v	f_b	F_v	F_b
10	10.00	64.6	20	30
15	6.67	23.7	20	30
20	5.00	16.2	20	30

Figure 12.23 Eccentrically loaded butt welds.

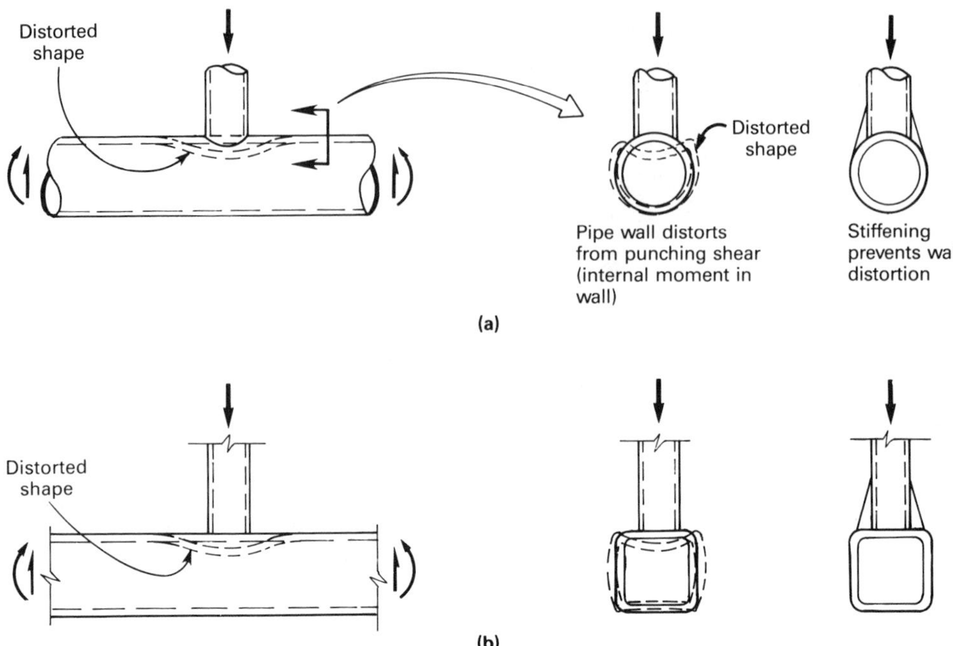

Figure 12.24 Tubular joints; wall bending with mismatched sizes: (a) pipes; (b) structural tubes.

12.9.13 Joints: Tubular Members

Joints J and K in Table 12.8 do not have counterparts in bolted joinery. The widespread use of pipes and rectangular tubes in structures followed the advent of welding and probably was not possible without it. Pipes and large round tubes are now used widely in offshore structures where their form reduces hydraulic drag in a wave train. Both round and rectangular tubes offer esthetic advantages in other exposed structures, as well as great torsional strength and stiffness and stability against buckling. The method of joinery in this case opens up great opportunities in member selections.

In both cases, it is necessary to choose the relative sizes of members to be connected carefully in order to avoid a problem general to tubular joints, that of punching shear. The problem in its simplest form can be seen in Fig. 12.24, where small tubes are shown attached perpendicular to large tubes. Transfer of load from the small tube to the large is accompanied by deformation and bending stress in the wall, threatening the geometric integrity of the large tube and its strength. One way to avoid the problem is to use tubes of equal or almost equal sizes. If that is not feasible, the tube wall bending can be avoided by stiffening the tube internally or externally, as shown in the figure (internal stiffening is usually feasible only for very large tubes).

Requirements for the design of tubular joints and their welding are given in AWS D1.1-80.[17] The AISC Specification offers no direct guidance.

12.10 BEAM END CONNECTIONS

12.10.1 Flexible and Rigid Connections

Beam end connections occur so often in all kinds of structures that they influence costs very strongly and have attracted a great deal of attention from design engineers and researchers. The result is a great variety of forms that can be executed safely and considerable difference of opinion about cost. To discuss end connections of beams, it is necessary to consider the range of assumptions made in frame analysis regarding these connections. The AISC Specification (Section 1.2) provides for three different types of framing, which relate to the end connections of beams to columns. By the AISC classification system*:

Type 1, or *rigid frame*, joints are sufficiently rigid to essentially prevent any change of angle between the members in the joint. Perhaps a better term would be *fully continuous joints*.

Type 2, or *simple framing*, uses beam end connections essentially free to rotate under gravity loading without mobilizing moment resistance to rotation. Connections are made to transfer the end shear only.

Type 3, or *semirigid framing*, is based on the assumption that the end connections have dependable and known moment capacity "intermediate . . . between Type 1 and Type 2."

The assumption of Type 1 joints is consistent with the elastic methods of statically indeterminate analysis where full continuity is often assumed at some or all joints. End moments emerge from analysis and may be anywhere from zero to and beyond the "fixed end moment" (which assumes zero rotation) and may sometimes reverse in direction, particularly when gravity and lateral loads are combined. Full continuity does not assume zero rotation of the joint (the fixed end condition), but equal rotation of all members at the joint. Fully continuous joints are difficult to achieve with bolted end connections, but can be approximated fairly closely if sufficiently stiff elements are used. Continuous welded joints are easier to achieve and much less bulky in form.

One positive way to achieve a Type 2 joint is with a single shear pin at the beam end. However, if connecting angles are made light enough to permit end rotation of the beam, it is possible, at much less cost, to transmit shear effectively with welded or bolted end connections of members designed for zero end moment.

*Three different framing types are used in Part 1 of the Specification, which relates to elastic design. Part 2, the section on Plastic Design, requires that all joints with minor exceptions be Type 1 joints.

The author has difficulty justifying semirigid joints based on the assumptions of the AISC Specification, which seem to require that a stipulated capacity be provided that may be less than the elastic moment that the joint is called on by analysis to resist. Semirigid joints are possible and presumably safe if connecting elements permit angle changes between the members of a predictable amount intermediate between simple beam end rotation and the zero angle change required for fully continuous joints. This requires a great deal of sophistication beyond the purposes of this book. Further discussion will deal with *continuous* joints (Type 1) and flexible joints (Type 2). It will deal with the nature of the joints, not the frame, since a frame may have a mixture of the two types.

12.10.2 Flexible Beam End Connections

Figure 12.25(a) shows a simple span beam supporting a uniformly distributed load and its deflected shape, which requires that end rotation θ take place without mobilizing any external resisting moment ($M_A = M_B = 0$). From familiar slope-deflection relations, it will be recalled that

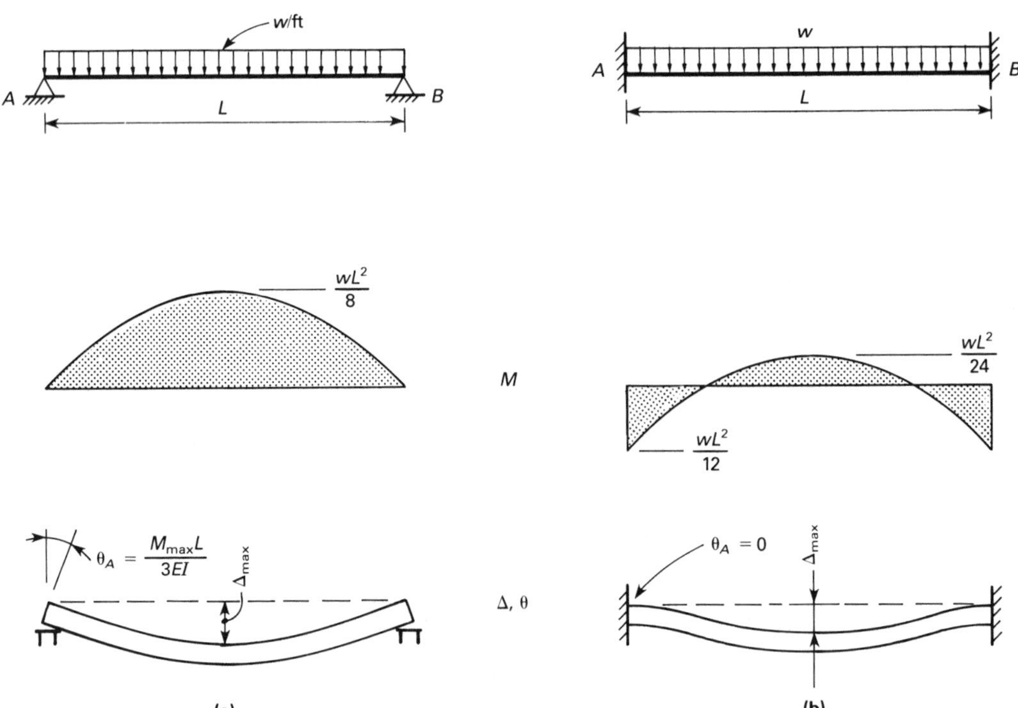

Figure 12.25 Rotation, simple and fixed spans: (a) simple span, uniform load; (b) fixed ends, uniform load.

$$\theta_A = \int_0^{L/2} \frac{M}{EI}\, ds$$

and is proportional to the M/EI area for half of the beam. For the common case shown,

$$\theta = \frac{2}{3} \cdot \frac{M_{\max}}{EI} \cdot \frac{L}{2} = \frac{M_{\max} L}{3EI}$$

$$\Delta = \frac{5}{48} \frac{M_{\max}}{EI} \cdot L^2$$

which, after some rearranging, leads to

$$\theta = \frac{48}{15} \frac{\Delta}{L}$$

Commonly, beams are proportioned to make

$$\frac{\Delta}{L} \leq \frac{1}{240}$$

which would make

$$\theta \leq \frac{48}{15(240)} = 0.01333 \text{ rad}$$

equivalent to an end slope of 0.16 inch per foot.

The fixed-end case shown in Fig. 12.25(b) has zero end rotation and at its end a negative moment equal in absolute magnitude to two-thirds of $M_{\max}$ for the simple span case. To reduce that end moment to zero requires the amazingly small rotation noted previously.*

Figure 12.26(a) shows an end connection of a beam to a column flange made with *clip* angles. If the distance from the bottom of the angles to the uppermost bolt is, say, 12 inches, the upper portion of the clip angles must displace slightly over 0.16 inch to provide full relief of moment. As can be seen from Figs. 12.26(b) and (c), a sufficiently light angle can do this if its outstanding leg yields in bending. Such yield is not alarming, since it stops when simple span rotation is reached. To proportion a connection of this type:

- Use thin angles.
- Use long outstanding legs.
- Use length of clip angles close to the minimum required to transfer the end shear through the weld.

*A corollary is that a very small relative rotation will invalidate an analysis based on full continuity.

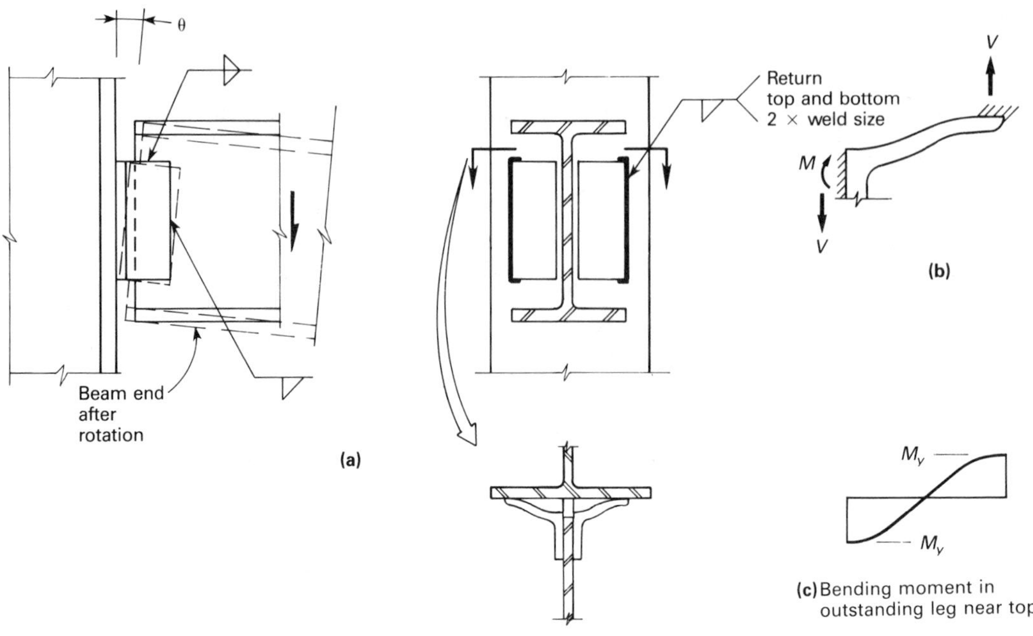

Figure 12.26 Flexible joint.

Making the connection stiffer may very well defeat the intent to provide flexibility. However, high shear strengths are achievable. Figure 12.27 shows three other types of flexible end connection. Tables IV, VI, and VIII, Manual, Part 4, give standard clip angle dimensions and corresponding shear capacities for the connections shown in Figs. 12.26 and 12.27.

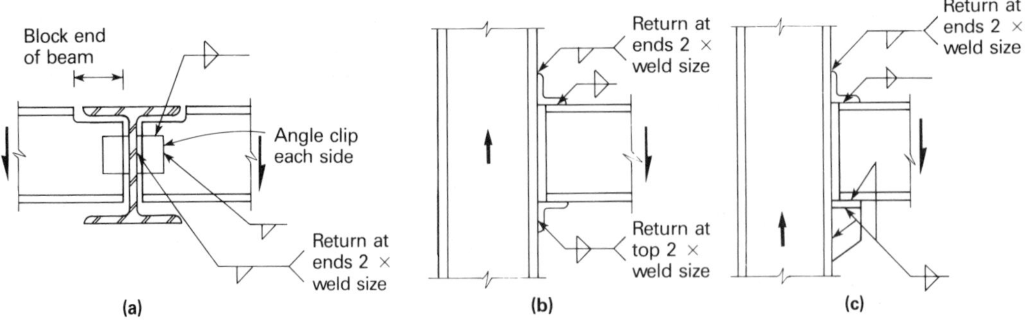

Figure 12.27 Flexible joints.

12.10.3 Example: Flexible Connection, Beam to Girder

In the analysis of a five-story tier building in Chapter 18, the following data applies to typical simple span beams (B1), not part of the primary ductile moment resisting frame.

$$\text{Load:}\quad D + L = 8'(0.2) = 1.6 \text{ k/ft}$$

$$\text{Seismic} = 0$$

$$\text{Span} = 25 \text{ ft}$$

$$\text{End moment} = 0$$

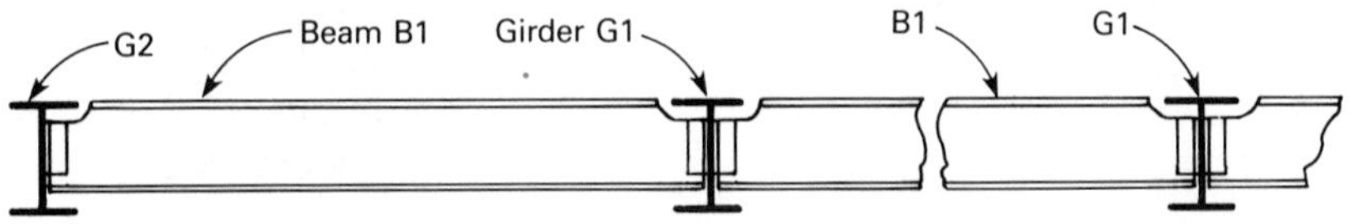

B1: W16 × 40; A36
G1 and G2: W18 × 97; A572, Gr. 50

There are 225 such identical beams in the building. The designer had the choice of making the beams continuous and using lighter beams because of reduced bending moment, or using flexible connections and heavier beams. He opted for simple spans, judging that the inexpensive simple connections would save more than the cost of the extra steel. (This decision is debatable.) Design a flexible connection to satisfy these conditions.

CALCULATIONS	*REMARKS*
(1) Calculate end shear:	
$V = \dfrac{1.6(25)}{2} = 20 \text{ k}$	From problem statement.
Allowable shear on W12 × 40:	
$V_{ALL} = 51.1 \text{ k} \gg 20$	Beam tables, Manual, Part 2.
Design connection for $V = 30$ k.	Allows for some overload.
(2) Sketch connection:	
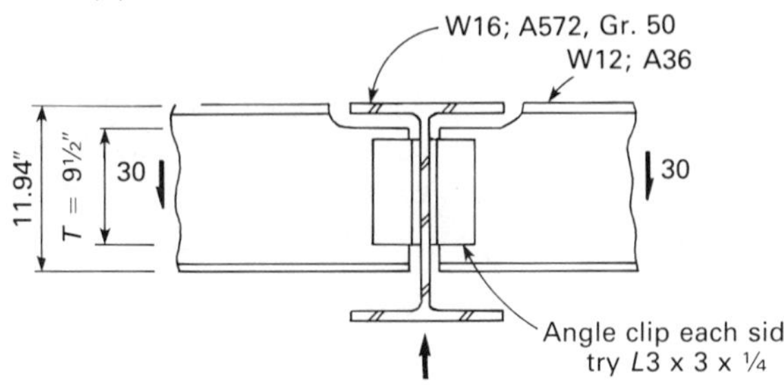	Dimensions from shape tables. T distance is flat part of web between fillets.

(3) Required weld:

(a) Maximum size fillet = 3/16″ E70XX electrodes, SMAW Allow $q = 3(0.9) = 2.7$ k/in. — AISC 1.17.3. Table 12.6.

(b) Weld to girder web, each L:

$$L \geq \frac{30}{2(2.7)} = 5.56''; \text{ Try } 6'' < 9\tfrac{1}{2}$$

(c) Weld to beam web:

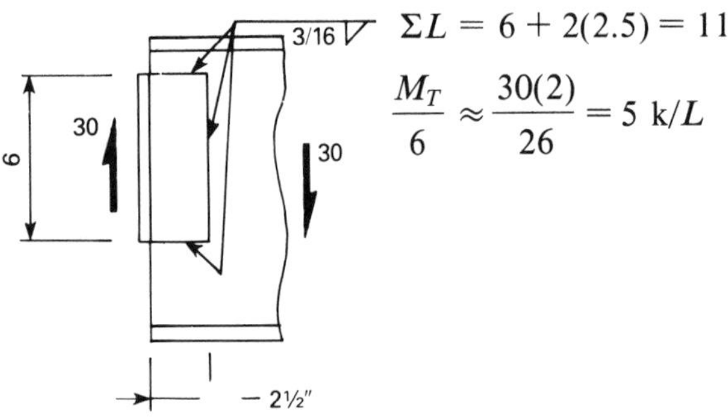

$$\Sigma L = 6 + 2(2.5) = 11$$

$$\frac{M_T}{6} \approx \frac{30(2)}{26} = 5 \text{ k}/L$$

Weld is checked for shear and torque. This is an approximation on the safe side, assuming all torque resisted by top and bottom welds.

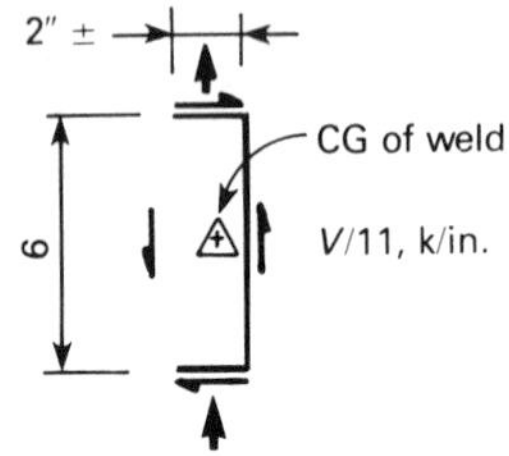

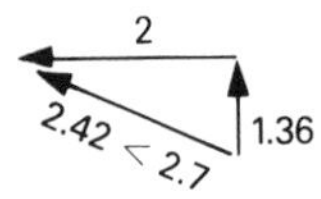

(4) Check outstanding leg of angle for flexibility.

See discussion, Section 12.10. $V_L = 5$ k as above; but V_{L1} cannot be larger than force to cause outstanding leg to yield.

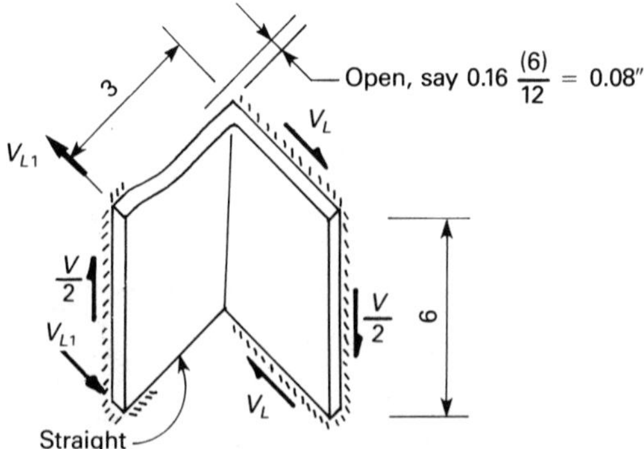

Required force at top to yield L:

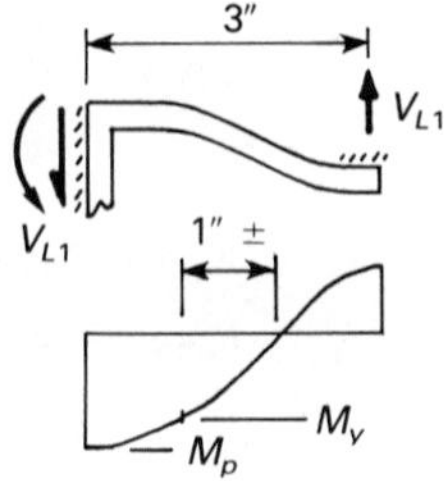

Calculation verifies that L can allow end of beam to rotate without offering significant moment resistance. Approximations made here seem sufficient for the purpose.

$F_y = 36$

$$S = \frac{bt^2}{6} \approx \frac{2(0.25)^2}{6} = 0.02$$

$M_y = F_y S \approx 36(0.02) = 0.72$ in. k

$V_{L1} \approx 0.72$ k

L yields at very low force. End rotation of beam permitted.

(5) Detail:

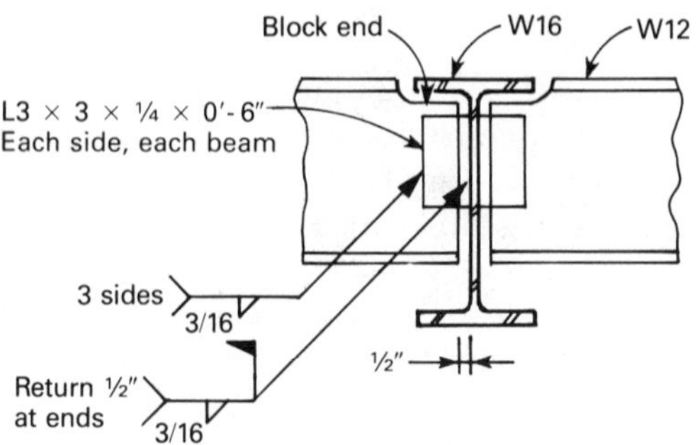

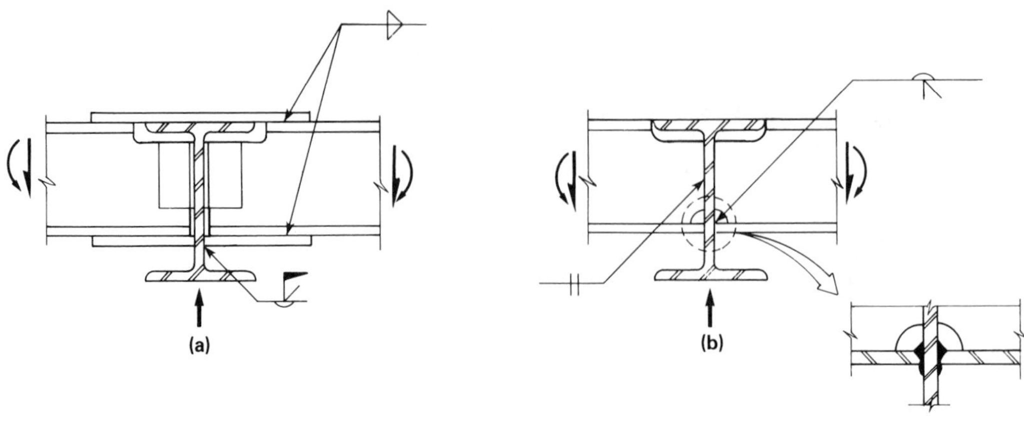

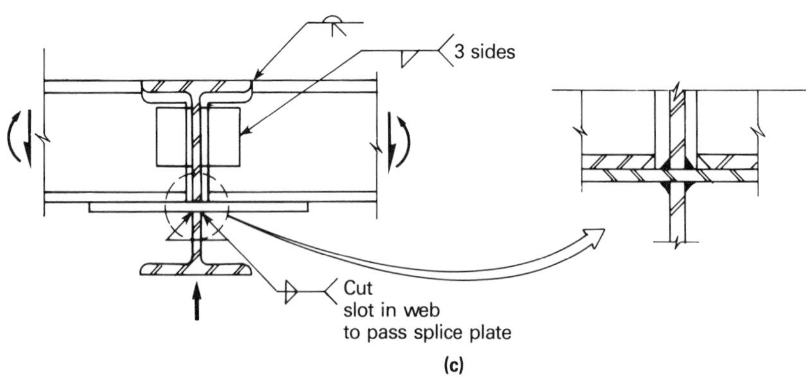

Figure 12.28 Continuous joints, beams to girder.

12.10.4 Rigid Beam End Connections

Rigid beam end connections are consistent with an analysis that assumes moment continuity through joints. We have already met two instances in Table 12.8, Types D(1) and D(2). Several more are shown in Fig. 12.28, covering instances where a beam is continuous through an intervening girder. In Fig. 12.28(c), where negative moment is indicated, the designer has decided to avoid the lamellar tearing problem by passing the tension flange splice plate through a slot cut in the web of the girder.

Many rigid connections have been devised to connect beams to columns, some of which are shown in Fig. 12.29, adapted from a booklet by the Steel Committee of

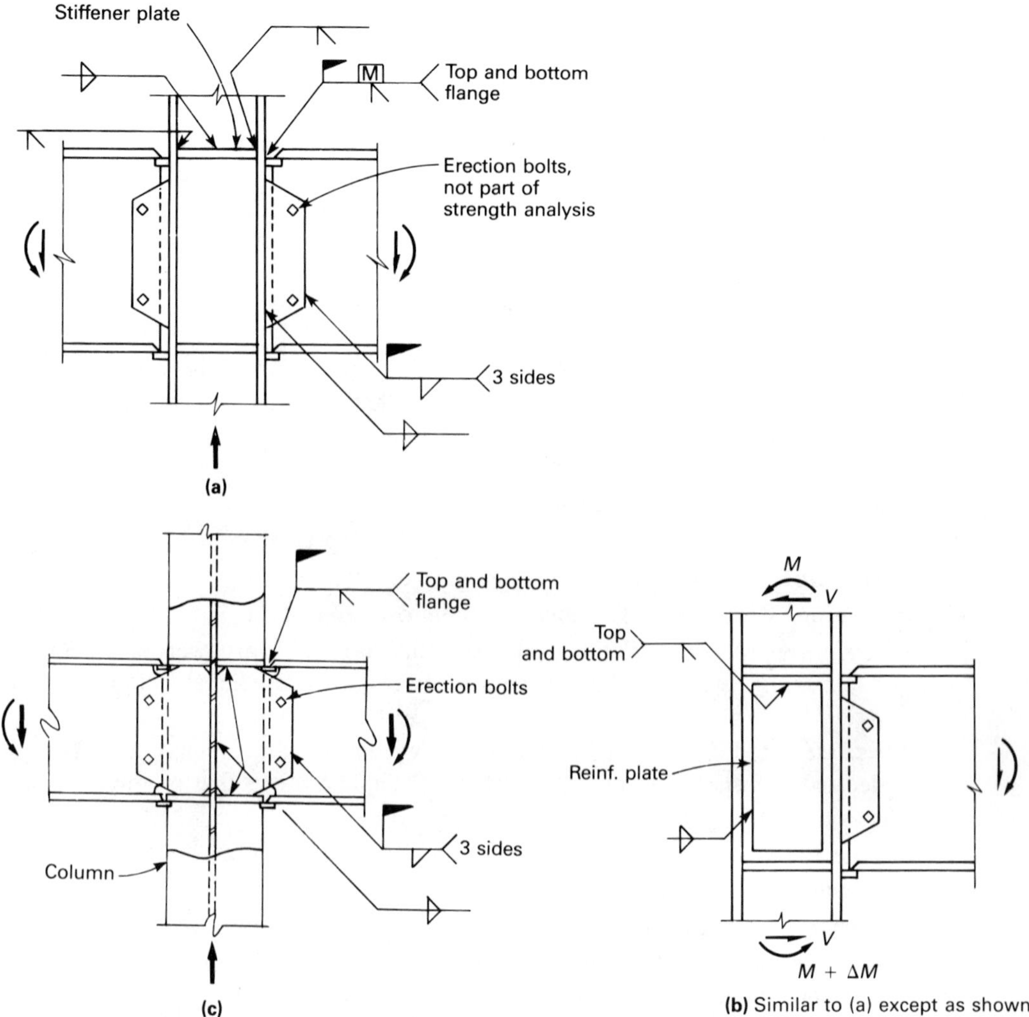

Figure 12.29 Continuous joints, beams to column.

California* in which the compilers give their opinions about relative cost of fabrication, a valuable service not repeated here. Calculations would be similar to those used in Example 12.9.7. A special problem arises in a one-sided connection such as Fig. 12.29(b), where the column web may be subjected to high shear stress within the connection. A reinforcing plate is indicated as one response to the high shear stress in accordance with AISC 1.15.5, an expensive procedure. Diagonal stiffeners are also effective and often less expensive.

*_Steel Connections, Details and Relative Costs_, Steel Committee of California, 1981 edition.

12.11 MIXED CONNECTIONS: WELDS AND BOLTS

The economics of joint design often indicate that connections made in the shop should be welded and those made in the field should be bolted. There are variants of all the connections shown in this chapter that allow for such a mix within the connections. Some will be shown in the next chapter after the design requirements for bolted connections are discussed.

12.12 GLOSSARY OF WELDING TERMS

Arc welding See *electric arc welding.*

Automatic welding Welding by processes that are mechanically and/or electronically controlled, with continuous feeding of electrodes, sometimes movable carriages for electrode reels and power source.

AWS American Welding Society.

AWS D1.1-80 1980 edition of *Structural Welding Code* of AWS.

Base metal The pieces of steel being welded.

Butt joint Connection of members without overlapping.

Electric arc welding Welding processes that use electric power as a source of fusion energy; weld metal is carried from electrode to base metal within an electric arc.

Electrode Rod or wire used to complete a welding circuit. In many processes, the electrode consists of weld metal with consumable coatings that produce shielding gas and flux.

Faying surfaces Surfaces that are in contact by overlapping.

Filler metal Weld metal.

Fitting up Preparing and aligning members prior to joining.

Flux Nonmetallic salts, often included in coating material on electrodes, which protects molten steel from oxidation and floats out impurities.

Gas (oxyacetylene) cutting Process for cutting steel by heating with an acetylene flame to red heat, followed by a jet of high-velocity oxygen.

Gas (oxyacetylene) welding Welding that uses a gas (usually oxygen and acetylene) for a source of fusion energy. Distinguish from gases used as shielding.

Hand welding Welding process not automatically controlled; usually SMAW or gas welding.

Heat-affected zone (HAZ) A zone within the base metal adjacent to the zone of fusion whose crystalline structure is altered by the welding process.

Hydrogen embrittlement Embrittlement caused by incorporation of hydrogen from water into filler metal or HAZ. Avoided by preheat or low hydrogen electrodes.

Interpass temperature Temperature of the base metal in the vicinity of the weld after a welding procedure has started.

Jig Means, often movable, for holding work in position for welding.

Lamellar tearing Cracking roughly parallel to the surface of steel when tension is applied perpendicular to the surface. May result from welding shrinkage in highly restrained welds or tension applied through attached members (Fig. 12.17).

Lap joint A joint made by overlapping members prior to welding (or bolting).

Low hydrogen electrode Electrode whose coating is formulated to reduce the formation of hydrogen from water vapor, avoiding hydrogen embrittlement.

Matching metal Weld metal accepted by AWS as capable of producing welds of at least equal strength and similar performance to the base metal to which it is matched for welding (Table 12.2).

Oxyacetylene welding (or cutting) See *gas welding, gas cutting.*

Pass The making of a continuous line of weld by one passage over the line. In some processes, such as SMAW, the maximum size of a single pass is limited. Large welds may be made by multiple passes over the same line of weld.

Peening Tapping, usually with a light, vibratory hammer, on surfaces of weld passes to provide some relief of internal stress.

Preheat Raising of temperature of base metal prior to welding. Often done with an oxyacetylene heating torch.

Prequalification Procedure or form of preparation accepted by AWS without further evidence of effectiveness, provided provisions of AWS D1.1 are followed.

Qualification Attaining of AWS acceptance of a weld procedure by evidence of tests following AWS qualification procedures.

Rod A short, round electrode, usually for hand welding.

Semiautomatic welding Welding done by an operator similar to hand welding but with a continuous electrode fed at a controlled rate.

Shielding Protection of the molten weld metal from the surrounding air, as by shielding gas emitted by the coating on an electrode.

Size of weld In a 45° fillet, one of the vertical legs; in groove welds, a specified thickness dimension or depth of groove. Zone of fusion may be larger than nominal "size." Critical strength dimension in a fillet weld is smaller than the "size."

Stress relieving Procedure to relieve internal stresses locked into a weldment from cooling shrinkage, often by heating in a furnace and allowing to cool at a controlled rate.

Stud Similar to bolt, but lacking either head or thread; in composite construction, welded to steel face as a shear connector.

Weld forms Forms are defined by the shape of the weld without including the zone of fusion (see Figs. 12.9 through 12.12 and Tables, Manual, Part 4).

Welding Uniting or fusing of two pieces of steel to form continuous crystalline structure. In usual structural practice, involves melting at the location of desired joining and addition of metal from an outside electrode.

Weldment An assemblage of steel parts joined by welding.

PRINCIPAL WELDING PROCESSES IN STRUCTURAL PRACTICE

See Section 12.3

Shielded metal arc welding (SMAW)
Submerged arc welding (SAW)
Gas metal arc welding (GMAW); sometimes (MIG)
Flux-cored arc welding (FCAW)
Electro gas
Electro slag

PROBLEMS

12.1. Design the joint used in Examples 8.9.1 and 9.8.5. The angles between chords and diagonals are 45°. Chord is ⌋⌊ 4 × 3 × ½ in. Use ⅜″ gusset.

12.2. Using members selected for Problem 9.9, design welded connections for the joints. Splice lower chord at L2, and upper chord at U2.

12.3. Design flexible connections for the following (you may use the AISC standard connections as a check, but do your own connection designs, and draw in detail):

	Steel	Beam	End Shear	Connect to
(a)	A36	W30 × 173	200	W36 × 300 girder (one W30 each side)
(b)	A588	W30 × 173	250	Flange of W14 × 145 column
(c)	A36	W12 × 40	40	Web of W8 × 31 (one W12 each side)
(d)	A572, Gr. 50	W18 × 35	70	W18 × 50 girder

12.4. Examine the hanger in Example 8.9.2. The hanger is to be connected to a W27 × 146 whose web is parallel to plates A and B.

(a) Connect plate A to plates B with welds (not as originally shown).

(b) Connect plates B to the W27 beam. Use P_{ALL} as found in the example. Consider at least two alternatives for the joint.

12.5. For the conditions of Example 12.10.3, you wish to make beam B1 continuous at its interior supporting girders G1. There are five spans total of B1. Dead load = 0.125 k/ft^2; live load = 0.075 k/ft^2; tributary width = 8′-0″.

(a) Select a size for B1 appropriate for a continuous member with continuous lateral support.

(b) Design a connection for B1 to G1.

(c) Design a connection for B1 to G2.

12.6. In Example 12.9.11 a 25-k horizontal force is added at the top of the bracket, acting away from the column. Redesign the weld pattern.

12.7. You wish to splice a W21 × 44 to a W21 × 62, both of A588 steel. The splice is to transfer the full moment capacity of the 44 lb. beam, assuming continuous lateral support and a shear of 50 k.

(a) Design a welded butt splice.

(b) Design a welded splice with splice plates.

12.8. Connect the flanges only to transmit the force and moment shown, applied simultaneously. Use a welded splice.

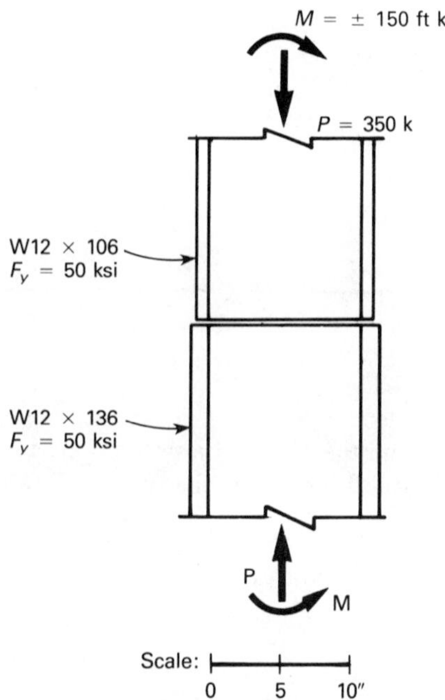

Problems 12.9 through 12.11 are suitable for major design reports to be done in conjunction with the case studies.

12.9. The main roof trusses in the Mill Building Case Study, Chapter 17, are *Fink trusses*, whose configuration is that of Alternative 1 shown on Calculation Sheet TR1 of that chapter. Design member loads for all truss members are shown in a table on Calculation Sheet TR7. For purposes of this problem, all truss joints are to be welded. Design all the joints of the truss. Steel is A36. Use *basic* allowable stresses; listed controlling member loads are based on dead plus live load.

12.10. Design all the rest of the connections for the mill building of Chapter 17, using welded joints.

12.11. Design typical welded connections for the tier building of Chapter 18.

13 Connections: Part 2

13.1 SCOPE OF CHAPTER 13

Recall from Fig. 12.1 and the discussion in Sections 12.1 and 12.2 that connections of steel members are usually made in one of four different ways:

With multiple mechanical connectors (bolts or rivets)
By direct bearing (limited to compression transfer)
By welds
By pins

Welding was discussed in detail in Chapter 12. This chapter will focus primarily on bolted and riveted joints, but will deal also with pin connections and several examples of mixed applications and special types of connections. We will also include some ideas about the "art" of connection design to ease fabrication and maintenance of structures. A glossary is provided at the end of the chapter to help with unfamiliar terms.

13.2 MECHANICAL CONNECTORS: RIVETS AND BOLTS

13.2.1 General

Examine Fig. 13.1. It shows a three-member joint in tension connected by 12 rivets. It could represent riveted or bolted joints in tension or compression and involving any

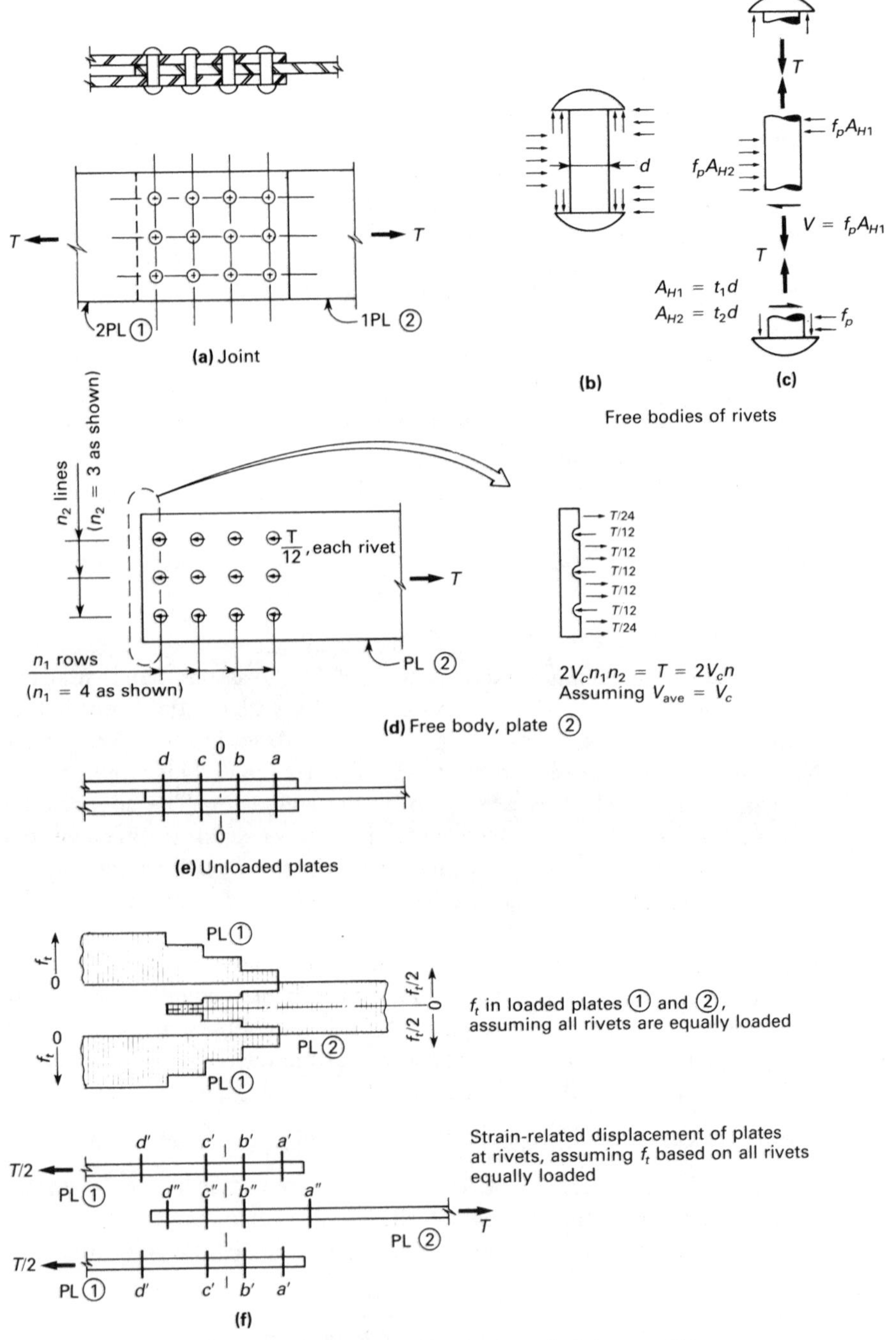

2PL ①
1PL ②
(a) Joint
(b)
(c)
Free bodies of rivets
f_pA_{H1}
f_pA_{H2}
$V = f_pA_{H1}$
$A_{H1} = t_1d$
$A_{H2} = t_2d$
n_2 lines
(n_2 = 3 as shown)
$\frac{T}{12}$, each rivet
n_1 rows
(n_1 = 4 as shown)
PL ②
T/24
T/12
$2V_cn_1n_2 = T = 2V_cn$
Assuming $V_{ave} = V_c$
(d) Free body, plate ②
(e) Unloaded plates
PL①
PL②
f_t in loaded plates ① and ②, assuming all rivets are equally loaded
Strain-related displacement of plates at rivets, assuming f_t based on all rivets equally loaded
T/2
(f)
Assumption that all rivets share load equally is incompatible with elastic action. At low force, T, $V_c \neq V_{ave}$. At high force, after ductile flow, $V_c \approx V_{ave} = T/n_1n_2$.

Figure 13.1 Three-member splice, riveted.

number of plates or similar elements ≥ 2. As a joint, the primary member tension is transferred by shear on the faying surfaces. (In this case, two surfaces lead to the term *double shear.*) Figure 13.1(c) shows that the connector is subjected to shear on the same plane. However, as can be seen in Figs. 13.1(b) and (c), there is a clamping action between the heads of the connector and the plates causing tension in the connector; there is an interchange of bearing stress (f_p) between the connector and the plate, which is the source of the shear in the connector. The mechanics of this type of joint are the same whether the longitudinal force to be transmitted results from axial load on a member or a couple of axial forces required to transmit bending moment.

In Fig. 13.1(d), plate 2 is shown with the axial tension balanced by the sum of the shear forces in the connectors. The assumption is made that the shears, V_c, on all connectors are equal; that is,

$$V_{\text{ave}} = V_c$$

and

$$T = 2N_c V_c \tag{13-1}$$

where $N_c = n_1 n_2$ = total number of connectors.

As can be seen by studying Fig. 13.1(f), the assumption that all connectors take equal shares of the total load is not compatible with elastic strains in the plates within the connection. Locations of high stress and strain in the outer plates coincide with locations of low stress and strain in the inner plate. This would require slip on the faying surface, which is prevented by the connector. The connectors take unequal shares of small loads. However, the shares tend to average out by ductile redistribution as loads approach failure. Failure of connections in shear may take place by shear or bearing in either the connectors or the connected material. See Fig. 13.2.

Following common practice, the procedures used herein for splices with multiple connectors will assume that a load is shared equally among a group of similar connectors in a splice. In the case of long splices of tension members, redistribution is incomplete; the value of a connector will be reduced to account for unequal connector shears. Equation (13-1) can be rewritten in more general form as

$$P = NV_c \tag{13-2}$$

where N equals the number of sheared connector planes (i.e., the number of connectors times the number of shear planes per connector). In another form, if VAL* = the allowable load on or *value* of a single connector, then

$$N_c \geq \frac{P}{VAL} \tag{13-3}$$

*The term *capacity* is commonly used. We avoid that term where possible since it may be confused with the *ultimate* capacity of the connector. VAL is the *allowable* load.

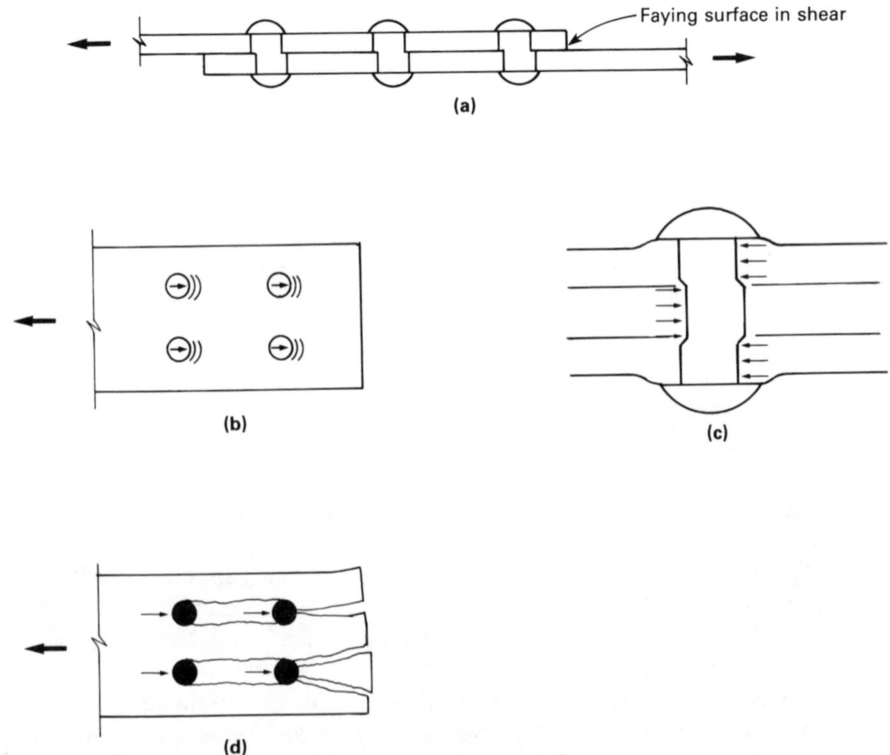

Figure 13.2 Modes of strain and failure, mechanical connectors: (a) shear on connectors; (b) bearing on plate; (c) bearing on connector; (d) tearing (shear) and splitting (cross tension) in plate.

can be used to determine the number of connectors required to transmit a force P. There are a number of different types and sizes of connectors in common use, each of which is assigned a different value.

In somewhat less common situations, such as is shown in Fig. 13.3, the primary action on a connector may be to resist tension [Fig. 13.3(a)] or a combination of shear and tension [Fig. 13.3(b)]. Connector values are different for these cases than for the shear case.

A series of tables at the beginning of Part 4, AISC Manual, lists allowable loads on the various types of connectors used under the AISC Specification. What follows looks at these types of connectors and their actions.

13.2.2 Classes of Connectors

Refer to Table 13.1. When classified by the form of the connector, there are two classes, *bolts* and *rivets*. These are illustrated in Fig. 13.4. The principal mechanical difference is in the mechanism used for developing the clamping action. These

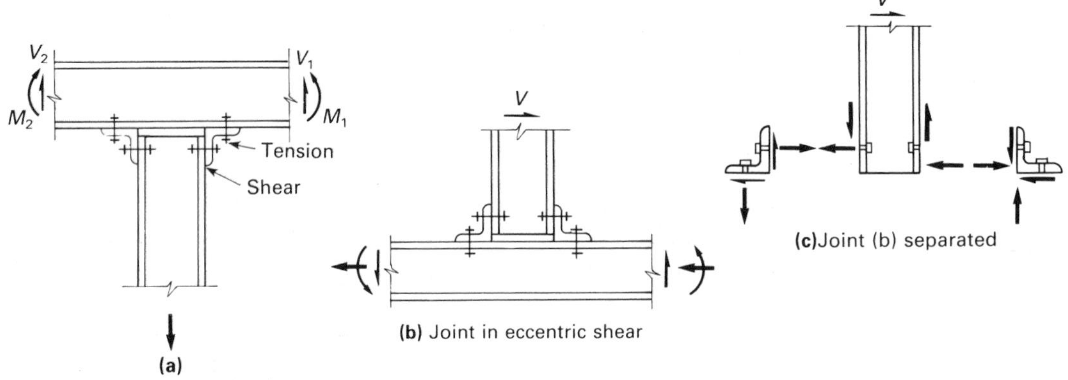

Figure 13.3 Connectors in shear, tension, and combined shear and tension.

mechanical differences lead to several other advantages for bolts over rivets. The method of tightening requires smaller crews and is less dangerous and less noisy. At one time, riveting was the principal method of joining structural steel. It is much less common today, having yielded to the advantages accruing to bolts and welds.

Another and perhaps more significant division in Table 13.1 is by clamping action and strength of connector (classes I, II, and III). A307 bolts and A502 rivets are made from *carbon steels,* with strength and stress–strain characteristics similar to those of A36 steel. As with A36 steel, they are amenable to cold working and cycles of heating and cooling without great change in mechanical properties. High-strength bolts (A325 and A490) are made from alloy steels of very high strengths, more closely resembling A514 steel. Their ratios of F_u/F_y are relatively low. Their mechanical properties depend on the controlled quenching and tempering used in their manufacture and would change if reheating were permitted in the field. They are applied by methods controlled by a special specification and are not intended for use except in the type of joints for which that specification was written.* However, used in such connections, they have very high connector values.

Bolts of A449 steel, in sizes larger than available in A325 or A490, are occasionally used in high-strength bolted connections. They are not covered under the specification governing the use of A325 and A490 bolts, but are permitted under the AISC Specification for buildings and should be used only by experienced designers familiar with their properties and the action of bolted joints.

*The collapse of an arena roof in Kansas City (1979) has been attributed by some to improper use of A490 bolts in a connection where the usual torquing was not possible. According to this theory (ENR magazine, August 7, 1979), fluctuating wind loads led to early fatigue fracture. The magazine cites a report by James L. Stratta, Structural Engineer, Menlo Park, California, to the Department of Public Works, Kansas City, Mo.

TABLE 13.1 MECHANICAL FASTENERS IN MOST COMMON USE IN CONNECTING STRUCTURAL STEEL

	Class by ASTM designation	Steel					Clamping action on joint	Stress type on connector	Class by clamping and strength
		Type[a]	Diameter in.	Stresses, ksi					
				Proof	F_y	F_u			
Bolts	A307	C	To 4	Not specified		60	Wrench tight; force not measured	Shear and bearing	I
	A325	C, QT	½ to 1	85	92	120	High measured torque leads to high, predictable clamping force	Friction or (bearing and shear)	II
			1⅛ to 1½	74	81	105			
	A490	Q	½ to 1½	120	—	150			
	A449	C, QT	1¾ to 3	55	58	90	Subject to designer decision; See AISC Spec. 1.16.1	Bearing and shear; friction possible	III
Rivets	A502-1	C		—	(28)[b]	(52)[b]	Clamping action of rivets results from cooling shrinkage; not predictable	Bearing and shear	I
	A502-2	C			(38)[b]	(68)[b]			

[a]C = carbon steel
Q = quenched
QT = quenched and tempered

[b]Strength not specified by ASTM. These values may be representative based on old specifications. See reference 29. Cold working during driving increases F_y.

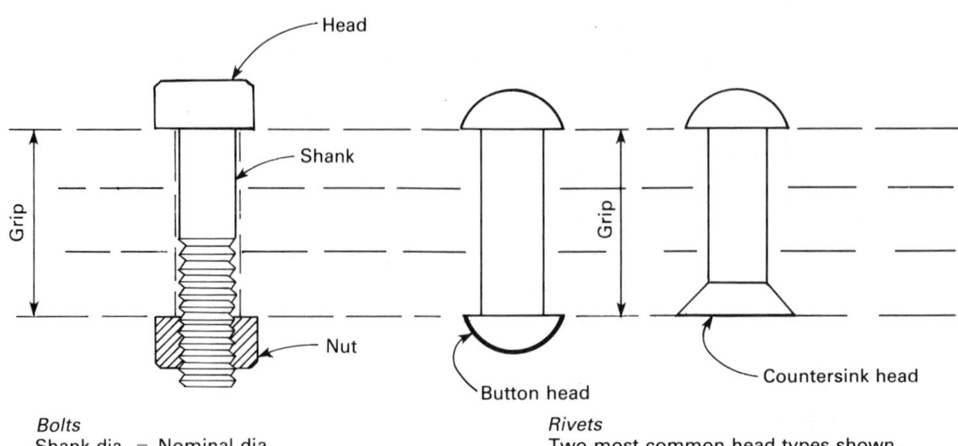

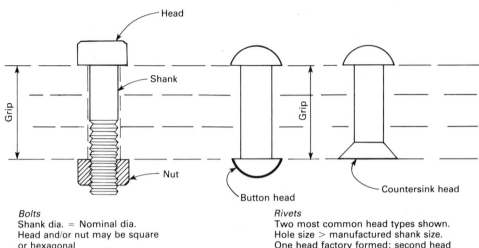

Bolts
Shank dia. = Nominal dia.
Head and/or nut may be square or hexagonal
Hole size > bolt size.
Installed at ambient temperature.
Clamping action achieved by applying torque to nut and head, advancing nut on thread. Torque must be enough to prevent future loosening of bolt.
Dimensions of all parts controlled to avoid failure of head or nut at lower force than threaded region of bolt.
Washers sometimes required below head and/or nut.
Construction noise level less than for rivets, higher than welding.
For *high-strength bolts,* torque must be measured as indicator of clamping force.

Rivets
Two most common head types shown.
Hole size > manufactured shank size.
One head factory formed; second head formed in place.
Usually heated (red) for driving.
Driving expands shank to hole size and forms head.
Clamping force results from axial shrinkage in cooling, inhibited by material gripped.
Clamping force not measured, but can be detected by frequency response to hammer tap.
Noise level very high during construction; construction hazard level also high.

Figure 13.4 Bolts and rivets.

13.2.3 Allowable Stresses

Allowable stresses in connectors are included in the AISC Specification in Section 1.5.2. Table 13.2 reproduces Table 1.5.2.1 of that Specification. Values are given for tension and shear (F_t and F_v). In some cases, F_v is different for *bearing-type* and *friction-type* connections, distinctions that apply to high-strength bolts (A325 and A490) only, which will be clarified later.

Among the notes in Table 13.2, several things should be mentioned. Note g refers to Section 1.5.6 of the Specification, which permits a 33% increase in allowable stresses for wind or seismic loading; the increase applies to values in the table. Notes a and d and their references exclude cycling tension loading on connectors, which might cause fatigue, except for A325 and A490 bolts. Cycling shear is permitted (see AISC B3.3) on connectors without penalty, although the

TABLE 13.2 ALLOWABLE STRESS ON FASTENERS, ksi

Description of Fasteners	Allowable Tension[g] (F_t)	Allowable Shear[g] (F_v)			
		Friction-type Connections[e,i]			Bearing-type Connections[i]
		Standard size Holes	Oversized and Short-slotted Holes	Long-slotted Holes	
A502, Grade 1, hot-driven rivets	23.0[a]				17.5[f]
A502, Grades 2 and 3, hot-driven rivets	29.0[a]				22.0[f]
A307 bolts	20.0[a]				10.0[b,f]
Threaded parts meeting the requirements of Sects. 1.4.1 and 1.4.4, and A449 bolts meeting the requirements of Sect. 1.4.4, when threads are not excluded from shear planes	$0.33F_u$[a,c,h]				$0.17F_u$[h]
Threaded parts meeting the requirements of Sects. 1.4.1 and 1.4.4, and A449 bolts meeting the requirements of Sect. 1.4.4, when threads are excluded from shear planes	$0.33F_u$[a,h]				$0.22F_u$[h]
A325 bolts, when threads are not excluded from shear planes	44.0[d]	17.5	15.0	12.5	21.0[f]
A325 bolts, when threads are excluded from shear planes	44.0[d]	17.5	15.0	12.5	30.0[f]
A490 bolts, when threads are not excluded from shear planes	54.0[d]	22.0	19.0	16.0	28.0[f]
A490 bolts, when threads are excluded from shear planes	54.0[d]	22.0	19.0	16.0	40.0[f]

[a] Static loading only.

[b] Threads permitted in shear planes.

[c] The tensile capacity of the threaded portion of an upset rod, based upon the cross-sectional area at its major thread diameter, A_b, shall be larger than the nominal body area of the rod before upsetting times $0.60F_y$.

[d] For A325 and A490 bolts subject to tensile fatigue loading, see Appendix B, Sect. B3.

[e] When specified by the designer, the allowable shear stress, F_v, for friction-type connections having special faying surface conditions may be increased to the applicable value given in Appendix E.

[f] When bearing-type connections used to splice tension members have a fastener pattern whose length, measured parallel to the line of force, exceeds 50 inches, tabulated values shall be reduced by 20 percent.

[g] See Sect. 1.5.6.

[h] See Appendix A, Table 2, for values for specific ASTM steel specifications.

[i] For limitations on use of oversized and slotted holes, see Sect. 1.23.4.

Reproduced with permission from AISC Specification, 1978, Table 1.5.2.1.

integrity of the connected materials must be checked. The allowable stresses will be further interpreted later in the discussion of each type of connector.

13.2.4 Rivets

Rivets are made from *carbon* steel, which permits heating, cooling, and hot and cold working under minimum control. There are two strength grades (A502-1 and A502-2, 3), both somewhat comparable to A36 steel. Values of F_t and F_v are slightly higher than for A36 steel (Table 13.2). Shear in rivets results from the action of *bearing-type connections* only; that is, the shear on the rivet results, as in Fig. 13.1, from bearing stresses interacting between the rivet and the connected part. Cooling of the hot rivet results in rivet tension and clamping pressure between the rivet head and the connected part on which it bears and between the various parts within the grip of the rivet. The magnitude of the clamping effect is not predictable.

Assuming that there is sufficient area in bearing, the value of a rivet is

$$VAL = F_v AN \tag{13-4}$$

where

F_v = 17.5 ksi or 22 ksi depending on the material

$A = \dfrac{\pi d^2}{4}$ = the area of cross section

N = number of planes in shear = 1 for a two-member connection, 2 for three members, etc.

Values are tabulated for rivets in single and double shear in the Manual, Part 4, Table I-D. Single shear values are in the range from 5.4 to 38.9 kips for rivet diameters in the range between $\frac{5}{8}$ and $1\frac{1}{2}$ inches.

To use the rivet for its full value, the bearing area between the rivet and the side of the hole must be sufficient. The allowable stress in bearing is

$$F_p = 1.5\ F_u\text{*} \qquad \text{AISC 1.5.1.5.3}$$

where F_u would be based on the material of the connected part. The example immediately below illustrates the basic procedure for rivets in shear.

13.2.5 Example: Riveted Splice, Single Shear

Nine 3/4″ diameter rivets, A502-1, are to be used in a two-member joint to transmit an axial tensile force T. The steel of the connected parts is A36. Assume that the spacing, edge, and end dimensions are sufficient so that the full rivet values may be

*The very large value of F_p relies on the fact that the material in bearing is confined by the limits of the hole.

used (see AISC Section 1.16 and later discussion herein). (a) What is the maximum permissible force T? (b) How thick does the plate have to be?

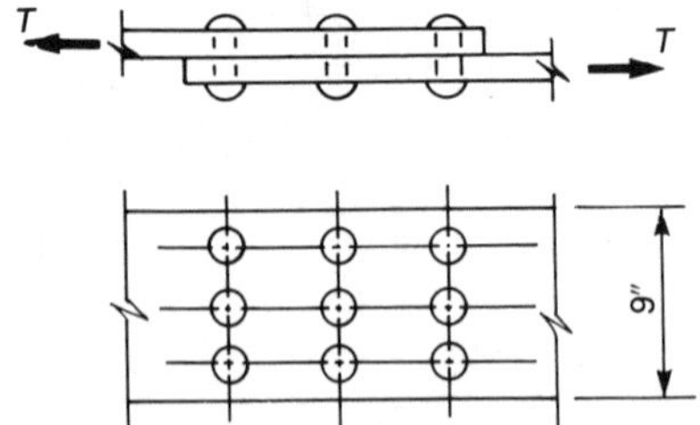

CALCULATIONS	REMARKS
(1) Rivets in single shear:	One shear plane.
shear value per rivet $= F_v AN$	F_v from Table 13.2. Length of pattern $\ll 50''$.
$= 17.5\left(\pi\dfrac{(0.75)^2}{4}\right)(1)$	
$= 7.73$ k	Agrees with Manual Table I-D.
If plates are thick enough:	
$T_{\max} = 9(7.73) = 69.6$ k	
(2) Size plate for $T = 69.6$ k:	
$A_g = 9t$	Procedure from Chapter 8; for A_e see AISC 1.14.2.2.
$A_e = 0.85t[9 - 3(.875)] = 5.42t$	
$F_t = 0.6F_y = 21.6$ ksi on A_g	
or $= 0.5F_u = 29.0$ ksi on A_e	
$T = 69.6 = 21.6(9)t = 194t$	
or $69.6 = 29.0(5.42)t = 157t < 194t$	Use smaller T.
$t \geq \dfrac{69.6}{157} = 0.44''$; try $7/16'' = 0.44''$	
(3) Check bearing on rivet:	
rivet value $= 7.73 = f_p(t)(d)$	
$f_p = \dfrac{7.73}{0.44(0.75)} = 23.4 \ll 1.5F_u = 87$ ksi	AISC 1.5.1.5.3
(4) $\therefore$ Use $t = 0.44$.	

13.2.6 A307 Bolts in Bearing and Shear

A307 bolts are similar to rivets with regard to strength and action within a joint. The material has an ultimate strength of 60 ksi; yield strength is not specified. When used to transmit loads by shear, they are used in *bearing-type* connections defined in the same way as previously for rivets. Joints with A307 bolts are clamped together by applying torque to the nuts, resulting in bolt tension and bearing at the head and nut and faying surfaces between parts of the joint. The torque is not measured and is limited by the relatively low value of F_u. The bolt tension and intersurface pressure are unknown.

Standard holes are $\frac{1}{16}$ inch larger than the diameter of the bolts. When the bolt patterns on two thicknesses of steel are superimposed for bolting (Fig. 13.5), some holes may match perfectly, while others may be slightly misaligned, although within acceptable dimensional tolerance. With proper workmanship it should be possible to enter all the bolts into the holes. However, as can be seen in the figure, if the force is large enough to overcome the friction on the faying surface, the joint will have to slip and some individual bolts distort before all the bolts bear on the sides of their holes. The ductile nature of the steel permits this readjustment. (The misalignment phenomenon is independent of the strain incompatibility phenomenon illustrated in Fig. 13.1.)

If member thicknesses are sufficient,

$$VAL = F_v AN \tag{13-4}$$

The area $A = (\pi d^2)/4$, where d is the shank diameter of the bolt. The value of F_v is set at the low value of 10 ksi (Table 13.2) to account for the loss of area, and hence of shear strength, at the location of threads, which may be in the plane being sheared.

The next example is very similar to the previous example, but illustrates the effect of double shear.

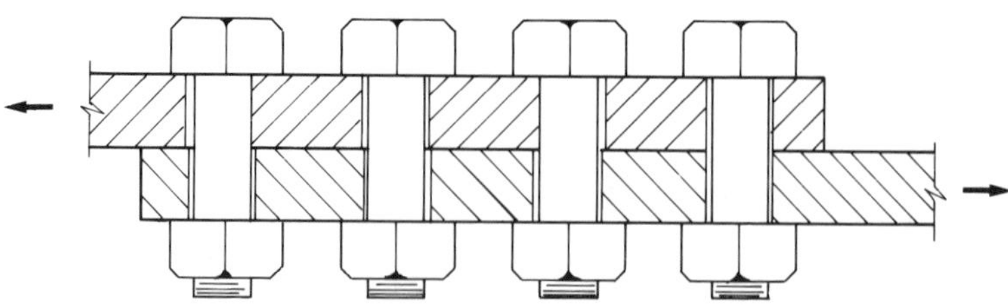

Figure 13.5 Mismatch of hole patterns (within tolerance) causes unequal shear.

13.2.7 Example: A307 Splice in Double Shear

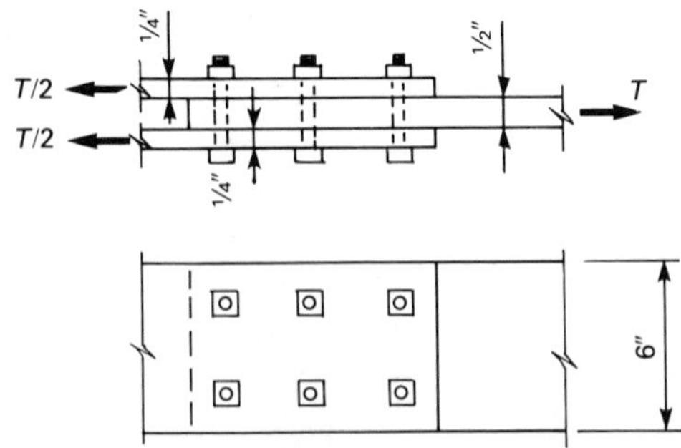

$P = 64$ kips. Steel is A36; bolts are A307. Find the required bolt size. Verify that the plate cross section is sufficient.

CALCULATIONS	REMARKS
(1) Force/bolt = 64/6 = 10.7 k. Requires	
$10.7 \le F_v A_b(2) = VAL$	A_b = cross-section area of bolt
$10.7 \le 10 \dfrac{\pi d_b^2}{4}(2)$	d_b = shank diameter
$d_b \ge \left[\dfrac{10.7(2)}{10\pi}\right]^{0.5} = 0.83''$	
Use 7/8" diameter bolts if bearing stress is OK.	Can be checked with Manual, Table I-D, Part 4.
(2) Check bearing on bolts:	
Inner plate: $f_p(0.875)(0.5) = \dfrac{46}{4} = 11.5$	
$f_p = 26.3 \text{ ksi} \ll 1.5\,F_u = 87$	AISC 1.5.1.5.3 (Check Manual, Part 4, Table I-E.)
Outer plates: same stress result	
(3) Check plate:	
$A_g F_y(0.6) = 6(0.5)(36)(0.6) = 64.8 > 64$	See Chapter 8 and AISC 1.5.1.1, 1.14.2.
$0.85 A_n F_u(0.5) = 0.85(6-2)(58)(0.5) = 98.6 > 64$	
(4) *Use 7/8" diameter bolts. Plate is OK.*	

13.2.8 High-Strength Bolts

General Discussion: High-strength (HS) bolts. In appearance, *high-strength bolts* and the joints that are made with them are hard to distinguish from A307 bolts and their joints; so special identifying markings are embossed on the heads and nuts of high-strength bolts, and inspection reports are required attesting that high-strength joints were assembled in accordance with specified procedures. In some ways, the design procedure for joints using high-strength bolts is also the same as for A307 bolts and rivets. A value is established for a single bolt based on its size, type of steel, and number of shear planes. The number of bolts required to transfer a force by shear perpendicular to the bolt axis is determined, as before, by Formulas (13-2) through (13-4).

However, in important ways, high-strength bolts and their joints constitute a connecting technology much more sophisticated and requiring more careful control of specification, joint preparation, and erection procedure than their relatively primitive cousins, the A307 bolts. The payoff for the more complicated construction constraints is the ability to transfer high forces with fewer bolts in more compact joints than can be done with A307 bolts. An added advantage derives from the requirement of high pretension on high-strength bolts. They become less susceptible to fatigue failure from external, cycling loads that cause axial tension in the bolts. Appendix B of the AISC Specification recommends only A325 or A490 bolts for such loads with more than 20,000 cycles. Consideration of tensile loads in bolts is reserved for later in this chapter, while we continue here to examine shearing action.

Our discussion of high-strength bolts here will of necessity be incomplete, but it is intended to clarify their basic action and make possible their use in joint design in accordance with the AISC Specification. The "Specification for Structural Joints Using ASTM A325 or A490 Bolts" and its "Commentary," both included in Part 5 of the AISC Manual, give more information and references.

ASTM A449 bolts are sometimes classified as high-strength bolts and may be used in similar ways. However, the rules for their use are not covered in the cited specification. In further discussion here, the use of the term *high-strength bolts* will mean A325 or A490 bolts.

Proof loading: Pretensioning. Note in Table 13.1 the very high tensile strength associated with high-strength bolts and the *proof* load, which is 70% to 80% of F_u. *Proof loading* is a nondestructive test done by applying a specified load high enough to reveal most manufacturing defects but not so high as to risk destruction of a sound bolt. It becomes very significant here because assembly specifications require that all high-strength bolts be tightened at assembly to 70% of their minimum ultimate strength, F_u, a requirement key to what follows (see Table 13.3).

Each bolt in a high-strength bolted joint is tightened sufficiently to cause a clamping force and corresponding bolt tension [similar to Figs. 13.1(b) and (c)] equal to 70% of the minimum bolt strength, F_uA_e, where A_e is the effective area at the threads. The tension may be measured by any one of three methods permitted by the Specification:

The iron workers are making up bolted joints with the aid of a pneumatic wrench. (Photos courtesy of American Bridge Division, United States Steel Corporation)

TABLE 13.3 MINIMUM TENSION ON BOLTS IN HIGH-STRENGTH BOLTED JOINTS

Nominal bolt diameter, in.	Minimum bolt tension, kips	
	A325 Bolts	A490 Bolts
½	12	15
⅝	19	24
¾	28	35
⅞	39	49
1	51	64
1⅛	56	80
1¼	71	102
1⅜	85	121
1½	103	148

From Table 3, Specification for Structural Joints Using High Strength Bolts, 1978.

- Turn-of-nut tightening.
- Calibrated wrench tightening.
- Tightening by use of a direct tension indicator.

The first two methods use torque as an indicator of tension. The third uses strain as the indicator. The very important result is that all parts of the joint are brought up tight and that the bearing force between parts is both high and predictable.

If these specified circumstances are realized, it becomes possible to design joints in shear on two different bases. The first is the *bearing-type* joint, which, as in the case of A307 bolts and rivets, acts by bearing between each bolt and the side of its hole and the consequent bolt shear. For reasons previously discussed (see Fig. 13.5), bearing-type joints experience small *slip* movements before mobilizing their full strength. The second, *friction type* of joint transfers the shearing load through friction at the faying surfaces; no slip occurs.

HS bolts in friction-type connections. Consider a surface between flat parts of steel members subjected to a known normal force and a force parallel to the surface. Such situations will be recalled from early studies of statics and are illustrated in Fig. 13.6 in the context of a three-member joint. The intersurface pressure, F_N, results from the clamping effect of the bolt heads and nuts, which itself is caused by tension in the bolts. The joint cannot slip until friction on the faying surfaces is overcome. If $2C_fN_c$ is the ultimate limit of F_N, then $2C_fN_cT_B/\text{SF}$ may be considered an allowable shear force. C_F is the static coefficient of friction, SF the safety factor against slip, and N_cT_B the sum of the tensions on all bolts within the joint. The lower limit of bolt tension is set by the requirements of the Specification

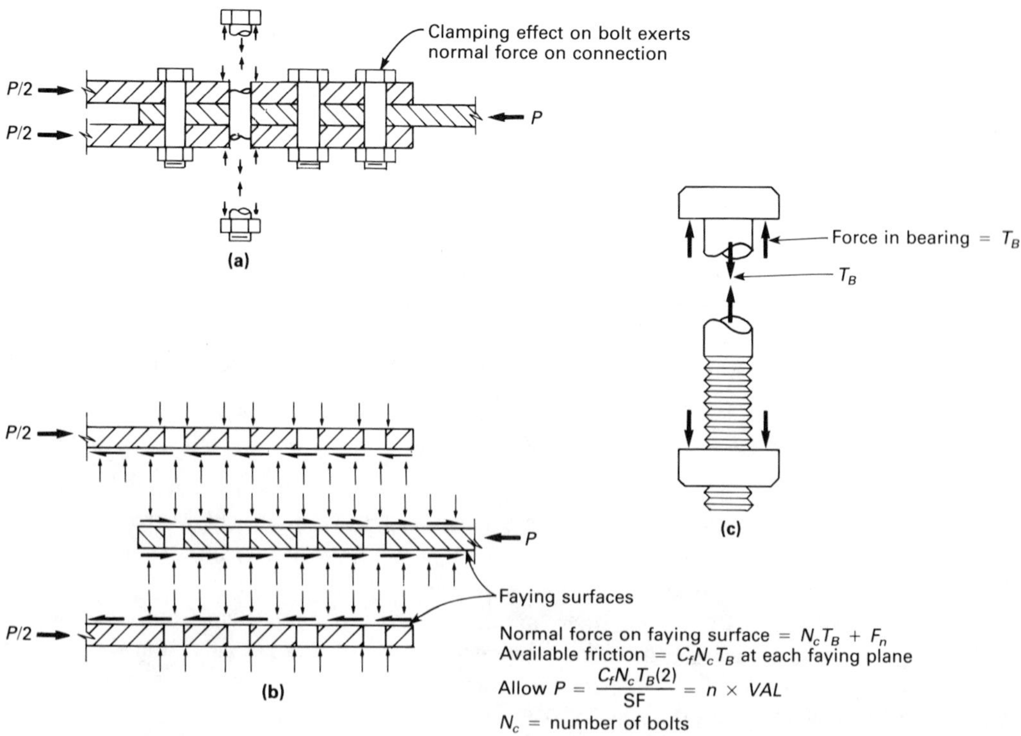

Figure 13.6 Three-member joint, friction bolts: (a) joint; (b) free body of plates; (c) free body of bolt.

and guaranteed by procedural specifications and inspection. Table 13.3 gives these values in kips. The coefficient of static friction of steel on steel may vary widely over the range $0 \leq C_f \leq 0.6$, depending on the roughness of the surfaces and the presence of intervening materials with lubrication-type effects, such as mill scale, oxides of iron, and paint, all of which reduce friction. Where bolts are used in friction-type connections, specifications require that faying surfaces be flat and free of contaminants, guaranteeing high friction.

HS bolts in bearing-type connections. The chief advantage of friction-type connections derives from their nonslip action. They can share loads with welds in the same joint since both act without slip (see Fig. 13.7 and AISC 1.15.10.). They permit holes larger than standard, although with some reduction in bolt values. Where hole sizes are standard (diameter + $\frac{1}{16}$ inch), an overloaded friction-type connection or a high-strength bolted connection without reliably high friction becomes a bearing-type connection. A small slip may occur similar to that in joints with A307 bolts or rivets. Bolt values, based on the shear/bearing strength of the bolts, are larger than those permitted in friction-type joints.

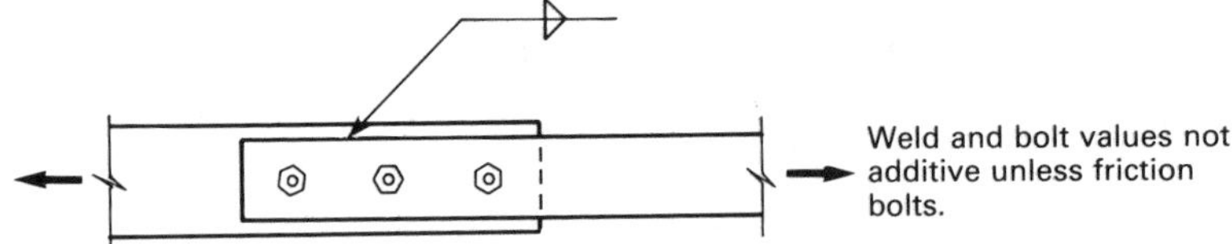

Figure 13.7 Mixed joints, bolts and welds sharing same load.

All the special constraints and problems associated with other bearing-type connections apply also to those using high-strength bolts. To use the full bolt value, the thickness of connected parts must be sufficient and holes spaced far enough apart so that bearing and shear can be resisted in the base materials. Minimum thickness is set by

$$VAL' = 1.5F_u\,dt \tag{13-5}$$

where VAL' is that portion of the bolt value that applies to the ply in question. Minimum spacing will be discussed later. Since bolt values for high-strength bolts are much greater than those for A307 bolts, high-strength bolts are much more likely to control the minimum thickness of connected parts.

As with A307 bolts, if there are threads in the sheared plane, the shear capacity of the bolt is reduced to that determined by the reduced area at the threads. The reduced capacity is recognized by reducing the allowable stress, F_v in the product F_vA, using for area the area of the shank.

However, a logical corollary to this procedure emerges if we ask the question, How strong is the bolt in shear if threads do not occur in the shear plane? The answer is obvious. To take advantage of the extra strength of the undisturbed shank area, two conditions are necessary. The designer must specify and the inspector certify that there are no threads in any plane of shear (see Fig. 13.8). Bolt manufacturers must be willing and able to supply bolts with specified lengths of both thread and bolt within

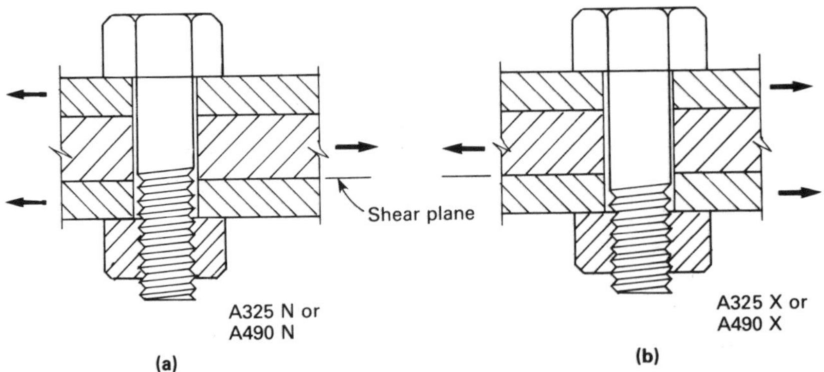

Figure 13.8 High-strength bolts, threads in and excluded from shear planes: (a) threads in shear plane; (b) no threads in shear plane.

narrow ranges. Both conditions are possible with high-strength bolts, and the gain is often worth the extra trouble.

For both A325 and A490 bolts, two sets of bolt values exist in bearing-type connections; one is based on the traditional condition where threads may or may not exist in a shear plane, the other on conditions where threads are excluded. Table 13.2 lists the values of F_v used for each case. In Part 4, Table I-D, of the Manual, these are converted to single and double shear values for different bolt diameters.

13.2.9 Example: HS Bearing Bolts to Develop Member

The capacity (allowable) of a W-shape member is 500 k. You wish to use A490 bolts of 3/4″ diameter in a bearing-type connection with threads permitted in the shear plane. How many bolts are required to develop the member in a splice such as shown? What is the minimum thickness of splice plate required to develop the bolt?

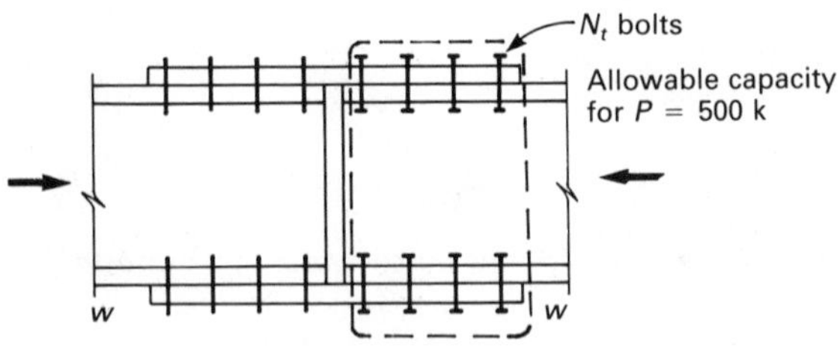

CALCULATIONS	*REMARKS*
(1) Bolts in single shear: Allowable shear per bolt is	
$$V_b = F_v A_b = 28\frac{(\pi)(0.75)^2}{4}$$	Table 13.2
$= 12.4$ k	May be verified, Manual, Part 4, Table I-D.
(2) Bolts to develop member:	
$$N_c = \frac{500}{12.4} = 40.3$$	
Use $4 \times 11 = 44$ bolts.	Min. no. $> N_c$ preserving biaxial symmetry in bolt pattern.
(3) Set $t_{\min}$ for plate by allowable bearing: $1.5F_u t d_b \geq 12.4$	AISC 1.5.1.5.3
$$t \geq \frac{12.4}{1.5(58)(0.75)} = 0.19''$$	Verify by Table I-E, Manual Part 4.
Use $t \geq$ *1/4″ to develop bolt.*	Plate size must also be determined for f_a.

13.2.10 Rivets and Bolts in Tension*

Direct tension and prying action. Recall Fig. 13.3 in which certain of the connectors are subjected to tension due to the nature of external loading. This effect can apply to any type of connector—in our context, rivets, A307 bolts, or high-strength bolts. It is a consequence of the requirements of external equilibrium and is separate from the pretension on connectors that results from rivet cooling or nut tightening and is internal in the joint.

Let us now look at the equilibrium problem for one of the connecting angles in Fig. 13.3(a), a free body of which is drawn in Fig. 13.9. It is apparent that equilibrium of moments around point O requires a force, C, at the toe of the angle. The magnitude of C depends on the dimensions of the angle, the *gage* distance g, and the distance e.

$$Ce = V_b g$$

and

$$T_b = V_b + C = V_b \left(1 + \frac{g}{e}\right)$$

where g/e may be equal to 2 or more, but can be minimized by the designer's choice of dimensions.

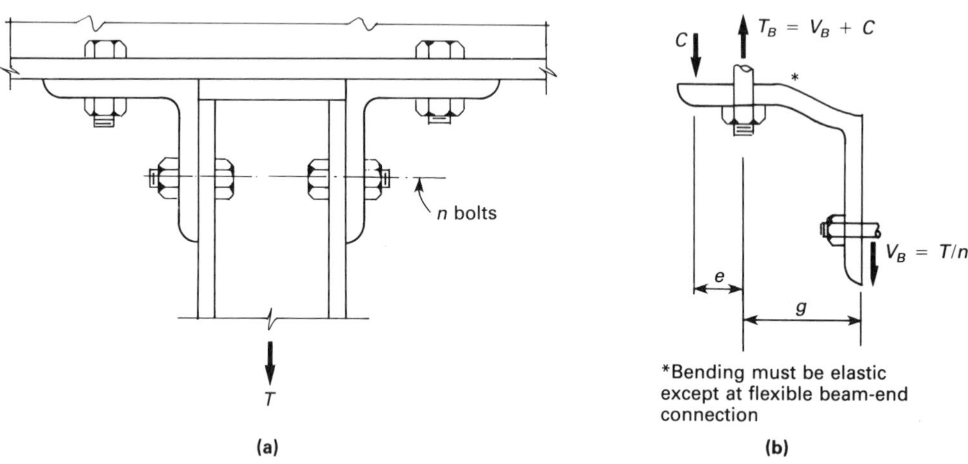

Figure 13.9 Prying effect, bolts in tension: (a) bolted hanger connection; (b) displaced form of L due to shear and prying.

*See AISC 1.5.2.1, 1.5.2.2.

The magnification of tension due to this *prying effect* can be quite severe and cannot be safely ignored. Values of F_t, listed in Table 13.2, are allowable stress. In the formula

$$T \leq F_t A \qquad (13\text{-}6)$$

as applied to connectors, the value of T must include the prying effect. A is calculated with the shank diameter.

Pretension and net tension; fatigue. Examine now a bolt (or rivet) that has been pretensioned to an initial tension, T_i, before an external force, T, is applied. What is the effect on tensile stress, f_t, when T is applied? Pretension has been applied by tightening a nut (or by cooling strain of a rivet) and is balanced by precompression of a much larger area of the connected parts (see Fig. 13.10). The length of the bolt within the grip has been increased by elastic strain and the thickness of the connected parts reduced elastically.* Since the tensile stress is many (10 or more) times the corresponding compressive stress, we can, without serious error, ignore the compressive strain and say

$$\Delta L \approx \frac{T_i L}{A_b E} \quad \text{or} \quad \frac{\Delta L}{L} = \varepsilon_i = \frac{T_i}{A_b E}$$

If strains from added tension are to be added to ε_i, the bolt must get longer than its grip, immediately relieving the compressive pressure between the connected parts

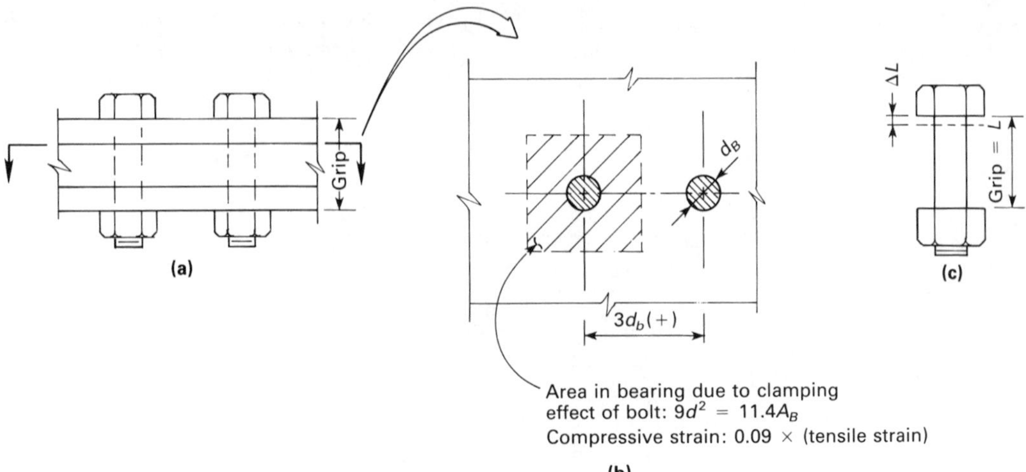

Figure 13.10 Bolt elongation and compression strain.

*Actually, the length within the grip remains unchanged, but the nut moves up the thread to take up the change ΔL.

(in other words, relieving the pretension). If $T \leq T_i$, strain, ε_i, is unchanged. *Therefore, stress in a pretensioned bolt is unchanged due to externally applied tension unless the external effects cause larger tensile forces than the initial pretension.*

Several consequences follow, both good and bad:

1. When high-strength bolts are pretensioned up to their proof load, $0.70F_u$, they do not suffer loss of safety or strength in resisting externally applied tension. For permissible levels of externally applied tension, there will be no change in tensile stress but some reduction in intersurface bearing pressure.
2. The pretensioning of rivets and A307 bolts, not measured and relatively low in magnitude, may be exceeded by externally applied tension at permissible levels, F_tA_b. Although not unsafe, since final $f_t < F_t$, slip within the now frictionless joint may occur at lower shearing force than otherwise.
3. Cycles of tensile stress caused by a repetitive externally applied tension cause a stress range, $f_{sr} = f_{max} - f_i$. The bolt threads are severe notches, corresponding to unfavorable categories for determining F_{sr} (AISC, Appendix B).
   a. If f_i is low, as in rivets and A307 bolts, f_{sr} may well exceed appropriate values of F_{sr}, leading to fatigue failure. The AISC Specification, Appendix B, gives no values for F_{sr} on such connectors and, in paragraph B3.2, recommends against their use.
   b. If $f_i >$ applied f_t, as would apply for A325 or A490 bolts, $f_{sr} = 0$. Such bolts are recommended by AISC for repetitive externally applied tensile loads. However, f_t is limited to F_t in Table 13.2, where $F_t = T/A$ and T includes the prying effect. If the prying effect is high and/or number of cycles high, values of F_t are reduced (AISC, Appendix B, paragraph B3.1) in accordance with Table 13.4.

TABLE 13.4 VARIATION OF F_t WITH CYCLING LOADS PER AISC SPECIFICATION, APPENDIX B (A325 OR A490 BOLTS)

No. of cycles	% Prying effect	F_t^*
≤20,000	All	Table (1.0)
20,000 to 500,000	≤10	Table (1.0)
	>10	Table (0.60)
500,000 to 2,000,000	≤5	Table (1.0)
	>5	Table (0.50)

*Table = F_t as given in Table 13.2; applies to externally applied tension, including its prying effect. Values are applied to shank area of bolts.

4. By reducing intersurface bearing pressure, externally applied tension reduces the friction available for *friction-type* high-strength bolted joints. As discussed previously, this does not reduce the ultimate strength of such connections if holes are of standard size. In other cases, or if the nonslip action of friction-type joints is important to the performance of the structure, it is necessary to reduce the allowable nominal shear stress (F_v) in proportion to the loss in normal intersurface pressure. The reduction factor provided in the AISC Specification is discussed in the next section.

13.2.11 Connectors in Combined Shear and Tension

Figures 13.3(b) and (c) showed one simple case of a connection where some bolts or rivets must resist shear and tension simultaneously. There are many such cases. It will be recalled that the principal tensile stress on an element subjected to simultaneous uniaxial tension and shear is larger than the tensile stress on the axis considered. Some mechanism is necessary to maintain safety in the presence of such combined effects. The method of the AISC Specification is to reduce the value of F_t for the effect of the shear, where tensile stress, $f_t \leq F_t$, is considered to act in the longitudinal direction of the connector. The adjusted values of F_t are given in AISC Specification, Table 1.6.3, reproduced herein as Table 13.5. The adjustments are linear, with F_t declining at a constant rate with increasing values of f_v. (*Note:* This is f_v, not F_v.) All equations are cut off at F_t values given in Table 13.2.

In friction-type connections, as long as there is no slip, the factor f_v has no direct meaning, since shear is resisted by surface friction. As discussed previously, externally applied tension reduces the clamping action of those bolts affected by it, thereby reducing the available frictional resistance. To account for this, the AISC Specification (Section 1.6.3) provides a reduction factor, which we can call C_{tf}.

TABLE 13.5 ALLOWABLE TENSION STRESS, F_t^*, FOR FASTENERS IN BEARING TYPE CONNECTIONS

Description of fastener	Threads *not* excluded from shear planes	Threads excluded from shear planes
Threaded parts A449 bolts over 1½-in. diameter	$0.43F_u - 1.8f_v \leq 0.33F_u$	$0.43F_u - 1.4f_v \leq 0.33F_u$
A325 bolts	$55 - 1.8f_v \leq 44$	$55 - 1.4f_v \leq 44$
A490 bolts	$68 - 1.8f_v \leq 54$	$68 - 1.4f_v \leq 54$
A502 Grade 1 rivets	$30 - 1.3f_v \leq 23$	
A502 Grades 2 and 3 rivets	$38 - 1.3f_v \leq 29$	
A307 bolts	$26 - 1.8f_v \leq 20$	

*For wind or seismic loading, multiply constants by 1.33. Do not increase coefficients of f_v.

From AISC Specification, 1978, Table 1.6.3.

$$F_v C_{tf} = \left(\frac{1 - f_t A_b}{T_b} \right) F_v \tag{13-7}$$

F_v is listed under *Friction-Type Connections* in Table 13.2. f_t is the tensile stress due to direct external load, T, averaged over all the bolts in the connection. T_b is the specified pretension load in the bolt. The value of a bolt, derived earlier as

$$VAL = F_v AN \tag{13-4}$$

becomes

$$VAL'' = C_{tf} F_v AN \tag{13-8}$$

13.2.12 Connector Values Summarized: Flow Chart and Tables

Flow Chart 13.1 gives a procedure for determining bolt and rivet values based on allowable stresses, type of connector, and type of stress applied. The AISC Manual,

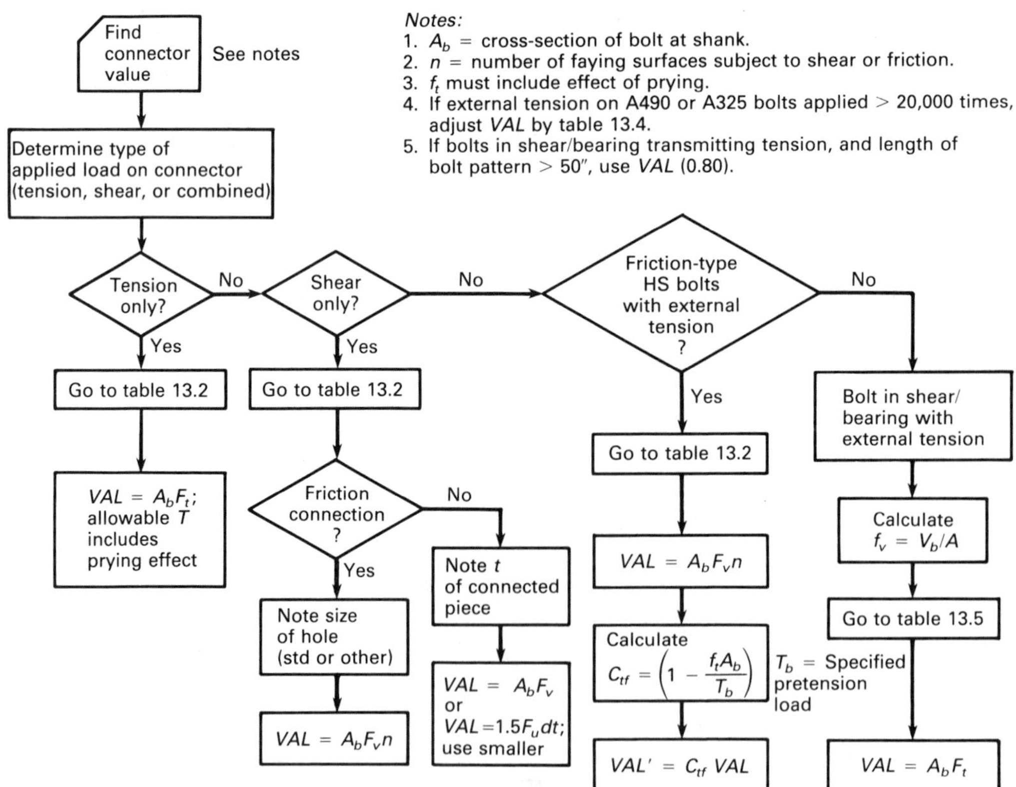

Flow Chart 13.1 Find connector value.

Part 4, Tables I-A, I-D, and I-E, gives connector values for tension, shear, and bearing by connector diameter. Tables I-B and I-C give information on the use of materials other than standard bolt or rivet materials for threaded tension connections.

13.3 DESIGN OF JOINTS WITH MECHANICAL FASTENERS

13.3.1 Joint Types

Table 13.6 illustrates a number of common types of bolted or riveted connections. They all lend themselves to use of any of the types of rivets and bolts discussed earlier. Wherever the applied force is concentric with the centroid of a group of connectors of given type and size, the simple equations (13-2) through (13-4) may be used to determine the number of connectors needed.

Two of the earliest decisions made in joint design are usually the determination of the type and size of connector. Some readers are no doubt asking themselves at this time why it is necessary to deal with so many types when one is clearly superior. After all, Why not the best? The best is clearly the strongest, isn't it? Or is it? If, for example, we standardized on A490 bolts of 1-inch diameter, spaced 3 inches on centers with threads excluded from the shear plane, we would very soon discover the connecting tail is in danger of wagging the structural dog, particularly if the structure is a light one and lightly loaded:

- If, as is common good practice, a criterion of two bolts minimum per connection were used, the minimum connection would have a capacity of $2(31.4) = 62.8$ k (AISC Table I-D).
- The minimum standard thickness of A36 connected steel consistent with the bolt value would be $\frac{7}{16}$ inch (AISC Table I-E), often more than necessary for gusset plates, very likely thicker than the webs or flanges of some of the shapes selected from design loads.
- There would not be enough room to attach the bolts or turn the nuts on angle legs less than about 3½ inches, or flanges of W shapes less than 7 inches wide, forcing uneconomical member selections (see Assembling Clearances Tables, Manual, Part 4).
- The designer's confidence in the quality of inspection would decline in proportion to the distance of the site from population and engineering centers and inversely with the number of similar connections involved in the project.

What is "best" for some structures is "overkill" for some others. There is a place for the full range of connector types (possibly excluding hot driven rivets)* and the full range of sizes.

*Since rivets are uncommon today, further discussion will refer to bolts. In most cases, the principles discussed apply also to riveted joints.

TABLE 13.6 ACTIONS OF SOME TYPICAL BOLTED OR RIVETED CONNECTIONS

Type	Form of connection	Action on members	Action on: Connectors	Action on: Splice material	Comments
A (1)	Lap joints; 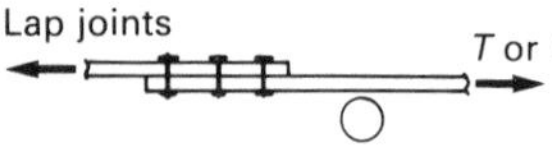 T or C	T or C Tension or compression	Single shear	NA	ⓐ See Fig. 13.1.
(2)	 T or C	↓	Double shear	NA	ⓑ Ends fitted to true plane perpendicular to longitudinal axis by grinding or milling.
	Butt joints;  T or C; ¼″ ± Gap	↓	Double shear	Tension or compression	ⓒ If reversal to tension possible, provide capacity in splice plates and connectors (AISC 1.1.5.8).
(3)	Butt joint, *W* or other shape; 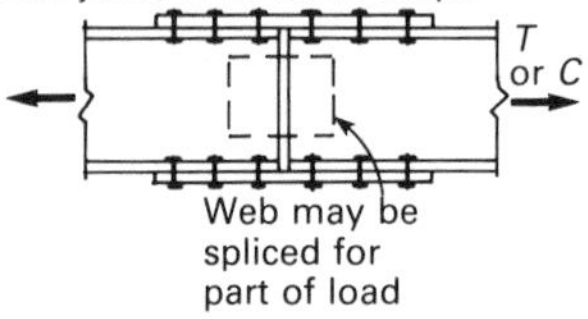 *T* or *C*; Web may be spliced for part of load	↓	Single shear on flange plates; if web splice, single or double shear ⓓ	Tension or compression ⓓ	ⓓ If truss members, provide capacity for 50% of computed load in connecting material (AISC 1.15.8).
B	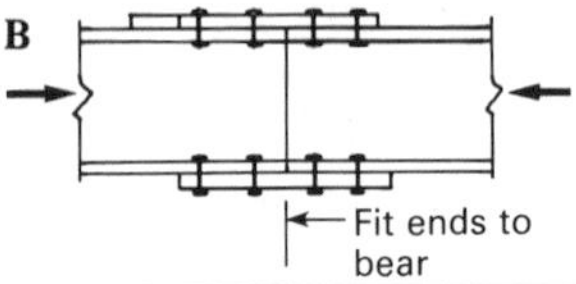 Fit ends to bear	Compression only; direct end bearing ⓑ	Positioning only or "incidental" shear ⓒ ⓓ	Positioning and incidental *T* or *C* ⓒ ⓓ	
C (1)	Lap joint; 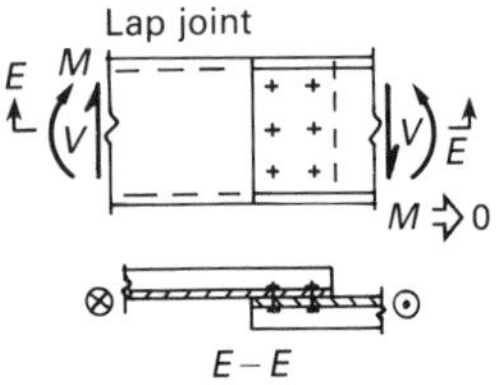 *E*, *M*, *V*; *M* → 0; *E*–*E*	Shear	Single shear	NA	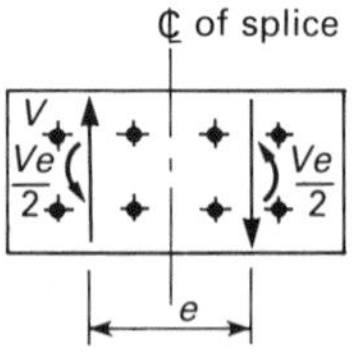 ⓔ ℄ of splice; *V*; *Ve*/2; *e*
(2)	Butt joint; 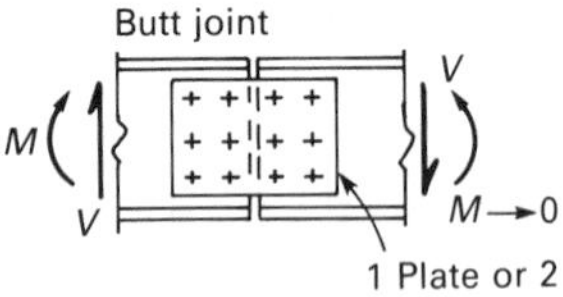 *M*, *V*; *M* → 0; 1 Plate or 2	↓	Single shear if 1 plate; double shear if 2 plates. Effect of load and torque ⓔ	Shear with some moment	ⓕ For web splice plates, see type C2. For flange splice plates, see type A(4). $M = Td = Cd = nf_vA_cd$

TABLE 13.6 *(continued)*

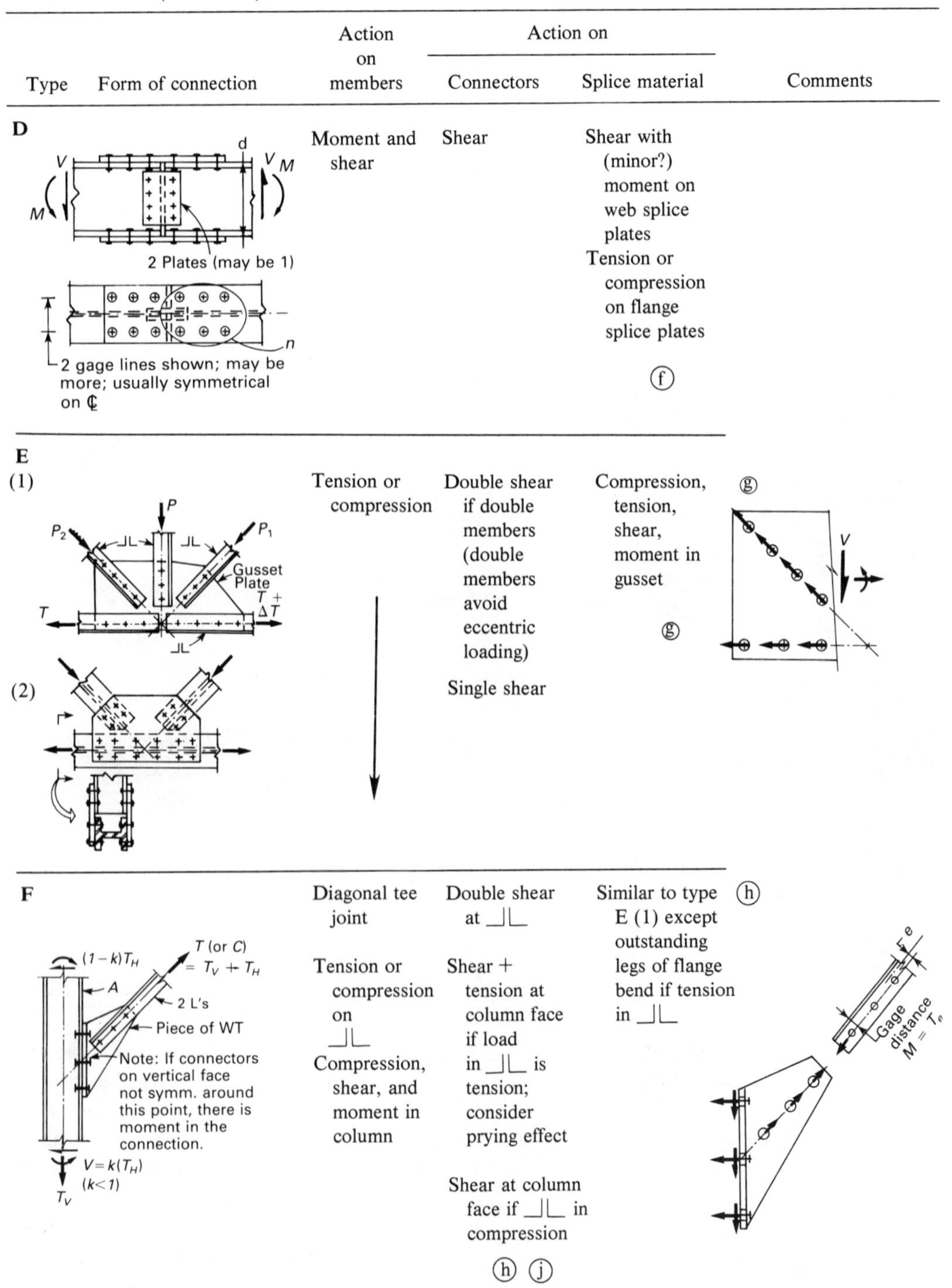

Type	Form of connection	Action on members	Action on: Connectors	Action on: Splice material	Comments
D	d; V; V; M; M; 2 Plates (may be 1); n; 2 gage lines shown; may be more; usually symmetrical on ℄	Moment and shear	Shear	Shear with (minor?) moment on web splice plates Tension or compression on flange splice plates (f)	
E (1)	P; P_2; P_1; Gusset Plate; T; $T + \Delta T$	Tension or compression	Double shear if double members (double members avoid eccentric loading)	Compression, tension, shear, moment in gusset (g)	(g) V
(2)		↓	Single shear		
F	$(1-k)T_H$; T (or C) $= T_V + T_H$; A; 2 L's; Piece of WT; Note: If connectors on vertical face not symm. around this point, there is moment in the connection.; $V = k(T_H)$ $(k<1)$; T_V	Diagonal tee joint Tension or compression on ⅃L Compression, shear, and moment in column	Double shear at ⅃L Shear + tension at column face if load in ⅃L is tension; consider prying effect Shear at column face if ⅃L in compression (h) (j)	Similar to type E (1) except outstanding legs of flange bend if tension in ⅃L	(h) e; Gage distance; $M = Te$

TABLE 13.6 *(continued)*

Type	Form of connection	Action on members	Action on Connectors	Action on Splice material	Comments
G	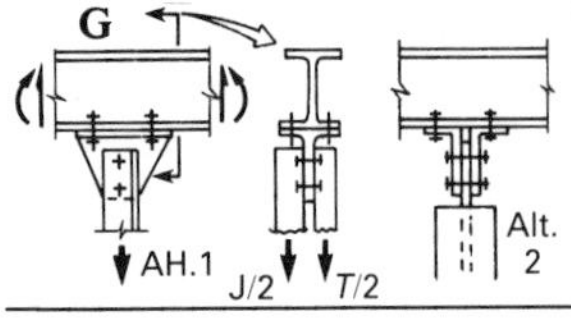 AH.1, J/2, T/2, Alt. 2	Tee joint in tension (compression similar)	Similar to F	Similar to F	(j) 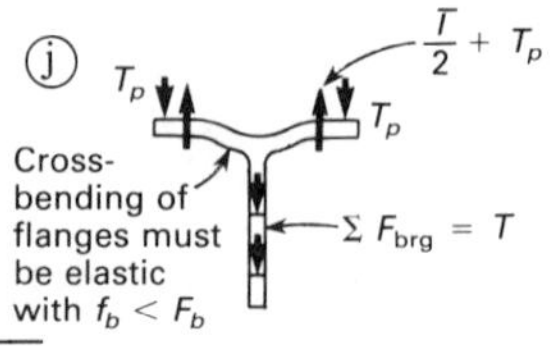 T_p; $\frac{T}{2} + T_p$; T_p; Cross-bending of flanges must be elastic with $f_b < F_b$; $\Sigma F_{brg} = T$
H (1)	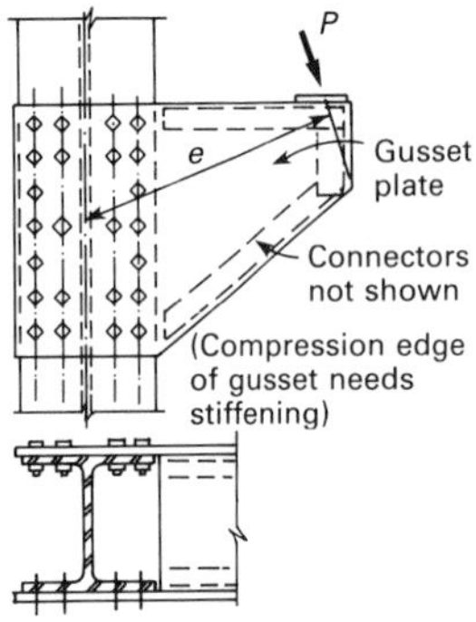 P; e; Gusset plate; Connectors not shown; (Compression edge of gusset needs stiffening)	Brackets	Combined shear from load and torque (k)	Shear, tension, compression, moment	(k) Eccentric action on connectors causes combined shear (see Fig. 13.13).
(2)	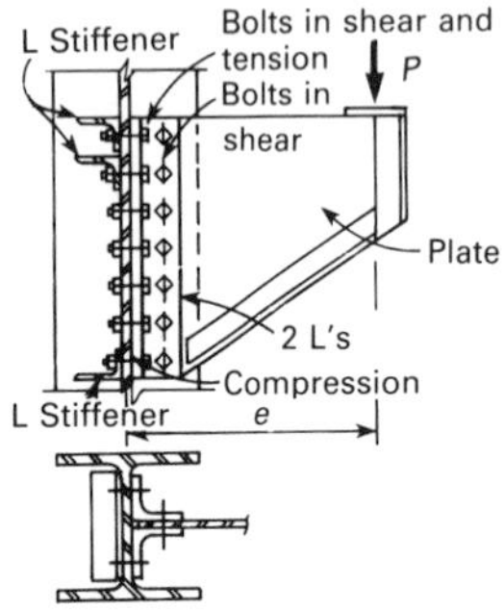 L Stiffener; Bolts in shear and tension; Bolts in shear; P; Plate; 2 L's; Compression; L Stiffener; e		See sketch (l)	Shear, tension, compression	(l) 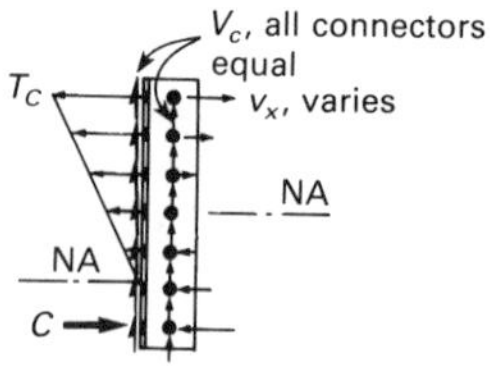 V_c, all connectors equal; v_x, varies; T_C; NA; NA; C

On any particular project, it is common and economical to select, early on, the type of connector and size that the lead designer judges to best suit the design. Sometimes it may turn out to be reasonable to use two sizes of the same type. It is seldom reasonable to make separate decisions for each joint. Having predetermined the type and size, the joint designer has then to decide quantity and arrangement of bolts, form of joint and necessary connecting plates, and so on.

The connections shown in Table 13.6 correspond in all cases with similar welded connections in Table 12.8. Some of the connections in Table 12.8, particularly groove welded butts, cannot be duplicated with bolts. For the most part, the discussion of Table 12.8 in Section 12.9 applies to the joints of Table 13.6 with the simple substitution of a value (VAL) in kips per connector for the weld value in kips per inch of weld. We will not repeat the earlier discussion but will point out some differences.

1. Bolted joint types C(1) and C(2) are somewhat less rigid than their counterparts made with welds and are therefore less likely to show distress from unanticipated bending moment at the joint. If bending moment tries to develop, its magnitude will be limited to that which will cause slip on the faying surfaces at the outer, most highly stressed bolts. A small angle develops between the axes of the two members, consistent with the action of a pinned joint.
2. In joints of types F and G, the gusset plate that was welded to the column or beam flange must be replaced with a gusset in the form of a *tee*, which can be bolted to the column or beam flange. This eliminates, in the case of tension members, the possibility of lamellar tearing, but requires consideration of cross-bending of the flanges and prying action on the bolts.
3. The gage lines (i.e., lines used in locating lines of bolt holes) are established on angle members, such as in joints of types E, F, and G, at distances from the heel of the angle established for convenience of punching and bolting. This usually differs slightly from the centroidal axis of the member. The resultant eccentricity causes some bending moment in the double angles. This is usually ignored in tension members. It may require checking in compression members.
4. Brackets of type H(1) are similar to their welded counterparts. However, the equations used to determine the bolt shears are slightly different in form from those used for a weld pattern. These will be discussed further later.
5. Brackets of type H(2) do not correspond directly to H(2) of Table 12.8, since that in Table 13.6 is attached to the web rather than the flanges of the column. This has two significant results.
   a. The column web is exposed to transverse bending between the flanges, which may require transverse stiffening to avoid overstress.
   b. The bolts need to develop horizontal components of resistance v_x on the legs attached to the bracket and T_c on the outstanding legs. Each is proportional to the distance from a neutral axis. The position of NA_2 can only be approximated. The magnitudes of T_c must include the prying effect.

Figures 13.11 and 13.12 should be self-explanatory.

13.3.2 Eccentric Load on a Pattern of Bolts

As in the case of an eccentrically loaded pattern of weld discussed in Section 12.9 of the previous chapter, we will present a traditional elastic solution to the problem of an eccentrically loaded pattern of bolts. This approach (see Fig. 13.13) assumes:

1. The effect of forces is shared equally among all the bolts.
2. The effect of torque is to cause shear in each bolt proportional to its distance

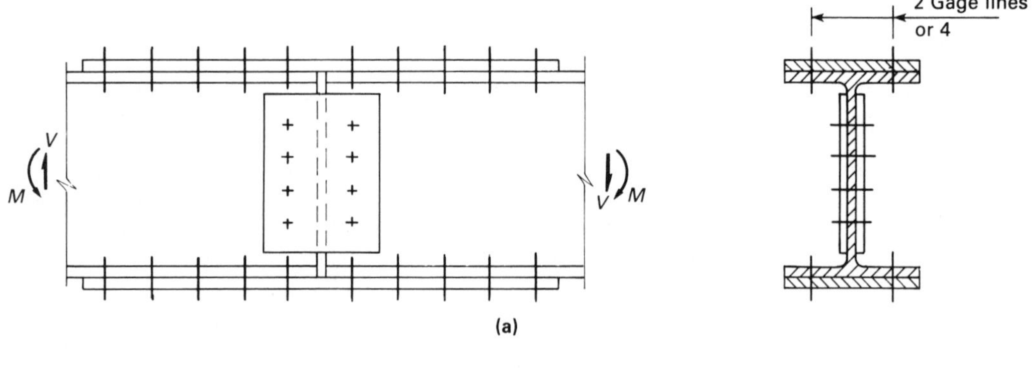

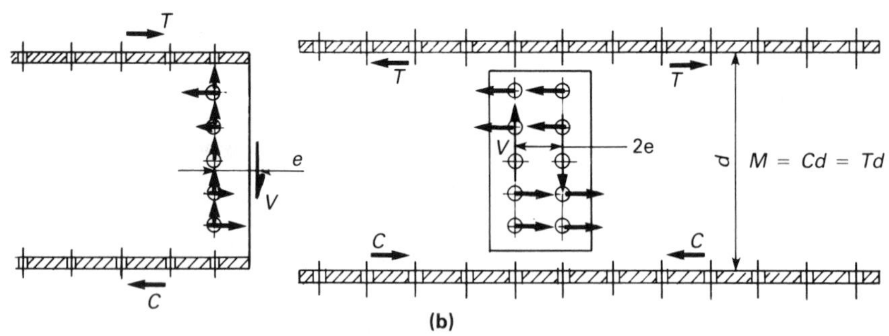

Figure 13.11 Beam splice, moment and shear: (a) joint; (b) equilibrium of parts.

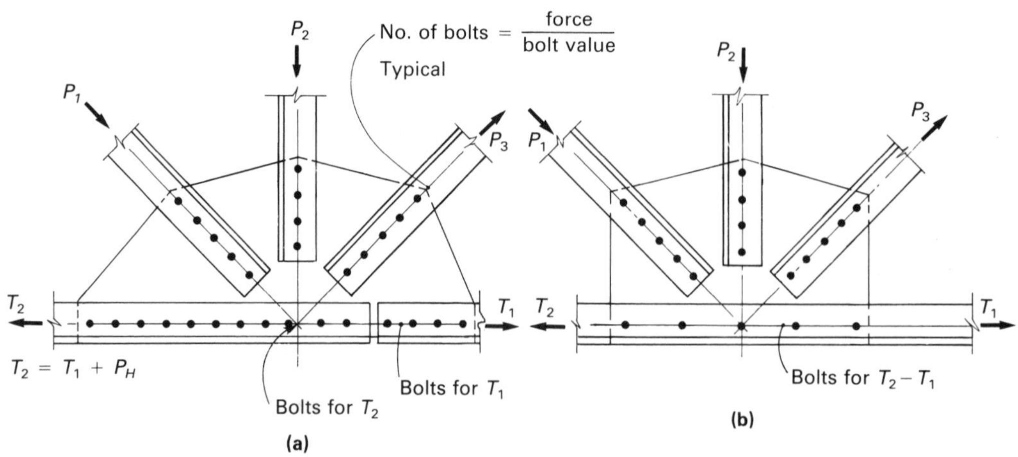

Figure 13.12 Gusset plated joint: (a) chord splice at joint; (b) chord continuous through joint.

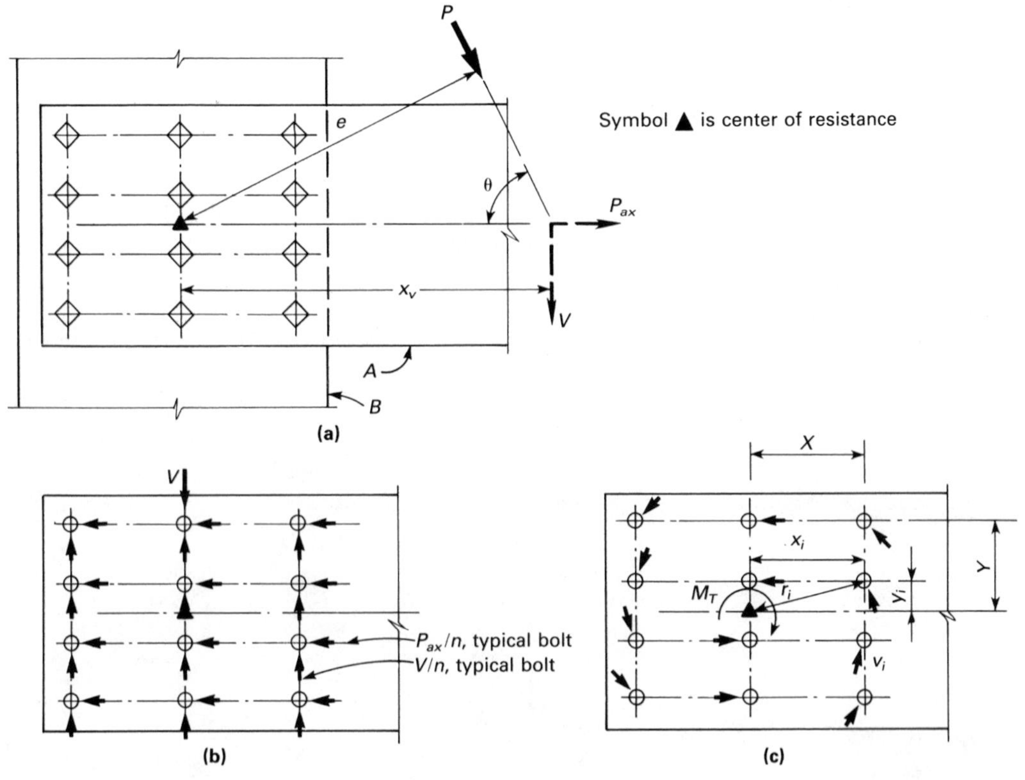

	Effect of	Bolt force coefficient		$J = \Sigma(x_i^2 + y_i^2)$
		x	y	
ith Bolt	*Forces*			
	P_{ax}	$\frac{P_{ax}}{n}$	0	
	V	0	$\frac{V}{n}$	
	Torque	$\frac{M_T y_i}{J}$	$\frac{M_T x_i}{J}$	$v_i = \left[\left(\frac{P_{ax}}{n} + \frac{M_T y_i}{J}\right)^2 + \left(\frac{V}{n} + \frac{M_T x_i}{J}\right)^2\right]^{1/2}$
Farthest bolt	*Forces*			
	P_{ax}	$\frac{p_{ax}}{n}$	0	
	V	0	$\frac{V}{n}$	
	Torque	$\frac{M_T Y}{J}$	$\frac{M_T X}{J}$	$v_{max} = \left[\left(\frac{P_{ax}}{n} + \frac{M_T Y}{J}\right)^2 + \left(\frac{V}{n} + \frac{M_T X}{J}\right)^2\right]^{1/2}$

(d)

Figure 13.13 Shear on bolt group, eccentrically applied load: (a) joint with eccentric load; (b) effect of forces (torque omitted); (c) effect of torque, $Pe = M_T = Vx_v = \Sigma v_i r_i = \Sigma(v_{ix} y_i + v_{iy} x_i)$; (d) shear in bolts.

from the center of resistance of the pattern, which is its geometric centroid.

3. The resistance to torque in each bolt acts perpendicular to a radius from the center of resistance and provides a resisting torque equal to the force in the bolt times that radius. The sum of all such resisting torques must equal the applied torque.
4. The shear in each bolt is the vector sum of the bolt shear resisting the force and that resisting the torque. The most heavily loaded bolt, which will be considered the critical bolt, is the one whose distance from the center of resistance is greatest.

This approach ignores some moment resistance in the bolts closer to the centroid, which would be available before ultimate failure. AISC Tables X through XVIII of the Manual, Part 4, are based on a less conservative ultimate strength approach, which is explained in the Manual.

For convenience of calculation, the equations in Fig. 13.13 are written in terms of x and y components of the applied and resisting forces:

$$P = P_{ax} \nrightarrow V^{*}$$

The resistance to P for each bolt equals

$$\frac{1}{n}(P_{ax} \nrightarrow V)$$

Resistance to the torque $P_e = M_T$ is

$$\Sigma\, r_i v_i = M_T$$

where the subscripts i refer to any ith bolt.

$$r_i = x_i \nrightarrow y_i$$

$$v_i = v_{ix} \nrightarrow v_{iy}$$

and

$$v_{ix} y_i \nrightarrow v_{iy} x_i = r_i v_i$$

The resisting force on the ith bolt

$$v_{Ti} = \frac{M_T}{J}(x_i \nrightarrow y_i)$$

where

$$J = I_x + I_y$$

$$I_x = \Sigma\, y_i^2(1) \quad \text{in.}^2 \times \text{bolts}$$

*The crossed arrow indicates vector addition.

$$I_y = \Sigma x_i^2(1) \qquad \text{in.}^2 \times \text{bolts}$$

$$J = \Sigma (x_i^2 + y_i^2) \quad \text{in.}^2 \times \text{bolts} \tag{13-9}$$

The dimensions of v_T are, then, kips/bolt if M_T is expressed in inch kips. The x and y components of v_i and v_{max} are each shown in Fig. 13.13 as the sum of a force effect and a torque effect. The bolt shears are calculated as

$$v_i = \sqrt{v_{xi}^2 + v_{yi}^2}$$

The equations for v_i and v_{max} are shown.

13.3.3 Hole Spacing and Detail Dimensions

In the examples that follow, we use a number of detail dimensions not fully explained until later. We will follow some commonly used detailing standards, which conform to the AISC Specification, but are also based on the needs of fabrication efficiency.

- Hole spacing along a line will be 3 inches center to center, useful for bolts up to 1-inch diameter.
- Hole spacing across a member will be based on Table 13.7.
- Edge distances, from center of hole, will conform to provisions and tables in AISC 1.16.
- Standard-sized holes (bolt diameter + $\frac{1}{16}$ inch) are used.

TABLE 13.7 COMMON GAGE DIMENSIONS (INCHES), USING STANDARD HOLES

	Leg (in.)	2	2½	3	3½	4	5	6	7	8
Angles or channel flanges	g	1⅛	1⅜	1¾	2	2½	3	3½	4	4½
	g_1	—	—	—	—	—	2	2¼	2½	3
	g_2	—	—	—	—	—	1¾	2½	3	3
	Max. dia.	⅝	¾″	⅞	1″	Check clearance and edge distance				

Symmetrical flanges	Flange width	3⅝	4±	5 to 5½	6	8 to 14	15
	g	2	2¼	3	3½	5½	5½
	Max. conn.	⅝	¾	¾			
	g_1	—	—	—	—	—	3

In all cases, check clearances for entering and tightening bolts or driving rivets; if gage spacing distance is less than 3 fastener diameters, stagger holes.

13.4 EXAMPLES: DESIGN OF JOINTS

13.4.1 Example: Column Butt Splice

Design a butt splice for a W8 × 24 connecting to the same size member and carrying an axial compression load of 70 k. Steel is ASTM A588. Type and size of connector and size of splice plate to be determined. We may assume that the controlling load for the splice is the 70-k load (but see AISC 1.15.7). An outline of the procedure can be seen in the numbered steps (1 through 5) of the solution.

CALCULATIONS	*REMARKS*
(1) $P = 70$ k, steel = A588	Given
$F_y = 50$ ksi, $F_u = 70$ ksi	Table 1, Manual, Part 1
$P/A = 70/7.08 = 9.89$ ksi	

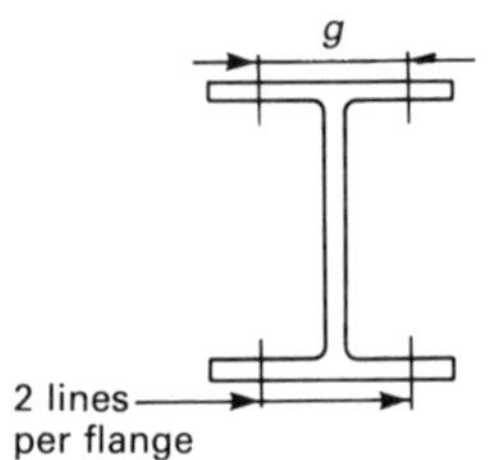

(2) Minimum number of bolts in flanges = 4

(3) Dimensions:

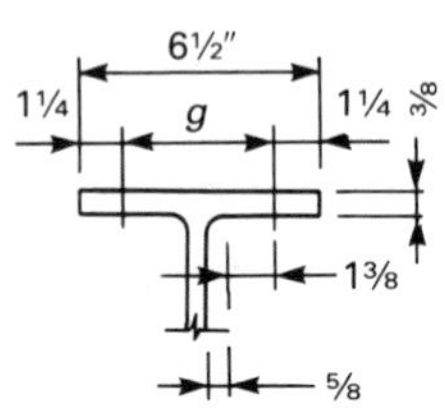

Shape Tables for dimensions. Table 13.7 gives $g = 3\frac{1}{2}$, but 4″ is better for bolt clearance. Long diameter of hexagonal head is $1\frac{5}{16}'' < 1\frac{3}{8}''$ (Manual, p. 4-136); $\frac{7}{8}''$ is possible, but tight.

Use 3/4″ diameter max:

$$A_b = 0.44$$

(4) Bolt values:

Type	F_v	Single shear value, kips
A307	10.0	4.4
A325: Bearing	21.0	9.24
Friction	17.5	7.70
A490: Bearing	28.0	12.32
Friction	22.0	9.68

F_v from Table 13.2. Values for bearing-type allow threads in shear plane. Compare results to Manual, Part 4, Table I-D.

Use 8 bolts, A325, bearing with threads in shear plane *or* 16 bolts, A307.

Either is acceptable
Joint with A325 bolts is more compact.

(5) Splice plate: $P = 70/2 = 35$ k. If plate $= 6.5 \times t$,

$$f_a = \frac{35}{6.5t} = \frac{5.38}{t}$$

Assuming $KL/r = 40$ for plate to avoid buckling and bolts at 3″ centers.

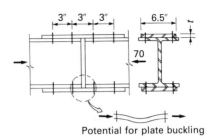

Potential for plate buckling

$$\frac{KL}{r} = \frac{0.65(3)\sqrt{12}}{t} = 40$$

$$t \gtrsim 0.17;$$

Then

$$F_a = 25.83$$

$$t \geq \frac{5.38}{25.83} = 0.21''$$

Use $t = 1/4''$ if develops bolt in bearing.

$$\text{value in bearing} = 1.5F_u td \qquad (13\text{-}5)$$

$$= 1.5(70)(0.25)(0.75)$$

$$= 19.7 \text{ k} \gg 9.24$$

Use 6½ × ¼ splice plates.

(6) Summary

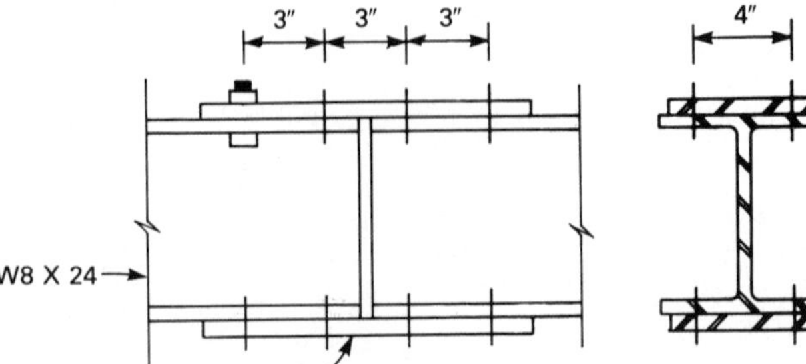

Symbol ϕ is bolt diameter. N indicates threads not excluded from shear plane.

13.4.2 Example: Beam Splice

Splice a W24 × 104 beam to a W24 × 68. At the location of the splice, $V_{max} = 110$ k and $M_{max} = 350$ ft k. Use $\frac{7}{8}''$ diameter A325 bolts in friction-type connections. Beam steel is ASTM A588. Capacity of members has been checked previously.

CALCULATIONS	*REMARKS*

(1) Sketch joint:

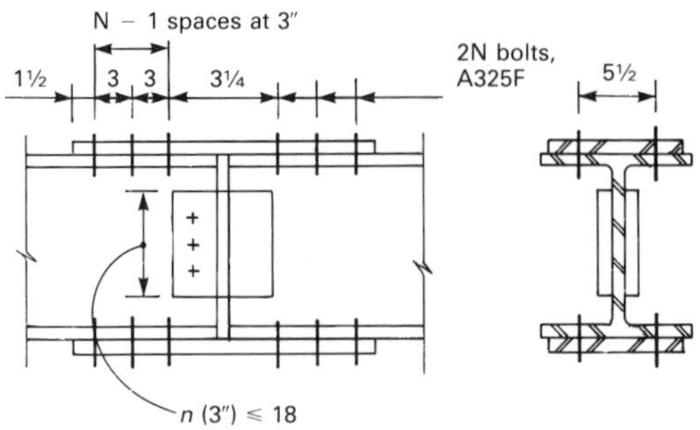

(2) Dimensions from Shape Tables:

Table 13.7 for $g = 5\frac{1}{2}$

	W24 × 104	W24 × 68
b_f	12.75	8.97
t_f	0.75	0.59
d	24.06	23.73
t_w	0.50	0.415
T	21	21

Shape Tables: Gage based on narrow flange, 2 gage lines only, fills not required.

Confirms 18″ limit on web holes.

(3) Bolts for flange splice for $M_{\max}$:

$$\frac{M}{d} = \frac{350(12)}{24} = 175$$

$$= \text{Shear on bolt group}$$

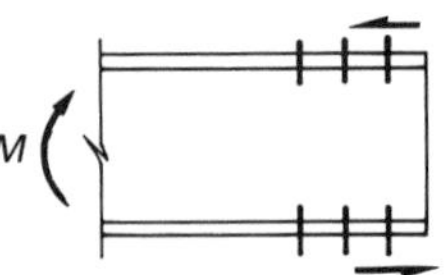

Connector value: for 7/8″ diameter A 325F, $V_c = 10.5$ k

Manual, Table I-D, Part 4

Check t_f for $V_c = 10.5$ k:

$$t_f d(1.5)F_u = 0.59(0.875)(1.5)(70)$$

$$= 54.2 \gg 10.5$$

Requires 175/10.5 = 16.7 bolts; use 18.

(4) Bolts for web splice for $V_{\max}$:

$$V = 110 \text{ k}$$

$$M_T = Pe = 110(1.625)$$

$$= 179 \text{ in. k}$$

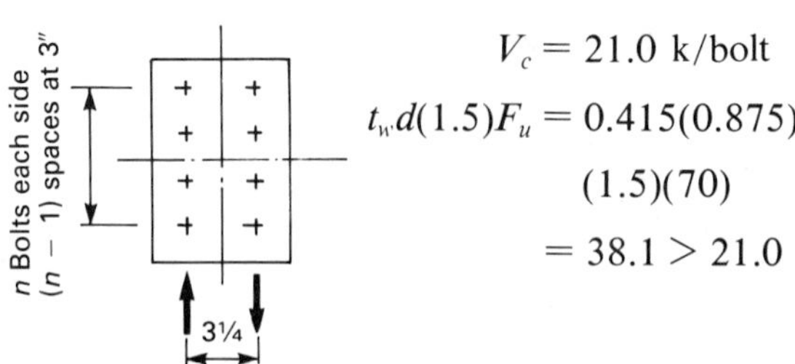

$V_c = 21.0$ k/bolt

Manual, Table I-D, Double Shear

$t_w d(1.5)F_u = 0.415(0.875)(1.5)(70)$

$= 38.1 > 21.0$

Spaces	Bolts	J	Y	V/n	$\dfrac{M_T Y}{J}$	$(v_b)_{max}$
2	3	18	3.0	36.7	29.8	47.3
3	4	45	4.5	27.5	17.9	32.8
4	5	90	6.0	22.0	11.9	25.0
5	6	158	7.5	18.3	8.5	20.1
6	7	252	9.0	15.7	6.4	17.0

Use 5 spaces, 6 bolts each side.

(12-10) with $P_{ax} = 0$; $X = 0$

$J = \Sigma y_i^2$

$$(v_b)_{max} = \left[\left(\frac{V}{n}\right)^2 + \left(\frac{M_T Y}{J}\right)^2\right]^{1/2}$$

(5) Size flange plates; use 9 × 5/8 PL

(6) Size web plates; use 2 at 6¼ × ¼

Reader may verify plate requirements, based on earlier chapters.

(7) Summary:

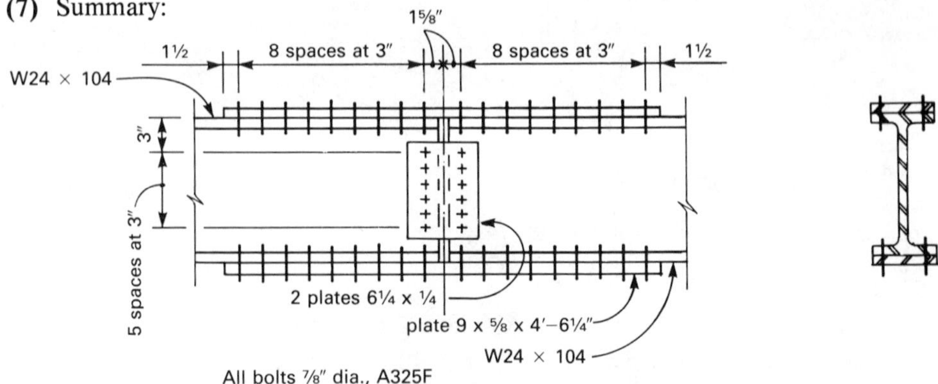

The moment splice is done separately from the web splice. Each is sufficient for the maximum value of its function. The splice plates almost duplicate the area of the portion they are splicing. If the beam was at its full moment capacity, the flange splice would require more area than the flange since some moment is resisted by the web of the beam. Similarly, if the full shear capacity of the beam is utilized, the shear area of the two web plates should equal $t_w d$, at least.

13.4.3 Example: Bracket Connection to Column

The bracket of Example 12.9.11 is to be connected to the W14 × 193 column with A325 bolts in shear/bearing. Gage lines shown on the column are based on Table 13.7. Choose a bolt size and design a bolt pattern.

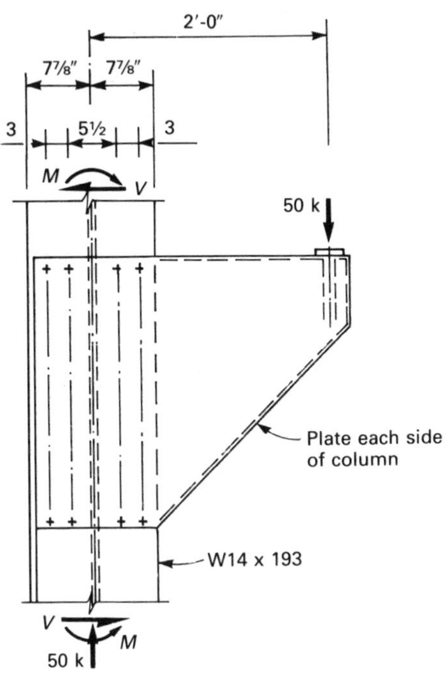

CALCULATIONS	REMARKS
(1) Shear and torque to one face: $$V = 25 \text{ k}$$ $$M_T = 25(24) = 600 \text{ in. k}$$	
(2) If outer bolts of pattern resist 90% of torque and load: $$(M_T)_{90} = 0.9(600) = 540 \text{ in. k}$$ $$V_{90} = 0.9(25) = 22.5 \text{ k}$$	A convenient approximation to simplify the calculation of J. The inner bolts are the least useful in resisting torque.
(3) Find v_{max} on corner bolt:	

(a) Vertical spaces, n	3	4	5	6
(b) No. outer bolts, n_1	12	14	16	18

(c) Σy_i^2, in.^{2}	171	324	540	828
(d) Σx_i^2, in.^{2}	295	361	427	493
(e) J, in.^{4}	466	685	967	1321
(f) Y, in.	4.5	6.0	7.5	9.0
(g) X, in.	5.75	5.75	5.75	5.75
(h) $(M_T)_{90}\ Y/J$, *kips* →	5.21	4.73	4.19	3.68
(i) $(M_T)_{90}\ X/J$, kips ↑	6.66	4.53	3.21	2.35
(j) V_{90}/n_1, *kips* ↑	1.88	1.61	1.41	1.25

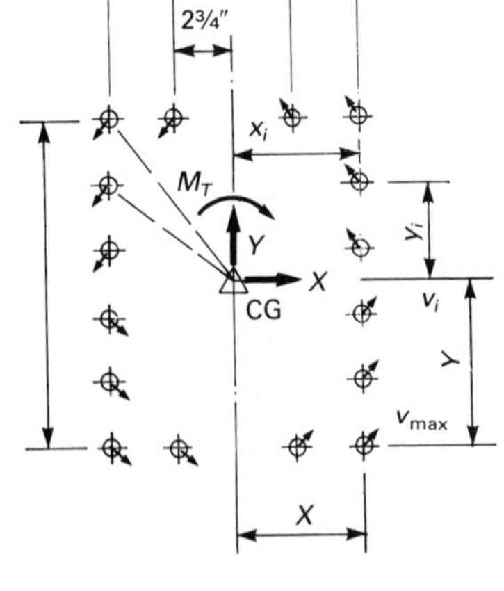

See. Fig. 13.13.

$$J = \Sigma(x_i^2 + y_i^2)$$

(k) v_{max}, kips	10.0	7.75	6.24	5.15	$[(h)^2 + (j+i)^2]^{1/2}$

(4) Select bolt size and number of vertical spaces. From Table, A325 N, SS, 3/4″ diameter, value = 9.3 k.

Table I-D, Manual
N = threads not excluded from shear plane
SS = single shear

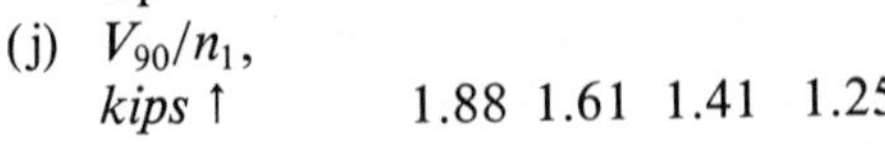

As above

$$J = 685$$
$$+ 2(1.5)^2 = \underline{\quad 5}$$
$$690$$

No. of bolts = 16

$$\frac{M_T Y}{J} = \frac{600(6)}{690} = 5.22$$

$$\frac{M_T X}{J} = \frac{600(5.75)}{690} = 5.00$$

$$\frac{V}{\text{No.}} = \frac{25}{16} = 1.56$$

$$v_{max} = 8.38 \text{ k} < 9.3$$

(5) *Use 3/4" diameter bolts, A325N in pattern as in part (4).*

Since torque was a large part of the problem, it was decided to place most of the bolts on the perimeter of the bolt pattern. The approximation of part (2) turned out to be useful and did simplify the calculation of J. Actually, the two inner bolts are not necessary by calculation since $(1.0/0.9)(7.75) < 9.3$. However, they help to bring the faying surfaces into contact.

13.5 BOLTED BEAM-END CONNECTIONS

Recall the discussion of flexible and rigid welded connections in Section 12.10. The same principles apply to bolted connections. The assumption that a beam end is free to rotate without encountering moment resistance requires flexible end connections. The assumption that the members are continuous at a rigid joint requires rigid connections.

Flexible connections can be made with bolts in ways very similar to those with welds. The connections in Fig. 13.14 resemble those in Figs. 12.26 and 12.27. The clip angles in Figs. 13.14(a) and (b) must be made light enough to "breathe" (i.e., to yield with little moment resistance). To permit bolting of the outstanding leg, it must be wider than its welded counterpart, and the top bolts must allow for some prying effect, in this case a small one.

There is a potential for shear failure in the *blocked* beam web that is shown in Fig. 13.14(b). Although little of the beam web is removed by the end block, the web can tear out on the line *abc*, which experiences a combination of shear and tension. The AISC requires a special calculation for this case, which treats the net area along *abc* as an area totally in shear and uses a special value for F_v.

$$F_v = 0.30F_u \qquad \text{(AISC 1.5.1.2.2)}$$

$$A_v = t_w(\overline{ab} + \overline{bc} - nd)$$

where n and d are the number and diameter of the holes. The allowable end shear is

$$V_{\text{ALL}} = F_v A_v = 0.3F_u t_w(\overline{ab} + \overline{bc} - nd) \qquad (13\text{-}10)$$

Figures 13.14(c) and (d) show two other devices to achieve flexibility at end connections.

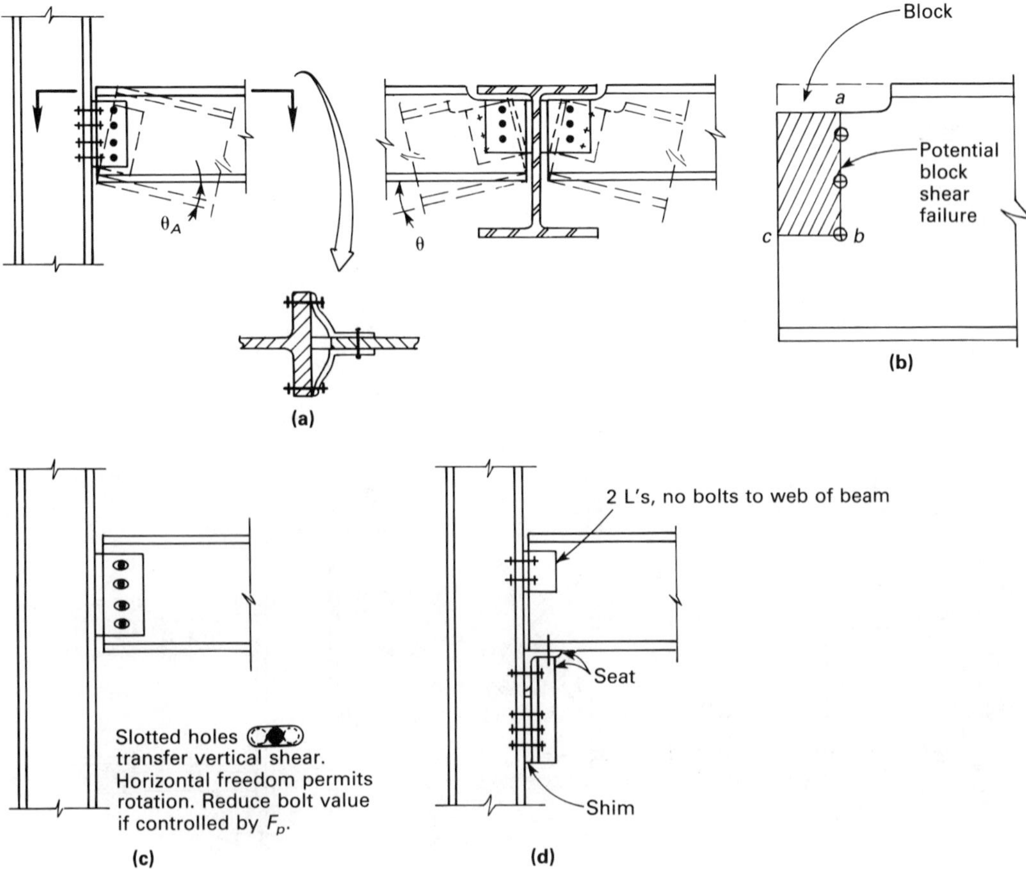

Figure 13.14 Flexible connections.

The AISC Manual, Part 4, has tables of standard connections for bolted angle clips similar to those for welded clips.

In trying to devise rigid beam connections, we soon confront the limitations of bolted joinery. It is necessary in this case to avoid the small relative rotations between members that we encouraged in the flexible connections. This is perfectly feasible in continuous beam splices such as in Figs. 13.11 and 13.15(a). However, it is much more difficult in beam to column connections such as in Figs. 13.15(b), (c), and (d). The connecting pieces must become very heavy or the joint itself very bulky, and even then these connections are suitable for relatively small values of moment. Transverse bending of outstanding legs of connecting pieces must be fully elastic to limit displacement. It would be instructive to compare these to the simplicity of the

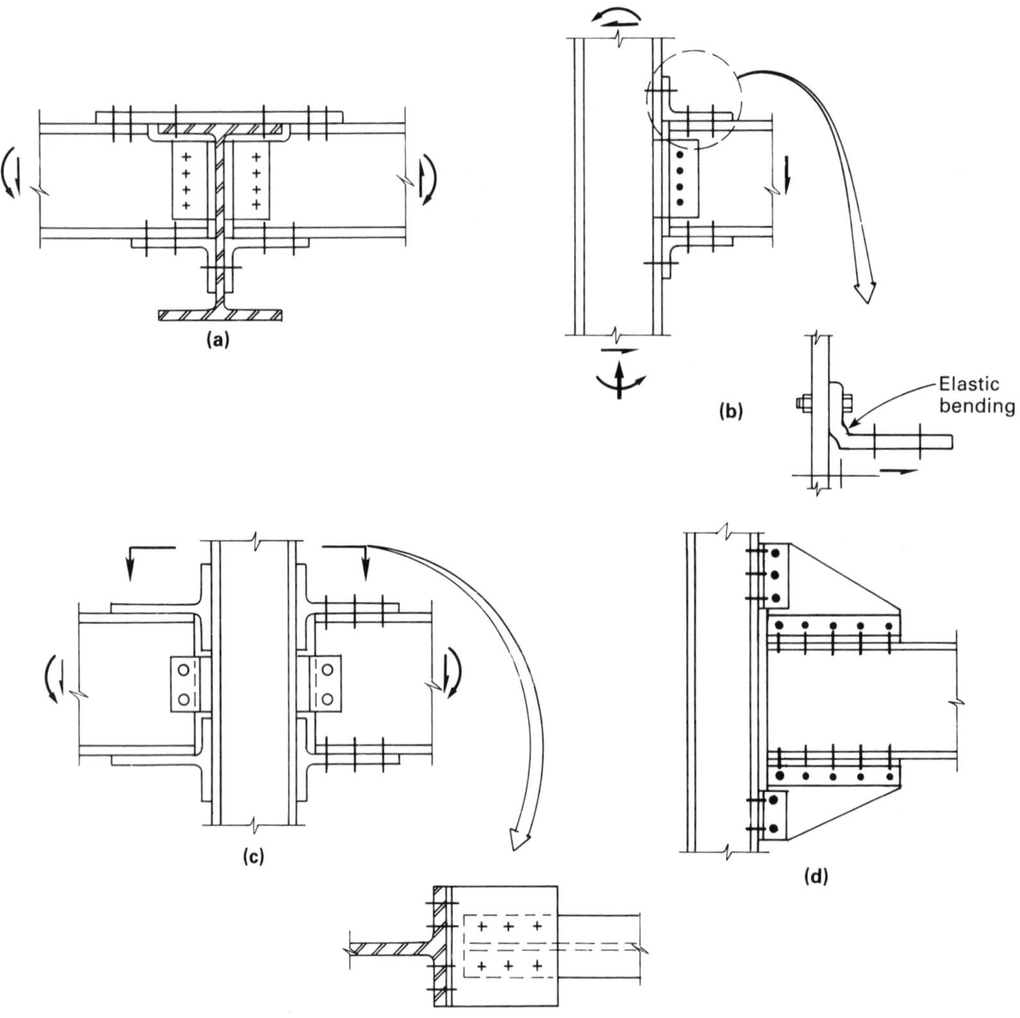

Figure 13.15 Rigid beam-end connections.

connections in Fig. 12.29, where the moment limitation is set by the strength of the members.*

We will not present an example of a flexible connection. However, the reader is invited to redo Example 12.10.3 with bolted clip angles. We will suggest a procedure to design a rigid connection for shear and moment similar to that in Fig. 13.15(c), but connected to the web of the column.

*Once again, we must caution the reader to take measures to avoid lamellar tearing of the column flanges.

13.6 DESIGN PROCEDURE: BOLTED RIGID BEAM—COLUMN CONNECTION

In the analysis of the same building as in Example 12.7.1, beams B2 are connected to columns and act as part of a ductile* moment-resisting frame for seismic (earthquake) loads. Joints must be rigid.

The following data apply to the connection of B2 to the interior columns at the third floor (data come from the calculations in Chapter 18).

Beam B2: W16 × 40
Column: W12 × 211
Steel: ASTM A572, Gr. 50

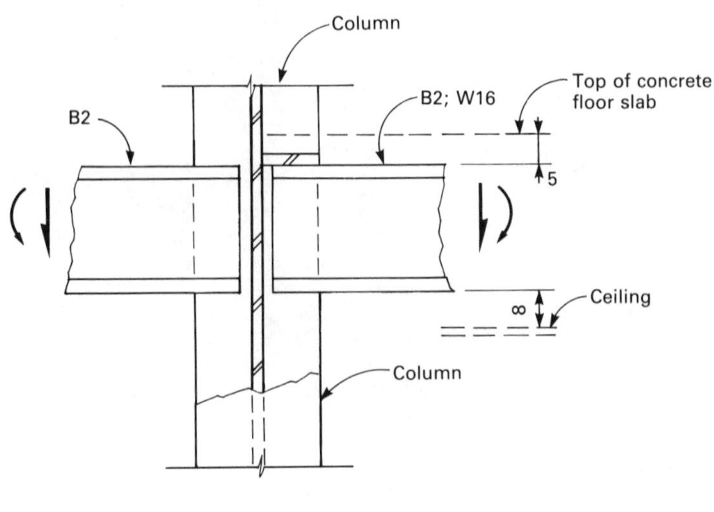

	$D+L$	EQ
End shear, kips	21.5	±5.1
End moment, ft kips	−101	±64

Design a rigid connection. Use A490X bolts.† (*Note:* The information given locating the top of concrete and ceiling is a constraint on joint design. The connection is not to intrude into the room.)

*A ductile moment-resisting frame is designed so that in a major earthquake its joints can yield and absorb large amounts of energy in the range of ductile strains without collapse. In less severe, "design" earthquakes, it is expected to act elastically. The forces and moments described represent elastic design conditions.

†In joints of this type, A490X bolts prove their value. Their high shear values permit the use of few bolts in compact joints, as compact as possible to make with bolts.

Procedure

1. Compare $D + L$ and $D + L + EQ$ loads to decide which controls. Allow $F(1.33)$ for $D + L + EQ$.
2. Sketch joint, using dimensions from Shape Tables.

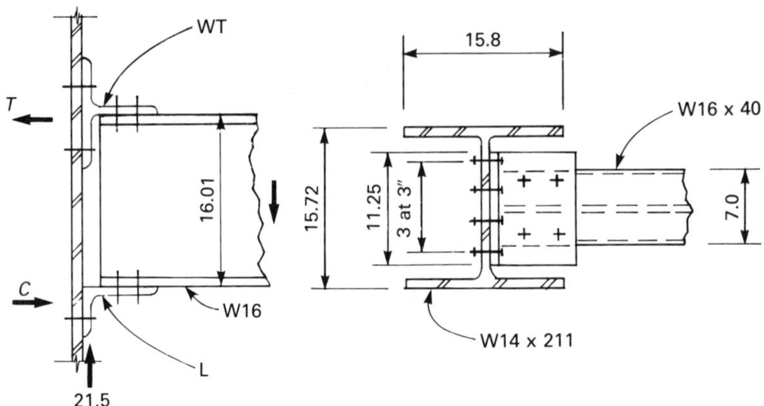

$$C = T = \frac{M}{d} = \frac{124(12)}{16} = 93 \text{ kips}$$

3. Examine form of joint for feasibility under given loads.
4. Select a size of A490X bolt, after dimensional study. List single shear value.
5. Select an upper WT. Analyze for force T.
   a. Number and location of bolts, web of WT to beam flange.
   b. Number and location of bolts, flange of tee to web of column.
   c. Elastic bending of flange of WT under pull by force T and reacting tension bolts with prying effect.

   This may require several iterations.
6. Select lower seat angle. Analyze to transfer force C.
7. Check bolts on vertical leg of seat L for design shear, V.
   a. If OK, use as is.
   b. If NG, consider separate web connection.
8. Draw entire joint to scale to verify dimensions and search for difficulties in assembly and function.
   a. If OK, stop.
   b. If NG, revise.

13.7 INCIDENTALS IN DESIGN OF BOLTED JOINTS

13.7.1 Long Grips

Up to now, our consideration of the action of connectors ignored the length of the grip, the space between heads and nuts occupied by connected material. The idealization of connector action in Fig. 13.1(b) omits from the free body the bending moment in the pin that is necessary to keep it in equilibrium. The connector acts like a beam; in the process it flexes and deforms, in proportion to M/EI. The section modulus of a solid round beam is $(\pi/4)R^3$, its moment of inertia $(\pi/4)R^4$; its shear resistance grows with its area, πR^2. The reader will appreciate the importance of the ratio grip/diameter to the flexural adequacy of the connector. What may not be quite as obvious is the role of pretension. In highly pretensioned high-strength bolts with head and nut bearing on thick connecting plates, the end conditions approach full fixity, with consequent reduction of moment in the bolt. Rivets and A307 bolts are somewhat less effective.

The problem of long grips arises in multimember (more than three) connections or joints with very thick parts. It is relatively uncommon. For most bolts or rivets, the ratio of grip/diameter is small. Based on performance tests on joints, the AISC (1.16.3 and C1.16.3) provides a simple rule for coping with it. Except with high-strength bolts, if the grip is more than five times the diameter, the calculated number of required bolts is increased at the rate of 1% for each 1/16 inch of excess length.

13.7.2 Washers

The clamping force of mechanical connectors is applied to the connected parts by the heads or nuts. The assistance of washers is necessary only in special cases.

1. Several cases are cited in the "Specification for Structural Joints Using A325 or A490 Bolts," where hardened washers are required. The reader is advised to refer to the joint specification. A similar provision applies to A449 bolts (AISC 1.16.1).
2. When a connected part has nonuniform thickness (e.g., flanges of channels or S beams), tapered washers are required with bolts so that the faces of head and nut are parallel (Fig. 13.16).

13.7.3 End and Other Cuts

In many of the figures of this chapter, a gap is shown between the end of a connected member and the member adjacent. That is not a defect of drafting. The gaps are intended. Where the connection is designed for full bearing, no gap is permitted. In

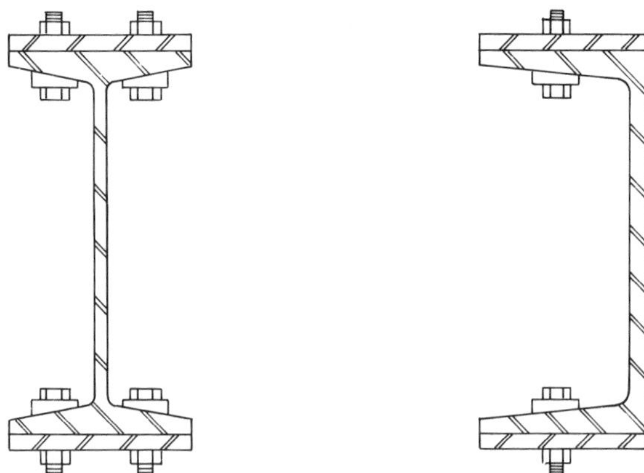

Figure 13.16 Use of tapered washers.

other cases, a gap of ¼ to ½ inch is advantageous to make for easy erection and compensate for mill tolerances on member sizes and inaccuracy of shop work.

Flange *blocks* have been seen in this chapter and are often seen on jobs. Figure 4.6 shows a number of other instances of cutouts in members. They are a common source of crack initiation unless large radii are used in reentrant (inside) corners, a problem that requires careful control by the designer.

13.7.4 Housekeeping

Careful design of joints can help the user of a building to maintain it free of unnecessary debris and with minimum corrosion. Details should be arranged so that rainwater is free to drain off, so that particles of dirt do not accumulate in troughs and corners, and so that all parts of the work are accessible for cleaning and painting.

13.7.5 Fills

Design requirements often lead to the need to connect members of different thicknesses or different depths. The column splice in Fig. 13.17 is a case in point. Beam and other splices are similar. In Fig. 13.17, two W14 columns are being spliced with bolts. Although nominally of the same depth, the actual depths differ by 0.76 inch. The gap must be filled so that faying surfaces can meet.

It is common, as done here, to make the fills total about ⅛ inch less than the nominal difference to ease the erector's problem in bringing the joint together. The theory is that tightening of the bolts will effectively close the small gap, a useful idea and reasonable if the thicknesses of parts are not too great and high-strength bolts are used.

The fill can cause a change in the action of the bolt, causing it to distort across the gap represented by a neutral fill. To avoid this, the fill plate itself must be made an

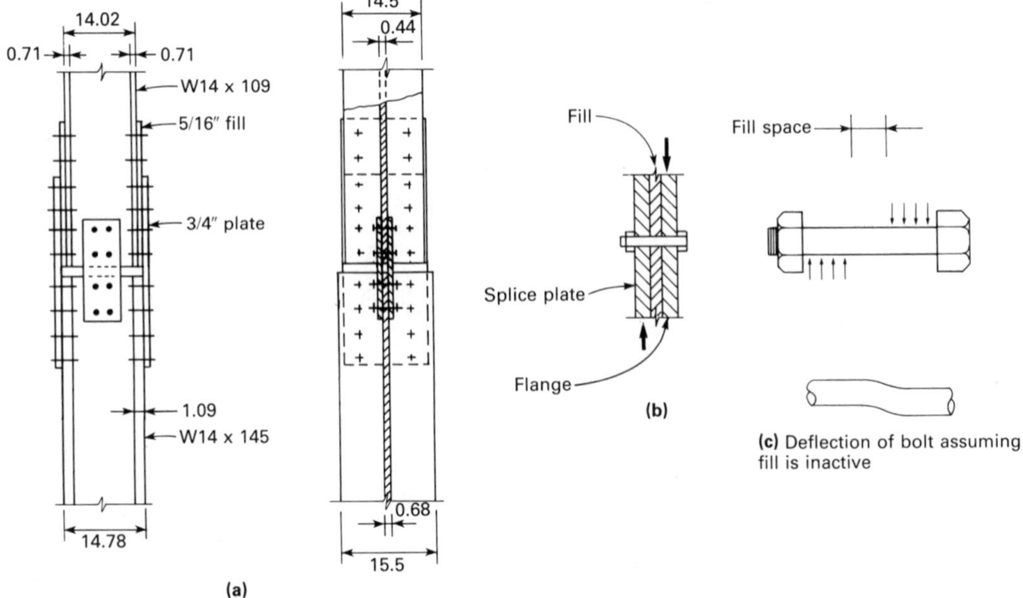

Figure 13.17 Column splice: W shapes with fills.

integral part of one of the connected parts. In the case shown, the 5/16-inch fill is bolted to the flange beyond the limits of the splice plate. The AISC Specification requires (1.15.6) that there be enough such extra bolts so that a proportionate share of the member stress is first transferred to the fill plate before the splice begins.

If $A_f = 14.5(.71) =$	10.30 in.^{2}	for the W14 × 90
and $\Delta A = 5/16\,(14) =$	$\underline{4.38}$	for the 5/16-inch fill
Combined A	14.68 in.^{2}	

of which the fill accounts for 30%.

If the stress in the column beyond the splice is, say, 18 ksi, then 0.3(10.3)(18) = 55.6 kips must be transferred by bolts to the fill plate before the splice begins. We have assumed that the web splice will carry its share of the total load that is in the web. Friction connections using high-strength bolts are excluded from AISC 1.15.6.

13.7.6 Holes: Sizes and Slots

Refer to Table 13.2. Values of F_v for *friction bolts* differ depending on the type of hole. AISC 1.2.3.4 gives more detail on the uses of the three types of hole and their dimensions. Bolts in *bearing-type connections* require *standard* holes whose diameter is

$$\text{hole diameter} = \text{bolt diameter} + 1/16 \text{ in.}$$

Friction bolts are permitted oversized or slotted holes. With such holes, washers are necessary if the specially sized hole is in an outer ply, and the design may not rely on the backup strength of the bolt in bearing after slip. Bolts in tension or high pretension with oversized or slotted holes need washers.

Oversized or slotted holes are useful for at least two purposes:

1. To speed up erection.
2. For slotted holes, to permit slip in one direction while transferring load in the perpendicular direction. See Fig. 13.14(c).

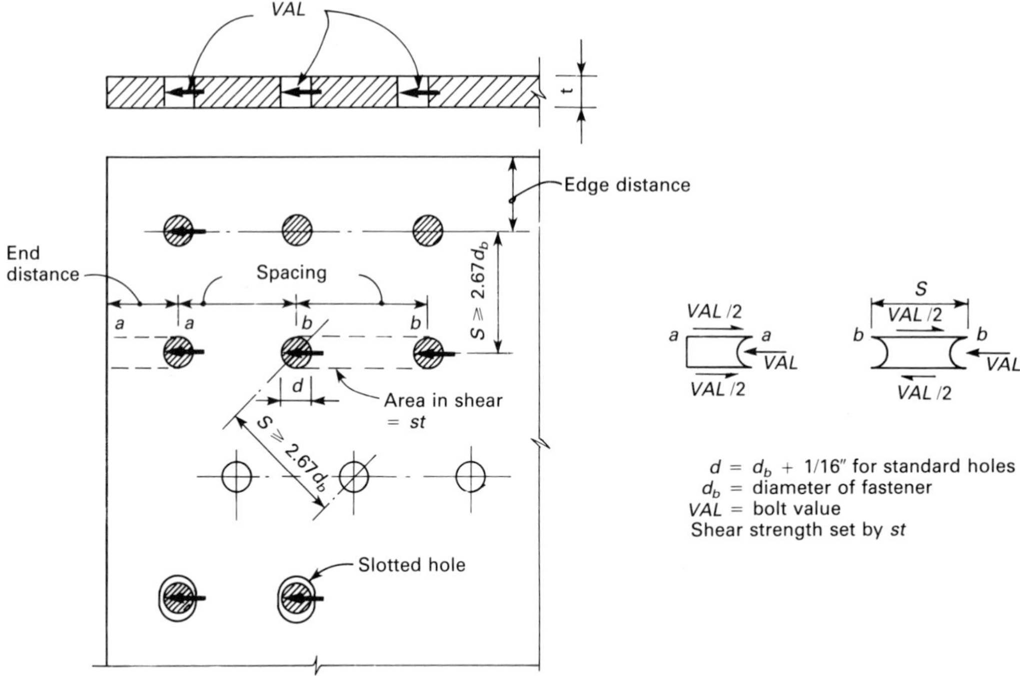

Min. $S = 2.76d_b$ for standard holes; increase per AISC 1.16.4.2 for oversized or slotted holes. May increase in direction of force.

Max. S may be set by need for corrosion protection on faying surfaces, watertightness as in tanks or plate buckling if plate is in compression.

Min. edge distance: see AISC Table 1.16.5.1; modify per AISC 1.16.5.4. Modify per AISC 1.16.5.2 or 1.16.5.3 based on direction of stress.

Max. edge distance $= 12 \times$ plate thickness $\leq 6''$.

Figure 13.18 Hole spacing (see also Fig. 13.2).

TABLE 13.8 TABLES OF FASTENER DATA IN AISC MANUAL, PART 4, 1978

Table no. (not in manual)	Page in 1978 manual	General title	Subjects and remarks
13.8.1	4-132	Threaded fasteners, assembling clearances	Minimum space for entering and clearing bolts
13.8.2	4-133	Rivets and threaded fasteners, field erection clearances	See title
13.8.3	4-135	Rivets and threaded fasteners	Dimensions and conventional symbols; usual gages for angles
13.8.4	4-136, 4-137	Threaded fasteners	Dimensions, bolt heads and nuts
13.8.5	4-138 through 4-141	Threaded fasteners	Weights, dimensions of threads
13.8.6	4-142 through 4-145	Clevises, turnbuckles, sleeve nuts, recessed pin nuts	Dimensions of these connecting devices

13.7.7 Spacing of Holes

The spacing of holes (see Fig. 13.18) is governed by a combination of constraints, some involving practical construction limitations, some stress flow as interpreted through tests. Bolts spaced closer than three bolt diameters on centers are difficult to tighten, since the long diameter of a nut is more than two bolt diameters and wrenches require additional room. An absolute minimum spacing is specified by AISC at 2.67 diameters. As is indicated in Table I-E, Part 4 of the AISC Manual, the most common spacing in a line of bolts is 3 inches. Automatic machines are often set for that spacing. Minimum spacing applies to the space between any two bolts, in line or not. Other practical limits on hole spacing and arrangement can be seen in the AISC Tables cited in our Table 13.8. The common gage dimensions in Table 13.7 are based on clearance requirements of many punching machines. They are not universally observed.

From the point of view of stress, requirements are set on minimum spacing between holes in the direction of the applied force. Not only must a connector be strong enough to develop, before failing, its *value* times the factor of safety; but the connected material between holes must not fail in bearing or shear at a lesser load.

Here we concern ourselves with the strength of the connected material, which in the case of high-strength bolts has much lower yield and rupture strength than that of the connectors. Figure 13.18 may be used with AISC 1.16.5 to interpret the requirements for the spacing of holes.

13.7.8 Procedure to Lay out Holes

A procedure to lay out a group of standard-sized holes for a flange splice, using one splice plate, would be as follows:

1. Determine the type of connector and diameter. — Usually a decision made early for the entire project.
2. Find the connector value: — F_v from Table 13.2; one shear plane for this case.

$$V_c = (1)\frac{\pi d^2}{4} F_v$$

3. Record flange dimensions and F_y, F_u. — Member has been selected previously. Use Shape Tables.
4. Check bearing capacity of flange at hole. — In bearing, $VAL = 1.5F_u\,dt$.
5. VAL in bearing $\geq V_c$?
   a. If yes, use V_c.
   b. If no, revise connector value.
6. On sketch of flange, determine gage dimensions for connectors: — See Tables 13.7 and 13.8. Check Fig. 13.18 and AISC 1.16.5. Check y with AISC Table 1.16.5.1. $x = (g/2) - k_1$ dimension of beam; $x \geq$ long head diameter of bolt $\times \frac{1}{2}$.

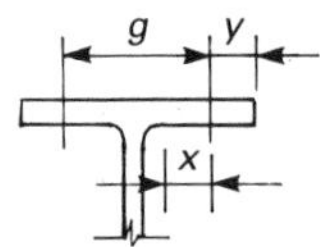

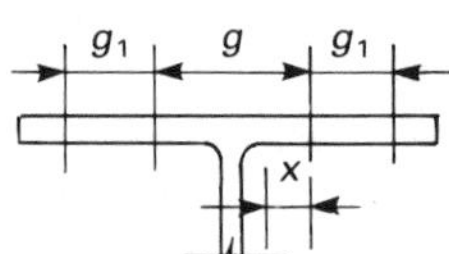

Revise connector size if necessary. If $g_1 < 2.67d_b$, establish staggered bolt pattern.

7. Tentatively, select spacing, s, of connectors in line of stress. — For example, use 3 diameters or 3 inches.
8. Check required s in direction of stress. Revise s if necessary in that direction. — AISC 1.16.4.2
9. Set edge distance. — AISC Table 1.16.5.1
10. Check required end distance in direction of stress. Adjust if necessary. — AISC 1.16.5.2

11. Calculate number of connectors required: — V is force to be transferred by this part of splice. $2n_R$ is required for a full splice plate.

$$n_R = \frac{V}{V_c} \quad \text{or} \quad = \frac{V}{V_c'}$$

12. Use even number $\geq n_R$.
13. Lay out holes on members to be spliced. — Allow for gap between members.
14. Select flange plate: — AISC Table 1.16.5.1
    a. Width $\geq \Sigma g + 2 \times$ minimum edge distance.
    b. Thickness $\geq VAL/1.5F_u d_b$.
    c. A_g (and A_e if tension) to suit requirements of F_t or F_a.
15. Lay out holes on splice plate to match holes in members. — Allow for gap between members.

If the two beams to be spliced are not identical, the lighter beam would usually control the spacing of holes. Use the same spacing on the heavier beam, adjusting the gage if necessary to clear a thicker web.

13.8 MIXED JOINTS: BOLTS AND WELDS

It has no doubt occurred to the reader that, since there are sometimes advantages accruing to welded joints and sometimes to bolted joints, there must be times when a mixture is advantageous. A mixed joint was shown in Fig. 12.1(a). Others will be found in the Mill Building Case Study (Chapter 17). In that case, the rationale, a common one, is that shop fabrication will be less expensive with welded joints, while field erection costs would favor bolts for joining large assemblies. Under the difficult conditions that exist in trying to hold large assemblies in position for joining high in the air, the steel erector finds it very convenient to have prepunched holes, which, once lined up within the tolerance of a hole diameter, can be brought into complete alignment by driving a tapered *drift pin*. When all holes at the periphery of two large assemblies are matched, the relative positions of the assemblies automatically conform to the planned position that was developed on templates (patterns) in the shop. Sometimes, even in joints designed as all-welded joints, a steel erector will find it helpful to have a few holes for *erection bolts* which can be used in positioning the assemblies in the field.

The joint shown in Fig. 12.1(a) presents no problems of compatibility, since each side of the joint can be designed separately. If, on the other hand, the designer wishes to use both bolts and welds to share load in the same connection, they are often not compatible. Figure 12.29 showed conditions where a designer has supplied some bolts for easy fit-up but intends to weld the joint to satisfy strength requirements.

Fabricated steel arrives at an upper floor of a building site. Prepunched holes are prepared for bolting the beams to the columns and to each other. The working surface is light gage steel decking which will later receive concrete as in Figure 7.3. (Photo courtesy of American Bridge Division, United States Steel Corporation)

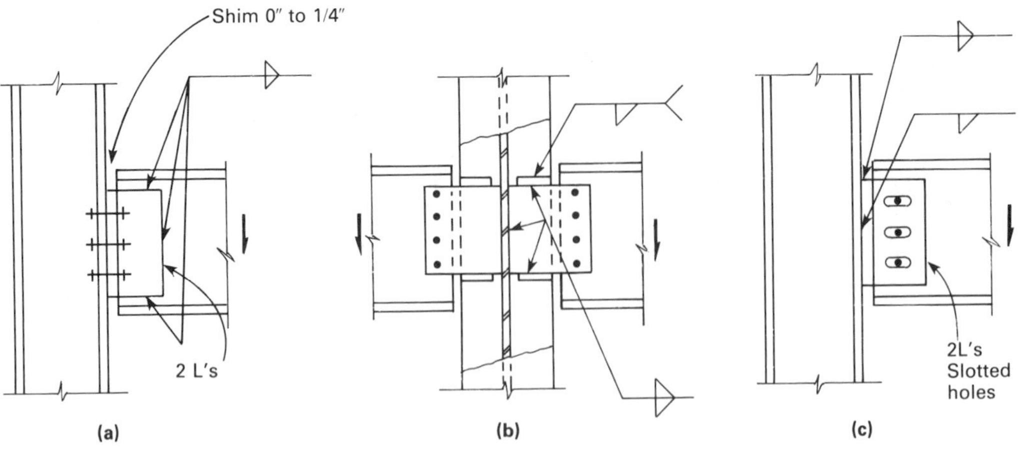

Figure 13.19 Flexible joints, mixed bolts and welds.

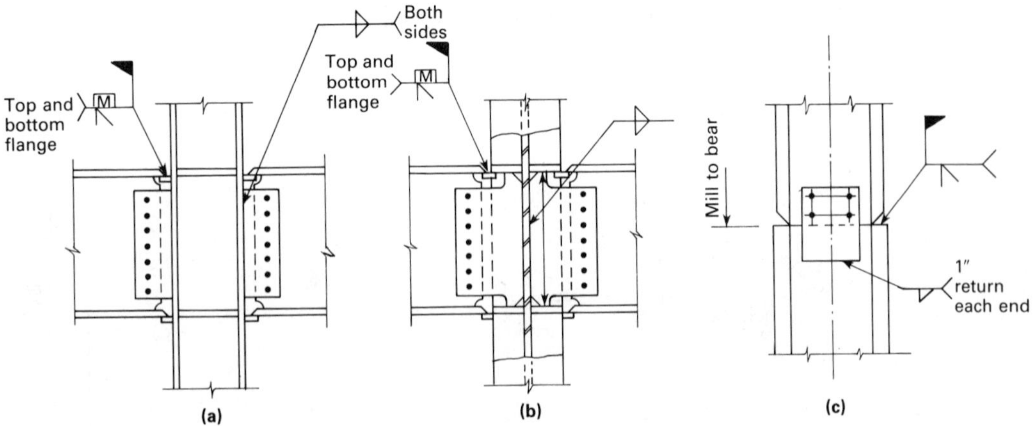

Figure 13.20 Rigid joints, mixed bolts and welds.

The temptation is to subtract the value of the bolts from the requirements of the weld on the assumption that they will "share" the load. The AISC (Section 1.15.10) permits such sharing only if the bolts are *high-strength* type in friction-type connections. Recall Fig. 13.8. In such a case the designer should realize that the reserve, second-stage strength of friction bolts in their ultimate bearing mode may not be available until after the weld has cracked.

Figures 13.19 and 13.20 illustrate a few cases of mixed usage of bolts and welds.

13.9 SOME SPECIAL JOINTS

13.9.1 Pinned Joints

We met pinned joints in Chapter 8 when we were considering the required sections of pin-connected members in tension. In Examples 8.9.3 and 8.9.4, it was not possible to choose the member cross sections themselves without at the same time finding the required pin sizes. Recall that AISC 1.14.5 makes the rules for proportioning pin-connected plates and their pins mutually dependent.

Pin connections are not limited to use with plates. They are the pure form of the theoretical pin-connected joint assumed in many problems in texts on strength of materials. They offer no resistance to rotation, but can transfer force in any direction perpendicular to the axis of the pin. Figure 13.21 shows several types of pinned joints, which can be seen in many structures. The complete joint usually also requires welding or bolting. In the case of Figs. 13.21(a) and (b), we see a recent evolution from pinned joints to supports on elastomeric pads. Elastomers can be selected with low enough compression moduli to allow end rotation while supporting high pressure. We will meet a number of pin-connected joints in the Stiffleg Derrick Case

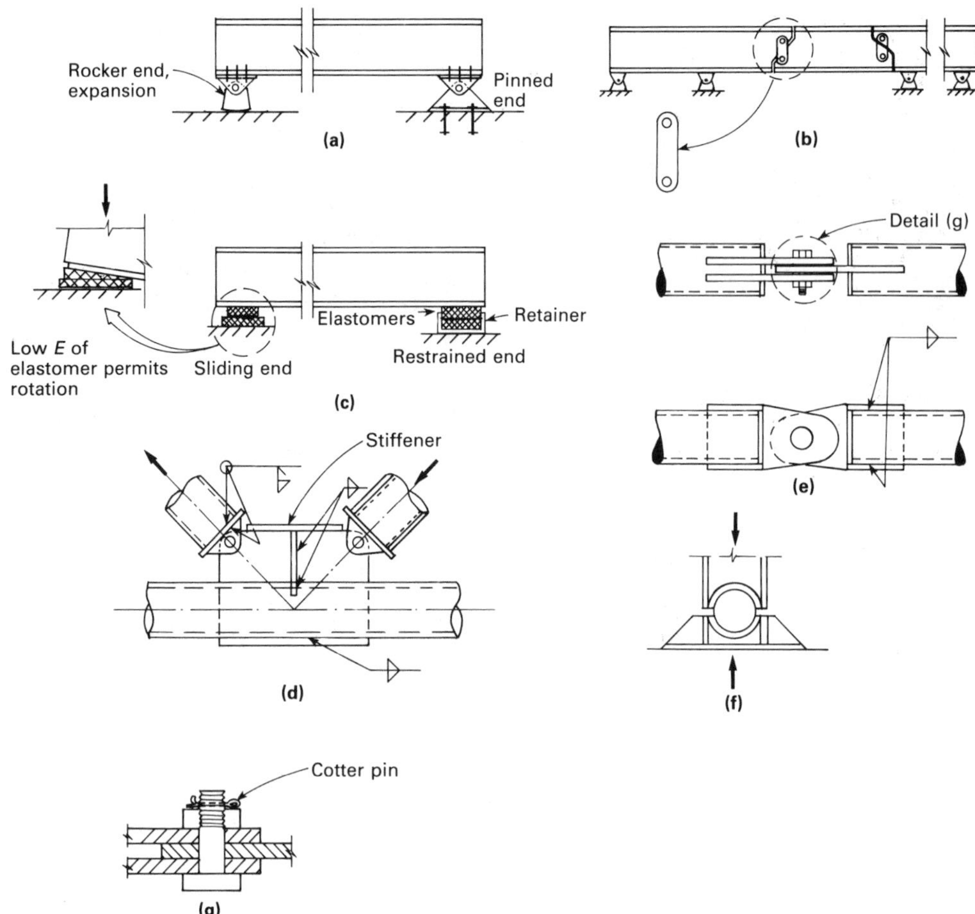

Figure 13.21 Pinned joints.

Study. In that system, the advantages of easy assembly and disassembly make pins advantageous for many nonmoving members, while the requirement to raise and lower the boom makes a heel pin a necessity.

Pinned joints are not normally brought up tight as are bolted joints, raising the possibility that nuts could vibrate off from vibration or working of the joint in use. Figure 13.21(g) shows one of a number of types of details used to keep a nut in place while avoiding undesirable tightening of the joint.

13.9.2 Column Bases

Figure 13.22 shows several conditions that often exist at the base of columns, where load must be transferred to the foundation, and base details that are suitable for these

C, large
Shear, small
Mill end for bearing
Friction resists shear
Drypack grout
Oversized holes for bolts
Excess shear transferred to concrete in bearing on side of rods

(a)

C, large or small
Shear, small
Alt. 1
Alt. 2

(b) Similar to (a) except as shown

Large C and/or small T
Shear, large
Fill-hole for grout
Drypack
Grout
Tube

(c) Similar to (a) except as shown

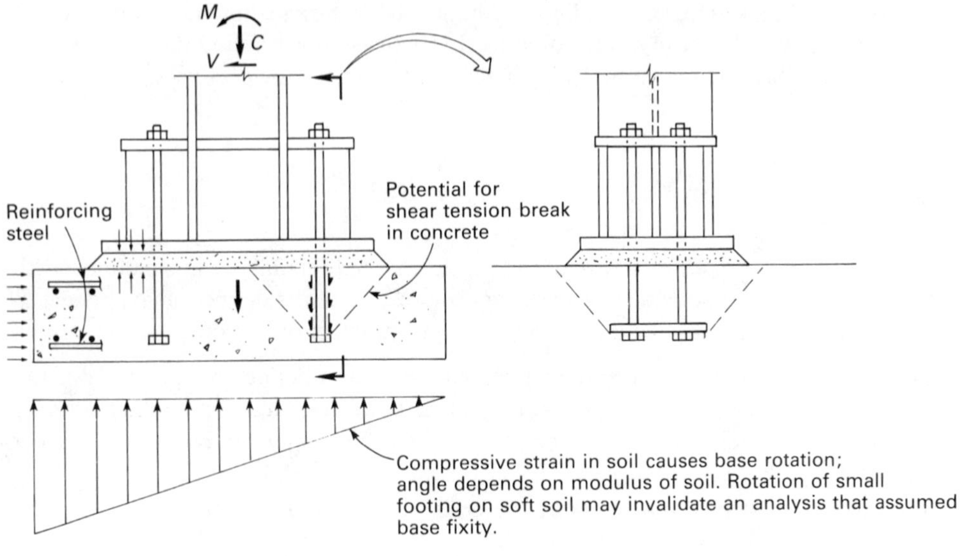

Reinforcing steel
Potential for shear tension break in concrete
Compressive strain in soil causes base rotation; angle depends on modulus of soil. Rotation of small footing on soft soil may invalidate an analysis that assumed base fixity.

(d)

Figure 13.22 Column base details.

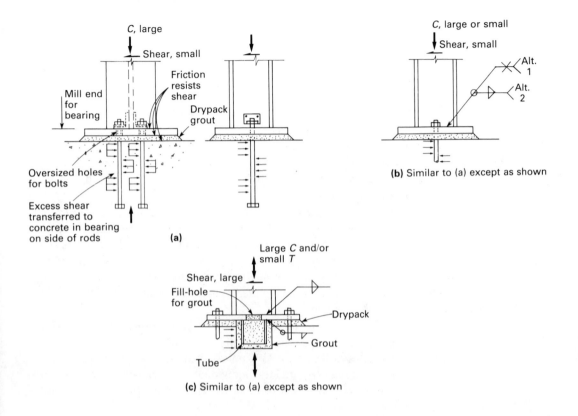

conditions. For the most part the reader can see the reasoning behind the details by studying the figures. A few comments here will help.

1. Anchor bolts are usually placed in oversized holes for ease of erection.
2. Base plates are necessary to transfer the load from steel with high capacity for compressive stress to concrete with much lower compressive strength. We met a similar problem in Chapter 10. See Fig. 10.15 and the accompanying discussion for base plate design. See AISC 1.5.5 for allowable bearing stress, F_p, on concrete or other masonry.
3. If uplift effect is high in Fig. 13.22(c), the column must be well anchored to a footing that weighs more, by a safety factor, than the maximum calculated uplift load. The calculated net uplift is usually the result of the dead load minus the uplift effect of any loads causing overturning. Most codes (e.g., UBC paragraph 2311[33]) establish a safety factor for such a calculation. The usual practice of assigning *conservative* (i.e., overestimated) values of dead load is *unsafe* in calculating net uplift.
4. When high base moments are found in analysis, as when ends are assumed *fixed*, the design of the foundation must prevent rotation. This may not be possible if the modulus of the soil is low, unless the foundation is supported on piles. With firm soil, it may require large individual footings with low maximum soil pressure or rigid combined or *mat* footings. The corner columns of the mill building (Chapter 17) exhibit net uplift that forces an increase in the weight of footings. The applicable calculation is in that study in the section on design for lateral forces. The tier building of Chapter 18 is designed on the assumption of full fixity at the base of the columns. In that case if it is found that full fixity is not feasible, the entire analysis, or at least that of the lower stories, would need revision.

13.10 GLOSSARY

The following are special terms used in this chapter relating to connections with mechanical fasteners. Generally understood terms or terms defined in Appendix B are omitted. Some terms here have different meanings in other contexts.

Bearing-type bolt (or ***connection)*** Bolt whose value is defined by the larger of the shear strength of the bolt or the bearing capacity of the plate–bolt contact surface. Also called *shear/bearing*. (Do not confuse with *body-bearing* bolts, which are a special type not discussed here.)

Butt Joint Joint made by bringing the ends of two pieces together before adding overlapping splice material.

Connector Bolt or rivet or other mechanical fastener.

Develop To have the same strength as . . . ; e.g., a connection may develop the strength of a connected member.

Double shear Shear on two plates of same connector.

Edge distance Distance from the center of a connector hole to an edge (Fig. 13.18).

End distance Edge distance in the direction opposite to applied load.

Fastener Connector.

Faying surfaces Overlapping surfaces.

Fill plate In a splice of two pieces of different thicknesses, a plate that compensates for the difference so that the splice plates can remain straight (Fig. 13.17).

Friction-type connection (or ***bolt)*** A connection whose strength is defined by the friction available on faying surfaces that are subjected to normal forces by tightening a group of HS bolts.

Gage Transverse distance between lines of fasteners or from the heel of an angle to a line of fasteners (Table 13.7).

Grip The length of a fastener between its heads.

Gusset plate (or other) A plate or other device to which several members are connected in a joint.

High-strength bolt (HS) A bolt made from steel conforming to ASTM-A325 or A490. Sometimes used for large-diameter bolts made of ASTM-A449 Steel.

Lap joint Joint made by overlapping two or more plates prior to fastening.

Machine bolt Term commonly used for A307 bolts.

Pin A steel cylinder, usually larger than bolts or rivets, used as a mechanical fastener.

Pretension As used with high-strength bolts, internal tension within the grip caused by tightening of the nut.

Rigid joint A connection joining two or more members in which the angles between member axes do not change under load. The joint as a whole may rotate.

Shank The unthreaded portion of the body of a bolt or pin.

Shear/bearing See *bearing-type bolt.*

Shim Fill.

Single shear Shear on one plane of a connector.

Spacing Distance between centers of fasteners. When used along a line of fasteners, sometimes called *pitch* (Fig. 13.18).

Splice Two-member connection.

Splice plate Plate used in a butt splice. The splice plate laps both members.

Value The allowable load on a fastener (sometimes imprecisely called *capacity*).

SYMBOLS

C_{tf} Coefficient reducing F_v when external tension is applied to a high-strength friction bolt. See Eq. (13-7).

F_N	Normal force
g, g_1, etc.	Gage distances
J	Polar moment of inertia
N	Number of shear planes in a joint
N_c	Number of connectors
HS bolts	High-strength bolts
S	Spacing
VAL	Value (of a fastener)
VAL'	Portion of value (of a fastener) determined by its value in bearing within one ply. May reduce the total value of the fastener. See Eq. (13-5).
VAL''	Modified shear value of a high-strength friction bolt when external tension is applied. See Eq. (13-8).
F, N, or X	When used with A325 or A490 bolts, "friction," "threads not excluded from shear plane," "threads excluded from shear plane."
ϕ	Often used for diameter.

PROBLEMS

Do the problems of Chapter 12, using bolts or rivets. Select the type of fastener in each case. Compare different types. In each case, be sure the members are wide enough for the type and size of fastener used. If not, revise the member or the fastener.

14 Beyond Rolled Sections: Plate Girders and Composite Girders

14.1 SCOPE OF CHAPTER 14

In earlier chapters, most member selection procedures led to a search for the most suitable among available rolled shapes. Members so selected are of constant cross section (prismatic). Often members are selected for the highest value of force or moment and are unnecessarily strong for most of their length, where forces and moments are less. This is most obvious in flexural members. In each of the four loaded beams shown in Fig. 14.1, the maximum moment and maximum shear exist at only one point. Only in Fig. 14.1(d) are the points of maximum shear and moment the same.

At all points where the shear or moment capacity is greater than required, a prismatic member may be considered to be wasteful of steel.* One is tempted to design members of constantly varying cross section in the search for minimum steel usage.† The temptation is usually resisted for at least one good reason. The use of prismatic sections, as rolled, will very often lead to *least cost design* in spite of the excess use of steel.

The cost of usual *mill-delivered* structural steels varies in a very small range when measured in dollars per pound. The cost of *fabrication* and *erection*, which may

**Caution:* This is true with respect to members selected for maximum stress. It may be less true if deflection is an important consideration.

†Further caution; in an indeterminate case, such as Fig. 14.1(d), varying the cross section will change the moment and shear diagrams.

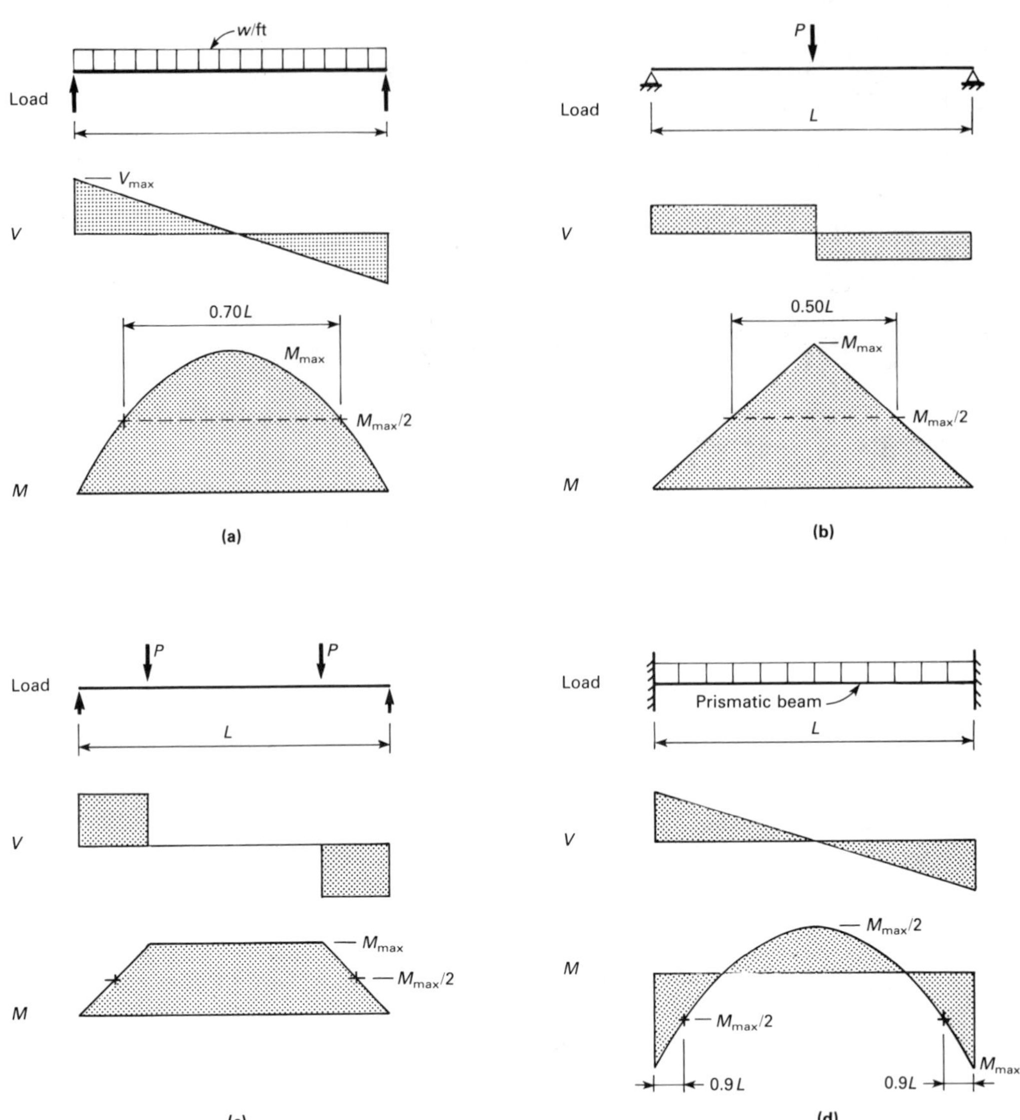

Figure 14.1 Loads and moments.

add more than 100% to the cost per pound, is very sensitive to the degree and complexity of required shop and field work. For most structural steels, it is insensitive to the type of steel. In this chapter, we will consider several simple modifications of rolled beams that sometimes lead to cost savings. In addition, we will go beyond the size and capacity range of rolled sections, where large girders are required that can only be made from plates. In such instances (*plate girders, hybrid girders, composite girders*), economical design will usually require cross sections varying in response to the shear and moment demands at different points.

14.2 SIMPLE METHODS TO EXTEND CAPACITY OF ROLLED BEAMS

For beams of usual span/depth ratios, a member selected to satisfy the maximum moment is often much stronger than necessary in shear. A designer may wish to select a beam of the proper depth but with a light web and then vary the moment capacity by adding cover plates to the flanges. In Figs. 14.2(a) and (b), cover plates are welded or bolted to the flanges to increase the moment capacity of a W shape, a procedure that can be followed for that part of the length where the excess capacity is helpful.

A cover plate equal in area to the flange of the W and fitted to both flanges will more than double the moment capacity, with no loss of capacity in shear. When would this be economical? There is no simple answer. However, consider Fig. 14.1. Assume we wish to select a rolled beam to satisfy $M/2$ and add flange reinforcing plates where moment is higher. It would probably be a futile gesture, since the augmented strength would be needed for almost the full length. A stronger prismatic member would be more economical. Even in Fig. 14.1(a), $M > 0.5M_{max}$ for over 70% of the span. The cost of attaching the doubler plates would probably overshadow the saving in weight. The possibility of net saving becomes much greater in Fig. 14.1(b) and very likely in Fig. 14.1(d), where $M > 0.5M_{max}$ for only 9% of the span length at each end. If used, doubler plates must be welded or bolted to the flanges. The requirements for welds of bolts may be derived from the shear flow equation,

$$q = \frac{VQ}{I}$$

using the methods of Chapters 12 and 13.

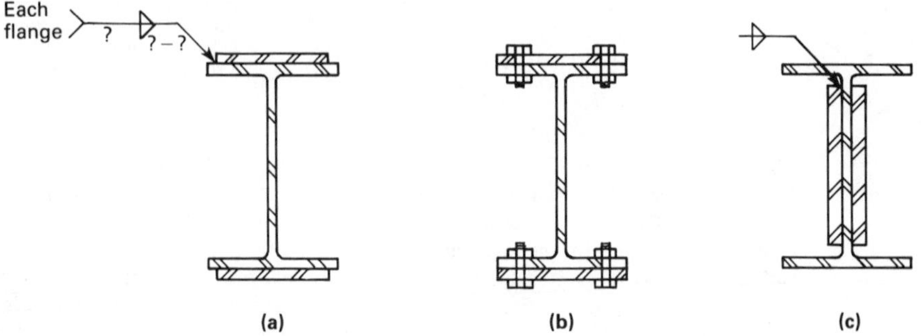

Figure 14.2 Augmenting moment or shear capacity by reinforcing plates.

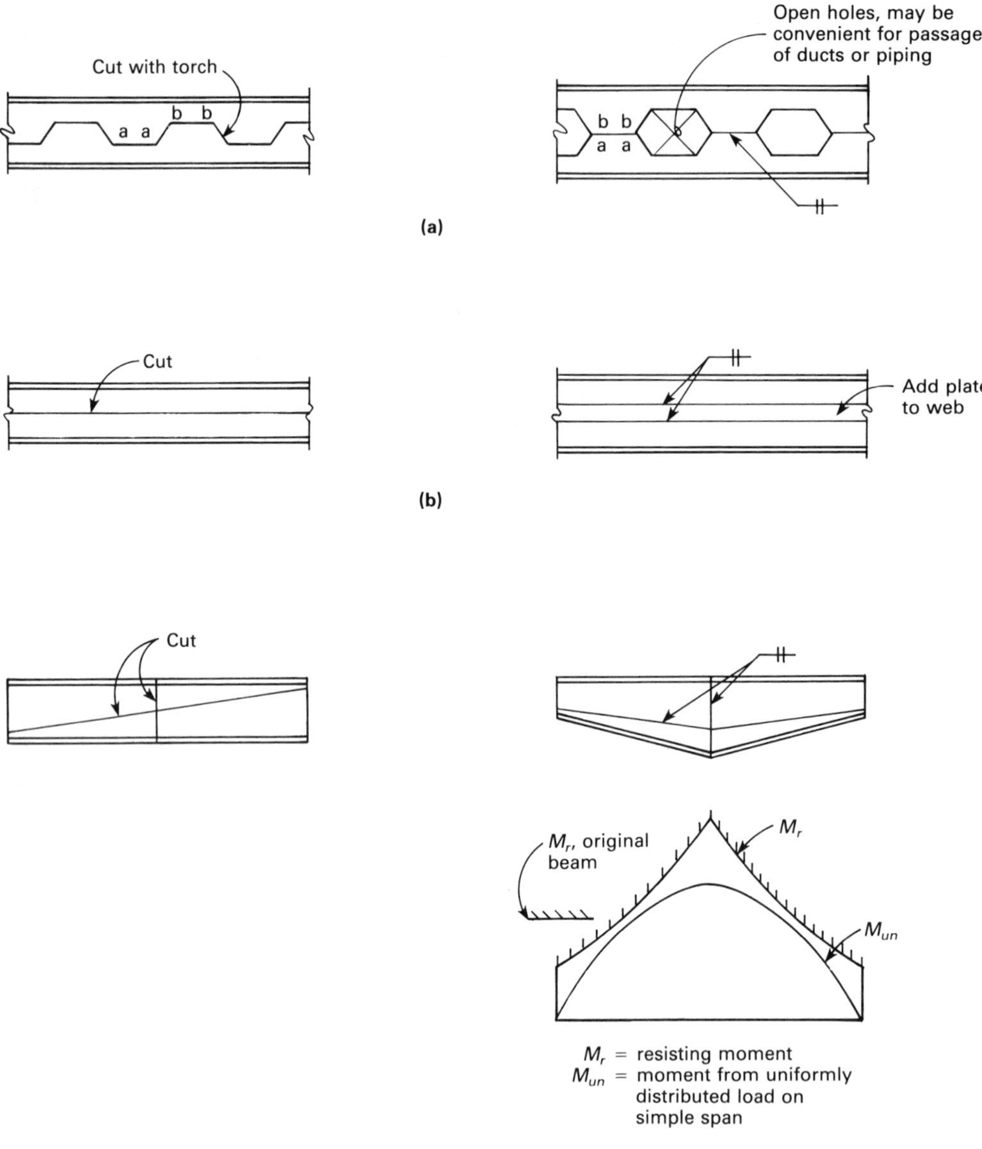

Figure 14.3 Several ways to augment moment capacity of rolled beams by reworking in shop: (a) The mill-supplied beam at left is cut as shown and welded together as at right, into a deeper beam with higher I and S. Shear strength is reduced. Possibility of web instability requires careful examination. (b) The mill-supplied beam at left is torch cut as shown and reassembled with an added plate as at right. I and S are increased. Shear strength may be increased or decreased depending on the final d_w/t_w ratio. (c) Mill-supplied beam at left is cut as shown and reassembled as at right. The resulting moment capacity is compared to the original M_r and to the requirements of a uniformly loaded simple span. Shear strength requires check.

If, on the other hand, a member selected for M_{max} is overstressed in shear for a short distance, the web may be reinforced as in Fig. 14.2(c). This may be found to be economical in a case such as Fig. 14.1(c), where moment demand is almost constant, but, over most of the length, the shear is much less than its maximum.

Several other ways are shown in Fig. 14.3, which, with simple shop fabrication procedures, extend the moment capacity of W shapes.

14.3 PLATE GIRDERS

14.3.1 Comparative Span Ranges

The largest W shape provided by U.S. steel mills is the W36 × 300. Its maximum moment capacity, by AISC standards, is 2220 ft kips in A36 steel, or 3050 ft kips if $F_y = 50$ ksi. If span/depth ratios are limited to 30, the maximum span for this shape would be 90 feet, which is also close to the practical limit for shipment by main highways. The deflections shown in Table 14.1 result from loading such beams uniformly to their maximum permissible moments. They are likely to be unacceptable* for buildings, and even less so for bridges.

TABLE 14.1 DEFLECTION OF W36 × 300

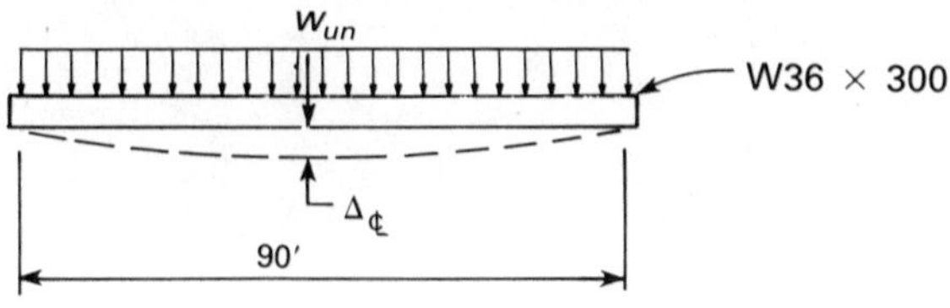

F_y, ksi	w_{un}, k/ft	M_{max}, ft k	$f_b = F_b = 0.66F_y$, ksi	$\Delta_{℄}$	$\frac{\Delta}{L}$
36	2.20	2220	24.0	5.50	1/196
50	3.05	3050	33.3	7.64	1/141

For most purposes, then, the limit of effective simple span usage of W shapes, as rolled, is likely to be somewhat less than 90 feet depending on the type of loading, the yield strength of the steel, and the effective limit of deflection. In the range of

$$60 \text{ feet} \leq \text{simple span} \leq 90 \text{ feet}$$

*The case is not clear unless we know how much of the deflection results from live load. A mill can be asked to camber (prebend) a 90-foot-long W36 in the range of 3 to 6 inches to compensate for dead load deflection (see Manual, Part 1).

TABLE 14.2 SPAN LIMITS FOR W SECTIONS AND PLATE GIRDERS

Nominal depth, d, (in.)	Span in feet limited by ratio L/d		Simple span in feet limited by ratio of deflection, Δ, to span (approx.)			
			$F_y = 36$ ksi		$F_y = 50$ ksi	
	$20d$	$30d$	300Δ	200Δ	300Δ	200Δ
W12	20	30	20	30	15	22
W18	30	45	31	46	21	32
W24	40	60	40	60	29	43
W30	50	75	49	73	36	54
W36	60	90	88	90	43	65
Plate girder d (in.)						
48	80		86	129	62	93
60	100		107	161	77	116
72	120		129	193	93	139
84	140		150	225	108	162
96	160		172	258	124	186

Sample Calculations

For *W shapes*, using AISC beam tables for maximum allowable load, W (kips) uniformly distributed over span, L (feet), and assuming compact sections and F_y as given in beam table:

$$\Delta = \frac{D_c L^2}{1000}, \text{ inches} \qquad (D_c \text{ from table})$$

If L' = span in inches = $12L$ and $L' = 300$

$$\frac{L'}{300} = \frac{D_c L L'}{12{,}000}$$

$$L = \frac{40}{D_c}$$

For *plate girder* with L = span in feet = $L'/12$; d = depth in inches, uniformly distributed load, simple span, n = 300 or 200:

$$\frac{L'}{n} = \Delta = \frac{5}{384}\frac{w(L)^4}{EI} = \frac{5}{48}\frac{M_{max}(L')^2}{EI} = \frac{5}{48}\frac{f_b S(L')^2}{EI} = \frac{5}{48}\frac{f_b^2(L')^2}{Ed}$$

If $F_b = 0.6F_y$ and $f_b = F_b$, after algebraic substitutions:

$$L = 19{,}333\frac{d}{nF_y}$$

the appropriate choice of steel section is often a plate girder. If the span exceeds 90 feet, the choice will almost inevitably be a plate girder or one of the other systems suitable for long spans, such as trusses and arches.

Table 14.2 shows simple span limits for members of various depths based on several different criteria of span/depth ratio and deflection. Deflection criteria and

TABLE 14.3 COMMON SPAN RANGES, VARIOUS SYSTEMS

System	Feet
W shapes, simple spans	0–90
Plate girders, simple spans	50–130
Composite girders, simple spans	50–150
Trusses	80–300
Arches or suspension systems	100–No set limit

span/depth ratios are intimately related. Plate girders are seldom less deep than one-twentieth of the span. With A36 steel and $F_b = 0.6F_y$, a depth of $L/20$ is almost consistent with total deflection of $L/300$. However, for high-strength steels a similar criterion for deflection would require deeper girders.

Table 14.3 is offered as a general guide to common span ranges for various structural systems using steel. Many exceptions exist. It should also be noted that structural steel competes with reinforced and prestressed concrete through most span ranges, with the exception of the longest steel suspension bridges.

The effective range of plate girder spans may be extended by continuity over multiple supports. We will limit further discussions here to the simple span case, which reveals much of what is special in plate girder design. Actual design of plate girders is not limited to simple spans.

14.3.2 Special Characteristics of Plate Girders

The change from the use of rolled beams to that of plate girders leads to several related changes in design approach. Some major changes include:

1. It may no longer be economical to use a constant cross section.
2. The economical depth is likely to be greater for a plate girder than for a rolled section.
3. The designer must be even more conscious than usual of the ratio $b_f/2t_f$.
4. Ratios d_w/t_w are likely to be high enough to threaten shear instability unless stiffening is provided.
5. Plate girders are unlikely to be classifiable as compact sections.
6. It may become advantageous to build plate girders of more than one kind of steel (*hybrid girders*) or of steel combined with other materials like concrete (*composite girders*).

We will expand on each of these in turn.

1. Nonprismatic sections are likely to be used in an attempt to save steel. As discussed earlier in this chapter, prismatic shapes are usually economical because

they require minimum shop fabrication to incorporate them into a design. When circumstances rule out rolled shapes, the extra fabrication cost required in building up the cross section from plates can be partially offset by reducing the total demand for steel. As can be seen in Fig. 14.4, this can be done in a number of different ways by tailoring the moment and shear capacity of the girder as closely as is feasible to the requirements of the moment and shear envelopes.

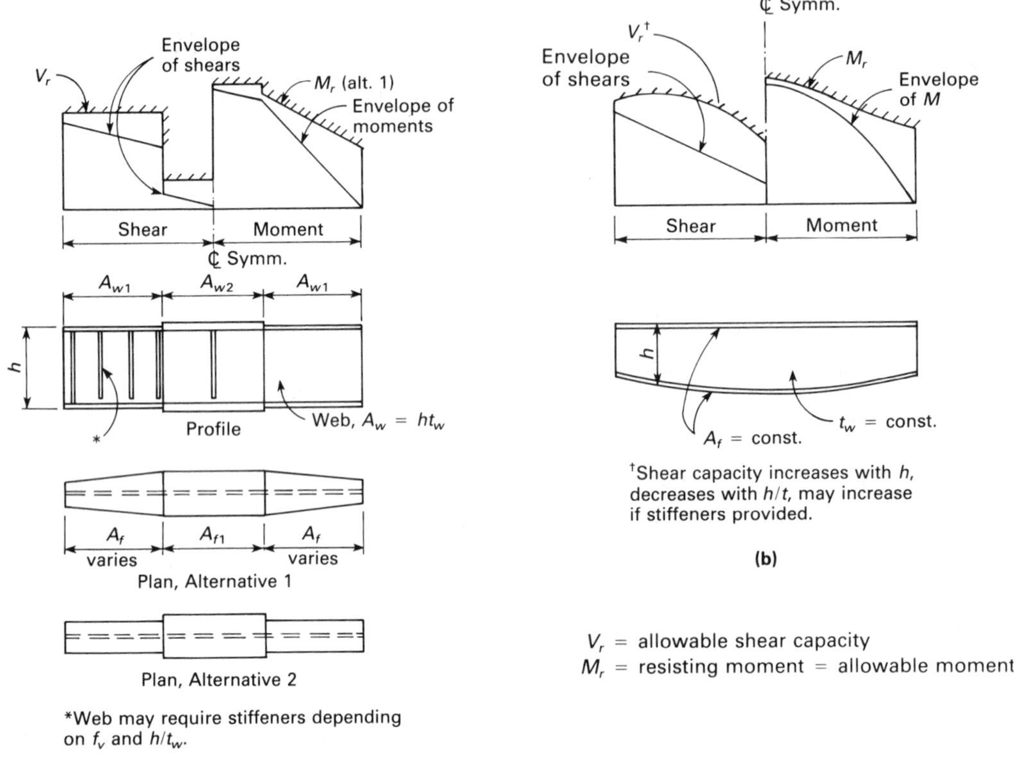

Figure 14.4 Moment and shear capacity of plate girders may be tailored to the moment and shear envelopes.

Since each change in cross section intended to save steel also requires additional work of fabrication, the number of such changes is usually limited. Each change used should save more in cost of steel than it adds in cost of fabrication.*

2. For a given total load W on a girder, deflection is a function of L^3/I. Deflection criteria, on the other hand, may be based on an absolute limit of deflection

*See article by J. C. Holesapple, "Civil Engineering," *ASCE*, November 1982.

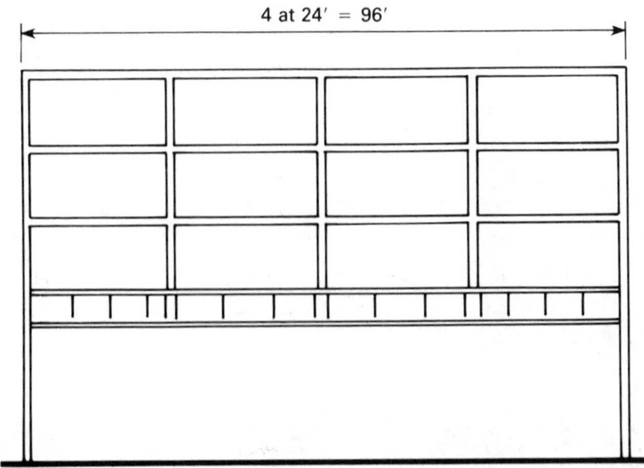

Figure 14.5 Plate girder allows column-free space in lower story of a building.

or the ratio of deflection to span. The former case may apply, for example in a bent of a multistory building such as is shown in Fig. 14.5, where a plate girder is used to create a large open space in the lower story. The girder spans the full 96 feet and carries the second floor dead and live loads as well as the column loads from the upper stories. Any live load deflection of the girder will change the bending moment and shear in the upper story girders and may also cause distress in nonstructural parts of the building. Both are strong reasons for an absolute limit on live load deflection of the girder. In the case of highway bridge girders, as will be seen in the case study of Chapter 20, desirable riding qualities for the deck lead to a live load deflection limit of $L/800$.

In the face of exponential growth of deflection with span, we seek exponential growth of moment of inertia. For a plate girder of I shape,

$$I \approx \frac{d^3 t_w}{12} + 2\left(\frac{d}{2}\right) b_f t_f \propto d^K$$

where $2 \leq K \leq 3$.
Since plate girders are used primarily for long spans, their required depth can be expected to grow at a faster rate than the span length.

3. Rolled sections are usually proportioned by steel producers to avoid reduction of allowable stress through local instability of the flanges. As can be seen by reference to the Shape Tables of the AISC Manual, all W shapes of depth 18 inches or more qualify as compact sections by the $b_f/2t_f$ criterion, F'_y, if $F_y \leq 58.5$ ksi. In almost all cases, this criterion is satisfied for all structural steels. On the other hand, the ratio $b_f/2t_f$ for plate girders must be established by the designer. By the criteria of AISC, the ratio for the compression flange must be set at

$$\frac{b_f}{2t_f} \leq \frac{95}{\sqrt{F_y}} \qquad \text{(AISC 1.9.1.2)}$$

unless, for some reason, it is desired to reduce the value of F_b below $0.60F_y$.

4. The desirability of deep sections leads to a tendency for uneconomically large web areas. If the ratio d_w/t_w were limited as it is in rolled sections, web areas would be very large, dominating the required section modulus and moment of inertia. This would result in very low web shear stresses and very small flanges, both being indications of inefficient use of steel for flexure.

The desirable (i.e., most efficient based on the strength of the steel) value of F_v is, by AISC standards,

$$F_v = 0.4F_y \qquad \text{(AISC 1.5.1.2.1)}$$

However, a web without stiffening starts to lose shear strength due to shear instability at ratios

$$\frac{d_w}{t_w} \approx 50$$

more or less, depending on the yield strength of the steel. Recall that pure shear on an element of the web of a girder can be converted to principal tension and compression on an element drawn at 45° to the original (Fig. 14.6). Compressive stress on a thin deep web may cause buckling. Once again, the proportions of rolled sections are

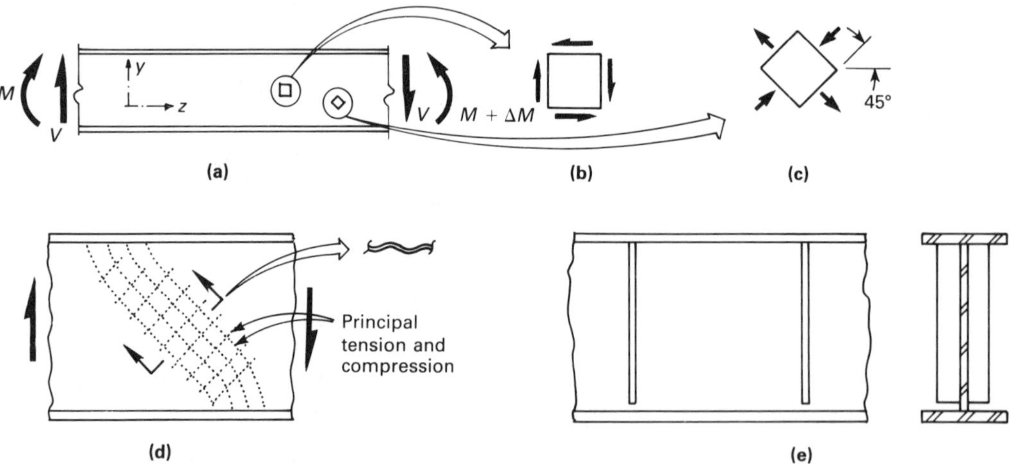

Figure 14.6 Stiffening to prevent web instability in shear: (a) Girder shown subjected to shear and moment. (b) Element of web as free body; if near neutral axis, the element displays pure shear if edges are parallel to y and z. (c) Similar element rotated 45° from y and z displays principal compression and tension. (d) Slender web will buckle in compression at low shear stress. (e) If stiffeners fitted to web, buckling is prevented. Depending on spacing of stiffeners and their stiffness, some or all of the inherent shear strength of the steel may be restored.

designed to guard against this, as can be verified by reference to the d/t_w columns in the Shape Tables. Plate girders are not so protected nor should they necessarily be for maximum economy. But F_v must be reduced when unfavorable proportions reduce shear strength. Compression buckling due to web shear can be inhibited and the reduction of F_v minimized by adding stiffeners to the web as shown in Fig. 14.6(e). The size and spacing of such stiffeners, as well as the ratio d_w/t_w, determine the shear strength of a plate girder. Much of the attention of the designer of a plate girder must be directed toward the proportioning of the web and its stiffeners.

5. One of the criteria set by the AISC for classification of a flexural member as "compact" is the ratio d_w/t_w (AISC 1.5.1.4.1). In the absence of axial load, the ratio is

$$\frac{d_w}{t_w} \leq \frac{640}{\sqrt{F_y}}$$

As discussed in paragraph 4, economical ratios of d_w/t_w tend to be high for plate girders. In the examples of this chapter and in Chapter 20, desirable ratios of d_w/t_w will tend to be two and more times the *compact* limit.

Compact section criteria also require that the connection of flanges to web be continuous, a natural consequence of the rolling process for mill-supplied girders. In bolted plate girders, this is an impossibility. It is possible in a welded plate girder to have a continuous weld from web to flanges. However, the weld requirement, based on shear flow,

$$q = \frac{VQ}{I}$$

is often low, and usually does not require continuous welding.*

As a consequence, compact sections are unusual in plate girder design. Allowable bending stress, F_{bx}, in noncompact sections is $0.6F_y$ or less.

6. The use of more than one type of steel in a girder is not feasible with rolled sections, but becomes so when parts are purchased as separate plates. For example, higher-strength steel may be used in regions of high stress and less expensive, say A36, steel where stresses are lower. Girders made of two steels are called *hybrid*.

A special application of the hybridizing approach is the use of a plate girder married to a concrete slab in such a way as to cause them to act as a combined section. A number of design advantages in such *composite* girders will be explored in Section 14.8.

14.3.3 Nature of Plate Girder Cross Sections

Figure 14.7(a) illustrates a bolted or riveted plate girder. The nature of the cross sections and details are based on the requirements inherent in that type of technology. Many large and heavily loaded riveted bridge and building girders built in the

*Fatigue or corrosion problems sometimes dictate continuous welding.

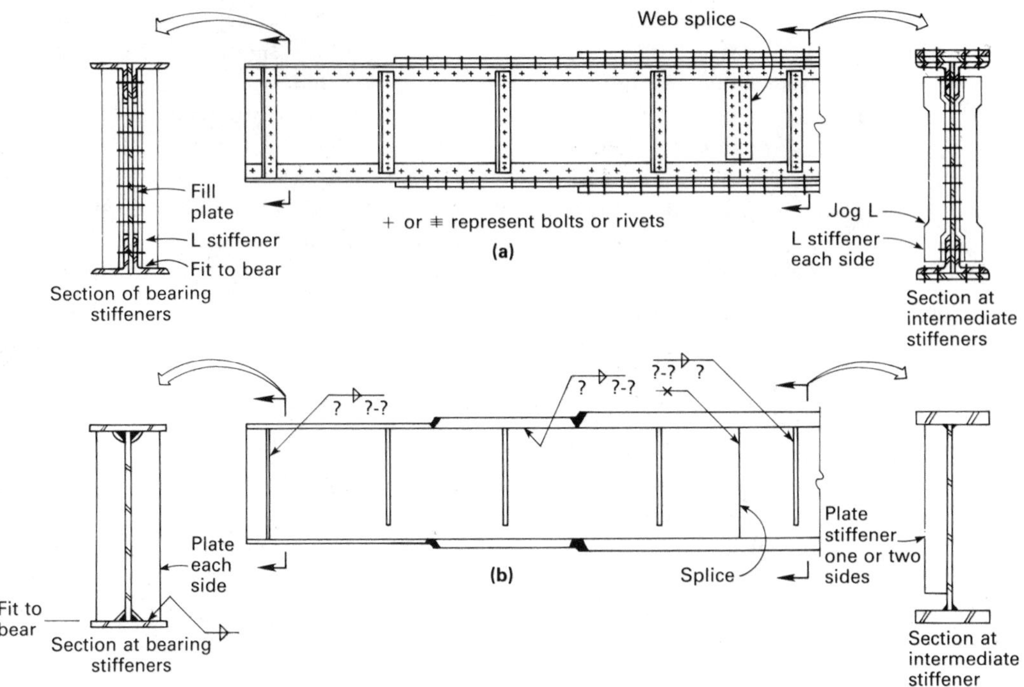

Figure 14.7 Plate girder, details: (a) bolted or riveted plate girder; (b) welded plate girder.

nineteenth and early twentieth centuries can still be seen in use today. Figure 14.7(b) illustrates the details of welded plate girders. By comparing the two figures, the reader may readily understand how, in this instance, the simplicity possible with welded details has caused the earlier riveted designs to be almost totally supplanted by welded ones.*

Some of the common demands of plate girder design can be seen on both figures:

1. Long lengths may require splices in both web and flanges.
2. Flange areas can be varied with the moment envelope diagram.
3. Web areas can be varied with the shear envelope by changing thickness at the web splices.
4. The capacity at each point along the length must be equal to or greater than the ordinates of the shear and moment envelope at that point.
5. Vertical web stiffeners are often needed for shear stability.

*There are still exceptions, not least of which are instances where bolted splices are used in the field to join large welded shop assemblies.

6. Bearing stiffeners are almost universally needed to transfer the end reactions to the supports. These are used to prevent local compression buckling (crippling) of the web.
7. Bearing stiffeners are sometimes needed within the span if high concentrated loads are applied at the top flange. This is discussed later.

Further discussion in this chapter will concentrate on welded plate girders. Except for details that derive from the use of mechanical fasteners, the principles used for welded girders apply to bolted or riveted girders.

14.4 DESIGNING PLATE GIRDERS USING THE AISC SPECIFICATION

14.4.1 Special Provisions

Section 1.10 of the AISC Specification covers most of the special requirements applying to plate girders. They can be summarized in the list of subsections, most of which will be discussed here.

1.10.1 Proportions
1.10.2 Web
1.10.3 Flanges
1.10.4 Flange Development
1.10.5 Stiffeners
1.10.6 Reduction in Flange Stress
1.10.7 Combined Shear and Tension Stress
1.10.8 Splices
1.10.9 Horizontal Forces
1.10.10 Web Crippling
1.10.11 Rotational Restraint at Supports

Although the concentration of Section 1.10 is on plate girders, all other requirements of the Specification not specifically changed by this section also apply to plate girders. Similarly, some provisions of Section 1.10 must be considered in connection with rolled beams.

14.4.2 Proportioning Plate Girder Sections: Flanges and Depth

When considering the required moment of inertia and section modulus, rivet or bolt holes may be ignored if the reduction of a flange area due to the holes is less than 15%, a provision that simplifies the calculation. The resisting moment of a plate girder, then, is

$$M_R = F_b(S_x),$$

where S_x is based on the gross area.

A convenient approximation of the section modulus of a deep I-shaped plate girder results from the fact that the clear distance, h, between flanges is almost equal to the overall depth, d. Then

$$I = \frac{h^3 t_w}{12} + 2A_f\left(\frac{d}{2}\right)^2$$

where $A_f =$ the area of one flange, and

$$I \approx \frac{h^3 t_w}{12} + \frac{h^2 A_f}{2} \tag{14-1}$$

$$S \approx \frac{h^2 t_w}{6} + A_f h \tag{14-2}$$

Then a first approximation of required flange area can be

$$A_f \approx \frac{M}{F_b h} - \frac{t_w h}{6} \tag{14-3}$$

the second term usually being very much smaller than the first. F_b may be assumed at $0.6F_y$ if the compression flange is to be laterally restrained. As will be seen later, a slightly lower assumption, based on AISC (1.10-5), would sometimes be closer.

Proportions b_f/t_f are limited by AISC 1.9. As in other instances, if the requirements of Section 1.9 are exceeded, F_b must be reduced.

14.4.3 Proportioning Plate Girder Webs (No Tension Field Action)

The starting point of a plate girder design is often a tentative determination of the depth of girder, which is likely to be controlled by a limit on deflection set by the design specification or the engineer's judgment regarding performance requirements. Table 14.2 is helpful in setting d. The depth of web $h \approx d$.

The thickness of the web must be sufficient so that the area ht satisfies the requirements of shear stress. However, F_v cannot be determined for deep, thin webs without considering the problems of shear buckling discussed earlier and other sources of potential buckling not yet mentioned. Therefore, the thickness of the web must be considered simultaneously with decisions about the nature of the stiffening.

In AISC 1.10.2, two absolute limits are set on the ratio h/t. They derive in part from the possibility of buckling of the web in compression due to the squeezing action shown in Fig. 14.8, where a portion of a girder is shown subjected to positive bending. The curvature results from the bending moment. As a consequence, the web is subjected to a compressive load required to satisfy equilibrium at the flange–web

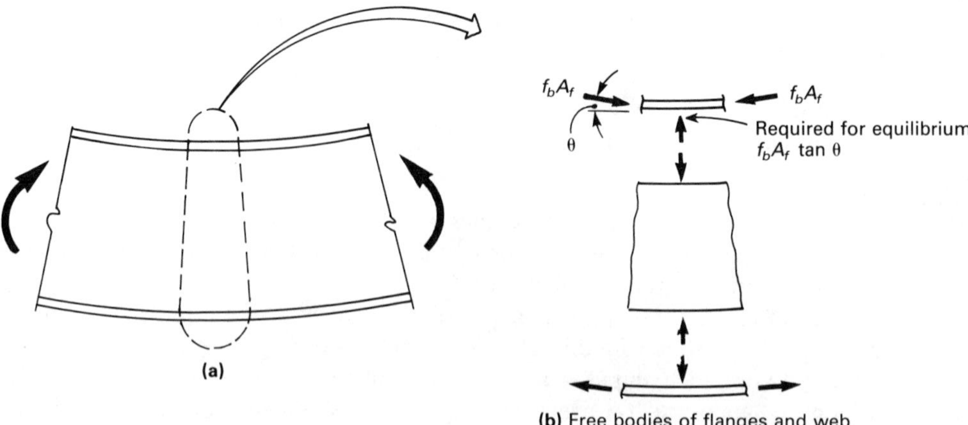

Figure 14.8 Transverse compression in web from bending moment.

contact. The curves of Fig. 14.9 show the limits of h/t plotted against yield stress. Note that the difference between the two curves becomes more significant for higher values of F_y. If there are no web stiffeners, the limit, set by AISC 1.10.2 and 1.10.5.3, is

$$\frac{h}{t} \leq \frac{14{,}000}{[F_y(F_y + 16.5)]^{1/2}} \leq 260 \tag{14-4}$$

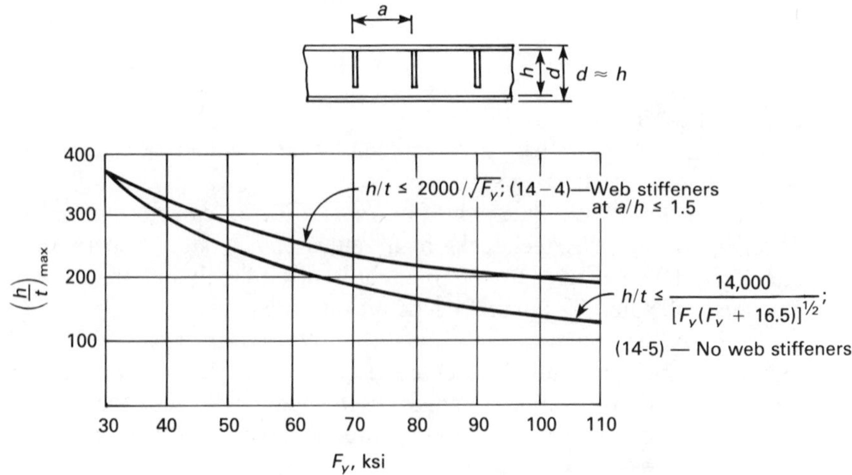

Figure 14.9 Maximum values of h/t for stiffened and unstiffened webs.

The limit is 322 for A36 steels and lower with higher yield strengths. If effective web stiffeners exist at a spacing a not greater than $1.5h$, h/t is allowed to increase. Under those conditions,

$$\frac{h}{t} \leq \frac{2000^*}{\sqrt{F_y}} \tag{14-5}$$

Engineering offices will sometimes set absolute lower limits of web thickness. For example, an engineer may reason that corrosion will cause greater relative loss of strength in thin webs than in thick ones. On that account, it is common office practice to set a lower limit of $\frac{1}{4}$ inch or $\frac{5}{16}$ inch for girders to be used under corrosive conditions. Consistent with this, an office may consider the *effective* thickness of a $\frac{5}{16}$-inch web to be $\frac{1}{4}$ inch and use the lower figure in calculating stress. If a welded plate girder must support high concentrated loads, the phenomenon of *crippling*, discussed later, may determine the minimum t.

The limits of h/t in formulas (14-4) and (14-5) establish the minimum permissible web thickness for a given depth of girder. They do not necessarily provide the thickness necessary to satisfy the shear. For that purpose, we need a value for F_v in the formula

$$f_v = \frac{V}{ht} \leq F_v$$

The maximum value of F_v is given in AISC 1.5.1.2.1 as

$$F_v = 0.4F_y$$

but is subject to reduction in accordance with Section 1.10.5.2, which in AISC equation (1.10-1) stipulates

$$F_v = \frac{F_y(C_v)}{2.89} \leq 0.4F_y$$

C depends on both h/t and a/h.

Shear stability, which is threatened by large values of h/t, can be restored by proper spacing of effective stiffeners.

C_v declines with large values of h/t and increases with reduced values of a/h. *Efficient* use of web steel seeks high values of F_v, but stiffeners are expensive to attach. High values of F_v are often unnecessary if depth and, therefore, web area are effectively dictated by flexural needs making shear stress low.

The AISC equations for C_v may be followed in Figs. 14.10 and 14.11 where they are written out and represented graphically. Values of C_v are found from two different equations in different ranges. Both equations include the variable k, which is independent of the steel used, and itself is governed by different functions of a/h in

*F_y is the yield stress of the flanges, a special requirement that becomes significant for hybrid girders.

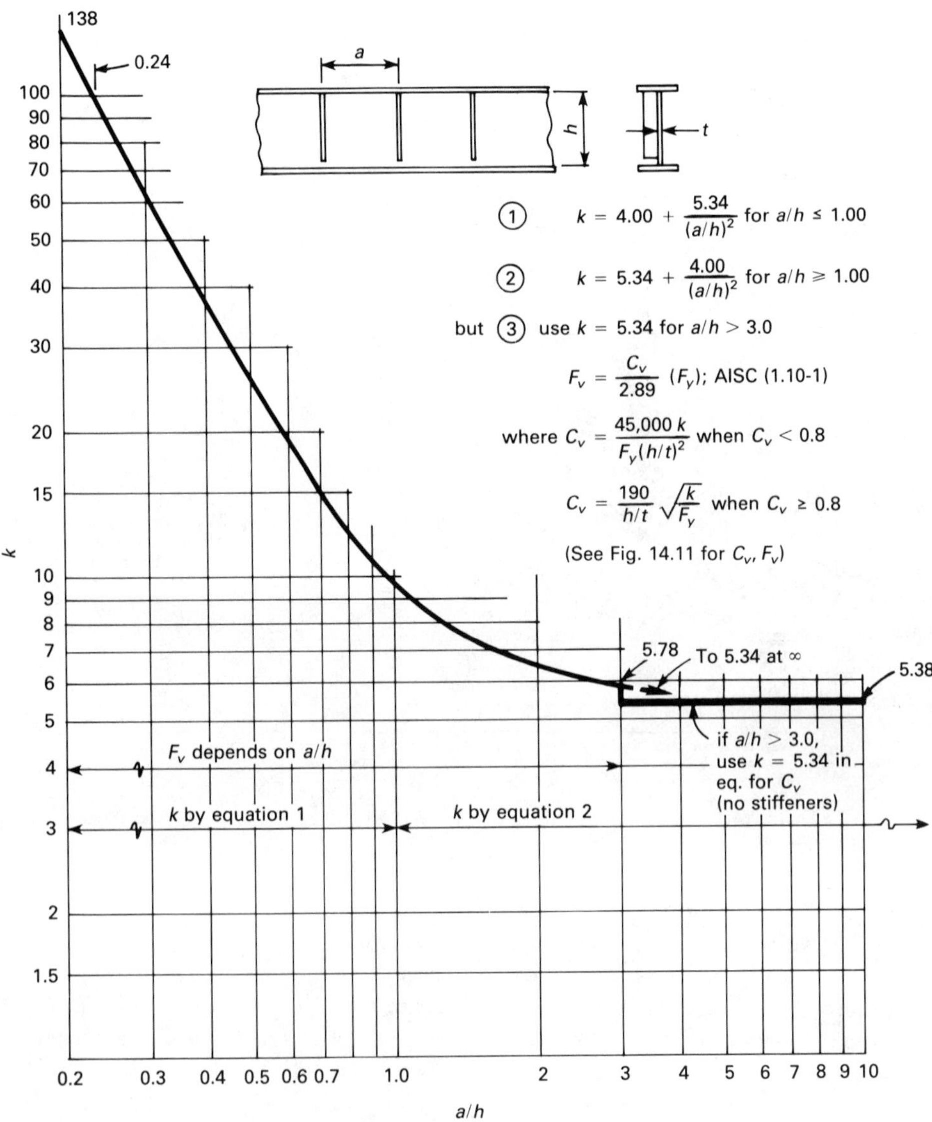

Figure 14.10 k versus a/h in formula (1.10-1), AISC Specification, 1978.

two ranges. The log–log plot of Fig. 14.10 shows the curve of k versus a/h in the interval for k between 0.2 and infinity. The two equations meet at $a/h = 1.0$. At $a/h = \infty$, $k = 5.34$, a value almost equal to that at $a/h = 3.0$. If stiffening is needed, AISC 1.10.5.3 stipulates that

$$\frac{a}{h} \leq 3 \text{ and } a/h \leq \left[\frac{260}{h/t}\right]^2 \tag{14-7}$$

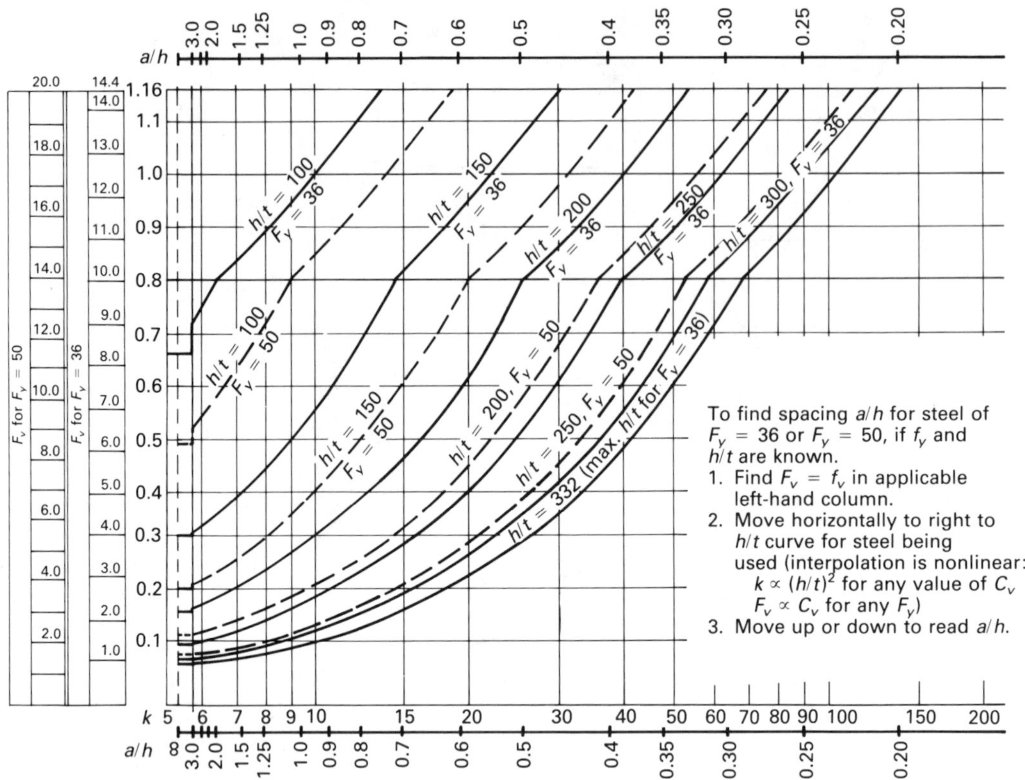

Figure 14.11 k (and a/h) versus C_v (and F_v) for values of h/t, AISC Specification, 1978, formula (1.10-1).

The value $k = 5.34$ is used for unstiffened webs.

In Fig. 14.11, k and the corresponding values of a/h are plotted against C_v, F_v, and h/t. F_v is proportional to F_y. Curves are shown for two common values of F_y, but may be adapted for others. The limit of C_v is set at 1.156, which in Formula (14-6) gives

$$F_v = 0.4F_y{}^*$$

Curves are plotted for several values of h/t in the range 50 to 300. Nonlinear interpolation may be used for other values of h/t.

AISC Tables 10-36 and 10-50 provide F_v directly for combinations of a/h and h/t and two values of F_y, 36 and 50. The author finds both the figures here and the

*See AISC 1.5.1.2.1.

AISC tables useful for different purposes. The range of a/h in Fig. 14.11 is greater than that of the tables. The tables reveal a very important fact not obvious from the figures. For large values of h/t, and up to the limit of h/t permitted for each steel by formulas (14-4) and (14-5), values of F_v are identical for different steels. This is consistent with other contexts in which allowable stress is limited by potential buckling.

14.4.4 Plate Girder Webs with Tension Field Action

In the preceding discussion and in Fig. 14.6, the problem of shear buckling was considered to be avoidable by a combination of reduced F_v and addition of stiffeners. It was shown that shear buckling is actually compression buckling on the compression diagonal of plate elements in shear.

Consider now the Pratt truss in Figs. 14.12(a) and (b), which must resist a shearing load and the consequent moments. The diagonals resist the shear by converting its effect to axial tension. This is only possible with the assistance of the compression struts and the chords, making equilibrium possible at the joints.

The behavior of a stiffened plate girder may be considered analogous to such a truss at shear loads that cause compression buckling. Although the compression diagonal of a shearing element is ineffective, the steel of the web can still resist tension in the direction of the tension diagonal. However, this requires that the stiffeners act as compression members. In Fig. 14.12(c), the analogy is drawn. The combination of numerous tension diagonals within each panel acts as a *tension field,* whose vertical component causes compression in the stiffeners. Under these circumstances, compression buckling may lead to small waviness in the web, but failure is not progressive as resistance is shifted to the new system.

If the designer decides to take advantage of tension field action, the AISC Specification allows values of F_v greater than those in Eq. (14-6). With tension field action and $C_v \leq 1.0$,

$$F_v = \frac{F_y}{2.89}\left[C_v + \frac{1 - C_v}{1.15\sqrt{1 + (a/h)^2}}\right] \tag{14-8}$$

This equation is identical to Eq. (14-6) except for the second term within the brackets. In deciding whether or not to take advantage of tension field action, the designer must judge whether or not the advantages are offset by the more expensive stiffeners that would be required.

A girder designed for tension field action requires (AISC 1.10.5.3) an end panel that has a low enough ratio, a/h, to satisfy Eq. (14-6). This panel may be considered to *anchor* the longitudinal component of the tension field. Similar anchorage is required if there are large holes in the web. Hybrid girders may not be designed with tension field action.

AISC Tables 11-36 and 11-50 give values of F_v with tension field action for various combinations of a/h and h/t and for F_y equal to either 36 or 50.

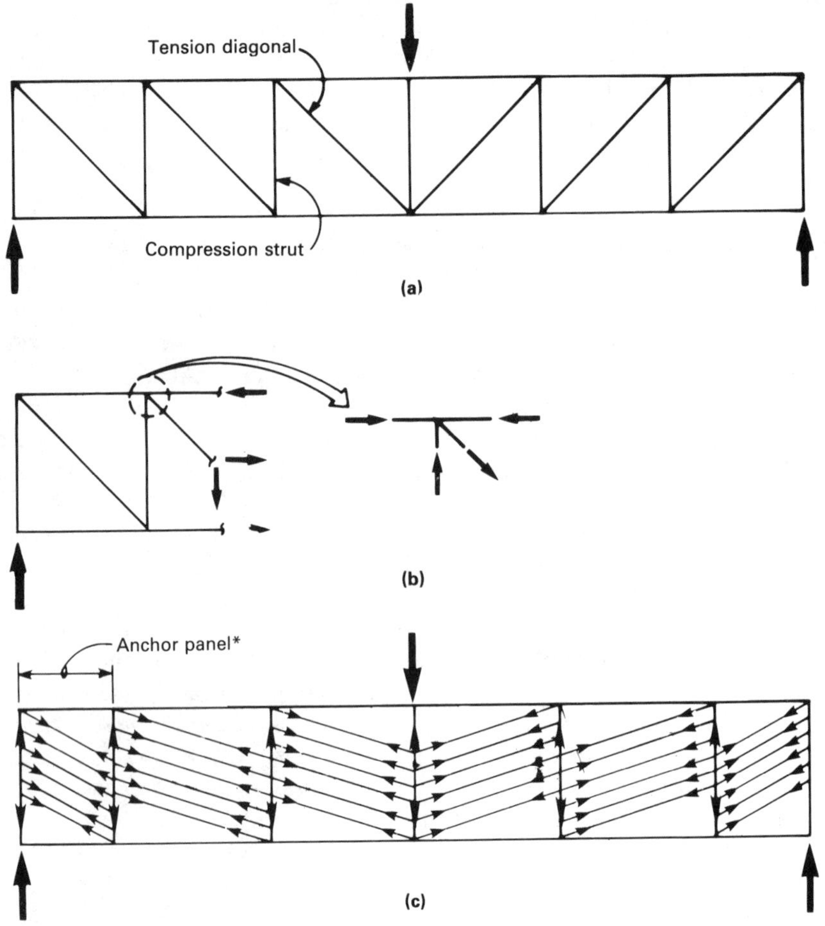

Figure 14.12 Tension field action, truss analogy: (a) Pratt truss, resistance to load causing shear and moment; (b) partial free bodies; (c) plate girder with tension field action analogous to truss.

14.4.5 Proportioning Intermediate Stiffeners (AISC 1.10.5.4)

Stiffeners used in girders *without tension field action* need only be stiff enough to prevent the initiation of web buckling. The criterion for selection is one of stiffness.

$$I_{st} \geq \left(\frac{h}{50}\right)^4 \tag{14-9}$$

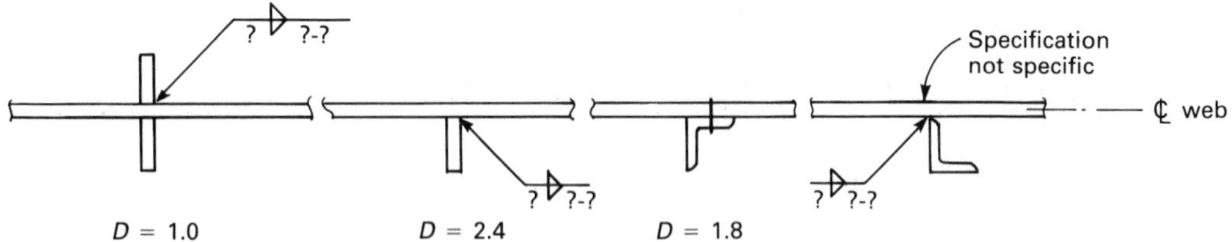

Figure 14.13 Values of D for various stiffeners in formula for required A_{ST} used with tension field action, formula (14-7), AISC 1.10-3.

I_{st} is the moment of inertia of the stiffener around an axis in the plane of the web. When used with a web in *tension field action*, the stiffener must also act as a compression strut. For such cases, a requirement of minimum area is added.

$$A_{st} \geq \frac{1 - C_v}{2} \left[\frac{a}{h} - \frac{(a/h)^2}{(1 + a/h)^2} \right] YDht \tag{14-10}$$

Y applies only when the material of the stiffener differs from that of the web.

$$Y = \frac{(F_y)_{\text{web}}}{(F_y)_{\text{stiff}}} \tag{14-11}$$

D relates, roughly, to the eccentricity of the stiffener with respect to the web. Types of stiffeners and corresponding values of D are illustrated in Fig. 14.13. When $f_v < F_v$, A_{st} may be reduced proportionately. AISC tables 11-36 and 11-50 include requirements for A_{st} in italics.

To assure effective interaction between the web and stiffeners, they must be connected for a shear flow at least equal to

$$f_{vs} \geq h \sqrt{\left(\frac{F_y}{340} \right)^3} \tag{14-12}$$

in kips/linear inch. This may be provided by rivets* or welds. Other issues of detail are also covered in AISC 1.10.5.4.

14.4.6 Bearing Stiffeners (AISC 1.10.5.1)

Bearing stiffeners are proportioned as parts of columns, as shown in Fig. 14.14. The total cross section consists of the stiffeners plus a limited portion of the web. The reaction or applied concentrated load must be transferred into the web to create web shear. At the bearing end of the stiffeners, firm bearing is required at

*The specification refers to "rivets." Bolts can provide equally effective shear transfer. Presumably, either is intended.

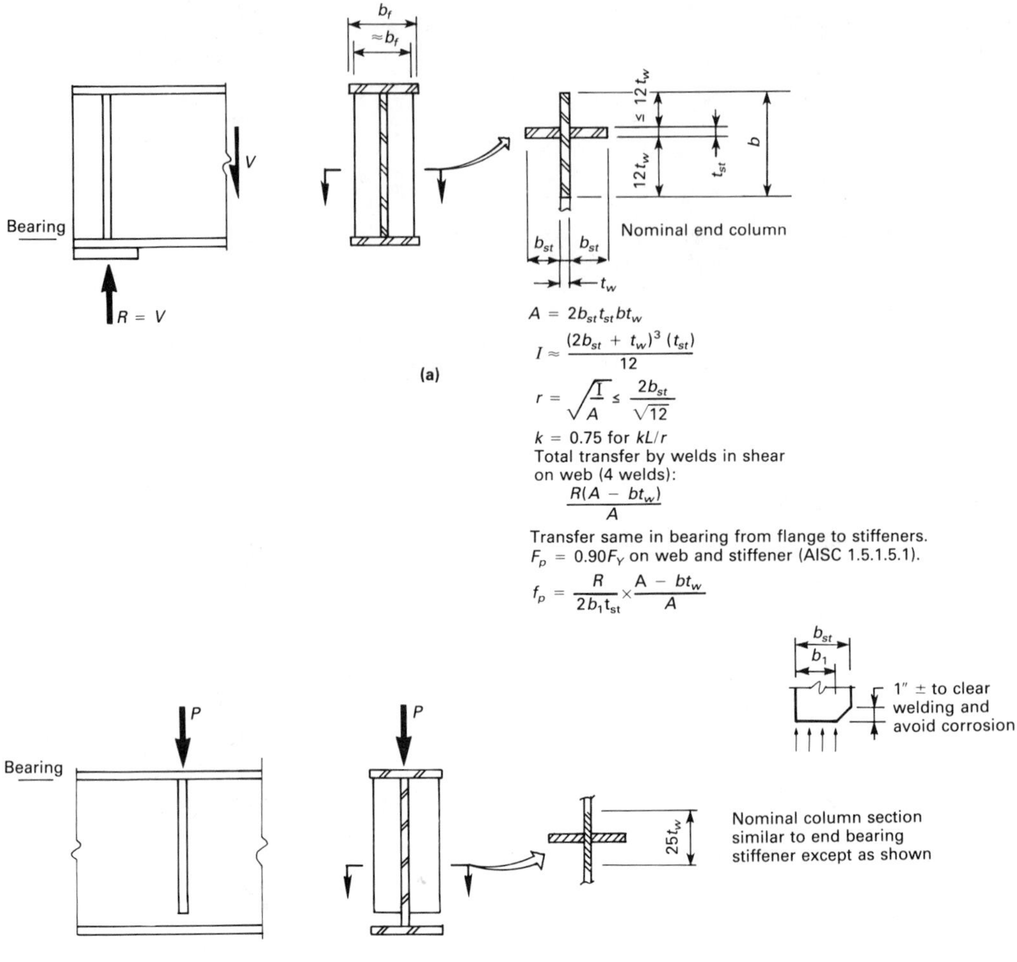

Figure 14.14 Bearing stiffeners: (a) end bearing stiffeners; (b) bearing stiffener at concentrated load within span.

$$f_p \leq 0.9F_y$$

to transfer part of the load to the column. The part of the load that is transferred to the ends of the stiffeners must be further transferred to the web through the bolts or welds connecting the stiffeners to the web. The value of K in the parameter KL/r is 0.75, in recognition of the fact that the stress in the column decreases progressively as the load is transferred to the web.

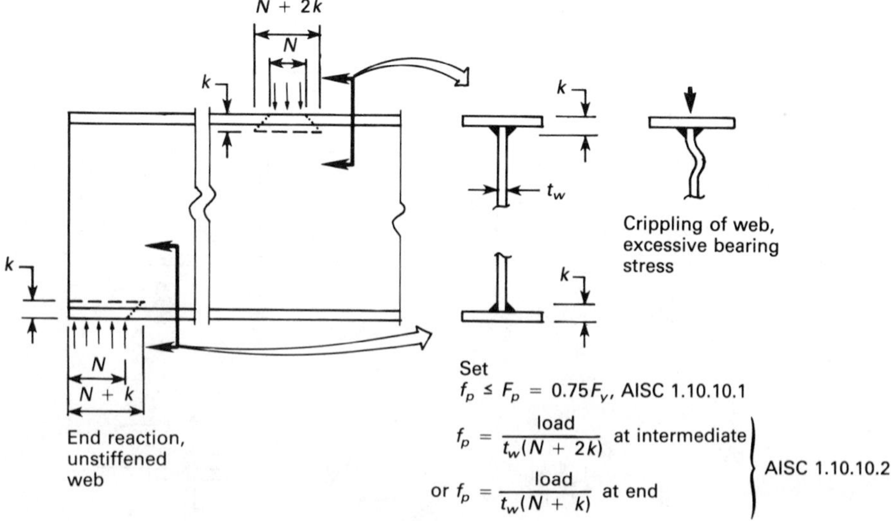

Figure 14.15 Crippling of web from excessive transverse bearing stress.

14.4.7 Crippling of Webs

The crippling problem in girder webs was examined earlier in Chapter 10 and illustrated in Fig. 10.17. It is similar in welded plate girders. The location of the critical value of bearing stress, f_p, as shown in Fig. 14.15(a) is at the toe of the weld from flange to web. If f_p is excessive, bearing stiffeners are required. They are usually necessary at reaction points [Fig. 14.14(a)] and may also be required, as in Fig. 14.14(b), at locations of large concentrated loads. Loads applied as in Figs. 14.16(a) and (b) so as to cause compression in the web in the direction of the load may cause web buckling. The buckling mode depends on whether or not the flange is restrained against rotation. See Figs. 14.16(c) and (d). Resistance to such buckling depends on a/h and h/t. The applicable AISC formulas are shown in the figure.

14.4.8 Reductions in Allowable Bending Stress (AISC 1.10.6, 1.10.7)

Values of F_b are established for plate girders in accordance with AISC 1.5.1.4. However, two different phenomena often lead to reductions in the allowable stress.

Very thin webs in regions of high compressive stress due to bending moment may experience some lateral buckling from the squeezing effect illustrated in Fig. 14.8. A small additional moment resistance must be supplied by the compression

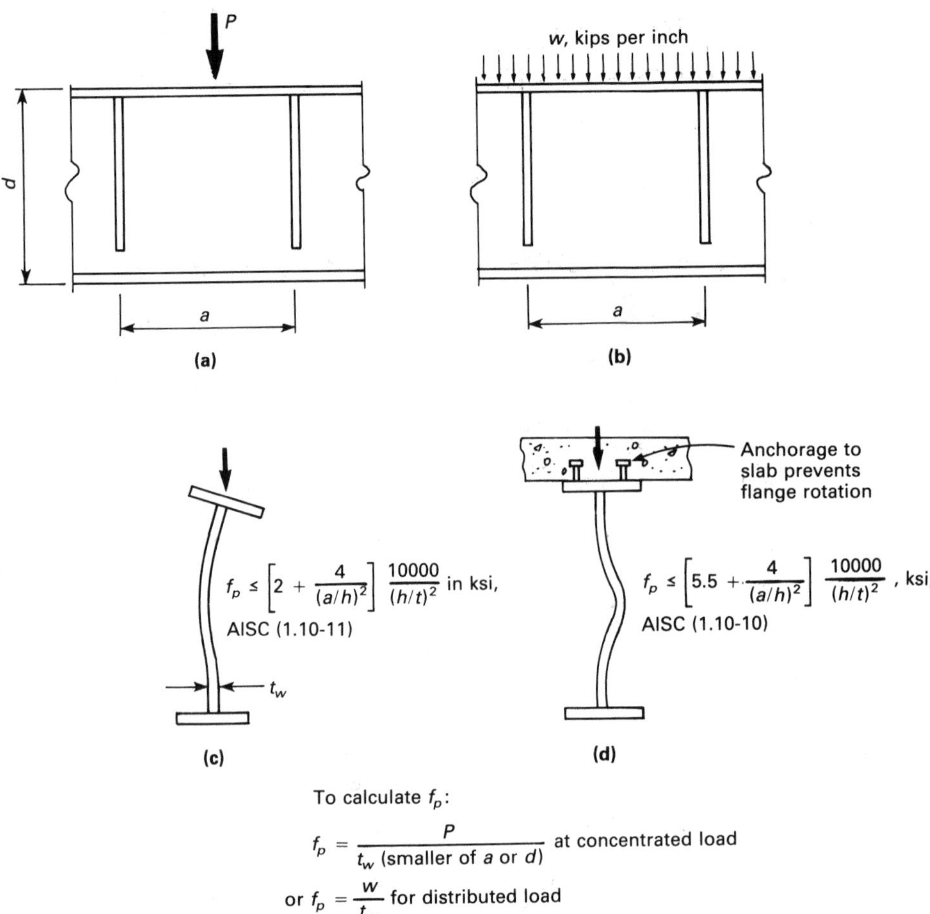

Figure 14.16 Buckling of web from excessive transverse compression.

flange to compensate for the loss of compression strength in the web. If $h/t \geq 760/\sqrt{F_b}$, the AISC stipulates a reduced value of allowable compressive bending stress [AISC Eq. (1.10-5)]:

$$F'_b = F_b \left[1.0 - 0.0005 \frac{A_w}{A_f} \left(\frac{h}{t} - \frac{760}{\sqrt{F_b}}\right)\right] \tag{14-13}$$

The reduction is usually small. A similar formula applicable to hybrid girders will be found later in Fig. 14.18.

When tension field action is required, the principal tension stress in the web near the tension flange may be considerably higher than the longitudinal tensile stress calculated by the usual formula,

$$f_b = \frac{My}{I}$$

To account for this without requiring direct evaluation of the principal stress, AISC formula (1.10-7) provides for a reduced value of allowable stress in tension. If the reduced value is called F_b'',

$$F_b'' = \left(0.825 - 0.375\,\frac{f_v}{F_v}\right)F_y \le 0.6F_y \qquad (14\text{-}14)$$

F_v is based on Eq. (14-5). Values vary between

$$0.45F_y \le F_b'' \le 0.6F_y$$

14.5 EXAMPLE: DESIGN OF PLATE GIRDER

In the second half of Chapter 20, a case study of a plate girder highway bridge design, a welded plate girder of 90-foot span is designed as one of three parallel girders carrying two lanes of traffic. The design conditions are established in the first half of the case study. For the particular girder, for dead plus live load plus impact,

M_{max} = 4065 ft kips at and near midspan of which live load plus impact account for 60%

V_{max} = 191 kips at ends.

The midspan cross section is reproduced in Fig. 14.17 with section characteristics and stresses. Considerations that led to the cross section are discussed in detail in Chapter 20 and will not be repeated here. Some comments are useful for the purposes of this chapter.

Although not usual practice in highway bridge design, it was found convenient to use the AISC Specification, familiar to readers of this book, in setting strength criteria. Live loads, impact, and deflection criteria were based on the AASHTO highway bridge specification[1].

Span/depth ratio:

$$\frac{L}{d} = \frac{90(12)}{88.5} = 12.12$$

Maximum deflection from live + impact loads = 1.32 inches, for a ratio

$$\frac{\Delta_{(L+I)}}{L} = \frac{1}{872} < \frac{1}{800} = \text{AASHTO limit}$$

The maximum value of f_b is low due to fatigue controls. The weight of the girder at midspan is 275 pounds/foot.

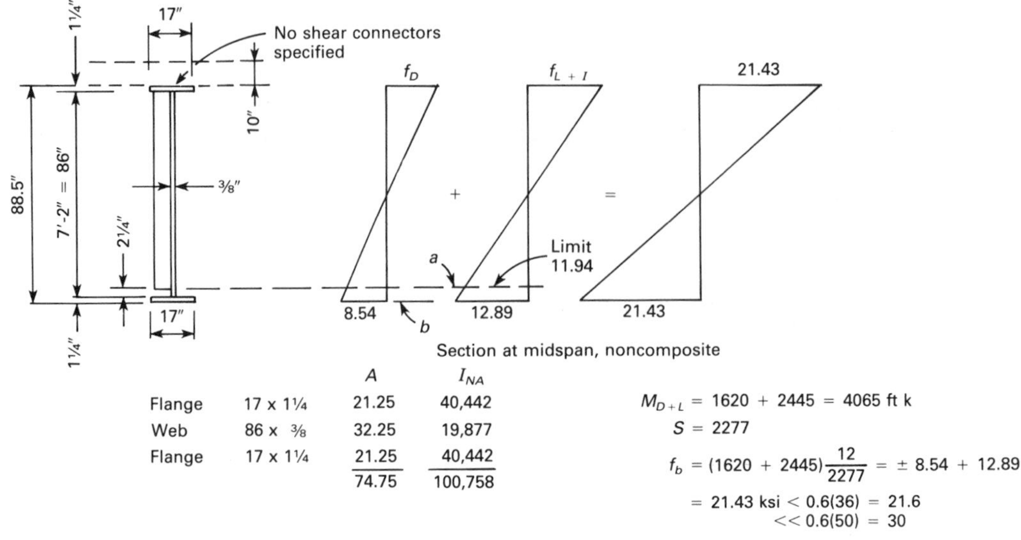

		A	I_{NA}
Flange	17 x 1¼	21.25	40,442
Web	86 x ⅜	32.25	19,877
Flange	17 x 1¼	21.25	40,442
		74.75	100,758

$M_{D+L} = 1620 + 2445 = 4065$ ft k

$S = 2277$

$f_b = (1620 + 2445)\dfrac{12}{2277} = \pm\ 8.54 + 12.89$

$= 21.43$ ksi $< 0.6(36) = 21.6$

$<< 0.6(50) = 30$

$V_{max} = 191$ kips

$(f_v)_{max} = \dfrac{191}{88.5(0.375)} = 5.76 \text{ ksi} << 0.6F_y$

At bottom of stiffener, level a, cond. 4, category F:

$f_{cr} = \dfrac{41}{44.25}(12.89) = 11.94 < 12.0$ OK

At soffit, level b, cond. 4, category B:

$f_{sr} = 12.89 < 16$ OK

Girder satisfactory for $F_y = 36$, but A588 used for reasons discussed in Chapter 20.

Figure 14.17 Noncomposite plate girder, from Case Study, Chapter 20.

In Section 14.3, we noted that the largest available rolled shape, W36 × 300, has a resisting moment of 3080 ft kips, based on compact section criteria and $F_y = 50$ ksi. If that moment resulted from loads similar to our plate girder, deflection would be over 7 inches, of which more than 4 would be from live load and impact. If the steel of a W36 × 300 used in that bridge were strong enough for a moment of 4065 ft kips, $\Delta_{(L+I)}$ would be over 6 inches.

The web area of the plate girder is constant at

$$A_w = 86 \times ⅜ = 32.25 \text{ in.}^2$$

and accounts for 45% of the total weight. Three percent of the weight is devoted to web stiffeners, saving several times that much in the weight of the girder.

By reducing the flange area in the quarter-span at each end, it was possible to reduce the total flange weight by 2500± pounds, 11% of the final weight of the girder. The cost saving would be a smaller percentage because of the added cost of welding the splices. Tension field action was considered for the design and rejected because it did not seem to offer cost savings.

Part 2 of the AISC Manual includes several illustrations of plate girder designs for buildings.

14.6 HYBRID GIRDERS

A similar cross section is the basis of Table 14.4, with resisting moment calculated using various steels and with *hybrids* made with A36 webs and stronger flange steels. Note that the resisting moments of the hybrid girders are almost the same as for those made completely with the stronger steels. The ratio used for h/t is close to the limit allowed by formula (14-2) for A514 steel ($F_y = 100$). The comparison applies if the same section is used for all cases. However, with the lower-strength steels, higher ratios of h/t are permitted and more efficient ratios of A_f/A_w are possible. On this account, the table does not properly compare design possibilities for the various separate steels and the hybrids. If loads are applied similarly, deflection can be expected to increase, for the single steel cases, in direct proportion to the maximum moment, and in almost the same proportion for the hybrids.

The AISC* defines a *hybrid girder* narrowly as one with stronger steel in the flanges than in the webs. Based on that definition, special provisions are applied to hybrid girders in the AISC Specification, most of which are summarized in Table 14.5. Of these, the only one that affects the comparison here is the determination of F'_b. The effect is small. Readers are reminded that the properties of A514 steel are based on the quenching and tempering applied in its manufacture and change with subsequent heating and cooling. Special care is necessary in welding that steel, covered in part by AWS D1.1.

The webs with lower-strength steel would have only slightly lower shear strength than those with stronger steel. However, the principal means for controlling

TABLE 14.4 COMPARATIVE MOMENT CAPACITY, GIRDERS OF SINGLE STEELS AND HYBRIDS

Steel, F_y, ksi		Area, in.^{2}					Stress, ksi		Resisting moment
Flanges	Web	$2A_f$	A_w	h/t	I, in.^{4}	S, in.^{3}	F_b	F'_b	M_r, ft k
36	36	45	50	128	100,934	2447	21.6	21.6	4405
50	50	45	50	128	100,934	2447	30.0	29.3	5975
100	100	45	50	128	100,934	2447	60.0	56.5	11521
50	36	45	50	128	100,934	2447	30.0	29.1	5934
100	36	45	50	128	100,934	2447	60.0	52.2	10644

$F_b = 0.6F_y$, using F_y of flange and continuous lateral restraint. For F'_b, see Table 14.5;

$$M_r = \frac{F'_b S}{12}$$

*Commentary, Section 1.10.1.

TABLE 14.5 SPECIAL PROVISIONS OF AISC SPECIFICATION RE HYBRID AND A514 GIRDERS[a]

AISC Section	Single material	Hybrid[b]
1.10.1	Proportion by gross I	Proportion by gross I, but f_a must be $\leq 0.15F_{y1}$; both flanges must have equal area and be of same steel
1.10.5.2	F_v established separately with and without tension field action	Tension field action not permitted
1.10.6	Reduction in flange stress: $F'_b \leq F_b \left[1.0 - 0.0005 \frac{Aw}{A_f}\left(\frac{h}{t} - \frac{760}{\sqrt{F_b}}\right)\right]$	Reduction in flange stress: $F'_b \leq F_b \left[\frac{12 + \left(\frac{Aw}{A_f}\right)(3\alpha - \alpha^3)}{12 + 2\left(\frac{Aw}{A_f}\right)}\right]$
1.10.7	Reduced (F_b) in tension with tension field action	Not applicable
1.5.1.4.1	Compact section requirements apply in determining F_b	Hybrid girders and A514 members qualify as compact
1.5.1.4.2	F_b for semicompact members	Hybrid and A514 girders excluded
1.5.1.4.3	F_b for minor axis bending, I and H shapes	A514 girders excluded
1.5.1.4.5	Part 2, "Compression," gives choice of satisfying (1.5–6a) or (1.5–6b), as applicable, or (1.5–7)	In (1.5–6a) and (1.5–6b), $F_y = F_{y1}$ (above); (1.5–7) does not apply
1.5.3	Welds: Allowable stresses, matching metals	Special provisions for joining different strength steels in AWS D1.1
1.17	AWS D1.1 incorporated into AISC Specification	Special provisions of AWS D1.1 apply to A514 steel

[a]If provisions are identical, no notation made.

[b]F_{y1} = yield stress of flanges
F_{y2} = yield of stress of web
$\alpha = F_{y2}/F_{y1}$

the shear strength would be by manipulation of stiffener spacing. Recall from the highway bridge example that the stiffeners accounted for a very small percentage of the weight of the girder. We will focus on the comparison of moment capacity. In the final column of Table 14.4, resisting moments are compared using the equation

$$M_R = \frac{S_x F'_b}{12}$$

with F'_b in all cases being based on F_y for the flange steel. Before accepting this simple comparison, we should examine the performance of the hybrid girder through its range of moment resistance. This is done in Fig. 14.18 for the girder under consideration. The principles are general.

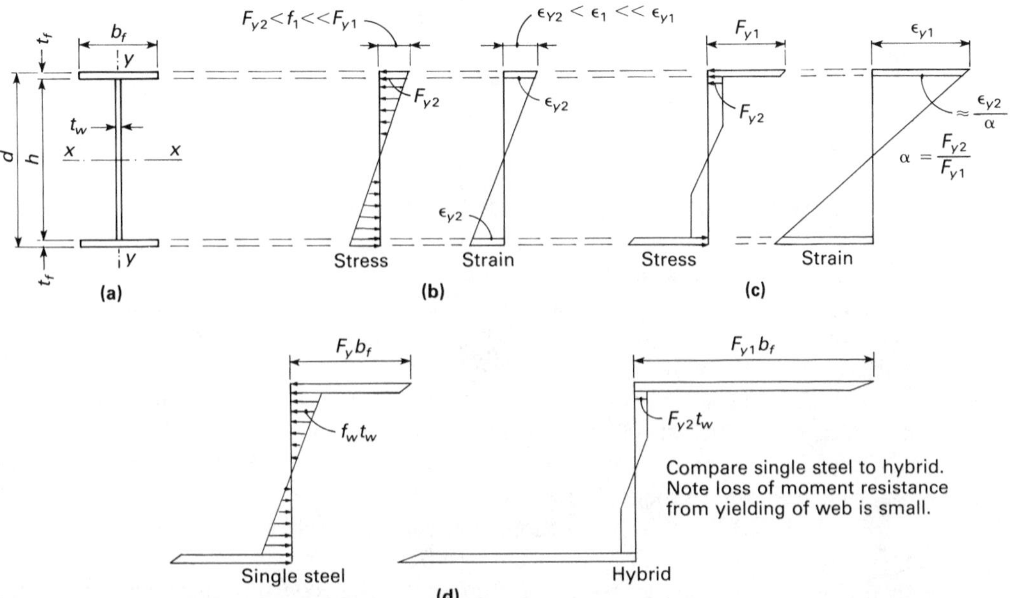

Figure 14.18 Flexure in hybrid girders: (a) cross section; (b) first yield in web; (c) first yield in flange; (d) force per unit depth, in flanges and web (Flanges account for most of M_r).

The beam in Fig. 14.18(a) is fully elastic up to the limit of condition (b), where the outer web fibers are at their yield stress and strain. The flange stresses are far below yield. Condition (c) represents the case of first yield in the flanges, the usual condition for establishing yield moment in a beam. Yield has progressed in the web, which contributes a very small additional amount to the *yield moment*. The flange contributes the bulk of the resisting moment. Strain in the outer fibers of the web is approximately $1/\alpha$ times its elastic strain, with α defined as in the figure. In Fig. 5.3, we found that the ductile range of strain was more than 11 times the elastic range.

Note, however, that the reserve of strength between the yield moment and fully plastic moment is very small, explaining why hybrid girders are excluded from consideration as compact sections. In the comparison of Fig. 14.18(d), we use the product of F times the web thickness or flange width as abscissas, highlighting the contribution of the relatively wide flange to the moment resistance. It is seen that the loss of resistance resulting from substitution of the lower-strength web in the hybrid is very small. At the working stress level, if F_b is based on the yield strength of the flanges, yield strain in the web is likely to involve a very small excursion into the ductile range.

The hybrid girders shown in Table 14.4 offer some saving of cost when compared to girders of the same cross section made completely with the higher-strength steel. In a design case, one would be more likely to compare different sections of different steels with hybrids, all with the same capacity, rather than the same section with the capacity varying with the steel. Problems are suggested at the end of the chapter where such a comparison can be made.

14.7 PLATE GIRDERS AND DESIGN OPPORTUNITIES

In adding cover plates to increase the capacity of rolled sections, we took one small step to expand the opportunities for design beyond the limits set by rolled shapes. When we moved to constant-depth plate girders, it was a big step. The range of feasible spans is increased far beyond the limits of rolled sections.

In Fig. 14.19 we point to other giant steps that are possible (and have been taken) once the bounds of single, prismatic rolled shapes are crossed. The reader may enjoy examining each of them. How do they arise from their moment and shear envelopes? from their function? from considerations of beauty? of efficiency?

Figure 14.20 shows cross sections that become possible with simple extension from the I shape, which rightly dominates the field of rolled shapes. The open double web girder is an obvious extension of a linear member resisting shear and moment. The box girder does that, but is also torsionally strong and stiff, usable for straight girders, but also for girders with curved longitudinal axes. Possible cross-sectional forms of box girders include all those shown here and many more.

The ship hull is the ultimate box girder, as is the airplane fuselage. Picture a ship in a rough sea. Its nonuniformly distributed weight is supported by the nonuniform and changing buoyancy of the wave in the direction of its longitudinal axis while it is being twisted by the component of the wave traversing its beam. Few forms other than the box girder could withstand this environment. Even fewer could be adapted for hydraulic shape, to accommodate the propulsion system, to allow the multiplicity of activities necessary at sea, comparable only to those of a small city.

14.8 COMPOSITE GIRDERS

14.8.1 Principles

The action of composite girders will be examined in the context of a specific girder. The AISC Specification, Section 1.11, covers a variety of issues of composite action, which we leave to the reader to examine when necessary. A number of references are cited in the Bibliography accompanying the AISC Commentary.

The highway bridge in the case study of Chapter 20 consists of a reinforced concrete deck slab supported by steel plate girders. The design methods of that study ignore the possibility of composite action. This has several effects.

1. If composite action actually exists, or can be reliably provided, the calculations

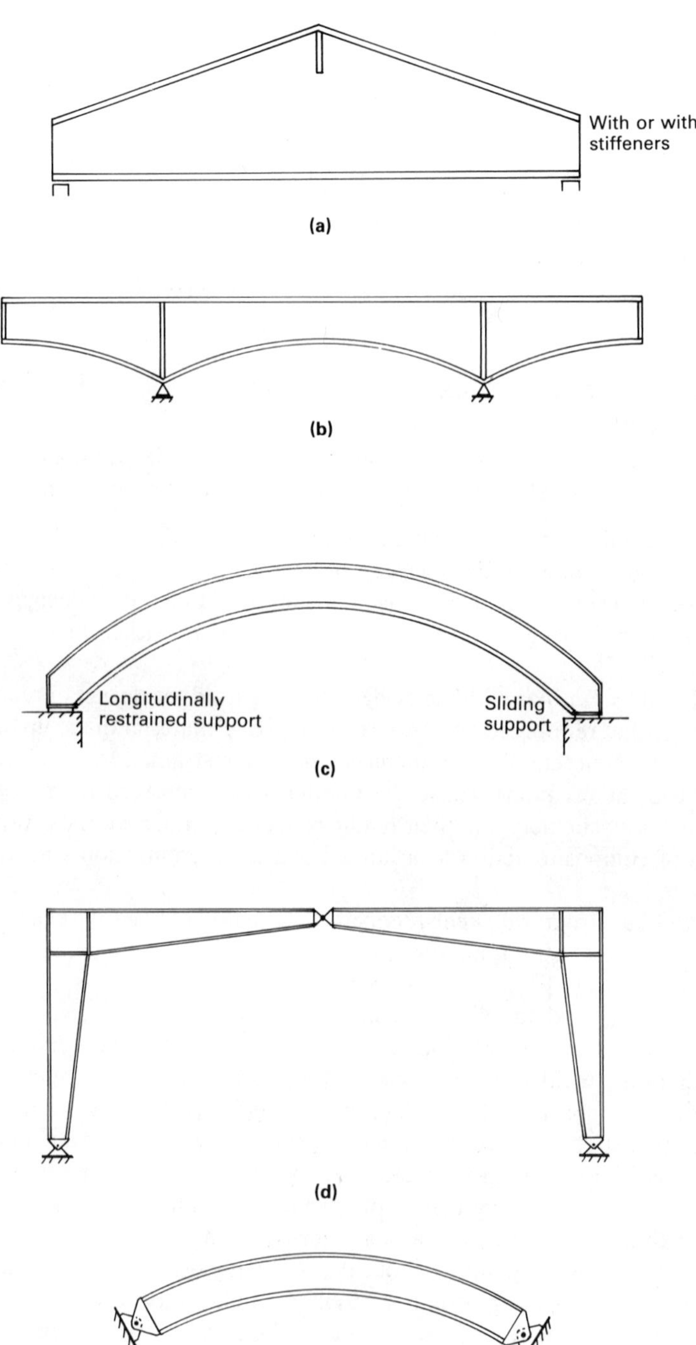

Figure 14.19 Opportunities offered by plate girders: (a) tapered girder; (b) continuous girder; (c) girder curved in profile; (d) three-hinged rigid frame arch; (e) two-hinged arch.

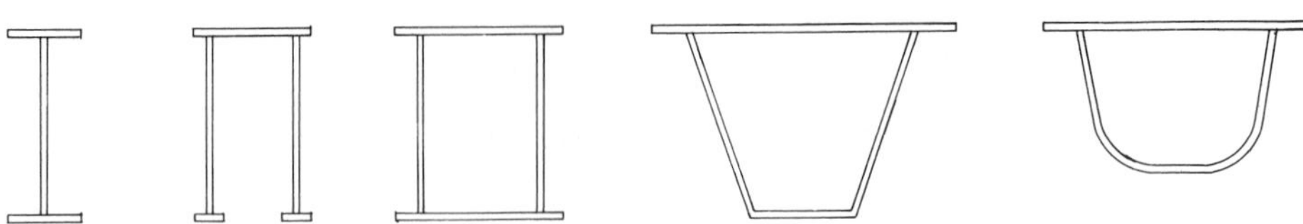

Figure 14.20 Plate girder cross sections.

underestimate the flexural strength of the girders and overestimate the deflections.

2. As a consequence, the cross section selected is deeper than it needs to be and uses more steel than necessary, with no offsetting saving in concrete.

If, then, composite action can be assured, it should be possible to effect savings in the design of plate girder highway bridges with concrete slabs. This is commonly done today in such designs. It is also done at times in building designs where concrete slabs may be considered to act compositely with either plate girders or rolled beams.

Readers familiar with the design methods used for design of reinforced concrete are aware that reinforced concrete is a composite material made up of a combination of steel and concrete, the two materials being constrained by bond to experience the same strain at any point. Those less familiar with reinforced concrete have, no doubt, studied composite action in their readings in strength of materials. We will review the nature of composite action to a limited degree in connection with the bridge girder design.

Unlike structural steels, concrete is neither linearly elastic nor, after an identifiable yield strain, ductile. The two materials can be compared by the stress–strain diagrams in Figs. 14.21(a) and (b). That of concrete has a constantly changing slope (tangent modulus, E_t), and, shortly after peaking, shows sudden, brittle failure. Its peak stress in compression (called f'_c under standard test conditions) is a small fraction of the yield stress of steels. It ruptures in tension at a much lower stress than f'_c. Its moduli are much lower than that of steel. While steel strain remains constant over time if stress is constant, concrete exhibits the characteristic of *creep* illustrated in Fig. 14.21(c) so that reinforced concrete structures show progressively increasing deflection under the influence of sustained loadings, for a long, theoretically infinite time. In spite of these characteristics, for short-time loading and within a stress and strain range up to about $0.50f'_c$, it is possible to consider concrete as an elastic material and to postulate a modulus of elasticity, E_c, as a function of the unit weight of the concrete and its f'_c. This is done by the American Concrete Institute (ACI 318-77)[2] and the formulas accepted by the AISC in the provisions of its specification relating to composite design.

$$E_c = w_c^{1.5}\, 33\sqrt{f'_c} \qquad \text{(ACI—8.5.1)}$$

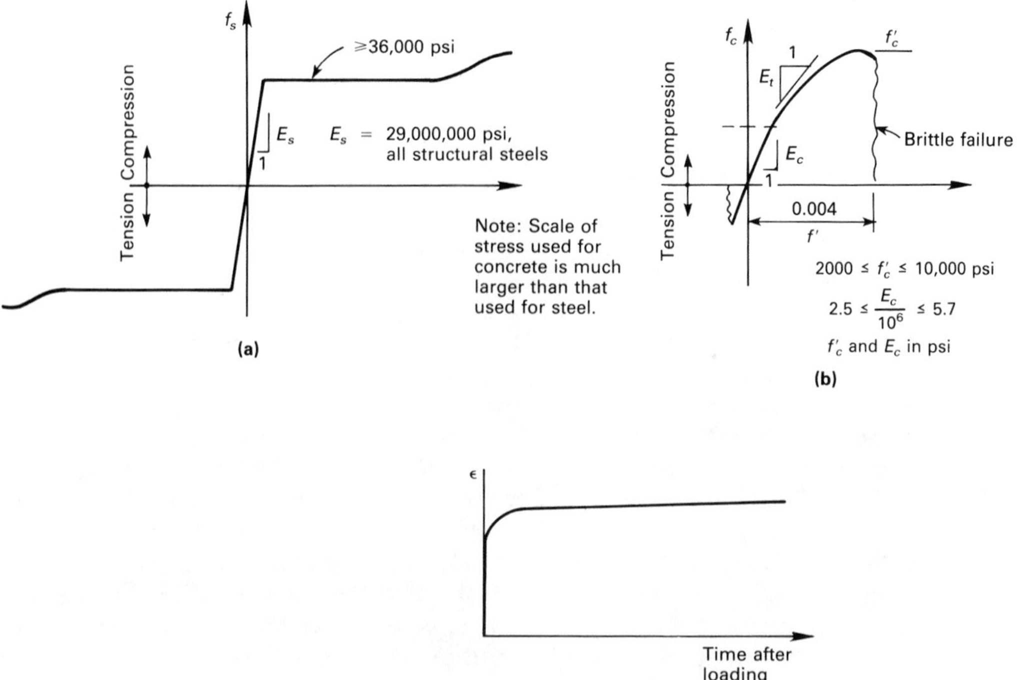

Figure 14.21 Stress-strain relations, steel and concrete: (a) steel; (b) concrete; (c) creep in concrete at constant stress.

E_c is the elastic modulus for concrete in pounds/square inch, w_c its unit weight in pounds/cubic foot, and f'_c and its square root are both in pounds/square inch. The range of w_c is

$$90 \leq w_c \leq 155$$

For "normal weight" concrete, at 145 lb/ft^3 ±, an approximation is used:

$$E_c = 57{,}000\sqrt{f'_c}$$

Commonly, f'_c is in the range between 3000 and 5000 psi, giving a range of E_c for normal weight concrete:

$$3.12 \times 10^6 \leq E_c \leq 4.03 \times 10^6$$

which may be compared to E for steel at 29×10^6 psi. If we define

$$n = \frac{E_s}{E_c}$$

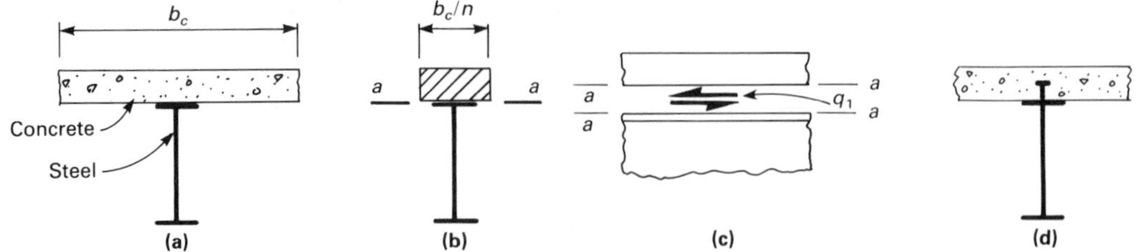

Figure 14.22 Composite section: (1) actual; (b) transformed; (c) shear flow at *a-a*; (d) connectors transfer shear flow.

$$9.3 \leq n \leq 7.2$$

n is usually approximated by the nearest whole number in recognition of the great amount of error in the formula for E_c when compared to actual performance.

If concrete and steel are bonded together so they strain equally, the stress in the steel will be n times the stress in the concrete. We can then, for purposes of analysis, transform the concrete into its equivalent in steel and consider the composite section as a member completely made of steel. Such a transformation is done in Fig. 14.22. At any distance above the top of steel, we replace a differential concrete element of width b_c by its equivalent in steel of width b_c/n. The transformed section has the elastic modulus of steel.

We now set the condition that at level *a-a* in Fig. 14.22 any strain in the top surface of the steel is matched by equal strain in the bottom surface of the concrete. If this were not so, deflections from flexure would take place as in Fig. 14.23(a). The girders and the slab would both be bent by the loads and would share the loads in proportion to their stiffnesses, EI. The contribution of the slab would be trivial.

By requiring that strains be equal, we establish the condition of Fig. 14.23(b), with a composite of the girder and slab much stiffer than the original girder resisting all the load. The corollary of this condition is that longitudinal shear flow must be resisted at plane *a-a* of Fig. 14.22, whose magnitude is

$$q = \frac{VQ}{I} \quad \text{kips/inch,}$$

the characteristic equation of longitudinal shear flow in a cross section. If we bury the upper flange steel in the concrete, bonding (adhesive) effects between the two materials are usually sufficient to force composite action. The same effect can also be achieved by one of a number of types of mechanical shear connectors welded to the steel and buried in the concrete. In either case the effect is analogous to the welds or bolts that connect the flange and web of a steel plate girder.

Consider now that the lower steel girder is one of several parallel girders placed between its end supports (Fig. 14.24) before the concrete deck is cast. It must be strong enough to resist, by itself, the loads caused by forming and casting of the

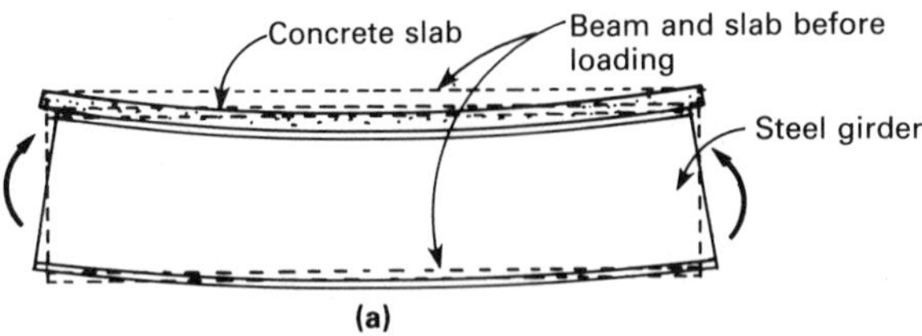

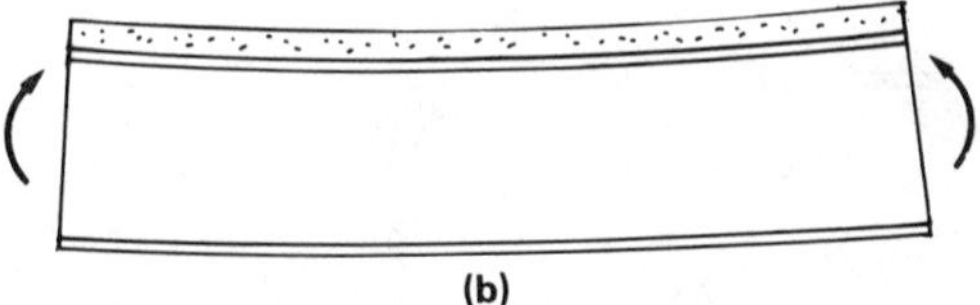

Figure 14.23 Girders subjected to moment: (a) noncomposite; (b) composite.

concrete and to hold those loads until the concrete has aged enough to harden and attain its required strength. From that time, any new loads, such as operating live loads, are resisted by the composite section. If the live loads are transitory, the concrete acts elastically and the transformed section based on the value n is reasonable.

Two stress–strain diagrams for longitudinal flexural stress in the transformed girder are illustrated in Fig. 14.24. If we use the terminology of the AISC Specification, Section 1.11:

S_s is the section modulus of the steel section "referred to its bottom flange."*

S_{tr} is the section modulus of the transformed section "referred to its bottom flange."

The initial loading causes stresses in the steel section. Loads added later cause stresses in the transformed section. The two may be superposed to evaluate the maximum stresses.

At the lower surface,

$$f_b = \frac{M_1}{S_s} + \frac{M_2}{S_{tr}}$$

*If $S_x = I/y_{max}$, and the X axis is not an axis of symmetry, there are two section moduli, one used as a measure of stress at the bottom, the other at the top. They are inversely proportional to the distances of bottom and top from the neutral axis. The term "referred to" may be interpreted as "yielding stress at" when used in the formula $f_{max} = M/S = M_c/I_{NA}$, where c is the distance from the neutral axis to the bottom and top from the neutral axis. The term "referred to" may be interpreted as "yielding stress at" top.

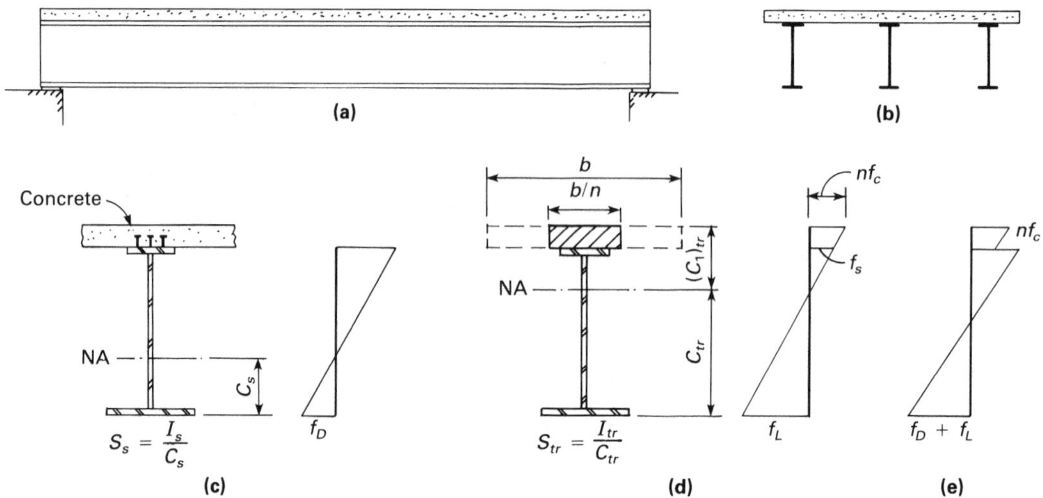

Figure 14.24 Stresses in composite girder: (a) profile; (b) cross section; (c) dead load of wet concrete and forms stresses steel section only; (d) live or other added loads stress transformed section; *transformed stress* is n times actual stress in concrete.

and must be $\leq F_b$. Similar addition gives stress at any other point in the steel. At the top surface of the transformed section,

$$(f_b)_{tr} = -\frac{M_2}{S_{tr}}\left(\frac{c_1}{c}\right)$$

and the real concrete stress

$$f_c = -\frac{(f_b)_{tr}}{n}\left(\frac{c_1}{c}\right)$$

If the allowable concrete stress, $F_c = 0.45f'_c$, f_{tr} must be $\leq 0.45f'_c n$. Vertical shear may be considered, on the safe side, to be resisted by the steel web, by the usual formula:

$$f_v = \frac{V}{d_s t_w}$$

AISC Section 1.11 permits averaging of horizontal shear flow; that is, the total number of connectors between the point of maximum moment and that of zero moment must be sufficient to satisfy the total shear flow within that length.

Values of connectors are treated similarly to bolt values. Commonly used connectors are listed in AISC Table 1.11.4, reproduced here as Table 14.6.

Since concrete strength in tension is low and not reliable, transformed sections are not used for negative moment. In such a case, the designer may either (1) use the

TABLE 14.6 ALLOWABLE LONGITUDINAL SHEAR LOADS ON SHEAR CONNECTORS IN COMPOSITE GIRDERS

Connector[b]	Specified Compressive Strength of Concrete (f'_c), ksi		
	3.0	3.5	≥4.0
½″ diam. x 2″ hooked or headed stud	5.1	5.5	5.9
⅝″ diam. x 2½″ hooked or headed stud	8.0	8.6	9.2
¾″ diam. x 3″ hooked or headed stud	11.5	12.5	13.3
⅞″ diam. x 3½″ hooked or headed stud	15.6	16.8	18.0
Channel C3 x 4.1	4.3w[c]	4.7w[c]	5.0w[c]
Channel C4 x 5.4	4.6w[c]	5.0w[c]	5.3w[c]
Channel C5 x 6.7	4.9w[c]	5.3w[c]	5.6w[c]

[a] Applicable only to concrete made with ASTM C33 aggregates.
[b] The allowable horizontal loads tabulated may also be used for studs longer than shown.
[c] w = length of channel, inches.

COEFFICIENTS FOR USE WITH CONCRETE MADE WITH C330 AGGREGATES

Specified Compressive Strength of Concrete (f'_c)	Air Dry Unit Weight of Concrete, pcf						
	90	95	100	105	110	115	120
≤4.0 ksi	0.73	0.76	0.78	0.81	0.83	0.86	0.88
≥5.0 ksi	0.82	0.85	0.87	0.91	0.93	0.96	0.99

Reprinted with permission from AISC Tables 1.11.4 and 1.11.4A.

strength of the steel section only, or (2) add reinforcing steel within the concrete slab, which, if properly bonded, may be considered part of the steel section.

The example that follows illustrates strength calculations for a composite section in shear and positive bending. The conditions are those of the highway bridge in Chapter 20. The section is compared with that in the case study where composite action was not considered.

14.8.2 Example: Composite Girder

A composite cross section is to be selected to satisfy the requirements of a 90-foot simple span, two-lane highway bridge designed for AASHTO HS-20 truck loading. The section is to be compared to a noncomposite section designed for the same conditions, which was met earlier in this chapter in Section 14.5 and Fig. 14.17.

The composite section of Fig. 14.25(a) is created after specifying shear connectors welded to the upper flange and cast into the concrete. The section shown emerged after several trials. It is not the only possible one, nor necessarily the best. It may be compared to that of Fig. 14.17(a).

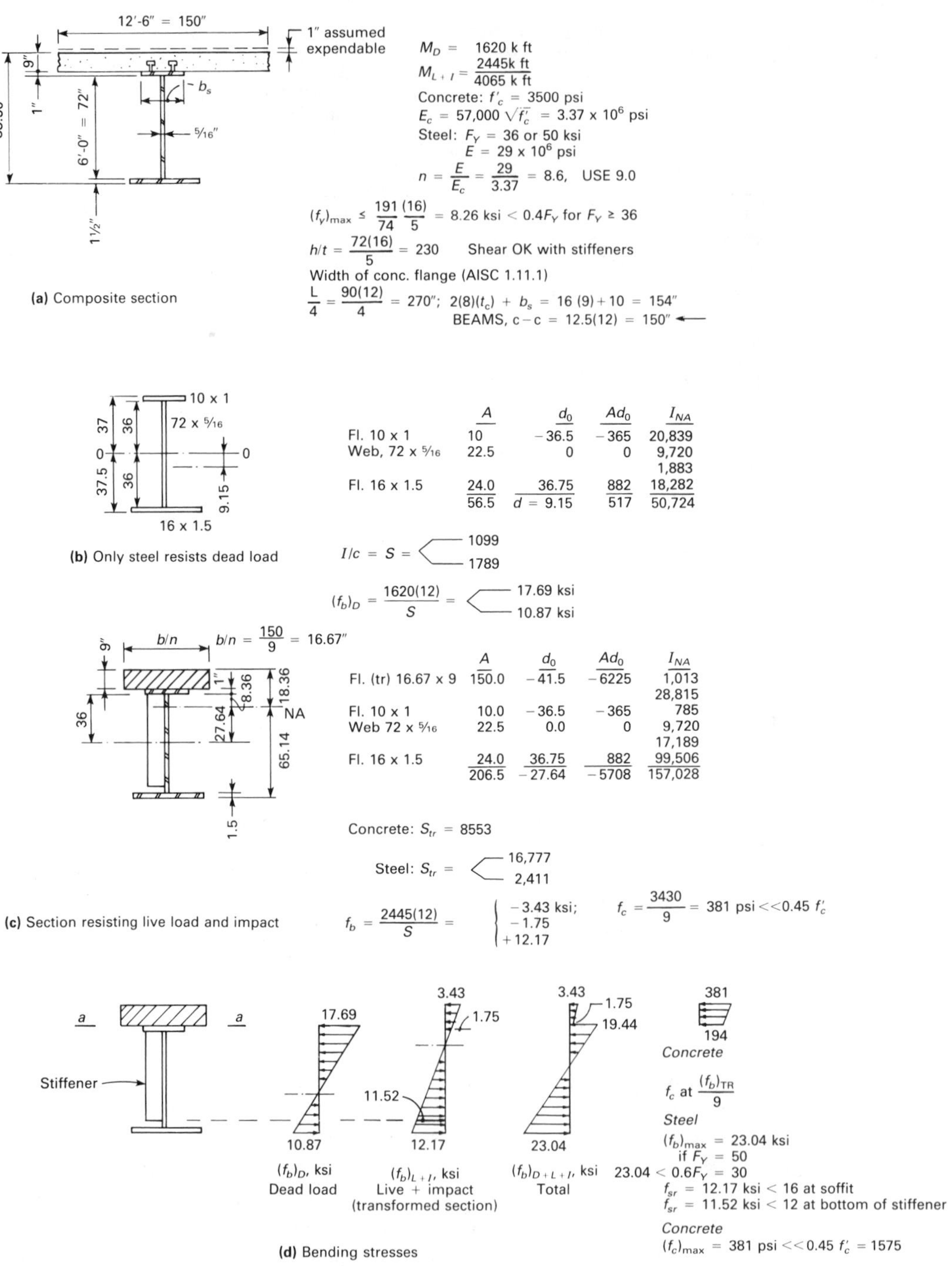

	A	d_0	Ad_0	I_{NA}
Fl. 10 x 1	10	−36.5	−365	20,839
Web, 72 x 5/16	22.5	0	0	9,720
				1,883
Fl. 16 x 1.5	24.0	36.75	882	18,282
	56.5	d = 9.15	517	50,724

	A	d_0	Ad_0	I_{NA}
Fl. (tr) 16.67 x 9	150.0	−41.5	−6225	1,013
				28,815
Fl. 10 x 1	10.0	−36.5	−365	785
Web 72 x 5/16	22.5	0.0	0	9,720
				17,189
Fl. 16 x 1.5	24.0	36.75	882	99,506
	206.5	−27.64	−5708	157,028

Figure 14.25 Composite plate girder: calculations.

1. The overall depth, including concrete, is somewhat less than that of the original steel section.
2. The moment of inertia of the transformed section is greater than that of the original steel section. Deflection from live load plus impact will be reduced.
3. The web is thinner than before, but h/t, at 230, is almost the same. The maximum shear stress is low enough to be acceptable with stiffeners.

The concrete is cast on forms supported by the girders. The girders at that time do not require temporary shoring supports, but do require lateral bracing for stability. The resisting section is that of Fig. 14.25(b). The weight of the concrete and resulting bending moments cause the dead load stresses shown, which become frozen into the steel cross section. (Later removal of forms reduces these stresses slightly; this reduction is ignored here.)

Traffic is admitted to the bridge only after the concrete has set, gaining strength and stiffness. Figure 14.25(c) shows the transformed section that resists moments resulting from live load and impact. It also shows the resulting stresses. Limits of F_{sr} control the increment of stress in the tension region.

In Fig. 14.25(d), the stresses of Figs. 14.25(b) and (c) are added for a resulting diagram of $(f_b)_{max}$ compared to an allowable value of

$$F_b = 0.6F_y$$

The weight of the steel cross section has been reduced by 24% near midspan when compared to that of Fig. 14.17. It would be reduced by a different percentage in regions of low moment (not calculated here).

Deflection has been reduced by the increase of moment of inertia to about two-thirds that of the noncomposite girder.

The requirements for shear connectors to transfer the longitudinal shear flow are determined in Fig. 14.26, based on the envelope of shears that is established in Chapter 20. Two alternatives are considered, either being permissible under Section 1.11 of the AISC Specification.

In alternative 1, the calculated longitudinal shear from live load plus impact is transferred from the concrete to the upper steel flange with stud connectors. The total number of connectors required is the number necessary to transfer the total horizontal shear force between points of zero and maximum moment.

The formulas of alternative 2 are intended to establish *full composite action.* The full change of longitudinal force ($f \times$ area) from zero to the maximum permitted ($F \times$ area) is developed in either the concrete or the steel by shear transfer between the points of zero moment and maximum moment. The smaller of the two possible values of V_h is used to determine the number of shear connectors required.

Although full composite action is not required, the decision is made to supply the additional studs required to establish it. The number of studs required is spaced evenly along the length of the girder. Although this is not completely consistent with the shear flow equation, the AISC Commentary cites tests which indicate that girder performance is not hurt. The AASHTO Specification[1] would require that shear

Shear flow at line $a-a$

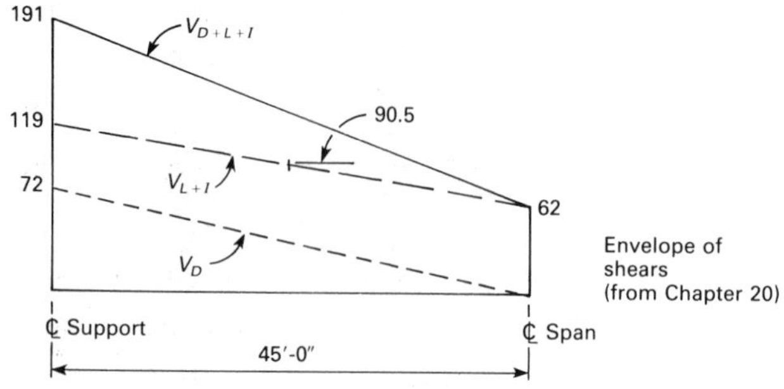

Alternative 1: Provide shear studs based on total calculated horizontal shear

AISC 1.11.2.2 would require recheck of effective section modulus, S_{eff}.

For live load and impact only

$$V_{\text{AVE}} = \frac{119+62}{2} = 90.5 \text{ kips}$$

Shear from dead load does not involve plane $a-a$.

$$q_{\text{AVE}} = \frac{V_{\text{AVE}}Q_{TR}}{I_{TR}}$$

At ℄ span:

$$Q_{TR} = A_{TR}\,(9.36 + 5.5) = 150\,(14.86) = 2229 \text{ in.}^3$$

$$I_{TR} = 157{,}028 \text{ in.}^4$$

Assuming Q/I is constant:

Assumption not exact since lower flange may be reduced in area near ends because of lower moment.

$$q = 90.5\left(\frac{2229}{157{,}028}\right) = 1.29 \text{ kips/inch}$$

$$V_h' = 1.29(45)(12) = 695 \text{ kips}$$

Total shear flow in 45′. Definition of V'_n in AISC 1.11.4 is not exactly the same as used here.

Using ¾″-dia. headed studs at 12.5 k/stud

AISC Table 1.11.4

Provide $\dfrac{695}{12.5} = 56$ studs in 45 ft

Alternative 2: Provide shear studs for full composite action

AISC 1.11.4

$$V_h = 0.85f_c'\,A_c/2 = 0.425(3.5)(150)(9) = 2008 \text{ k}$$

AISC (1.11-3). Use actual area $= A_c$

or

$$V_h = \frac{A_sF_Y}{2} = \frac{56.5(50)}{2} = 1412 \text{ k}$$

AISC (1.11-4). Select smaller shear per AISC 1.11.4.

On this basis and with ¾″ ϕ studs

provide $\dfrac{1412}{12.5} = 113$ studs.

Use alternative 2:
Spacing studs evenly in threes
38.5 spaces at 14″ = 45′ (12)

The cost of additional studs required for alt. 2 is justified by added ultimate strength.

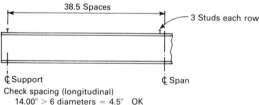

38.5 Spaces
3 Studs each row
℄ Support
℄ Span

Check spacing (longitudinal)
14.00″ > 6 diameters = 4.5″ OK

AISC 1.11.4, last paragraph.

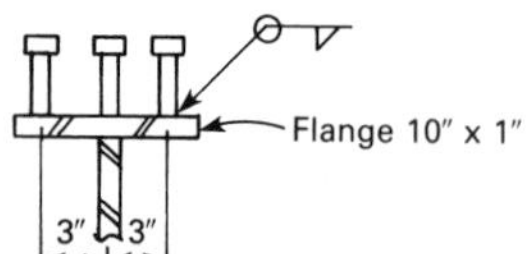

Spacing (transverse); 4 dia. = 3″

Figure 14.26 Calculation of shear connector requirements.

connector spacing reflect the shear flow at each point. The same total number of connectors would be spaced closer together where shear is high and farther apart where shear is low.

PROBLEMS

Except for Problems 14.1 and 14.2, the problems here are long ones, suitable for major assignments.

14.1. Your design provides for a W27 × 84 beam of A36 steel, which has already been ordered from the mill. Final checking shows that the moment for 10 ft of the beam's length is 470 ft k, which is too much for that beam. You wish to use the original beam, but reinforce it as necessary for the full moment. Make the necessary design change assuming shear stress will not be a problem.

(a) By reinforcing the flanges only (Fig. 14.2).

(b) By one of the methods of Fig. 14.3.

(c) If the maximum shear is 120 k, is your solution for part (b) satisfactory?

14.2. Design a hybrid cross section using A36 and A514 steel, suitable for a moment of 11,500 ft k. Compare to the section with $F_y = 100$ in Table 14.4. You are not limited to 80″ web depth nor to ⅝″ web thickness.

14.3. After studying the design of the 90′ span plate girder in Chapter 20, design a composite girder for the same loads. The steel cross section should be suitable for the total dead load and the composite section for all loads. You should indicate where temporary lateral support is necessary for the top flange of the steel girder, but you do not have to design the lateral bracing. The steel design should be complete, including variations of cross section, stiffening if needed, splices, and shear connectors. The complete design of the reinforced concrete deck is properly a part of this problem, but may be omitted if you wish to concentrate on the steel design only. In that case, use the concrete thickness indicated in Chapter 20 and $f'_c = 3000$ psi.

14.4. The plate girder shown in Fig. 14.5 is to be designed for the following loads at each interior column.

Dead: 4 floors at 30′ × 24′ × 115lb/ft^2 = 331 k

Live: 4 floors at 30′ × 24′ × 35lb/ft^2 = 101 k

Deflection from live load shall not exceed 1.0″. You are to describe a camber diagram so the fabricator can compensate for the dead load deflection. (See calculation sheet 31 for the bridge girder in Chapter 20.) You should consider steels of A36 or A572, Gr. 50, and select the more appropriate. Assume the end columns are W14 × 370 and the inner columns W14 × 90. All column webs are parallel to the girder web. You are to connect your girder with bolts to the end columns. The interior columns are to have base plates, which are to be bolted to the girder. Design the base plates and other details necessary to transfer the column loads. Your design should include all the details necessary for a complete girder.

14.5. Design a tapered girder such as in Fig. 14.19(a) for an industrial building roof. Controlling loads:

Dead load: 15lb/ft^2

Snow load: 30lb/ft^2

Span is 80′ and girders are to be spaced 20′ on centers. Steel is ASTM A36. The slope of the top flange should be set by you. Compare your result to the mill building truss designed in Chapter 17.

15

Some Odds and Ends

15.1 WHAT ELSE IS THERE?

The opening of Part II of this book, in Section 7.1, set the task of fashioning some tools of design specific to structural steel. Most of that is now behind us. The sufficiency of that work may be measured in part by the fact that the material of Chapters 7 through 15, joined to the analytic procedures and basic grounding in mechanics, which is clearly outside the scope of this book, is sufficient to design the four common and representative structures covered in Part III. Space is not available here for much more. But we must use a little space to indicate that there is much more in a field where engineers constantly seek new and better ways to solve traditional problems and ways to solve new ones. We will touch on just a few that are closely related to the materials in the rest of the book. The engineering literature, both books and current journals, contains many more.

15.2 BEYOND SECTION 1.9: SEMISTABLE ELEMENTS

Recall from Chapters 9 and 10 that the potential for instability due to buckling arises in two different ways: from the slenderness ratios (KL/r, ld/A_f, or l/r_T) of compression members or compression flanges between points of lateral restraint, and from the ratios b/t or d/t of elements of cross sections. In dealing with the former, procedures of the AISC Specification and others reduce the allowable stress on the members under consideration to compensate for the loss in strength from slenderness; the effect is to seek, as closely as possible, a constant factor of safety. A different approach is usually taken with respect to element stability. The elements themselves are made stiff enough so that the question of instability from that source does not arise. Table 9.3, based on Section 1.9 of the AISC Specification, gives limiting ratios of significant width to thickness. When these limits are adhered to, as they are for most rolled sections, members may be used without further concern for element instability—a simplifying approach, but one which ignores the strength that remains when the limits of AISC 1.9 are exceeded. For such cases, the AISC Specification provides procedures for adjusted calculations in its Appendix C. The Commentary to AISC Section 1.9 explains the basis of the procedures in part and refers to sources from which they are derived. We will limit ourselves to some comments on how the procedures are applied.

The procedures of Appendix C provide three different bases on which formulas in the body of the Specification are to be adjusted:

1. Reduction in allowable stress.
2. Reduction in the cross-sectional area to be considered.
3. Adjustment of the value C_c (see Chapter 9).

All three apply to axially loaded compression members. Number 2 also applies to members in flexure. Since buckling of unstiffened elements involves the entire outstanding width of the element, allowable stresses on such elements are reduced, recalling the reductions of F_a and F_b for KL/r, ld/A_f, and l/r_T. The nature of buckling of stiffened elements leaves a portion of the element, close to the stiffeners, still effective for compression stress. For such elements the procedures provide formulas for computing the effective width of the elements.

To accomplish the adjustments, four new variables, Q_s, Q_a, b_e, and C_c', are defined in Appendix C.

1. Q_s applies to unstiffened compression elements of a cross section whose values of b/t exceed the limits of Section 1.9.1.1 (see Table 9.3). Values of Q_s are less than 1.0 and decline with increasing values of b/t in accordance with Formulas (C2-1) through (C2-6), which may be found in Appendix C.
2. b_e is defined as a "reduced effective width" applying to stiffened compression elements of the cross section. Its values decline with increasing b/t, but increase with decreasing stress, f. The governing formulas, (C3-1) and C3-2), are in Appendix C.
3. Q_a = effective area/actual area, where the effective area is $\Sigma (b - b_e)t$.
4. $$C_c' = \sqrt{\frac{2\pi^2 E}{Q_s Q_a F_a}} = \frac{C_c}{\sqrt{Q_s Q_a}} \qquad \text{(see Chapter 9 for } C_c\text{).}$$

Allowable axial stresses, F_a, for members with unstiffened or stiffened compression elements are given by

$$F_a = \frac{Q_s Q_a \left[1 - \dfrac{(kl/r)^2}{2(C_c')^2}\right] F_y}{\dfrac{5}{3} + \dfrac{3}{8}\dfrac{Kl/r}{C_c'} - \dfrac{1}{8}\left(\dfrac{Kl/r}{C_c'}\right)^3} \qquad \text{(AISC C5-1)}$$

which the reader will recognize as a modified form of AISC (1.5-1) (see Chapter 9). When $Kl/r > C_c'$, the formulas of AISC Section 1.5 apply. For circular tubes with $D/t > 3300/F_y$, f_a is governed by the smaller of the values in AISC 1.5.1.3, or

$$F_a = \frac{662}{D/t} + 0.40F_y \qquad \text{(AISC C3-3)}$$

The reduced effective width, b_e, is also used in flexural calculations to find effective values of the flexural properties of cross sections. I, S, r_T, and A_f are computed, using b_e in place of b. Formulas for allowable stress, F_b, are unchanged from those found in AISC 1.5.1.4. However, it should be clear that the effective resisting moment is reduced.

The interaction formulas of AISC 1.6 are also adjusted to account for the reduced axial and flexural capacities.

15.3 COLD-FORMED STEEL STRUCTURAL MEMBERS

A child who makes a paper airplane (Fig. 15.1) may think it has created a miracle when it folds a floppy piece of paper and it becomes stiff enough to remain rigid as it glides through the air. Engineers recognize that the miracle is the creation of a beam, with depth, moment of inertia (therefore section modulus), and shear rigidity. An extension of the same principle is familiar to us all in the corrugated cardboard box that the child finds convenient for a roof over his playhouse [Fig. 15.1(b)].

Thin sheets of steel cold formed in similar manner into stiff, folded shapes are used to form the roof and side walls of the mill building in Chapter 17 and are used widely in steel buildings as *cladding* [Fig. 15.1(c)]. Pressed into C and double C form (recall Fig. 7.3), light gage* steel can be made to resemble and act like individual beam and column framing members, often used in standard designs for factory-built† trailers, mobile homes, warehouses, and the like. Cold-formed decks and sides are also common in roofs and sides of buildings where the primary frame may be made with rolled structural steel, reinforced concrete, or other materials. In the tier building

**Gage* in this case measures thickness. See Table of Sheet Metal Gages in AISC Manual, Part 6. Higher gage numbers apply to thinner sheets.

† We differentiate between factories, which produce many identical pieces, and fabricating shops, which make a limited number for custom designs. The dividing line between the two is not always clear.

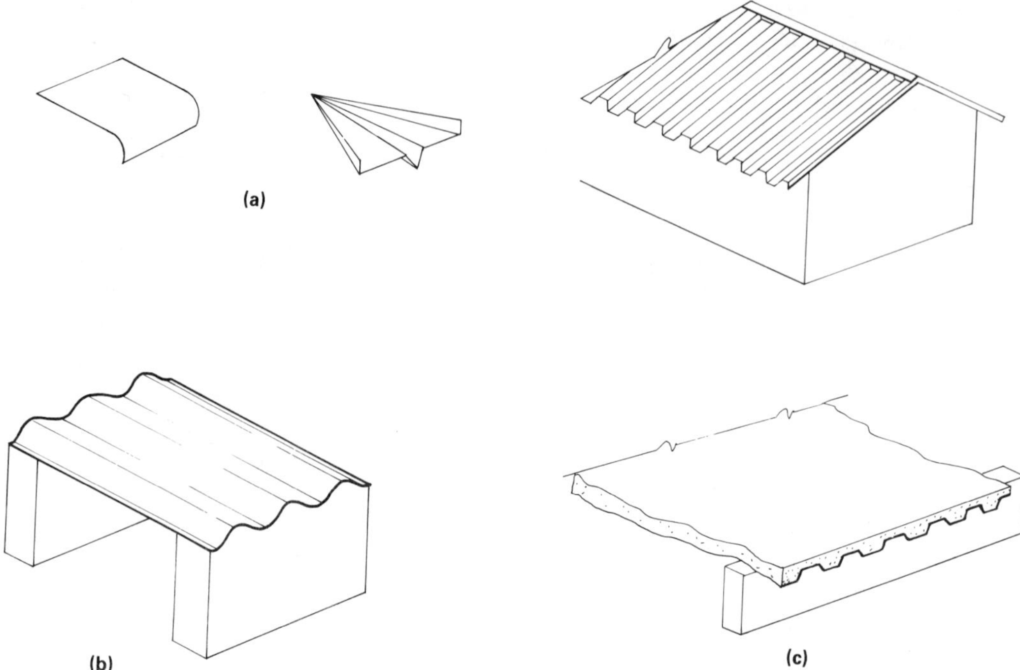

Figure 15.1 Stiffness and strength from folded forms: (a) a flimsy sheet of paper becomes a beam when folded, which a child can fly like an airplane; (b) pressed into corrugations, the paper becomes stiff enough to span as the roof of the child's play house; (c) thin sheets of steel pressed into stiff forms become roofs of buildings and parts of composite floors.

of Chapter 18, metal deck is used to *form* (act as a mold for) concrete when it is poured as a paste, which ultimately hardens and works compositely with the metal deck in much the same way as the composite girders of Chapter 14.

Cold-formed members are made by pressing thin sheets of steel between the halves of heavy dies without addition of heat. The resulting uniform-thickness members have depth of cross section, cross-sectional area, moment of inertia, and section modulus, all of which can be calculated by the usual methods. The wide, thin surfaces between folds have large ratios of width/thickness. It should be no surprise that the compression elements of the cross sections tend to be unstable, and measures similar to the special ones in AISC Specification, Appendix C, become basic to the proportioning of light-gage, cold-formed members. Allowable stresses and effective section properties depend on b/t.

The most commonly used specification governing the use of such members in structures is issued and periodically updated by the American Iron and Steel Institute.[12] See also the discussion in reference 53. The form of the AISI Specification resembles the AISC Specification, already familiar to readers of this book, with increased emphasis on element stability. The steels most commonly used are among those listed in Table 5.1 and can be identified by reference to the "Principal Uses" column. Welding procedures are somewhat special because of the low energy required to melt the light-gage materials. Suppliers issue tables of section properties for the products they supply, including properties calculated for the full cross section and those based on the effective parts of the cross section as reduced due to stability considerations.

Cold-formed deck and wall sections are used as flexural members, but they also cover large areas. Properly joined to supporting members, they can also act as diaphragms (see Chapter 2) in the horizontal force resisting system. This is an option that the designer of the mill building of Chapter 17 chooses to forego, for reasons explained therein. However, if it had been used, considerable savings would have resulted in the members of the structural frame. Effective diaphragm action is a problem of shear transfer, controlled not only by the area being sheared but by the stability of the elements of what may degenerate to a very flimsy structure if the geometry of the folded section is allowed to change. *Allowable* values of shear are often derived from full-scale tests of the diaphragms, followed by reports to regulatory bodies and agreement if the evidence is persuasive. For example, manufacturers submit data to the International Conference of Building Officials supporting proposed allowable shears for use with the Uniform Building Code.[33] If approved, a notice to that effect is issued by ICBO, which may be published with the manufacturer's promotional literature.

15.4 WEB-TAPERED MEMBERS

The bulk of this book, as of the AISC Specification, deals with members of constant cross section (prismatic). In Chapter 14, the line was crossed in the case of plate girders, where the usual reasons for using prismatic members do not apply. In the design of the boom and back legs of the stiffleg derrick in Chapter 19, reasons emerge that cause the designer to turn to Timoshenko and Gere[59] for guidance in modifying the KL/r of end-tapered compression members. For symmetrical, linearly tapered members, such as were seen in Fig. 6.2, the AISC offers guidance in Appendix D of its design Specification. Modified formulas are substituted for those in the body of the Specification, which apply to prismatic members. Where compression is axial, Eqs. (1.5-1) and (1.5-2) for F_a are modified to account for adjusted values of KL/r, based on the actual r and the sharpness of taper. Formulas for F_b (1.5-6a, 1.5-6b, and 1.5-7 in the body of the Specification) are modified by factors that substitute for the tapered beam, an *equivalent* beam with a changed length and a cross section equal to that of the smaller end. The interaction formula of Section 1.6 of the AISC Specification is also modified. The AISC Commentary to Appendix D provides some explanation of the revised formulas and references on which they were based. It also provides charts for determining the *modified lengths* of members in frames such as those in Fig. 6.2.

15.5 ARCHES

In Fig. 2.5 and the accompanying discussion, it was pointed out that the arch as a form can do the same job as that done in other ways by the several systems of Fig. 2.4. The arch was a central feature of early Roman architecture and has been found in somewhat different forms in the architecture of many other cultures of that and later periods, which relied on the compressive strength of masonry to span forces across large openings. With the advent of materials such as steel capable of resisting both tension and compression on parts of the same cross section, it became possible to free the arch from the constraint of masonry, which can only permit very small bending moments. Nevertheless, a *funicular arch*, one whose profile is based on the *funicular polygon* of its main loads, can be designed for much less bending moment than a girder or a truss. The elegance of well-designed arches, which continue to gratify the esthetic sense of people, is in no small part based on its *natural* form, which evolves directly from its loads.

A loaded weightless cable such as is shown in Fig. 15.2 must take the shape of a funicular polygon—a polygon of pure internal tension, with the cable in equilibrium under the action of the loads and their reactions. It must do so because the cable can only resist tension. It will be recognized that, at any point along the cable, the vertical component of the cable tension is the transverse shear on the system, and its horizontal component multiplied by the lever arm h [Fig. 15.2(b)] is the moment on the system. The change in moment within the length of any straight segment of the cable is equal to the shear times the length of the segment or, by direct proportion, the horizontal component times the change in h. These are the same relationships that are used in Chapter 10 to determine the shear and moment on a beam and the diagrams that represent them.

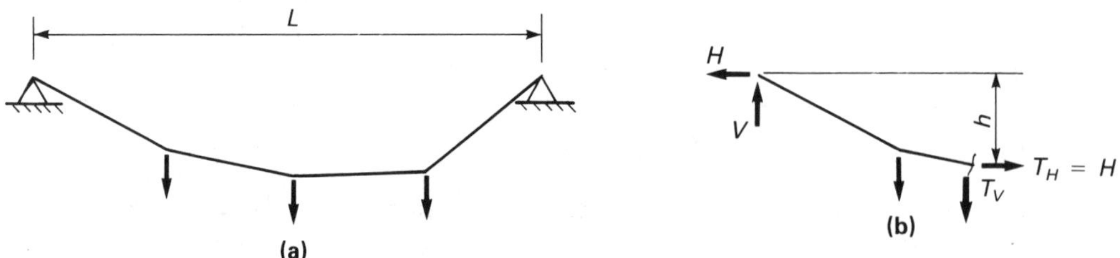

Figure 15.2 Funicular system, loaded cable: (a) equilibrium shape of cable is a *funicular* system of internal tension; (b) T_V = shear in system, $T_H h$ = moment in system.

If the shape of the funicular polygon is *frozen* and turned upside down, the tension system of the cable may be replaced by an arch, each segment of which is in axial compression (see Fig. 15.3). The shape of a funicular arch is the shape of the moment diagram of the straight, simple-span beam between the same supports subject

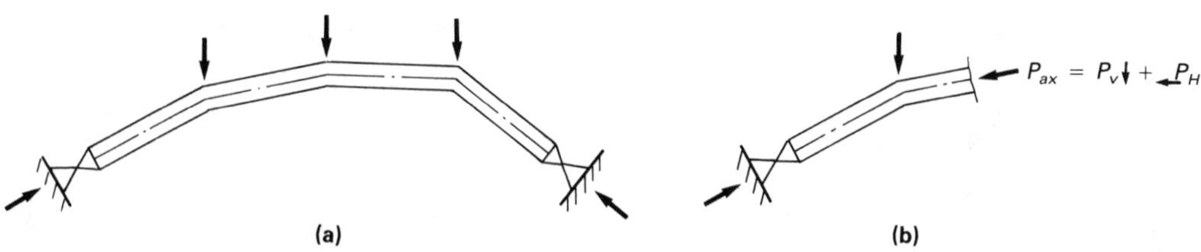

Figure 15.3 Funicular arch for same load system as Fig. 15.2.

to the same loads and drawn to an appropriate scale. However, the funicular arch has no bending moment or shear.

A parabolic arch, then, has no bending moment if its load is uniformly distributed along its projected length, a good approximation of the dead load case for many structures. For long spans, where dead loads dominate the loading system, parabolic arches approach the ideal, compressive shape* (see Fig. 15.4). Moment and shear arise only from loads whose funicular polygon is not in the shape of the arch. Their magnitudes are determined only by the deviation between the two curves (see Fig. 15.5) and are much smaller than the moments and shears on the equivalent

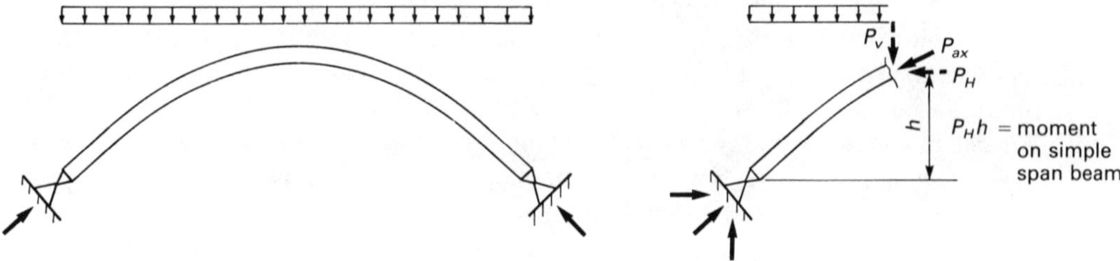

Figure 15.4 Parabolic arch is funicular form for uniformly distributed load.

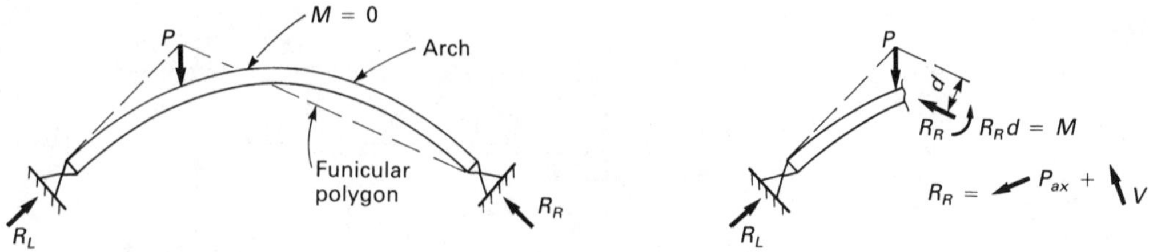

Figure 15.5 Concentrated load on parabolic arch causes axial load, shear, and bending moment.

*If the rise of the arch is large compared to its span and the load uniformly distributed along the curve itself, the ideal shape would be a catenary, the curve assumed by a cable under the action of its own weight. For low rise, the catenary and parabola are almost identical.

beam. The resulting arch can be much less stiff and more slender than the equivalent beam, although it can only be used if space is available for its profile and its ends can be restrained from separating. Its capacity is governed by Kl/r, as with any compression member.

Parallel arches lend themselves to long-span linear structures, such as bridges. Crossing each other at a common center, two arches may be used to form a vaulted roof and several a dome (see Fig. 15.6).

Figure 15.6 Crossed arches form vaults and domes.

15.6 ECCENTRICALLY BRACED FRAMES

Lateral-load resistance used in the Tier Building Case Study, Chapter 18, is derived from two systems, one a shear wall system similar to Fig. 2.11(a), the other a Vierendeel frame like Fig. 2.11(c). To explain the selection, it is noted in the study that the moment-resisting Vierendeel frame can be made ductile for energy absorption through damped displacement during a severe earthquake, but that movement of such a frame in windstorms may be objectionable. The more rigid *shear walls* act to resist the more common events; however, partial failure of the shear walls leads to sufficient displacement to actuate the ductile moment-resisting system. Similar dual systems are also found, utilizing braced frames such as Fig. 2.11(b) in place of the shear walls.

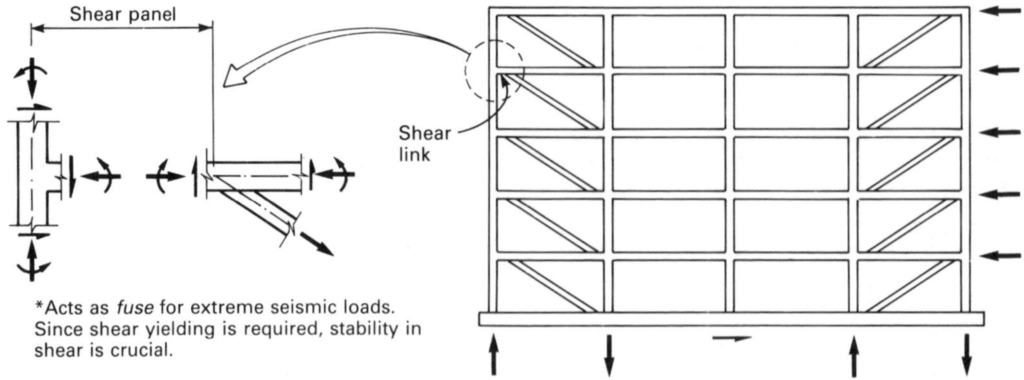
Shear panel
Shear link
*Acts as *fuse* for extreme seismic loads. Since shear yielding is required, stability in shear is crucial.

Figure 15.7 Eccentrically braced frame.

Historically, engineers designing braced frames have taken considerable care to be sure that the axes of all members are concurrent, avoiding bending moments in members which would arise from eccentricity. This was the approach emphasized in this book in discussing truss and frame members in Chapters 8 and 9 and their connections in Chapters 12 and 13. The bracing of the mill building in Chapter 17 is designed on that basis. In parts of the stiffleg derrick design in Chapter 19, it becomes impossible to avoid eccentricity at the joints, forcing the designer reluctantly to design heavier members.

Recent work by Popov and others[45,46,57] has led to a new approach to the dual system concept, with deliberate use of eccentrically connected bracing. One variant of the system is illustrated in Fig. 15.7. Bracing is connected to a beam at each floor at a short distance from the column. When the bracing is activated by horizontal forces, the vertical component of its force causes shear to travel across the short link to the column, with some associated bending. Unlike many other design instances discussed earlier, the shear tends to dominate the design problem for the link, requiring special detailing for shear strength and shear stability. At usual force levels, the rigidity of the system is similar to that of a concentrically joined braced frame. However, in the extreme case, the short link provides ductile energy resistance in the plastic range of shear and moment resistance. Eccentrically braced frames have been found attractive enough in performance tests and cost to be adopted by a number of design engineers within a very few years.

Part III
Case Studies

16
Introduction to Part III

16.1 CASE STUDIES IN THIS BOOK

16.1.1 Purpose, Selection, Format

In the structure of this book, Part I represents context, and Part II provides some tools; Part III illustrates product, which in a real sense is the fruit of engineering study and activity.

In Part 3, we will present four design problems and, for each, a solution. In none of the cases is the solution the only possible one; nor is it necessarily the best. It is certainly not the best solution for all times and all places. The Eiffel Tower in Paris and the Space Needle in Seattle (Fig. 16.1) solved almost identical problems, each in its own magnificent way. Although both are built of structural steel, the solutions are markedly different. Put aside the fact that the problem itself, a theme structure for an international exposition, demanded "different" solutions for public relations purposes; the solution in each case is a product of its time and its place. Minasian in Seattle was not rejecting Eiffel's design approach as much as he was applying his creative talent to the problem in a new context. Among other things, the new context contained a steel industry capable of producing large-sized structural sections, the availability of versatile and powerful hoisting equipment, and the imperative for smaller numbers of members, which results from the current ratio of material to assembly costs.

Figure 16.1 Eiffel Tower and Space Needle. (Photos courtesy of French Government Tourist Office and Seattle-King County Convention and Visitors Bureau)

There are no Eiffel Towers or Space Needles among the four case studies presented. These are much more prosaic, run-of-the-mill design problems of the sort that large numbers of engineers have to solve month after month. However, there are universals in design process that apply equally to these cases and to the major structures that the fledgling engineer may dream of designing.

What we have is four different types of structures, each representing a major arena for structural design in which steel is often the material of choice. Each case studied comes from the background of practice of the authors. In that sense, they are truly case studies. However, none are actual designs of specific structures in exactly the way they were originally done. Just as fiction often illuminates life more effectively than straight reporting, we believe the doctored form of these real-life designs will serve the educational purposes of this book more effectively than a straight recounting of the designs. Readers will best serve their own educational purposes if they accept our suggestion to try different approaches to the same problems after they have understood the thought processes of the designers as presented here.

16.1.2 The Selections

We have attempted to select design problems that are simple enough to be followed by readers of limited experience while they are still in the process of mastering the other parts of this book and books on structural analysis. However, anxious that the case studies be of sufficient scope so that important elements of the engineering design process are grappled with and finally grasped, we have avoided problems so simple that they do not reveal the nature of engineering thinking. As a child learns through contact with a world that is always more complex than its current understanding, an engineer finds, in the course of executing a design, that he must sometimes learn new things from the problem itself and be guided by surrogate parents—previous designs, books on engineering theory, senior engineers. The form of the case studies is intended to help engineering novitiates to reach beyond their present stage of mastery. The reader should not wait to start on the case studies until after everything else is learned. Rather use the broader problems presented here to help learn in a more effective way the materials of Parts I and II, which, we are convinced, are fully grasped only when applied to complete design problems.

The studies of the mill building and the stiffleg derrick, each in its own way, address simple, essentially statically determinate systems, and can be profitably attacked at a relatively early stage. The tier building study would be easier to follow for readers with at least some background in the stiffness issues central to indeterminate analysis. For the plate girder bridge it is necessary to abandon the safe harbor of rolled sections selected from manufacturers' lists and use, in a less structured way, the underlying principles of material science, fabrication technology, and the like, in the creation of tailor-made sections. The reader should understand or be prepared to study something about shear stability and joinery before getting too deep into this study.

16.1.3 Calculations and Discussion: Format for Study

Each case study is presented in the form of a set of calculations accompanied by discussion. The calculation set is the most usual form used by structural designers to create and justify designs that are ultimately presented in the form of drawings and specifications and the other contract documents discussed in Chapter 4. The set of calculations is often submitted to reviewing bodies, either public agencies or the client's staff or advisors, whose proper function is not to second guess the designer, but to be sure that the work is thoroughly thought out, documented, and consistent with purposes of the reviewing agency or department, which may not be the primary priorities of the designer.

A structure is designed for specific purposes, to resist specific loads and to satisfy technical and legal constraints as they exist at the time of the design. Purposes, loads, and constraints are likely to change during the life of the structure. Most of the older buildings in San Francisco have had to be strengthened to account for advances in knowledge about seismic resistance. Most U.S. cities have had old buildings altered as their uses changed: from one type of industry to another; from warehousing to offices or residences; from private to institutional use. The Brooklyn Bridge, designed for pedestrians, trams, and horse carts, now carries automobiles and trucks. The collapse of the Tacoma Narrows Bridge in 1940 triggered major engineering studies[20] on the aerodynamics of structures, which led not only to changed design approaches on new structures, but to alterations in a number of major bridges that were already in place.

The calculations are a record of the designer's thinking at the time of the design, the problem and constraints as they were seen at that time. Along with other design documents, if complete and clear, they provide a sound base line for alteration of the structure when necessary.

For the designer and reviewer, the calculations are intended to be self-contained and self-explanatory. Ideally, they are, although the results of many meetings, design conferences, and so on, may not be recorded in them. The calculation sets as presented here are edited so that design concepts are presented more completely than they might otherwise be, and some details are omitted. The accompanying discussions are intended to explain the designers' thinking, particularly for the less experienced reader for whom the process is still new.

16.2 THE FOUR CASE STUDIES IN BRIEF

16.2.1 Common Threads

Certain things are common to the four case studies. Each is a relatively small and simple example of a large class of steel structures, where the basic approach presented can be extended, with adjustment for the specifics of the problems, to many large and small examples of the class. Each is a *framed structure* in the sense that its primary elements are linear; they are assemblages of members each of which has a longitudinal axis much longer than its cross-section dimensions. Framed structures of this sort may be distinguished from surface structures such as plates and shells, where resistance to loads depends on the shape of the surface rather than of the cross section. The boundaries are not sharp and the two classes of structure overlap in the roofing and siding of the mill building and in the deck and deep webs of the plate girder bridge.

Each represents a type of structure where experience of the profession has been sufficient so that most of the necessary theory is at the relatively simple level of the rest of this book; so much of it has been set down in generally accepted Codes, Specifications, and procedures. All utilize the AISC Specification and design aids in the AISC Manual, or, where a more broadly applicable code is necessary, the *Uniform Building Code*. This is done uniformly to dovetail with the work of the first two parts of the book and in spite of the fact that highway bridges are usually designed with other specifications. Generally accepted specifications do not, to our knowledge, exist for derricks, and the *Uniform Building Code* is only one of the widely used codes in the United States. The principles used are transferrable, and there seems to be little purpose in confusing the reader at this stage with a multiplicity of codes and specifications, which for the most part cover the same material in slightly different words or form.

While every attempt has been made to make the calculations correct and the designs valid, none of the results are intended for direct transfer to design of other structures; nor is any case study complete enough to be used without further work and further checking for structures closely resembling these. Such use is beyond the purposes of this book and is not authorized.

16.2.2 The Mill Building, a Light Industrial Building

The setting is in light industry and warehousing, although the larger arena is the very broad field of steel industrial structures. The client needs a one-story building of about 16,000 square feet (medium size). Column-free area is desirable, fire protection requirements minimal. Plant activities are supported on the floor and do not affect the steel frame, except that a light traveling hoist must be supported from the roof structure.

Foundation conditions are poor, leading to desirability of statically determinate design. Snow loads are at a level common to large areas without deep mountain snow packs. Winds are fairly severe, seismic problems less so. The design utilizes repetitive parallel bents with 80-foot-long trusses on steel columns. Walls and roof are clad in light-gage steel. Wind forces are resisted by bracing systems that demand a great deal of design attention.

16.2.3 The Tier Building

A five-story office building covering 12,000 square feet per floor is a fairly modest example of a very large number of fire-protected multistory commercial and residential buildings in urban settings designed by structural engineers as part of architect-led teams. Principles and design methods can apply with little change to regularly framed tier buildings of larger area and more stories, although the validity of the "portal" method of analysis may be questionable if there were many more stories.

The location is one of severe seismic hazard, an important factor in establishing not only the design loads but the framing system, which has two stages of resistance to seismic loads: rigid shear walls and a less rigid ductile moment-resisting frame. The structural steel frame utilizes W shapes for beams, girders, and columns, with floor diaphragms of light-gage steel and concrete, and continuous (moment-resisting) joints at all frame connections.

The structure is statically indeterminate with a high degree of redundancy. Calculations presented are made for the preliminary stage of design (see Chapter 4), are approximate, and have the limited purpose of establishing basic member sizes, the space needs of the structural frame, tonnage of steel, and the structural budget. More precise analysis and (presumably minor) design refinements are left for the final design stage, as is the detailing of joints.

Loads are standard dead and floor live loads from the building code; seismic loads, also based on the code, are of magnitudes determined not only by the mass of the structure but by the framing system itself. Limitation of story drift is a design criterion.

16.2.4 The Stiffleg Derrick

The stiffleg derrick is one of the simplest types of structure used as the structural part of a major hoisting system in construction. The principles discussed apply (with additions) to a great variety of cranes and derricks used in construction and general industry. The structural form is a combination of a simple three-dimensional truss fixed in position combined with a plane truss of varying geometry, which is used to move lifted loads. The analysis is dominated by the large lifted design load, its movements, and its accelerations. Members are large latticed ones, heavy W shapes, a large-diameter tube, wire ropes, multipart blocks, and incidental parts to complete the system.

The structure, broken down into parts, must be transportable from site to site to site over highways. As a piece of construction equipment, it requires machinery and moving parts, making close collaboration necessary between structural and mechanical engineers.

16.2.5 The Plate Girder Bridge

The problem is to develop a design system for two-lane simple-span highway bridges applicable to a range of spans from 60 to 120 feet. The span range lends itself to design with two or more parallel plate girders (see Chapter 14) and concrete deck. Two different arrangements are considered and two alternative types of steel. The basic characteristics of the system are determined. Plate girders are designed in detail for a 90-foot span to establish calculation procedures that can be used, with computer assistance, to design a family of similar structures covering the assigned range.

Live loads are moving heavy truck loads in standardized form as established by the American Association of State Highway and Transportation Officials. Dead load is a much greater part of total load than in the other case studies, becoming even more so as the spans get longer. The depths of girders are beyond the range of mill-supplied sizes. Designed to be manufactured from plates, they are made nonprismatic (of varying cross-sectional dimensions) for economy and weight saving. The deep thin webs are potentially unstable in shear and require stiffening to avoid shear buckling. Live load deflection criteria are severe to preserve the riding quality of the roadway. Repetitive loadings require consideration of fatigue. The nature of the system makes it stable against wind and seismic effects without special measures.

Similar designs, utilizing composite action of steel and concrete, would probably lead to some cost savings. The design presented should be easier to follow for engineers unfamiliar with concrete theory. The methods of Chapter 14 can be used to adapt the design to composite form.

16.3 PROCEDURE FOR CALCULATION

16.3.1 Structure of the Calculations

In a generalized form, all the calculations follow the sequence of Flow Chart 16.1. However, the details phase is omitted from calculations in this book. The ideas of Chapters 12 and 13 apply, and some problems in that chapter derive from these case studies. This sequence is, we believe, ideal after the fact for purposes of information retrieval and evaluation of the results. It is a less than perfect representation of the order (or lack of order) in which creative design proceeds.

Many designs come to the designer predetermined (by someone in authority) to be examples of systems selected in advance. The reasons are many: prior investigation validating the system for the type of problem; past success in similar projects; capitalized investment in particular types of fabrication facilities; relationship to other on-site structures. We hesitate to mention the (sometimes less justifiable) reasons of limitations of fee, minimization of office costs, mental inertia. Whatever the reason, such a problem requires, for the first two phases, merely the recording of the problem and the system. There is not much virtue, in the commonly expressed opinion of senior engineers frustrated with their juniors, in "reinventing the wheel" for each design. The design work is then limited to quantification and selection; the calculations can be made and recorded almost exactly as presented. Many engineers learn to define design in these limited terms, failing to realize that the total process is much less mechanical.

If engineering seeks the most effective solution to a problem in the total and often unique combination of circumstances, the design process must be less highly structured, much more iterative. It proceeds backward and forward through various stages of conceptualization, quantification, testing, revision of concept, requantification, and so on. Sometimes, quantitative results lead to reexamination of the design

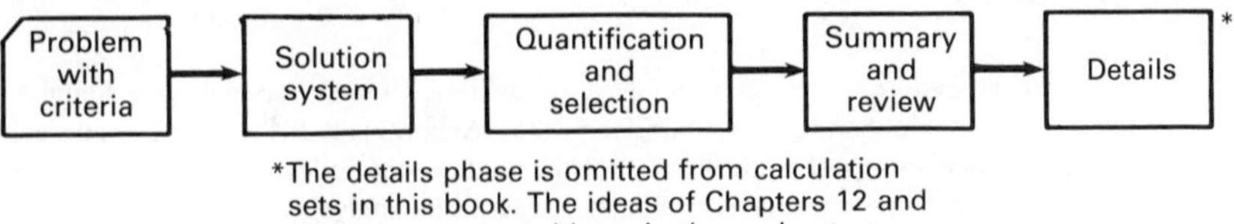

*The details phase is omitted from calculation sets in this book. The ideas of Chapters 12 and 13 apply; some problems in those chapters derive from these case studies.

Flow Chart 16.1 Typical structure of a set of calculations.

criteria and constraints that were originally conceived as part of the problem presented and therefore sacrosanct.*

The sequence of the flow chart remains an effective way to present the calculations and to judge them after they are complete. However, during the process of design, strict adherence to that sequence may unnecessarily constrict thought and prevent forward movement. When groping for an efficient structural system, the designer, possibly on expandable scratch sheets, makes calculations and draws sketches to see what sizes of members are implied and what type of connections. These are early tests of the system, best made before the system is frozen. Early decisions about the type of steel may need to be reexamined if unhappy surprises occur in member selection. The process of joint detailing often reveals the wisdom of revising member sizes for easier fabrication.

Nevertheless, to avoid chaos, the sequence later to be used in presentation can be effectively used as a skeleton on which to hang the various parts of the design process as they are developed. Iterations take place within this structure as the engineer, at each stage, asks himself the yes or no (or maybe) questions on the basis of which he decides between reopening an earlier decision or proceeding to the next stage.

This requires a certain freedom in the organization of calculations, which is gained by devices we have adopted from some very effective engineers. Instead of writing a sequence of calculation sheets numbered in the order in which they were made, divide the set in advance into separate sections, each to be numbered serially. Afterthoughts and revisions can then be fitted into the place where they reveal the ultimate logic, rather than the relatively unimportant time sequence. In the *mill building* and *stiffleg derrick* cases, sections are separated into separate, open-ended numerical series distinguished by identifying letters. In the *highway bridge* study, a blank block of numbers is introduced between the two main sections of a relatively short set. The *tier building study*, limited to the preliminary design phase, also has relatively few sheets; in this case there is a single numerical sequence, a fact that we regretted several times during the course of preparation and review.

We do not suggest that the final calculation set be limited to the final decisions. If they are to reveal the thinking process to both the designer, after the passage of time, and to others, the comparisons and iterations that led to choice should often remain, although in a very long set some may best be relegated to an appendix or separate file.

*In the early days of ocean offshore oil drilling, Cooper was part of a team designing an offshore tower for what was then thought of as deep water. According to the (imperfect?) perception of Cooper as a middle-level engineer not central to the critical action, the following took place. The project was almost complete when some unexpected difficulty arose in the analysis of the interaction of some closely linked planes of bracing. The problem was solvable quantitatively but brought to the fore some fundamental questions about the structural system. The decision, rather than fight through to a solution on the original basis, was to abandon the system and start over on a new basis. Both client and engineering consultant have had reason, in the years that followed, to be happy about that decision, which could only have been made because both client and engineer perceived engineering and economics in creative ways. (It would not have been possible with the usual fee structures under which engineers function.)

16.3.2 Sections of a Calculation Set

Each set opens with a Table of Contents, indispensable as an organizing and administrative tool, doubly so because of the sectional subdivisions used. There should be no question left in any receiver's mind that the numbers 12 through 20 of the highway bridge set were not inadvertently omitted from the set received, but in fact do not exist. It should be easy for a reviewer to find the section on roof design in the mill building, for a drafter to find the truss arrangement and member sizes without plowing through all the calculations.

The first section in each case develops and records information general to the design: the purpose, people, organizations, loads, codes, specifications, constraints, structural system, and arrangement. Collected in one place, this information is easily available to designer and others for use with the other sections.

Other sections are divided according to the logic of the design. In the tier building, selections are made first for vertical loads and then tested and revised as necessary for lateral loads. Calculations are divided in the same manner. If the selections of steel members were more thoroughly controlled by the seismic loading, as in a building without shear walls or of many more stories, the sequence might very well be reversed. The stiffleg derrick calculations demand a different type of partitioning related to this different type of problem.

A summary section is used in all the sets except the stiffleg derrick. Commonly omitted, we recommend it for a number of reasons. For the designer, it is the (almost*) last chance to test the wisdom of individual decisions made earlier—tests for consistency, omissions, opportunities for simplification, efficiency. The validity of the design may be tested by a calculation of tonnage of steel and/or pounds per square foot of steel. These results, compared to initial assumptions, help lead to a decision about whether further refinement of the calculations is necessary. The same results, compared to those for similar structures, may reveal basic mistakes in calculation or inefficiency of the design system; the comparison may also reinforce the designer's confidence, and that of his supervisor or client, in the work.

The summary is used to budget structural costs, as a tool for the drafter in presenting the results, as a tool for both designer and drafter in deciding what joint and other details require elaboration, and as a reference for the designer or other engineers in checking the design drawings and the materials take-offs and estimates made from them.

16.3.3 Methods of Analysis

There is no way to design an engineered structure without analyzing it. However, the primary purpose of these case studies and this book is to teach design, of which analysis is only one necessary part. The selection of a structural system requires at least a qualitative understanding of the response of the system to loads. However,

*The stage of drawing review is another such chance.

once the system is selected the designer usually has a choice among a number of mathematical methods for modeling and calculating the system. He can usually also choose between hand and computer versions of the models used.

We had to decide between seeking the most elegant or the most sophisticated models we could find or the models that would be most easily followed by our readers, many of whom may not yet be thoroughly grounded in complex methods of analysis. We chose the latter route, which we believe keeps the focus where it will be most useful in learning design—to learn how the tools of analysis are combined with all the other ingredients to achieve an integrated design.

The methods of the mill building are for the most part the simple methods applicable to statically determinate structures. Since the designer at times deals with a number of similar members, he uses tables frequently as a useful and time-saving methodological approach, which we recommend, and which is the form of output for many computer-aided solutions. It becomes necessary in a number of cases to deal with interaction in two-way bending and bending with axial load. The methods of Chapter 11 are sufficient for the cases involved. They can be mastered before or along with the case study. Perhaps the most important methodological issue is one that is met here and repeated in all the other studies, a systematic approach to separation of separate loading effects and their reintegration in different ways in search of design controls. This implies elastic methods and superposable cases. It also explains why in all the studies the designers use unit load or influence line approaches to analysis. Coefficients resulting from unit load solutions are easily multiplied by separate magnitudes for each type of load and combined in all ways necessary to find the worst effects.

Graphics is a fundamental tool of both thinking and communication for structural engineers, among themselves and with others. The calculations include sketches liberally for both purposes. They are most effective if done at least roughly to scale. In the case of the stiffleg derrick, most members sizes and connecting details emerge only after careful scale drawing of the movable parts in many positions.

As mentioned earlier, the framing of the tier building is highly indeterminate and the method of analysis used is approximate. Although approximate, the portal method can be expected to yield fairly close approximations for regular frames of the type and size involved. It is specifically sanctioned even for final design by some codes, although it is a fairly simple matter with computer programs available today to get more precise analyses. Used as it is in the case study, the portal method helps to establish member sizes, a prerequisite for any precise method of analysis to be used in final design. For those who are unfamiliar with indeterminate analysis, the approximate methods and the discussion should clarify some of the relationships among relative stiffness, shear, and moment, which are at the heart of indeterminate analysis. With these relationships understood qualitatively, it should be possible to follow the quantitative approximations and later move on to precision. Once again, the designers here tabulate much of the work when similar problems appear repetitively.

The stiffleg derrick uses another statically determinate frame. The influence line approach is a powerful one, particularly because the primary loads are moving loads, and a central problem of analysis involves the search for critical positions of the load. The designer is forced to extend the interaction ideas of Chapter 11 when she recognizes that the latticed boom is both a compression and flexural member, with flexure resisted by truss action; the equations of AISC Section 1.6 cannot be applied directly, but the interactive effects covered in equations (1.6-1) cannot properly be ignored.

In the highway bridge, the problem is limited to that of a simple span system. The loads again are moving loads. In this case, also, the spans are variable. The influence approach to analysis becomes useful again. With influence effects established, much of the rest of the work is systematized and tabulated and is amenable to programming for those who want to use the results to solve the problem of the entire family of designs.

16.3.4 Tables and Design Aids

In an abstract sense, every design should be examined on its individual merits and developed from its basics, an often-expressed idea of great validity, within limits. The engineer should be able to, and sometimes must, calculate the moment of inertia, section modulus, and radius of gyration of an I-shaped section from its dimensions. To ignore the Shape Tables where they are conveniently listed for rolled sections is to create unnecessary work, but also to make almost impossible the kind of comparisons that are central issues in design choice. To prove each time a simple-span beam is subjected to a uniformly distributed load that $M_{max} = WL/8$ and $\Delta_{max} = (5/384)(Wl^3/EI)$ is similarly unnecessary, since the standard beam cases of the AISC Manual have that information and more in a form for easy use and easy comparison of cases. So, also, do the beam tables and the column tables and figures and charts in this and other books. The engineer who retains the ability to re-create the proofs will find it most useful for the more unusual cases whose results are not tabulated. The wheel needs no reinvention as long as its properties and its limitations are understood.

The calculations in the case studies draw directly from the aids in the AISC Manual where they apply to the problem exactly. They are also found to be useful starting points when the tabulated problem is part of a more complex problem. Thus the column tables are easily usable to get the compression capacity of an axially loaded member. They clearly overstate the compression capacity of a beam column, but, with judicious adjustment, are found useful in sizing the columns of both the mill building and the tier building. Design aids are particularly useful when decisions about the system are made, requiring early, impressionistic ideas of size and shape to be subjected later to detailed check and refinement.

16.3.5 Loads

Much of the problem of design of a structure revolves around the definition of loads and determination of those combinations of loads that can be expected during the life of a structure and that will most severely test it. Some general issues of loading and factors of safety were discussed in Chapter 7 and met in several different ways in the chapters of Part II. In the case studies here, we meet the same issues in a number of specific, possibly more complex ways; it will be useful to dwell somewhat more on these same matters here.

In Table 7.1, loads were divided into vertical and horizontal and then subdivided into different types of each. All the loads in the studied structures come from this list. Dead and live loads apply to all, but live loads are prescribed differently for each. Wind loads are pertinent to all, but control parts of the design quantitatively* only for the mill building and the derrick. Seismic loads are controlling for the tier building. In the highway bridge design, wind and seismic loads are resisted primarily within the concrete deck, although their transfer to the foundations (details not developed here) would involve the structural steel. For the two buildings and the bridge, both design loads and load combinations are prescribed by a code or specification, relieving the designer of the necessity to do, possibly less well, what has already been formalized from the collective experience of the profession. However, the mill building designer judges it necessary to account for wind in a somewhat more severe way than required by code. The derrick designer is the initial prescriber of operating loads, setting from them requirements for the purchase of operating machinery and limits on the user.

Dead loads pose a special problem in that they cannot really be known until the design is done. The designer must make an educated guess in order to start the quantitative design, risking the necessity to redesign if the estimate turns out to be very high or even slightly low. Unnecessary cost may go with unduly high dead load estimates. Loss of safety usually goes with unrealistically low estimates of dead load, but, as in the wind analysis in the mill building, may result from overestimate. The impact of error in estimating dead load is much less in the buildings and much more in the bridge, where the ratio of dead to live load is large.

Live loads for the mill building consist of roof snow load and hoist load. The snow load is prescribed by a building official, but a design decision about the shape of the roof is aimed at preventing more severe loads from drifting snow. The hoist load is a moving live load selected from operating requirements by the designer and affecting different parts of the structure as it moves from place to place. Live loads used for the tier building are the minimum floor live loads prescribed, based on ample precedent, by the *Uniform Building Code*. These are usually considered sufficient except for special cases, such as fireproof files, computer floors, and lead-shielded x-ray rooms. Those used for the bridge are the standardized form of moving loads from heavy trucks incorporated into the highway bridge design specification by AASHTO. The derrick is designed for moving live loads, which are in fact its reason for being. In all

*Both wind and seismic loads require complete resisting systems within the total structural system. Once the system is established, quantification of effects is sometimes unnecessary for one if the other is clearly more severe.

cases of moving live load, both horizontal and vertical accelerations of the moving masses increase the live load effects. They are applied in the form of static equivalents, as percentages of the live load, called *impact* for the vertical accelerations and *longitudinal* or *lateral* for horizontal accelerations. Allowable stresses are not adjusted when such accelerations are considered.

Temperature stresses are discussed in the mill building design but not quantified. The Code offers no guidance, but the engineering and physics literature does. In the highway bridge they are avoided by the simple expedient of providing an expansion joint at one end so that temperature strain is not resisted, and hence causes no stress.

Wind loads affect the design of the mill building very significantly. A highly complex dynamic problem based on the acceleration of air and water particles in an air stream of infinitely varying velocities, wind loads have been standardized for building design purposes in many codes. The 1982 version of the *Uniform Building Code*, used in the mill building design, has the most complex and probably the most satisfactory wind load requirements in the history of that Code. Suction loads are seen to be as significant as pressure loads, a fact recognized through wind tunnel studies as well as theoretical studies and dramatized by the experience of structures in tornadoes and blast. Calculation of wind effects is complicated by the arbitrary division into elements and primary frame. Elements are assigned higher load intensities than the frame so that when the loads on elements are transferred to the supporting frame they are reduced. This reflects the physical fact that the wind effects are not uniformly distributed as they are calculated, but makes for some ambiguity and uncertainty at the boundaries.

Seismic loads become important to the design of the tier building. Again, a static surrogate is used for a highly complex dynamic load. The UBC and many engineers consider the static equivalent approach satisfactory for *regular** buildings and acceptable at the present state of the art for many other structures. However, the current provisions are under comprehensive review at this writing and may be revised soon in the UBC. Other dynamic and semidynamic approaches are available. They do not yet correct all the deficiencies of the UBC static approach and are applied at this writing mostly to unusually large, complex, or important structures. The static Code requirements themselves have been progressively changing through various editions of the UBC. Seismic loads are unique in building design in that structural engineers recognize that the design loads, for a variety of reasons, are so much lower than reasonably expectable ones that "failure" in an extreme earthquake is recognized as a possibility in design. The resisting system of the tier building is designed to convert the form of possible failure from one that threatens life to one that damages materials.

**Regular* in this context applies to framing systems, generally repetitive from floor to floor without large, abrupt changes in mass or stiffness within the stories or between stories.

16.3.6 Load Combinations

Few failures of structures occur from the application of single loads. Rather, they occur from combinations, sometimes unanticipated, of the separate effects. Combinations are prescribed as design requirements by building and bridge codes. The additive effects of live and dead loads are obvious, but some controlling conditions for the tier building only emerge when live load is on some spans and not on others. The overturning problem is most dangerous when gravity live loads are not in place. It is found in the mill building that truss members designed for large tension force from dead plus live load are threatened with buckling in compression if the design wind occurs in a summer storm when the snow is not conveniently ballasting the roof system. An appropriate design decision follows.

Safety might be served by adding the worst possible effects of all separate loads. But the probability of simultaneous occurrence of a 100-year wind and 100-year earthquake is so low that no code we know asks that the wind and seismic loads be combined. The requirements of the UBC, for the two buildings, those of AASHTO for the highway bridge, and the advice of AISC for the stiffleg derrick, all both guide and constrain the designers in their choice of combinations for calculation. In the derrick mast, the controlling combination is obscure and emerges only after considerable searching. The seismic loadings used, in conformance to the UBC, for the tier building design, are applied only in the direction of the principal building axes, a fact that simplifies the calculations but ignores the physical probabilities. This may be no more arbitrary a decision than the design seismic loads themselves.

Although maximum values of all separate loads are not prescribed, design combinations do include combinations of reduced values. When combinations that include wind (mill building) or seismic loads (tier building) are applied in accordance with UBC, allowable stresses increase by one-third. (Arithmetically, the equivalent is sometimes done by multiplying the loads by 0.75.)

In all the studies, the designers make it a point to calculate and record the separate effects of the loads before combining them. This is a mathematical necessity to satisfy the interaction formulas of AISC 1.6 (see Chapter 11). It also serves other design purposes. By maintaining the separate effects and seeing their relative magnitudes, the designers are able to make judgments about the most effective way to adjust member or system properties during the iterations leading to design decisions. It informs the designer's decision about whether or not to redesign when assumed dead load turns out to include a degree of error. It reduces the amount of calculation involved when mistakes or errors are corrected. Other advantages will be seen in each case study and by readers who apply this principle in their work.

17 Case Study: Light Industrial Mill Building

17.1 THE LIGHT INDUSTRIAL BUILDING

Light industry is very widely dispersed in a technological society or a society in the developing stages of technological transition. It demands a very large number of relatively small buildings housing many diverse manufacturing and warehousing functions. They are often located in uncongested areas with relatively low land values, where vehicular or railroad access is easy. If the manufacturing process is such that the building is not usually subject to uncontrollable fires and/or a localized fire is not likely to spread to other adjacent properties, the legal or insurance requirements for fire-resistive construction are often minimal. Many such buildings are of one-story construction, often of structural steel.

The steel itself does not support combustion. However, it is not considered fire resistant, since steel strength degenerates with temperature. Steel structures offer advantages of easy fabrication and erection, and easy alteration with changing needs. The light industrial building (often called *mill building*) is particularly suited for our first case study. As in the case here, it can often be of simple design, which can be followed by readers early in their exposure to structural steel design. Statically determinate methods of analysis are sufficient and suitable for many structures of this type, and are often desirable since much light industry is located in areas offering poor foundation conditions. In our case, many of the members are axially loaded and can be selected by methods covered in the early chapters of Part II. Most flexural members are relatively simple.

The reader who has not yet read Chapters 10 and 11 will find it useful to read them while studying the mill building design. The discussion that accompanies the calculations should help with qualitative ideas even for those not yet ready for the quantitative ones; they can, if necessary, be sharpened later. Material from Chapter 16 will help in understanding the purpose and form of the calculations and the loads. Many of the design ideas, while not limited to mill buildings, are derived from the special needs of such structures and the specific conditions of this one.

17.2 GENERAL CONSIDERATIONS IN THE DESIGN (G Sheets)

17.2.1 Table of Contents (Sheet G1)

The set of calculations opens with a table of contents, which is convenient in following the work. The calculations are in seven parts represented by separate serially numbered sheets. G sheets assemble in one place the basic conceptual thinking and design constraints. Data included in this series are used as reference in developing other parts of the design. The design system is established for the entire structure. Subsystems for *roof framing* (R), *main trusses* (TR), *lateral resisting systems* (L) and *main columns* (C) can be best understood as parts of the total system shown in the G series. The *summary* sheets (S) assemble the results of the member selections. They make it possible to estimate the weight of steel used and, through that, to estimate costs. We have also listed, but did not include, a D series, which would be used to develop the design of connections and other details that must be examined before the design can be considered fully validated. For reasons discussed in Chapter 16, each number series is open ended.

We have not included the foundation design, which would not be of structural steel. However, foundation conditions will affect the design of the steel superstructure.

The designer is identified as Millman, acting for Coopchen Engineers on behalf of their client, XYZ Manufacturing Company. None of these names are real.

17.2.2 General Requirements, Criteria, Framing Alternatives (Sheets G2 through G4)

The design work starts with an examination of the specific requirements of the client and the constraints set by location and environment. The requirements of total column free area, dimensions, access, and interior hoists have been established by an earlier study of the needs of the manufacturing process and use of space. These will each be important in the selection of the structural system. The clearance requirement

Project : LIGHT INDUSTRIAL BUILDING
Subject : JOB TITLE & CONTENTS

SHEET G1 OF

STRUCTURAL CALCULATIONS

LIGHT INDUSTRIAL BUILDING

FOR

XYZ MANUFACTURING CO.
MILLTOWN, COLO.

CONTENTS

FOUNDATION DESIGN NOT INCLUDED
MONORAIL DESIGN NOT INCLUDED

COOPCHEN ENGINEERS
AUGUST 1982
BY MILLMAN

and "column-free" stipulation will each rule out some less expensive alternative designs. Neither should be adopted lightly. However, one important fact of industrial plant design is that saving of structural cost at the expense of optimum operational cost is likely to represent poor economy.

The local constraints of weather, location, soil, and Code will influence the design very strongly. Seismic hazard in Zone 1 is minimal. However, the wind exposure cited will require serious attention to the forces caused by wind. Roof live loads (snow) of 30 pounds/square foot are characteristic of large parts of the United States. They are more demanding than the generally snow-free West Coast, but less demanding than structures in remote high mountain areas such as California's Sierra Nevada or the Rockies. Experience with problems of structures has led to tightening up of both snow and wind requirements in many areas in recent years. The 1982 edition of the *Uniform Building Code* includes a more stringent and more completely elaborated section on wind requirements than earlier editions.*

The selection of materials is common enough so that there is little justification offered for the choices. Actually, there is some question in our minds about whether A36 steel was the best choice. The reader might find it useful, after going through this design, to test the choice. Millman's reasoning was, in part, thus:

Most individual members will be long and slender. The selection of compression members is likely to be based more on the requirements of stiffness (KL/r) than of the yield strength of the material. However, not all members are controlled by minimum stiffness. Those not so controlled would be lighter with higher yield steel at a small increase in unit price. Assuming one steel for the entire structure, it is not certain at this stage which would result in lower cost.

We will discuss the selection of material for the roof deck and siding when Millman checks the possible sections.

The *Uniform Building Code* is only one of several possible choices for this type of structure. The actual choice is based on local legislation and the designer would be obliged to research that. Although requirements tend to be similar, they are not identical. The engineer is obliged to use local requirements as at least minimum standards. The UBC is used here, as well as in the next case study. The AISC Specification is used for steel design because the reader has already become familiar with it. We know of no serious disagreement between the applicable chapter, No. 27, of the 1982 UBC and the 1978 AISC Specification, at least in the matters addressed in this design. That is not always true. The responsible designer has an obligation to check conformance if conformance to a set of specifications is certified.

The five-part listing of load combinations is more detailed in the 1982 UBC than in previous editions, and Millman has further subdivided each numbered section,

*If this design had been done in the 1920s or 1930s, the wind bracing system would very likely have been less fully developed and less strong. Perhaps, as in some buildings we have been asked to alter, they would have been conceptually incomplete. Even if Millman had been guided by the 1979 UBC, the individual members would have been somewhat lighter. We believe that the more stringent new requirements are justified.

Project : LIGHT INDUSTRIAL BUILDING
Subject : GENERAL REQUIREMENTS AND CRITERIA

SHEET G2 OF

CLIENT REQUIREMENTS

ONE STORY STRUCTURE, LIGHT INDUSTRIAL USE - MANUFACTURING PROCESSES WITHOUT HIGHLY FLAMMABLE LIQUIDS OR GASES. INCOMBUSTIBLE MATERIAL REQUIRED BUT RESISTIVITY TO SPREAD OF FIRE NOT REQUIRED EXCEPT SPRINKLERS.

ENCLOSED AREA 80'± X 200'± = approx. 16,000 SQ. FT.

CLEARANCES

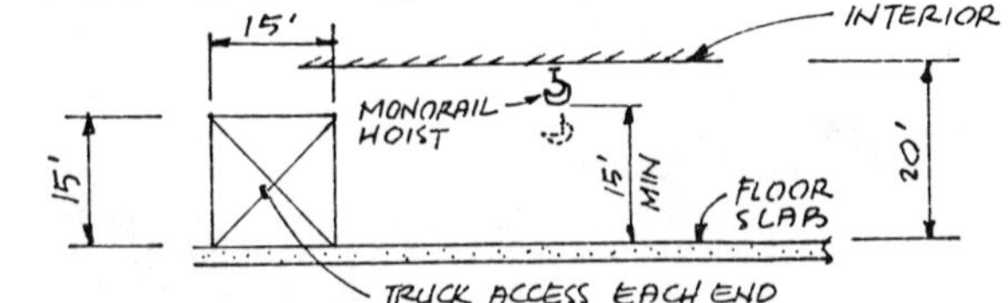

INTERIOR SPACE TO BE COLUMN-FREE

TRAVELLING HOIST ON MONORAIL TO SERVICE OPERATING STATIONS ON BOTH SIDES OF THE BUILDING - 2 TON CAPACITY

TRUCK ACCESS BOTH ENDS OF BUILDING

ALLOW FOR WINDOWS AT SIDES AND ENDS, AND SKYLIGHTS IN ROOF

CONSTRAINTS - LOCAL CONDITIONS

URBAN SETTING - OUTSIDE DOWNTOWN FIRE ZONE. NO STATUTORY REQUIREMENT FOR FIRE RESISTIVE CONSTRUCTION

SNOW - BASIC SNOW LOAD = 30 psf ON HORIZONTAL PROJECTION

FLAT SITE - WIND EXPOSURE "B" BY UBC STANDARDS (1982)
BASIC WIND SPEED - 80 mph PER FIG. 4 - UBC

SEISMIC - ZONE #1 - PER SEISMIC MAP, FIG. 1, UBC

ACCESS - BY PUBLIC HIGHWAYS AND STREETS

SOIL IS ALLUVIAL TO DEPTH OF 40'±, MIX OF CLAYEY SANDS AND SANDY CLAYS OF MEDIUM DENSITY. SOILS ENGINEER REPORTS PROBABILITY OF 1/2" TO 1" SETTLEMENT UNDER EFFECT OF SHALLOW FOUNDATION LOADS ON ORDER OF 1500 psf. SETTLEMENT MAY VARY FOR DIFFERENT FOOTINGS. DIFFERENTIAL SETTLEMENT UP TO 1/2". FROST PENETRATION ≤ 3.0 FEET.

GOVERNING CODE - UNIFORM BUILDING CODE - 1982, WITH LOCAL AMENDMENTS GOVERNING FIRE ZONING, UTILITIES, ETC. (CHAPTER 27, UBC, IS IN GENERAL, IDENTICAL TO THE 1978 "SPECIFICATION FOR DESIGN, FABRICATION AND ERECTION OF STRUCTURAL STEEL FOR BUILDINGS" OF AISC. REFERENCES TO STEEL DESIGN SPECIFICATIONS WILL BE BY AISC NUMBERS AND FORMULAS UNLESS NOTED.)

Project : LIGHT INDUSTRIAL BUILDING
Subject : GENERAL REQUIREMENTS AND CRITERIA

SHEET G3 OF

MATERIALS

ROOF DECK AND SIDING - COLD FORMED LIGHT GAGE SHEET AND STRIP; ASTM A570, Gr. D OR E, OR ASTM A607

STRUCTURAL STEEL ASTM A36

BOLTS ASTM A307

WELDING RODS - SERIES E60. CONFORM TO AWS STRUCTURAL WELDING CODE - D1.1-80 FOR QUALIFICATION OF WELDERS AND QUALITY CONTROL OF WELDED JOINTS.

PAINTING OF STEEL - CORROSION RESISTANT PRIME COAT REQUIRED. FINISH COAT TO BE SELECTED IN CONSULTATION WITH CLIENT. NORMAL INDUSTRIAL QUALITY PAINT SYSTEM TO BE SELECTED BASED ON "GUIDE TO SHOP PAINTING OF STRUCTURAL STEEL" BY STEEL STRUCTURES PAINTING COUNCIL AND AISC.

FOUNDATIONS - REINFORCED CONCRETE SHALLOW FOOTINGS. (FOUNDATION DESIGN AND CALCULATIONS NOT INCLUDED IN THIS SET OF CALCULATIONS.)

DESIGN CRITERIA

GENERAL - AS REQUIRED BY LOCAL ORDINANCE - UNIFORM BUILDING CODE (UBC) - 1982. SEISMIC ZONE AND WIND EXPOSURE AS STATED ABOVE.

STRUCTURAL STEEL DESIGN - AISC "SPECIFICATION FOR THE DESIGN, FABRICATION AND ERECTION OF STRUCTURAL STEEL FOR BUILDINGS" -1978

COLD FORMED STEEL DESIGN - "SPECIFICATION FOR THE DESIGN OF COLD-FORMED STEEL STRUCTURAL MEMBERS" -1980, BY AMERICAN IRON AND STEEL INSTITUTE (AISI)

LOAD COMBINATIONS (BASED ON AND ADAPTED FROM UBC Sec. 2303(d) and (f) where F = BASIC ALLOWABLE STRESSES

1. a. DEAD PLUS ROOF SNOW* (LIVE) LOAD	F(1.00)
b. DEAD + CRANE + 0.75(SNOW)	F(1.00)
2. a. DEAD + WIND	F(1.33)
b. DEAD + SEISMIC	F(1.33)
c. 0.75(DEAD) + WIND	F(1.33)
d. 0.75(DEAD) + SEISMIC	F(1.33)
3. DEAD + CRANE + WIND + 0.50(SNOW)	F(1.33)
4. a. DEAD + SNOW + 0.50(WIND)	F(1.33)
b. DEAD + CRANE + 0.75(SNOW) + 0.50(WIND)	F(1.33)

5. DEAD PLUS SNOW PLUS SEISMIC NOT REQUIRED SINCE SNOW LOAD $\leq$ 30 psf

* APPLY SNOW LOAD TO FULL ROOF OR HALF ROOF FOR WORST EFFECT

Project : LIGHT INDUSTRIAL BUILDING
Subject : GENERAL REQUIREMENTS AND CRITERIA

SHEET G4 OF

CONSIDER POSSIBLE FRAMING SYSTEMS

A. CONTINUOUS FRAMING OR STATICALLY DETERMINATE?
DIFFERENTIAL SETTLEMENT MAY INVALIDATE STIFFNESS ANALYSIS OF CONTINUOUS FRAME. USE DETERMINATE SYSTEM, PINNED JOINTS (OR EFFECTIVELY "PINNED").

B. FLAT ROOF OR SLOPING ROOF
PRESCRIBED SNOW LOAD = 30 psf - FAIRLY HIGH. WINDS ARE HIGH. ALTHOUGH WE ARE NOT REQUIRED BY UBC TO CONSIDER FULL WIND AND SNOW SIMULTANEOUSLY, SNOW IS MORE LIKELY TO DRIFT IN HIGH WINDS, CAUSING DEEP, HEAVY AREAS, PARTICULARLY AT ANY VERTICAL SURFACES ON ROOF

STEEP ROOF PREFERRED, PARTICULARLY CONSIDERING THAT THE ROOF FRAMING WILL BE LIGHT AND EXTRA SNOW LOAD MAY CAUSE SIGNIFICANT, UNANTICIPATED STRESSES.

C. POSSIBLE DETERMINATE SYSTEMS - ROOF SUPPORT

1. WELDED TAPERED GIRDERS

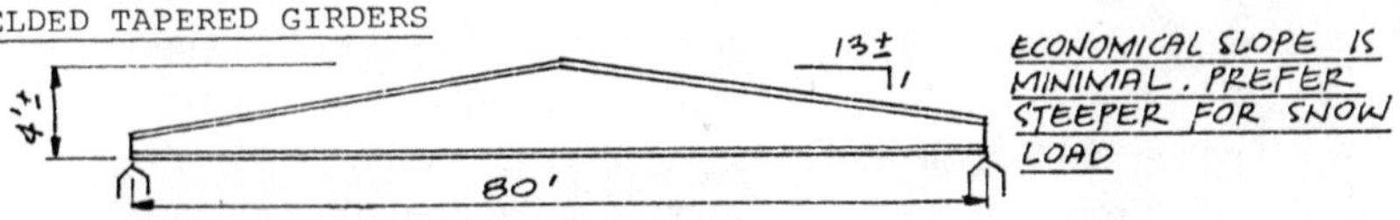

2. WELDED TAPERED BEAMS - 3 HINGED ARCH

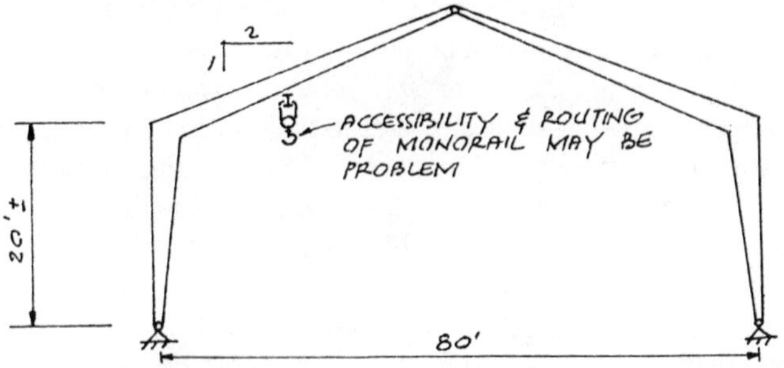

3. ROOF TRUSSES - HORIZONTAL LOWER CHORD - SLOPING UPPER CHORD

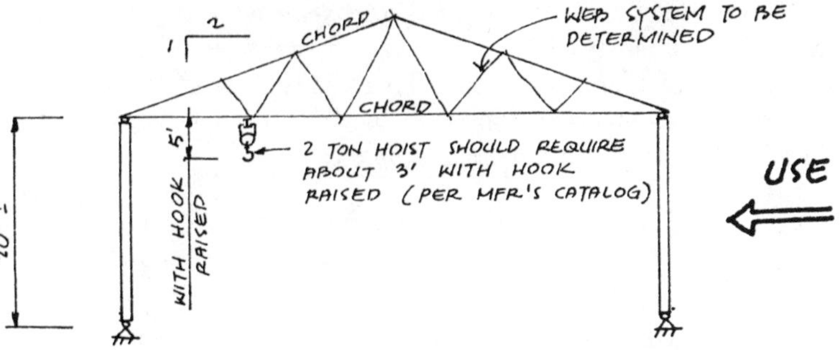

largely to reflect the type of *live loads* applicable to this design. *Floor live loads* are transmitted directly through the floor slab to the ground and do not affect the steel frame. The *crane* (hoist) *load* may add stress to the same members stressed by the snow. However, in Combination 1b, consistent with UBC requirements, the crane load is to be combined with only 75% of the *snow load*. One could reason that the greatest snow accumulation on the roof would occur when the plant is not operating. Wind and seismic requirements must both be checked. Since wind causes uplift (suction) on the roof, one critical wind loading on the roof structure is likely to occur when the only gravity load is the *dead load* (Combination 2c). The lightweight roof that is foreseen for this structure may blow away if not securely anchored to the foundation through the steel frame. In general, the load combinations stipulated by the UBC represent standardized forms of real load combinations that may be expected to occur during the life of the structure. Short-term stresses from wind or earthquake are permitted by Code criteria to be one-third higher than the basic operating stresses. Crane loads and their impact will be treated as basic loads, but are not expected to occur often enough to fatigue the structure.

On sheet G4, the designer considers alternative approaches to framing systems. As noted on sheet G2, the soil at foundation depth is a relatively uncompact alluvium, which may settle different amounts at different columns. Continuous rigid frame structures subjected to differential settlement are subjected to stresses that arise from the resistance of the frame to the resulting distortions. On the other hand, several types of statically determinate structure with *hinged* joints can accommodate to expected small foundation movements without distress.

Among the common types of statically determinate transverse frame considered, the designer is led in several stages to the choice of a sloping roof to minimize the accumulation of snow and a horizontal lower member to support the specified monorail hoist. Trusses will be used as shown at the bottom of sheet G4, supported by columns *hinged* at top and bottom. Although it is not common in light frames of this sort to use expensively detailed true pin joints, a number of simple types of connections can be used that offer little rotational restraint. In addition, the form of the end joints of the truss will result in little moment continuity between truss and column.

Examination of the form of the selected bent, as sketched, will reveal the possibility of collapse of the roof support structure; the columns could drift sideways uninhibited by their pinned ends. To avoid such an eventuality, it will be necessary in the framing arrangement shown on sheet G5 to introduce bracing against translation in both the transverse and longitudinal directions.

17.2.3 Arrangement of Framing (Sheet G5)

On G5, we see in several views the entire structural system. It is shown here as it finally evolved from the detailed calculations. In a slightly different form, it was set down by Millman immediately after the final decision on the previous sheet was

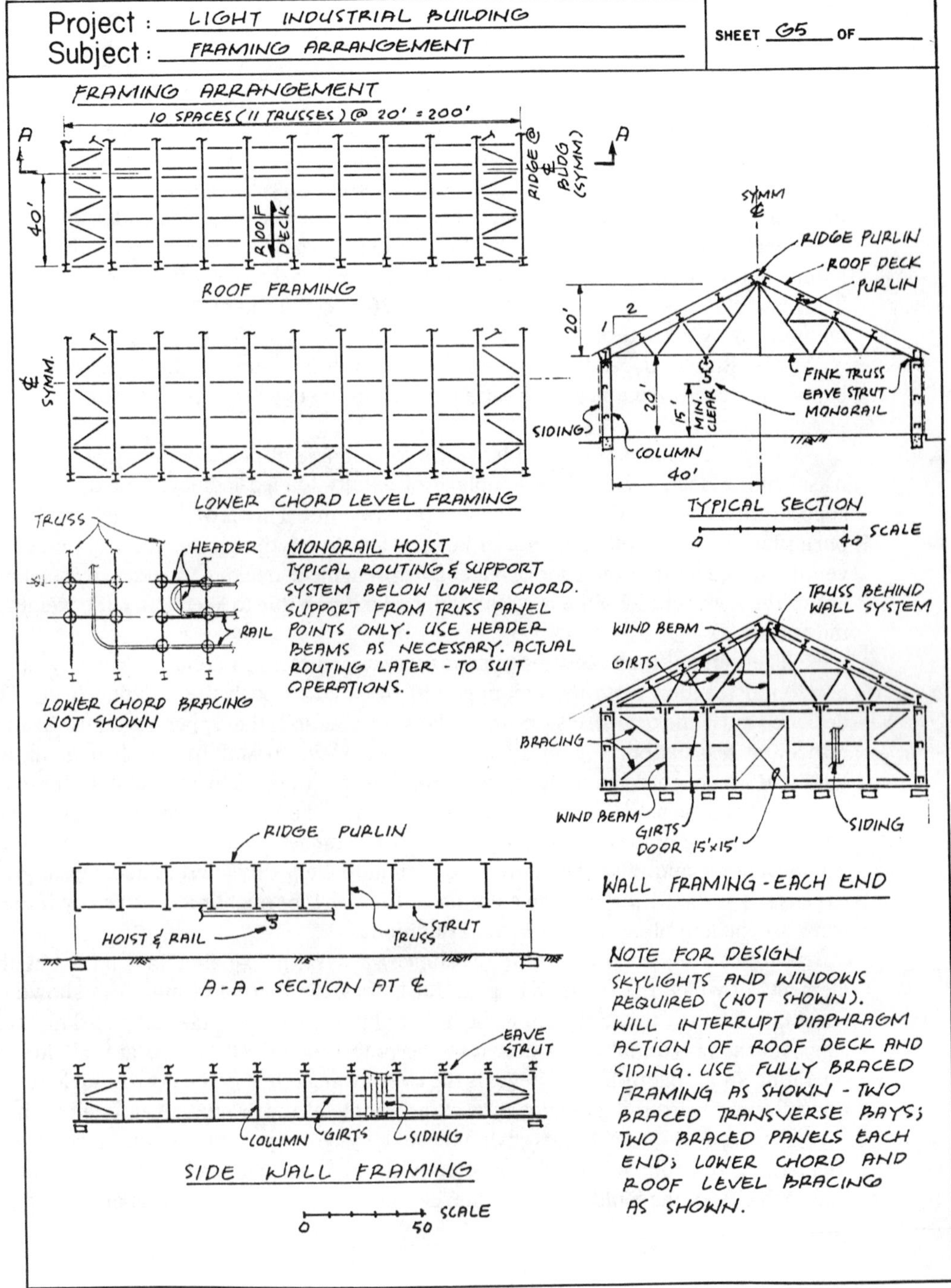
Project : LIGHT INDUSTRIAL BUILDING
Subject : FRAMING ARRANGEMENT
SHEET G5 OF
FRAMING ARRANGEMENT
10 SPACES (11 TRUSSES) @ 20' = 200'
A
A
40'
ROOF DECK
RIDGE @ ℄ BLDG (SYMM)
ROOF FRAMING
℄ SYMM.
LOWER CHORD LEVEL FRAMING
SYMM ℄
RIDGE PURLIN
ROOF DECK
PURLIN
20'
1
2
20'
15' MIN. CLEAR
FINK TRUSS
EAVE STRUT
MONORAIL
SIDING
COLUMN
40'
TYPICAL SECTION
0
40
SCALE
TRUSS
HEADER
RAIL
LOWER CHORD BRACING NOT SHOWN
MONORAIL HOIST
TYPICAL ROUTING & SUPPORT SYSTEM BELOW LOWER CHORD. SUPPORT FROM TRUSS PANEL POINTS ONLY. USE HEADER BEAMS AS NECESSARY. ACTUAL ROUTING LATER - TO SUIT OPERATIONS.
TRUSS BEHIND WALL SYSTEM
WIND BEAM
GIRTS
BRACING
WIND BEAM
GIRTS
DOOR 15'x15'
SIDING
WALL FRAMING - EACH END
RIDGE PURLIN
STRUT
TRUSS
HOIST & RAIL
A-A - SECTION AT ℄
NOTE FOR DESIGN
SKYLIGHTS AND WINDOWS REQUIRED (NOT SHOWN). WILL INTERRUPT DIAPHRAGM ACTION OF ROOF DECK AND SIDING. USE FULLY BRACED FRAMING AS SHOWN - TWO BRACED TRANSVERSE BAYS; TWO BRACED PANELS EACH END; LOWER CHORD AND ROOF LEVEL BRACING AS SHOWN.
EAVE STRUT
COLUMN
GIRTS
SIDING
SIDE WALL FRAMING
0
50
SCALE

made. In both versions, it was intended to show a complete system of stable geometry, defining the elements necessary to keep out the weather, keep clear the space needed for intended activities, provide required access, and maintain the structural integrity of the system itself.

The roof and side walls are clothed in continuous deck and wall surfaces, which, as they keep out the snow and wind, must flex between supporting beams called purlins and girts. The purlins span between roof trusses, which occur at regular intervals of 20 feet and themselves span the 80 feet between side walls. Reactions from each truss are carried to the foundation by end columns, as axially loaded members that must also act flexurally to resist the horizontal reactions from the girts. The entire space, 20 feet high by 80 by 260 feet, is kept free for the intended manufacturing activities.

The structural concept permits the use of hoists hung from single (monorail) beams that can be routed just about anywhere in the enclosed space to suit operating needs.

All direct connections to a truss are to be made at panel points (joints), so the truss can be designed as an assemblage of axially loaded members, flexing only to support their own small weight between panel points. This proviso, which applies to purlin locations as well as monorail beams (and to anything else, such as pipes and ventilation ducts that the user or mechanical engineer decides to route in the space above the lower chord of the trusses), will make it possible to keep down the weight of the trusses.

The vertical load resisting system is now complete in concept, but any upper horizontal reaction from the columns will threaten to push the building over. The designer must make a decision about how to stabilize the upper structure against horizontal translation. Figure 17.1 shows two ways to stabilize each bent individually. Both are common in this type of building. Neither is acceptable in this case for reasons that appear on the figure. The intrusion of the knee braces on the clearance envelope would, among other things, restrict the range of operation of the monorail. The designer could raise the entire roof system to clear this interference, but at great expense. The cantilever column solution of Fig. 17.1(b) might be satisfactory if there were no concern about the poor soil conditions.

Millman prefers for lateral resistance, a system like that in Fig. 17.2, the elements of which the reader can pick out among the myriad of members shown on sheet G5. The key to the system is the horizontal truss at lower chord level. This truss transfers the horizontal loads shown to the braced end bents, which are outside the operating area. Figure 17.2(c) reveals the overturning effect of the horizontal loads on this system. Overturning creates a resisting couple at the ground level, one of whose forces causes uplift that must be ballasted by gravity loads of the structure or foundation. In Millman's layout on sheet G5, there are two such horizontal trusses, one each side of the building centerline, connected by the lower chords of the vertical

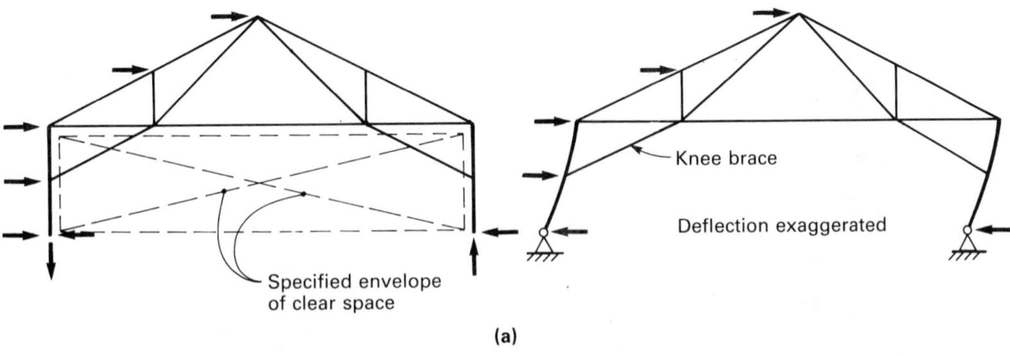

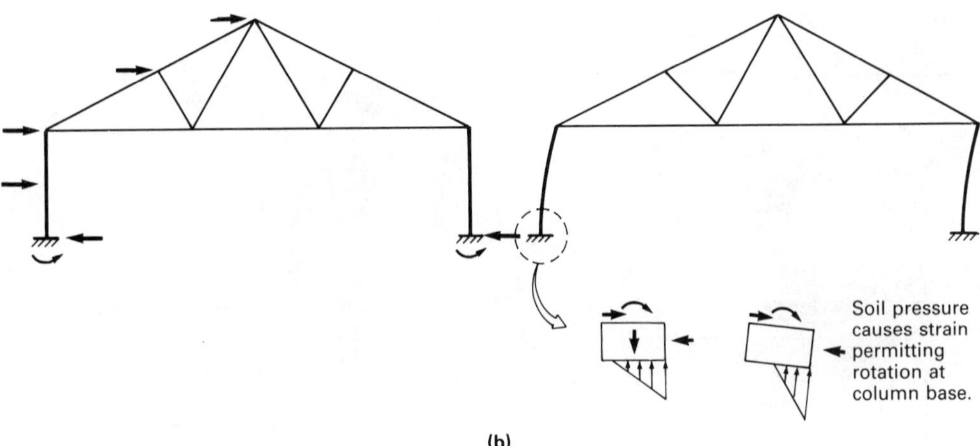

Figure 17.1 Possible lateral support systems at each transverse bent: (a) System stable with flexing columns. Knee braces intrude on interior space. (b) System stable if column bases can be effectively fixed. Poor soil may allow base rotation; ability to fix bases questionable.

trusses so that they each resist half of the lateral forces on the upper part of the building.

From the discussion of stability in Chapter 2, the reader will recognize that the structure is still unstable longitudinally. Millman remedies this by a system similar to that in Fig. 17.3, which is subjected to longitudinal forces. At each end of the building, this system has a bent transverse truss in the planes of the roof and a horizontal transverse truss at the level of the lower chords of the vertical trusses. In

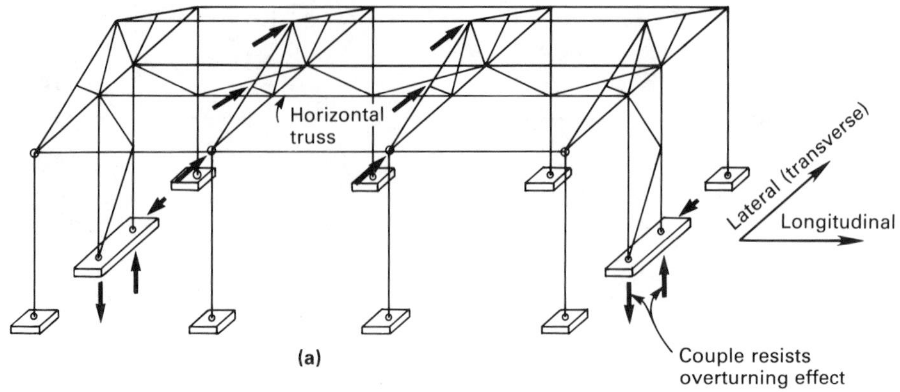

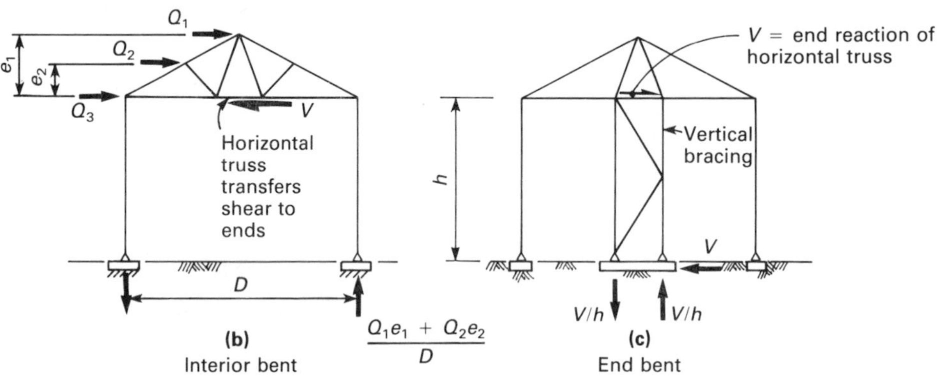

Figure 17.2 Lateral bracing system.

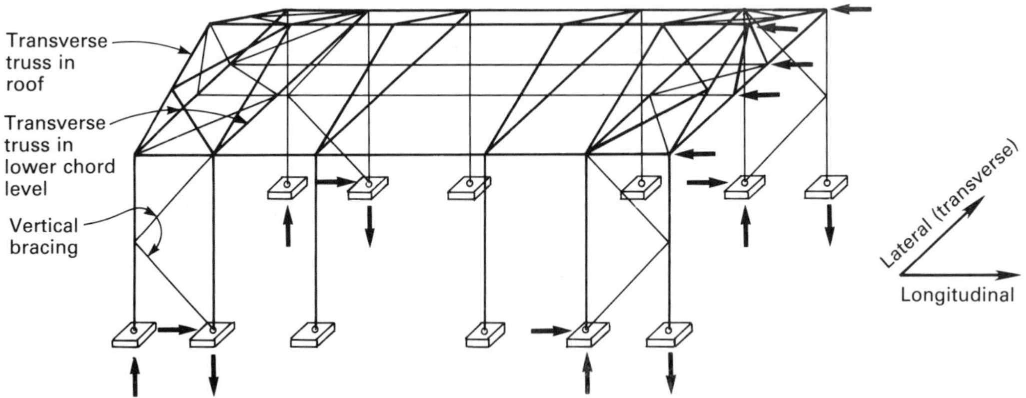

Figure 17.3 Longitudinal bracing system.

each case, longitudinal *collectors* between the two trusses make them share the total longitudinal force between the two ends. The end reactions of these trusses introduce horizontal shear and overturning moment to vertical bracing trusses at each side wall. Sheet G5 has such a longitudinal force resisting system.

The three subsystems, vertical load resistance, lateral load resistance, and longitudinal load resistance, are parts of the total building system, all necessary for stability and force resistance. In this type of structure, the analysis of each subsystem for its forces is independent of the others, although many of the members are common to the three subsystems. The building at any one time can be subjected to combinations of forces affecting all three resisting subsystems. By separating the vertical, lateral, and longitudinal force components, their effects can be found in each resisting subsystem and then superposed in finding internal member forces and reactions. That will be Millman's procedure in the analysis and design of this building.

The reader may have noticed, in the design criteria on sheet G3, that temperature effects are not mentioned. They are not unimportant. Steel is a good conductor of heat, and steel structures of this type will grow and shrink, often following closely the cycles of outside temperature. If the temperature at this site fluctuates in the range −20° to 100°F, and the building is built in a season of 60° weather, the superstructure of a 200-foot-long building will grow in the summer:

$$\Delta L = C_e \cdot L \cdot \Delta T = 6.5(200)(12)(40) \times 10^{-6} = 0.624 \text{ in.}$$

where $C_e = 6.5 \times 10^{-6}$ in./in./°F

In the winter it would have to shrink twice that amount, 1.25 inches. Assuming that the foundation were dimensionally unchanged while the superstructure experiences these movements, a rigid, nondeterminate bracing system such as is drawn on sheet G5 would act to prevent the changes in length. To reduce ΔL to zero would introduce stresses in the longitudinal members:

$$\frac{\Delta PL}{AE} = \Delta f_a \left(\frac{L}{E}\right) = -\Delta L$$

causing wintertime tensile stress

$$\Delta f_t = \frac{1.25(29{,}000)}{200(12)} = +\ 15 \text{ ksi!!}$$

and summertime

$$\Delta f_t = -7.5 \text{ ksi}$$

Fortunately, things are seldom as bad as this sounds. The assumption of the previous paragraph about dimensions at the foundation is somewhat unreal. An operating plant tends to be at fairly constant temperature. The bolted connections of a bracing system may slip slightly under high load, offering some relief of longitudinal stress. Millman follows a common rule of thumb that, up to a length of 200 feet, temperature effects are acceptable. In general, we urge caution with regard to temperature effects, but for purposes of this design we will concur in Millman's judgment.

The Fink truss that is shown in the typical section on sheet G5, and which will be used in the typical *bent* of the building, is one of several forms of truss available to support steep sloping, symmetrical roofs.* The virtues of the Fink truss should become apparent when Millman does the roof truss design.

A tentative partial layout also appears on the framing arrangement sheet covering the monorail hoist. The monorail system will not be designed in this set of calculations. In all likelihood it will be designed by the hoist supplier or the mechanical engineers who lay out the operating portions of the plant. What we find here is the assumptions regarding that system, which are used in the structural analysis and which must be observed when the monorail is designed. The key assumption is that all connections from the monorails to the structure must be made at truss panel points to avoid bending in members designed for axial load. This may at times require secondary *header* beams between trusses. The same constraint applies to any other loads the plant operators might wish to support with the roof trusses.

We can now see a complete structure in concept, although Millman has still to make the first numerical calculation directly related to it. The design cannot be properly calculated until a framing concept is complete. However, the concept itself must be validated and sometimes revised in response to what is revealed by the numerical work. Sheet G5 at this time is not identical to its appearance before the calculations were made. It now reflects some relatively minor changes that were made when the numbers led the designer to them. Earlier forms, presumably, are filed among many other rejects perhaps useful in the future as memory joggers or as ideas for future designs. The numerical part of the *general* section is contained in sheets G6 through G8.

17.2.4 Loads: Sheets G6 through G8

The data assembled on sheets G6 through G8 will be used as part of a reference file, to be referred to as needed in other sections of the calculations. The designer here assembles information on loads to be used in design, separated into vertical and

*According to Johnson, Bryan, Turneaure,[34] Fink was a nineteenth-century railroad engineer who originally developed the form of this truss for railroad bridges. It later became common for building roofs.

horizontal loads, and further separated as to their source and direction. Throughout the calculations, Millman will find it convenient to deal with combinations of separate loads rather than to lump them all together. Reference to the multiplicity of load combinations on sheet G3 should help to explain this. The source of most of the loads other than the dead load is either Chapter 23 of the UBC (Appendix A to this book) or the AISC Specification.

The vertical loads shown on sheet G6 are divided into dead load, roof live load (snow), and crane load, which in this case is a moving live load suspended from the lower chords of the trusses. Some critical design controls are found when live load is maximum, others when it is removed or redistributed.

To establish the magnitude of the dead loads, it is necessary, at this stage, to make some "educated guesses," which are subject to recheck and revision after actual weights are determined. To avoid much time-consuming recalculation, it is usually a good idea to guess slightly on the high side if possible. Some small understress in members is usually acceptable. However, overestimating dead load can sometimes hide potential instability. In the analysis of wind effects on the trusses, it will be found that overestimating the dead load would hide what could be a serious problem of stress reversal, which is important in the selection of some of the members.

A load table is started by setting down the magnitude of dead load effects separately for different parts of the structure. The roof deck supports only itself and the attached insulation. The purlins support themselves plus the roof deck and insulation. The trusses support themselves in addition to the purlins and their loads. The columns support it all. The listings are done on the basis of pounds per square foot of horizontal projection. Magnitudes are based in part on preliminary calculations, as is shown for the truss, in part on manufacturers' catalogs or reference to data from previous jobs done by this designer or from other engineers willing to share the results of their experience. In any case, the later stage of recheck and recalculation will protect against the consequences of incorrect early assumptions. The dead load figures on sheet G6 are slightly revised from earlier assumptions; the calculation set now reflects the loads on the current G6.

The basic roof *live* load for buildings in the Milltown area is 30 pounds/square foot, established on the basis of local weather records and included among local building regulations. This would represent approximately 3 feet of newly fallen snow or a somewhat smaller thickness of late winter snow which may have been compacted somewhat by cycles of partial melting and refreezing before runoff. The slope of the roof and absence of valleys that might accumulate drifted snow are chosen to allay any worry the designer might have about higher load intensities. The Code [paragraph 2305(b)] permits a slight reduction in live load based on the slope of the roof, which the designer elects not to use.

Since the initial line of defense against snow load is the roof deck, its effects are carried without change through the whole system. Later calculations will apply the snow load to either one or both slopes of the roof, searching for possible design controls from each condition.

Project : LIGHT INDUSTRIAL BUILDING
Subject : LOADS VERTICAL
SHEET 66 OF ____

VERTICAL LOADS (#/FT² OF HORIZONTAL PROJECTION)

DEAD LOADS	TO ROOF DECK	TO PURLIN	TO TRUSS	TO COLUMN
ROOF DECK W/ INSULATION (ASSUME 2.5 + 1)	3.5	3.5	3.5	3.5
PURLINS, SAY 20#/LF @ 10' CTRS (INCL. SAG RODS & CONNECTIONS)	–	2.0	2.0	2.0
TRUSS* - ALLOW (SUBJECT TO CHECK)	–	–	3.0	3.0
COLUMNS, SAY $\frac{40\#/FT \times 20' \text{ OF HEIGHT}}{40' \times 20'}$		–	-	1.0
BRACING - SAY	–	–	1.5	1.5
MECHANICAL SYSTEMS, ALLOW	–	–	2.0	2.0
TOTALS, DEAD	3.5	5.5	12.0	13.0 + WT. OF SIDING

LIVE LOAD = SNOW ON ROOF @ 30#/FT²

BASIC SNOW LOAD = 30 PSF
POSSIBLE REDUCTION PER UBC 2305 (d)

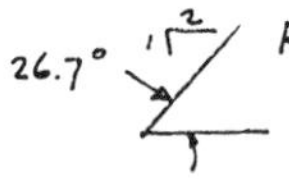

$R_s = (\frac{30}{40} - \frac{1}{2})6.7 = 1.68$ PSF

30 - 1.68 = 28.32
USE 30 PSF

	TO ROOF DECK	TO PURLIN	TO TRUSS	TO COLUMN
LIVE (SNOW)	30.0	30.0	30.0	30.0
DEAD + SNOW	33.5	35.5	42.0	43 + WT. OF SIDING

"CRANE" (AT ANY ONE LOWER CHORD JOINT OF A TRUSS)

USE MONORAIL HOIST FOR "CRANE" LOAD

DEAD WT.	=	600 #	(CATALOG)
LIFT	=	4000	(SPECIFIED)
IMPACT @ 10%	=	400	(AISC 1.3.3)
TOTAL:		5000 #	
LATERAL @ 20%	=	1000 #	(AISC 1.3.3)
LONG'L @ 10%	=	500 #	(AISC 1.3.3)

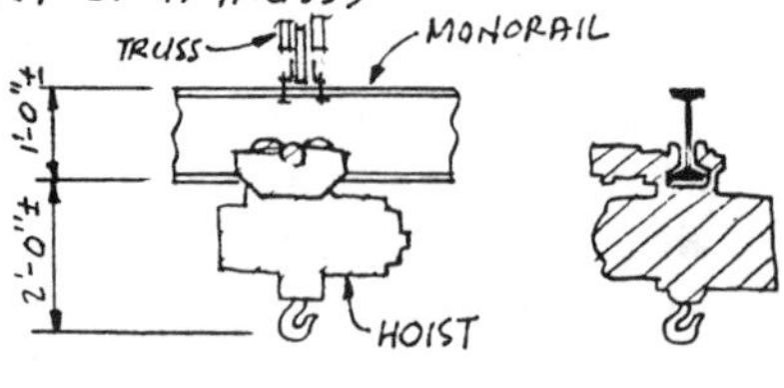

* ESTIMATE WEIGHT OF TRUSS

ASSUME $w_{(D+L)} = 20'(15+30) = 900$ #/LF

REACTION $= .9(\frac{80'}{2}) = 36$ k

AT HEEL JOINT

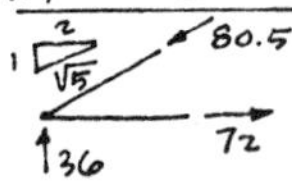

IF CHORD STRESSES @ 18 ksi

A = 4.0 ; LOWER CHORD
4.5 ; UPPER CHORD

WEIGHT @ A (490/144)

LOWER CHORD	= 13.6 (1)	= 13.6
UPPER CHORD	= 15.31 (√5/2)	= 17.1
WEB SAY	=	12.3
DETAILS ALLOW		5.0

} 48 #/LF
CONVERT TO PSF : $\frac{48}{20} = 2.4$; USE 3

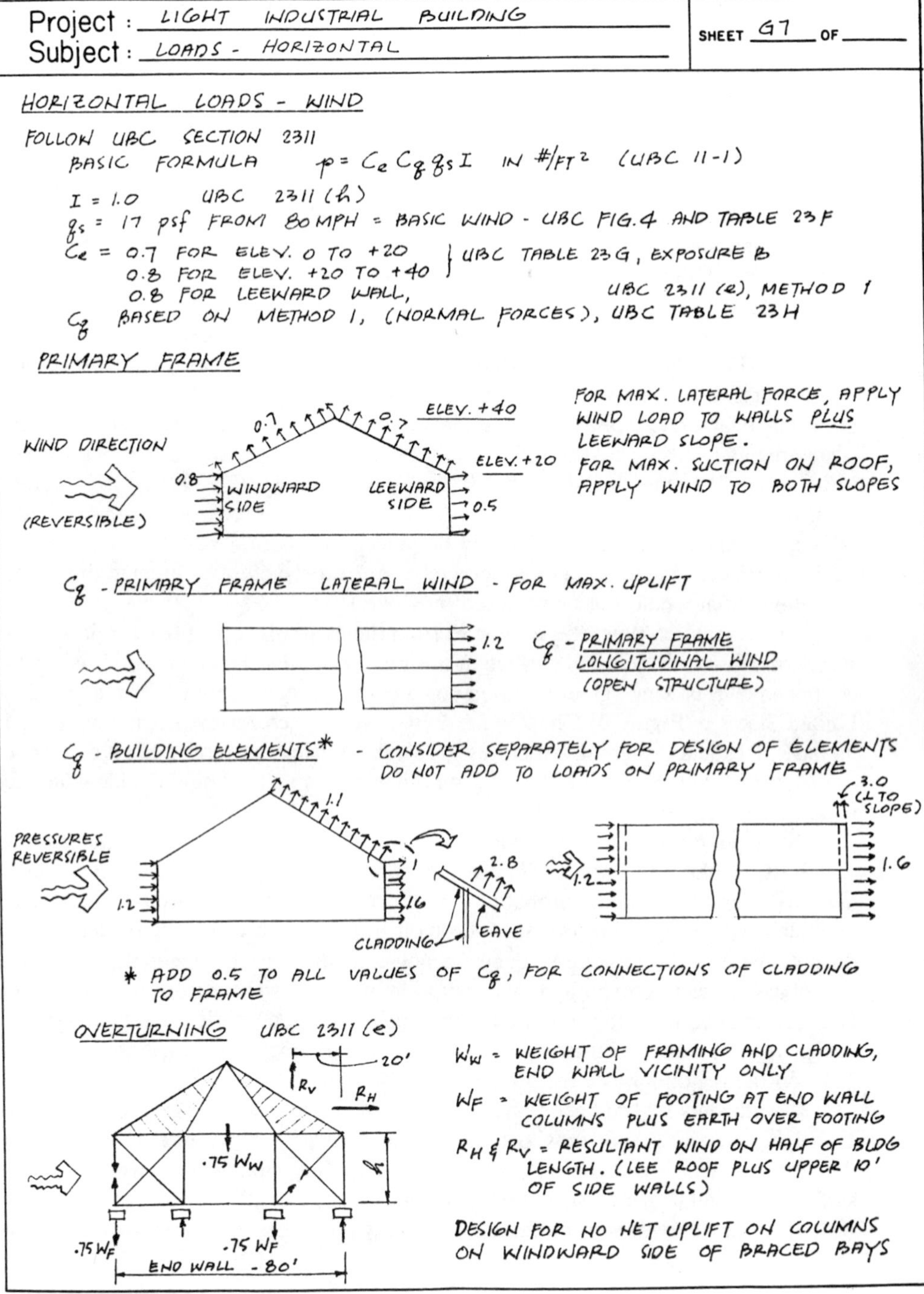

Project : LIGHT INDUSTRIAL BUILDING
Subject : LOADS - HORIZONTAL
SHEET G7 OF

HORIZONTAL LOADS - WIND

FOLLOW UBC SECTION 2311

BASIC FORMULA $p = C_e C_q q_s I$ IN #/FT² (UBC 11-1)

$I = 1.0$ UBC 2311 (h)

q_s = 17 psf FROM 80 MPH = BASIC WIND - UBC FIG. 4 AND TABLE 23 F

C_e = 0.7 FOR ELEV. 0 TO +20 } UBC TABLE 23 G, EXPOSURE B

0.8 FOR ELEV. +20 TO +40

0.8 FOR LEEWARD WALL, UBC 2311 (e), METHOD 1

C_q BASED ON METHOD 1, (NORMAL FORCES), UBC TABLE 23 H

PRIMARY FRAME

FOR MAX. LATERAL FORCE, APPLY WIND LOAD TO WALLS PLUS LEEWARD SLOPE.
FOR MAX. SUCTION ON ROOF, APPLY WIND TO BOTH SLOPES

C_q - PRIMARY FRAME LATERAL WIND - FOR MAX. UPLIFT

C_q - PRIMARY FRAME LONGITUDINAL WIND (OPEN STRUCTURE)

C_q - BUILDING ELEMENTS* - CONSIDER SEPARATELY FOR DESIGN OF ELEMENTS DO NOT ADD TO LOADS ON PRIMARY FRAME

* ADD 0.5 TO ALL VALUES OF C_q, FOR CONNECTIONS OF CLADDING TO FRAME

OVERTURNING UBC 2311 (e)

W_W = WEIGHT OF FRAMING AND CLADDING, END WALL VICINITY ONLY

W_F = WEIGHT OF FOOTING AT END WALL COLUMNS PLUS EARTH OVER FOOTING

R_H & R_V = RESULTANT WIND ON HALF OF BLDG LENGTH. (LEE ROOF PLUS UPPER 10' OF SIDE WALLS)

DESIGN FOR NO NET UPLIFT ON COLUMNS ON WINDWARD SIDE OF BRACED BAYS

The *crane* load does not really arise, by strict definition, from a crane but from a monorail-supported hoist that may travel on rails to many places in the building. It consists of the specified lift of 2 tons (4000 lb) plus associated effects. The dead weight of the hoist is treated as a live load, because, as a design control on any part of the structure, it is transitory in nature. The vertical impact and lateral and longitudinal loads all result from acceleration of the hoist and lifted loads in their travels. The magnitude used for dead weight is based on data from a manufacturer and may be somewhat in error for the actual hoist used—not likely to be a major problem in this case. The acceleration effects are standardized in the referenced section of the AISC Specification. Operationally, this equipment is not expected to deviate much from the norm.

Horizontal loads from wind on sheets G7 and G8 are based on a formula in Section 2311 of the 1982 UBC, which describes wind pressure as the product of four coefficients, three of which reflect physical variables, the fourth a social one. The importance factor, I, represents the social urgency of guaranteeing the survival of this structure in a windstorm—the stake of the community as opposed to that of the owner in its survival. Values higher than 1.0 are assigned, in UBC 2311(h), to structures critical to a community's disaster response network or that are used for the assembly of large numbers of people who could be placed at risk in a structural failure. (An engineer may choose to use a factor greater than 1 for structures critical to the client's operations; that would not be required by Code.)

The stagnation pressure, q_s, comes from the hydraulic equations for pressure of standard density air on a flat surface in a wind stream. The basic pressure parameter contained in q_s is wind velocity, based on velocity contours mapped for the mainland United States in Figure 4, Chapter 23, UBC. Notes accompanying the map explain its basis. Readers designing structures in regions of irregular terrain are well advised to take seriously the caution of note 5 on the wind map and should decide when and whether special wind criteria should be applied.

Stagnation pressure varies with the square of wind velocity, as do the values of q_s in UBC Table 23F. Values of q_s are applied to a location and represent the unusual *design* wind with an annual probability of 0.02 (*50-year wind*). Usual winds at a site are much less strong. Since pressures drop off with the square of velocity, usual winds do not threaten the structure. They do not constitute a significant design control except insofar as the manufacturing operation must be shielded from them by siding. Design winds do not occur often enough to be considered a fatigue problem by most designers. The *basic* stagnation pressure on our building is 17 psf, based, in Table 23F, on the regional basic wind velocity of 80 mph.

Coefficients C_e are based on variations to the basic wind pressure and depend on height, exposure, and transitory gusts. The designer (or building official) assigns the less restrictive category of exposure (B) to this site, which, while flat, has sufficient irregularities of structures and other results of urban development in the wind approach corridors to mitigate the worst wind effects. C_e varies with elevation,

Project : LIGHT INDUSTRIAL BUILDING
Subject : LOADS - HORIZONTAL

SHEET G8 OF ____

HORIZONTAL LOADS - WIND (CONT.)

SUMMARY - WIND PRESSURES, p, (#/FT²)

		Ce	Cq	qs	I	p	DIRECTION	APPLICABILITY
PRIMARY FRAME	LATERAL WIND							
	SIDE WALLS : 0' TO +20'							
	WINDWARD	0.7	0.8	17	1.0	9.52	IN →	TOTAL FRAME INCLUDING BRACING
	LEEWARD	0.7	0.5	↓	↓	5.95	OUT →	
	ROOF : +20' TO +40'							
	WINDWARD AND/OR LEEWARD	0.8	0.7	↓	↓	9.52	↖ ↗ (roof, 2:1)	
	LONGITUDINAL WIND							
	LEEWARD WALL	0.8	1.2	↓	↓	16.32		
ELEMENTS	LATERAL WIND							
	WINDWARD WALL	0.7	1.2	↓	↓	14.3	IN →	SIDING, WINDOWS, GIRTS, (COLS?)
	LEEWARD WALL	0.7	1.6	↓	↓	19.0	OUT →	" " " "
	ROOF	0.8	1.1	↓	↓	15.0	↖ ↗ OUT	ROOF DECK, PURLINS
	EAVE OVERHANG	0.7	2.8	↓	↓	33.3	↑ UP	ROOF DECK, EAVE PURLIN
	LONGITUDINAL WIND							
	WINDWARD WALL	0.8	1.2	↓	↓	16.3	IN →	SIDING, WINDOWS, GIRTS, WIND BMS
	LEEWARD WALL	0.8	1.6	↓	↓	21.8	OUT →	" " " "
	END OVERHANG	0.8	3.0	↓	↓	40.8	↖ ↑ UP	ROOF DECK, PURLIN OVERHANGS

TOTAL WIND SHEAR (OMIT WIND ON LOWER 10' OF WALL)

LATERAL -

WALLS ; $\frac{(9.52+5.95)(10')(202')}{1000} = 31.25k$

ROOF : $\frac{9.52(45)(202)}{1000}\frac{(21)}{45} = 40.38k$

TOTAL: 71.63k

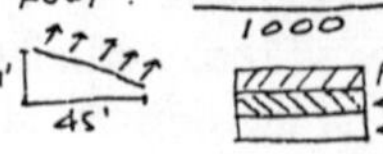

21'
45'
ROOF
← WALLS - UPPER 10'
← OMIT LOWER 10'

LONGITUDINAL

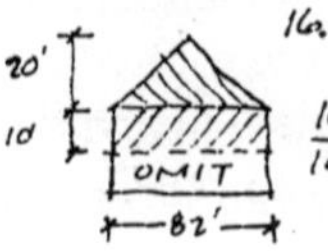

20'
10'
OMIT
82'

$\frac{16.32(10')(82')}{1000} = 13.38k$

$\frac{16.32}{1000}\left(\frac{20'}{2}\right)(82') = 13.38k$

TOTAL = 26.76 k

HORIZONTAL LOADS - SEISMIC

UBC SECTION 2312 ; ZONE 1 ; BOX SYSTEM

V = ZIKCSW (12-1)

Z = .188	(2312-C)
I = 1.0	(TABLE 23-K)
K = 1.33	(TABLE 23-I)
$C = 1/15\sqrt{T} \leq 0.12$	(12-2)
(T NOT ESTABLISHED)	
S = 1.5	

BUT CS ≤ 0.14 < .12(1.5) = .18 ; USE CS = 0.14

$W = 13\,psf\,\frac{(82')(202')}{1000} = 215k$ ROOF D.L.

$4\,psf\,\frac{(10')(202')(2)}{1000} = 16k$ OMITS LOWER 10'

$4\,psf\,\frac{(10+10_{AVE})(82)(2)}{1000} = 13k$

244 k

TOTAL SHEAR, W = .188(1.33)(1)(.14)(244) = 8.54 KIPS ≪ WIND SHEAR

DESIGN FOR WIND LOADS

being greater at higher elevations where frictional drag is less effective in slowing the wind.

Coefficients C_q help the designer in interpreting the localized intensities of pressure as they apply to various parts of a structure. The designer of a structure of this sort may choose, in calculating the effects of wind on the *primary frame*, between factors based on two different calculation methods, the second of which has the virtue of greater simplicity. Millman elects the perhaps more complex *normal force method*, which appears to be a more rational approach more closely related to theory and tests.*

The diagrams on sheet G7 interpret the code requirements for C_q, as they apply to the primary frame and to various building elements. The more severe criteria applied to elements recognize localized effects of pressure variation at different parts of the building. The coefficients used assume an *open structure*, since opening of the large end doors, and possibly windows, will sometimes invite the wind inside, causing, for example, the sum of pressure and suction effects to be felt by the same exterior wall. The *open-structure* category may be too severe for this structure, since the large truck access doors are not likely to be open in a severe storm. However, the designer wishes to permit large window areas, for which impact-resisting glazing may be costly (see note 1, Table 23H, UBC). The reader may find it interesting to trace the structural cost of this decision through the various parts of the design. In Table 23H, UBC, method 1 gives $C_q = 0.7$ for both windward and leeward roofs, leading to no net horizontal force on the roof. Method 2 gives a large horizontal force. While Millman recognizes that method 1 was derived from wind tunnel tests, he is reluctant to accept completely a zero net force. By setting up a subcase with wind suction on only half of the roof, he will achieve what he considers a safer control.

The table at the top of sheet G8 summarizes the values of the coefficients and design pressures, p, as they will be applied to various parts of the design. The total wind shear force on the building is also evaluated for later comparison to the design seismic force, which is calculated on the same page.

UBC equation (12-1) is used to calculate the total seismic shear. It appears in Section 2312 of the Code and consists of a series of coefficients, each of which is multiplied by the building weight, W, resulting in a total horizontal force from earthquake accelerations expressed as a percentage of the gravity force on the structure. Each coefficient is defined in UBC 2312. The very low *zone factor*, Z, for this region is the key to the low seismic shear calculated. The weight of the total structure must be estimated. The earlier estimate of dead load on sheet G6 is close enough for this purpose. However, the side and end walls were omitted from that

*The *Transactions of the American Society of Civil Engineers* report many excellent studies of wind effects on structures, whose conclusions are reflected in the 1982 UBC. The Metal Building Manufacturers' Association, also proceeding from test data, publishes a Code[42] that uses, for buildings of the shape of this one, somewhat lower pressures.

estimate. They are added at this time. The total includes only half of the weight of the walls, since seismic force initiating in the lower half will be transmitted directly to the foundation; insofar as the steel frame is concerned, only the girts and columns will feel its effect. The upper half of the walls contribute seismic loads to the upper chord level bracing system.

The seismic load turns out to be much less than the wind load and will not be pursued further in these calculations. Seismic effects will exist on individual interior members that are shielded from the wind. However, the completely intertied bracing system already established on sheet G5 will provide ample protection against these effects.

Selection of members may now begin. It will be done, as is common in buildings, from the roof down.

17.3 MEMBER DESIGN: VERTICAL LOADS

17.3.1 Roof Framing and Roof Deck: Sheet R1

The roof is to be clad in light-gage steel decking. Light-gage steel can be made stiff enough to support light loads in flexure by pressing it into a variety of shapes. Millman's roof deck will use such steel folded to shapes like those shown on sheet R1. The design rules developed by the AISI* are similar to those published by the AISC for heavy steel shapes. However, light-gage designs are more often affected by the ratios b/t of the individual flat elements; b/t values tend to be large since steel thickness is usually small. A table of thickness versus sheet metal gage is included in the AISC Manual and referenced in the calculation. The thickness of the section ultimately selected is 0.0299 inch. For the upper flange, a stiffened compression element, $b/t = 5.62/0.0299 = 188$. If we were to follow the AISC Specification (which we are not recommending here), Section 1.9.2 would set a limit of $253/\sqrt{F_y}$, which equals, for this 40-ksi steel, 40.0. Beyond this value of b/t, element buckling controls allowable stress, triggering the use of Appendix C. This requires the use of an *effective* cross section based on the stable b/t ratios of AISC Section 1.9. The approach of the AISI Specification is similar.

A number of manufacturers market similar roof and floor sections and publish engineering properties of these sections based on the AISI criteria. In the search for the proper combination of strength and stiffness, the designer copies some of the material supplied by one manufacturer. The sections are identified by proprietary numbers. However, a competitive supplier usually has no difficulty in showing which of their sections are equivalent.

*American Iron and Steel Institute.[12]

Project: LIGHT INDUSTRIAL BUILDING
Subject: ROOF DECK & SUPPORTS
SHEET R1 OF ____

ROOF DECK

w; 11.18'; 2; 1; √5; 10' = 120" = ℓ_H

DESIGN FOR 12" WIDTH

$w_{(D+L)} = 3.5 + 30 = 33.5$ #/LF/FT OF WIDTH

SINCE ROOFDECKS MAY BE INTERRUPTED FOR SKYLIGHTS OR VENTS, CONTINUITY OVER PURLINS MAY NOT EXIST. DESIGN FOR SIMPLE SPAN.

$M_{\mathbb{C}}$ @ $\frac{w\ell_H^2}{8}(12) = 525 + 4500 = 5025$ "#/LF OF WIDTH

FOLLOWING AISI SPECIFICATION - 1980

ASTM A 570 STEEL; $F_Y \geq 40$ ksi;

$F_{bx} = .60\,F_Y \geq 24$ ksi ON EFFECTIVE S_x AS LIMITED BY b/t RATIOS

FROM H.H. ROBERTSON CO. TABLES:

	DECK SECTION & GAGE	WEIGHT #/FT²	OVERALL DEPTH IN.	PER FT OF WIDTH I_x IN.⁴	$+S_x$ IN³	$-S_x$ IN³
#3 DECK (2 1/8", 3 5/8", 2 3/8", 1 1/2" NOM)	3-22	1.8	1.530	0.18	0.20	0.22
	3-20	2.2	1.536	0.23	0.27	0.27
	3-18	2.9	1.548	0.34	0.40	0.38
	3-16	3.5	1.560	0.44	0.51	0.48
#21 DECK (2 1/8", 5 5/8", 2 3/8", 3" NOM.)	21-22	2.1	3.030	0.67	0.39	0.47
	21-20	2.6	3.036	0.85	0.50	0.58

TRY #3 DECK: $S_R = \frac{5025}{24{,}000} = 0.21$ TRY 3-20

CHECK DEFLECTION $\Delta = \frac{ML^2}{EI}\left(\frac{5}{48}\right) = \frac{5025(11.18)^2(12)^2}{29\times10^6(.23)}\left(\frac{5}{48}\right) = 1.41" = L/95$ N.G.

TO REDUCE DEFLECTION, GO TO SECTION 21 - 3" NOMINAL DEPTH

#21-22; $I_x = 0.67$ $\Delta = \frac{.23}{.67}(1.41) = 0.48 = L/277$ OK

FOR SECTION 21-22 CHECK BENDING OF SHEET ELEMENT DUE TO WIND SUCTION

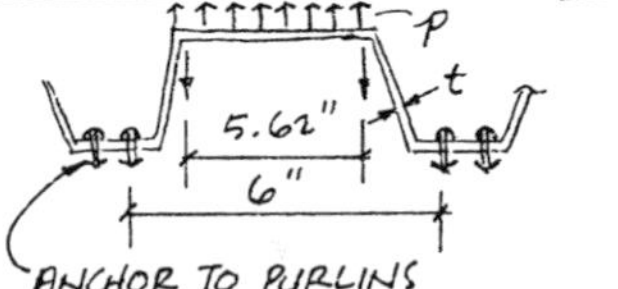

$p = \frac{15}{144}$ #/IN² (SHT. G8)

$t = .0299"$ (AISC "GAGE" TABLE MANUAL, p. 6-3)

$M \approx \frac{p\ell^2}{10} = \frac{15}{144}\frac{(6)^2}{10} = 0.38$ IN #/INCH

$f_b = \frac{M}{S} = \frac{.38(6)}{(.0299)^2(1)} = 2517$ psi $\ll .6F_Y$

BENDING OK BUT MUST SECURE TO ALL PURLINS AT SIDES OF TROUGH

USE H.H. ROBERTSON Co. SECTION 21-22 (OR EQUIVALENT BY OTHERS)

STEEL MUST HAVE $F_Y \geq 40$ ksi (ASTM A570 OR EQUIVALENT)

GALVANIZE PER ASTM A525-79 CLASS 40

CONTINUITY OK FOR MULTIPLE SPANS. SINGLE SPAN OK

ANCHOR TO PURLINS AT ALL TROUGHS FOR WIND SUCTION

SPECIAL ANCHORAGE AT EAVE OVERHANGS

Loads (dead, live, and wind) and span length are taken from the *G* sheets. After comparing the properties of the sections considered, the choice of a 3-inch-deep section is made in preference to an adequately strong 1½-inch-deep section, which would deflect excessively under the snow load. Being thinner, the 3-inch-deep section also turns out to be slightly lighter in pounds per square foot of deck than the 1½-inch section, a happy result. The designer is somewhat concerned about the local bending stress caused by wind suction on the thin, flat flange which must span between the screws that will anchor the deck to the purlins. This check being satisfied by calculation at the bottom of the sheet, the selection is recorded.

A historical note is pertinent here. A short time ago, a span of 11.2 feet would have been considered outlandish for a light-gage steel deck. The corrugated steel roofs that are still seen in thousands of farm and industrial buildings scattered around the United States and other nations were formed to sections ½ inch in depth and could span snow and wind loads no farther than 6 feet, more or less. The designer of a Fink truss such as this one would have had to use more closely spaced purlins and decide between the alternatives of introducing bending moments in the top chord or further subdividing the truss in order to receive all purlin loads at panel points. If the table of sheet G6 turns out to be correct, the purlins and trusses account for almost half of the total structural steel demand of the structure. Each would have been substantially heavier. The modern light-gage deck uses the same principles as the older corrugated roofing, and it extends them to establish, for many uses, more effective sections.

17.3.2 Typical Purlins: Sheet R2

In approaching the selection of the typical purlins (P1 in the sketch on sheet R1), we exercise a privilege not really available to Millman as designer of the building. To save space, we refer on sheet R2 to a previous calculation of an earlier chapter in which a purlin is selected for almost identical conditions. That calculation is for a beam in two-way bending spanning 20 feet around its strong axis and 6.67 feet in the perpendicular direction.

If the reader, anxious to study the design of axially loaded members, has arrived at this point before absorbing the ideas of Chapter 11, he or she may wish to simply note the results of the next few sheets of calculations and move immediately to the truss design that follows in the *TR* sheets and is discussed in Section 17.4.

The typical purlin turns out to be a W8 × 18. Its weight, converted to pounds per square foot of projected roof area, is 1.8 psf, slightly less than the 2.0 psf assumed on sheet G6. Sag rods will add an almost trivial amount to the average weight. The ridge purlin, P2, with special functions that make it heavier than the typical purlin, will increase the average.

Project: LIGHT INDUSTRIAL BUILDING
Subject: ROOF DECK & SUPPORTS

SHEET R2 OF

PURLINS

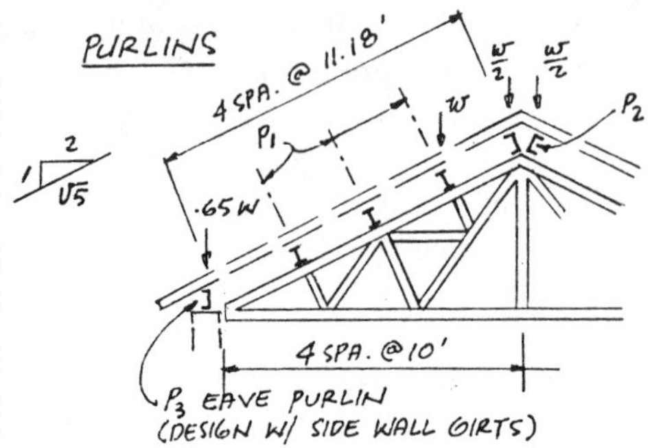

SPAN = 20'
SHEET G3 FOR LOAD COMBINATIONS,
G6 FOR LOADS

LOADING 1A - DEAD + LIVE

$w_{(D+L)} = 10'(5.5 + 30)$
$= 55 + 300 = 355$ #/LF
$= \searrow 317.5 + \swarrow 159$

LOADING 2A

DEAD + WIND

$w = 10'(5.5\downarrow + 15\nwarrow)$
$= \nwarrow 100 + \swarrow 25$, #/LF

P_1 (NOTE TO READERS

EXAMPLE 11.2.3, THIS BOOK, SHOWS CALCULATIONS FOR A PURLIN DESIGNED FOR ALMOST IDENTICAL CONDITIONS TO THOSE INCLUDED HERE AS LOADING 1a. THE BENDING OF THE PURLIN IS BIAXIAL. SELECTION IS CONTROLLED BY LOADING 1a. WE CAN THEREFORE USE THE SECTION CHOSEN IN THAT EXAMPLE WHICH IS:

W8 × 18

LOADING 2A MAKES IT NECESSARY TO CONNECT THE PURLIN TO THE TRUSS FOR 100 (10') = 1000# UPLIFT AT THE END OF THE PURLIN.

WE WILL NOT REPEAT THOSE CALCULATIONS HERE, ALTHOUGH WE RECOMMEND THAT, IN AN ACTUAL JOB, COMPLETE DOCUMENTATION SHOULD BE PROVIDED IN A SET OF CALCULATIONS.)

17.3.3 Ridge Purlins: Sheets R3 through R5

The problem of the ridge purlin is of particular interest since it is one of the few cases in the building design where a flexural member is "designed" in the sense that it is created from a number of parts rather than simply selected from a list of available sections. The trusses, which are also such designed flexural systems, consist of assemblies of axially loaded members. The individual parts of the ridge purlins as well as the composite, resist two-way bending plus axial load, combinations that can be considered by extending the ideas of Chapter 11.

The geometry of the situation, shown on sheet R3, makes it convenient to use channels as the primary members of this composite, rather than the W shapes used for the typical purlins. By lacing the top flanges together with diagonal bars, a horizontal truss is created capable of resisting horizontal bending forces and spanning their effects the full 20 feet between trusses. As a bonus, the close spacing of the lacing bars reduces the effective length, l, between lateral supports of the compression flange, increasing the allowable compressive stresses, even though the channel flanges are narrower than those of the typical W8 purlin.

Here, we see for the first time the effect of applying an unbalanced live load as required by UBC Section 2305(C). It is realistic to expect unbalanced snow load on a roof where, for example, the snow on the sunny south side might melt off sooner than that on the north side. In designing the typical purlins, the gravity loads were converted to components in the strong and weak directions of the members. The weak-way spans were shortened by introducing intermediate supports in the form of rods that act in tension in the upslope direction. The sag rods accumulate such tension from the reactions of three purlins below the ridge, so two rods are pulling on the ridge purlin in the downslope direction on each side. As long as the dead and live loads are balanced, the net effect is two concentrated vertical loads causing bending around the horizontal transverse axis of the composite section. If the sag rod pulls are unbalanced, there is also a net horizontal force that must be spanned the full distance between trusses. For this function, we have the horizontal truss created by the laced top chords.

The force analysis requires, on sheet R3, reconversion of the sag rod pulls to vertical and horizontal components. The pull at the ridge is the accumulated interior reactions from weak-way bending of three downslope purlins (see Example 11.2.3). Analysis is done first for the effect of unbalanced load, and then for that of balanced load. In the first part we see the unbalanced horizontal load combined with the related vertical load. The channel sections are selected for the vertical load moments. The horizontal load analysis of the laced beam is done as for a truss. The lacing and end tie plates are chosen by the methods of AISC 1.18, which is intended for a laced column with an arbitrary, low shear, but is suitable for this case. The initial selection of channels is found insufficient for combined effects and is changed in favor of heavier channels.

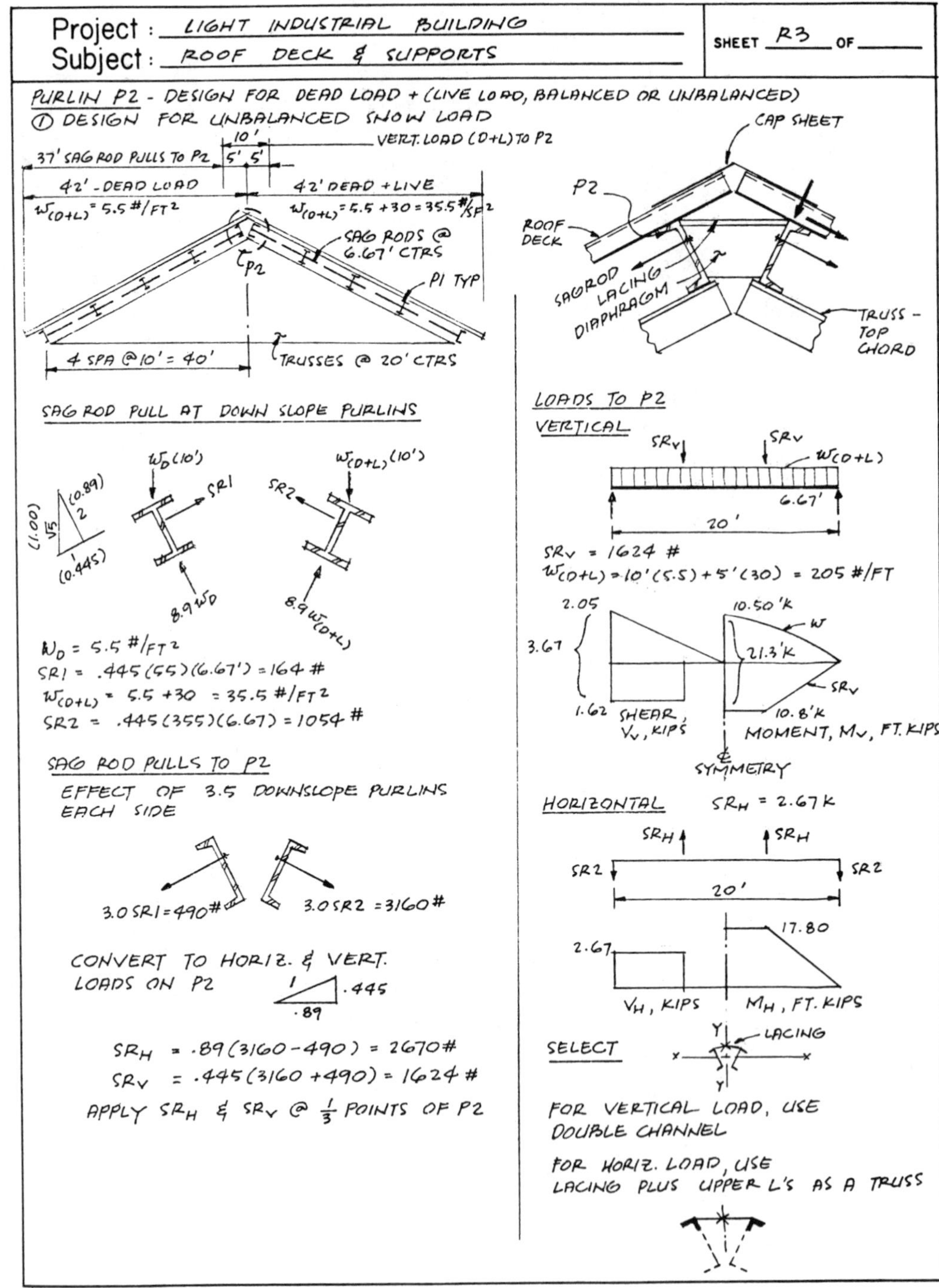
Project : LIGHT INDUSTRIAL BUILDING
Subject : ROOF DECK & SUPPORTS
SHEET R3 OF
PURLIN P2 - DESIGN FOR DEAD LOAD + (LIVE LOAD, BALANCED OR UNBALANCED)
① DESIGN FOR UNBALANCED SNOW LOAD
10'
VERT. LOAD (D+L) TO P2
37' SAG ROD PULLS TO P2
5' 5'
42' - DEAD LOAD
42' DEAD + LIVE
W(D+L) = 5.5 #/FT²
W(D+L) = 5.5 + 30 = 35.5 #/SF²
SAG RODS @ 6.67' CTRS
P2
P1 TYP
4 SPA @ 10' = 40'
TRUSSES @ 20' CTRS
CAP SHEET
P2
ROOF DECK
SAG ROD
LACING
DIAPHRAGM
TRUSS - TOP CHORD
SAG ROD PULL AT DOWN SLOPE PURLINS
WD (10')
W(D+L) (10')
SR1
SR2
(1.00) √5
(0.89) 2
1 (0.445)
8.9 WD
8.9 W(D+L)
WD = 5.5 #/FT²
SR1 = .445 (55)(6.67') = 164 #
W(D+L) = 5.5 + 30 = 35.5 #/FT²
SR2 = .445 (355)(6.67) = 1054 #
SAG ROD PULLS TO P2
EFFECT OF 3.5 DOWNSLOPE PURLINS EACH SIDE
3.0 SR1 = 490#
3.0 SR2 = 3160 #
CONVERT TO HORIZ. & VERT. LOADS ON P2
1
.445
.89
SRH = .89 (3160 - 490) = 2670#
SRV = .445 (3160 + 490) = 1624 #
APPLY SRH & SRV @ 1/3 POINTS OF P2
LOADS TO P2
VERTICAL
SRV
SRV
W(D+L)
6.67'
20'
SRV = 1624 #
W(D+L) = 10' (5.5) + 5' (30) = 205 #/FT
2.05
3.67
1.62
SHEAR, VV, KIPS
10.50 'K
W
21.3 'K
SRV
10.8 'K
MOMENT, MV, FT. KIPS
℄ SYMMETRY
HORIZONTAL SRH = 2.67 K
SRH
SRH
SR2
SR2
20'
17.80
2.67
VH, KIPS
MH, FT. KIPS
SELECT
Y
LACING
X
X
Y
FOR VERTICAL LOAD, USE DOUBLE CHANNEL
FOR HORIZ. LOAD, USE LACING PLUS UPPER L'S AS A TRUSS

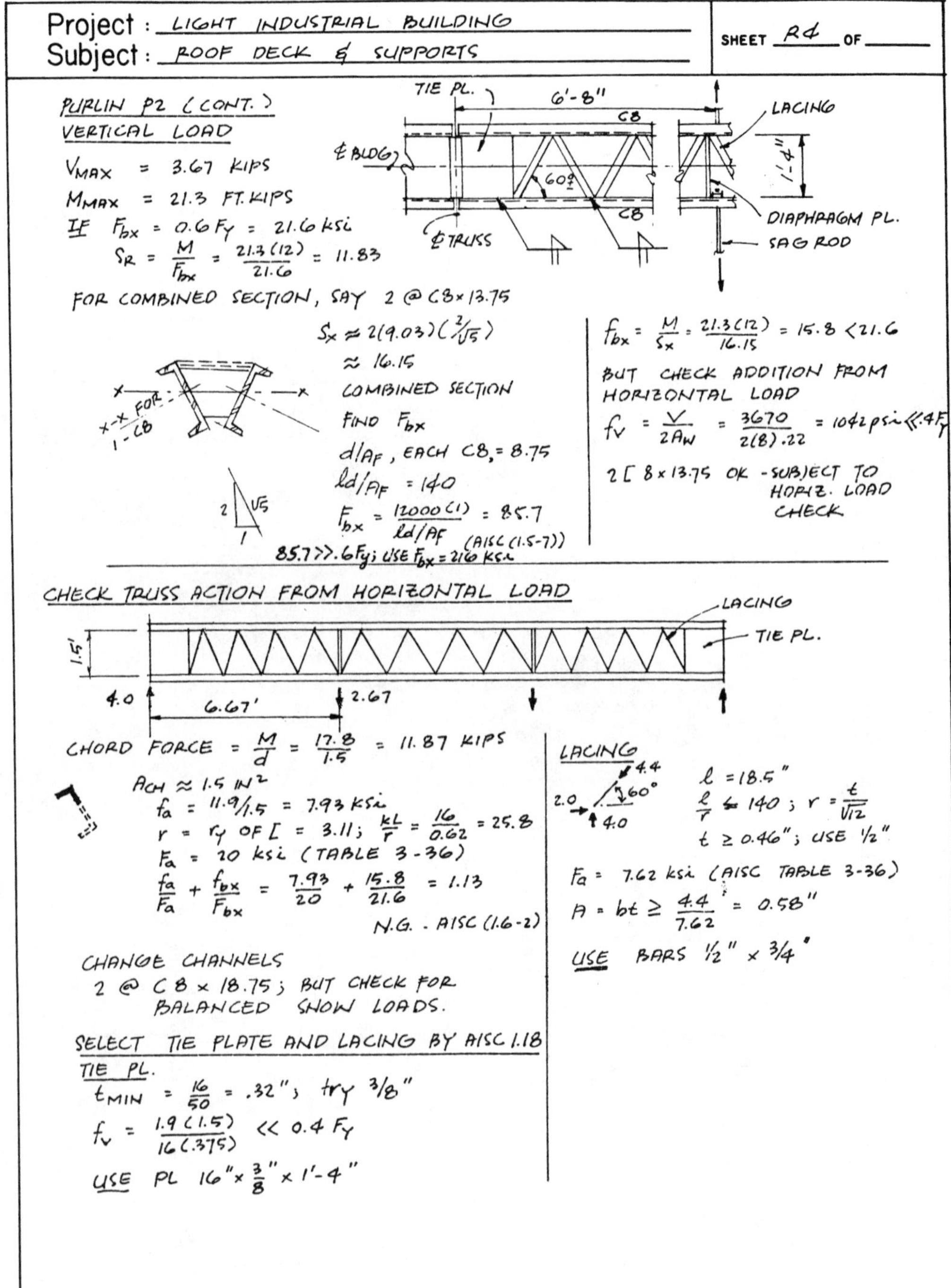

Project : LIGHT INDUSTRIAL BUILDING
Subject : ROOF DECK & SUPPORTS
SHEET R4 OF
PURLIN P2 (CONT.)
VERTICAL LOAD
VMAX = 3.67 KIPS
MMAX = 21.3 FT.KIPS
IF Fbx = 0.6 FY = 21.6 ksi
SR = M/Fbx = 21.3(12)/21.6 = 11.83
FOR COMBINED SECTION, SAY 2 @ C8×13.75
TIE PL.
6'-8"
C8
LACING
℄ BLDG
60°
1'-4"
℄ TRUSS
DIAPHRAGM PL.
SAG ROD
Sx ≈ 2(9.03)(2/√5)
≈ 16.15
COMBINED SECTION
FIND Fbx
d/AF, EACH C8, = 8.75
ℓd/AF = 140
Fbx = 12000(1)/(ℓd/AF) = 85.7
(AISC (1.5-7))
85.7 >> .6Fy; USE Fbx = 21.6 ksi
x-x FOR 1-C8
fbx = M/Sx = 21.3(12)/16.15 = 15.8 < 21.6
BUT CHECK ADDITION FROM HORIZONTAL LOAD
fv = V/2AW = 3670/2(8).22 = 1042 psi << .4Fy
2 [8×13.75 OK - SUBJECT TO HORIZ. LOAD CHECK
CHECK TRUSS ACTION FROM HORIZONTAL LOAD
LACING
TIE PL.
1.5'
4.0
6.67'
2.67
CHORD FORCE = M/d = 17.8/1.5 = 11.87 KIPS
ACH ≈ 1.5 IN²
fa = 11.9/1.5 = 7.93 ksi
r = ry OF [= 3.11; kL/r = 16/0.62 = 25.8
Fa = 20 ksi (TABLE 3-36)
fa/Fa + fbx/Fbx = 7.93/20 + 15.8/21.6 = 1.13
N.G. - AISC (1.6-2)
CHANGE CHANNELS
2 @ C8 × 18.75; BUT CHECK FOR BALANCED SNOW LOADS.
SELECT TIE PLATE AND LACING BY AISC 1.18
TIE PL.
tMIN = 16/50 = .32"; try 3/8"
fv = 1.9(1.5)/16(.375) << 0.4 FY
USE PL 16"×3/8"×1'-4"
LACING
4.4
60°
2.0
4.0
ℓ = 18.5"
ℓ/r ≤ 140; r = t/√12
t ≥ 0.46"; USE 1/2"
Fa = 7.62 ksi (AISC TABLE 3-36)
A = bt ≥ 4.4/7.62 = 0.58"
USE BARS 1/2" × 3/4"

Project : LIGHT INDUSTRIAL BUILDING
Subject : ROOF DECK & SUPPORTS

SHEET R5 OF ____

PURLIN P2 (CONT.)

② CHECK FOR BALANCED SNOW LOAD

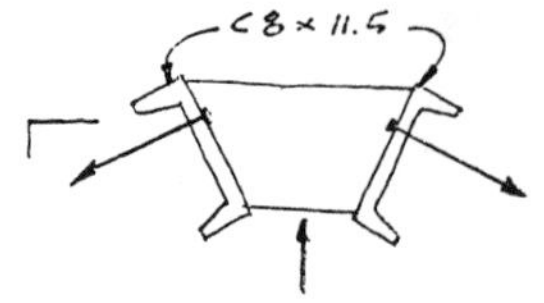

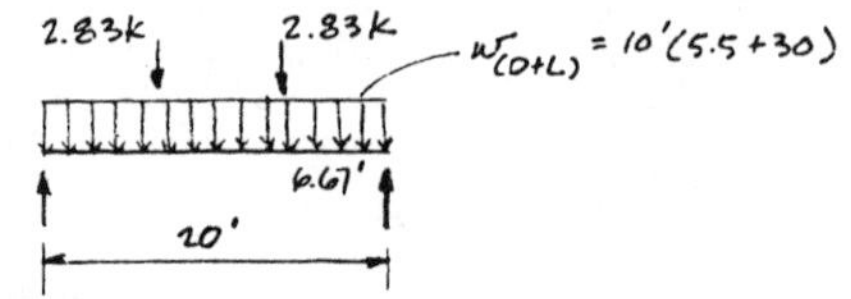

M_x = 18.9 + 17.8 = 36.7 FT.K
V = 2.8 + 3.6 = 6.4K
S_x = 2(11)(2)/$\sqrt{5}$ = 19.7
$f_{bx} = \frac{36.7(12)}{19.7}$ = 22.4 > 21.6 ≐ F_{bx} ; NG.

CHANGE TO 2 @ C9 × 15.0
S_x = 4(11.3)/$\sqrt{5}$ = 20.2
f_{bx} = 21.8 ksi ≈ 21.6 SAY OK
COPE ENDS 8"

USE 2 @ C9 × 15.0 - COPE ENDS
TIE PL'S EA. END - 16" × 3/8" × 1'-4
LACING AT 60°, 1/2" × 3/4" BAR
3/8" DIAPHRAGM PLATES AT 1/3 POINTS
WT PER FOOT :
2(18.75) = 37.5
LACING:
2(1/2×3/4)(3.04) = 2.28#/LF = 2.3
TIE PL'S
$\frac{2(1.3)^2(15.3\ \#/FT^2)}{20'}$ = 2.6

42.5 #/FT

SAG RODS

T = 3160 # AS ABOVE
ALLOW F_t = .5 F_u ON A_e ; F_t = 29 ksi (AISC 1.5.1)
$A_R = \frac{3.16}{29}$ = 0.11 IN²
USE 1/2" φ RODS ; A_e = .14 (MANUAL P 4-141)

However, even that selection is overstressed for the maximum vertical load, which occurs with the balanced roof live load. The third selection, C9 × 15, is lighter than the second, but deeper than the typical purlins, and requires shop cutting at its ends. It is a reluctant choice.

17.3.4 Sag Rods: Sheet R5

The sag rods are selected on sheet R5 to resist their accumulated tensile loads. They are threaded rods, sized for stress on their net area by the methods of Chapter 8.

17.3.5 The Diaphragm Alternative

Roof decks of the type being used, if continuous from eave to ridge and properly connected, are capable of acting as diaphragms, which are continuous resisting surfaces in their own planes. If approached that way, several advantages would emerge. They would intercept components of load in the X direction of the purlins, eliminating bending about the Y axis and the need for sag rods (except perhaps as a temporary erection device). The purlins would be sized for unidirectional bending with allowable stress, F_b, maximized since lateral support would be continuous. Since the diaphragm could resist in-plane shear, the need for diagonals in the roof system would disappear. The deep girders formed by the light-gage deck as web and the eave and ridge purlins as flanges would resist the effects of unbalanced live load, reducing much of the special demand on the design of the ridge purlins. The potential savings would be considerable.

Diaphragm design, to be effective, must be executed completely and in detail. Each interruption in the continuity of a diaphragm creates complications requiring careful engineering analysis and special, often expensive detailing. The designer in this case wishes to be free to add skylight and vent openings in patterns not yet determined and has decided to forego the savings possible with use of the roof as a diaphragm. Some engineers would take the opposite tack. We leave it to the reader to form his own opinion after calculating the differences in steel usage, construction complexity, and cost.

17.4 TYPICAL TRUSS: ANALYSIS AND DESIGN (TR SERIES)

17.4.1 Geometry, Member Types, Joint Types: Sheet TR1

Some of the advantages of the Fink truss result from its simplicity, a fact not readily apparent the first time it is seen. It becomes apparent in part when Millman examines the geometry of the typical roof truss for our mill building. The Fink truss is *compound*, consisting of two identical trusses, arranged symmetrically around a

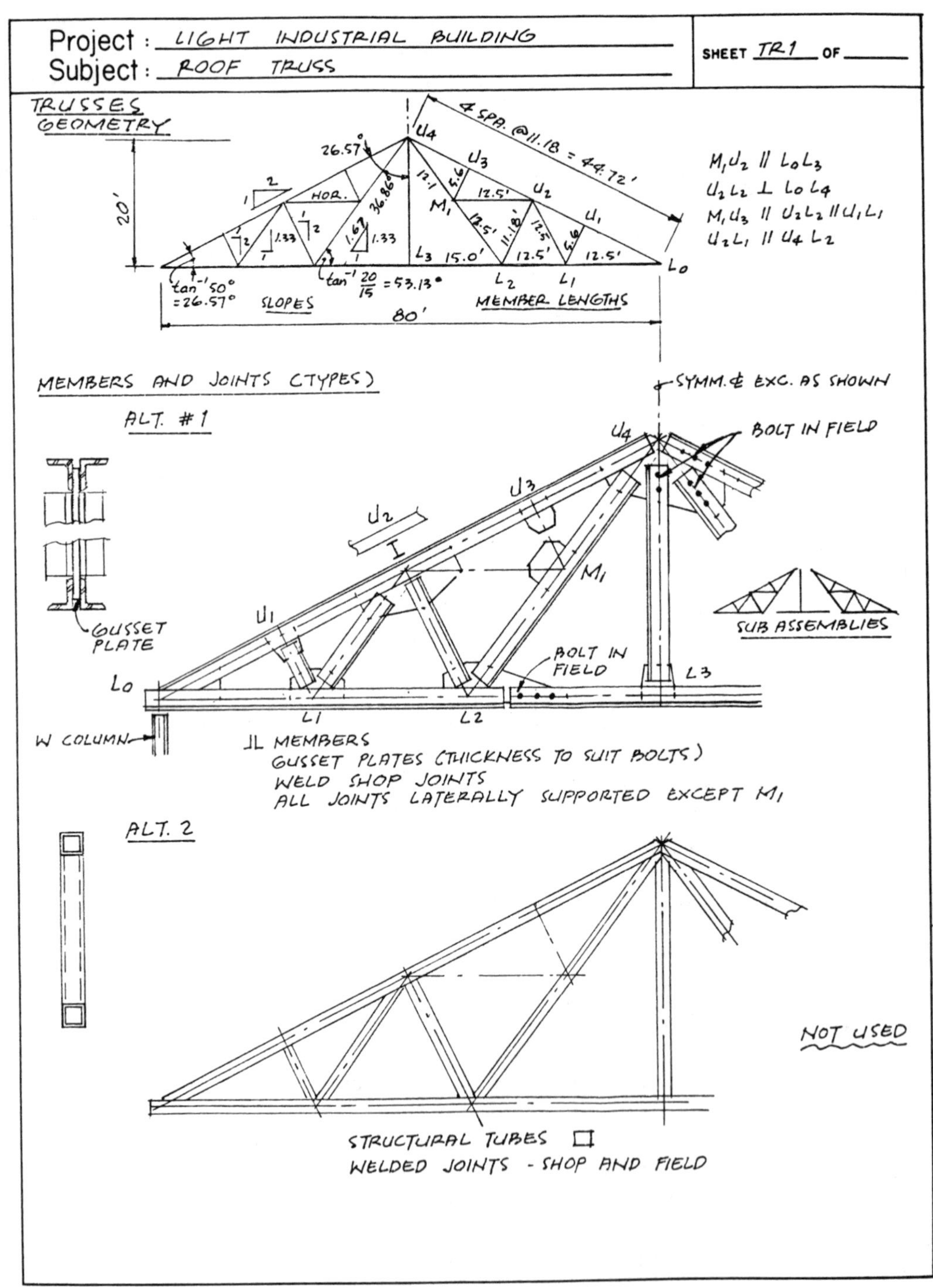

Project : LIGHT INDUSTRIAL BUILDING
Subject : ROOF TRUSS
SHEET TR1 OF
TRUSSES GEOMETRY
4 SPA. @ 11.18 = 44.72'
26.57°
36.86°
20'
HOR.
tan⁻¹ 50° = 26.57°
tan⁻¹ 20/15 = 53.13°
SLOPES
MEMBER LENGTHS
80'
$M_1U_2 \parallel L_0L_3$
$U_2L_2 \perp L_0L_4$
$M_1U_3 \parallel U_2L_2 \parallel U_1L_1$
$U_2L_1 \parallel U_4L_2$
MEMBERS AND JOINTS (TYPES)
ALT. #1
SYMM. ℄ EXC. AS SHOWN
BOLT IN FIELD
GUSSET PLATE
SUB ASSEMBLIES
BOLT IN FIELD
W COLUMN
JL MEMBERS
GUSSET PLATES (THICKNESS TO SUIT BOLTS)
WELD SHOP JOINTS
ALL JOINTS LATERALLY SUPPORTED EXCEPT M1
ALT. 2
NOT USED
STRUCTURAL TUBES
WELDED JOINTS - SHOP AND FIELD

vertical centerline, the separate trusses being joined at the ridge and tied at the bottom. Each component truss is symmetrical around its own centerline, which is normal to the slope of the roof, is divided into four panels of equal length measured parallel to the roof, and consists of triangles similar to both the main compound truss and its primary subtrusses. The angle established by the designer between the roof and the horizontal lower chord is repeated in most of the subfigures. Thus, the designer and steel detailer have to contend with only a very small number of angles of intersection and member lengths, repeated often within the system—a boon to both. The two side trusses can be shipped to the erection site in stable subassemblies of manageable size, easy to ship and to handle during erection.

The designer calculates and records angles, slopes, and lengths of members, and then proceeds to study possible types of members and joints, a procedure that requires sketches to reveal the physical implications of choices. The two alternatives shown do not exhaust the possibilities for a Fink truss, but each could be appropriate to the scale of span and loads of this building. Longer spans or heavier loads, for example, might demand W shapes.

The range of available angles is quite large. Their use in pairs permits light, stable members, as well as fairly heavy, strong ones. Used with gusset plates, they allow for simple joints with most angle ends cut at 90° to their axes. Either bolts or welds can be used to connect the parts of this type of joint. It then becomes possible to use welds for shop-assembled joints and bolts for those assembled in the field. In Millman's opinion, perhaps verified by consulting with some fabricators, each of these choices is economical.

Structural tubes are also available in a number of light, stable sections. If the various members have equal or nearly equal widths, they can be easily and effectively joined by welding without the necessity for gusset plates. Bolted connections, both within the truss and for purlins and bracing, are somewhat more of a problem, leading to the designer's opinion that use of tubes would be best, with welded connections in both shop and field (see Chapter 12).

For reasons that may or may not be valid, but are certainly backed by many precedents in similar buildings, the decision is made to proceed with alternative 1.

17.4.2 Unit Load Analysis: Sheets TR2 and TR3

The temptation is strong at this point to rush into a truss analysis based on the actual forces already fairly well known from the work on the G sheets. Millman decides that more time will be saved and better decisions will emerge from a more indirect procedure: first analyzing the truss for the effect of unit loads and then applying them, by direct proportion, to the effects of the actual loads.

Actually, Millman has available in office files the results of a unit load analysis done for a Fink truss of similar geometry. These results are independent of span length and can be applied to this and any truss where all dimensions are in proportion. Member loads are recorded on sheet TR2 for four load cases, two being for balanced

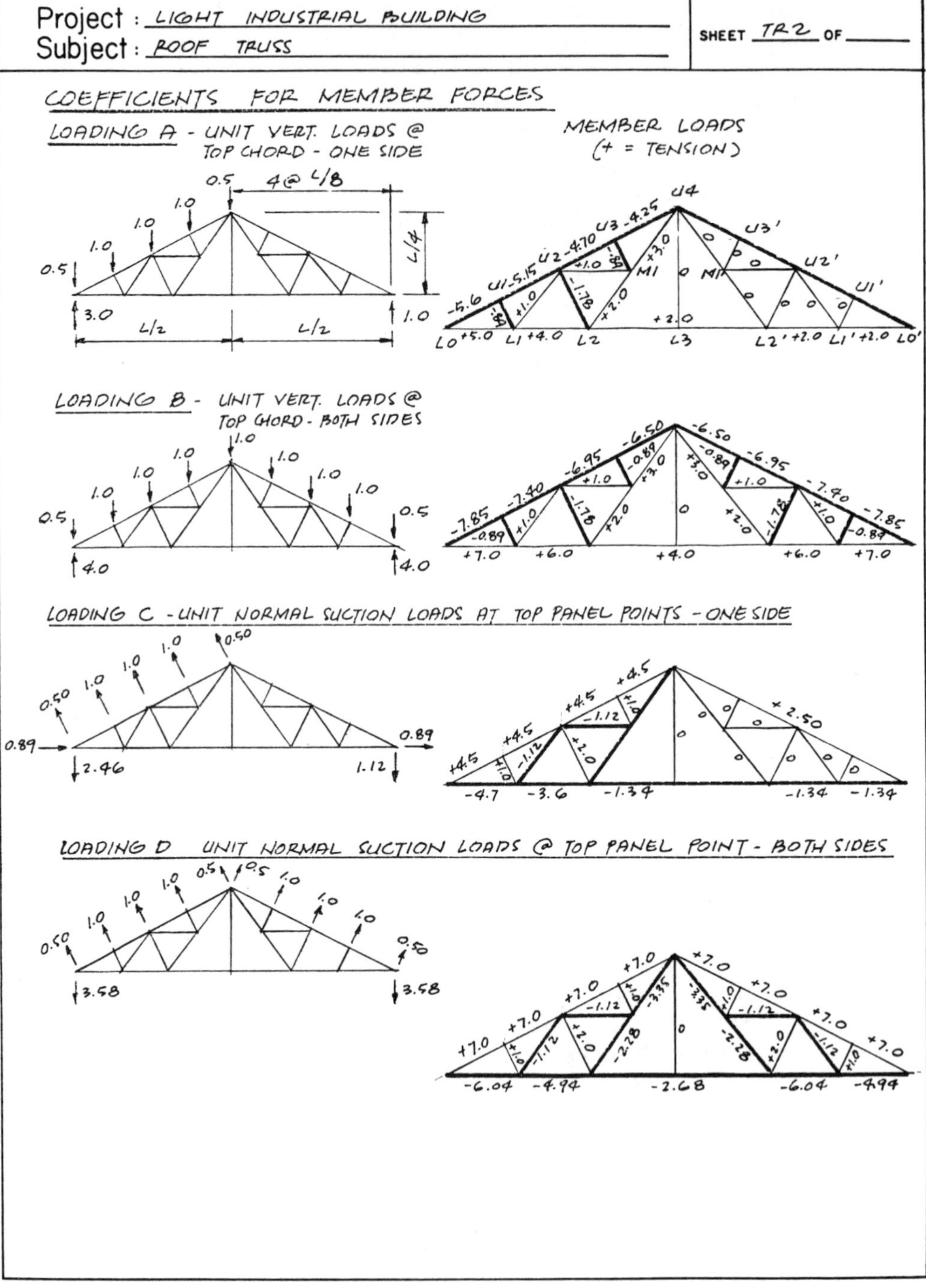
Project : LIGHT INDUSTRIAL BUILDING
Subject : ROOF TRUSS
SHEET TR2 OF
COEFFICIENTS FOR MEMBER FORCES
LOADING A - UNIT VERT. LOADS @ TOP CHORD - ONE SIDE
MEMBER LOADS (+ = TENSION)
4@ L/8
L/4
L/2
L/2
LOADING B - UNIT VERT. LOADS @ TOP CHORD - BOTH SIDES
LOADING C - UNIT NORMAL SUCTION LOADS AT TOP PANEL POINTS - ONE SIDE
LOADING D UNIT NORMAL SUCTION LOADS @ TOP PANEL POINT - BOTH SIDES

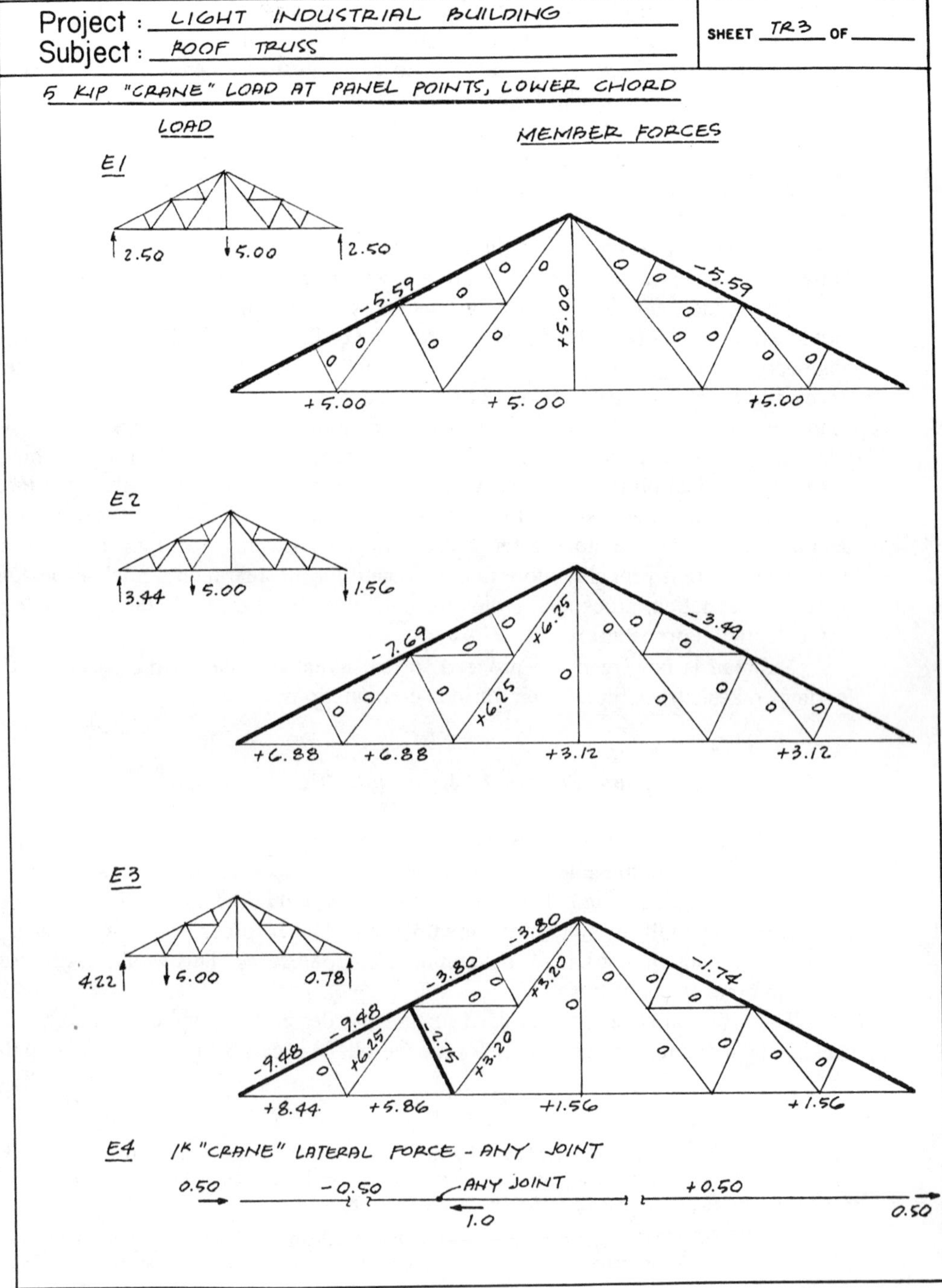
Project : LIGHT INDUSTRIAL BUILDING
Subject : ROOF TRUSS
SHEET TR3 OF
5 KIP "CRANE" LOAD AT PANEL POINTS, LOWER CHORD
LOAD
MEMBER FORCES
E1
2.50
5.00
2.50
-5.59
-5.59
+5.00
+5.00
+5.00
+5.00
E2
3.44
5.00
1.56
-7.69
-3.49
+6.25
+6.25
+6.88
+6.88
+3.12
+3.12
E3
4.22
5.00
0.78
-3.80
-3.80
-9.48
-9.48
-1.74
+6.25
-2.75
+3.20
+3.20
+8.44
+5.86
+1.56
+1.56
E4
1K "CRANE" LATERAL FORCE - ANY JOINT
0.50
-0.50
ANY JOINT
+0.50
1.0
0.50

roof loads, the other two with loads on only half of the roof. Neither of the unbalanced load cases gives coefficients higher than the corresponding balanced load case. However, loadings C and D give member forces opposite in sign to A and B. Wind loads will be based on C and D.

In each of the three cases on sheet TR3, a single load is applied to one panel point. This will represent the effect of the *crane* load, which has only one value of 5 kips, but may appear at any lower joint. The member forces entered are five times those that would result from a unit load. The effects of single joint loads are of interest since they reveal one of the consequences of the geometry of a Fink truss. Only the very closest web members are affected by a load at a single joint.

The reader can easily check the member force coefficients resulting from the unit load analyses. Methods of analysis are those ordinarily studied in a first course in engineering statics—the method of joints or that of sections. The coefficients were developed by a graphical version* of the method of joints. Its results include some scaling error,† and therefore include few significant figures in the numbers. The reader who finds, when checking, that his calculator provides more accurate answers should not be too critical on that account. Considering, among other things, the very large uncertainty about the level of loads assigned throughout the work, the increased accuracy does not lead to greater validity in the answers. Accuracy to the third significant figure is probably more than sufficient for most structural analysis as long as the mathematical procedure avoids magnification of error, a caveat applicable to any calculation procedure.

Millman is now ready to proceed to the actual member loads, based on the loadings on this truss, in all their various combinations.

17.4.3 Member Loads: Sheets TR4 through TR6

Member loads for actual design load cases are entered on the truss on sheets TR4 and TR5. For each case, the magnitudes of joint loads are determined from data in the *G* sheets and multiplied by member load coefficients found in the unit load analysis. As in the unit load analysis, none of the members had higher forces from unsymmetrical live or wind loads than it has for the symmetrical load cases. This would not be true for some other types of truss.

The table on sheet TR6 is used to find the design force for each member by seeking the maximum member force for the various load combinations specified in the

*This and other graphical procedures of analysis are described in detail in many older texts and structural design handbooks. They are especially simple to apply to truss problems if the engineer is proficient with the use of drafting instruments and careful about scale.

†We use the engineering distinction between *errors*, which are small and result from limitations of accuracy in arithmetic or graphical procedures, and *mistakes*, which result from incorrect procedure or theory.

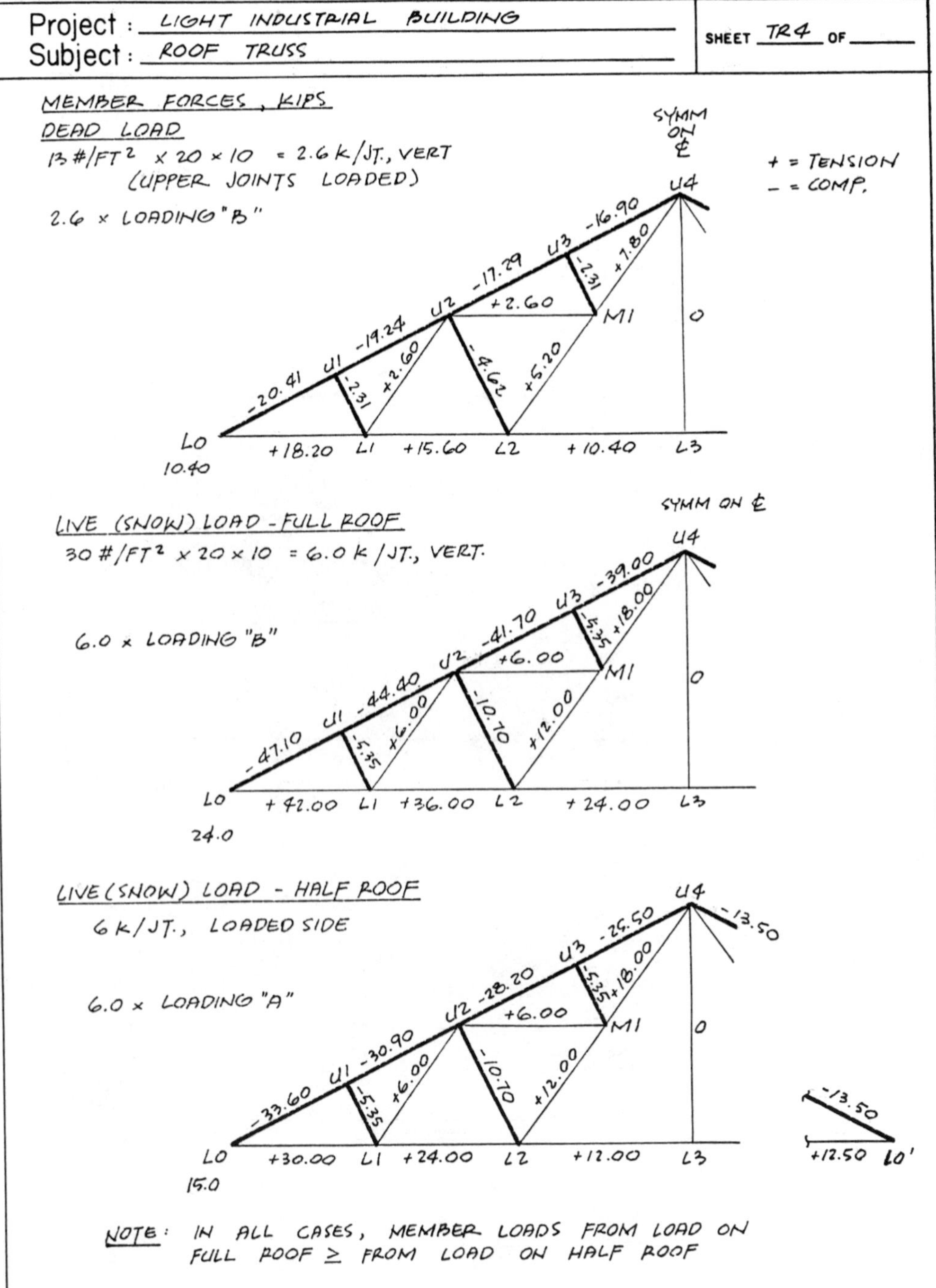
Project : LIGHT INDUSTRIAL BUILDING
Subject : ROOF TRUSS
SHEET TR4 OF
MEMBER FORCES, KIPS
DEAD LOAD
13 #/FT² × 20 × 10 = 2.6 K/JT., VERT
(UPPER JOINTS LOADED)
2.6 × LOADING "B"
SYMM ON ℄
+ = TENSION
− = COMP.
U4
-16.90
U3
+1.80
-2.31
-17.29
U2
+2.60
M1
0
-19.24
U1
+2.60
-4.62
+5.20
-20.41
-2.31
L0
10.40
+18.20
L1
+15.60
L2
+10.40
L3
SYMM ON ℄
LIVE (SNOW) LOAD - FULL ROOF
30 #/FT² × 20 × 10 = 6.0 K/JT., VERT.
6.0 × LOADING "B"
U4
-39.00
U3
+18.00
-5.35
-41.70
U2
+6.00
M1
0
-44.40
U1
+6.00
-10.70
+12.00
-47.10
-5.35
L0
24.0
+42.00
L1
+36.00
L2
+24.00
L3
LIVE (SNOW) LOAD - HALF ROOF
6 K/JT., LOADED SIDE
6.0 × LOADING "A"
U4
-13.50
-25.50
U3
+18.00
-5.35
-28.20
U2
+6.00
M1
0
-30.90
U1
+6.00
-10.70
+12.00
-33.60
-5.35
L0
15.0
+30.00
L1
+24.00
L2
+12.00
L3
-13.50
+12.50
L0'
NOTE: IN ALL CASES, MEMBER LOADS FROM LOAD ON FULL ROOF ≥ FROM LOAD ON HALF ROOF

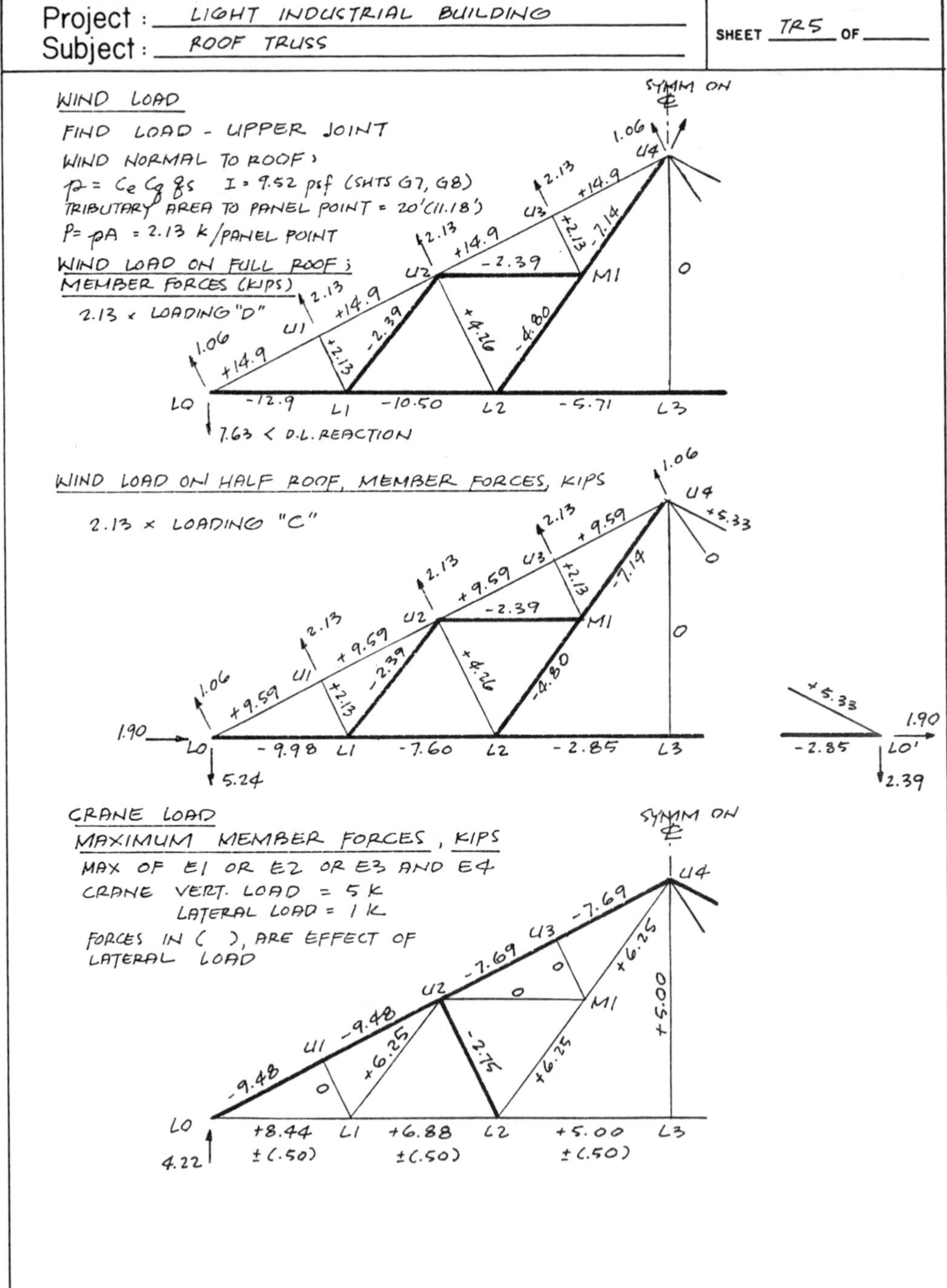
Project : LIGHT INDUSTRIAL BUILDING
Subject : ROOF TRUSS
SHEET TR5 OF
WIND LOAD
FIND LOAD - UPPER JOINT
WIND NORMAL TO ROOF ;
p = Ce Cq qs I = 9.52 psf (SHTS G7, G8)
TRIBUTARY AREA TO PANEL POINT = 20'(11.18')
P = pA = 2.13 k/PANEL POINT
WIND LOAD ON FULL ROOF ;
MEMBER FORCES (KIPS)
2.13 × LOADING "D"
SYMM ON ℄
7.63 < D.L. REACTION
WIND LOAD ON HALF ROOF, MEMBER FORCES, KIPS
2.13 × LOADING "C"
CRANE LOAD
MAXIMUM MEMBER FORCES, KIPS
MAX OF E1 OR E2 OR E3 AND E4
CRANE VERT. LOAD = 5 K
LATERAL LOAD = 1 K
FORCES IN ( ), ARE EFFECT OF LATERAL LOAD
SYMM ON ℄
L0 L1 L2 L3 U1 U2 U3 U4 M1
+8.44 ±(.50)
+6.88 ±(.50)
+5.00 ±(.50)
4.22

Project: LIGHT INDUSTRIAL BUILDING
Subject: ROOF TRUSS
SHEET TR 6 OF ____

MEMBER FORCES

	MEMBER	FORCES (KIPS)										
		D	L (SNOW)	CR	W	COMBINATION (SEE BELOW)						
						1a	1b	2a	2c	3	4a	4b
CHORDS	L0 - L1	+18.2	+42.0	+8.9	-12.9	+60.2	+58.6	CLEARLY NOT CONTROLLING FOR ANY MEMBER BY INSPECTION	+0.75	CLEARLY NOT CONTROLLING - BY INSPECTION	LESS THAN COMBINATION 1A - THROUGHOUT	CLEARLY NOT CONTROLLING - BY INSPECTION
	L1 - L2	+15.6	+36.0	+7.4	-10.5	+51.6	+50.0		+1.20			
	L2 - L3	+10.4	+24.0	+5.5	-5.7	+34.4	+33.9		+2.10			
	L0 - U1	-20.4	-47.1	-9.5	+14.9	-67.5	-65.2		-0.40			
	U1 - U2	-19.2	-44.4	-9.5	↓	-63.6	-62.0		+0.50			
	U2 - U3	-17.3	-41.7	-7.7	↓	-59.1	-56.4		+1.93			
	U3 - U4	-16.9	-39.0	-7.7	↓	-55.9	-53.9		+2.23			
WEB	U1 - L1	-2.31	-5.35	0	+2.13	-7.66	-6.32		+0.40			
	U2 - L2	-4.62	-10.70	-2.75	+4.26	-15.3	-15.4		+0.80			
	U3 - M1	-2.31	-5.35	0	+2.13	-7.66	-6.32		+0.40			
	L1 - U2	+2.60	+6.00	+6.25	-2.39	+8.60	+13.35		-0.44			
	U2 - M1	+2.60	+6.00	0	-2.39	+8.60	+7.10		-0.44			
	L2 - M1	+5.20	+12.00	+6.25	-4.80	+17.2	+20.5		-0.88			
	M1 - U4	+7.80	+18.00	+6.25	-7.14	+25.6	+27.4		-0.88			
	L3 - U4	0	0	+5.0	0	0	+5.0		0			

LOAD COMBINATIONS - FROM SHT. G3

#	LOADS	FACTOR FOR F
1 a	D + L	1.00
b	D + 0.75L + CR	1.00
2 a	(D + W) 0.75	1.33 × 0.75 = 1.00
b	D + SEISMIC (DON'T USE)	
c	(.75D + W) 0.75	1.33 × 0.75 = 1.00
d	.75D + SEISMIC (DON'T USE)	
3	(D + 0.50L + CR + W) 0.75	1.33 × 0.75 = 1.00
4 a	(D + L + 0.50W) (0.75)	1.33 (0.75) = 1.00
b	(D + 0.75L + CR + 0.5W) .75	1.33 (0.75) = 1.00
5	D + L + SEISMIC	(DON'T USE)

NOTES:

1. FOR DIRECT COMPARISON OF MEMBER LOADS FROM DIFFERENT CASES, LOADS FOR WHICH F(1.33) IS PERMITTED ARE MULTIPLIED BY 0.75. 1.33 × 0.75 = 1.00
2. CONTROLLING MEMBER LOADS UNDERLINED
3. RESULTS OF CASE 2C SHOW POTENTIAL REVERSAL OF STRESSES IN ALL MEMBERS WITH SLIGHT INCREASE OF CALCULATED WIND. LIMIT L/r OF ALL MEMBERS ≤ 200

Code and previously noted on sheet G3. To make the comparisons easier, instead of multiplying allowable stresses by 1.33 for combinations 2, 3, and 4, as permitted by the UBC and by AISC 1.5.6, the member loads are multiplied by $1/1.33 = 0.75$.

In the table, the columns for most of the *combinations* are not filled in because, after inspection, it is clear that those combinations will yield forces much lower than combinations 1a and 1b. Most members are controlled by combination 1a, a few web members by the local effect of the hoist loads in 1b. Combination 2c reveals an important possible source of danger to the structure. The member forces are almost zero, indicating that a small increase in wind load would cause stress reversals. The magnitude of wind load is debatable in itself, and it is quite clear that the 100-year wind will be more severe than the design basis 50-year wind. To prevent dangerous instability of the members and truss in an extreme wind, Millman decides to set the L/r* limit for all members at 200, the limit set in the AISC Specification for compression members. Although not required by the Code, this is very inexpensive disaster insurance for the client.† Millman is now ready to select members.

17.4.4 Member Selection: Sheet TR7

The selection of each member is done on the basis of procedures elaborated in Chapters 8 and 9. The table used is a helpful way of systematizing the work. Besides saving calculation space (a trivial advantage), it has other purposes. It makes it possible for the designer to check the individual selections with others for consistency, thereby avoiding mistakes and complications in fabrication. It also makes the necessary checking process much easier and more effective. The completed table is also used to calculate the weight of the truss and compare it to the original assumptions, a necessary step in deciding whether or not the dead load calculation has to be redone, possibly requiring changes in member design loads and sizes. In this case, the results in pounds per square foot validated the earlier rough calculation on sheet G6. The table will also be used later as a source of data for joint design.**

*$K = 1.0$ for all members in a usual truss (see AISC 1.8.2 and Chapter 9).

†It will turn out later, in the design of the wind bracing system, that the *tension* lower chord of the vertical truss can become a compression member from the design wind.

**Although we have not included Millman's D sheets in the calculation set presented here, similar conditions are included in examples and problems in joint design in Chapters 12 and 13.

Project : LIGHT INDUSTRIAL BUILDING
Subject : ROOF TRUSS

SHEET TR7 OF ______

MEMBER SELECTION - DOUBLE ANGLES - SHOP WELDING; FIELD BOLTING

	MEMBER	LENGTH C-C, FT (IN PLANE / ⊥ PLANE)	DESIGN LOAD, KIPS	SIZE OF DBL ∠ & AREA (JL = L.L. b-b; ⊥ = S.L. b-b)	r_x / r_y	LARGER OF $(L/r)_x \le 200$	LARGER OF $(L/r)_y \le 200$	STRESSES*, ksi F_t (+)	F_a (−)	f_t (+)	f_a (−)	WEIGHT (1/2 OF TRUSS) #/FT	#/MEMB.
CHORD	L0-L2	2@12.5 / 2@12.5	+60.2	JL 3½×2½×¼, A_g = 2.88	1.12 / 1.09		138	21.6		20.9		9.80	245
	L2-L2'	2@15.0 / 2@15.0	+34.4	JL 3×2½×3/16, A_g = 1.99, A_e = 1.66	.954 / 1.12	189		21.6 ON A_g		17.3		6.77	203/2
	L0-U4	4@11.18 / 4@11.18	−67.5	JL 5×3½×3/8, A = 6.09	1.60 / 1.46		92		14.0		11.1	20.8	930
WEB MEMBERS	U1-L1	5.6 / 5.6	−7.66	JL 2×2×3/16, A = 1.43	.617 / .98	109			11.81		5.36	4.88	27
	U2-L2	11.2 / 11.2	−15.3	JL 3×2½×¼, A = 2.63	.945 / 1.13	142			7.41		5.82	9.0	101
	U3-M1	5.6 / 5.6	−7.66	SAME AS U1-L1								4.88	27
	L1-U2	12.5 / 12.5	+13.4	JL 2½×2×3/16, A_g = 1.62, A_e = 1.79	.79 / .92	190		21.6 ON A_g		8.27		5.5	69
	U2-M1	12.5 / 12.5	+8.6	USE SAME AS L1-U2								5.5	69
	L2-U4	2@12.5 / 25.0	+27.4 / −0.9	⊥ 3½×3×¼, A_g = 3.13, A_e = 2.69	.914 / 1.65	164	182	21.6 ON A_g	4.51	8.75	0.30	10.8	270
	L3-U4	20.0 / 20.0	+5.0	JL 4×3×¼, A_g = 3.38, A_e = 2.94	1.28 / 1.29		188	21.6 ON A_g		1.48		11.6	232/2

*F_a FROM TABLE 3-36 AISC SPEC.

F_t = 0.6 F_y = 21.6 KSI ON GROSS SECTION
or 0.5 F_u = 29.0 ksi ON EFFECTIVE NET AREA (BOLTED) MEMBERS ONLY)

FOR FIELD BOLTED CONNECTIONS OF TENSION MEMBERS (JOINTS U4, L2, L2') - SEE AISC 1.14.2.2

USING 3/4" ∅ BOLTS IN 1 LINE

$A_n = A_g - 2(.875)t$

$A_e = A_n C_t$, WHERE C_t = 0.85 IF 3 BOLTS OR MORE;
C_t = 0.75 IF 2 BOLTS (WEB MEMB. ONLY)

DIMENSIONS FOR MIN. BOLTED LEG OF JL

3" MIN; 1"; 2"; ℄ 3/4" ∅ BOLT; 3/8" GUSSET PL.

VALUE = 8.8 K. IN DBL SHEAR (MANUAL p 4-5)

SEE SPEC. TABLE 1.16.5.1 & MANUAL p. 4-136 FOR BOLT DIMENSIONS.

WEIGHT OF TRUSS

SUBTOTAL, 1/2 TRUSS	1,955
— " — 1 TRUSS	3,910
ADD DETAILS - 8%	310
TOTAL WT. 1 TRUSS	4,220
TRUSS WT., #/FT²	2.64

ASSUMED WT. OF TRUSS = 3.0 #/FT² OF BUILDING - SAY OK

FOR COMPRESSION $b/t \le 12.67$

L'S USED

- 5×3½×3/8 ; b/t ≈ 11
- 3½×2½×¼ - TENSION ONLY
- 3½×3×¼ ; b/t = 12
- 3½×2½×3/16 ; TENSION ONLY
- 2×2×3/16 ; b/t = 8
- 3×2½×¼ ; b/t = 10
- 2½×2×3/16 - TENSION ONLY
- 4×3×¼ - TENSION ONLY

17.5 WIND-RESISTING SYSTEM (L SERIES)

17.5.1 Loading: Elements, Frame

The L series of sheets is used to size members in the wind-resisting system. The basis of loading is taken from the table previously developed on sheet G9, in which, it will be recalled, different wind pressures were stipulated for the design of elements of the structure and for the primary frame, the former being more severe than the latter. This makes for some inconsistencies in the mathematical model, which may sometimes confuse, but should not prevent, reasonable analysis. Since the definitions of primary frame and elements are not precise, there is some uncertainty as to the dividing line between them.

Millman includes as *elements* the siding, girts, and *wind beams* in the end walls. These are addressed on sheets L1 through L7. In the *primary frame* are included the horizontal bracing system, which consists in part of members previously selected for other purposes and doing double duty in this system (purlins, truss chords). It also includes the diagonals in the vertical braced bents and the columns with which they interact, although the columns will be selected later, primarily for other purposes. Elements of the end wall are designed for the total pressure, from wind and lee sides, as required for open structures.

17.5.2 Siding: Sheet L1

Pressure diagrams for the end wall appear on sheet L1 with a framing elevation on which girts and wind beams are superimposed on the typical truss behind them. Siding will be light-gage steel functioning in very similar manner to the roof deck (refer to sheet R1). The calculation is made for the longest vertical span between girts, considering, for the worst case, simple span action. The same maximum span exists in the side walls; the same siding will be used. Once again, as for most cases with large span/depth ratios, finding a strong enough section is no problem, but deflection is. The designer considers the same two sections as were considered for the roof deck. Both are possible choices. The number 21 type (3-inch depth) deflects less than the number 3. There are no generally accepted criteria for deflection applicable to this case. Although $L/163$ seems to be a lot, Millman decides that it will not be damaging and chooses the number 3 section for other reasons. Unlike the roof deck, the siding is at a level where local damage from plant activities is highly likely for the number 21 section, which is of very light gage and has wider expanses of flat sheet between the stiffening webs.

17.5.3 Girts and Sag Rods, End Walls: Sheets L2 through L4

The girts are horizontal beams loaded by the end reactions of the siding in a manner analogous to the way the roof deck reactions load the purlins. Clearly, they require

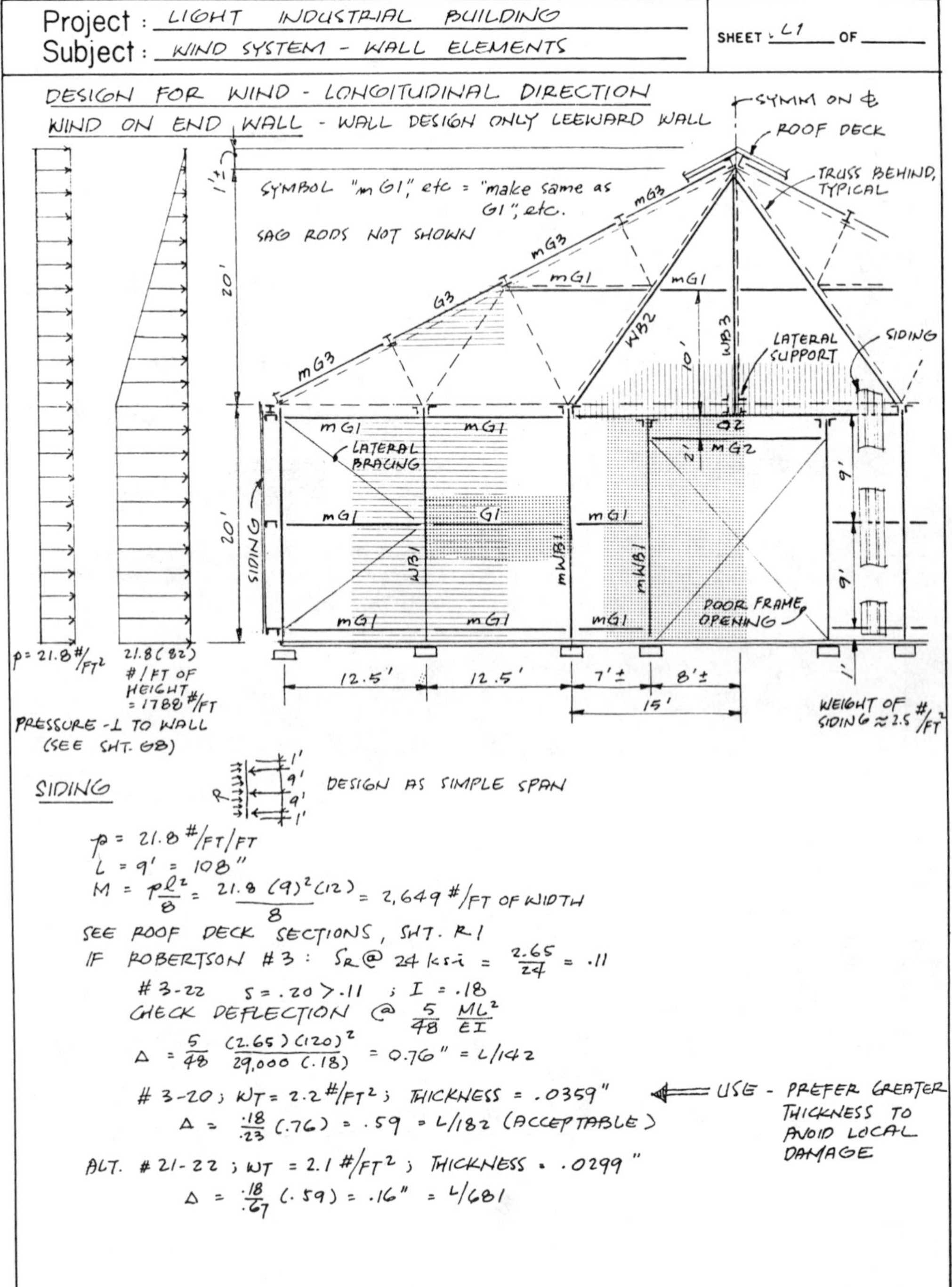

SIDING

DESIGN AS SIMPLE SPAN

$p = 21.8$ #/FT/FT

$L = 9' = 108''$

$M = \frac{pl^2}{8} = \frac{21.8\,(9)^2(12)}{8} = 2{,}649$ #/FT OF WIDTH

SEE ROOF DECK SECTIONS, SHT. R1

IF ROBERTSON #3 : S_R @ 24 ksi $= \frac{2.65}{24} = .11$

#3-22 $S = .20 > .11$; $I = .18$

CHECK DEFLECTION @ $\frac{5}{48}\frac{ML^2}{EI}$

$\Delta = \frac{5}{48}\frac{(2.65)(120)^2}{29{,}000\,(.18)} = 0.76'' = L/142$

#3-20 ; $W_T = 2.2$ #/FT² ; THICKNESS = .0359" ⟸ USE - PREFER GREATER THICKNESS TO AVOID LOCAL DAMAGE

$\Delta = \frac{.18}{.23}(.76) = .59 = L/182$ (ACCEPTABLE)

ALT. #21-22 ; WT = 2.1 #/FT² ; THICKNESS = .0299"

$\Delta = \frac{.18}{.67}(.59) = .16'' = L/681$

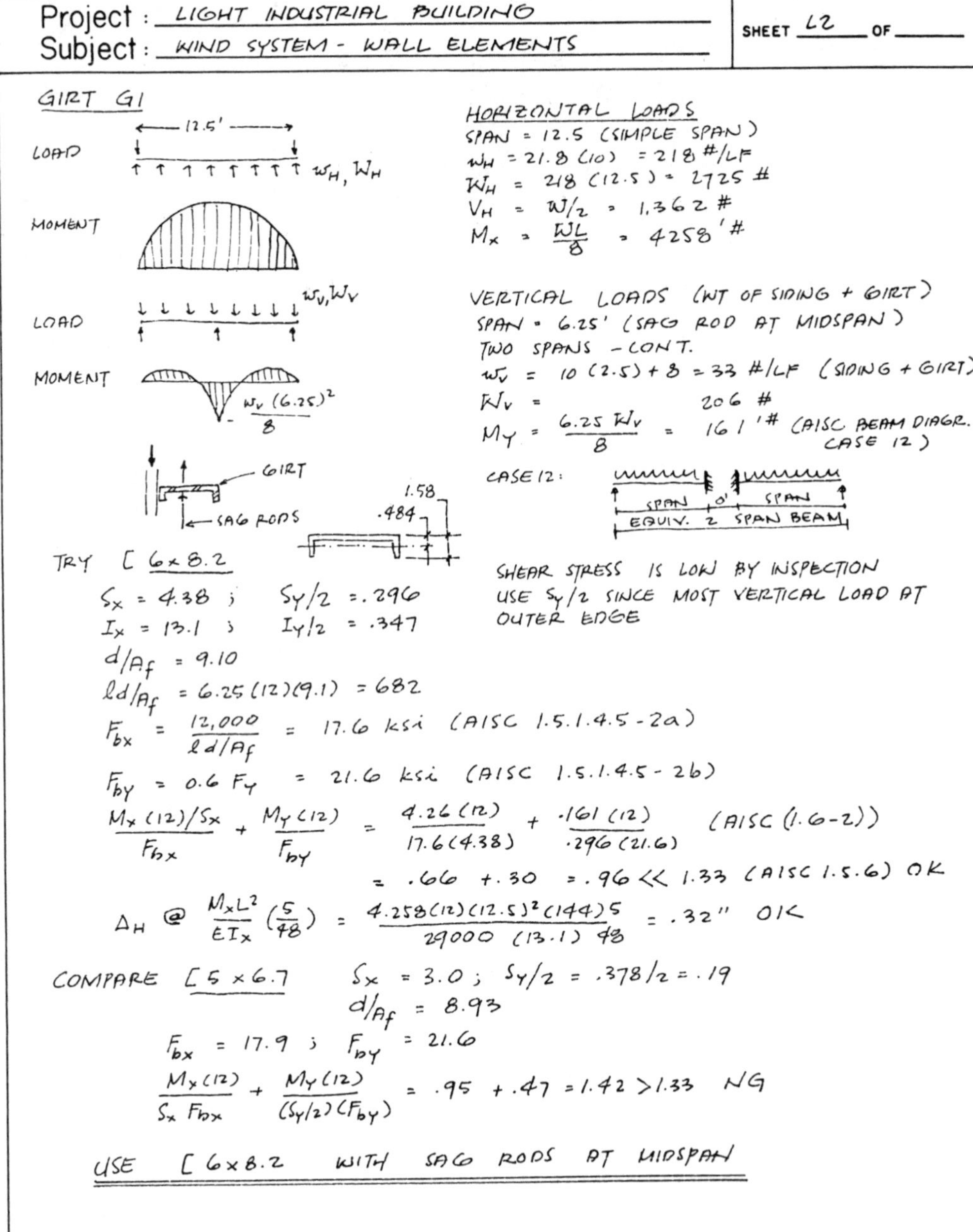

Project : LIGHT INDUSTRIAL BUILDING

Subject : WIND SYSTEM - WALL ELEMENTS

SHEET L2 OF

GIRT G1

LOAD — 12.5' — w_H, W_H

MOMENT

LOAD — w_v, W_v

MOMENT — $\frac{w_v (6.25)^2}{8}$

GIRT

SAG RODS

1.58

.484

HORIZONTAL LOADS

SPAN = 12.5 (SIMPLE SPAN)

$w_H = 21.8\,(10) = 218\ \#/LF$

$W_H = 218\,(12.5) = 2725\ \#$

$V_H = W/2 = 1{,}362\ \#$

$M_x = \frac{WL}{8} = 4258\ '\#$

VERTICAL LOADS (WT OF SIDING + GIRT)

SPAN = 6.25' (SAG ROD AT MIDSPAN)

TWO SPANS - CONT.

$w_v = 10\,(2.5) + 8 = 33\ \#/LF$ (SIDING + GIRT)

$W_v = 206\ \#$

$M_Y = \frac{6.25\,W_v}{8} = 161\ '\#$ (AISC BEAM DIAGR. CASE 12)

CASE 12: SPAN, 0', SPAN — EQUIV. 2 SPAN BEAM

TRY [6×8.2

$S_x = 4.38$; $S_Y/2 = .296$

$I_x = 13.1$; $I_Y/2 = .347$

$d/A_f = 9.10$

$\ell d/A_f = 6.25\,(12)(9.1) = 682$

$F_{bx} = \frac{12{,}000}{\ell d/A_f} = 17.6$ ksi (AISC 1.5.1.4.5-2a)

$F_{bY} = 0.6\,F_Y = 21.6$ ksi (AISC 1.5.1.4.5-2b)

SHEAR STRESS IS LOW BY INSPECTION

USE $S_Y/2$ SINCE MOST VERTICAL LOAD AT OUTER EDGE

$$\frac{M_x\,(12)/S_x}{F_{bx}} + \frac{M_Y\,(12)}{F_{bY}} = \frac{4.26\,(12)}{17.6\,(4.38)} + \frac{.161\,(12)}{.296\,(21.6)} \quad \text{(AISC (1.6-2))}$$

$$= .66 + .30 = .96 \ll 1.33 \quad \text{(AISC 1.5.6) OK}$$

$$\Delta_H \;@\; \frac{M_x L^2}{E I_x}\left(\frac{5}{48}\right) = \frac{4.258(12)(12.5)^2(144)5}{29000\,(13.1)\,48} = .32'' \quad \text{OK}$$

COMPARE [5×6.7 $S_x = 3.0$; $S_Y/2 = .378/2 = .19$

$d/A_f = 8.93$

$F_{bx} = 17.9$; $F_{bY} = 21.6$

$$\frac{M_x\,(12)}{S_x F_{bx}} + \frac{M_Y\,(12)}{(S_Y/2)(F_{bY})} = .95 + .47 = 1.42 > 1.33 \quad \text{NG}$$

USE [6×8.2 WITH SAG RODS AT MIDSPAN

Project : LIGHT INDUSTRIAL BUILDING
Subject : WIND SYSTEM - WALL ELEMENTS

SHEET L3 OF ____

GIRT G2

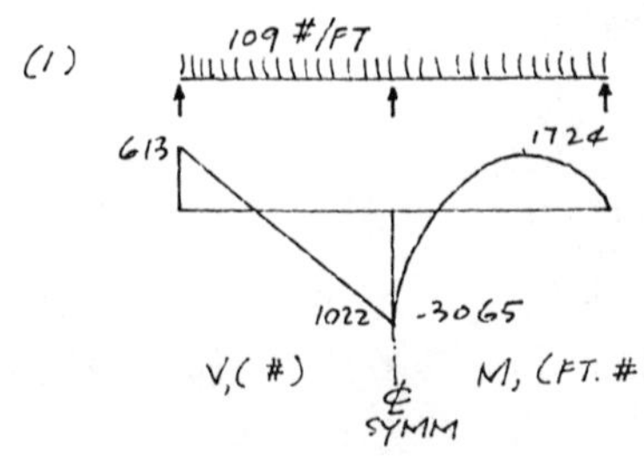

HORIZONTAL LOADING

$w = 5'(21.8\ \#/\square) = 109\ \#/FT$

$Q \leq (8' + \frac{7'}{2})\ 9'(21.8) = 2060\ \#$

(1) BY SYMMETRY, USE AISC BEAM DIAGRAMS PART 2, # 12.

$w = 109\ \#/FT$

$L = 15'$

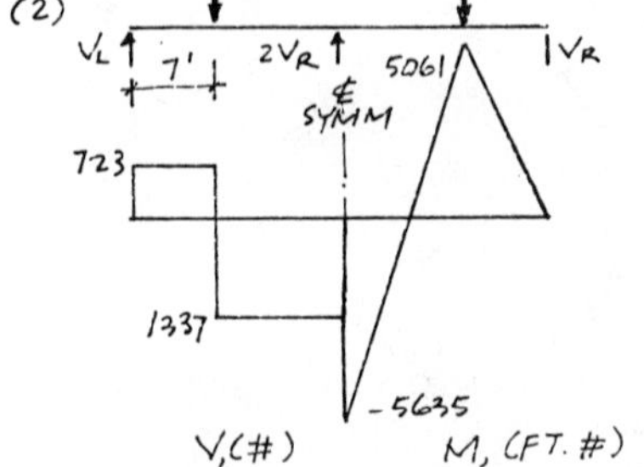

(2) AISC BEAM DIAG. CASE 14

a
a
L
0'
L

$L = 15'$; $a = 7'$

$V_L = 723\ \#$ $V_R = 1337\ \#$

$M_1 = 723(7) = 5061\ FT\ \#$

$M_2 = 5061 - 8(1337) = -5635\ '\#$

(1) + (2)

$M_{MAX} = -(3065 + 5635) = -8700\ '\#$

$V_{MAX} = 1022 + 1337 = 2{,}359\ \#$

IF C6 × 8.2 (AS G1)

$f_v = \frac{2359}{6(.2)} = 1966 \ll F_v$

(CONT. NEXT SHEET)

Project: LIGHT INDUSTRIAL BUILDING
Subject: WIND SYSTEM - WALL ELEMENTS
SHEET L4 OF ____

GIRT G2 (CONT.)

VERTICAL LOADING ; SIDING + BEAM

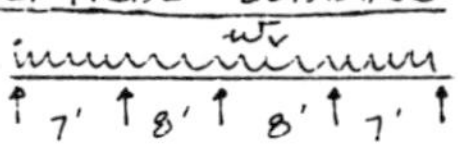

$w_V \approx 8'(2.5) + 8\ \#/FT = 28\ \#/FT$

$M_{MAX} \approx \frac{w\ell^2}{10} = \frac{28(8)^2}{10} = 179\ '\#$

SHEAR VERY LOW - NOT CONTROLLING

TRY C6×8.2 (SAME SIZE AS G1)

PROPERTIES: SHT. L2

$\frac{\ell d}{A_f} = 8(12)(9.10) = 874$

$F_{bx} = \frac{12,000}{874} = 13.7$ ksi (AISC (1.5-7))

$F_{by} = 21.6$ ksi (AISC 1.5.1.4.5 -2b)

$\frac{M_x(12)/S_x}{F_{bx}} + \frac{M_Y(12)/.5S_Y}{F_{bY}} = \frac{8.7(12)}{4.38(13.7)} + \frac{.179(12)}{.296(21.6)} = 1.74 + 0.34 = 2.08 >> 1.33$ NG

TRY C8×11.5

$S_x = 8.14$; $S_{Y/2} = .39$; $d/A_f = 9.08$

$\ell d/A_f = 96(9.08) = 872 \longrightarrow F_{bx} = 13.8$; $F_{bY} = 21.6$

$\frac{M_x(12)}{S_x(F_{bx})} + \frac{M_Y(12)}{.5S_Y(F_{bY})} = 0.93 + 0.25 = 1.18 < 1.33$ OK

USE C8×11.5

GIRT G3

HORIZONTAL LOADS

$w_H \approx 4(21.8) = 87\ \#/FT$

[6×8.2 OK BY COMPARISON WITH G1

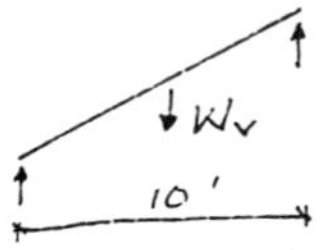

VERTICAL LOADS - SAG ROD PULL + BEAM

$W_v \leq 30' \times 6.5' \times 3\ \#/FT^2 = 585\ \#$

$M \leq \frac{PL}{4} = 1462\ '\#$

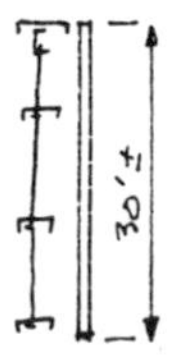

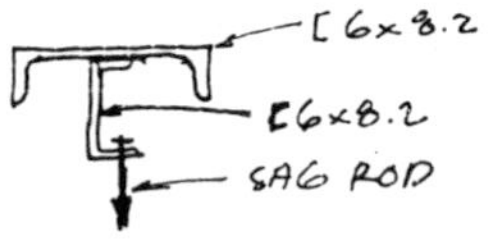

[6×8.2

$S_x = 4.38$

$f_{bx} = \frac{1462(12)}{4.38(1000)} = 4.01\ ksi << F_{bx}$

LIGHTER SECTIONS OK, BUT USE AS SHOWN TO AVOID ADDITIONAL SECTIONS ON JOB.

SAG RODS

MAX. LOAD $\leq 585\ \#$ << ROOF SAG RODS

USE 1/2" ϕ THREADED RODS AS FOR ROOF

sections with the strong axis of cross section vertical. Although the design wind is strong, the force effects on the girts are fairly light, particularly considering the effect of increased allowable stresses. The girts do not participate in the bracing system and are often, as in this case, set out from the main frame. As with the purlins, although for slightly different reasons, they are loaded in the direction of both axes, the vertical loads being their own weight and that of the siding. Light channels are commonly used, of sufficient depth for the normal wind load, the member x axis being oriented vertically so its strong bending direction is horizontal. As with the purlins, weak-axis bending moments are reduced by the inexpensive expedient of sag rods.

Designers often try to settle on one size of girt to use for the entire building on the theory that small savings in the weight of individual nontypical girts are overshadowed by the extra expense of ordering and handling more different sections. The calculations are similar to those used for the purlins. There are a number of different spans and loads per linear foot, but the calculations and results are similar. The calculation for girt G1 establishes the typical size. Member G2 is somewhat more complicated than the typical girts since it is a two span beam horizontally, being supported in the middle by the centerline wind beam. Since both the beam and loads are symmetrical, and therefore the rotation at the interior support is zero, Millman is able to use, for analysis, a standard fixed-end beam case among the Beam Diagrams of the AISC Manual. Girt G2 turns out to require a heavier and deeper section than the others. Millman makes an exception in this case and uses a nontypical section.

Members G3 are also special since they must carry the vertical loads from the sag rods as well as their share of wind load. To avoid new sections here, the designer simply combines two of the C6's as shown on sheet L4.

The same size rod is selected for sag rods on sheet L4 as was used for the roof (sheet R5), although the loads are considerably less. Millman agrees with the view of many engineers that ½ inch is the smallest diameter of threaded rod that should be permitted in an industrial application. In any case, the sag rods are very light and the difference in cost of steel between ½ inch diameter and ⅜ inch would be slight, while the cost of placing the rods in the structure would be practically the same for either.

17.5.4 Wind Beams: Sheet L5

We choose to define as *wind beams* vertical members that are sometimes called wind columns, since Millman has specified that, in this building, they not be permitted to act as columns. The lower tier of wind beams span vertically from floor level to lower chord level. Those in the upper tier span from lower chord to roof. Sliding joints are specified at the upper end of members WB1 to prevent transfer of axial loads from the roof support system to the wind beams. A typical truss and columns are provided in the end bents, and the wall is considered an appendage. The designer shares the philosophy of many that one should not permit a member to have to resist loads that were not investigated in its design; unexpected consequences sometimes result.

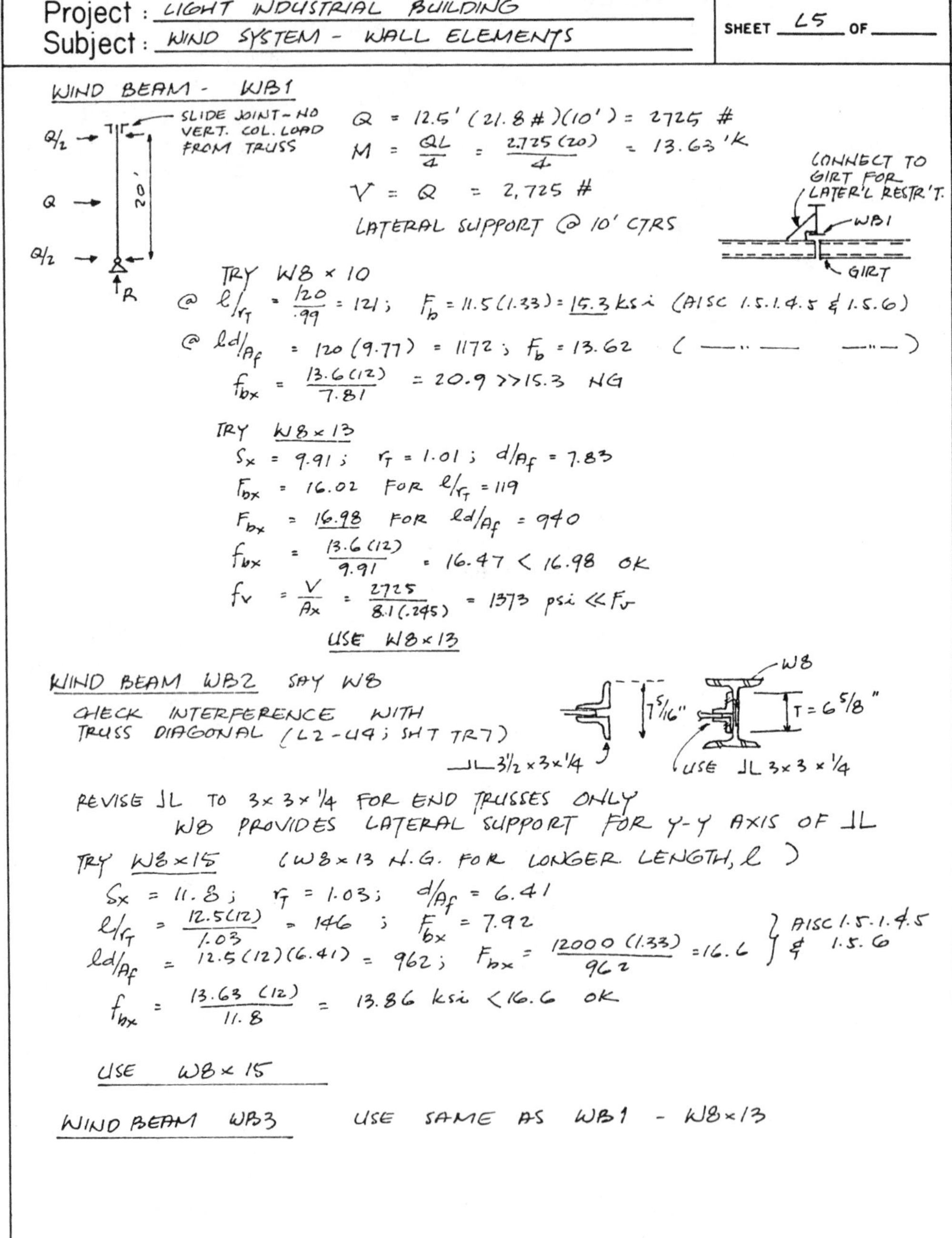

Project: LIGHT INDUSTRIAL BUILDING

Subject: WIND SYSTEM - WALL ELEMENTS

SHEET L5 OF

WIND BEAM - WB1

SLIDE JOINT - NO VERT. COL. LOAD FROM TRUSS

Q/2, Q, Q/2, 20', R

$Q = 12.5'(21.8\#)(10') = 2725\ \#$

$M = \frac{QL}{4} = \frac{2725(20)}{4} = 13.63\ 'K$

$V = Q = 2,725\ \#$

LATERAL SUPPORT @ 10' CTRS

CONNECT TO GIRT FOR LATER'L RESTR'T.

WB1

GIRT

TRY W8 × 10

@ $\ell/r_T = \frac{120}{.99} = 121$; $F_b = 11.5(1.33) = 15.3$ ksi (AISC 1.5.1.4.5 & 1.5.6)

@ $\ell d/A_f = 120(9.77) = 1172$; $F_b = 13.62$ (— " — — " —)

$f_{bx} = \frac{13.6(12)}{7.81} = 20.9 >> 15.3$ NG

TRY W8 × 13

$S_x = 9.91$; $r_T = 1.01$; $d/A_f = 7.83$

$F_{bx} = 16.02$ FOR $\ell/r_T = 119$

$F_{bx} = 16.98$ FOR $\ell d/A_f = 940$

$f_{bx} = \frac{13.6(12)}{9.91} = 16.47 < 16.98$ OK

$f_v = \frac{V}{A_x} = \frac{2725}{8.1(.245)} = 1373$ psi $<< F_v$

USE W8 × 13

WIND BEAM WB2 SAY W8

CHECK INTERFERENCE WITH TRUSS DIAGONAL (L2-U4; SHT TR7)

$7\frac{5}{16}$" ; W8 ; T = $6\frac{5}{8}$"

JL $3\frac{1}{2} \times 3 \times \frac{1}{4}$; USE JL $3 \times 3 \times \frac{1}{4}$

REVISE JL TO $3 \times 3 \times \frac{1}{4}$ FOR END TRUSSES ONLY

W8 PROVIDES LATERAL SUPPORT FOR Y-Y AXIS OF JL

TRY W8 × 15 (W8 × 13 N.G. FOR LONGER LENGTH, ℓ)

$S_x = 11.8$; $r_T = 1.03$; $d/A_f = 6.41$

$\ell/r_T = \frac{12.5(12)}{1.03} = 146$; $F_{bx} = 7.92$

$\ell d/A_f = 12.5(12)(6.41) = 962$; $F_{bx} = \frac{12000(1.33)}{962} = 16.6$ } AISC 1.5.1.4.5 & 1.5.6

$f_{bx} = \frac{13.63(12)}{11.8} = 13.86$ ksi < 16.6 OK

USE W8 × 15

WIND BEAM WB3 USE SAME AS WB1 - W8 × 13

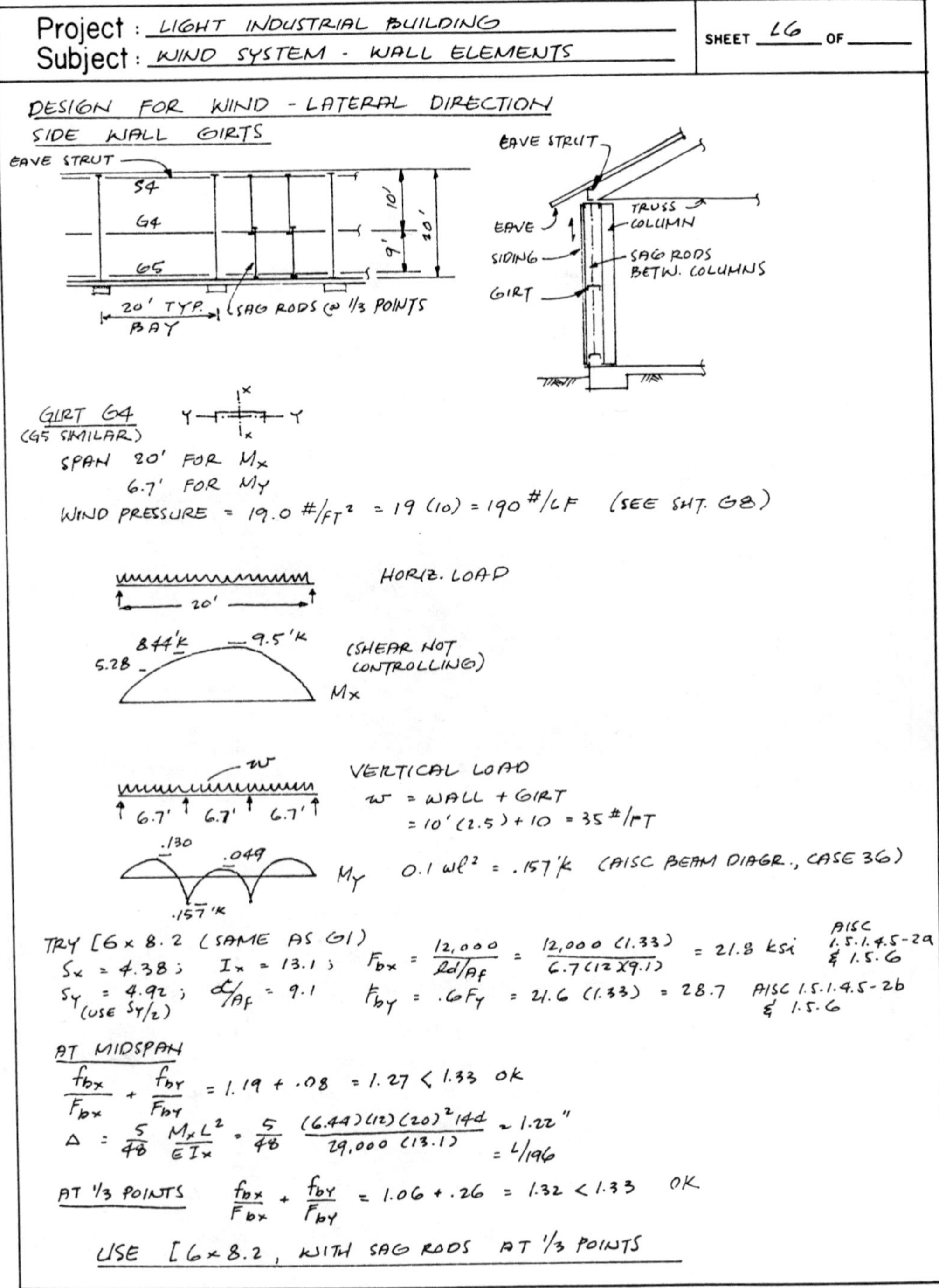
Project : LIGHT INDUSTRIAL BUILDING
Subject : WIND SYSTEM - WALL ELEMENTS
SHEET L6 OF
DESIGN FOR WIND - LATERAL DIRECTION
SIDE WALL GIRTS
EAVE STRUT
S4
G4
G5
10'
9'
20'
20' TYP. BAY
SAG RODS @ 1/3 POINTS
EAVE STRUT
TRUSS
COLUMN
EAVE
SIDING
SAG RODS BETW. COLUMNS
GIRT
GIRT G4
(G5 SIMILAR)
SPAN 20' FOR Mx
6.7' FOR My
WIND PRESSURE = 19.0 #/FT² = 19 (10) = 190 #/LF (SEE SHT. G8)
HORIZ. LOAD
20'
8.44'k
9.5'k
5.28
(SHEAR NOT CONTROLLING)
Mx
w
VERTICAL LOAD
6.7' 6.7' 6.7'
w = WALL + GIRT
= 10' (2.5) + 10 = 35 #/FT
.130
.049
.157'k
My
0.1 wl² = .157'k (AISC BEAM DIAGR., CASE 36)
TRY [6 x 8.2 (SAME AS G1)
Sx = 4.38; Ix = 13.1; Fbx = 12,000 / (ld/Af) = 12,000 (1.33) / 6.7(12)(9.1) = 21.8 ksi AISC 1.5.1.4.5-2a & 1.5.6
Sy = 4.92; d/Af = 9.1 Fby = .6Fy = 21.6 (1.33) = 28.7 AISC 1.5.1.4.5-2b & 1.5.6
(USE Sy/2)
AT MIDSPAN
fbx/Fbx + fby/Fby = 1.19 + .08 = 1.27 < 1.33 OK
Δ = 5/48 MxL²/EIx = 5/48 (6.44)(12)(20)²144 / 29,000 (13.1) = 1.22" = L/196
AT 1/3 POINTS fbx/Fbx + fby/Fby = 1.06 + .26 = 1.32 < 1.33 OK
USE [6 x 8.2, WITH SAG RODS AT 1/3 POINTS

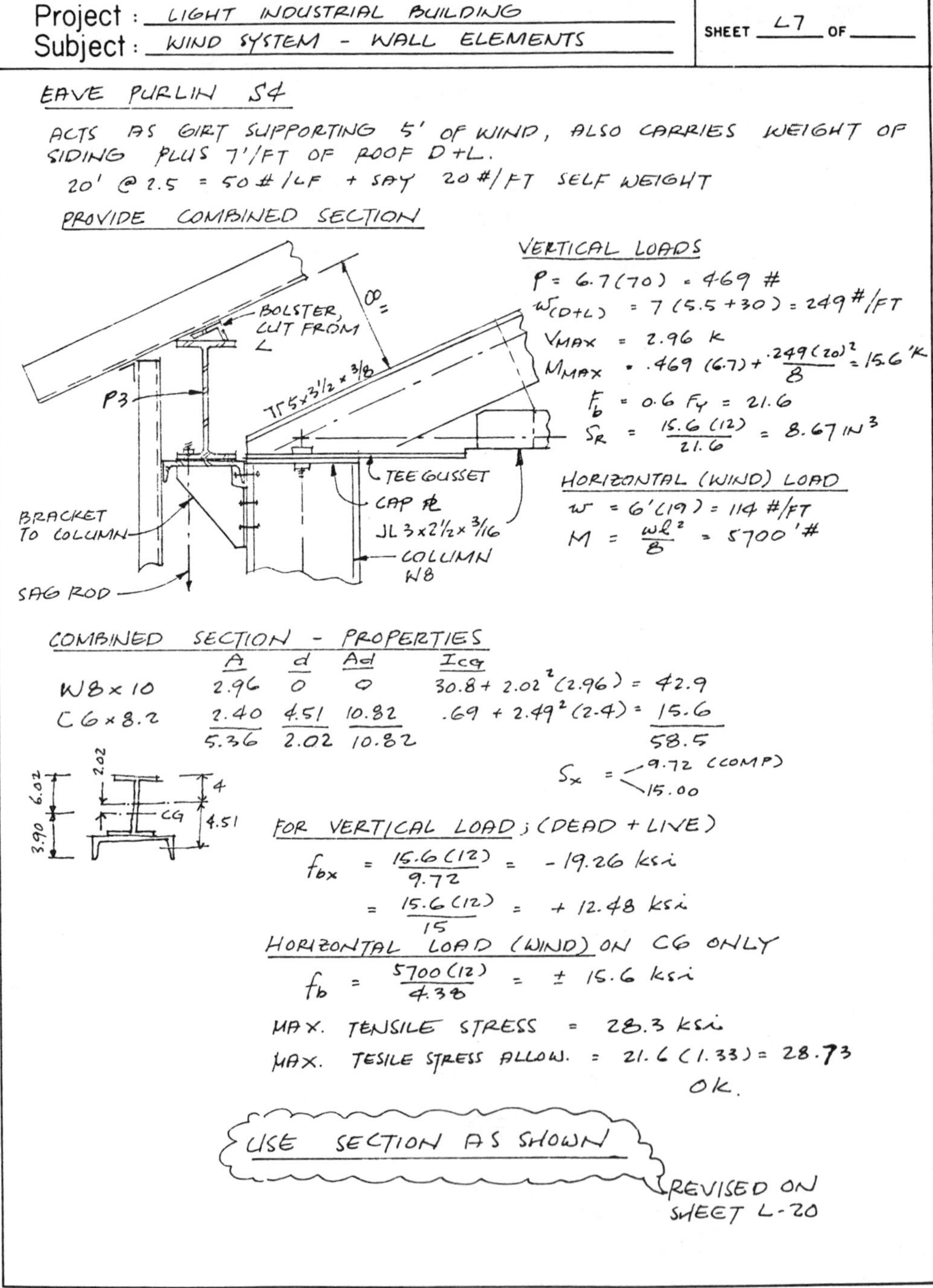
Project : LIGHT INDUSTRIAL BUILDING
Subject : WIND SYSTEM - WALL ELEMENTS
SHEET L7 OF
EAVE PURLIN S4
ACTS AS GIRT SUPPORTING 5' OF WIND, ALSO CARRIES WEIGHT OF SIDING PLUS 7'/FT OF ROOF D+L.
20' @ 2.5 = 50 #/LF + SAY 20 #/FT SELF WEIGHT
PROVIDE COMBINED SECTION
BOLSTER, CUT FROM L
8"
P3
TT 5x3½x3/8
TEE GUSSET
CAP PL
JL 3x2½x3/16
BRACKET TO COLUMN
COLUMN W8
SAG ROD
VERTICAL LOADS
P = 6.7(70) = 469 #
w(D+L) = 7 (5.5 + 30) = 249 #/FT
VMAX = 2.96 K
MMAX = .469 (6.7) + .249 (20)² / 8 = 15.6 'K
Fb = 0.6 Fy = 21.6
SR = 15.6 (12) / 21.6 = 8.67 IN³
HORIZONTAL (WIND) LOAD
w = 6'(19) = 114 #/FT
M = wl² / 8 = 5700 '#
COMBINED SECTION - PROPERTIES
A d Ad Icg
W8x10 2.96 0 0 30.8 + 2.02²(2.96) = 42.9
C6x8.2 2.40 4.51 10.82 .69 + 2.49²(2.4) = 15.6
5.36 2.02 10.82 58.5
2.02
6.02
3.90
4
4.51
CG
Sx = 9.72 (COMP) / 15.00
FOR VERTICAL LOAD; (DEAD + LIVE)
fbx = 15.6 (12) / 9.72 = -19.26 ksi
= 15.6 (12) / 15 = +12.48 ksi
HORIZONTAL LOAD (WIND) ON C6 ONLY
fb = 5700 (12) / 4.38 = ± 15.6 ksi
MAX. TENSILE STRESS = 28.3 ksi
MAX. TESILE STRESS ALLOW. = 21.6 (1.33) = 28.73
OK.
USE SECTION AS SHOWN
REVISED ON SHEET L-20

An alternative approach would be to use these members as columns and not provide a truss at all here, designing a special vertical load system for the end bays. The approach used here has the virtue that it more easily permits extension of the building, a common need during the life of an industrial building. Members WB1 are sized on sheet L5 from a simple beam calculation such as is discussed in Chapter 10. Loads Q come from the end reactions of the girts that frame into the wind beam.

Members WB2 and WB3 occupy space immediately alongside members of the vertical truss. In the case of diagonal L2–U4, there is a slight geometric interference problem. The $7\frac{5}{16}$-inch space occupied by the outstanding legs of the angle is larger than the available flat space on the web of the W against which it must fit. Taking advantage of the extra stiffness provided by attaching the double angle to the W, the designer changes to a smaller angle for that member in the end trusses.

17.5.5 Girts, Sidewalls: Sheets L6 and L7

As can be seen on sheet L6, the sidewall wind system is simpler than the end wall system. The two lower girts span farther than G1, but the wind pressure from sheet G8 is less. With the bigger total span, two sag rods are needed per bay. The calculation for wind pressure effects is the familiar one for a simple span uniformly loaded. The vertical load is taken as the weight of 10 square feet/foot of siding plus the beam itself as a continuous beam over two interior supports. For this common indeterminate case, an AISC Beam Diagram gives moment and shear coefficients. The typical girt selected for the end wall is checked for two locations on girt G4, where the interaction formula could become critical. Maximum utilization is almost all the 133% allowed for wind conditions.

The eave purlin, P3 (sheet L7), is like G3 in that it carries the sag rod pulls from the weight of the full height of siding. In addition, it supports about 7 square feet/foot of roof dead plus live load. A study of the geometry leads to the selection of a combination section, which is then checked for the combined vertical and horizontal load and seems satisfactory. However, a note indicates that later analysis for different conditions forced a revision.

17.5.6 Wind on Primary Frame, Force Analysis: Sheets L11 through L15

The analysis for wind effects on the primary frame is done separately, first for longitudinal wind and then for lateral wind. Since many members are common to both systems, member selection is delayed until both analyses are complete. The wind pressures used are taken from the first half of the table on sheet G8.

The longitudinal load system is described in the upper diagram on sheet L11 as a group of forces applied to the joints of the bracing system. Each force to the bracing system is then evaluated as the product of the pressure and its tributary area and the total checked by multiplying the pressure times the total area. The forces at ground level go directly to the foundation. After deducting the lower forces, the total force on sheet L11 compares fairly well with the longitudinal wind shear found earlier on sheet G8.

The force analysis is done on sheet L12, first for the roof level transverse truss, and then for the lower chord level truss and finally for the vertical bracing. Half of each total input force causes shear on the truss, the other half being transmitted to the far end through the longitudinal struts. (Recall the total system on sheet G5.) The method of analysis for either transverse truss is simple and straightforward since the truss chords are parallel. It can be compared to analogous concepts in beam analysis where it is recognized that shear is the source of change in moment. In these parallel chord trusses, the shear is resisted by the diagonals alone, the moment by the chords alone. Analysis starts with a diagram of panel shear and a geometric diagram showing the proportions of the right triangle with the diagonal as hypotenuse. All transverse shear must be carried by a diagonal whose member force is equal to the shear times the diagonal coefficient of the right triangle. Chord forces accumulate starting from the end supports. At each joint, the change in chord force from each diagonal at a joint is the shear times the coefficient of the right triangle in the direction of the chord.

The total of the two end reactions of the transverse trusses plus two locally applied loads are used to load the longitudinal bracing. The bracing panel experiences both shear and overturning moment. Recall the discussion of Fig. 17.2 and 17.3. Overturning effects cause uplift at one column, which must be resisted by gravity effects of dead load. UBC 2311(e) requires an excess of such ballasting load for safety.

Analysis of the transverse resisting system is done by a more complex version of the same method, which is first studied on sheet L13 with unit joint forces. To understand this analysis, the reader should have some familiarity with stiffness concepts for analysis of statically indeterminate systems and also moment distribution procedures. Again, with the help of the connecting members, this time the lower chords of the vertical trusses, each longitudinal truss must resist half of a total load of two per joint. The horizontal system consists of four trusses joined at their ends into a rectangular frame. It is statically indeterminate and has many members. An exact analysis would be long and would require initial assumption of unknown member sizes subject to later revision and reanalysis. An acceptable approximation is found through study of an analogous rigid frame with solid webs. The deflected shape shows that, due to the antisymmetrical loading on the frame, contraflexure (zero moment) exists at the two points on the frame on the centerline of the building. That makes it possible to proceed to analyze half of the frame. At the bulding centerline the half-frame is supported by longitudinal reactions supplied by the other half-frame and has no moment. The deflected shape shows two additional points of contraflexure along the longitudinal element, whose positions are unknown. The reactions cannot be found without an indeterminate analysis.

To solve the indeterminate system, stiffnesses must be assigned to the legs of the frame. In the analogous frame, member stiffnesses are proportional to L/I. In a truss, the equivalent stiffness can be based (approximately) on L/Ad^2 of the chords. If the

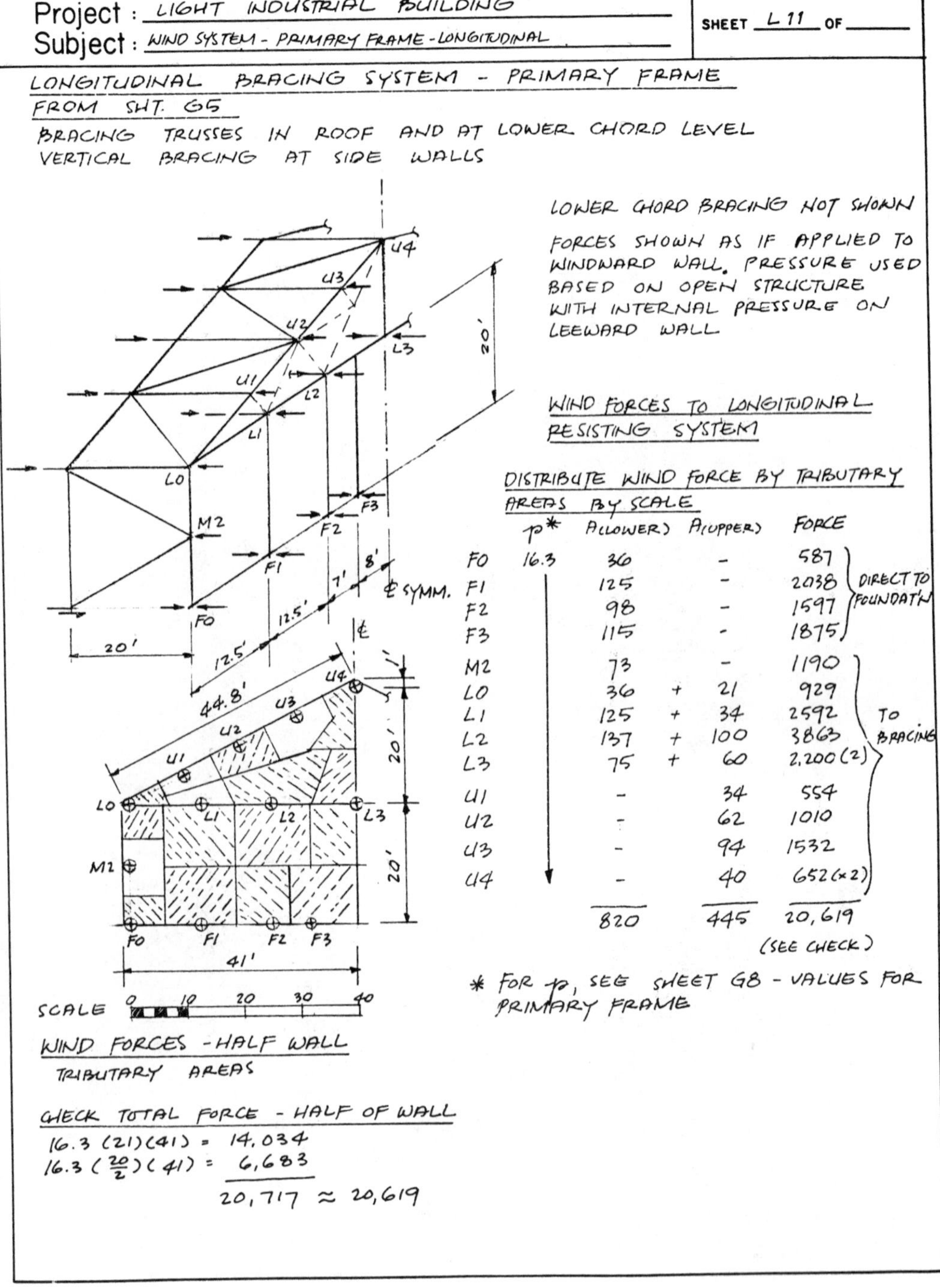

Project : LIGHT INDUSTRIAL BUILDING

Subject : WIND SYSTEM - PRIMARY FRAME - LONGITUDINAL

SHEET L11 OF

LONGITUDINAL BRACING SYSTEM - PRIMARY FRAME

FROM SHT. G5

BRACING TRUSSES IN ROOF AND AT LOWER CHORD LEVEL

VERTICAL BRACING AT SIDE WALLS

LOWER CHORD BRACING NOT SHOWN

FORCES SHOWN AS IF APPLIED TO WINDWARD WALL. PRESSURE USED BASED ON OPEN STRUCTURE WITH INTERNAL PRESSURE ON LEEWARD WALL

WIND FORCES - HALF WALL

TRIBUTARY AREAS

WIND FORCES TO LONGITUDINAL RESISTING SYSTEM

DISTRIBUTE WIND FORCE BY TRIBUTARY AREAS BY SCALE

	p*	A(LOWER)	A(UPPER)	FORCE	
F0	16.3	36	-	587	DIRECT TO FOUNDAT'N
F1	↓	125	-	2038	
F2		98	-	1597	
F3		115	-	1875	
M2		73	-	1190	TO BRACING
L0		36 +	21	929	
L1		125 +	34	2592	
L2		137 +	100	3863	
L3		75 +	60	2,200 (2)	
U1		-	34	554	
U2		-	62	1010	
U3		-	94	1532	
U4		-	40	652 (×2)	
		820	445	20,619	

(SEE CHECK)

* FOR p, SEE SHEET G8 - VALUES FOR PRIMARY FRAME

CHECK TOTAL FORCE - HALF OF WALL

16.3 (21)(41) = 14,034

$16.3\,(\frac{20}{2})(41) =$ 6,683

20,717 ≈ 20,619

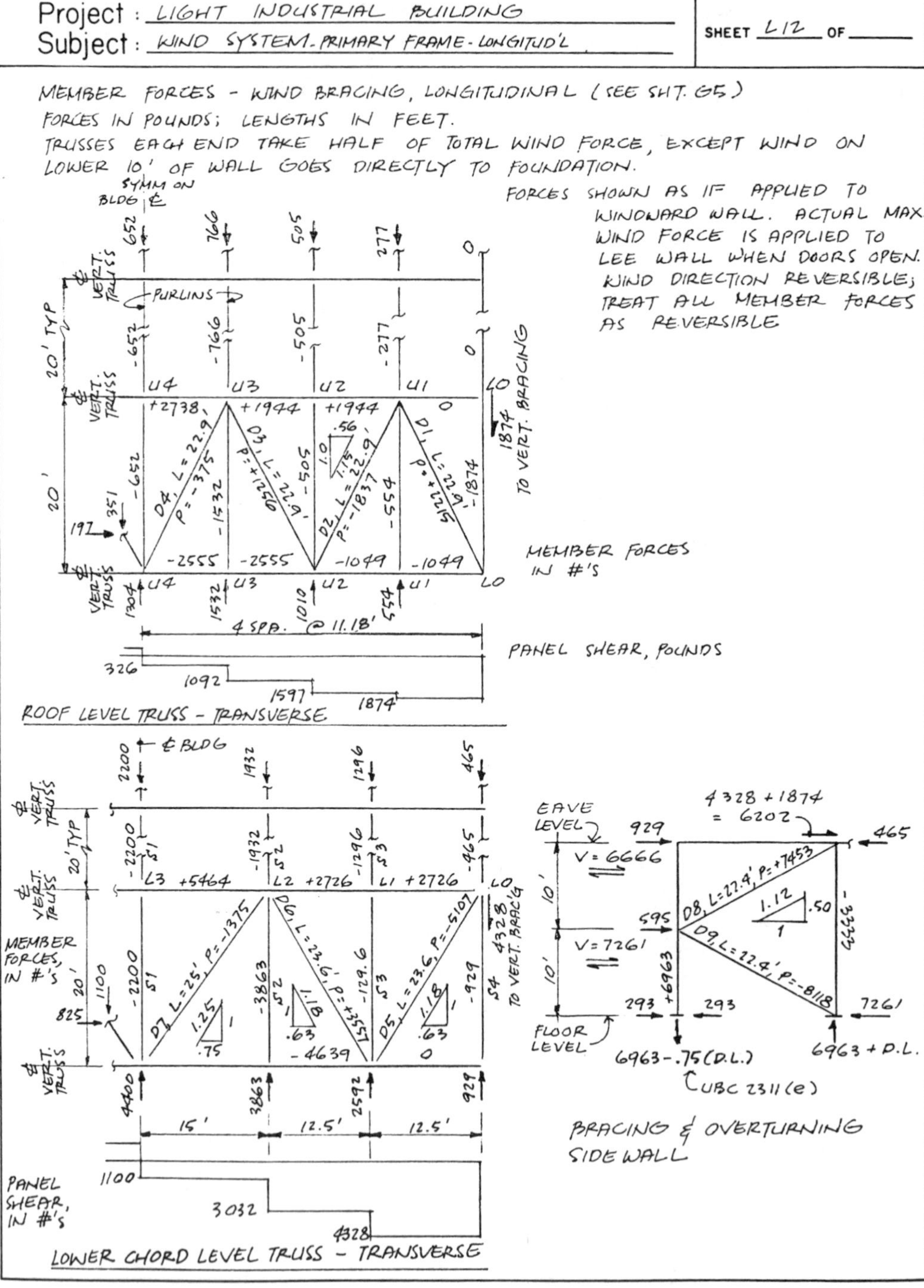
Project : LIGHT INDUSTRIAL BUILDING
Subject : WIND SYSTEM - PRIMARY FRAME - LONGITUD'L
SHEET L12 OF
MEMBER FORCES - WIND BRACING, LONGITUDINAL (SEE SHT. G5)
FORCES IN POUNDS; LENGTHS IN FEET.
TRUSSES EACH END TAKE HALF OF TOTAL WIND FORCE, EXCEPT WIND ON LOWER 10' OF WALL GOES DIRECTLY TO FOUNDATION.
FORCES SHOWN AS IF APPLIED TO WINDWARD WALL. ACTUAL MAX. WIND FORCE IS APPLIED TO LEE WALL WHEN DOORS OPEN. WIND DIRECTION REVERSIBLE; TREAT ALL MEMBER FORCES AS REVERSIBLE
MEMBER FORCES IN #'S
PANEL SHEAR, POUNDS
ROOF LEVEL TRUSS - TRANSVERSE
LOWER CHORD LEVEL TRUSS - TRANSVERSE
PANEL SHEAR, IN #'S
BRACING & OVERTURNING SIDE WALL
EAVE LEVEL
FLOOR LEVEL
V = 6666
V = 7261
4328 + 1874 = 6202
6963 - .75 (D.L.)
6963 + D.L.
UBC 2311 (e)

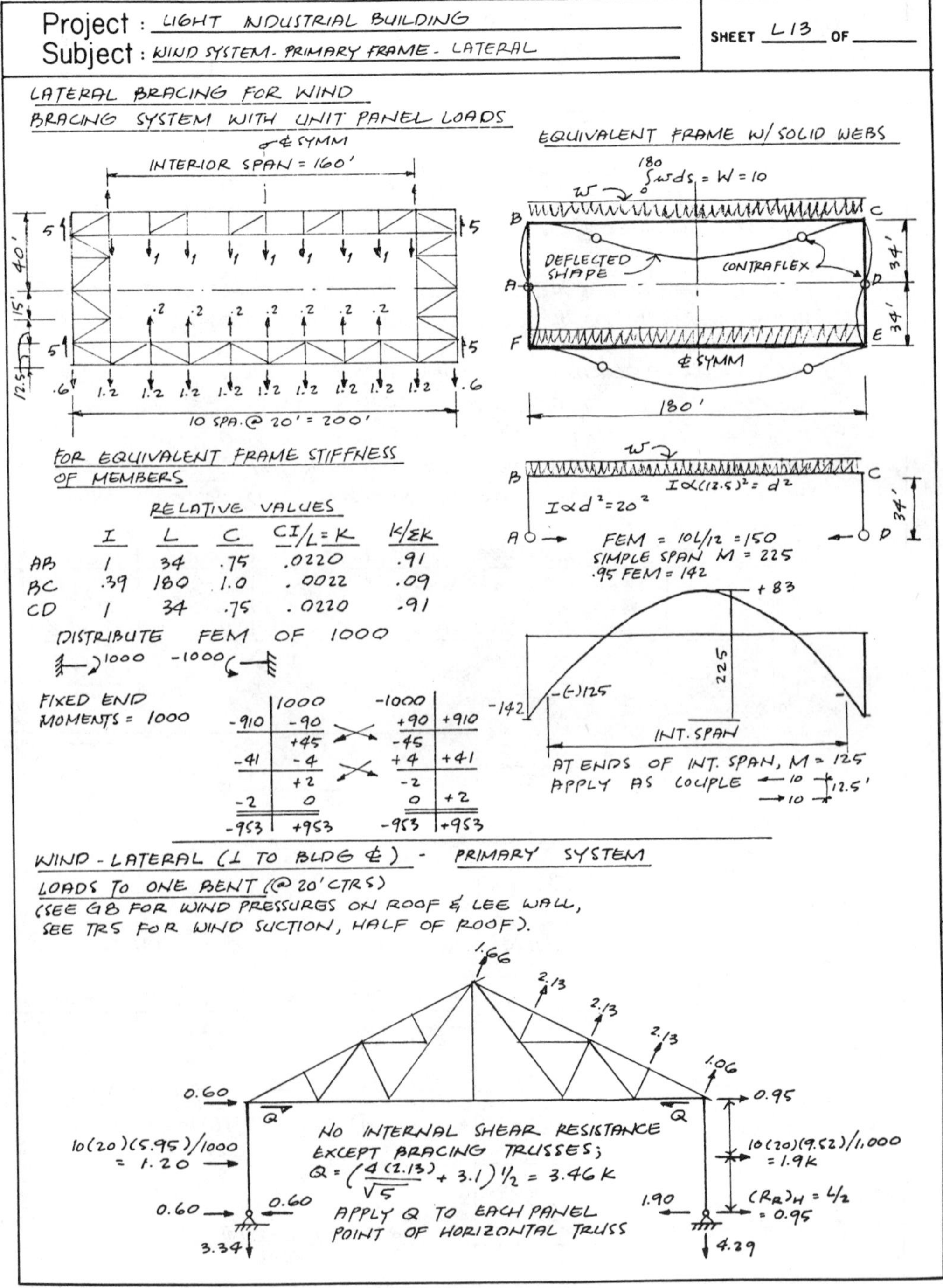
Project : LIGHT INDUSTRIAL BUILDING
Subject : WIND SYSTEM - PRIMARY FRAME - LATERAL
SHEET L13 OF
LATERAL BRACING FOR WIND
BRACING SYSTEM WITH UNIT PANEL LOADS
℄ SYMM
INTERIOR SPAN = 160'
10 SPA. @ 20' = 200'
EQUIVALENT FRAME W/ SOLID WEBS
∫ w ds = W = 10
DEFLECTED SHAPE
CONTRAFLEX
℄ SYMM
180'
FOR EQUIVALENT FRAME STIFFNESS OF MEMBERS
RELATIVE VALUES
I L C CI/L = K K/ΣK
AB 1 34 .75 .0220 .91
BC .39 180 1.0 .0022 .09
CD 1 34 .75 .0220 .91
DISTRIBUTE FEM OF 1000
FIXED END MOMENTS = 1000
-953 +953 -953 +953
I ∝ d² = 20²
I ∝ (12.5)² = d²
FEM = 10L/12 = 150
SIMPLE SPAN M = 225
.95 FEM = 142
+83
225
-142
(-)125
INT. SPAN
AT ENDS OF INT. SPAN, M = 125
APPLY AS COUPLE
WIND - LATERAL (⊥ TO BLDG ℄) - PRIMARY SYSTEM
LOADS TO ONE BENT (@ 20' CTRS)
(SEE G8 FOR WIND PRESSURES ON ROOF & LEE WALL,
SEE TR5 FOR WIND SUCTION, HALF OF ROOF).
NO INTERNAL SHEAR RESISTANCE
EXCEPT BRACING TRUSSES;
Q = (4(2.13)/√5 + 3.1) 1/2 = 3.46 K
APPLY Q TO EACH PANEL
POINT OF HORIZONTAL TRUSS
10(20)(5.95)/1000 = 1.20
10(20)(9.52)/1,000 = 1.9k
(R_R)_H = 1.9/2 = 0.95

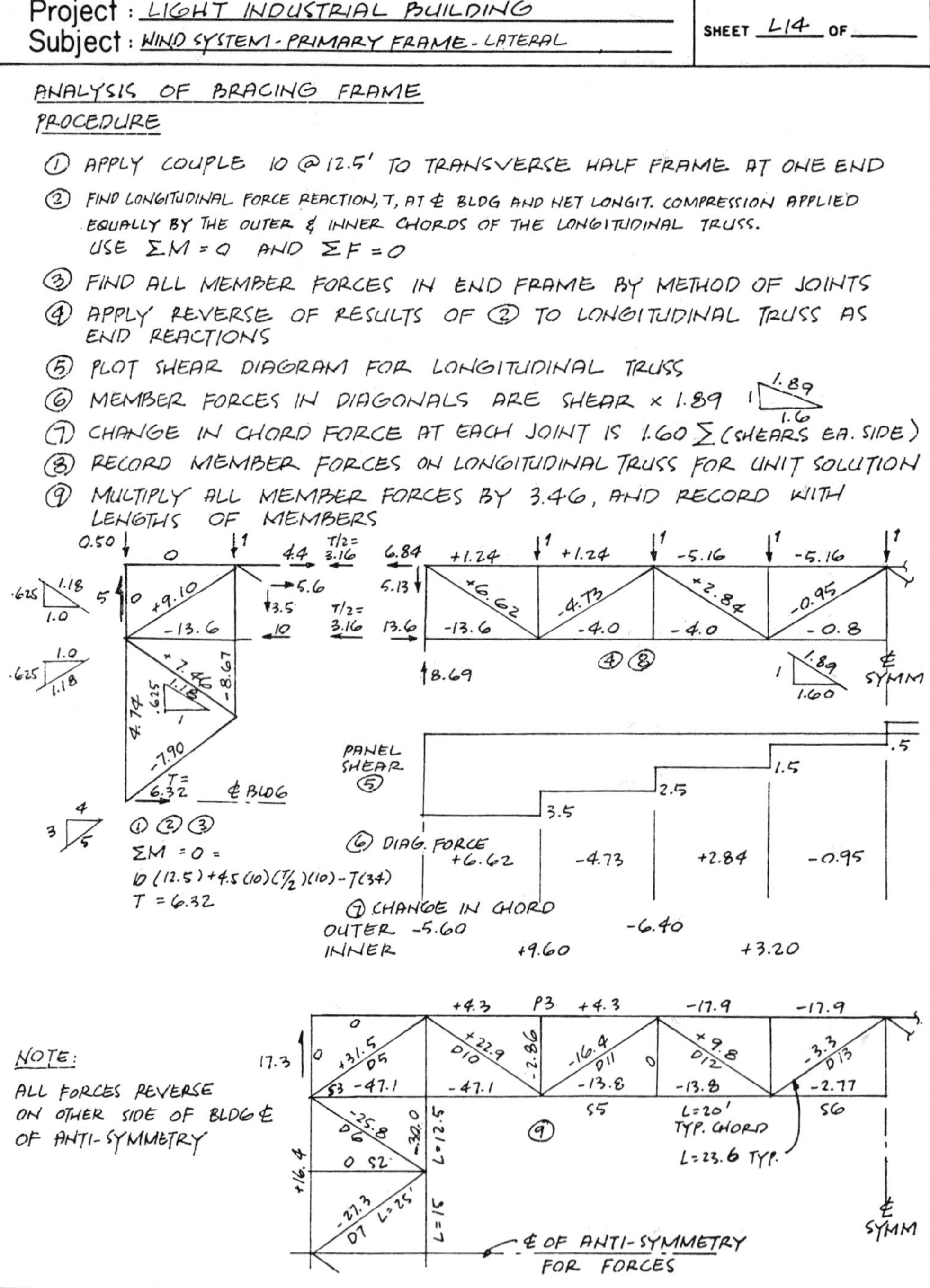

Project : LIGHT INDUSTRIAL BUILDING
Subject : WIND SYSTEM - PRIMARY FRAME - LATERAL
SHEET L14 OF
ANALYSIS OF BRACING FRAME
PROCEDURE
① APPLY COUPLE 10 @ 12.5' TO TRANSVERSE HALF FRAME AT ONE END
② FIND LONGITUDINAL FORCE REACTION, T, AT ℄ BLDG AND NET LONGIT. COMPRESSION APPLIED EQUALLY BY THE OUTER & INNER CHORDS OF THE LONGITUDINAL TRUSS.
USE ΣM = 0 AND ΣF = 0
③ FIND ALL MEMBER FORCES IN END FRAME BY METHOD OF JOINTS
④ APPLY REVERSE OF RESULTS OF ② TO LONGITUDINAL TRUSS AS END REACTIONS
⑤ PLOT SHEAR DIAGRAM FOR LONGITUDINAL TRUSS
⑥ MEMBER FORCES IN DIAGONALS ARE SHEAR × 1.89
⑦ CHANGE IN CHORD FORCE AT EACH JOINT IS 1.60 Σ (SHEARS EA. SIDE)
⑧ RECORD MEMBER FORCES ON LONGITUDINAL TRUSS FOR UNIT SOLUTION
⑨ MULTIPLY ALL MEMBER FORCES BY 3.46, AND RECORD WITH LENGTHS OF MEMBERS
T =
6.32
℄ BLDG
① ② ③
ΣM = 0 =
10 (12.5) + 4.5 (10)(T/2)(10) − T(34)
T = 6.32
④ ⑧
℄ SYMM
PANEL SHEAR ⑤
⑥ DIAG. FORCE
+6.62
-4.73
+2.84
-0.95
⑦ CHANGE IN CHORD
OUTER -5.60
-6.40
INNER
+9.60
+3.20
NOTE:
ALL FORCES REVERSE ON OTHER SIDE OF BLDG ℄ OF ANTI-SYMMETRY
⑨
L=20'
TYP. CHORD
L=23.6 TYP.
℄ OF ANTI-SYMMETRY FOR FORCES
℄ SYMM

Project : LIGHT INDUSTRIAL BUILDING

Subject : WIND SYSTEM - PRIMARY FRAME - LATERAL

SHEET L15 OF ______

END WALL BRACING AND OVERTURNING FROM LATERAL WIND

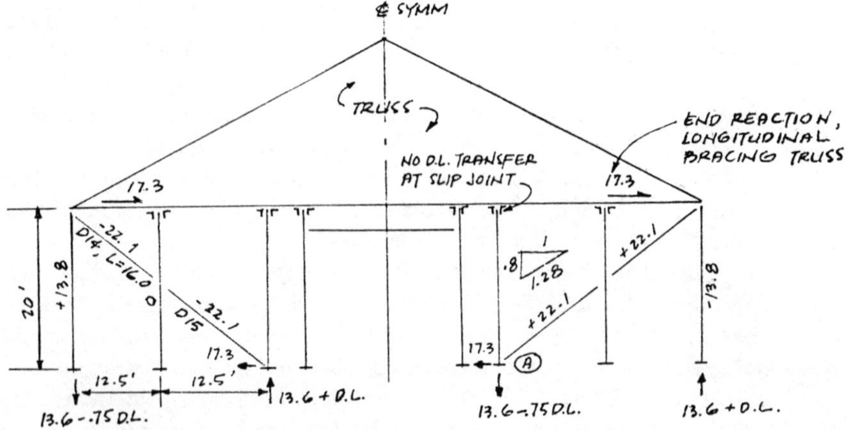

AT POINT (A), WE REQUIRE $\frac{13.6}{.75}$ = 18.1 KIPS OF DEAD LOAD TO BALLAST AGAINST UPLIFT

LOCAL WEIGHT OF END WALL IS VERY SMALL

TRUSS REACTION FROM D.L. GOES TO CORNER COLUMN

PROVIDE BALLAST IN WEIGHT OF FOUNDATION + SOIL OVER (UBC 2311(e))

CHECK EFFECT OF WIND BRACING SYSTEMS ON DESIGN OF CHORDS OF VERTICAL TRUSSES.

MEMBERS OF TRUSS WERE SELECTED ON SHT. TR7 FROM LOADS TABULATED ON SHT. TR6. ALL CONTROLLING LOADS WERE FROM DEAD + LIVE LOAD CASES. IF LOAD FROM WIND BRACING EFFECTS ARE ADDED TO A FULLY LOADED MEMBER, STRESSES MAY BE INCREASED 33% (AISC 1.5.6). IF MEMBER UTILIZATION < 100% FROM D+L, TOTAL WITH WIND MAY BE 133%.

TOP CHORD - MAX. BRACING LOAD = ± 2.7 KIPS (SHT. L12)
DESIGN LOAD FOR D+L = −67.5 (SHT. TR7) OK.

LOWER CHORD - MAX. BRACING LOAD = ± 30 k (SHT L12 & L14)
DESIGN LOAD FOR L0-L2 = D+L = 18.2 + 42.0
MEMBER IS ⅃L 3½ × 2½ × ¼ ; A_g = 2.88 ; KL/r = 138
FOR D+W ; P = +48.2, f_t = 16.74 ksi < 21.6 (1.33) OK
OR P = −11.8 ksi ; f_a = −4.10 ; F_a = −7.84 ; OK

NO CHANGE REQUIRED IN TRUSS DESIGN

areas of the two chords are approximately equal, the stiffnesses of the legs of the frame are proportional to L/d^2. These relative stiffnesses are used in the analogous frame. In this case, the end members are very much stiffer than the longitudinal member, the condition for almost complete end fixity for the longitudinal members.

A simple application of moment distribution procedure leads to a distributed corner moment and moment diagram for the equivalent frame. On that diagram the moment at the position of the end of the interior truss span can be calculated or, closely enough, scaled from the diagram. That moment, converted to a couple, can be applied to the now statically determinate, interior span and transverse truss. This will be done on sheet L14. But first Millman stops to evaluate the real wind forces, Q, to be applied at each panel point, the result of normal forces on the leeward roof and the upper half of the forces on both side walls.

The determinate analysis on sheet L14 is done according to a procedure set down there. It is first done for unit loads. The solution of the end truss provides an end reaction that for equilibrium must be balanced by a force interacting between the transverse and longitudinal trusses at the same location as the moment couple found previously. The rest of the unit analysis is done by methods discussed earlier. Resulting member forces are multiplied by Q and assigned to the members in the final diagram on sheet L14.

Before proceeding to the selection of the members specific to the horizontal force resisting frames, Millman stops to look at the effects of the wind force analysis on the chords of the vertical trusses that are parts of these frames. New forces are added to the previous truss analysis, but do not show any need to change member sizes. They do cause the lower *tension* chord to reverse to compression, further justifying an earlier decision to limit KL/r to ≤ 200.

17.5.7 Wind on Primary Frames, Member Selection: Sheets L16 through L20

To start the member selection process, for the wind frames, Millman assembles in a table on sheet L16 all the member loads from the previous analyses and the lengths of the members. All lengths are long and most forces low, indicating a need for long slender members, close to the permissible limits of KL/r. All the members are axially loaded, but bend from their own weight over long spans. They must be stiff enough and strong enough around their horizontal transverse axes to avoid high bending stress and excessive sag. Some general tentative decisions are made about the type of members to be selected and the procedure for checking them. The choice is between double angles and W shapes. The decision to use W8's in the horizontal frame is unusual and would probably not have been made if the controlling forces were lower. To consider the wisdom of this decision, the reader might examine the column tables in the Manual for compression capacity of double angles and W8's, keeping in mind that all capacities can be increased by one-third by AISC 1.5.6, but must be reduced by self-weight bending requirements. The bending problem is much worse for the

Project : LIGHT INDUSTRIAL BUILDING
Subject : WIND SYSTEM - PRIMARY FRAME - LONGITUDINAL & LATERAL

SHEET L16 OF ______

MEMBER	LENGTH (FT)	REQ'D r_{MIN}, IN.	LOCATION (LEVELS)	MAX (-) FORCE AXIAL (KIPS)	WIND DIRECTION FOR MAX. F
DIAGONALS					
D1 THRU D4	22.9	1.37	ROOF	2.2	LONGITUD'L
D5, D6	23.6	1.42	LOWER CHORD	31.5	LATERAL
D7	25.0	1.50	LOWER CHORD	27.3	LATERAL
D8, D9	22.4	1.34	SIDE WALL	8.1	LONGITUD'L
D10, D11	23.6	1.42	LOWER CHORD	22.9	LATERAL
D12, D13	23.6	1.42	LOWER CHORD	9.8	LATERAL
D14, D15	16.0	0.96	END WALLS	22.1	LATERAL
STRUTS					
PURLINS P1	20.0	1.20	ROOF	1.9	LONGITUD'L
P3	20.0	1.20	EAVE STRUT	17.9	LATERAL
S1, S2	20.0	1.20	LOWER CHORD	3.9	LONGITUD'L
S3	20.0	1.20	LOWER CHORD	47.1	LATERAL
S5	20.0	1.20	LOWER CHORD	13.8	LATERAL
S6	20.0	1.20	LOWER CHORD	28	LATERAL

USE W SHAPES, WEB VERTICAL
TRY ALL W8 FOR LOWER CHORD LEVEL
ROOF & SIDE WALL DIAGONALS, TRY ⫫L's
SELECT FROM COLUMN TABLES OR TABLE 3-36
CHECK SELF-WEIGHT BENDING - STRESS AND DEFLECTION

$M_{MAX} = \frac{w\ell^2 (12)}{8}$; w IN #/FT, M IN INCH KIPS

$\Delta = M_{MAX} (5/48)(L^2/EI)$; E = 29,000 ksi

CHECK $\frac{f_a}{F_a} + \frac{f_{bx}}{F_{bx}} \leq 1.33$ - AISC (1.6-2) AND 1.5.6

F_a FROM TABLE 3-36

F_b FOR LARGER OF AISC [(1.5-6a) OR (1.5-6b)] OR (1.5-7)

DEPENDING ON ℓ/r_T — FOR $\ell d/A_f$

Project : LIGHT INDUSTRIAL BUILDING
Subject : WIND SYSTEM - PRIMARY FRAME - LONGITUDINAL & LATERAL

SHEET L 17 OF ____

ROOF LEVEL

MEMBER SELECTION - WIND BRACING, LONGITUDINAL

PURLINS P1 AS STRUTS

W8 × 18 PREVIOUSLY SELECTED - CHECK FOR WIND EFFECT

$A = 5.26\ IN^2;\ r_x = 3.43;\ r_Y = 1.23;\ L = 20.0'$

$P_{MAX} = 1532\ \#$

$KL/r_Y = \frac{(1)(20)(12)}{1.23} = 195$ AT LOWER FLANGE

$\frac{195}{3} = 65$ AT UPPER FLANGE

USE A/2 AND BASE F_a ON $\frac{kl}{r} = 65$; $F_a = 16.94$

$f_a = 0.29$ ksi

$\frac{f_a}{F_a} = 0.02$. SINCE $.02 \ll 0.33$, PURLINS OK FOR WIND ACTION (F_a MAY BE INCREASED 33% FOR LOADS INCLUDING WIND)

DIAGONALS D1 THROUGH D4 (PLANE OF ROOF)

$P_{MAX} = 2.22$ KIPS

$L = 22.9'$ BUT USE 22' BETWEEN CONNECTIONS ; K = 1

$\rightarrow$ 2.22

$r_{MIN}\ @\ \frac{L}{200} = \frac{22.0(2)}{200} = 1.32$

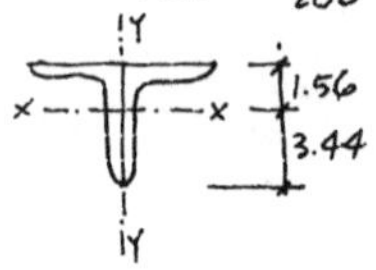

TRY $5 \times 3\frac{1}{2} \times \frac{1}{4}$* (0" BACK TO BACK)

(section sketch: Y–Y and X–X axes; 1.56, 3.44)

$A = 4.12$; WT = 14 #/FT

$I_x = 10.78$; $S_x = 3.14\left(\frac{3.44}{1.56}\right)$ / 3.14 = 1.42

(IN DBL ANGLE TABLES - MANUAL, PART I - EXTRAPOLATE FOR ⫪ 5×3½×¼)

$r_x = 1.62;\ r_Y = 1.33$

$\frac{KL}{r} = \frac{22.0(12)}{1.33} = 199 < 200$

$F_a = 3.77$ ksi (TABLE 3-36, PRIMARY)

$\frac{f_a}{F_a} = \frac{2.22}{4.12\,(3.77)} = 0.14 \ll 1.33$

(LIGHTER ⫫L's DO NOT CONFORM TO L/r REQUIREMENTS)

USE ⫪ $5 \times 3\frac{1}{2} \times \frac{1}{4}$

(IF NOT AVAILABLE, USE 5/16" THICKNESS)

* STEEL MANUAL SAYS L 5×3½×¼ MAY NOT BE AVAILABLE

CHECK SELF WEIGHT BENDING & SAG

FOR ⫪ $5 \times 3\frac{1}{2} \times \frac{1}{4}$

14 #/FT, 22.9'

$M = \frac{14(22.9)^2}{8} = 918\ '\#$

$f_{bx} = \frac{918\,(12)}{S_x} = -1590$ psi / $+3507$ psi

$F_{bx} = \frac{12{,}000}{ld/A_f}$ (AISC 1.5-7)

$= \frac{12{,}000\,(7)(.25)}{22.9(12)(5)} = 15.28$ ksi

$\frac{f_a}{F_a} + \frac{f_{bx}}{F_{bx}} = .14 + \frac{1.6}{15.28} = .14 + .10 = .24 < .33$ OK

SAG, $\Delta = \frac{5}{48}\frac{ML^2}{EI_x}$

$= \frac{5}{48}\ \frac{(918)(12)(22.9)^2(144)}{29 \times 10^6\ (10.78)}$

$= 0.28$ INCH $= L/991$

OK

Project : LIGHT INDUSTRIAL BUILDING
Subject : WIND SYSTEM - PRIMARY FRAME - LONGITUDINAL & LATERAL

SHEET L18 OF ____

D5, D6

$P = -31.5$, $L = 23.6'$
FROM COL. TABLE, TRY W8×24
$P_{ALLOW} = 35$
$\frac{P}{P_{ALL}} = \frac{f_a}{F_a} = 0.90$

BENDING - SELF WEIGHT
$M_{MAX} = 20.88$ IN.K.
$\Delta = 0.70''$ FOR $I = 82.8$
$\Delta/L = 1/405$
$r_T = 1.76$; $d/A_f = 3.05$
$\ell/r_T = 161$; $\ell d/A_f = 864$
$F_{bx} = 6.56$ ksi OR 13.89
$f_{bx} = \frac{M}{S} = \frac{20.88}{20.9} = 1$ ksi
$\frac{f_{bx}}{F_{bx}} = 0.07$

$\frac{f_a}{F_a} + \frac{f_{bx}}{F_{bx}} = 0.90 + 0.07 = 0.97 \ll 1.33$ OK. USE W8×24

D7

$P = -27.3$ k ; $L = 25.0$
TRY W8×24 FROM COL. TABLE
$P_{ALL} = 30$ K
$r_x = 3.42$; $r_y = 1.61$; $r_T = 1.76$; $d/A_f = 3.05$
$\frac{f_a}{F_a} = \frac{27.3}{30} = 0.91$

BENDING - W8×24
$I = 82.8$; $S_x = 20.9$
$M_{MAX} = 23.44$ IN.K
$\Delta = 0.88$
$\Delta/L = 1/341$; OK
$\ell/r_T = 171$; $\ell d/A_f = 1830$
$F_{bx} = 5.81$ OR 6.56 ksi
$f_{bx} = \frac{23.44}{20.9} = 1.12$ ksi
$\frac{f_{bx}}{F_{bx}} = 0.17$

$\frac{f_a}{F_a} + \frac{f_{bx}}{F_{bx}} = 0.91 + 0.17 = 1.08 \ll 1.33$
OK. USE W8×24

D8, D9

$P = -8.1$ K ; $L = 22.4'$
FROM COL. TABLES
TRY ⊤⊤ 5 × 3½ × 5/16 ; WT = 17.4 #/FT
$P_{ALLOW} = 22.2$; SAY OK
$\frac{f_a}{F_a} = 0.37$

1.59
3.41
5

BENDING, SELF-WEIGHT

10 / 20 / 22.4

$M_{MAX} = \frac{17.4(22.4)(20)/12}{1{,}000} \cdot \frac{}{8} = 11.7$ IN K

$\Delta = 11.7\left(\frac{5}{48}\right)\frac{(22.4)^2(144)}{29{,}000\,(13.2)} = 0.23''$

$\Delta/L = 1/1169$; $S = \begin{cases} 8.30 \\ 3.87 \end{cases}$

$f_{bx} = \frac{-11.7}{8.3} = -1.41$ ksi

$F_{bx} = \frac{12{,}000}{\ell d/A_f} = \frac{12{,}000(7(.375))}{22.4(12)(5)}$ (AISC 1.5-1)

$= 23.44$; USE $21.6 = .6F_y$

$\frac{f_{bx}}{F_{bx}} = 0.07$

$\frac{f_a}{F_a} + \frac{f_{bx}}{F_{bx}} = .37 + 0.07 = .44 < 1.33$ OK.

USE ⊤⊤ 5 × 3½ × 5/16 ; LONG LEGS VERTICAL

Project : LIGHT INDUSTRIAL BUILDING

Subject : WIND SYSTEM - PRIMARY FRAME - LONGITUDINAL & LATERAL

SHEET L19 OF ______

D10, D11 / D12, D13

$L = 23.6'$; $P = -18.9\,K$

USE SAME AS D5, D6; CONTROLLED BY $KL/r \le 200$

W8×24

D14, D15

$P = -22.1\,K$; $L = 16.0$

TRY ⫪4×3×1/4 (COL. TABLE)

$P_{ALL} = 16$

$\dfrac{fa}{Fa} = \dfrac{22.1}{16} = 1.38 > 1.33$ NG

TRY ⫪4×3×5/16

$\dfrac{fa}{Fa} = \dfrac{4}{5}(1.38) = 1.11$

BENDING - SELF WEIGHT ⫪4×3×1/4

$WT = 11.6\ \#/FT$

$I = 5.54$

$S_x = \begin{cases} 4.45 \\ 2.00 \end{cases}$

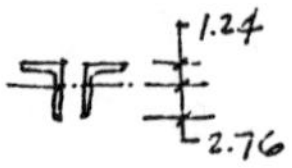

$M_{MAX} = \dfrac{11.6\,(16)}{1{,}000}\,\dfrac{(12.5)}{8}\,(12) = 3.48$ IN.K

$\Delta = 0.80$

$\Delta/L = 2400$

$f_{bx} = -0.78$ ksi (TOP)

$F_{bx} = \dfrac{12{,}000}{\ell d/A_f} = \dfrac{12000\,(6)(25)}{16\,(12)(4)} = 23.44$ ksi

USE 21.6 ksi = .6 F_y

$\dfrac{f_{bx}}{F_{bx}} = 0.04$; USE .05 FOR ⫪4×3×5/16

$\dfrac{fa}{Fa} + \dfrac{f_{bx}}{F_{bx}} = 1.11 + 0.05 = 1.16 < 1.33$ OK

USE ⫪4×3×5/16; LONG LEGS VERTICAL

STRUTS S1, S2, S5, S6

$3.9 \le P \le 13.8$; $L = 20'$

$r_{MIN} = 1.20$

TRY W8×18; $r_Y = 1.23$

$A = 5.26$

$I = 61.9$; $S_x = 15.2$

$KL/r = 195$; $F_a = 3.93$ ksi (PRIMARY)

ALLOW

$\dfrac{fa}{Fa} = 1.33 - \dfrac{f_{bx}}{F_{bx}} = 1.26$

THEN $fa \le 1.26\,(3.93) = 4.95$ ksi

ALLOW $P = 4.95\,A = 26.0$ KIPS

USE W8×18

BENDING - SELF WEIGHT

W8×18

$M_{MAX} = 10.8$ IN.K

$\Delta = 0.04''$

$r_T = 1.39$; $d/A_f = 4.70$

$\ell/r_T = 173$; $\ell d/A_f = 1128$

$F_{bx} = 5.68$ OR 10.64

$f_{bx} = \dfrac{10.8}{15.2} = 0.71$

$\dfrac{f_{bx}}{F_{bx}} = \dfrac{.71}{10.64} = 0.07$

STRUT S3

$P = -47.1$; $L = 20'$

USE FROM COL. TABLE: W8×24; $P_{ALL} = 48\,k$

$\dfrac{fa}{Fa} = 0.9 \ll 1.33$; OK FOR SELF WEIGHT BY COMPARISON TO ABOVE

Project : LIGHT INDUSTRIAL BUILDING
Subject : WIND SYSTEM - PRIMARY FRAME - LONGITUDINAL & LATERAL

SHEET L20 OF ____

P3 (EAVE STRUT)

FROM SHT. L6, L7, THIS IS A COMBINED SECTION, CARRYING THE WEIGHT OF SIDING PLUS RESISTING LATERAL WIND

W8×10
C6×8.2
A = 5.83

FROM SHT. L7
MAX. TENSILE STRESS = 23.3; ALLOW 23.7
ADD P = ± 17.9 FROM LAT'L BRACING
P/A = ± 2.54 ksi
⇒ USE W 8 × 18 AND C6×8.2

OVERTURNING FROM LONGITUDINAL WIND

EFFECT ON BRACED BAYS

END RF TRUSS
ROOF TRUSS
929
465 #
595
293
293
7261
6963 − .75(D.L.)
6963 + D.L. → LOAD TO FOOTING

FORCES FROM SHT. L7

6.96 K WIND < COLUMN LIVE LOAD
WIND O.T. NOT CONTROLLING

DEAD LOAD

40
20
40
10

ROOF TRUSS; ½ BAY = 5.5 K
(SEE SHT. TR4)

WALL, $\frac{1400 FT^2 \times 2.5}{1000}$ = 3.5

FOUNDATION
SAY ≥ 6 C.F. CONCR. @ 150 = 0.9
9.9

6.96 − .75 (9.9) = − 0.47 K

NO NET UPLIFT, BUT MAKE FOOTING HEAVIER TO SATISFY UBC 2311-(e)

angles than for the W's. After checking all the possible member weights, the reader might disagree with Millman's decision to use W8's. But we suggest that the decision to use all members of one type is economical from the point of view of fabrication costs.

Selections are done on sheets L17 through L20. Little additional comment is necessary regarding most selections. The capacity of purlins P1 to take on their additional function as struts is confirmed. The addition of the frame load to P3 causes the required member to become heavier than the one selected earlier.

17.6 TYPICAL COLUMNS: SHEET C1

Typical columns, at the ends of all vertical trusses, are all beam-columns. Bending arises from the reaction of the mid-height girt. The member is selected by methods of Chapter 11, using loads resulting from previous work.

17.7 SUMMARY: SHEETS S1 THROUGH S2

Two sheets at the end of the calculations summarize the work. The first of these, sheet S1, shows line diagrams of the whole structure. Several functions are served:

1. By drawing the entire system and entering member sizes, the designer can often find members that were not considered in the calculations.
2. The designer can see how the members relate to each other and judge whether the system appears both rational and economical.
3. He can make some ex post facto decisions to simplify the design, such as reducing the total number of different member sizes with similar functions.
4. He can simplify the work of the drafter who has to convert the results to contract drawings and his own work in checking the drafter's drawings.

The weight take-off on sheet S2 also has multiple uses.

1. To compare the results with the assumptions made for dead load in the early stages. If the actual structure is heavier than assumed, recalculation is necessary, possibly leading to redesign. If the structure is much lighter than assumed, the designer should judge whether or not savings might result from recalculation.
2. To judge the contributions of various parts of the system to total steel usage. Ideas may suggest themselves to save on individual members by a revised approach to the system. The value of pursuing such ideas depends to a great degree on how large the savings might be in the total design picture.
3. To be able to estimate and report to the client the probable cost of the steel frame.

Project: LIGHT INDUSTRIAL BUILDING Subject: MAIN COLUMNS	SHEET C1 OF ____

COLUMNS AT ENDS OF TRUSSES

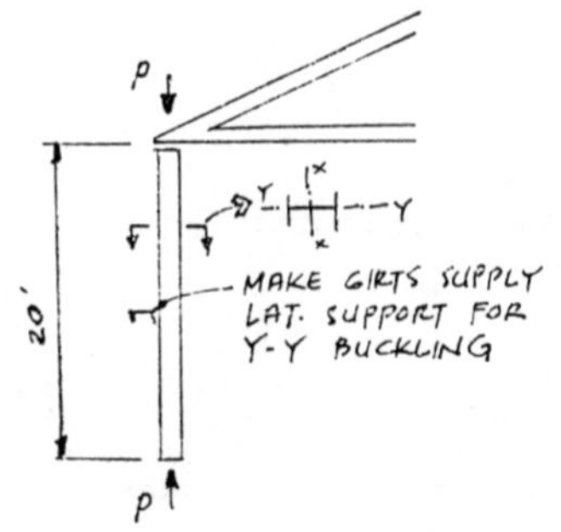

FROM TRUSS CALC'S

$P_{(D+L)} = 10.4 + 24 = 34.4$ kips

$L_x = 20(12) = 240"$; $L_Y = 10(12) = 120"$

BENDING (WIND LOAD)

$Q = 10'\frac{(19)(20)}{1000} = 3.8k$ (SHT. G8)

$M_x = \frac{QL}{4} = 19$ FT. K

TRY W8×18 AND RESTRAIN LATERALLY BY CONNECTING GIRTS TO BOTH FLANGES

$A = 5.26$; $S_x = 15.2$; $r_x = 3.43$; $r_Y = 1.23$; $r_T = 1.39$; $d/A_f = 4.70$

$\frac{kl}{r_x} = \frac{1(240)}{3.43} = 70$; $\frac{kL}{r_Y} = \frac{1(120)}{1.23} = 98$

$F_a = 13.23$; $F_{bx} = 17.8$ FOR $\frac{1.20}{1.39} = \frac{l}{r_T} = 86$ (AISC 1.5-6a)

OR $\underline{21.3}$ @ $12{,}000/ld/A_f$ (AISC 1.5-7)

$F'_{ex} = 30.48$ (AISC TABLE 9)

$f_a = 6.54$ ksi ; $f_a/F_a = 0.49$ OK

$f_{bx} = \frac{19(12)}{15.2} = 15.0$ ksi ; $\frac{f_{bx}}{F_{bx}} = 0.70$ OK

COMBINATION 4a (SHT G3)

$$\frac{f_a}{.6F_y} + \frac{f_{bx}}{2F_{bx}} = 0.30 + 0.35 = .65 \ll 1.33 \quad \text{(AISC 1.6-2)}$$

$$\frac{f_a}{F_a} + \frac{f_{bx}\ (1)}{2F_{bx}(1 - f_a/F'_{ex})} = 0.49 + 0.45 = 0.94 \ll 1.33 \quad \text{OK}$$

USE W8×18

(W8×15 IS LIGHTER, BUT UNFAVORABLE r_Y DUE TO NARROW FLANGE LEADS TO OVERSTRESS)

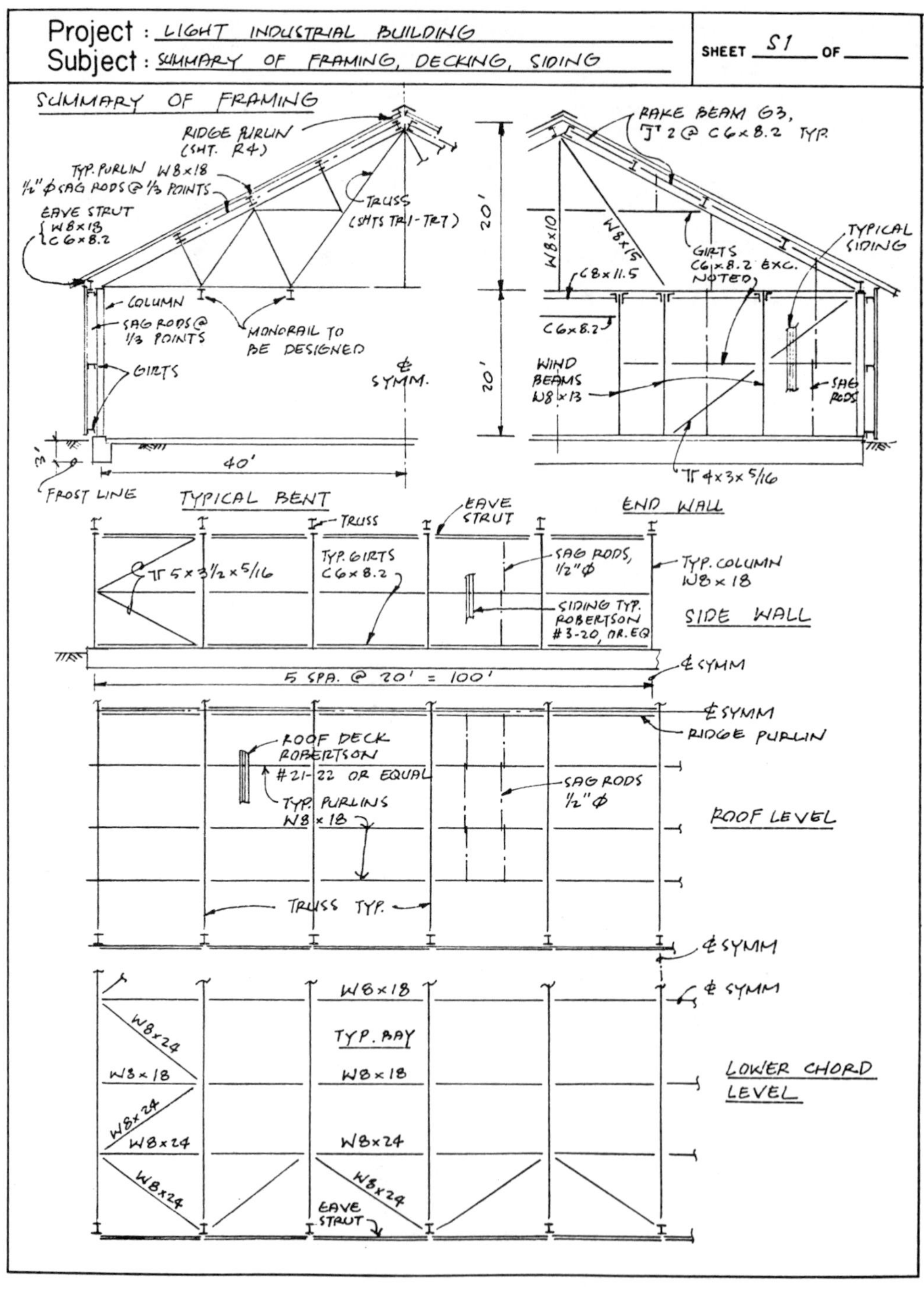
Project : LIGHT INDUSTRIAL BUILDING
Subject : SUMMARY OF FRAMING, DECKING, SIDING
SHEET S1 OF
SUMMARY OF FRAMING
RIDGE PURLIN (SHT. R4)
TYP. PURLIN W8×18
½" ϕ SAG RODS @ ⅓ POINTS
EAVE STRUT W8×18 C6×8.2
TRUSS (SHTS TR1-TR7)
COLUMN
SAG RODS @ ⅓ POINTS
MONORAIL TO BE DESIGNED
GIRTS
℄ SYMM.
FROST LINE
40'
TYPICAL BENT
RAKE BEAM G3,][2 @ C6×8.2 TYP.
20'
20'
W8×10
W8×15
C8×11.5
GIRTS C6×8.2 EXC. NOTED
TYPICAL SIDING
C6×8.2
WIND BEAMS W8×13
SAG RODS
ㄒㄒ 4×3×5/16
END WALL
TRUSS
EAVE STRUT
ㄒㄒ 5×3½×5/16
TYP. GIRTS C6×8.2
SAG RODS, ½" ϕ
TYP. COLUMN W8×18
SIDING TYP. ROBERTSON #3-20, OR. EQ
SIDE WALL
5 SPA. @ 20' = 100'
℄ SYMM
℄ SYMM
RIDGE PURLIN
ROOF DECK ROBERTSON #21-22 OR EQUAL
TYP. PURLINS W8×18
SAG RODS ½" ϕ
ROOF LEVEL
TRUSS TYP.
℄ SYMM
℄ SYMM
W8×18
W8×24
TYP. BAY
W8×18
W8×18
LOWER CHORD LEVEL
W8×24
W8×24
W8×24
W8×24
W8×24
EAVE STRUT

Project : LIGHT INDUSTRIAL BUILDING
Subject : SUMMARY OF FRAMING, DECKING, SIDING

SHEET S2 OF ______

TAKEOFF (WEIGHT ESTIMATE)

FRAMING MEMBER		SHAPE	LENGTH	WT/FT	WEIGHT, EACH, #	QTY	WEIGHT, KIPS	REFERENCE & REMARKS
TRUSSES			80'		3.910	11	43.00	TR1; DET. WT NOT INCL.
COLUMNS		W8×18	21'	18	378	22	8.32	C1
PURLINS	P1	W8×18	20'	18	360	48	17.28	R2
			21'	18	378	12	4.54	INCLUDES END OVERH'G
	P2	{2@C9×15	20'	42.5	850	8	6.80	R4
		+LACING	21'	42.5	892	2	1.18	WITH END OVERHANG
	P3	{W8×18	20'	26.2	524	16	8.38	L20
		C6×8.2	21'	26.2	550	4	2.20	
GIRTS, SIDE		[6×8.2	20'	8.2	164	40	6.56	L6
END		[6×8.2	500LF	8.2	-	-	4.10	L2
		[8×11.5	15'	11.5	172	2	0.34	L2
RAKE][		2@ C6×8.2	11.2'	16.4	184	16	2.94	L4
DIAGONALS								
ROOF		TT5×3½×¼	22.4	14.0	314	16	5.02	L12, L17
LOWER CHORD		W8×24	25.0	24	600	4	2.40	L14, L18
		W8×24	23.6	24	566	24	13.59	L14, L18, L19
SIDES		TT5×3½×5/16	22.4	14.0	314	8	2.51	L13, L18
ENDS		TT4×3×5/16	16.0	11.6	186	8	1.49	L14, L19
STRUTS	LOWER CHORD	W8×24	20.0	24	480	20	9.6	L19
		W8×18	20.0	18	360	40	14.4	L19
WIND BEAMS (END WALLS)		W8×13	288	13	-	-	3.74	L1, L5
		W8×15	25	15	375	4	1.50	L1, L5
SAG RODS		½"ϕ	55'(2)	.70	-	20	1.54	L4

SUBTOTAL WEIGHT: 162.03
ADD 8% + DETAIL 12.97
175 KIPS. = 87.5 TONS

WT. PER SQ.FT. BLDG. = 175,000/200(80) = 10.94 #/FT²

TAKEOFF - ROOF DECK & SIDING

ROOF DECK # 21-22 : 92' × 202' = 18504 FT² @ 2.1 #/FT² = 39,000 # *

SIDING # 3-20 : 2 @ 20' × 202'
2 @ 20' × 82'
2 @ 20/2 × 82

13,000 FT² @ 2.2 #/FT² = 28,600 #

TOTAL: 67,600 #

* DOES NOT INCLUDE WT. OF INSULATION

17.8 CLOSING REMARKS

The first of the case studies has also been the first opportunity we have to look at the design of a complete structural steel system rather than an individual member. The problem itself is a common one, similar to many faced daily by many structural designers. No great breakthroughs are made; nor did we present a brilliant new system that might result in the outmoding of all previous buildings with similar uses. Such things happen from time to time, but for most of us not often enough to make them the usual part of our experience. Few brand-new design systems are developed by engineers who have not first mastered the more prosaic design problems and seen how they are solved by systematically combining understanding of the total system with that of its parts.

If the reader has noted nothing else in the course of this calculation and discussion, he or she should note: (1) decisions on each subsystem were affected by decisions on all others; (2) for each stage and for each member there were always alternative answers from which the designer chose one, often by judgment only imperfectly backed by calculation; and (3) at times, the designer was impelled by his own opinions to provide more than was required by calculation, even if the latter conformed strictly to the Code.

There are a number of judgment calls made by the designer that may not be the wisest ones. We have noted several points at which different decisions might lead to a more satisfactory total design. Readers may have noticed other possibilities. They are invited to pursue some of those suggested after this chapter, or others that occur to them. Our purpose is not to present the only or best solution; in that we would certainly fail. If readers understand the process we presented, we have succeeded. If, with the help of this case study, they can improve the design, we and they have succeeded even more.

PROBLEMS

The following problems are suggested for term papers for students or design investigations for other readers. They vary in complexity and each may be approached at different levels of detail. The common thread is tracing of the implications of changed criteria or changed decisions on the design of this type of building. The work may stand on its own or the results may be compared to the design of the case study on one or more of a number of bases, some of which are interdependent:

- Cost
- Simplicity of erection

- Usefulness for adoption as a prototype system for a number of similar types of buildings of varying spans
- Operational utility for the client company

Depending on the time available and the depth to which this exercise is to be investigated, you may make "reasonable" assumptions about the various sources and magnitudes of costs, or you may research from available estimating data or local fabricators and erectors the costs of various types of materials, joints, erection problems, and so on. If costs of alternatives are to be compared to the case study design, you will first have to establish base-line costs for that design.

When investigating alternative design ideas, "winning" does not mean proving that the alternative is better; rather it is in finding out whether or not it constitutes an improvement and sometimes under what circumstances that is so.

It is often useful for at least two people to collaborate on one of these problems. Depending on what other work and reading has been done before attempting the problem, some of the suggested problems may require more research than you are prepared to do at this time. If so, do it later or choose an alternative closely related to the case study presentation.

17.1. Using the same criteria for design that were used in the case study, vary one or more of the system decisions that were made. Trace the implications of that change by appropriate calculations leading to a revised design. Make judgments as to whether the revised design is:

Clearly superior or clearly less good?

Superior under some conditions and less good under others, describing the conditions?

A stand-off?

Some things that could be considered as system revisions.

(a) Use rectangular tubes for all members.

(b) Provide basic lateral resistance at each bent by changing the form of truss. For example,

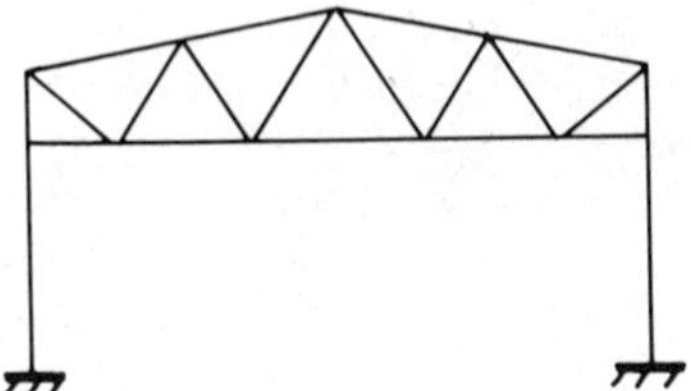

column continuous over lower chord

Any one of a number of web systems may be used. For example,

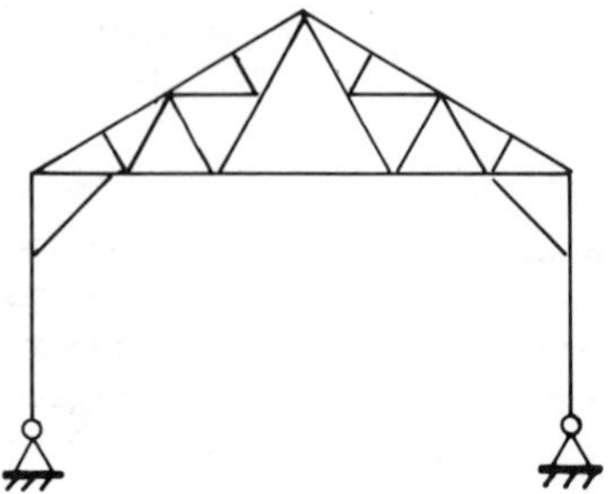

column continuous at knee brace.

In this approach, it is possible to (1) maintain the previously specified 20 ft. clearance under the lower chord; or (2) investigate how valid that requirement is, and how costly it is to retain it; advise the project engineer of your opinion on whether to allow the knee braces to intrude on the specified clearance envelope.

(c) Treat the roof and side walls as diaphragms, allowing a limited number of window and skylight openings in locations you control as part of your system arrangement. Don't attempt this alternative until you have studied light-gage steel more deeply than was covered in the case study. If you do it, consider what bracing members may still be necessary or useful.

(d) Using the same general arrangement, provide purlins at 5.59-ft centers along the roof instead of 11.18 ft. Revise the design to suit.

(e) How would the decision to treat the building as a closed structure affect the results of the analysis? This should be examined in terms of loading changes and design changes.

(f) Select double-angle members for all the horizontal bracing frame, with due regard to methods of connection and the stress effects of bending and bending stability.

17.2. If the operational requirement of a monorail hoist did not exist, how would you design this structure?

17.3. If the supporting soil was dense, frost resistant, and not subject to differential settlement, how would you change the design, if at all?

17.4. Assuming the mill price of ASTM A588 steel is 10% more than that of A36 steel, would it be advantageous to use A588 steel for all or part of the design? If so, what parts?

17.5. For the case study as presented, prepare, in the form of *D* series sheets, details of all or some of the principal connections.

17.6. If columns were permitted in the interior of the building, how would you design it?

18
Case Study: Preliminary Design of Tier Building

18.1 INTRODUCTION

The case study of a steel frame tier building is presented with a dual purpose in mind. It is representative of a large class of steel frame buildings of small to medium height often used for office, small commercial, and residential occupancies. Although we use a five-story building here, the original from which this study is derived has eight stories. The approach to the structural system can be extended beyond this building to a large class of structures.

Of equal importance, we use this case study to focus on the very important *preliminary design* stage during which many central decisions must be made with great bearing on the functional and financial success of the ultimate design. The function of the preliminary design stage in the design process is discussed in Chapter 4. We suggest that readers refer to that chapter at this time. The ideas are extended here.

An office building falls within the class of structure where, customarily, an architectural firm is chosen by an *owner* to provide a design. The architects retain the services of other design consultants to provide specific expertise in, among other things, the various branches of engineering: civil, structural, mechanical, electrical, and sometimes others. The architects retain overall design responsibility. The other professionals obligate themselves to provide sound, economical systems, within their own disciplines, which fulfill the design program worked out between architects and owner.

In this case study, the structural engineers, Coopchen Engineers, make a set of engineering calculations resulting in the choice of a structural steel framing system and tentative selection of the members to be used in that system. The calculations should be able to stand on their own as a self-contained document. However, we will, in parallel discussion, comment on the details that may not be obvious to students and relatively inexperienced engineers. The final two sheets of the calculations summarize the findings of the earlier work in a form which focuses on the questions that most concern the architect.

1. What is the arrangement of the structural frame?
2. Is it compatible with the space usage the architect has in mind?
3. What space is to be reserved for structural members?
4. How much will it cost to build the structural frame? What is included and what excluded in calculating the cost?

For purposes of this book, the calculations are limited to those required to evaluate the structural steel portion of the structural system. The total preliminary structural design would also include the concrete floor and roof slabs, the reinforced concrete shear walls, which are the primary system for resisting wind and seismic loads, and, possibly, the "nonstructural" precast concrete exterior wall elements. While supported on the steel frame, these must be prevented from interfering with the deflections that the structural system must experience in order to perform its function.

The calculations start with an initial definition by the architects of space arrangement, functional requirements, treatment of facade, and finishes. This has emerged from earlier conceptual studies, which presumably included concepts of supporting frame supplied by the same structural design team. At this point the architects have reexamined and modified the earlier design concepts and now present a building arrangement for a preliminary design.

The problem for each engineering discipline is to develop, in preliminary form, systems that will solve the problems of that discipline within the constraints set by others. This is an iterative process, coordinated through the achitects, with each discipline conceiving of solutions and adjusting as necessary to the requirements of others, sometimes through several cycles. A good integrated solution that is feasible for all offers optimum economy for the total project, not necessarily for each discipline; it achieves the objectives of the total design as defined by the architect and owner.

The preliminary design stage offers great opportunity for creative thinking. Solutions are fluid at this stage. Decisions made at this time will have profound effect, and in fact may determine whether or not the entire project is viable (i.e., whether project objectives can be achieved within the budget available). This phase is also crucial in the engineer's business management. Judicious use of the time of key staff in preliminaries will leave the bulk of the design budget still available for the stage of final design. Failure to pay sufficient attention to preliminary design could result in costly reworking at a later stage without additional compensation.

The level of detail should be sufficient to define the basic problem, but not more. At the completion of this phase, most of the basic parameters of design should be frozen, to be revised further only under extreme circumstances.

The following purposes are to be served:

1. Review structural feasibility of the proposed architectural layout.
   a. Consider and develop an approach to a structural frame.
   b. Choose structural materials and system.
   c. Establish space requirements for framing coordinated with other requirements of total design.
   d. Define loads*:
      (1) Loads required by Code, both vertical and horizontal.
      (2) Special loads, if any, required by design objectives.
   e. Define all reasonably expected combinations of loads and, where appropriate, adjustments of allowable stresses or safety factors applicable to each combination.
   f. Make calculations to establish preliminary sizes closely enough to develop total tonnage and preliminary structural budget.
   g. Compare structural budget to costs of similar earlier buildings as index to adequacy and economy of design approach.
2. Report to architect:
   a. Structural system selected.
   b. Reasons for choice.
   c. Space required.
   d. Potential cost.
3. Be prepared to respond to questions about:
   a. Feasibility of alternative systems.
   b. Consequences of special requirements by architect and other engineering disciplines.
4. After review by all concerned, and adjustment where necessary, *freeze* the basic design, particularly the elements that affect the work of the architect and mechanical and electrical engineers.

When all preliminary work is complete, when there is agreement on the basic design by all concerned, when authorized to proceed by the architect, then and then only, the structural engineer should proceed to the final design phase, which includes final calculations, development of details, preparation of drawings and specifications,

*See discussions, Chapter 7 and 16.

and so on. ASCE Manual 45[16] suggests "up to" (i.e., not more than) 40% of the design fee be assigned to the preliminary phase. Actually, it should take considerably less than 40% of the total person-hours of engineering time, but they are critical hours expended by the most qualified personnel.

Examination of the calculations will reveal several instances where Coopchen chose not to take advantage of the opportunity to reduce the magnitude and effect of applied loads based on the results of calculation. This does not indicate unwillingness to rework the numbers. Rather, it is a deliberate attempt to retain an extra degree of conservatism until, after careful final design calculations, the engineer is convinced that all special problems of design and cost have surfaced. Most experienced engineers, after having been burned by the results of early optimism, use one or another device to do the same thing. It is not difficult to "sell" reduced sizes in final design, as long as they don't complicate other people's work. Increased costs are much harder to explain.

For similar reasons, there are some cases of rounding off of numbers that may be distressing particularly to students weaned on electronic calculators. Again, the purpose of this exercise is to define a suitable system, the space it occupies, and the cost to produce it. Some uncertainty is inevitable, some later revision to be expected; extreme precision is unwarranted.

The order of the calculation sheets follows only roughly the sequence in which the actual calculations were made. The reader is referred to Chapter 16 for some thoughts on ordering of calculations.

In various places throughout the calculations, notations are found regarding matters to be considered in final design. Final design may be done much later in time than preliminary and very likely by another designer, possibly junior to the one who did the preliminary. It should be possible to follow, through the calculations, the thinking of the designer. These notations can be very helpful if available to the ultimate designer.

18.2 TITLE, CONTENTS, GENERAL ARRANGEMENT

For reasons discussed in Chapter 16, the opening sheet of the calculations is a Table of Contents and the next five sheets record general design information that will be used as reference in the later sheets. Sheet 2 lists the purpose, participants, and general design constraints. It is noted that necessary soils information is not yet known, a reminder, for final design stage, to reexamine the assumptions regarding frame performance that depend on soil characteristics and foundation design. The location of the building in earthquake probability zone 4 triggers a series of *Uniform Building Code** requirements for the design.

*See Appendix A.

Project : PRELIMINARY DESIGN - TIER BUILDING Subject : CALCULATIONS - CONTENTS	SHEET 1 OF 24

PRELIMINARY CALCULATIONS FOR 5-STORY OFFICE BUILDING

CONTENTS

Project : PRELIMINARY DESIGN - TIER BUILDING Subject : GENERAL INFORMATION	SHEET 2 OF 24

CALCULATIONS FOR PRELIMINARY STUDY - PRIMARY STRUCTURAL FRAME

5 STORY OFFICE BUILDING
311 Q STREET
TIMBUCTOO, CA.*

* ARCHITECT: DESIGN ASSOCIATES
STRUC. ENGR: COOP CHEN ENGINEERS
MECH. ENGR: ENVIRONMENTAL SYSTEMS CO.
ELEC. ENGR: " " "

SOILS & GEOLOGICAL INVESTIGATION TO BE DONE BY EARTHLAB INC.*
ADVANCE INFO INDICATES SPREAD FOOTINGS WILL BE FEASIBLE

GOVERNING CODES :
- UNIFORM BUILDING CODE (UBC) - 1982
- STRUCT. STEEL PER AISC SPEC. - 1978 (BUT NOT TO CONFLICT WITH UBC)
- SEISMIC ZONE 4 PER UBC ZONE MAP.

FIRE REQUIREMENTS (PER ARCHITECT - BASED ON ZONE USE & OCCUPIED AREA)

- TYPE II BLDG (FIRERESISTIVE) - DETAILS, CH. 19 - UBC - ALLOWS STEEL, CONCR, MASONRY

FIRE PROTECTION (TABLE 17A - UBC)

EXT. WALLS - 4 HRS EXC. ALLOW REDUCTION AT STREET & BY DISTANCE FROM PROPERTY LINE (UBC #1903a)

STRUCT. FRAME - 2 HRS
FLOORS - 2 HRS
ROOF - 1 HR

ARCHITECTS' SCHEMATIC DRAWINGS SET DESIRED COLUMN ARRANGEMENT, STORY HEIGHTS. NEED CHECKS (STRUCT'AL) ON:

FRAMING MATERIAL
TYPE OF VERTICAL & LATERAL RESISTING SYSTEM
SIZE OF COLUMNS & COLUMN FIREPROOFING
DEPTH REQUIREMENTS FOR BEAMS, GIRDERS, & STRUCTURAL FLRS
PRELIMINARY BUDGET - STRUCTURAL FRAME

* NAMES SHOWN ARE FICTITIOUS ENTITIES

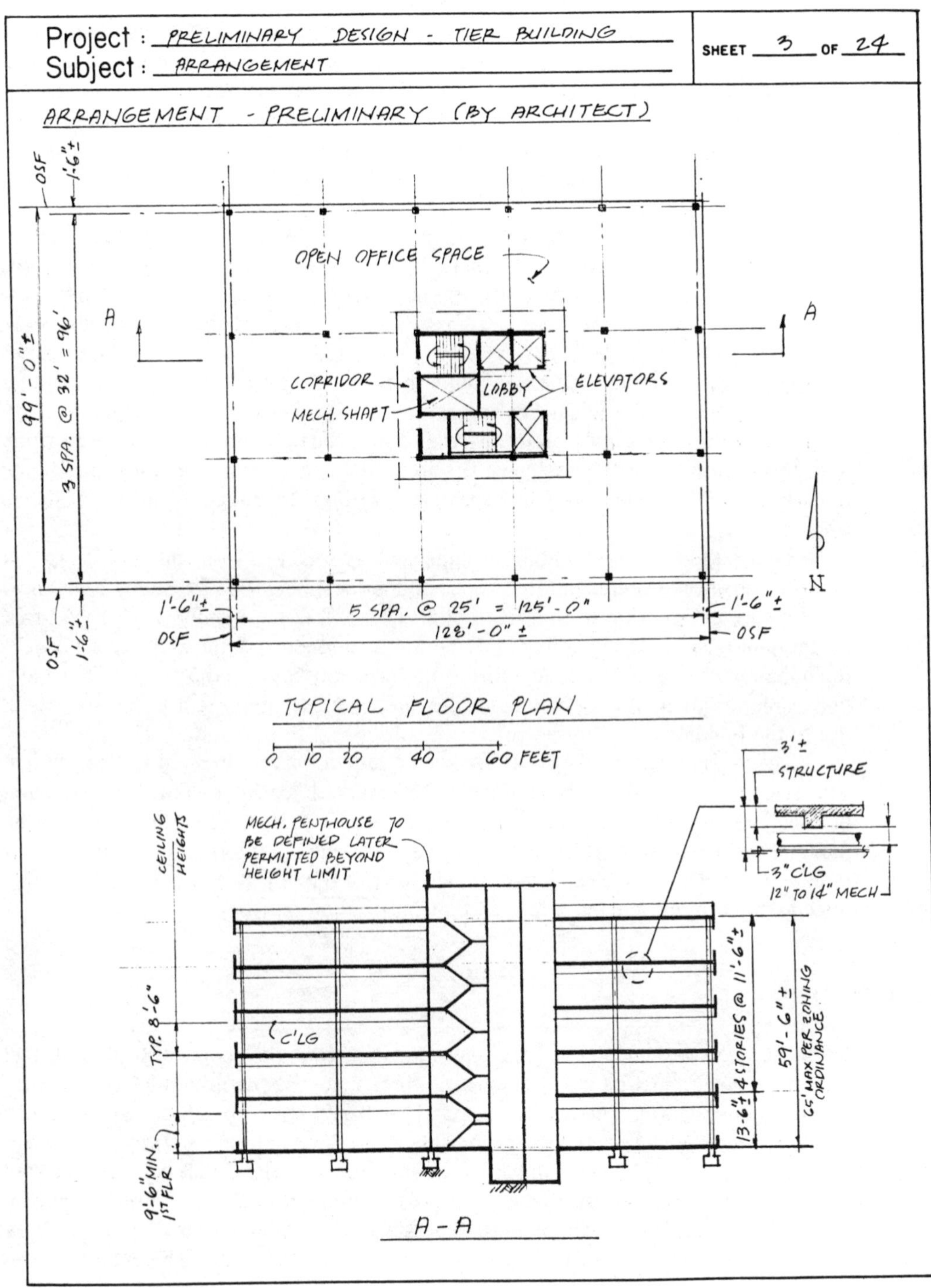
Project : PRELIMINARY DESIGN - TIER BUILDING
Subject : ARRANGEMENT
SHEET 3 OF 24
ARRANGEMENT - PRELIMINARY (BY ARCHITECT)
OPEN OFFICE SPACE
CORRIDOR
MECH. SHAFT
LOBBY
ELEVATORS
A
A
OSF
1'-6"±
99'-0"±
3 SPA. @ 32' = 96'
1'-6"±
OSF
5 SPA. @ 25' = 125'-0"
128'-0"±
N
TYPICAL FLOOR PLAN
0 10 20 40 60 FEET
3'±
STRUCTURE
3" C'LG
12" TO 14" MECH
MECH. PENTHOUSE TO BE DEFINED LATER PERMITTED BEYOND HEIGHT LIMIT
CEILING HEIGHTS
TYP. 8'-6"
C'LG
9'-6" MIN. 1ST FLR
13-6"± 4 STORIES @ 11'-6"±
59'-6"±
65' MAX PER ZONING ORDINANCE
A - A

Fire protection requirements are crucial to structural steel design, since steel loses strength rapidly when heated. The requirements are usually stated by the architects, but must be understood and considered by the engineers in conceiving the structural frame. The designation by "hours" represents the time during which the protected steel can be exposed to fire of "standard" intensity before it effectively loses its ability to do its structural job. Protection is provided by other materials that insulate the steel from the heat of the fire.

Type II is a designation used in the UBC to cover a class of fire-resistive building. Fireproof buildings are not really achievable. We actually seek various levels of fire-resistive construction. Municipalities often require fire-resistive construction in downtown commercial areas. Fire-resistance requirements in hours are found in tables in the UBC, based on type of building, fire zone, area, and type of occupancy. They define the time in hours that it would take for a structural or other part to become unsafe when heat representing a "standard" fire is applied to it. Similarly, for fire separation partitions, they define the time during which the partition is expected to prevent the spread of fire from one space to an adjoining one. Some materials such as steel require special coverings to function in fire-resistive systems.

Information on the building arrangement as received from the architects and recorded on sheet 3 points out to Coopchen that local height limits, set by ordinance, require careful attention to the depth of structural framing members. If the 65-foot height limit is exceeded by a five-story building as conceived, the architects will face the unhappy choice of losing an entire story, probably making the project financially unfeasible, or giving up the ceiling heights they consider important to the successful use of the building.

The mechanical penthouse is noted as a functional requirement of final design. The size, layout, and construction will be reserved for later. This is a common problem, adding to the structural engineer's uncertainty. It stems in this case from the fact that the mechanical engineers do not yet have enough information to define their requirements. The structural engineers will have to find a way of working around this uncertainty without leaving a hole in the construction budget.

18.3 MATERIALS

On sheet 4, the designers record the reasoning that led to the choice of structural steel framing. Choice must be made among the three materials permitted for type II fire-resistive construction. The choice ultimately is based not so much on calculation as on judgment. The engineers feel responsible to produce not only a structure that can be made to conform, by the numbers, to requirements of the Code, but one that they can believe in and defend as able to give satisfactory performance. Their reasoning on sheet 4 is strongly influenced by their experience with the materials in previous work and "bias" based on observation as to what degree of reliability can be expected from

Project : PRELIMINARY DESIGN - TIER BUILDING Subject : MATERIALS	SHEET 4 OF 24

CHOICE OF MATERIALS - BASIC FRAME

CHOICES ARE LIMITED BY FIRE REQUIREMENTS TO STEEL, CONCRETE OR MASONRY. IF MASONRY, SEISMIC REQUIREMENTS DEMAND REINFORCING.

ARCHITECT PLANS HIGHLY SCULPTURED WINDOW OPENINGS IN FACADE. THIS SHOULD BE DONE IN PRECAST CONCRETE PANELS; - OK FOR FACING, BUT CONNECTION DETAILS PROBABLY UNSUITABLE FOR SHEAR WALLS.
CENTRAL ELEVATOR - STAIR - MECHANICAL CORE CAN BE BOUNDED BY SHEAR WALLS WITHOUT DISTURBING BUILDING FUNCTIONS.

UNIT MASONRY - EXPERIENCE OF THIS OFFICE MAKES IT RELUCTANT TO USE REINF. MASONRY IN LARGE STRUCTURES. IN THIS SIZE, MAYBE MARGINAL, BUT ARCHITECT'S PLANS FOR WINDOW OPENINGS PROBABLY WON'T WORK WITH UNIT MASONRY. REJECT UNLESS FACADE DESIGN IS RECONSIDERED.

EITHER CONCRETE OR STEEL OK FOR VERTICAL SYSTEM. SOME COST EDGE GOES TO CONCRETE FOR THIS SIZE BUILDING.
HOWEVER:
1. 32' SPAN CAN PROBABLY BE HANDLED WITH LESS DEPTH IN STEEL (CAN BE CHECKED);
2. USING A DUCTILE MOMENT RESISTING FRAME AS BACK-UP FOR A PRIMARY SHEAR WALL SYSTEM, PERMITS REDUCTION OF "K" (SEE UBC FORMULA 12-1, PAR. 2312 d, AND TABLE 23-I) FROM 1.33 TO 0.80, A 40% REDUCTION IN DESIGN LATERAL FORCE FOR EARTHQUAKE. THIS OFFICE BELIEVES DUCTILE FRAMES ADD TO SAFETY. DUCTILE MOMENT FRAME PROBABLY EASIER TO DETAIL IN STEEL, ALSO LIGHTER, LEADING TO LESS SEISMIC LOAD;
3. A STEEL FRAME CAN BE ERECTED QUICKLY ON SITE, OPENING A LOT OF WORK EARLY FOR MECHANICAL AND FINISHING TRADES; REDUCING TOTAL TIME OF CONSTRUCTION.
4. THIS OFFICE HAS MORE CONFIDENCE IN THE DUCTILE PERFORMANCE OF STRUCTURAL STEEL THAN THAT OF CONCRETE.

∴ RECOMMEND STRUCTURAL STEEL FRAME

DEVELOP PRELIMINARY CALCULATIONS FOR STRUCTURAL STEEL. IF RESULTS LOOK GOOD, RECOMMEND TO ARCHITECT.

a structure designed in each material. Such judgments vary among qualified engineers.

Unit masonry made of reinforced brick or concrete block is becoming increasingly popular for buildings of this size, partly based on favorable costs. However, many engineers feel that the lower costs are achieved in part by poor quality control in the field. They avoid that material unless they can be assured of quality control based on careful materials testing and close field inspection. If Coopchen were more receptive to unit masonry, they might point out the potential cost saving to the architects and see if they want to reconsider the use of precast concrete wall panels, which are planned for the exterior walls. Precast concrete walls, in vogue in recent years in office construction, offer interesting opportunities to architects in the treatment of the facade.

The real choice in the opinion of these designers is between reinforced concrete and structural steel, either of which can lead to an acceptable design. The cost edge would probably go to reinforced concrete at this time. (Relative costs are influenced by changeable market considerations.) Many engineers would make that choice. But the designers list secondary, but not trivial, arguments to buttress their underlying judgment which is in sheet 4. They prefer to have a ductile moment-resisting frame to help resist extreme seismic events, because of their convictions about structurally safe systems. Their convictions do not arise from the reduced K factor permitted by UBC (argument 2); but the reduced K factor helps to make the system they believe in economically competitive.

Ductile moment frames grow naturally from the use of steel, which is an elastoplastic material capable of absorbing relatively large amounts of energy after initial yield without loss of capacity to support loads. Use of reinforced concrete in ductile moment frames seems, at first blush, alien to the brittle nature of concrete. In recent years, analysis, testing, and long argument have led to a new class of reinforced concrete in which the concrete is sufficiently confined by surrounding reinforcing steel so that it exhibits ductile performance. Many now accept the validity of such design. Some engineers are still not convinced. Some others are uncertain and prefer to stay with the material they feel has proven itself to the point where they can sleep comfortably when they use it.

Coopchen's engineers prefer structural steel and will proceed on that basis, but are prepared to reexamine the choice if major cost problems surface in the preliminary design phase.*

18.4 LOADS

Sheets 5 and 6 assemble the loads to be used in analyzing the structural system, dividing them into *vertical* and *lateral* loads. The vertical loads are themselves divided into dead loads, representing the weight of materials permanently in place, and live

*We are not intending to "sell" the choice of structural steel. However, it is a common enough choice so that it could legitimately be made here. If this were a book on reinforced concrete, we feel an adequate design could be presented using that material.

Project : PRELIMINARY DESIGN - TIER BUILDING
Subject : VERTICAL LOADS

SHEET 5 OF 24

LOAD TABLES - TYPICAL FLOOR* - (#/ft²) - VERTICAL LOADS

	TO SLAB	TO BEAMS	TO GIRDERS	TO COLUMNS
DEAD LOAD (① FINISH ②)				
SLAB ① OR ② 6"± (CONC. METAL DECK)	85	85	85	85
BEAMS, SAY 64#/L.F. @ 8' CTRS	-	8	8	8
GIRDERS, SAY 100#/L.F. @ 25' CTRS	-	-	4	4
COLUMNS (25' × 32' = 800 ф TRIB.) 125 # × 12' HT/800 ; USE	-	-	-	2
FIREPROOFING (CHECK SYSTEM)	-	2	4	5
MECHANICAL, ALLOW	-	5	5	5
CEILING	5	5	5	5
SUBTOTAL	90	105	111	114
ADD PARTITIONS : SAY EACH WAY AT 15' CTRS AVERAGE & 8.5' HT. $2\left(8.5' \times \frac{10\ \#/ф}{15'}\right) = 12\ \#/ф$ FLOOR				
USE LOAD PER UBC 2304(d)	20	20	20	20
TOTAL DEAD LOAD	110	125	131	134
LIVE LOAD (TABLE 23-A, UBC WITH REDUCTIONS PER PAR. 2306)				
OFFICE AREAS 50#/ф OR 2000#				
PUBLIC EXITS 100 #/ф USE FOR PRELIM. DESIGN	100	75	50	50
TOTAL DEAD + LIVE	210	200	181	184

* AT THIS STAGE CONSIDER ROOF SAME AS TYPICAL FLOOR. REDUCE LOADS & LIGHTEN ROOF FRAMING IN FINAL DESIGN. HOWEVER, SAVINGS WILL BE OFFSET BY COST OF PENTHOUSE.

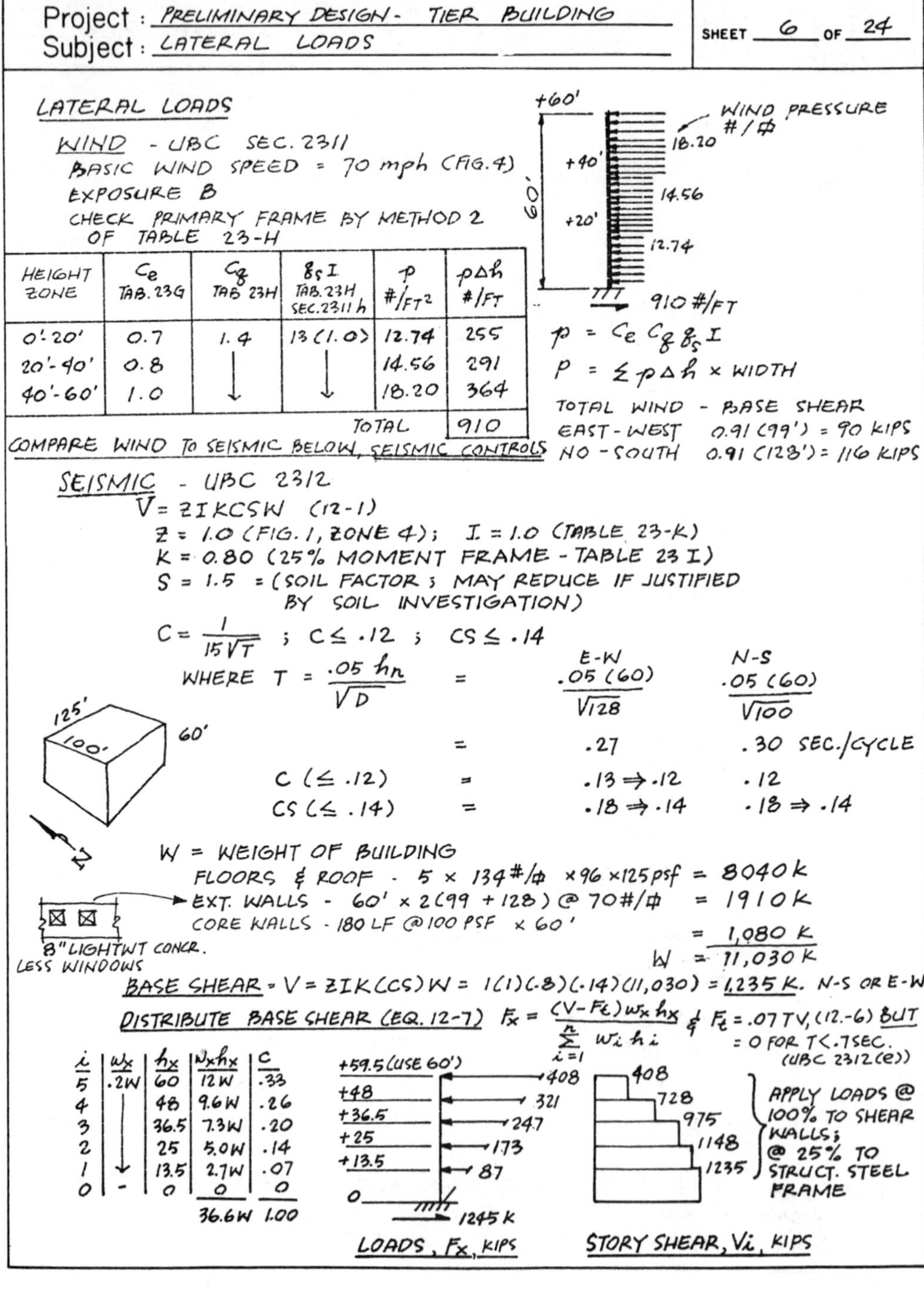

Project : PRELIMINARY DESIGN - TIER BUILDING
Subject : LATERAL LOADS
SHEET 6 OF 24

LATERAL LOADS

WIND - UBC SEC. 2311
BASIC WIND SPEED = 70 mph (FIG. 4)
EXPOSURE B
CHECK PRIMARY FRAME BY METHOD 2 OF TABLE 23-H

HEIGHT ZONE	C_e TAB. 23G	C_q TAB 23H	$q_s I$ TAB. 23H SEC. 2311 h	p #/FT²	$p\Delta h$ #/FT
0'-20'	0.7	1.4	13 (1.0)	12.74	255
20'-40'	0.8	↓	↓	14.56	291
40'-60'	1.0	↓	↓	18.20	364
				TOTAL	910

$p = C_e C_q q_s I$

$P = \Sigma p \Delta h \times$ WIDTH

TOTAL WIND - BASE SHEAR
EAST-WEST 0.91 (99') = 90 KIPS
NO-SOUTH 0.91 (128') = 116 KIPS

COMPARE WIND TO SEISMIC BELOW, SEISMIC CONTROLS

SEISMIC - UBC 2312

$V = ZIKCSW$ (12-1)

$Z = 1.0$ (FIG. 1, ZONE 4); $I = 1.0$ (TABLE 23-K)

$K = 0.80$ (25% MOMENT FRAME - TABLE 23 I)

$S = 1.5$ = (SOIL FACTOR; MAY REDUCE IF JUSTIFIED BY SOIL INVESTIGATION)

$C = \dfrac{1}{15\sqrt{T}}$; $C \le .12$; $CS \le .14$

		E-W	N-S
WHERE $T = \dfrac{.05\, h_n}{\sqrt{D}}$	=	$\dfrac{.05\,(60)}{\sqrt{128}}$	$\dfrac{.05\,(60)}{\sqrt{100}}$
	=	.27	.30 SEC./CYCLE
$C\ (\le .12)$	=	.13 ⇒ .12	.12
$CS\ (\le .14)$	=	.18 ⇒ .14	.18 ⇒ .14

W = WEIGHT OF BUILDING
FLOORS & ROOF - 5 × 134 #/□' × 96 × 125 psf = 8040 K
EXT. WALLS - 60' × 2 (99 + 128) @ 70 #/□' = 1910 K
CORE WALLS - 180 LF @ 100 PSF × 60' = 1,080 K
W = 11,030 K

BASE SHEAR = $V = ZIK(CS)W = 1(1)(.8)(.14)(11{,}030) = 1{,}235$ K. N-S OR E-W

DISTRIBUTE BASE SHEAR (EQ. 12-7) $F_x = \dfrac{(V - F_t)\, w_x h_x}{\sum_{i=1}^{n} w_i h_i}$ & $F_t = .07\,TV$, (12.-6) BUT = 0 FOR T < .7 SEC. (UBC 2312(e))

i	w_x	h_x	$w_x h_x$	c
5	.2W	60	12W	.33
4	↓	48	9.6W	.26
3		36.5	7.3W	.20
2		25	5.0W	.14
1	↓	13.5	2.7W	.07
0	-	0	0	0
			36.6W	1.00

loads, which are transitory and sometimes movable loads resulting from use of the building. Dead and live loads are listed separately and will be carried forward as separate parts to be used in various load combinations.

The vertical load table is set up on the basis of pounds per square foot (psf) of floor area. In assembling the table, it is necessary to project ahead to a framing arrangement as shown in sheet 7 and estimate the anticipated unit weights of members. The weights chosen are subject to later correction. Reference to beam capacity tables in the AISC Manual, Part 2, can reduce the degree of error resulting from these early arbitrary selections. Weights of linear elements such as beams are converted to pounds per square foot by dividing weight per linear foot by the spacing between them. The dead loads get progressively larger as we proceed through the hierarchy of members supporting other members. Slabs support only themselves and other loads applied directly to them. Beams support all the slab loads plus their own weight and other loads directly applied to the beam. Girders are loaded by the beams and their own weight. Columns must support all loads within a *tributary bay*, in this case 25 by 32 feet; in fact, due to the continuity of the beams and girders, some columns may usurp some of the load within the area tributary to others. Column loads accumulate from story to story, with a column at any one level carrying its part of all stories above.

The UBC (paragraph 2304d) requires, in office occupancies, the addition of a *dead load* of 20 psf on floors to account for partitions between office spaces, which may be subject to many relocations during the life of the building. The actual weight is very uncertain, but as shown by the thumbnail calculations accompanying the listing of that load, it is not likely to exceed 20 psf. Note that partitions are shown that weigh 10 psf of partition area, but the resulting weight is converted to pounds per square foot of floor.

Live loads are taken from the UBC table of required minimum live loads to be used in the design of buildings based on type of occupancy. The engineers have no reason in this building to provide for heavier loads.

While most of the floor is subject to 50-psf office loading, those parts used as public corridors and emergency escape routes require 100 psf. For present purposes, 100 psf is used as live load for design of the entire floor slab. UBC Section 2306 allows reduction of live load effects on members whose loading derives from large tributary areas. This recognizes the fact that the intensity of live load may be relatively high over a small area but is likely to be smaller, on the average, when the loaded area is larger. Live loads in the table reflect this type of reduction, although at this stage the designers do not use the total reductions permitted. They may in the final design process.

For the preliminary stage, the roof framing will be treated like a floor. The excess weight resulting will offset the weight of penthouse construction, which is being omitted. For this lowland California location, final design will be based on 20-psf roof live load. In snow country (see mill building study), the roof live load would usually be higher.

The structural arrangement plan and sections that follow on sheet 7 are used in establishing the seismic loads on sheet 6. The engineers plan to resist horizontal forces by transferring them to the ground through the action of a stiff concrete *shear wall* tower.* They will provide a much less stiff second line of resistance to extreme earthquake effects by building the steel frame with continuous joints between columns and beams and with members and joints chosen for their ability to mobilize ductile moment resistance. This reserve defense would only act after the stiffer shear walls had cracked, but it would help prevent collapse in an extreme event. Lines of resistance for such rigid frames exist in each direction on column lines. The concrete slabs, which act as very deep girders (diaphragms) in resisting horizontal forces, will tend to equalize the participation of the various parallel rigid frames.

Beams between column lines (B1, sheet 7) cannot help in this rigid frame resisting system. The girders (G1 and G2, sheet 7) will clearly have very small torsional stiffness (see Chapter 11). The joint rotations between beams B1 and their supporting girder will also be very small. The torsional moment transferred by G1 or G2 to the columns will be trivial compared to the flexural moments transferred to the columns from B2 and B3.

The magnitude of lateral forces from both wind and earthquake are calculated and compared on sheet 6. The term *lateral loads* describes horizontal forces applied in either the longitudinal or transverse direction of the building. As used here, both wind and seismic (earthquake) forces are applied in accordance with Code requirements. Chapter 16 contains discussion on these provisions, which are under continuing review by engineers and Code bodies.

The base shear, which represents the total of all horizontal loads to be transferred from the building to the ground, is very much higher in this case for seismic loads than for wind. Although that is not always the case, it is not surprising here, since this is a relatively heavy building with a high ratio of volume to surface exposed to wind. Seismic loads are functions of mass; in the pseudostatic approach used in the Code, they are expressed as the product of building weight and accelerations, which are percentages of the acceleration of gravity.

The calculation of base shear and its conversion to forces applied at floor levels is based on attempts by engineers in writing the UBC to convert a very complex dynamic problem into a somewhat less complex pseudostatic problem by coefficients of force applied to dead weight. The coefficients reflect location, importance of the structure in a public disaster-response system, the type of structural system, and the effect of soil–structure interaction on building response. The K value of 0.8 reflects the decision, on sheet 4, to use the steel frame as a second stage defense against extreme seismic forces. When $K = 0.8$ in evaluating base shear, the ductile moment-resisting frame is expected to be designed elastically for 25% of that shear. At the same time, the stiffer shear wall resisting system is designed for 100% of the same shear. The shear wall design is not included here, but would be included in the total design.

*See Chapter 2.

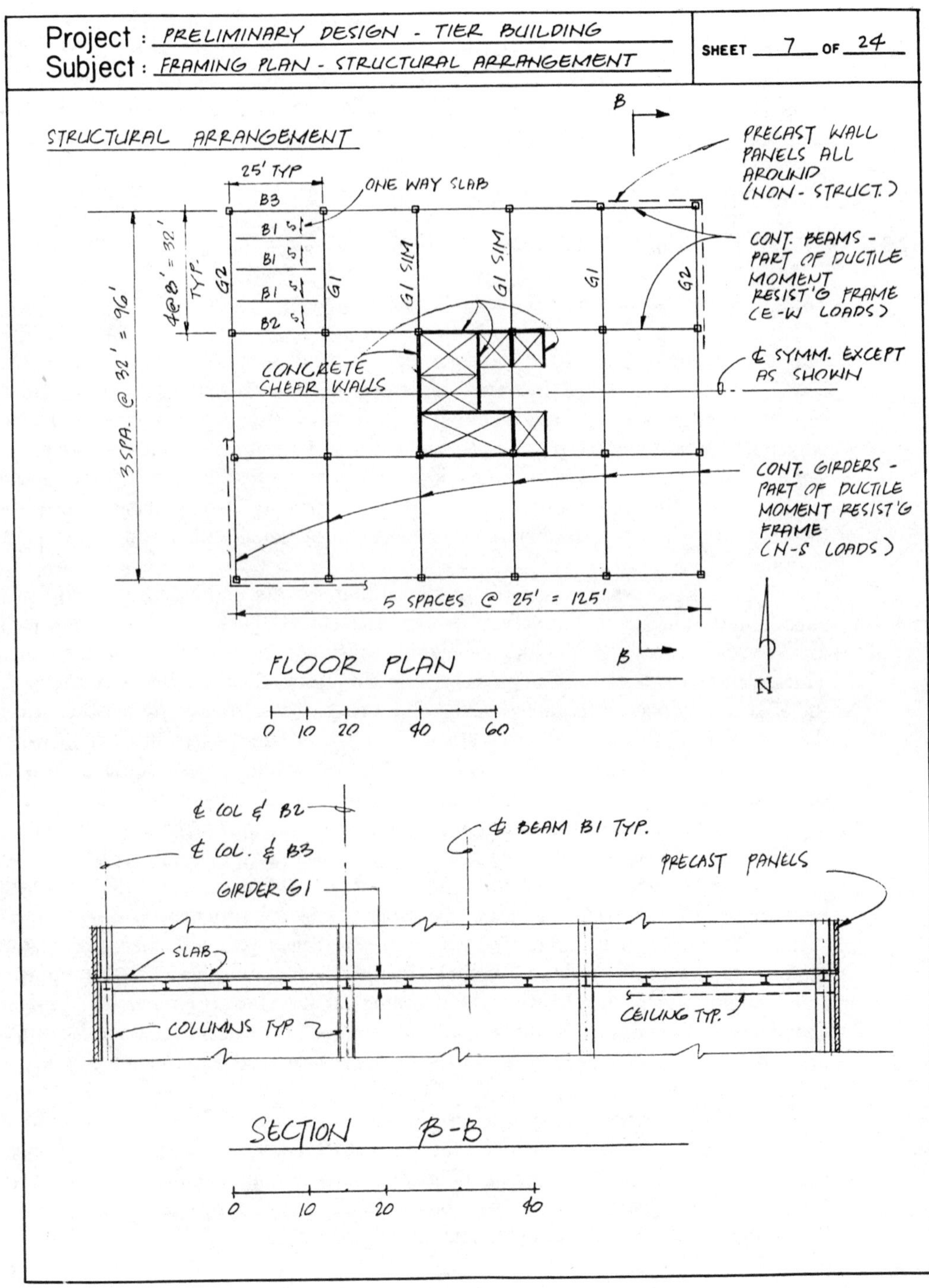
Project : PRELIMINARY DESIGN - TIER BUILDING
Subject : FRAMING PLAN - STRUCTURAL ARRANGEMENT
SHEET 7 OF 24
STRUCTURAL ARRANGEMENT
25' TYP
ONE WAY SLAB
B3
B1
B1
B1
B2
G2
G1
G1 SIM
G1 SIM
G1
G2
4@8' = 32' TYP.
3 SPA. @ 32' = 96'
PRECAST WALL PANELS ALL AROUND (NON - STRUCT.)
CONT. BEAMS - PART OF DUCTILE MOMENT RESIST'G FRAME (E-W LOADS)
℄ SYMM. EXCEPT AS SHOWN
CONCRETE SHEAR WALLS
CONT. GIRDERS - PART OF DUCTILE MOMENT RESIST'G FRAME (N-S LOADS)
5 SPACES @ 25' = 125'
B
N
FLOOR PLAN
0 10 20 40 60
℄ COL & B2
℄ COL. & B3
℄ BEAM B1 TYP.
GIRDER G1
PRECAST PANELS
SLAB
COLUMNS TYP
CEILING TYP.
SECTION B-B
0 10 20 40

Note that the weight of exterior and core walls, which were not included in floor dead loads, must be included in the evaluation of building weight for base shear. Note also that the application of UBC formula 12–7 leads us to apply relatively high proportions of the base shear to the upper stories although the distribution of mass is uniform with height. This is in keeping with the approximate distribution of the "real" dynamic forces.

18.5 TENTATIVE MEMBER SELECTION: VERTICAL LOADS

18.5.1 Arrangement and General Considerations

Sheets 7 through 11 are used for tentative selection of members, based on vertical loads (i.e., dead and live loads). The effects of lateral loads will be considered separately later, and selected member sizes rechecked and possibly revised. Some engineers might prefer to consider all load combinations at the same time. We find the separation effective as a way to develop a feel for the action of this frame and the relative importance of different effects. Later revisions are likely and can be done best by manipulating the parameters that have the greatest effect on the resulting tonnage.

The designers are aware of two things that are not expressed explicitly in the calculations. The specification permits increases of one-third in allowable stresses for load combinations that include the highly transient effects of wind or earthquake. Thus, a member sized for dead plus live load automatically has a one-third reserve for the effects of added earthquake loads, still leaving a safety factor greater than 1.0 for the action of these transient loads at the level prescribed for design by the Code.

Second, the designers will be sizing the steel frame to resist only 25% of the specified seismic force, while also making the shear walls capable of resisting 100%. Thus, they do not expect the lateral problem to dominate in establishing member sizes as it might if the rigid steel frame had to resist 100% of the seismic loads.

A *framing arrangement plan* (sheet 7) opens this section, intended to identify, by symbol, all members that are to be sized. Note the progression of transfers of vertical load effects from member to member. Slabs (S) will span north–south, directly supporting floor loads. Beams (B1, etc.), spanning east–west, support the slab, transferring all the loads that originated on the slab to girders (G1 and G2) spanning north–south to columns (C). The columns must accumulate all loads from all floors, plus whatever may come to them directly, and carry them down to the foundation.

In drawing the framing plan as shown, the designers have chosen among a number of possible arrangements. The one chosen was based, in part, on earlier indications that the depth of members may have to be minimized in order to keep the five-story building within the height limit. Beams B1 will be treated as simple spans to facilitate erection. By spanning them in the 25-foot rather than 32-foot direction, their

Project: PRELIMINARY DESIGN - TIER BUILDING

Subject: MEMBER SELECTION - VERTICAL LOADS

SHEET 8 OF 24

SLAB (CONSIDER 2 ALTERNATIVES FOR THICKNESS ONLY, DETAILED DESIGN LATER)

ALT. 1 - SOLID SLAB

FINISH, h

PER ACI CODE # 318-77, TABLE 9-5 (a)

$\frac{\ell}{h} \le 24$ WILL YIELD ACCEPTABLE DEFLECTION

$h \ge \frac{8(12)}{24} = 4''$ ASSUME 5"

WEIGHT/FT²	NORMAL WT @ 150 PCF	LTWT CONC @ 110 PCF
5" SLAB	62.50	45.8
ADD 10# FINISH	10	10
	73 #/FT³	56#/FT³

USED 85 ON SHT. 5; SAY OK BUT MAY REDUCE IN FINAL DESIGN.

ALT. 2 COMPOSITE METAL DECK & CONCRETE

FROM ROBERTSON CO. TABLES - OTHER SUPPLIERS SIMILAR.

FINISH, DECK, t, CONC, h

FOR 2 HR FIRE SEPARATION

U.L. # 840 - UKX DECK (1½" DEPTH) + 3¼" CONCR., LTWT.

TABLES GIVE PERMISSIBLE SUPERIMPOSED LOADS VS. SPAN

SPAN = 8'; ALLOW. SUPER. LOAD = 390 PSF

SUPERIMPOSED LOAD ≤ 100 #/□ LIVE
10 #/□ FINISH
110 #/□ OK PER TABLE

COMPARE TO ACI TABLE 9.5 (a)

@ $h = 1.5'' + 3.25''$, $\frac{\ell}{h} = \frac{8(12)}{4.75} = 20.21 < 24$ OK

WOULD LEAD TO

FINISH
3¼" LTWT CONC.
ROBERTSON QL-UKX 20 GA. OR EQUIV.

TOTAL D.L. = 41 #/□ + FINISH ≪ 85 ASSUMED
OK FOR NOW, REVISE LATER

BEAMS

B1 (@ 8' CTRS)

w_{D+L}, 25'

$w_{D+L} = 8'(125 + 75) = 1000 + 600 = 1600$ #/LF

$W_{D+L} = 25\,w = 25 + 15 = 40$ K

$M_{D+L} = \frac{W\ell}{8} = 78 + 47 = 125^{k'}$

FROM AISC - BEAM TABLES p. 2-29, 2-30
A36 STL, CONT. LAT. SUPPORT (FR. SLAB)

$W = 40k = \frac{W_c}{L}$

$W_c \ge 25(40) = 1000K'$

SECTION	ALLOWABLE W_c	SHEAR, V	DEFLECTION D_c	Δ_{D+L}	Δ_{D+L}/L	Δ_L/L
W12×50	1040	51	2.0	1.20	1/250	1/666
W14×43	1000	60.4	1.8	1.13	1/265	1/708
W16×40	1040	70.8	1.6	0.96	1/312	1/832

$V_{D+L} = \frac{W_{D+L}}{2} = 20 \ll V_{ALL}$

ACTUAL $\Delta_{D+L} = \frac{D_c L^2}{1000} \times \frac{1000}{W_c} = \frac{625\,D_c}{W_c}$

$\Delta_L = \frac{15}{40}(\Delta_{D+L})$

1ST CHOICE IS W16×40
W14×43 OR W12×50 OK IF NEC'Y FOR MECH. SPACE NEEDS

POTENTIAL ADDED COST TO USE W12×50 = 10#/LF = 1.25#/□
IF AT \$0.50/LB, ADDS \$0.625/□
ADDED COST = 0.625 (96×125) 5 FLRS ≈ \$37,500

depth can be kept smaller. The girder will have to span in the long direction, but, being continuous, their controlling moments (therefore required section modulus) and deflections will be less than the simple span case. Thus, a good balance is expected between required depth of beams and girders. This is a judgment call. Considerations may arise during preliminary or final design, either structural or nonstructural, that may cause reevaluation of this choice. For present purposes, Coopchen proceeds on this basis.

18.5.2 Slabs

The slab calculation is introduced here primarily as a check on the assumptions of the load table, sheet 5. Since the slab is not part of the structural steel frame, the authors omit detailed calculations for it from this set. However, note, on the load table, that the assumed slab weight of 85 psf is almost half of the total load to be supported by the beams and columns.

Checking the implications of using a reinforced concrete slab, it is discovered that a 5-inch slab would be structurally acceptable. After allowing 10 psf for additional dead load of applied finish materials, a 5-inch slab would not weigh more than 73 psf. If the extra unit cost of lightweight concrete is justifiable, the slab would weigh even less. The choice of finish materials will ultimately be decided by the architect and may vary between zero and 15 psf, although it is not very likely to exceed the assumed 10 psf. Note also that the original assumption of 6-inch slab depth seems excessive for the beam spacing shown, giving the designers the option to increase the spacing between beams or reduce the thickness of slab. The concrete slab is satisfactory without other protection as a 2-hour fire barrier, although the supporting beams will still need fire protection.

Another possibility for the floor slab is a composite of light-gage steel deck with concrete. From data supplied by manufacturers of the steel deck, which the designers know has been approved by the International Conference of Building Officials (the issuers of the UBC), they find:

1. Several types of available metal deck configurations with 3¼ inches of lightweight concrete are satisfactory as a 2-hour fire separation.
2. A particular configuration of deck is satisfactory by both stress and deflection considerations for the loading we have.

The deck section noted is supplied by the H. H. Robertson Company and is identified under their numbering system. Equivalent sections bearing different numbers can be bought from others, and final contract documents would probably call out the required performance, leaving the choice of supplier to the building contractor.

The combination of metal deck and lightweight concrete turns out to be lighter than the reinforced concrete deck of normal-weight concrete and even lighter than the deck of reinforced lightweight concrete. Recall also that one of the reasons for using structural steel (see sheet 4) was to save construction time. The use of metal deck makes formwork and temporary supporting posts unnecessary while working on the floor slab, so the floor below can be made available much sooner for the use of mechanical and finishing tradespeople. So the designers note that the composite alternative will probably be best, but are unwilling to be fully committed until cost comparisons are made at a later date.

The alert reader may now be concerned that the designers have shown, by two reasonable alternative choices, that the assumption of the weight of floor slab in the vertical load table, sheet 5, was unnecessarily heavy, but show no inclination to revise the calculations which relied on that assumption. They are indeed tempted. If, for example, the assumed slab weight of 85 psf were reduced to 75, which is still likely to turn out somewhat high, this alone would represent about a 5% reduction in loading on all members, with some corresponding reduction in sizes required. The effect on seismic base shear would also be a 5% reduction. For the 25% of shear used in steel frame design, the corresponding reduction is from 309 to 271 kips, a 12% change.

The designers prefer to note the probability that the slab weight can be reduced in later calculations, continuing the calculations on the original basis at this time, and building up a reserve for unanticipated requirements that may be discovered in final design.

18.5.3 Beams and Girders

The selections of beams and girders can be made by procedures discussed in Part II of this book. It is economical to choose standard rolled W sections with webs vertical so that dead and live loads act through the vertical axis of symmetry and bending takes place around the *x-x* axis in the Shape Tables (Section 1 of the AISC Manual). This can be referred to as *strong-way* bending.

Calculations actually show, for B1, three alternative selections from the beam tables in Part 2 of the AISC Manual. Coefficients in the tables are used to determine the capacity of each shape listed, in kips of load uniformly distributed, by allowable stresses and span. Capacities can be found in bending and shear and the corresponding deflections calculated. Each of these can be derived very easily from formulas in Part II of this book; it is suggested that inexperienced people do such calculations in addition until they understand fully the use of the tables. The designers use the tables since it makes it possible to see several alternatives simultaneously. Since A36 steel is proposed, the tables used are those for $F_y = 36$ ksi. The members must satisfy separately the section requirements for maximum shear and maximum moment, which in this case occur at different locations on the members. It turns out that the beams selected for moment requirements have much more capacity in shear than is needed. That is often, but not always, true.

Note that the list shows a preferred size, W16 $\times$ 40, and acceptable alternatives in shallower, but heavier sizes, which the designers are prepared to substitute if the building height requirements demand it. The shallower members are more costly and deflect more under load. The effect in each case is noted for ready reference if needed.

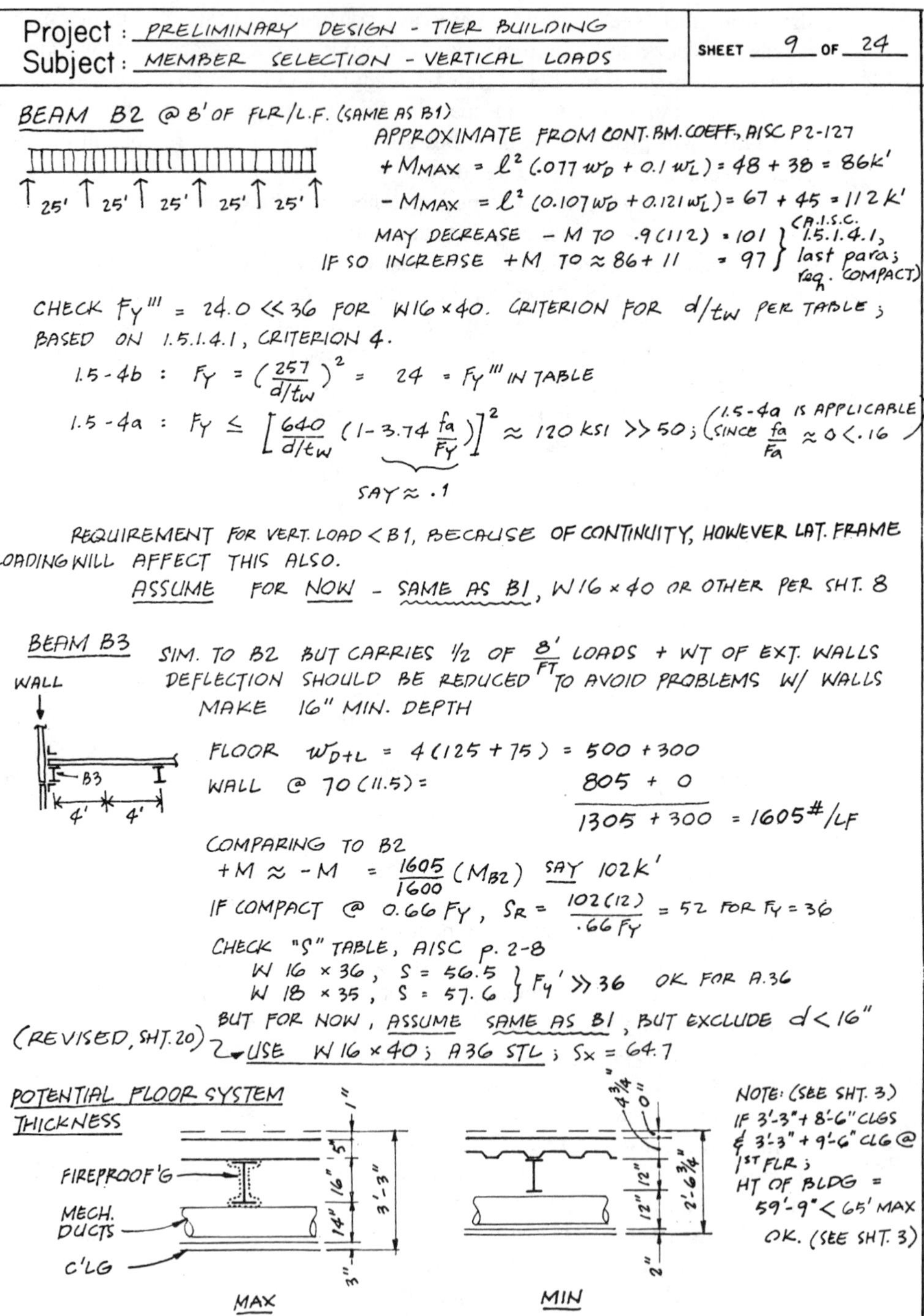

Project: PRELIMINARY DESIGN - TIER BUILDING
Subject: MEMBER SELECTION - VERTICAL LOADS

SHEET 9 OF 24

BEAM B2 @ 8' OF FLR/L.F. (SAME AS B1)

25' 25' 25' 25' 25'

APPROXIMATE FROM CONT. BM. COEFF., AISC P2-127

$+M_{MAX} = \ell^2 (.077 w_D + 0.1 w_L) = 48 + 38 = 86k'$

$-M_{MAX} = \ell^2 (0.107 w_D + 0.121 w_L) = 67 + 45 = 112k'$

MAY DECREASE $-M$ TO $.9(112) = 101$
IF SO INCREASE $+M$ TO $\approx 86 + 11 = 97$
(A.I.S.C. 1.5.1.4.1, last para; req. COMPACT)

CHECK $F_Y''' = 24.0 \ll 36$ FOR W16×40. CRITERION FOR d/t_w PER TABLE; BASED ON 1.5.1.4.1, CRITERION 4.

1.5-4b : $F_Y = \left(\frac{257}{d/t_w}\right)^2 = 24 = F_Y'''$ IN TABLE

1.5-4a : $F_Y \le \left[\frac{640}{d/t_w}\left(1 - 3.74\frac{f_a}{F_Y}\right)\right]^2 \approx 120$ KSI $\gg 50$; (1.5-4a IS APPLICABLE SINCE $\frac{f_a}{F_a} \approx 0 < .16$)

SAY ≈ .1

REQUIREMENT FOR VERT. LOAD < B1, BECAUSE OF CONTINUITY, HOWEVER LAT. FRAME LOADING WILL AFFECT THIS ALSO.

ASSUME FOR NOW - SAME AS B1, W16×40 OR OTHER PER SHT. 8

BEAM B3 SIM. TO B2 BUT CARRIES ½ OF 8'/FT LOADS + WT OF EXT. WALLS
DEFLECTION SHOULD BE REDUCED TO AVOID PROBLEMS W/ WALLS
MAKE 16" MIN. DEPTH

FLOOR $w_{D+L} = 4(125 + 75) = 500 + 300$
WALL @ 70(11.5) = 805 + 0
1305 + 300 = 1605#/LF

COMPARING TO B2

$+M \approx -M = \frac{1605}{1600}(M_{B2})$ SAY 102k'

IF COMPACT @ $0.66 F_Y$, $S_R = \frac{102(12)}{.66 F_Y} = 52$ FOR $F_Y = 36$

CHECK "S" TABLE, AISC p. 2-8
W 16 × 36, S = 56.5
W 18 × 35, S = 57.6
$F_y' \gg 36$ OK FOR A.36

BUT FOR NOW, ASSUME SAME AS B1, BUT EXCLUDE $d < 16"$

(REVISED, SHT. 20) USE W16×40; A36 STL; $S_x = 64.7$

POTENTIAL FLOOR SYSTEM THICKNESS

NOTE: (SEE SHT. 3)
IF 3'-3" + 8'-6" CLGS & 3'-3" + 9'-6" CLG @ 1ST FLR;
HT OF BLDG = 59'-9" < 65' MAX
OK. (SEE SHT. 3)

In comparing costs, the designers use a figure in cost per pound of steel somewhat lower than will be used later in calculating the cost of the total frame. A lower than average figure is properly applicable when one is adding only weight per foot of purchased member and not increasing other costs proportionately.

For the case of B2, a continuous member, there are no equivalent tables for automatic selection of beam sizes. The member is statically indeterminate, and analysis will ultimately depend on the relative stiffnesses of the beams and columns (not yet known) and careful examination of the implications of shifting live loads. These will be considered in the final design stage.

The designers decide to make their first selection based on moments and shears that result from a simpler case of four equal continuous spans on knife-edge supports. Results of analysis of this type of beam are shown in the AISC Manual, Part 2. Coefficients are shown to be applied to wl for shear and wl^2 for moment. (Readers accustomed to seeing positive moments plotted above the base line and negative below will have to stop and notice that the opposite convention is used here.)

Case 39 (AISC) shows loading applied to all spans. This is the actual case for dead load. Cases 37 and 38 are intended to find both the maximum negative and maximum positive live load moment that occur for different combinations of loaded and unloaded spans. Since the designers have kept separate track of dead and live loads, it is a simple procedure to apply the dead load wl to case 39 and the live load to either 37 or 38 to get the maxima. They then add the results of case 39 and the larger values from 37 or 38, being sure that the values added apply, in each case, to the same location.

Not surprisingly, the maximum negative moment turns out to be greater than the positive, and both are less than the simple span moment that applies to beam B1. We would expect the maximum shear to be somewhat higher, due to continuity, than found in B1. However, looking back at B1, we remember that shear is not likely to control the selection here. The designers consider the same shapes listed for beam B1 in order to minimize the number of different sizes used in the building. The fact that moments from dead plus live load are smaller gives some reserve for the added moments from seismic effects to be considered later. Recall that B1 is not part of the moment-resisting frame.

However, a way is sought to justify more nearly equal negative and positive moments from dead and live load. Specification Section 1.5.1.4.1 offers such a justification. Provided the section is compact and the controlling moment is not on a cantilever, the AISC Specification permits an arbitrary 10% reduction of design negative moment in a continuous beam provided the positive design moment is correspondingly increased. As discussed in Chapter 10, this is a device by which the Specification recognizes the plastic reserve moment capability of compact sections.

The designers examine the compact section criteria as reflected in the values of F_y' and F_y''' in the Shape Tables of the AISC Manual. These values were previously listed when selecting the shapes considered for B1. F_y' and F_y''' as listed there are intended to show the maximum values of F_y for which the section may be considered compact when judged by the width–thickness ratios of flange and web, respectively.

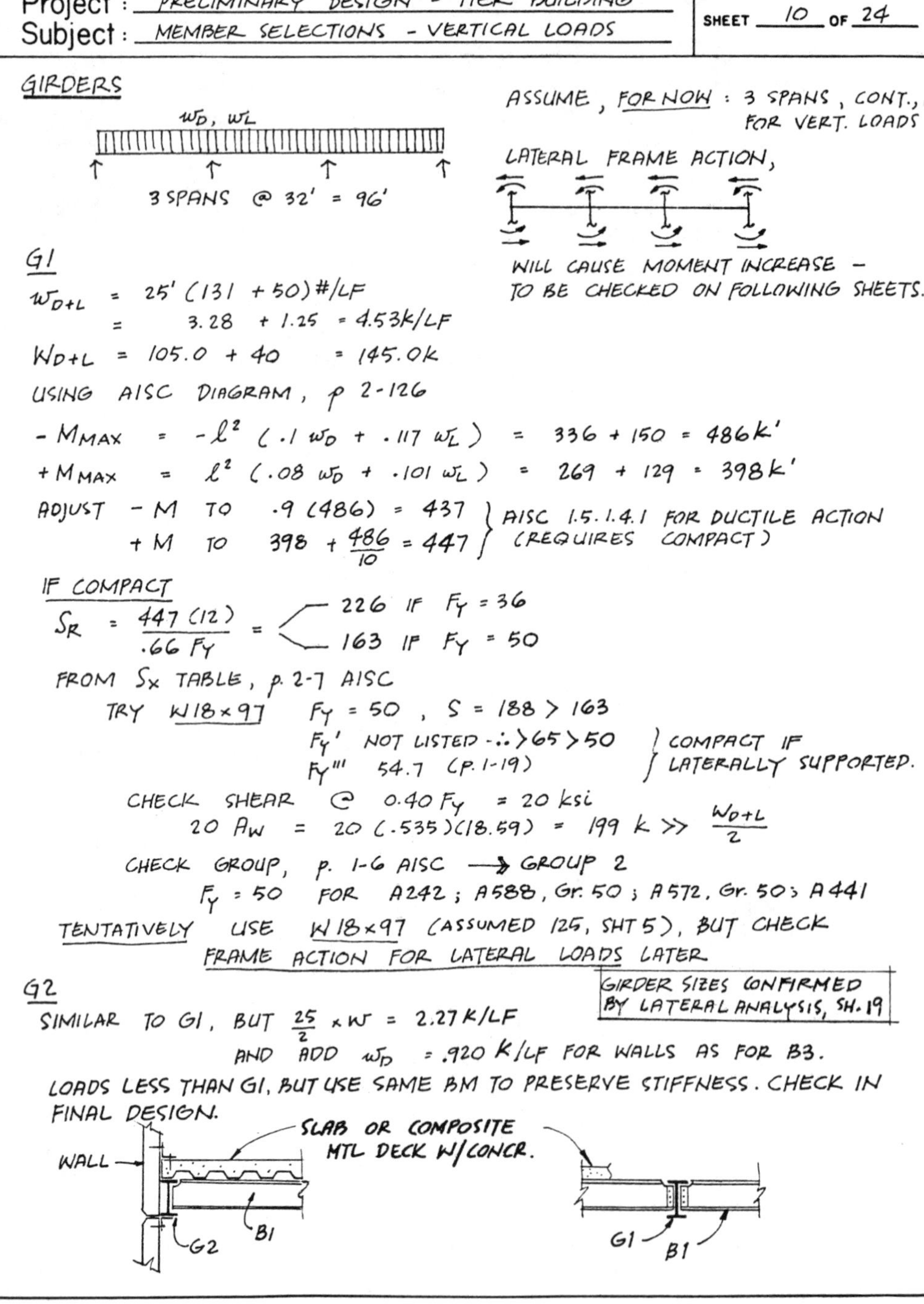

Project: PRELIMINARY DESIGN - TIER BUILDING
Subject: MEMBER SELECTIONS - VERTICAL LOADS
SHEET 10 OF 24

GIRDERS

w_D, w_L

3 SPANS @ 32' = 96'

ASSUME, FOR NOW: 3 SPANS, CONT., FOR VERT. LOADS

LATERAL FRAME ACTION, WILL CAUSE MOMENT INCREASE – TO BE CHECKED ON FOLLOWING SHEETS.

G1

w_{D+L} = 25' (131 + 50) #/LF
= 3.28 + 1.25 = 4.53 k/LF

W_{D+L} = 105.0 + 40 = 145.0 k

USING AISC DIAGRAM, p 2-126

$-M_{MAX} = -\ell^2 (.1\, w_D + .117\, w_L) = 336 + 150 = 486\,k'$

$+M_{MAX} = \ell^2 (.08\, w_D + .101\, w_L) = 269 + 129 = 398\,k'$

ADJUST $-M$ TO $.9\,(486) = 437$
$+M$ TO $398 + \frac{486}{10} = 447$
} AISC 1.5.1.4.1 FOR DUCTILE ACTION (REQUIRES COMPACT)

IF COMPACT

$S_R = \frac{447\,(12)}{.66\,F_Y}$ = 226 IF $F_Y = 36$; 163 IF $F_Y = 50$

FROM S_x TABLE, p. 2-7 AISC

TRY W18×97 $F_Y = 50$, $S = 188 > 163$
F_Y' NOT LISTED $\therefore > 65 > 50$
F_Y''' 54.7 (p. 1-19)
} COMPACT IF LATERALLY SUPPORTED.

CHECK SHEAR @ $0.40\,F_Y = 20$ ksi
$20\,A_W = 20\,(.535)(18.59) = 199\,k \gg \frac{W_{D+L}}{2}$

CHECK GROUP, p. 1-6 AISC → GROUP 2
$F_Y = 50$ FOR A242; A588, Gr. 50; A572, Gr. 50; A441

TENTATIVELY USE W18×97 (ASSUMED 125, SHT 5), BUT CHECK FRAME ACTION FOR LATERAL LOADS LATER

G2

GIRDER SIZES CONFIRMED BY LATERAL ANALYSIS, SH. 19

SIMILAR TO G1, BUT $\frac{25}{2} \times w = 2.27$ k/LF
AND ADD $w_D = .920$ k/LF FOR WALLS AS FOR B3.

LOADS LESS THAN G1, BUT USE SAME BM TO PRESERVE STIFFNESS. CHECK IN FINAL DESIGN.

These are criteria 2 and 4 under 1.5.1.4.1, AISC Specifications. The W16 × 40 appears to be unacceptable when judged by the tabulated value of F_y'''. However, that value is based on formula 1.5-4b, which assumes, in combination with the bending moment, more axial load than is expected here. A check by formula 1.5-4a permits the designer to treat the W16 × 40 as compact.

The adjusted positive and negative moments are almost equal. The choice, tentatively, is the same section as for B1.

Beam B3 has a very similar function in the lateral force resisting frame to that of B2. It carries only half as much floor load, but it must also carry the weight of exterior wall panels one story in height. The results lead to selection of the same shape as used for B1 and B2. In this case, in order to minimize deflection, which can cause difficulties with the wall panels, the designers rule out the shallower selections in favor of the W16. This will be studied further in the final design stage. A note is appended to this calculation, indicating that a later calculation sheet led to a revision of this selection.

The designers now stop long enough to consider the implication of their beam sizes on the available space between ceiling and floor. Two simple sketches illustrate this.

Selection of girder G1 is done in a manner similar to beam B2, using, from the AISC Manual, coefficients for moment in the case of a continuous beam of three equal spans.* The reader may have noticed that the loading is expressed in pounds per lineal foot, whereas the Framing Arrangement Plan, sheet 7, indicates that G1 actually receives concentrated loads from two beams, B1, every 8 feet. B2 and B3 also contribute concentrated loads to the columns, but they do not contribute to moment and shear in G1. The choice of uniformly distributed load is recognized as an approximation. The reader may wish to test this approximation in part by examining the table on page 2-113 of the AISC Manual. Compare the cases of $n = \infty$ and $n = 4$. Keep in mind that the value of P used with $n = \infty$ is the entire uniformly distributed load. With $n = 4$, $P =$ one-fourth as much.

The obvious question is, Why not use the real load case directly? However, solutions on page 2-113 apply to cases that are only roughly approximated by our continuous girder. The designers find it easier to trace the separate live and dead load effects with the three span cases on page 2-126. The results are considered sufficiently accurate for preliminary design.

To keep member sizes down, the designers find it useful here to switch to higher-strength steel. Since the girders are continuous, higher stresses will not, for the span/depth ratio applicable, lead to excessive deflections. The designers expect that the unit price of the higher-strength steels will be only slightly higher than for A36. They will recheck market conditions at the time of final design. Several types of possible

*Shear is as important a criterion for beam selection as moment. In this case, for $L/d \approx 24$ and the applicable load pattern, the designers expect a beam selected for moment to be adequate in shear. Readers may wish to verify this.

steels are noted, with $F_y = 50$ ksi. The reasoning in the choice of G2 is similar to that of B3.

Note on this sheet:

1. The designers sketched the source of girder movements resulting from horizontal shear in columns, the effect of seismic action. This points to a check to be made later.
2. A notation indicates that the seismic checks confirmed the size selected.
3. The designers, in thinking ahead, sketch the type of connections they have in mind for beams to girders and for wall panels to girders. Although not yet ready to design the connections, they know that selected members will not prove out if they cannot be connected in a way consistent with the assumptions used in analysis.
4. Calculations seem to allow a smaller size for G2 than G1. The tentative decision is to use the same size. Excessive deflection of G2 may cause trouble with the precast wall panels, particularly in an earthquake. Further checks will be made later in these calculations. Before final design selection, it will be necessary to study the joints of the precast panels.

18.5.4 Columns

Typical interior columns support the uniformly distributed floor load from a *tributary area* of 25 by 32 feet (see Plan, Sheet 7). However, the columns also act as part of rigid frames resisting seismic forces. In addition, they must resist moments and shears resulting from unbalanced live load on the continuous beams and girders.

From Chapter 9, readers are aware that the allowable stresses in columns are inversely proportional to $(KL/r)^2$, where r is the smaller radius of gyration, L is the column length, and K is a factor that depends, first, on the relative rotational stiffnesses of the girders and columns meeting at a joint and, second, on the nature of resistance to sidesway in the total framing system.

In our case, as long as the shear walls are undamaged, they act as horizontal restraints for sidesway and $K < 1.0$. However, since the design requirement that the steel frame must resist 25% of the seismic forces assumes damage to the shear walls, the designers assume that sidesway resistance must come from the frame itself. K then becomes an unknown quantity > 1.0.

The value $K = 1.2$ is chosen subject to later refinement. This is an unconservative choice. It is the value recommended by AISC for columns with fixed ends subject to translation of one end with respect to its original position (see AISC, Commentary to Specification, Table C1.8.1). The column ends are not fully fixed in this building. However, these will be short columns, with relatively small ratios of L/r, say approximately $150/4 < 40$. In this region, allowable stresses, F_a, are not extremely sensitive to variations of K, as can be seen in AISC Specification, Appendix A, Table 3.

Project: PRELIMINARY DESIGN - TIER BUILDING
Subject: MEMBER SELECTIONS - VERTICAL LOADS

SHEET 11 OF 24

COLUMNS CHOOSE SIZES FOR VERTICAL LOADS, USING $K = 1.2$
THEN CHECK FOR LATERAL FRAME ACTION & ADJUST IF NECESSA'Y
REVIEW K VALUE LATER

FOR VERT. LOADS, $K = 1.2$, $L = 13.5$ @ 1ST STORY
11.5 @ OTHER STORIES

INTERIOR COLUMNS

$w_{D+L} = 134 + 50 = 184$ #/□ (SHT. 5)

$W_{D+L} = 25(32)(134+50) = 107.2 + 40 = 147$ k/STORY

MULTIPLY BY 6/5 FOR BEAM CONTINUITY = 176k/STORY
(SEE AISC p. 2-126, 127)

ELEV		LOADS IN STORY
+60	ROOF	
		176
+48	5	
		352
+36.5	4 (ASSUME SPLICE)	
		528
+25	3	
		704
+13.5	2	
		880 KIPS
0	1	

LOWER COLUMN $P = 880^k$ @ 1ST STORY
ASSUME W/ SEISMIC BENDING
$P_{EQUIV.} = 1050$ k, $F_y = 50$
$KL = 13.5(1.2) = 16.2'$

PREFER → W14 × 145 $P = 1050$ (COMPACT FOR $L_c = 13.9'$)
ALT. W12 × 170 $P = 1130$ (COMPACT FOR $L_c = 11.3'$; OK)

LATERAL EFFECT
W14×145 REVISED SHT-17
W14×109 CONFIRMED SH. 18

UPPER COLUMN $P = 528^k$ @ 3RD STORY
ASSUME W/ BENDING FROM LAT. LOADS
$P_{EQUIV.} = 700$ k.
AISC COL. TABLES CIRCA 3-19
$KL = (11.5)(1.2) = 13.8$; $F_y = 50$
W 14 × 99 (SEMI-COMPACT)$^{N.G.}$ $P = 730 > 700$
W 14 × 109 (COMPACT, $L_c = 13.1'$) $P = 806 \gg 700$
ALT. W 12 × 106 (COMPACT, $L_c = 10.9'$) $P = 744 > 700$ OK

ALL GROUPS OK FOR A572, GR. 50
AISC P.1-5, 1-6 $F_y = 50$

CHECK SPACE REQUIREMENTS FOR COLUMNS

FIREPROOFING & FINISH
W14×145
20"±
$d = 14\frac{3}{4}"$
$b = 15\frac{1}{2}"$
20"±

IF ARCH'T FINDS b×d EXCESSIVE FOR W14 × 145, COULD TEST W12's GIVEN AS "ALT" ABOVE. HOWEVER CHECK, AT FINAL DESIGN, THAT FAILURE IN COLUMNS OCCURS @ HIGHER LOAD COMBINATIONS THAN FAILURE IN BEAMS.

EXTERIOR COLUMNS FLOOR LOAD << THAN LOAD FOR INTERIOR COLS. BUT WALL LOAD ADDED. PARTICIPATION IN SEISMIC RESISTING SYSTEM SIMILAR TO INT. COLS.
USE SAME AS INTERIOR COLUMNS (MAY REDUCE IN FINAL DESIGN).

For closer estimates of K, we could refer to chart C1.8.2 in the Commentary to the AISC Specifications, which derives K as a function of the relative stiffnesses of columns and girders framing in to each end of the column. The designers do not have firm data to do this at this time, and reserve this for later check. As it turns out, the later check will require some changes in column weights.

Columns will be chosen from those standard W sections that are rolled specifically for use as columns. These are W10, W12, and W14 sections with flange widths approximately equal to member depth. They have smaller ratios of r_x/r_y than sections usually used for beams. After being protected by fireproofing, they appear as square columns if in the middle of a floor and not enclosed by partitions.

Column loads per story are initially based on the tabulated (sheet 5) dead plus live load per square foot multiplied by 25- by 32-foot tributary area. However, recognizing that the beams and girders are continuous, we can expect at least some amplification of load on interior columns. The designers add 20%, which is probably well on the safe side. Column loads accumulate story by story from top to bottom.

A practical matter arises. How many individual members does it take to make a five-story series of columns, totaling slightly over 60 feet in length? Since the column load increases with each floor, the column section should ideally be changed at the same intervals. However, each change requires a costly splice, offsetting the saving in steel. The decision is made to choose a section suitable for the first-story loads and make it continuous into the third story. The continuing member will be suitable for the third-story loads and will be more than sufficient to use up to the roof.

A splice will be used close to mid-height of the third story. This is close to the location of zero bending moment from horizontal loads, minimizing the cost and maximizing the safety of the splice. An actual pinned splice would not be acceptable, since that would violate the assumed KL/r.

In choosing the column sections, the designers keep in mind that the column must resist, in addition to the axial effects of dead plus live load, bending moments due primarily to horizontal shears. This will ultimately require an interaction analysis, using Section 1.6 of the AISC Specification. However, data for this interaction being insufficient at this time, an arbitrary addition is tentatively made to the calculated axial loads to account for the effects of bending moment. The magnitude of this tentative addition is not critical since the real moment effect will be calculated later and the column sizes adjusted as necessary.

Having converted the problem of axial load plus bending into an *equivalent* axial load problem, the designers can use methods of selection suitable to an axially loaded member (see Chapter 9). They decide to use the Column Tables in Section 3 of the AISC Manual, which give permissible axial loads on various column sections as a function of KL and for two values of F_y, 36 and 50 ksi. The tables represent solutions to formulas in Section 1.5.1.3 of the Specification, arranged so one can make useful comparisons to help in selection.

The designers start with the lower-story column, with $KL \approx 16$ feet. Examining the allowable load on W14 columns, they could use W14 × 145 if $F_y = 50$, whereas if $F_y = 36$ the weight/foot would have to increase to 193 pounds. The ratio of 193/145 = 1.33 is much greater than the ratio of costs for the two types of steel.

For similar reasons, W12 columns turn out to be much heavier than W14. So they note a preference for the W14, and, using its actual dimensions plus an allowance for fireproofing, they record the required floor space of 20 by 20 inches. If this unduly complicates the architectural space problem, they are prepared to consider W12 sections further.

The selection of shape and F_y is followed by a check of availability. In Tables 1 and 2 of Part 1, AISC Manual, it can be seen that W14 × 145 is a group 3 shape, which can be found, with $F_y = 50$, in Steel of ASTM A572, grade 50.

The upper columns are examined in a similar manner, trying for W14's in order to simplify the third-story splices. KL is now 14 feet ± and, with $F_y = 50$, W14 × 99 appears to be the lightest suitable column. However, the table warns us that this section is *noncompact** with respect to flange dimensions, an important consideration, since the axial load being considered contains a fictitious part substituting for bending moment from frame action. Since compact sections are required for ductile moment-resisting frames, the designers choose between the lightest compact W14 and the lightest suitable W12, which turn out to be very close to equal in weight. Section moduli of the W14 × 109 are larger than those of W12 × 106, leading to preference for the W14.

The tentative decision to use the same columns on exterior column lines as interior follows reasoning similar to that used for the beams. In addition, the designers note that the requirements of lateral force resistance are likely to require strength in the exterior columns similar to the interior. A later notation on this sheet indicates that the selection of lower columns made here was overoptimistic. The check for lateral force effects on sheets 17 and 18 was accompanied by a reevaluation of K values, which turned out to be significant.

18.6 LATERAL LOAD ANALYSIS: EFFECTS ON MEMBER SELECTIONS

18.6.1 General Approach

The effects of lateral loads are addressed in sheets 12 through 22. We have already noted on sheet 6 that seismic (earthquake) forces are much more significant than wind forces. The probability of an extreme wind and an extremely severe earthquake occurring at the same time is remote enough so that the UBC does not require consideration of their combined effects. The designers will investigate the effects of combining dead plus live loads with seismically induced forces.

The lateral resisting system consists of a series of rigid frame *bents* in the planes of the columns. A bent consists of columns and either beams or girders, depending on the plane of the bent. Bents resist horizontal forces in their own planes. The system of bents (six acting north–south and four acting east–west) is interconnected through a

*A check of AISC (1.5-5a) would show the member to be *semicompact*.

Project : PRELIMINARY DESIGN - TIER BUILDING
Subject : LATERAL LOAD ANALYSIS

SHEET 12 OF 24

PRELIMINARY LATERAL ANALYSIS

SHEAR WALLS (CONCRETE) WILL BE DESIGNED FOR 100% OF THE HORIZONTAL FORCES AND RESULTING STORY SHEARS IDENTIFIED ON SHEET 6

STRUCTURAL STEEL WILL BE DESIGNED FOR 25% OF THE SAME FORCES AND SHEARS.

ARRANGEMENT OF STEEL FRAME IS QUITE SYMMETRICAL ON BOTH PRINCIPAL AXES OF THE BUILDING. DISTRIBUTION OF MASSES IS SYMMETRICAL WITH RESPECT TO EAST-WEST ℄ AND ALMOST SO WITH RESPECT TO NORTH-SOUTH ℄. UBC (2312-(e)5) REQUIRES APPLICATION OF TORSIONAL MOMENT DUE TO ECCENTRICITY OF 0.05 × MAX. BUILDING DIMENSION D BETWEEN CENTER OF RESISTANCE AND CENTER OF MASS. USE .05DV = T FOR PRELIMINARY ANALYSIS

RESPONSE OF FRAMES TO LATERAL FORCES MUST BE BASED ON A STIFFNESS ANALYSIS; i.e. MEMBERS OF A GROUP OF FRAMES RESISTING IN ANY DIRECTION OF FORCE WILL RESIST IN PROPORTION TO THEIR RELATIVE STIFFNESSES AS MEASURED BY FORCE PER UNIT OF DISPLACEMENT.

FOR THIS STAGE OF ANALYSIS, THE "PORTAL METHOD" WILL BE CONSIDERED TO DESCRIBE A SATISFACTORY APPROXIMATION OF THE STIFFNESS CHARACTERISTICS OF THIS RELATIVELY SHORT REGULAR BUILDING. ASSUMPTIONS OF THE PORTAL METHOD:

1. IN A SINGLE PLANE FRAME, TOTAL STORY SHEAR IS DISTRIBUTED TO ALL COLUMNS IN PROPORTIONS OF 1.0 TO EACH INTERIOR COLUMN AND 0.5 TO EACH EXTERIOR COLUMN
2. POINTS OF CONTRAFLEXURE (M = 0) ARE ASSUMED AT MIDSPAN OF EACH BEAM OR GIRDER IN THE FRAME AND AT MID-HEIGHT OF EACH COLUMN IN A STORY.

WITH THESE ASSUMPTIONS, THE FRAME IS MADE STATICALLY DETERMINATE.

STORY DRIFT (RELATIVE TRANSLATION OF COLUMN TOP TO BOTTOM WITHIN A STORY) MUST BE ≤ 0.005 × STORY HEIGHT. DRIFT IS CALCULATED IN ACCORDANCE WITH UBC 2312-(h).

ALL ALLOWABLE STRESSES ARE INCREASED BY 1/3 FOR LOAD COMBINATIONS WHICH INCLUDE SEISMIC LOADS. SEISMIC LOADS MUST BE COMBINED WITH DEAD LOAD ALONE OR W/ DEAD + LIVE FOR MAX. EFFECT.

very stiff diaphragm at each floor, each diaphragm being a floor slab acting as a very deep girder resisting forces in the plane of the floor and distributing their effects to the bents that act as their supports. The rigid diaphragms limit the difference in deflection between parallel bents so that, in the absence of torsion around a vertical axis of the building, they all deflect equal distances. Since we have symmetrical steel construction in both principal building directions, the center of resistance of the steel space frame is at the geometric center of the building. Joints within a bent are continuous.

Eccentricity between the seismic forces, which act through the mass center of the building, and the center of resistance causes torsional (twisting) effects on the building. The primary mechanism for resisting such torsion in this doubly symmetric system is by pairs of coupled, equal and opposite forces in the planes of the bents. The designer sets down some basic information and criteria for seismic design on sheet 12.

The rigid frames are statically indeterminate structures and any strength analysis must be based on analysis of the deflections of the frame. Forces causing horizontal shear in the building will be resisted internally by horizontal shears in the columns. Column shears cause column moments that must be resisted by girder moments. Girder moments lead to girder shears. Shear, and therefore moment, is attracted to each of a group of columns in proportion to its stiffness EI/L, modified by the stiffnesses of other members, such as girders that connect to the joints at the top and bottom of a column within a story. Having chosen tentative sizes, a deflection analysis is possible. The *portal method* chosen on sheet 12 is a widely used approximate method of analysis, useful for simple, regular rigid frame bents, particularly in relatively short, squat, multistory frames. Most texts on statically indeterminate analysis, including reference 32, describe the method. To analyze a single bent does not require preselection of members. However, since in our building parallel east–west bents do not have equally stiff columns, we need to examine the effects of member size in our analysis.

The plan and sections on sheet 13 show the lines of bents and proposed orientation of column axes with respect to the building axes. All columns within a story are the same. Half of the columns have their stiffer direction resisting east–west shear forces; the other half are stiffest for north–south shears. The designers are seeking relatively equal deflections in both directions. Further calculations will be required before they know whether that has been achieved.

Since the relative stiffnesses of the columns are roughly proportional to their moments of inertia, and there are two moments of inertia affecting each direction of bents, the designers record the ratio I_y/I_x for the columns being considered.

Horizontal forces are applied to the frames in each direction, representing one-fourth of the forces found on sheet 6 for the building. The corresponding story shears are recorded. Torsion around vertical axes is found by code requirements by establishing the minimum eccentricity required by the UBC between the center of mass and center of resistance. The resulting torsional resisting shears are approxi-

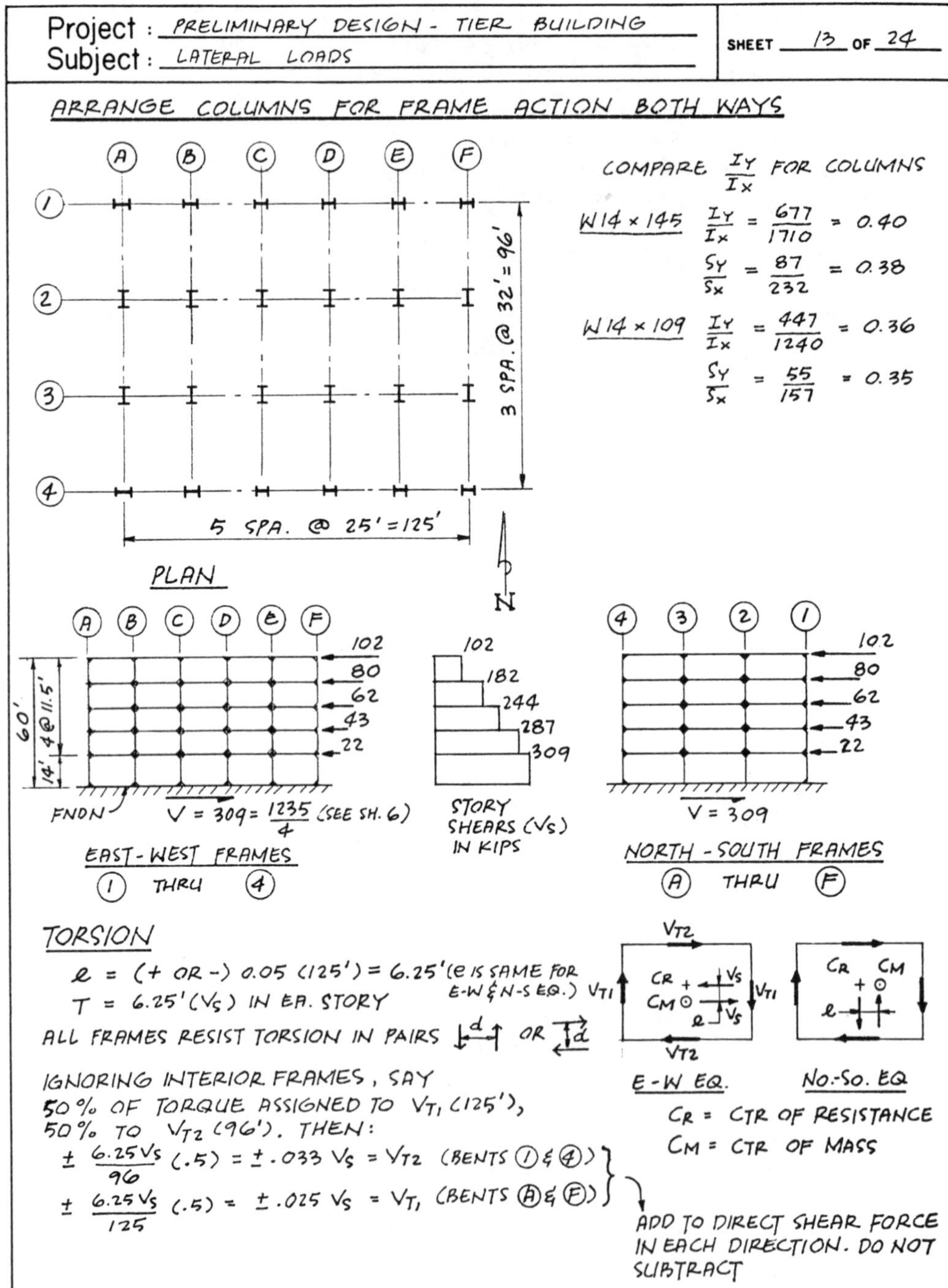
Project : PRELIMINARY DESIGN - TIER BUILDING
Subject : LATERAL LOADS
SHEET 13 OF 24
ARRANGE COLUMNS FOR FRAME ACTION BOTH WAYS
A B C D E F
1 2 3 4
3 SPA. @ 32' = 96'
5 SPA. @ 25' = 125'
PLAN
N
COMPARE IY/IX FOR COLUMNS
W14 × 145 IY/IX = 677/1710 = 0.40
SY/SX = 87/232 = 0.38
W14 × 109 IY/IX = 447/1240 = 0.36
SY/SX = 55/157 = 0.35
102
80
62
43
22
60'
14' 4@11.5'
FNDN
V = 309 = 1235/4 (SEE SH. 6)
EAST-WEST FRAMES
1 THRU 4
102
182
244
287
309
STORY SHEARS (VS) IN KIPS
V = 309
NORTH-SOUTH FRAMES
A THRU F
TORSION
e = (+ OR -) 0.05 (125') = 6.25' (e IS SAME FOR E-W & N-S EQ.)
T = 6.25'(VS) IN EA. STORY
ALL FRAMES RESIST TORSION IN PAIRS
OR
d
IGNORING INTERIOR FRAMES, SAY
50% OF TORQUE ASSIGNED TO VT1 (125'),
50% TO VT2 (96'). THEN:
± 6.25VS/96 (.5) = ±.033 VS = VT2 (BENTS 1 & 4)
± 6.25VS/125 (.5) = ±.025 VS = VT1 (BENTS A & F)
ADD TO DIRECT SHEAR FORCE IN EACH DIRECTION. DO NOT SUBTRACT
VT2
VT1
CR
CM
VS
e
E-W EQ.
NO.-SO. EQ
CR = CTR OF RESISTANCE
CM = CTR OF MASS

mated, on the safe side, by assuming coupled shearing forces in pairs of exterior bents only. Frames closer to the center of rotation will also resist torsion, but they will deflect less and therefore contribute smaller resisting torques. The results, V_{T2} and V_{T1}, will be added to shears resulting from loads through the center of resistance to represent the effects of seismic action on the most heavily loaded frames.

18.6.2 Distribution of Story Shears to Columns

On sheet 14, story shears are distributed to the columns within a story by the portal method. The bents considered are in east–west planes. All columns within a bent have equal I; all beams spans are equal. The portal method assumptions of ½ : 1 for shear distribution to exterior and interior columns apply. The sum of all coefficients of column shear at each half-story level must equal 1.00. Points of zero moment (*contraflexure*) are assumed at the mid-story heights of columns and the midspan points of beams and girders.

There are four east–west bents. However, columns on lines 1 and 4 bend about their stiff axes; those on lines 2 and 3 about their minor axes. It is necessary to seek the ratio of shear forces resisted by the exterior bents (1 and 4) to those on the interior bents (2 and 3) under conditions of equal lateral translation.

If all bents were equally stiff (force/unit deflection), they would share the shear load equally. If the beam stiffnesses were very much larger than the column stiffnesses,

$$\frac{I_B/L_B}{I_c/I_c} = \frac{K_B}{K_c} \rightarrow \infty$$

the shear would be shared in the ratio of column stiffnesses, $I_x/I_y = 2.5:1$. The actual case lies between the two extremes. For present purposes, the designers approximate the stiffness ratio as 1.5:1 so that exterior bents each take 30% of total shear and interior bents 20% each. With the added effect of torsional shears, these become 33% and 21%.

Coefficients shown in part (c) of this sheet combine the results of parts (a) and (b). Thirty-three percent of V_i or V_j is assigned to each of bents 1 and 4, 21% to each of bents 2 and 3. These percentages are then distributed to columns within each bent in the proportions shown in part (a).

In the case of the north–south bents (sheet 15), it is observed that all six are identical and therefore equally stiff. Each resists one-sixth of the total story shear and what we now recognize are small added shears due to torsion. The method of distribution of shear to bents and to columns is the same as was used for east–west bents.

A table of column moments completes this sheet. These are moments at column–girder joints. Consistent with the portal method assumption, column end moments are column shear times half of the story height. The larger values of M_x and

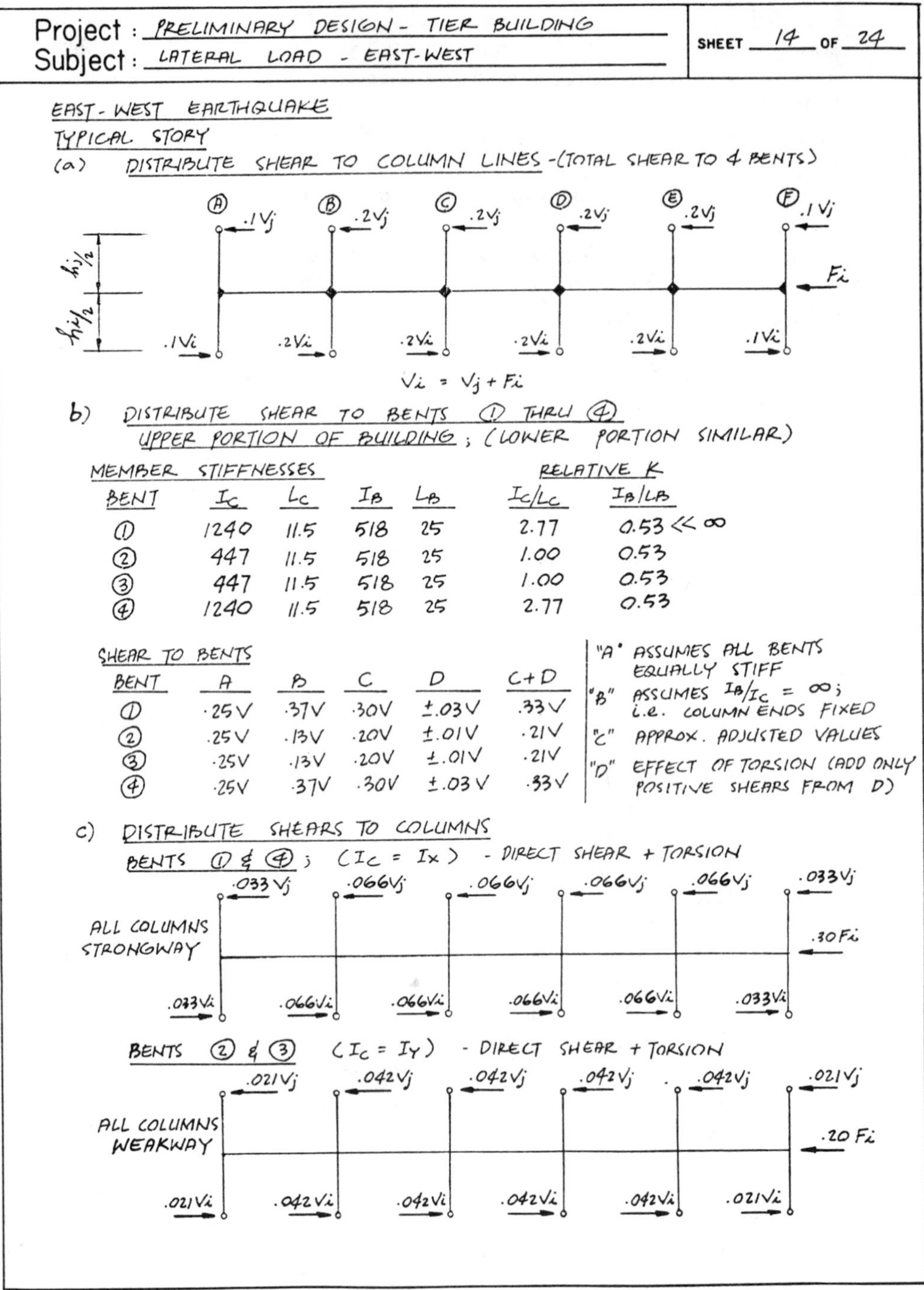
Project : PRELIMINARY DESIGN - TIER BUILDING
Subject : LATERAL LOAD - EAST-WEST
SHEET 14 OF 24
EAST-WEST EARTHQUAKE
TYPICAL STORY
(a) DISTRIBUTE SHEAR TO COLUMN LINES -(TOTAL SHEAR TO 4 BENTS)
A .1Vj
B .2Vj
C .2Vj
D .2Vj
E .2Vj
F .1Vj
Fi
hi/2
hi/2
.1Vi
.2Vi
.2Vi
.2Vi
.2Vi
.1Vi
Vi = Vj + Fi
b) DISTRIBUTE SHEAR TO BENTS 1 THRU 4
UPPER PORTION OF BUILDING ; (LOWER PORTION SIMILAR)
MEMBER STIFFNESSES
RELATIVE K
BENT Ic Lc IB LB Ic/Lc IB/LB
1 1240 11.5 518 25 2.77 0.53 << ∞
2 447 11.5 518 25 1.00 0.53
3 447 11.5 518 25 1.00 0.53
4 1240 11.5 518 25 2.77 0.53
SHEAR TO BENTS
BENT A B C D C+D
1 .25V .37V .30V ±.03V .33V
2 .25V .13V .20V ±.01V .21V
3 .25V .13V .20V ±.01V .21V
4 .25V .37V .30V ±.03V .33V
"A" ASSUMES ALL BENTS EQUALLY STIFF
"B" ASSUMES IB/Ic = ∞; i.e. COLUMN ENDS FIXED
"C" APPROX. ADJUSTED VALUES
"D" EFFECT OF TORSION (ADD ONLY POSITIVE SHEARS FROM D)
c) DISTRIBUTE SHEARS TO COLUMNS
BENTS 1 & 4 ; (Ic = Ix) - DIRECT SHEAR + TORSION
.033Vj
.066Vj
.066Vj
.066Vj
.066Vj
.033Vj
ALL COLUMNS STRONGWAY
.30Fi
.033Vi
.066Vi
.066Vi
.066Vi
.066Vi
.033Vi
BENTS 2 & 3 (Ic = IY) - DIRECT SHEAR + TORSION
.021Vj
.042Vj
.042Vj
.042Vj
.042Vj
.021Vj
ALL COLUMNS WEAKWAY
.20Fi
.021Vi
.042Vi
.042Vi
.042Vi
.042Vi
.021Vi

Project : PRELIMINARY DESIGN - TIER BUILDING
Subject : LATERAL LOADS - NORTH-SOUTH

SHEET 15 OF 24

NORTH-SOUTH EARTHQUAKE - DISTRIBUTE SHEAR TO COLUMNS

TYPICAL STORY

SINCE SIX EQUAL BENTS (Ⓐ THRU Ⓕ), ASSIGN $\frac{V}{6}$ TO EACH BENT = .1667 V

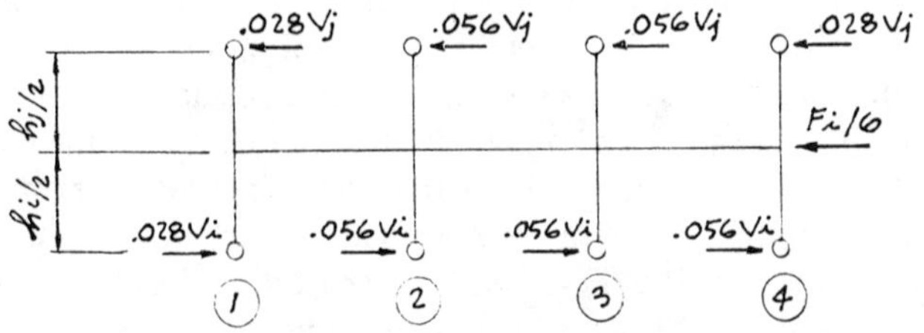

ADD TORSION - MAX. TO BENTS A & F = .025 Vs

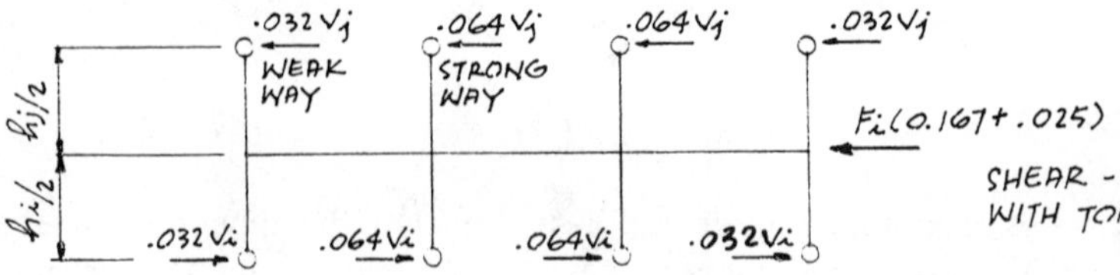

MAX. MOMENTS - EAST-WEST OR NORTH-SOUTH E.Q.

	E-W	N-S	USE
M_x	$.066\, V_i \frac{h_i}{2}$	$.064\, V_i \frac{h_i}{2}$	$.033\, V_i h_i$
M_y	$.042\, V_i \frac{h_i}{2}$	$.032\, V_i \frac{h_i}{2}$	$.021\, V_i h_i$

M_y will be used in examining the columns for the combined effects of axial load plus bending. Since the UBC does not require east–west and north–south earthquake effects to be applied simultaneously, axial loads will be combined with either M_x or M_y in seeking the worst effect.

18.6.3 Lateral Effects on Column Sizes

Sheets 16 through 18 test the earlier (sheet 11) assumptions regarding interaction of axial loads and moments in the columns. The critical cases are the first and third story. The test is that of Section 1.6, AISC Specification. Since the axial load is high, the applicable formulas are (1.6-1a) and (1.6-1b). The sum of utilization effects is permitted to be ≤ 1.33 rather than 1.0. At the same time, a somewhat closer and more conservative evaluation of the factor K is made for use in estimating Kl/r of the columns. The nomograph in AISC Figure C1.8.2 (Fig. 9.17, this book) is intended to yield values of K for unbraced frames such as this one.

The values of

$$G = \frac{\Sigma\, I_c/L_c}{\Sigma\, I_g/L_g}$$

are the ratios of column to girder stiffness at each of the upper and lower joints. Large values of these ratios give large values of K, indicating that the girders are not stiff enough to *fix* the ends of the columns against rotation. The original assumption (sheet 11) of $K = 1.2$ appears to be too low. After examining the range of column and girder stiffnesses based on the previous work, a (still tentative) larger value is chosen.

Results in the first story force a substantial increase in the weight of the column section resulting from the application of AISC formula (1.6-1a) to the condition of axial load combined with bending about the Y axis. The ratio f_a/F'_{ey} is quite large, indicating substantial loss of bending capacity in the presence of potential instability due to axial loads. The rapid decline of the parameter F'_{ey} is quite large with increasing values of Kl/r_y and can be examined in AISC Specification, Table 9. The formula for F'_e in Specification Section 1.6.1 can be recognized as the Euler column formula with the usual safety factor, but applied for potential buckling in the plane of bending.

Coefficients found for the third-story columns turn out to be below 1.33 for vertical load combined with earthquake. The column was originally chosen to be more than adequate for dead plus live loads. It can be expected to handle, without difficulty, the relatively small additional moments resulting from unbalanced live loads. Verification of this is left for the more complete analysis to be done in the final design stage, at which time the value of K will be more closely evaluated.

Project : PRELIMINARY DESIGN - TIER BUILDING
Subject : LATERAL LOADS - COLUMNS

SHEET 16 OF 24

1ST STORY COLUMNS - COMBINED AXIAL LOAD + BENDING

CHECK W 14 × 145 (SHT. 11); $h_i = 14.0'$ TO FOUNDATION

F_Y = 50 ksi ; COMPACT BY SECTION PROPERTIES

L_c = 13.9 > CLEAR DISTANCE < 14.0'

A = 42.7 in²
I_x = 1710 in⁴
S_x = 232 in³
r_x = 6.33 in

I_Y = 677 in⁴
S_Y = 87.3 in³
r_Y = 3.98 in

APPROX. "K"
FOLLOW AISC C1.8.2
LOWER END FIXED: $G_B \to 1$ *
UPPER END

COLUMN - I_x = 1710
I_Y = 677
$L_{AV} \approx 13$, USE 14

BEAMS (W16×40)
I_x = 659 ; L = 25

GIRDERS (W18×97)
I_x = 1750 ; L = 32

$G_A \lesssim 5$; USE K = 1.7

$$\frac{KL}{r_x} = \frac{1.7(14)(12)}{6.33} = 45$$

$$\frac{KL}{r_Y} = \frac{1.7(14)(12)}{3.98} = 72$$

1) AXIAL + M_X

P_{D+L} (FR. SH. 11) = 880K
$M_X = .033\, V_i h_i = .033(309)(14) = 143$ K-FT

OR

2) AXIAL + M_Y

P_{D+L} = 880K
$M_Y = .021\, V_i h_i$ = 91 K-FT

[NOTE: M_X & M_Y COMBINE IN SOME COLUMNS DUE TO DIRECT SHEAR & TORSION. SINCE TORSION EFFECT IS SMALL, IGNORE COMBINATION AT THIS TIME.]

FROM AISC 1.6.1

f_a = 20.6 F_a = 20.6 ksi (TABLE 3-50)
f_{bx} = 7.40 F_{bx} = .66(50) = 33.0 (1.5.1.4.1)
f_{by} = 12.51 F_{by} = .75(50) = 37.5 (1.5.1.4.3)
$C_{mx} = C_{my}$ = 0.85 $\frac{f_a}{F'_{ex}}$ = .28
F'_{ex} = 73.7 (TABLE 9)
F'_{ey} = 28.8 $\frac{f_a}{F'_{ey}}$ = .72

$C_{MX} = C_{MY}$ = 0.85

(CONT'D SHT. 17)

* ASSUMPTION OF G_B = 1.0 ASSUMES ALMOST FULL FIXITY AT TOP OF FOOTING. FOUNDATION DESIGN MUST PROVIDE FIXITY OR FIRST STORY ANALYSIS MUST BE BASED ON DEGREE OF FIXITY PROVIDED BY FOOTINGS. COLUMN MOMENTS WILL INCREASE IF FOOTING NOT FIXED. RE-CHECK IN FINAL DESIGN STAGE.

Project : PRELIMINARY DESIGN - TIER BUILDING
Subject : LATERAL LOADS - COLUMNS

SHEET 17 OF 24

1ST STORY CONT'D.

FOR D+L+EQ, ALLOW UTILIZATION FACTOR ≤ 1.33

1) P + Mx (SECTION 1.6 AISC SPEC)

(1.6-1a) $f_a/F_a + \dfrac{C_{mx} f_{bx}}{(1 - f_a/F'_{ex}) F_{bx}} + 0 = 1.00 + 0.26 = 1.26 < 1.33$ OK

(1.6-1b) $f_a/.6F_y + f_{bx}/F_{bx} + 0 = .69 + .22 = 0.91 \ll 1.33$ OK

2) P + My

(1.6-1a) $f_a/F_a + \dfrac{C_{my} f_{by}}{(1 - f_a/F'_{ey}) F_{by}} = 1.00 + .88 = 1.88 \gg 1.33$ NG ⟵

(1.6-1b) $f_a/.6F_y + f_{by}/F_{by} = .69 + .33 = 1.02 \approx 1.33$ OK

INCREASE COL. SIZE IN PROPORTION $\geq \dfrac{1.88}{1.33} = 1.41$; $1.41 \times 145 = 204$

TRY W 14 × 211 $A_{211}/A_{145} = 1.46$; $S_{y(211)}/S_{y(145)} = 130/87.3 = 1.49$

FOR P + My, FORMULA 1.6-1a

$$\frac{f_a}{F_a} + \frac{C_{my} f_{by}}{(1 - f_a/F'_{ey}) F_{by}} = .68 + .59 = 1.27 < 1.33 \text{ OK}$$

USE W14 × 211

Project : PRELIMINARY DESIGN - TIER BUILDING
Subject : LATERAL LOADS - COLUMNS
SHEET 18 OF 24

CHECK 3RD STORY COLUMNS (SHEET 11)

STORY SHEAR = 244 K ; (SHEET 13)

AXIAL LOAD = 528 K (SHT. 11)

FROM SHT. 11, W 14 × 109 - COMPACT SECTION TO $L_c = 13.1 > h_i = 11.5$

$A = 32.0 \text{ in}^2$

$I_x = 1240 \text{ in}^4$

$S_x = 173 \text{ in}^3$

$r_x = 6.22 \text{ in}$

$I_y = 447 \text{ in}^4$

$S_y = 61.2 \text{ in}^3$

$r_y = 3.73 \text{ in}$

$$\frac{KL}{r_x} = \frac{1.7(11.5)(12)}{6.22} = 37.7$$

$$\frac{KL}{r_y} = \frac{1.7(11.5)(12)}{3.73} = 62.9$$

1) $\begin{cases} P = 528K \\ M_x = .033(244)(11.5) = 93 K' \end{cases}$

OR

2) $\begin{cases} P = 528K \\ M_y = .021(244)(11.5) = 59 K' \end{cases}$

$f_a = 16.5$ $\quad F_a = 22.2$ (TABLE 3-50)

$f_{bx} = 6.45$ $\quad F_{bx} = .66 F_y = 33$ ksi

$f_{by} = 11.6$ $\quad F_{by} = .75(50) = 37.5$ ksi

$\left.\begin{matrix} F'_{ex} = 105 \\ F'_{ey} = 38 \end{matrix}\right\}$ TABLE 9

$\frac{f_a}{F'_{ex}} = .16 \qquad \frac{f_a}{F'_{ey}} = .43$

$C_{mx} = C_{my} = 0.85$

1) P + Mx

1.6 - 1a $\quad \dfrac{f_a}{F_a} + \dfrac{C_{mx} f_{bx}}{\left(1 - \frac{f_a}{F'_{ex}}\right) F_{bx}} = 0.74 + 0.20 = .94 < 1.33$ OK

1.6 - 1b $\quad \dfrac{f_a}{.6F_y} + \dfrac{f_{bx}}{F_{bx}} = 0.55 + 0.20 = 0.75 \ll 1.33$ OK

2) P + My

1.6 - 1a $\quad \dfrac{f_a}{F_a} + \dfrac{C_{my} f_{by}}{\left(1 - \frac{f_a}{F'_{ey}}\right) F_{by}} = 0.74 + 0.46 = 1.20 < 1.33$ OK

1.6 - 1b $\quad \dfrac{f_a}{.6F_y} + \dfrac{f_{by}}{F_{by}} = 0.55 + 0.31 = 0.86 \ll 1.33$ OK

VERIFIES USE OF W14 × 109 FOR UPPER STORY COLUMNS.
RECHECK "K" AT FINAL DESIGN.

Project : PRELIMINARY DESIGN - TIER BUILDING
Subject : LATERAL LOADS - GIRDERS

SHEET 19 OF 24

CHECK LATERAL MOMENT EFFECT ON GIRDERS - NO-SO EARTHQUAKE

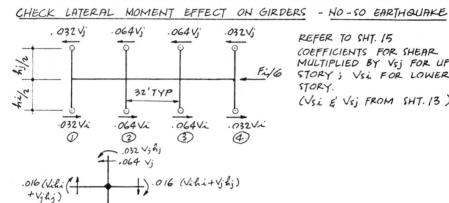

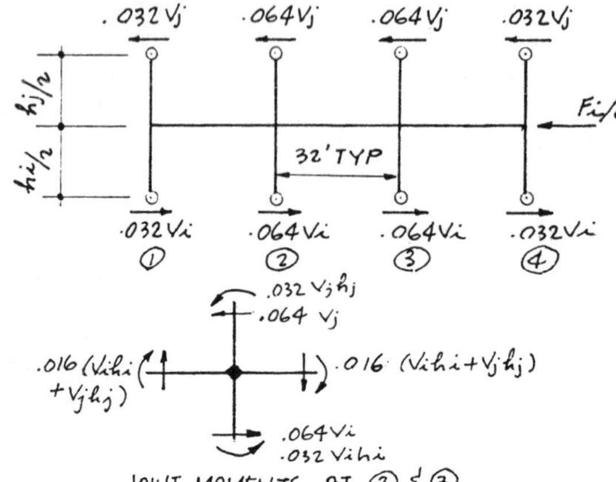

REFER TO SHT. 15
COEFFICIENTS FOR SHEAR MULTIPLIED BY V_{sj} FOR UPPER STORY ; V_{si} FOR LOWER STORY.
(V_{si} & V_{sj} FROM SHT. 13)

INPUT MOMENT FR. HORIZ. SHEAR = $.016\ (V_i h_i + V_j h_j) = M_G$

GIRDER MOMENTS - SEISMIC

FLOOR	V_i	V_j	h_i	h_j	M_G
5 (ROOF SIM.)	182	102	11.5	11.5	52
4	244	182	11.5	11.5	78
3	287	244	11.5	11.5	98
2	309	287	14	11.5	122

FROM SHT. 10

$(-)\ M_{D+L} = 437^{K'}$ TYPICAL FLOOR

$M_{(D+L+EQ)} \leq 122 + 437 = 559^{K'}$

$$f_{bx} = \frac{M}{S} = \frac{559(12)}{188} = 35.7 \text{ ksi}$$

ALLOW

$F_{bx}\ (1.33) = .66\ (50)(1.33) = 43.9 \text{ ksi}$

∴ W 18 × 97 OK ALL FLOORS

18.6.4 Lateral Effects on Beam and Girders

The continuous beams and girders on bent lines, originally selected for vertical loads, are tested for combined bending effects of dead, live, and seismic loads on sheets 19 and 20. Shears and bending moments arise from previous lateral analyses on sheets 13 through 15. It is clear that shear is not high enough to control the selection. But there are significant bending moments that, near the joints, add to the negative moments from vertical loads. The Code stipulates increases in allowable stress of one-third, so there is automatically a reserve for seismic increase of moment equal to one-third of the design moment for dead plus live load. If the beam was not fully stressed due to dead plus live loads, the reserve is even greater. For combined effects, $M_{allow} = F_b\,(1.33)(S_x) = 0.88S_x$, since the section is compact.

Bending moments from seismic loads are calculated for each floor and combined with moments previously calculated for dead plus live load. The reserve capacity is sufficient for all north–south girders. It turns out to be insufficient for the east–west beams in the lower three floors. The decision is made to change the steel of beams B2 and B3 to a type with $F_y = 50$. Although not strictly necessary, this will be done for all floors.

A flag on this sheet notes that these beams required further revision when, on a later sheet, the effects of story drift were considered.

18.6.5 Check of Story Drift

The limits of acceptable *story drift* are prescribed in the UBC in 2314(h). This limit is intended to avoid two potential problems:

1. Excessive deflection would cause additional bending moments in the columns as the axial column loads (not shown on the diagrams on sheets 21 and 22) become eccentric with their reactions at the base of the columns. This is often referred to as the $P\Delta$ *effect*.
2. Excessive drift would cause difficulty with the joints in the precast wall panels. The panels themselves cannot distort to follow the steel frame. Joints between them are deliberately made with small gaps, filled with soft expansion material or otherwise sealed against the weather. This allows the steel frame to distort without causing these nonstructural members to resist effects for which they were not designed. Corresponding requirements for the details of the precast wall panels are found in UBC 2312(j)3.C.

The calculated story drift is elastic, based on the values of story shear previously assigned. However, the concept of ductile moment-resisting frames implies that under extreme seismic conditions the frame may distort plastically. This

Project : PRELIMINARY DESIGN - TIER BUILDING
Subject : LATERAL LOADS - BEAMS

SHEET 20 OF 24

CHECK SEISMIC EFFECT ON BEAMS (EAST-WEST EQ)

BEAM B3 (REFER SHT 14, SHT 7) (B2 SIM., BUT V & M LOWER)

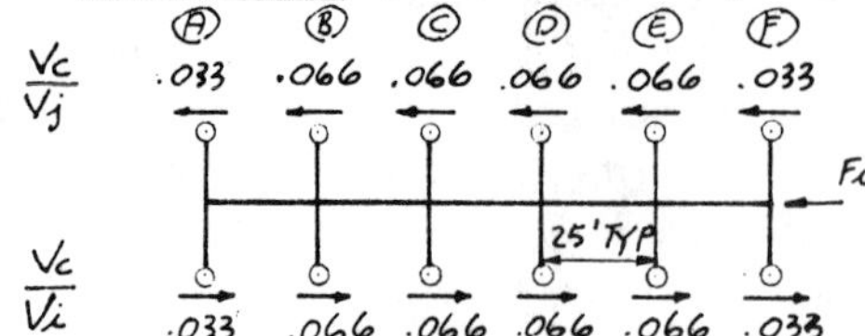

REFER TO SHT 13 FOR STORY SHEARS.

$\text{MAX. } M_{BM} = \frac{.066}{2(2)}(V_i h_i + V_j h_j)$

BEAM MOMENTS - SEISMIC

FLOOR	V_i	V_j	h_i	h_j	M_{BM}
5 (ROOF SIM.)	182	102	11.5	11.5	54
4	244	182	11.5	11.5	81
3	287	244	11.5	11.5	101
2	309	287	14.0	11.5	126

FROM SHT. 9, W16×40 - (S_x = 64.7; COMPACT, A36 STEEL)

$-M_{D+L} \approx 101\,K'$

M_{ALLOW} @ $F_b S$ WHERE $(F_b)_{MAX} = .66 F_Y (1.33) = 31.6\text{ksi}$

$M_{ALL.} = 31.6 \frac{(64.7)}{12} = 170$; $170 - 101 = 69$; OK 5TH + ROOF, N.G. OTHERS

TRY SAME SECTION WITH F_Y = 50 ksi

$M_{ALL} = 50(.66)(1.33)\frac{(64.7)}{12} = 237^{K'} > 101 + 126 = 227 = M_{MAX}$

USE W16 × 40 - ASTM A572 - GR. 50, ALL FLOORS

REVISED FOR DRIFT CONTROL, SEE SHEET 22.

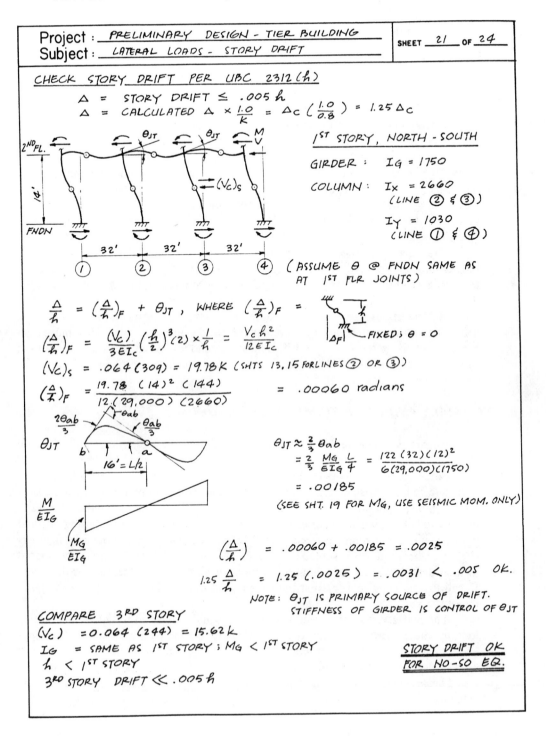

Project: PRELIMINARY DESIGN - TIER BUILDING
Subject: LATERAL LOADS - STORY DRIFT
SHEET 21 OF 24

CHECK STORY DRIFT PER UBC 2312 (h)

Δ = STORY DRIFT $\leq .005\,h$

Δ = CALCULATED $\Delta \times \frac{1.0}{K} = \Delta_C \left(\frac{1.0}{0.8}\right) = 1.25\,\Delta_C$

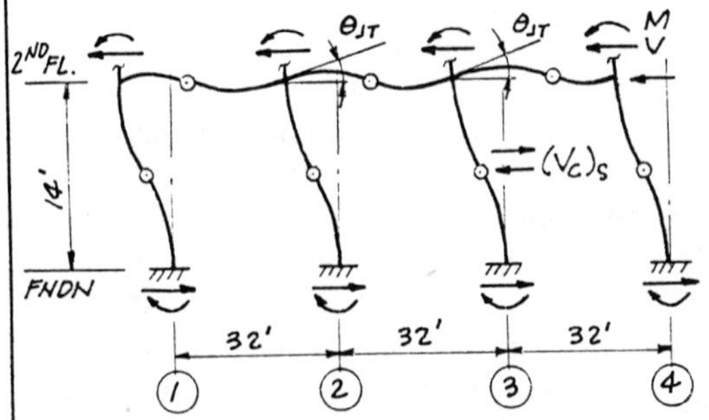

1ST STORY, NORTH - SOUTH

GIRDER: $I_G = 1750$

COLUMN: $I_X = 2660$ (LINE ② & ③)

$I_Y = 1030$ (LINE ① & ④)

(ASSUME θ @ FNDN SAME AS AT 1ST FLR JOINTS)

$\frac{\Delta}{h} = \left(\frac{\Delta}{h}\right)_F + \theta_{JT}$, WHERE $\left(\frac{\Delta}{h}\right)_F$ =

$\left(\frac{\Delta}{h}\right)_F = \frac{(V_c)}{3EI_c}\left(\frac{h}{2}\right)^3 (2) \times \frac{1}{h} = \frac{V_c h^2}{12 E I_c}$

$(V_c)_s = .064\,(309) = 19.78K$ (SHTS 13, 15 FOR LINES ② OR ③)

$\left(\frac{\Delta}{h}\right)_F = \frac{19.78\,(14)^2\,(144)}{12\,(29{,}000)\,(2660)} = .00060$ radians

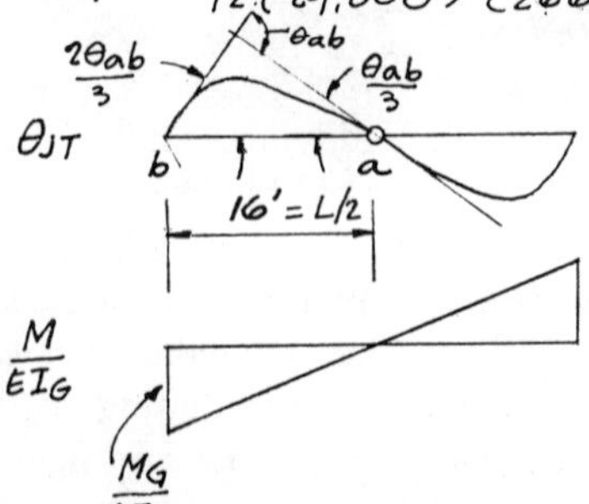

$\theta_{JT} \approx \frac{2}{3}\theta_{ab}$

$= \frac{2}{3}\,\frac{M_G}{EI_G}\,\frac{L}{4} = \frac{122\,(32)(12)^2}{6(29{,}000)(1750)}$

$= .00185$

(SEE SHT. 19 FOR M_G, USE SEISMIC MOM. ONLY)

$\left(\frac{\Delta}{h}\right) = .00060 + .00185 = .0025$

$1.25\,\frac{\Delta}{h} = 1.25\,(.0025) = .0031 < .005$ OK.

NOTE: θ_{JT} IS PRIMARY SOURCE OF DRIFT. STIFFNESS OF GIRDER IS CONTROL OF θ_{JT}

COMPARE 3RD STORY

$(V_c) = 0.064\,(244) = 15.62\,k$

I_G = SAME AS 1ST STORY; M_G < 1ST STORY

h < 1ST STORY

3RD STORY DRIFT $\ll .005\,h$

STORY DRIFT OK FOR NO-SO EQ.

apparent contradiction is partially resolved by the Code requirement that the limit of

$$\frac{\Delta}{h} \leq 0.005$$

be satisfied with 125% of the calculated elastic story drift.

The calculated drift, Δ, is made up of two parts, the effect of translation with joints fixed against rotation, plus the effect of joint rotation. The relative contribution of the two parts depends on the relative stiffness (I/L) of the columns, whose moments rotate the joints, and the girders or beams that resist those moments. Recall that these are the same relative stiffnesses considered in the nomograph of column K values cited on calculation sheet 16. Variation in K results from variation in the drift restraint of the system.

The calculation is based on the moment area approach to analysis of statically indeterminate structures. In this case, the designers are interested to note (sheet 21) that, because of the relatively flexible girders, joint rotation contributes the bulk of the displacement. This is even more true for the east–west direction for which the beams are even less stiff relative to the columns than are the north–south girders. Once again, the north–south bents appear adequate, but the east–west bents drift more than is permitted. Since the source of drift is mostly due to low beam stiffness, the solution used is to increase the I of the beams; the columns require no change.

18.7 SUMMARY AND STRUCTURAL STEEL BUDGET

The designers summarize the results of the calculations in readily accessible form on sheets 23 and 24. A calculation of total steel tonnage is now possible from a *take-off* of quantities of all members of the structural frame. Since only main members have been selected, the designers add an allowance for *details*, intended to cover gussets, angles, bolts, and other connecting pieces, as well as some secondary, special-purpose members, which experience tells them will become necessary during the final design state. Total tonnage is converted to pounds per square foot. The figure is slightly higher than was assumed on sheet 5. The difference is small enough compared to total dead load that recalculation of member sizes is not necessary at this time. Recall the earlier discovery that the weight of slab assumed on sheet 5 was higher than necessary.

The budget estimate is recorded in dollars and converted to dollars per square foot, in which form it is useful for comparison to available data on similar buildings and for early projections of structural costs on future proposed projects. The unit cost used in dollars per ton is often derived from publications that keep current records based on published data from many projects. In some engineering companies an estimating department follows actual costs on their own projects and/or derives costs

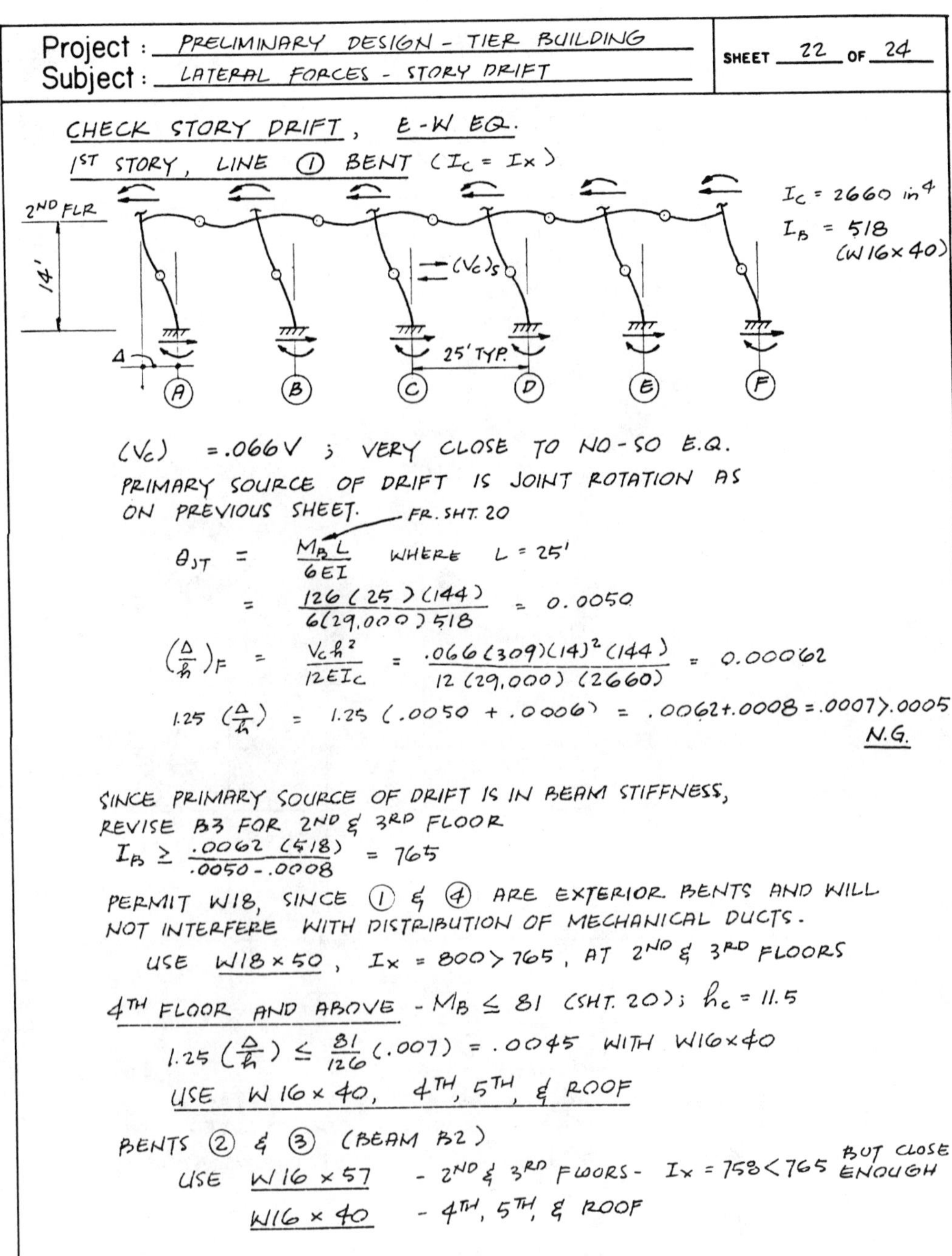

Project: PRELIMINARY DESIGN - TIER BUILDING

Subject: LATERAL FORCES - STORY DRIFT

SHEET 22 OF 24

CHECK STORY DRIFT, E-W EQ.

1ST STORY, LINE ① BENT ($I_c = I_x$)

$(V_c) = .066V$; VERY CLOSE TO NO-SO E.Q.

PRIMARY SOURCE OF DRIFT IS JOINT ROTATION AS ON PREVIOUS SHEET. — FR. SHT. 20

$$\theta_{JT} = \frac{M_B L}{6EI} \quad \text{WHERE } L = 25'$$

$$= \frac{126(25)(144)}{6(29{,}000)518} = 0.0050$$

$$\left(\frac{\Delta}{h}\right)_F = \frac{V_c h^2}{12EI_c} = \frac{.066(309)(14)^2(144)}{12(29{,}000)(2660)} = 0.00062$$

$$1.25\left(\frac{\Delta}{h}\right) = 1.25(.0050 + .0006) = .0062 + .0008 = .0007 > .0005$$

N.G.

SINCE PRIMARY SOURCE OF DRIFT IS IN BEAM STIFFNESS, REVISE B3 FOR 2ND & 3RD FLOOR

$$I_B \geq \frac{.0062(518)}{.0050 - .0008} = 765$$

PERMIT W18, SINCE ① & ④ ARE EXTERIOR BENTS AND WILL NOT INTERFERE WITH DISTRIBUTION OF MECHANICAL DUCTS.

USE W18×50, $I_x = 800 > 765$, AT 2ND & 3RD FLOORS

4TH FLOOR AND ABOVE - $M_B \leq 81$ (SHT. 20); $h_c = 11.5$

$$1.25\left(\frac{\Delta}{h}\right) \leq \frac{81}{126}(.007) = .0045 \quad \text{WITH W16×40}$$

USE W16×40, 4TH, 5TH, & ROOF

BENTS ② & ③ (BEAM B2)

USE W16×57 - 2ND & 3RD FLOORS - $I_x = 758 < 765$ BUT CLOSE ENOUGH

W16×40 - 4TH, 5TH, & ROOF

Project : PRELIMINARY DESIGN - TIER BUILDING
Subject : SUMMARY
SHEET 23 OF 24

SUMMARY OF PRELIMINARY DESIGN - STRUCTURAL STEEL FRAME

FOR TRANSMISSION TO ARCHITECT FOR BUDGETING AND ARRANGEMENT.

COSTS DERIVED DO NOT REPRESENT TOTAL STRUCTURAL COSTS

INCLUDES COST TO GEN'L CONTRACTOR + CONTR. O.H. & P.
— " — STRUCTURAL STEEL FRAME
— " — METAL DECK IN COMPOSITE SLABS

EXCLUDES (OF STRUCTURAL COSTS):
CONCRETE IN DECK SLABS
PRECAST WALLS
FOUNDATIONS
DESIGN, TESTING & INSPECTION FEES

COSTS ARE AS OF JAN. 1981. ESCALATE TO TIME OF CONTRACT.

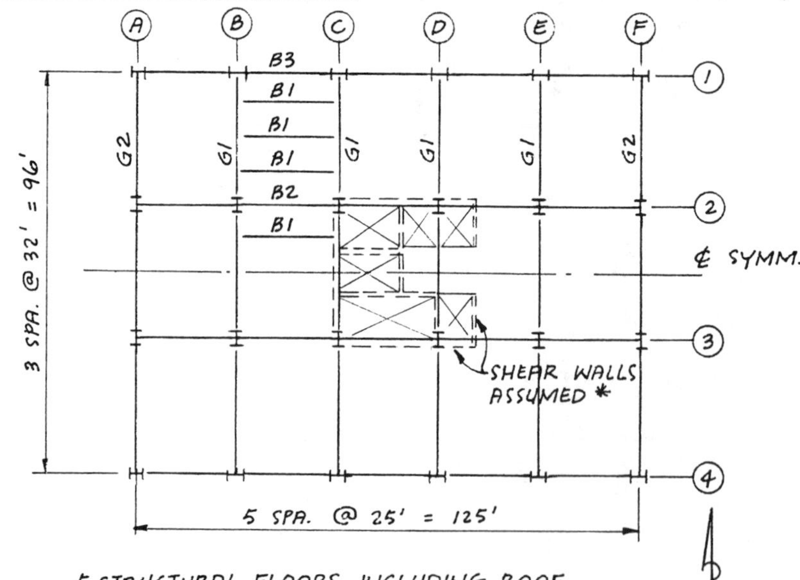

N

5 STRUCTURAL FLOORS INCLUDING ROOF
EACH FLOOR 99' × 128' = 12,672 ☐'
5 USABLE FLOORS INCLUDE 4 STRUCTURAL FLRS
PLUS CONC. SLAB ON GRADE @ STREET LEVEL
12,672 ☐' × 5 = 63,800 ☐' LESS ELEVATORS, STAIRS,
MECH. SPACE

* DESIGN REQUIRES SHEAR WALLS. IF ELIMINATED, STEEL DESIGN MUST BE HEAVIER.

Project : PRELIMINARY DESIGN - TIER BUILDING
Subject : SUMMARY

SHEET 24 OF 24

STEEL FRAMING SUMMARY (CONT)

MEMBER	LOCATION	SHAPE	STEEL	LENGTH (FT)	Q'TY	WEIGHT (LBS)	WEIGHT TOTALS
BEAMS							
B1 (SH. 8)	ALL FLRS	W16×40	A36	25.0	225	225,000	
B2 (SH. 20)	4TH, 5TH, RF	W16×40	A572 GR 50	25'×5	6	30,000	
	2ND, 3RD	W16×57	↓	↓	4	28,500	
B3 (SH. 20)	4TH, 5TH, RF	W16×40	↓	↓	6	30,000	
	2ND, 3RD	W18×50	↓	↓	4	25,000	338,500
GIRDERS G1 & G2 (SH. 10)	ALL FLRS	W18×97	↓	32'×3	30	297,360	297,500
COLUMNS ALL (SH. 11)	ROOF TO FLR 3½	W14×109	↓	28'	24	73,250	
	FLR 3½ TO FOUND'N	W14×211	↓	32'	24	162,050	236,000

SUBTOTAL: 872,000±
ADD 10% FOR DETAILS: 87,000
TOTAL : 959,000
SAY 480 TONS

STR. STL/SQ. FT = $\frac{960,000}{5(99)(128)}$ = 15.15 #/□
(COMPARE TO SHT. 5, BEAMS + GIRDERS + COLUMNS ASSUMED = 14 #/□ - SAY OK, BUT REDUCTION IS PROBABLE IN FINAL DESIGN.)

BUDGET COST OF STRUCTURAL STEEL

480 TONS @ $ 1600 / TON* = $ 768,000
= $ 12.12/□

METAL DECK

1½" MTL DECK QL - UKX (BY ROBERTSON) OR EQUAL.
99' × 128' × 5 FLRS = 63,360 □
@ 1.00 /SQ. FT = $ 63,360

MINIMUM SPACE REQ'D - STRUCTURE AT FLOORS

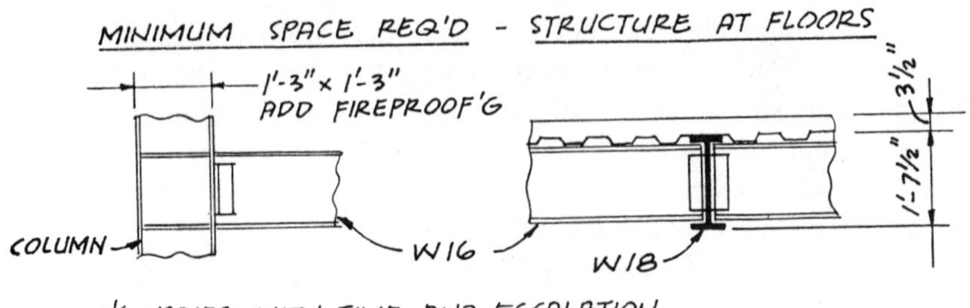

* VARIES WITH TIME AND ESCALATION

from their own analyses of costs of labor, materials, and equipment. In either case, such costs vary over time and require periodic updating.

The designers are careful to note that contractor overhead and profit is included and that the budget covers specific items of work and not others. Ambiguity in presenting costs can lead to very damaging misinterpretations.

In a form dependent on the business practices of the engineering company, the material in the summary will be transmitted as part of the preliminary design report to the architects. If the architects present proposals for structural revisions based on nonstructural needs, the engineers can, by reference to the body of the calculations, quickly ascertain the implications of such proposed changes so they can be weighed against alternative changes by others. Decisions can then be made based on total needs of the project.

18.8 RECAPITULATION OF THE PROCESS

The process of preliminary design is now complete, and the required report is made. If the designers have chosen wisely, and if the original problem statement remains unchanged, they now have a sound framing system that will, if proved out in review by the architect, live well with other systems on the project. Assuming concurrence by others, they are ready to proceed to the final design stage.

It would be instructive at this stage to consider what has been done. The process may be described, grossly, by Flow Chart 18.1. Between conceptual and final design, the preliminary design required seven stages plus two key decision points, each of which could result in return to an earlier stage. The reader may find it useful to develop his or her own subcharts for each stage.

As charted, the process appears to be a linear sequence. Each later stage depended on completion of the preceding stage. A path was followed that turned out to be clear and effective, with minimum need for internal reworking.

However, charting runs the risk of oversimplification. The fact is the designers worked in a more complex and less mechanical way.

1. In stage 2, the designers drew on their own experience, that of their associates, information in the technical literature, and the specifics of the problem to make an informed judgment as to the appropriate primary materials.
2. While proceeding with their choice of steel, they still reserved the option of changing that choice if later stages revealed that it was not the best.
3. When choosing steel in stage 2, the designers already had in mind the system that was formalized in stage 3. Very likely, early exploratory sketches, made during stage 2, were held for use later.
4. Stage 3 also required anticipation of stage 4.
5. In choosing the sequence 4 to 5 as shown, the designers were anticipating that stage 5 would not require major revision of members selected in stage 4. With

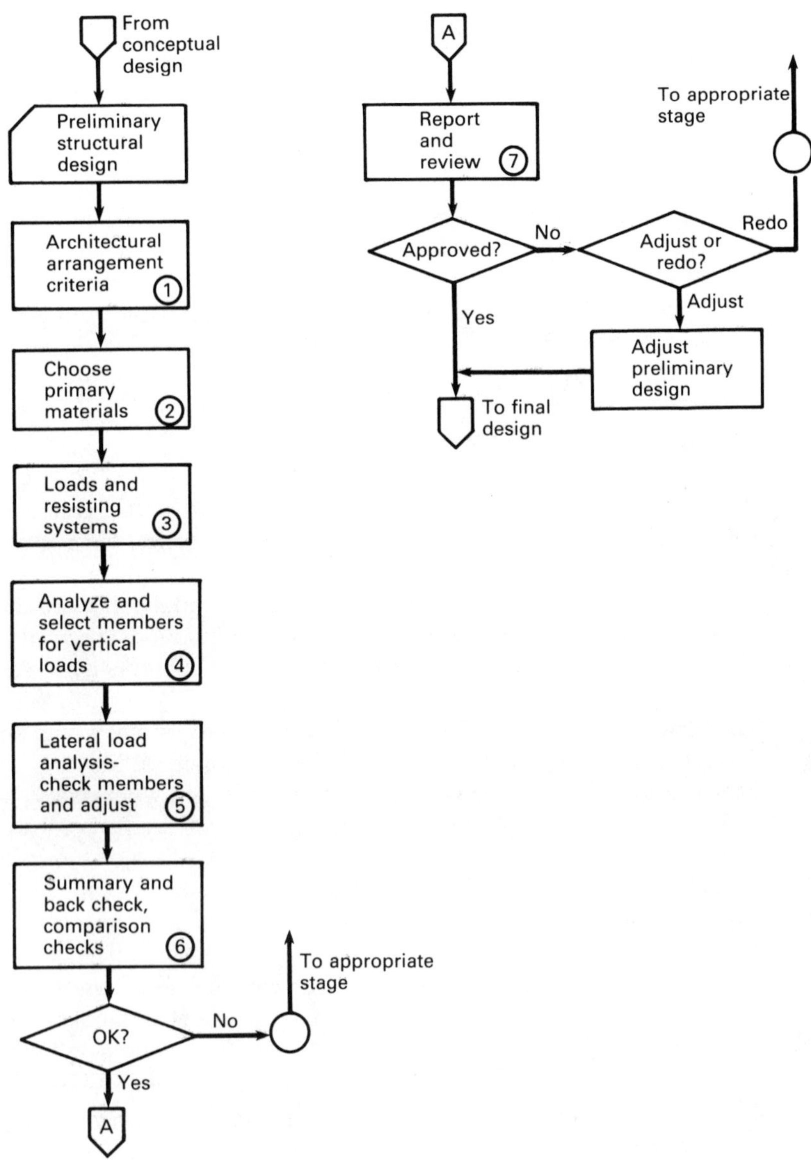

Flow Chart 18.1 Preliminary structural design, tier building.

some exception this turned out to be a reasonable assumption, in part because the lateral loads assigned to the steel frame were only 25% of the total. Such anticipation can be based on experience with similar structures or advance, rough calculation of utilization factors.

6. At stage 7, they faced a moment of truth. Even if internally sound and well documented, the resulting design is unacceptable if it unduly complicates the work of others. The structural designers must defend their solution, but must also be open to persuasion and negotiation in the interests of the success of the total project. There are other moments of truth to come. If the decision to proceed is based on a preliminary design defective in its relationship to space and the functional needs of others, the need to go back may assert itself in much more damaging ways later.

The reader will find it useful at this point to reread the material in Section 18.1 of this case study.

18.9 LOOKING FORWARD: FINAL DESIGN

Assuming that the preliminary design emerges from the architect's and owner's process of review and coordination without major change, there is much remaining to be done. Some of it we will discuss next.

The final calculations will be more precise and much more complete, suitable for review by all interested parties and for use in connection with possible future alterations. It is hoped that final design will result in a slightly lower structural steel budget.

The connection details, each a design problem in itself, will be developed and subjected to mathematical analysis. Where standard details are used, which have been previously calculated or tested, new analysis is not necessary. But the designers will verify that the requirements of this structure are consistent with the basis of the standard details.

The connections will be examined for consistency with the mathematical model used in analyzing the frame. For example:

1. The connections, if any, between the concrete shear walls and the steel frame must be such that failure of the shear walls in an earthquake does not destroy the steel members or connections.
2. If the foundations are unable to provide the assumed degree of rotational fixity at the base of the columns, the assumptions must be revised and the calculation adjusted.

Connections will have to be developed between the precast walls and the steel frame. These are critical connections, the failure of which could result in extreme danger to people in the street. Code requirements have become very severe for such connections. But the designers will also have to study building motions to be sure that the connections can actually work when needed.

The joint spaces between precast panels will be examined to be sure that they do not close completely from deformation of the steel frame, invalidating the structural analysis and endangering the panel-to-frame connections.

Engineers responsible for steel design will ascertain that potential foundation settlements will not invalidate the frame analysis.

Secondary framing will be provided for a host of items specified by architects, mechanical engineers, and others, which must be supported on the frame. The effects of such items on the frame analysis must be considered.

The final steel design, along with other items of structural design, will be expressed in a series of drawings and construction specifications. These are to be included in a larger set of drawings, specifications, and other requirements intended to be part of a construction contract between the owner and a builder.

PROBLEMS

The following problems are suitable for term papers or major student assignments or design investigations by other readers. The preamble opening the section on problems in Chapter 17 applies to these problems also. Keep in mind that any change in the type or arrangement of the structural system may affect many members as the effects are traced through the system.

18.1. Make members B1 continuous. What effect does this have on the seismic resistance? on the weight of the structure? on the cost? You may want to discuss the cost of continuous versus flexible joints with a local fabricator/erector.

18.2. Design some or all of the member connections for the frame of the case study. (Chapters 12 and 13 have examples and problems relating to this.)

18.3. Design the precast concrete panels of the exterior walls and their connections to the frame. (This should not be done unless you have some familiarity with reinforced concrete construction.) Recheck the supporting beams and girders, B3 and G2, and revise as appropriate.

18.4. Refine the preliminary design of the case study with computer assistance. Start with the arrangement and member sizes resulting from the preliminary design, and analyze the structure, using one of the standard programs available for analysis of structural frames. Check both the adequacy and efficiency of the preliminary system and revise as appropriate. More than one iteration may be necessary to get the best results.

18.5. Write a report:

(a) Describe and critically examine the design approach used in the case study.

(b) Examine qualitatively one or more of the alternative approaches suggested in Problem 18.7.

(c) Predict the effects of the alternatives considered on tonnage, cost of construction, and structural adequacy. This should not be done with detailed calculations, but it will be necessary to analyze roughly the changes in loads and type of response of the alternative system(s).

18.6. Using the same arrangement and criteria as in the case study, trace the effect of increasing the number of stories to one or more of seven, nine, eleven, thirteen. If different people do each of these, it should be possible to combine data and examine the effect of height on weight of steel in pounds per square foot and the relative costs of different parts of the frame.

18.7. Redesign the building frame of the case study using one or more of the alternative systems given in the following table. The results for different systems may be compared to each other and to that of the case study. Ideally, this work should be divided among a number of people. Resulting data may be used to "compare" the systems.

Vertical resisting system	Horizontal Resisting System East–West	Horizontal Resisting System North–South	Use concrete core walls?
(a) Slabs beams, girders, columns	Brace exterior bents	Brace exterior bents	No
(b) Same	Same as (a)	Ductile moment-resisting frame	No
(c) Same	Ductile moment-resisting frame	Same as (b)	No
(d) Same	Eccentric bracing in exterior bents; see Chapter 15	Eccentric bracing in exterior bents	No
(e) Same	Ductile moment-resisting frames, exterior bents only	Ductile moment-resisting frames, exterior bents only	No

19
Case Study: Stiffleg Derrick

19.1 INTRODUCTION

The *stiffleg derrick* offers some unusual opportunities to learn about design in structural steel. It is a relatively simple structural system of few parts, much of which is familiar to students and engineers who have analyzed simple force systems in courses in statics and dynamics. Yet the optimization of the system requires careful and detailed analysis of connecting details and of the effects of moving and changing loads. It also requires creative thinking linking design choices to the requirements of use. Unlike many building structures, the extreme design loads on a derrick are likely to be experienced many times during the life of the system. In fact, some imprudent construction superintendents will knowingly overload the system, responding to the pressures of construction economics and eating into the safety factor which they know the design engineer provides in establishing load limits for the derrick.

The stiffleg derrick is a form of hoisting equipment. Its uses are found in construction and industrial settings where the requirement is to lift, move, and place loads within a limited area. Anchored to a stiff base, it operates within three quadrants of what may be a circle of radius 100 feet more or less; the actual limits are set by the length of the boom. Mounted on a platform, it can move on wheels or *skids* in a prescribed path, extending its range linearly. Fixed to a barge deck, its range is limited only by the extent of a waterway. An early developed, simple system, which

A stiffleg derrick is mounted atop a tower on the deck of a barge in a California waterway. The derrick configuration is similar to the one discussed in this chapter.

evolved through the ingenuity of practical workers in construction, logging, quarrying, and similar occupations, stiffleg derricks have been optimized through engineering analysis to rugged, relatively inexpensive machines offered by manufacturers who specialize in hoisting equipment.

Recent development of a wide range of more sophisticated and often more mobile types of hoisting systems has narrowed the field of application of the stiffleg derrick. However, stiffleg derricks continue in use, and in the right combination of circumstances they are still the choice of many construction superintendents and industrial managers. The author has been part of teams that applied this technology to a variety of uses in bridge, shoreline, and offshore construction. In each case, the need was for a specialized, rugged piece of equipment, capable of handling heavy loads (50 to 100 tons and more). The equipment was to be operated on a single mounting for a long enough time to justify the high cost of moving onto a site and erecting the derrick. Conditions did not require over-the-road capability for the assembled rig; conversely, the lack of road width limitations highlighted the advantages of the wide, stable base.

The case study will focus primarily on an *erection* derrick, used to handle heavy loads in relatively slow operation. Stifflegs are also used, if more rarely, in *duty cycle* operation, where the problem is to move smaller loads, such as buckets of gravel, many times each hour in a repetitive, rapid operation. Such use introduces different problems: high operating speed, high accelerations, fatigue. We will limit ourselves to the more slow-acting erection derrick, where the number of cycles of high loads is usually below the fatigue range.

The primary action of the principal structural system is to resist the effects of lifted loads in the form of axial forces on a group of straight, pin-connected members. Once the axial forces are identified, members can be tentatively selected by the methods of Chapters 8 and 9. However, it will be discovered that the requirements of the "real" system make it necessary, and sometimes even advantageous, to consider the effects of bending moments on final member selections. Chapters 10 and 11 will help in addressing bending problems and combined effects.

19.2 CASE STUDY PROCEDURE

We will present a set of design calculations and sketches for the design problem as commonly addressed by engineers charged with stiffleg derrick design. The calculations will be accompanied by comments and discussions intended to reveal the considerations a designer has in mind, which are only partially revealed by the calculations. The experienced derrick designer may need no more than the calculations, which are intended to be complete enough to stand on their own. However, for the less experienced, the discussion will be helpful.

A generalized process analysis for the total system is presented in Flow Chart 19.1. There are two principal branches, one mechanical, one structural, which diverge after the choice of a total engineering system is made. The design paths along the two branches are separate, taking advantage of the specialized knowledge of civil and mechanical engineers. However, there is constant interaction between them. The loads on the structural frame are dependent on the operating characteristics (accelerations, type of braking system, and so on), as well as the lifted loads and the system geometry. Some structural components involve active, moving joints requiring

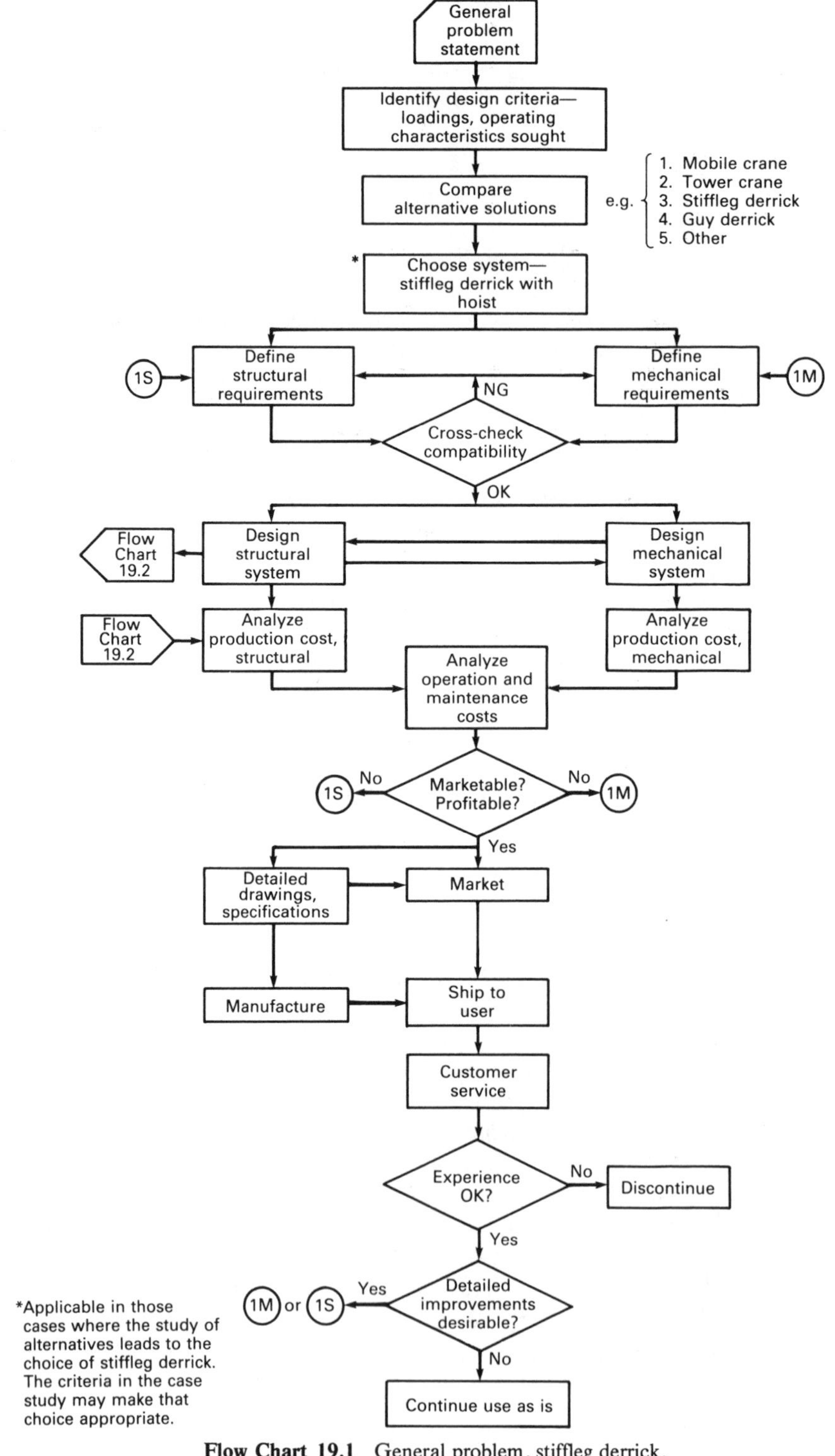

Flow Chart 19.1 General problem, stiffleg derrick.

mechanical bearings and lubricating systems. *Running ropes*, part of the structural system, are wound on *blocks* (assemblages of sheaves), the proper choice of which demands mechanical engineering knowledge. There is, probably, although not shown in the figure, an electrical engineering branch, which may be more or less important depending on the detailed criteria. Lighting is probably needed. Control systems for overload may be needed, as may also be operating dashboards indicating loads, positions, and other pertinent information. However, the beauty of a stiffleg derrick is often in its very simplicity and lack of sophisticated add-ons. Properly operated, it is a dependable piece of equipment that is unlikely to act up or shut down due to complexities in design that are justified in other settings.

Flow Chart 19.2 covers the part of the total problem addressed in the included calculations, which cover the main elements of the structural system. These calculations are not complete. However, within the space available, it is felt that they illustrate the nature of the structural system, design and analysis methods appropriate to it, and a number of engineering considerations arising from the problem. The calculations end at the point where member selections have been made for all the main members in the system. In the four parallel branches of Flow Chart 19.2, this leaves the designers with the following situation:

1. The *boom* section has been chosen, and they are ready for design of details and connections.
2. The members of the stiffleg frame have been selected. They are ready to prove out and if necessary revise their selections based on consultation with the mechanical department.
3. The *running ropes* and *bails* have been selected and the desired characteristics of multisheave blocks identified. The structural engineers are ready for input from the mechanical engineers on the feasibility of their tentative choices. If desirable, they are prepared to revise their selections.
4. They are prepared to proceed with the further processes involved in developing design details for the structural members and their attachments to mechanical members. This stage can be expected to lead to revision and refinement of initial selections, resulting in a workable design consisting of parts and details all of which are compatible with each other, and which they hope will survive the later stages of review and coordination shown after the return to Flow Chart 19.1.

As the reader can see by reference to the flow charts, there will still be much to do before the design can be considered complete. At any of the later stages, it may be desirable to reconsider earlier decisions, making appropriate revisions and refinements. Important in this process is user feedback, which can be sought during the marketing stage and again after the derrick has been put into service. Construction people have a great deal of knowledge and strong opinions about what makes a piece of equipment desirable. Engineers ignore these opinions at their peril.

A caveat. The calculations and decisions here are intended to illustrate design process. Many engineers will do things differently. Many who specialize in derricks

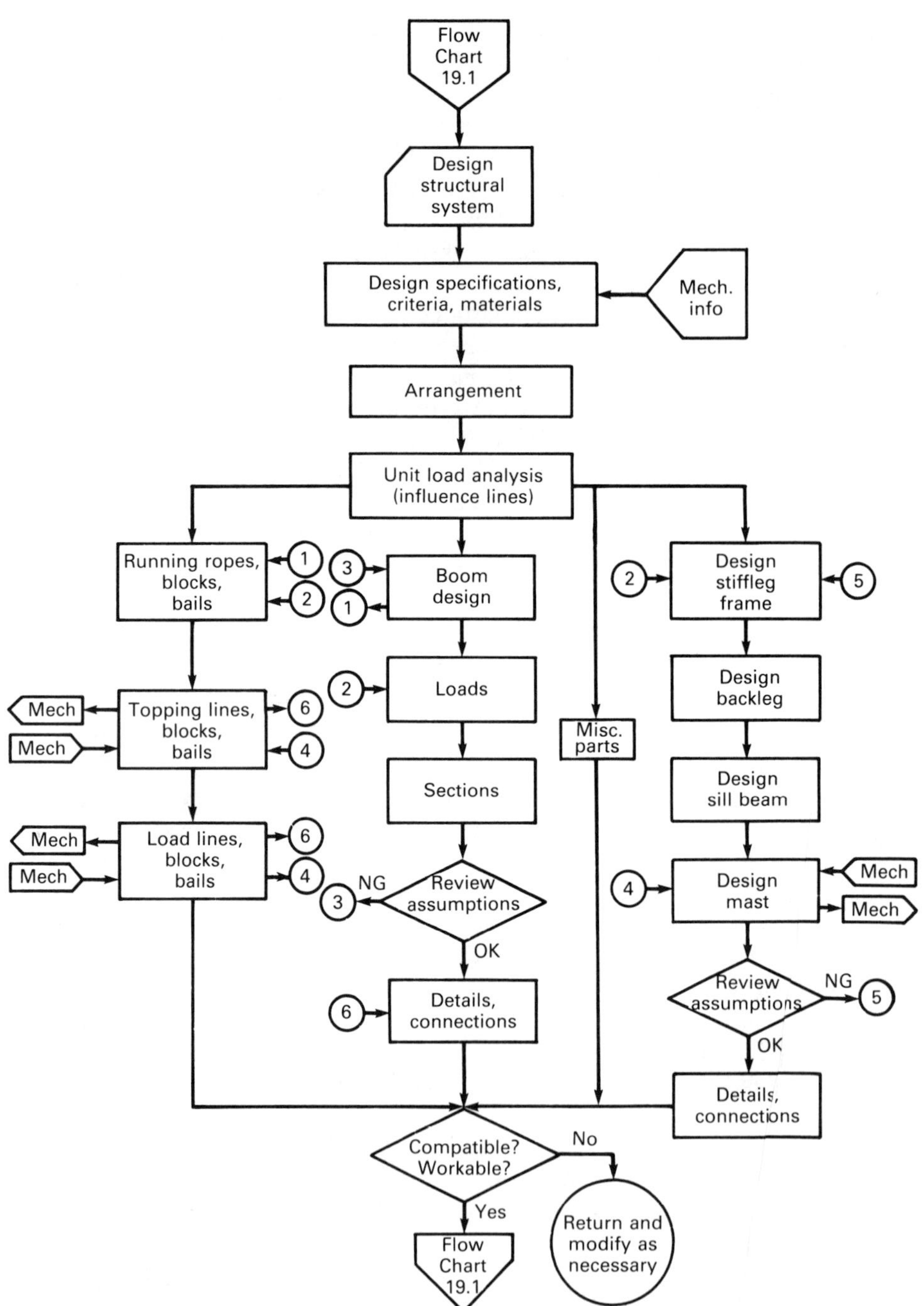

Flow Chart 19.2 Design structural system (not complete).

may be aware of much better design choices. We suggest readers use this material for its limited purpose. Before doing an actual design, use all available sources of wisdom—including your own.

Many of the terms used in the calculations and discussion are specialized *trade* terms that are unfamiliar to many readers or that are defined in special ways in the construction industry. A glossary of specialized terms will be found in Section 19.7. The definitions reflect usages with which we are familiar. Local usages and those in some industries may differ.

To our knowledge, there is no generally accepted and comprehensive specification governing design of derricks in U.S. practice. The American National Standards Institute publishes ANSI-B30.6, "Safety Code for Derricks,"[14] with general safety requirements. The AISC has a more detailed proposed specification[6] for design of guy and stiffleg derricks. It generally parallels the "steel building" specification, which is referred to extensively in this book. The calculations are based primarily on the requirements of the building specification. For problems derived from the special nature of derrick structures, we have been guided to some degree by the proposed derrick specification, as well as other sources.

19.3 STATEMENT OF PROBLEM

The problem (refer to Flow Chart 19.1) is to design a piece of hoisting equipment with specific characteristics.

1. Its primary use will be in construction or similar activities, handling heavy loads with relatively small accelerations (low *impact*).
2. It shall be mountable on a fixed base at specified locations for long enough periods to justify field assembly.
3. Its parts must be movable over highways and construction roads, and such that they may be assembled into a complete system on site, with minimum use of equipment other than its own parts.
4. Operated by highly skilled personnel, it should require a minimum of sophisticated and sensitive control equipment for safe operation and minimum maintenance.
5. Its primary *range of operation* (positions of main hook) must be within a 100-foot radius from a central vertical axis and may be limited to an angle of approximately 125° to each side of a vertical plane of symmetry including that axis.
6. The range of vertical movement of the main hook, while an operating consideration, is not considered a controlling one, since this range can be extended when necessary by mounting the equipment on a raised platform or tower.

7. Its hoisting machinery may be mountable on separate foundations outside the operating range of the main hook. The hoist must be set far enough away from the derrick so that running ropes can approach the hoist drums at small *fleet angles* (see glossary).
8. It shall have the capacity to lift, move, and lower up to 60 tons within a radius between 20± and 40 feet from its central axis.
9. It will have lifting capacities within radial ranges, $40 \leq R \leq 100\pm$ feet, which shall be determined consistent with the system designed as above for 60 tons at 40-foot radius.
10. It may be desirable to extend its *reach* somewhat beyond the 100-foot radius, lifting light loads on an auxiliary, light *jib* and hook.

A number of different types of equipment might serve these requirements. After comparison of alternatives (not discussed here), the decision is made to design a stiffleg derrick. Preliminary structural and mechanical design criteria are written and cross-checked and departmental assignments made for the next stage of design. We will discuss the structural part of this phase, following Flow Chart 19.2 and the calculations leading to selection of the principal members, as done by Kathie Lifta* of the structural group.

19.4 STRUCTURAL CALCULATIONS: GENERAL

The calculation set is introduced, on sheet C1, by a Table of Contents listing its principal subdivisions. The arrangement and numbering of the calculations is consistent with discussions in Chapter 16.

Sheets G1 and G2 list the major criteria for structural design. Parts A and B are in large part the criteria discussed in the problem statement of Section 19.3, with a few details added. In part C, listing of department assignments helps to clarify for the structural designers the scope of their work and the interactions needed with the mechanical engineers. There will also be some interaction required with electrical engineers, but this is of relatively minor nature and it is not noted here. The project engineer will have a more complete breakdown of assignments representing detailed planning for the entire design problem. Of the assignments listed for the structural department, further calculations cover all the main parts except the bull wheel and boom stays. They also omit a number of additional tasks listed in the *MS* sheets. These are all necessary parts of the system and must be included in a full set of calculations. They are reluctantly omitted here due to space constraints.

*Fictitious name.

Project : STIFFLEG DERRICK	SHEET C1 OF
Subject : CONTENTS	

STRUCTURAL CALCULATIONS FOR
PROTOTYPE DESIGN FOR
STIFFLEG DERRICK

TABLE OF CONTENTS

COOPCHEN ENGINEERS
STRUCTURAL GROUP
BY KATHIE LIFTA
MARCH, 1983

Project : STIFFLEG DERRICK Subject : GENERAL STRUCTURAL DESIGN CRITERIA & GENERAL ARRANGEMENT	SHEET G1 OF ____

GENERAL STRUCTURAL DESIGN REQUIREMENTS

A. PRIMARY REQUIREMENTS

1. LIFT, HANDLE & PLACE 60 TON MAX. LOADS WITHIN A REGION:
   a. 20' ≤ RADIUS ≤ 40' FROM CENTER OF MAST
   b. WITHIN A HORIZONTAL ANGLE OF 250°± LIMITED BY A 90° STIFFLEG FRAME
2. MAIN BOOM IS TO HAVE A REACH OF 100±FT. RADIUS
3. MUST PERMIT ASSEMBLY AND DISASSEMBLY FOR ROAD TRANSPORT:
   a. NO PIECE LONGER THAN 60'
   b. NO PIECE HEAVIER THAN 20 TONS
   c. NO PIECE BULKIER THAN PERMITTED FOR ROAD TRANSPORT WITHOUT SPECIAL WIDE LOAD PERMITS
4. STIFFLEG FOUNDATIONS TO BE AT MAX. 40' CENTERS
5. GEOMETRIC AND LOAD REQUIREMENTS FOR FOUNDATION DESIGN TO BE SUPPLIED FOR USERS

B. SUPPLEMENTARY REQUIREMENTS

1. ESTABLISH LIFTING CAPACITIES FOR REGION 40' ≤ RADIUS OF REACH ≤ 100'
2. DESIGN A JIB** TO EXTEND REACH TO 115 FT. RADIUS WITH CAPACITY:
   a. LIMITED TO SINGLE PART OF LINE
   b. LIMITED BY STRENGTH OF OTHER ELEMENTS AS CONTROLLED BY USE WITH MAIN BOOM

C. DEPARTMENTAL ASSIGNMENTS

STRUCTURAL	MECHANICAL	MECH. & STRUCT.
* BOOM	LOAD BLOCKS & SHEAVES	MAST STEP
* STIFFLEG FRAME *- BACKLEGS *- SILL BEAM *- MAST - PRIMARY STRUCTURE	TOPPING BLOCKS & SHEAVES DEFLECTION SHEAVES & SNATCH BLOCKS HOIST	MAST HEAD
* TOPPING AND LOAD BAILS	BEARINGS & LUBRICATION SYSTEMS	
* TOPPING AND LOAD LINES	OPERATING MANUAL	
BULL WHEEL		
BOOM STAYS		
FOUNDATION REQUIREMENTS		

** NOT INCLUDED IN THIS SET OF CALCULATIONS

* PARTIAL DESIGN CALCULATIONS ARE INCLUDED IN THIS CASE STUDY

Parts D and E list design specifications and advisory references that the designers use in the development of the design. The AISC building specification listed will be of great value insofar as it represents the state of the art of structural design with steel, and similarly for the AWS welding code. However, even more than in usual building design practice, Lifta is on her own in defining the nature of the loads and other aspects of the design problem. She will get help from the advisory references listed, but must be prepared to answer for her decisions on their merits.

In part F, the designer chooses the steels to be used, listing different materials specifications for different types of members. The reader may refer to Chapter 5 of this book or to the ASTM Specifications themselves for more information on these steels. Several important considerations enter into the choices made.

1. There is a premium on weight saving, making high-strength steels desirable. This arises in at least two ways.
   a. The weight of the boom and of other parts at and near the boom tip act through their resultant effects as loads applied at the boom tip. For any given hoist capacity and strength of running ropes, this leads to reduction in available capacity for *pay load*. Conversely, for a given pay load capacity, heavier parts require greater capacity in hoists and/or running ropes. (See calculation sheet G4 for insight into this problem, which will be discussed in more detail later.)
   b. In transporting derrick parts to job sites, weight may affect transportation feasibility as well as costs.
2. It is desirable to choose steels whose properties are not sensitive to change due to *detail* additions by field personnel. Any alterations made without full knowledge of the design requirements is to be discouraged. However, it is not uncommon for field people to make attachments to derrick parts for purposes the designers may not be aware of. (We are aware of at least one major failure directly attributable to welding on an "exotic" steel by improperly instructed personnel.) For this reason, Lifta rejects the potential weight savings from use of ASTM-A514 steel.
3. Superior corrosion resistance is a consideration for steel that will be used outdoors and may be outdoors for long periods without renewal of protective paints.

The basic choices for shapes, plates, bars, and tubes are high-strength steels (F_y = 50 ksi) with corrosion resistance improved by alloying elements, and with other desirable mechanical properties (see Chapter 5 and Table 5.1). Both are listed as acceptable for use with the AISC Specification (AISC 1.4.1), as are the steels listed

Project : STIFFLEG DERRICK
Subject : GENERAL STR'AL DESIGN CRITERIA & GENERAL ARRANGEMENT

SHEET G2 OF

D. STRUCTURAL DESIGN SPECIFICATIONS

1. AISC "SPECIFICATION FOR THE DESIGN, FABRICATION AND ERECTION OF STRUCTURAL STEEL FOR BUILDINGS" (EFFECTIVE NOV. 1, 1978)
2. AWS - STRUCTURAL WELDING CODE D1.1 - 80

E. GENERAL INFORMATIONAL AND ADVISORY REFERENCES

1. AISC "MANUAL OF STEEL CONSTRUCTION" - 8TH EDITION - 1980 (CALL "AISC MANUAL")
2. AISC "GUIDE FOR THE ANALYSIS OF GUY AND STIFFLEG DERRICKS" - 1974 (CALL "AISC DERRICK GUIDE")
3. ANSI STANDARD B30.6 - "SAFETY CODE FOR DERRICKS" ADOPTED BY DEP'T OF LABOR - OSHA

F. STRUCTURAL MATERIALS (UNLESS OTHERWISE NOTED FOR SPECIALLY SELECTED ELEMENTS)

SPECIFIED BY ASTM NUMBER UNLESS OTHERWISE NOTED

1. STEEL SHAPES, PLATES, BARS — A 588 - Gr. 50
2. TUBES AND PIPES — A 618 - Gr. 3
3. STEEL CASTINGS - MEDIUM STRENGTH — A 27 - Gr. 65-35
   - HIGH STRENGTH — A 148 - Gr. 80-50
4. STEEL FORGINGS — A 688
5. WELDING RODS - SUITABLE FOR PARENT STEEL PER AWS SPECIFICATION
6. BOLTS — A 307, GALV.
7. WIRE ROPES - "IMPROVED PLOW STEEL" PER ANSI STD. B30.6 GALVANIZING PERMITTED FOR STANDING ROPES ONLY

G. SIGN CONVENTIONS

UNLESS OTHERWISE NOTED:
FOR MEMBER FORCES, + = TENSION
FOR REACTIONS + = ↑ VERTICAL REACTION
→ HORIZONTAL REACTION
(REACTIONS ARE FORCES APPLIED BY THE FOUNDATION TO THE DERRICK)

for castings and forgings. The decision to use *A307* bolts* reflects the designer's opinion that she cannot be certain of enforcement of the detailed field conditions necessary for safe and reliable use of high-strength bolts. Galvanizing (zinc coating) is specified for these bolts for corrosion resistance.

Wire rope is not listed under acceptable materials in the AISC Building Specification. However, it is a necessary material for this type of hoisting system, with a long history of successful use (see Chapter 8). The ANSI Standard listed is an authoritative source for safe use of wire rope in construction derricks. Galvanizing will be used for corrosion protection of standing ropes; in such use, galvanizing can be expected to be effective, and the attendant 10% loss of strength can be compensated by increased size. For running ropes, galvanizing is neither desirable nor effective.

Sheet G3 is used to define the stiffleg derrick, its parts and operating range. The three-dimensional stiffleg frame (mast, back legs, and sill beams), when connected to a foundation, forms a stiff, stable space truss capable of supporting the effects of remote hook loads when they are transmitted to the top and bottom of the mast by the boom and topping line. The mast is also an element in a trusslike plane system of mast–topping line–boom, which receives hook loads transmitted through the multipart load line blocks. To allow for angular movement (swinging), the mast must be able to rotate around its own central axis. When the mast rotates, the boom, topping line, and load lines rotate with it.

A multidrum hoist is used to operate the system. Cable from the *topping line drum* is fed through the mast into the *boom falls*. When reeled in and out on the drum, this cable shortens and lengthens the topping line, raising and lowering the boom tip as the boom pivots around a pin at its lower end. Cable from the *load line drum* is fed, after several bends, over the boom tip and into the multipart *load falls*, which is attached to the boom tip at its top and to the hook at its lower end. Winding cable in or out on this drum raises or lowers the hook and the load suspended from it. A relatively short length of cable of fixed length is wound around the *slewing drum* and the *bull wheel*. When the drum is turned clockwise or counterclockwise, the *slewing ropes* and *bull wheel* rotate the mast on its axis either clockwise or counterclockwise.

As in ideal trusses, most of the effects of the lifted loads are resisted by axial member loads, and ultimately by their reactive effects at the three foundation points at the corners of the stiffleg frame. Figure 19.1, familiar to students of engineering statics, illustrates the nature of one plane in the idealized system. As will be seen later, bending effects resulting from the operation of the system modify the ideal truss and must be considered in member selection.

In the three-dimensional sketch, the designer has added a jib, or boom extension, with a single part of line shown attached to a light hook. This line would be

*See Chapter 13.

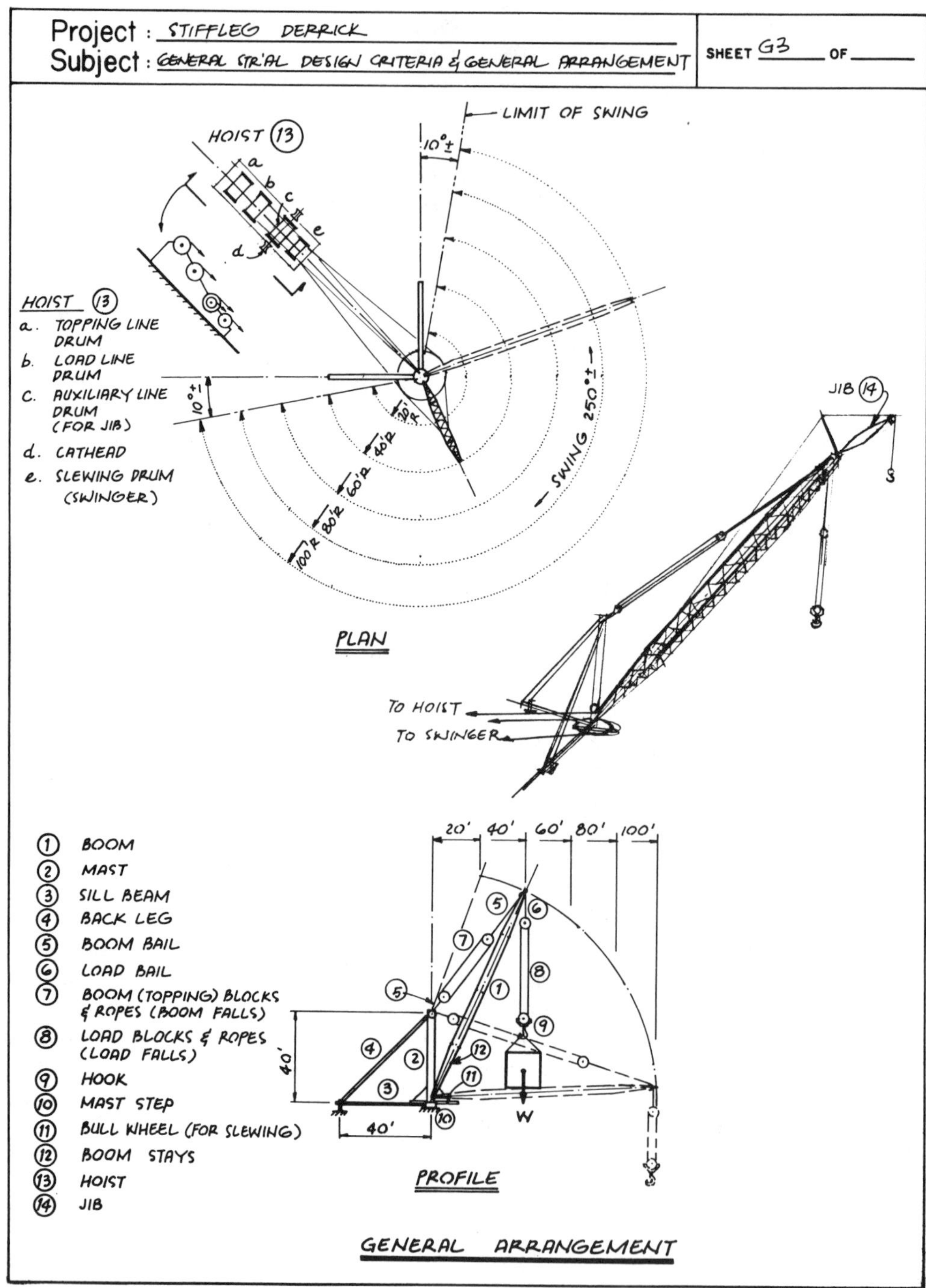
Project : STIFFLEG DERRICK
Subject : GENERAL STR'AL DESIGN CRITERIA & GENERAL ARRANGEMENT
SHEET G3 OF
HOIST (13)
LIMIT OF SWING
10°±
HOIST (13)
a. TOPPING LINE DRUM
b. LOAD LINE DRUM
c. AUXILIARY LINE DRUM (FOR JIB)
d. CATHEAD
e. SLEWING DRUM (SWINGER)
SWING 250°±
20'R
40'R
60'R
80'R
100'R
PLAN
JIB (14)
TO HOIST
TO SWINGER
① BOOM
② MAST
③ SILL BEAM
④ BACK LEG
⑤ BOOM BAIL
⑥ LOAD BAIL
⑦ BOOM (TOPPING) BLOCKS & ROPES (BOOM FALLS)
⑧ LOAD BLOCKS & ROPES (LOAD FALLS)
⑨ HOOK
⑩ MAST STEP
⑪ BULL WHEEL (FOR SLEWING)
⑫ BOOM STAYS
⑬ HOIST
⑭ JIB
20' 40' 60' 80' 100'
40'
40'
W
PROFILE
GENERAL ARRANGEMENT

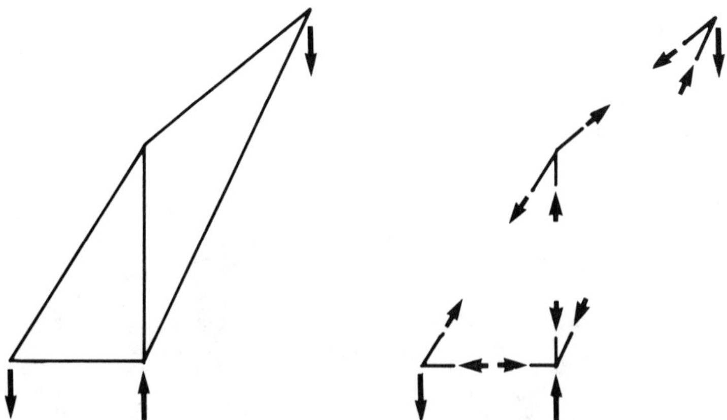

Figure 19.1 Simple derrick with joint free bodies.

controlled by a fourth drum at the hoist. Dimensions have been selected for the boom length, the mast height, and length of sill beams.

On sheet G4, the designer extends the basic requirements for load capacity, converting them into load criteria to be used in detailed analysis. As in most structures, design loads are first divided into dead and live loads. Since the live loads are movable, the associated accelerating forces (*impact*) must be considered, as must the effects of wind.

Each member has its own dead weights associated with it and requiring no special consideration here. As will be shown in the influence analysis to follow, loads applied at the hook (force vectors on a line through the boom tip and the hook) affect all members in the system. Examination of the sketch reveals that some parts of the dead load act effectively as hook loads.

It is immediately obvious that the weights of the *hook, load blocks, load ropes*, and *bail* hang from the boom tip on the same line as the lifted load vector. The distributed weight of the boom is supported at the lower end by the mast step and at the upper end by the topping line. If the resultant of the distributed weight is applied, half at each end, the upper end acts as an added hook load,

$$\frac{W_B}{2} = \frac{1}{2}(w_B L_B)$$

where w_B is weight per foot measured in the direction of the boom length, L_B. For similar reasons, half the weight of the topping system is considered as *effective hook load*.

The lifted load hangs from the hook and is clearly classified as live load. As a moving load, it is subject to vertical acceleration at the beginning and end of any vertical travel. In AISC Specification 1.3.3, this effect is referred to as *impact* and is

Project : STIFFLEG DERRICK
Subject : LOADS & LOAD COMBINATIONS

SHEET 64 OF

DEFINE LOADS

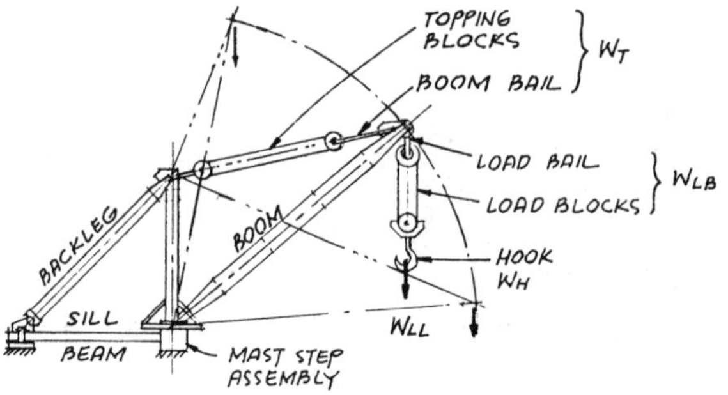

LOAD AT HOOK* (EFFECTIVE)

DEAD LOAD 'D'

$\frac{W_B}{2} = \frac{1}{2}$ OF WEIGHT OF BOOM

$\frac{W_T}{2} = \frac{1}{2}$ OF (TOPPING BLOCKS + BOOM BAILS)

W_{LB} = WEIGHT OF LOAD BLOCKS + BAILS

W_H = WEIGHT OF HOOK

LIVE LOAD 'L1'

W_{LL} = WEIGHT OF LIFTED LOAD

+

I_V = IMPACT (ACCELERATION) OF VERTICAL LOAD $= .20\left(W_{LL} + W_H + \frac{W_{LB}}{2}\right)$

LIVE LOAD 'L2'

W_{LL} +

$I_V = .03\,(W_{LL} + W_H + W_{LB} + W_B/2 + W_T/2)$

I_H = SLEWING = HORIZONTAL LOAD ⊥ BOOM $= .02\,(W_{LL} + W_H + W_{LB} + W_B/2 + W_T/2)$

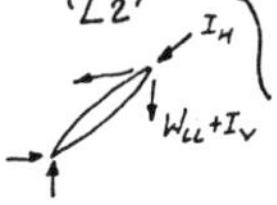

LIVE LOAD 'L3'

OPERATING WIND

H_W = HORIZ. FORCE ⊥ TO BOOM @ HOOK = WIND ON BOOM + WIND ON LIVE LOAD.

$H_{W_B} = C_d$(1#/SQ.FT.) × EXPOSED AREA OF ($\frac{1}{2}$ OF BOOM + BOOM BAIL)

$H_{W_{LL}} = .02\,W_{LL}$

C_d = 1 FOR ROUND MEMBERS
= 2 FOR OTHERS

+

L1

HOOK LOAD COMBINATIONS

I = D + L1

II = D + L2

III = D + L3

DESIGN FOR WORST OF LOADS, I, II, OR III

USE ALLOWABLE STRESSES PER AISC BLDG CODE WITHOUT INCREASE FOR WIND, IMPACT, OR SLEWING.

SINCE L3 INCLUDES L1, CONSIDER ONLY COMBINATIONS II & III

* SEE ALSO AISC "GUIDE FOR ANALYSIS OF GUY AND STIFFLEG DERRICK" - 1974. LOADS LISTED HERE NOT IDENTICAL.

treated as a percentage impact of live load. No direct guidance is given for the percentage appropriate to this case. The rate of such accelerations will be determined in part by the skill and prudence of the operating crew. However, the designer has some degree of control in specifying the starting and stopping characteristics of the hoist. Tests on stiffleg derricks reported and discussed in the AISC Derrick Guide* lead to an AISC recommendation of 20% impact factor for stiffleg derricks in construction service. Lifta follows this recommendation and will inform the mechanical engineers of the hoist characteristics needed to validate it. However, she notes that the parts subjected to accelerated motion include, in addition to the lifted load, the hook and lower load block. One condition of live load, $L1$, is defined as the lifted load plus 20% of the sum of that load plus the other accelerated parts.

The operation of *slewing* (swinging of the load through a horizontal angle) involves horizontal acceleration perpendicular to the boom in starting and stopping the swing plus minor vertical acceleration associated with vibration. The designer follows the recommendation of the AISC Derrick Guide, using simultaneous horizontal and vertical impact forces applied at the boom tip. Each such force is equal to 3% of the parts in slewing motion, load, blocks, hook, and boom. Live load $L2$ combines these impact forces with the lifted load.

Unlike most structures, it is not necessary to provide for the effects of unusual wind storms during derrick operation.† In fact, it usually becomes impossible to operate the derrick in high winds. A small operating wind force is recommended by the AISC Derrick Guide, $0.02W_{LL}$, applied as a horizontal force at the boom tip perpendicular to the boom. Such a force results when the wind force on the load moves the hanging ropes and hook out of plumb by 2%. The boom itself will simultaneously experience a small wind force evaluated as

$$F_H = C_d \times 1 \times (\text{net exposed area as seen in profile})$$

The effect at the boom tip is the force shown in $L3$, which also includes $W_{LL} + I_v$. C_d is a shape factor dependent on the type of members the boom will present to the wind stream.

Design will consider the worst effects from the combination of dead loads with either of $L1$, $L2$, or $L3$. In building or bridge design, it is customary, when combining design wind loads with other loads, to permit an increase in allowable stresses of 33%. In this case, the small *operating wind* being considered occurs frequently and is associated with the basic operating conditions of the derrick. No such increase will be permitted.

*Reference 6.

†Procedures for protecting the derrick during a major storm are necessary.

19.5 UNIT LOAD ANALYSIS

As a hook load is moved through the operating range of the derrick, its effects on the various parts of the system vary. Resulting member loads are functions of the magnitude of the hook load, the vertical angle of the boom, and the horizontal angle between the boom and the stiffleg frame. To determine the maximum (controlling) effect on each member, it will be necessary to define such a function for each member. As will be seen, the controlling position of the load is not the same for all members.

The designer chooses to use an influence line approach. A unit load will be placed at the hook, and as it is moved through the range of boom vertical and horizontal angles, its effects on each member will be found. This will provide her with influence curves. When multiplied by the actual hook load, the ordinates of the influence curves lead to member loads and reactions.

This is done on calculation sheets I1 through I6. Only axial effects are considered, assuming ideal joints and concurrent member forces. The modifying effects of joint members will be introduced later.

Sheets I1 and I2 examine the effect of the vertical angle of the boom on the boom and topping line. The geometry is studied on sheet I1. Since the users are interested, in load handling, in distance from mast centerline (*reach*), the geometry is plotted for the upper and lower extremes of reach, 20 and 100 feet, and for intervals between. Vertical angles are noted for each position. Two problems become apparent when the reach is small. The upper topping line block gets close to the boom, and they will hit each other if the boom is brought in much closer. The angle at which such interference takes place depends on the size of the boom and topping line blocks and the position of the upper topping line block at the high boom position. The line gets shorter as the boom is raised. Since the bails are of fixed length, the variation takes place in the length of the running ropes, determining the distance between the blocks. The blocks should not be allowed to come too close to each other since that will cause chafing of the ropes in the grooves as the sheaves turn. Twenty feet may not be the actual minimum reach, but it is not far from it. In addition, if the boom were raised much further, it would tend to become unstable and might try, in the presence of minor disturbing effects, to kick back in the direction of the mast.

On sheet I2, a vector analysis leads to coefficients for boom and topping line forces at various reaches. A series of force triangles is drawn that duplicate at a different scale the triangles formed by the mast, boom, and topping lines. The load of 1 is parallel to the mast axis and the other vectors parallel, respectively, to the boom and topping line. The result is constant boom compression of 2.5 (mast height/boom length = 2.5), and topping line tension that varies with the length of the line. The coefficients are determined by scale and their variations plotted against reach. The designer also notes that the boom will experience simple span bending due to its own

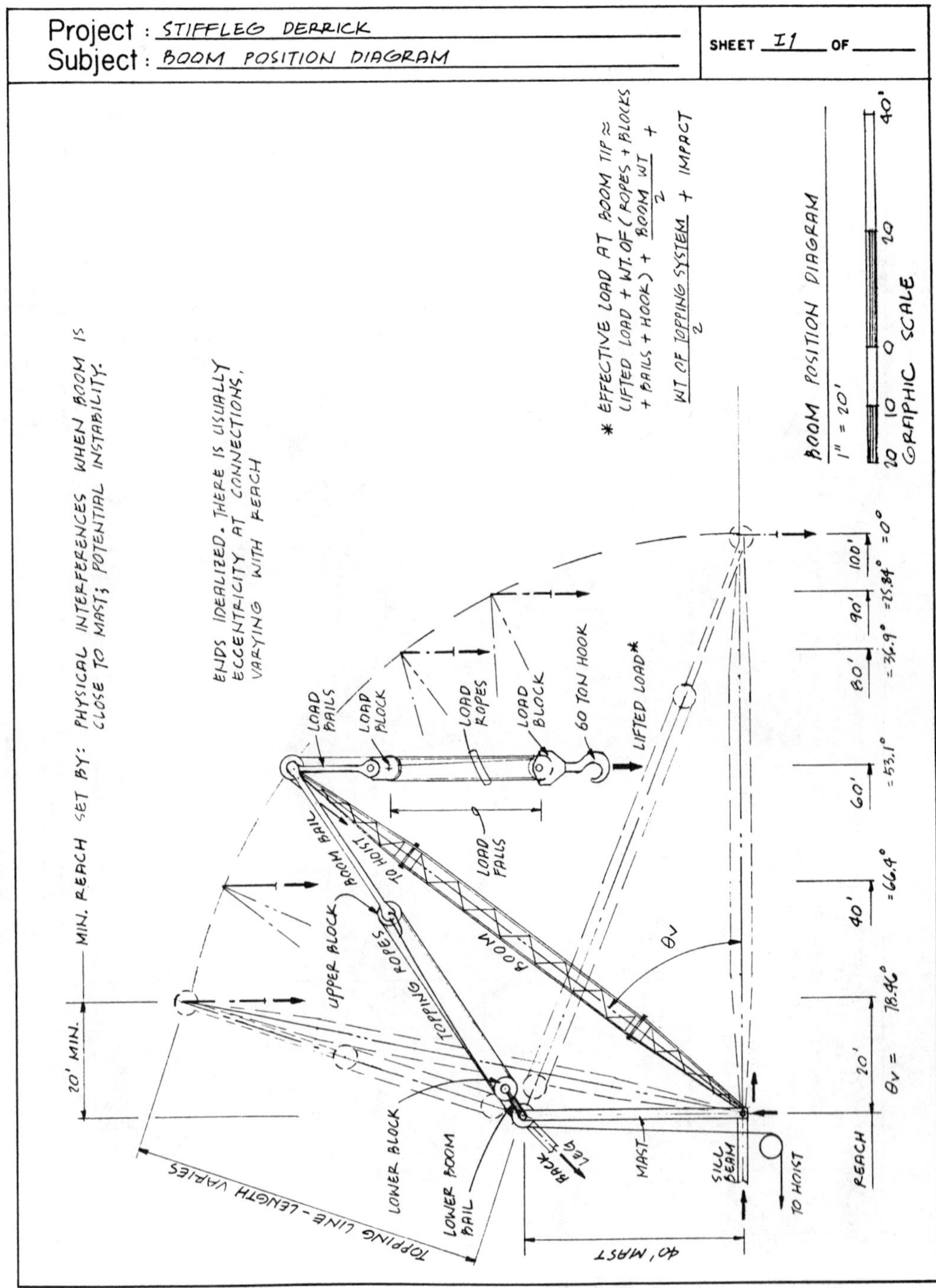

Project : STIFFLEG DERRICK
Subject : BOOM POSITION DIAGRAM
SHEET I1 OF
MIN. REACH SET BY: PHYSICAL INTERFERENCES WHEN BOOM IS CLOSE TO MAST; POTENTIAL INSTABILITY.
ENDS IDEALIZED. THERE IS USUALLY ECCENTRICITY AT CONNECTIONS, VARYING WITH REACH
* EFFECTIVE LOAD AT BOOM TIP ≈ LIFTED LOAD + WT. OF (ROPES + BLOCKS + BAILS + HOOK) + BOOM WT/2 + WT OF TOPPING SYSTEM/2 + IMPACT
BOOM POSITION DIAGRAM
1" = 20'
20 10 0 20 40'
GRAPHIC SCALE
LOAD BAILS
LOAD BLOCK
LOAD ROPES
LOAD BLOCK
60 TON HOOK
LIFTED LOAD*
LOAD FALLS
TO HOIST
BOOM BAIL
UPPER BLOCK
TOPPING ROPES
BOOM
θv
20' MIN.
LOWER BLOCK
LOWER BOOM BAIL
BACK LEG
MAST
SILL BEAM
TO HOIST
TOPPING LINE - LENGTH VARIES
40' MAST
REACH 20' 40' 60' 80' 90' 100'
θv = 78.46° = 66.4° = 53.1° = 36.9° = 25.84° = 0°

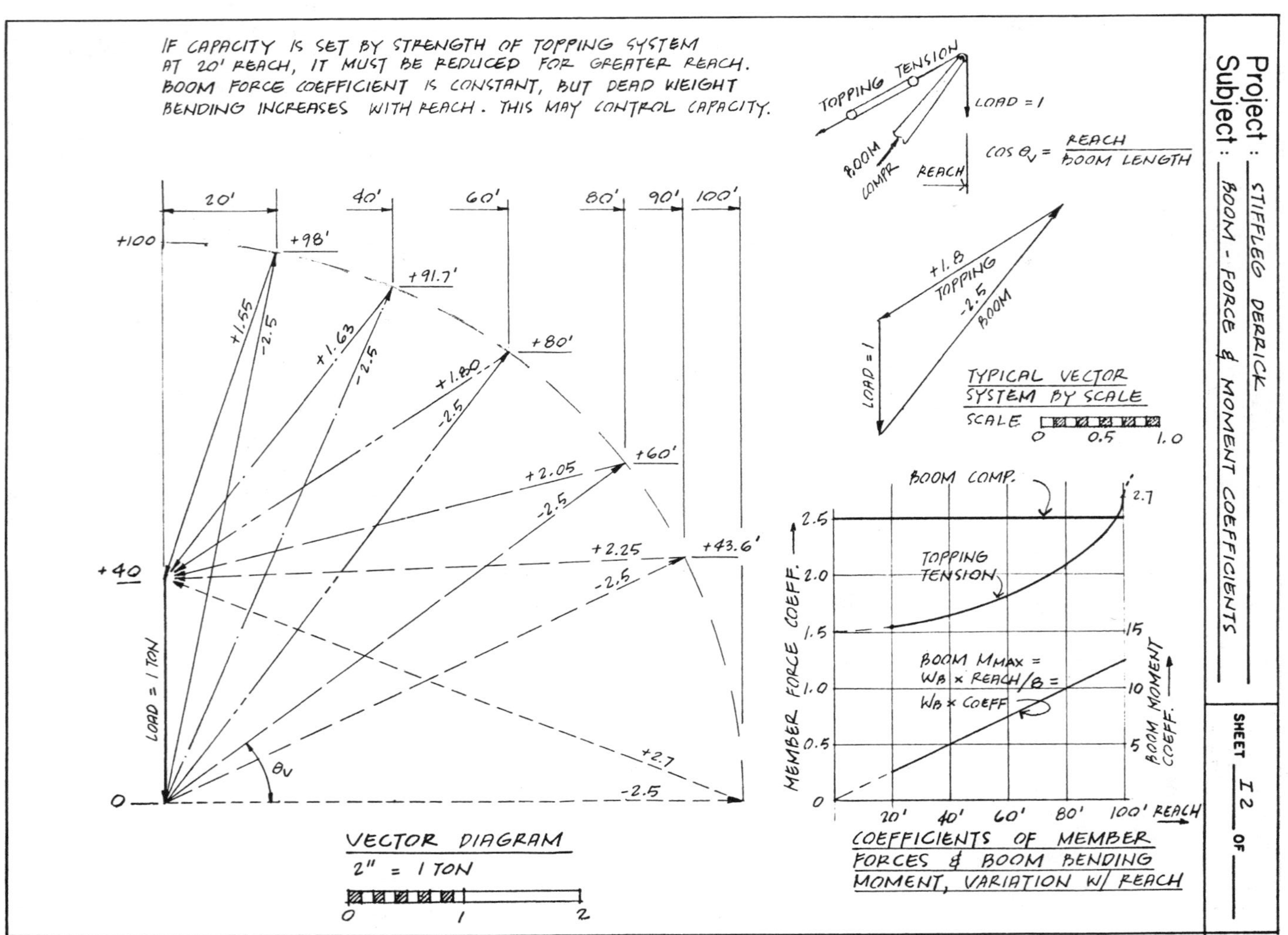
Project : STIFFLEG DERRICK
Subject : BOOM - FORCE & MOMENT COEFFICIENTS
SHEET I2 OF
IF CAPACITY IS SET BY STRENGTH OF TOPPING SYSTEM AT 20' REACH, IT MUST BE REDUCED FOR GREATER REACH. BOOM FORCE COEFFICIENT IS CONSTANT, BUT DEAD WEIGHT BENDING INCREASES WITH REACH. THIS MAY CONTROL CAPACITY.
TOPPING TENSION
BOOM COMPR
LOAD = 1
REACH
COS θV = REACH / BOOM LENGTH
+1.8 TOPPING
-2.5 BOOM
LOAD = 1
TYPICAL VECTOR SYSTEM BY SCALE
SCALE 0 0.5 1.0
20' 40' 60' 80' 90' 100'
+100
+98'
+91.7'
+80'
+60'
+43.6'
+1.55
-2.5
+1.63
+1.80
+2.05
+2.25
+2.7
+40
LOAD = 1 TON
θV
0
VECTOR DIAGRAM
2" = 1 TON
BOOM COMP.
TOPPING TENSION
2.7
BOOM MMAX = WB × REACH/8 = WB × COEFF
MEMBER FORCE COEFF.
BOOM MOMENT COEFF.
20' 40' 60' 80' 100' REACH
COEFFICIENTS OF MEMBER FORCES & BOOM BENDING MOMENT, VARIATION W/ REACH

weight (W_B). The maximum moment at midlength will vary linearly with reach. This effect will be small when lifting design loads at close reach, but may have a significant effect on the derrick capacity when the boom angle is small.

Note that the vertical angle affects the coefficients studied so far, but they are independent of horizontal angle (swing).

Sheets I3 through I6 extend the influence analysis to the members of the stiffleg frame and the reactions. Here it becomes necessary to consider both vertical and horizontal angles. This becomes apparent on sheet I3 where equations are written for influence reactions at the foot of mast and outer ends of sill beams. The resulting curves show the effect of varying θ_H over a 270° range between the xz and yz planes, which are the planes of the two stifflegs. To find reactions, the coefficients, reach, and angle are used in the equations below the curves. Reactions can be either positive (thrust) or negative (uplift), an important fact to be transmitted to the user who has to provide foundations. The angular positions for maximum reactions are different for reactions M, $B1$, and $B2$.

On sheet I4, using information from sheets I2 and I3, influence relationships are developed giving, for various values of reach and θ_H and a hook load of 1, the values of axial loads in the members of the stiffleg frame. Sheets I5 and I6 are used to convert the information on the previous two sheets to the more revealing form of curves. Note, on these sheets, that the angle $\theta_H = 135°$ is a position of symmetry for the force in and reaction at the mast, while the effects on back leg, sill beam, and R_B show symmetry about $\theta_H = 0$ or 180°. If the range of θ_H were extended to 360°, the symmetry would be complete. For convenience, a new angle, θ'_H, is defined:

$$\theta'_H = \theta_H - 135°$$

whose origin is the center of swing. The curves for $R = 40$ feet are drawn in bold lines, since they will be used in the next stage of design, the selection of members. The other curves will be useful in determining the permissible hook load at different radii (reaches).

The ordinates of the curves are influence coefficients. When multiplied by the magnitude of the hook load, they yield the effect of that load on members or reactions.

Project : STIFFLEG DERRICK
Subject : UNIT LOADS AT HOOK - REACTIONS VS R & θ_H

SHEET 13 OF ___

VARIATION OF REACTIONS WITH REACH, R, AND ANGULAR POSITION, θ; UNIT LOAD AT HOOK

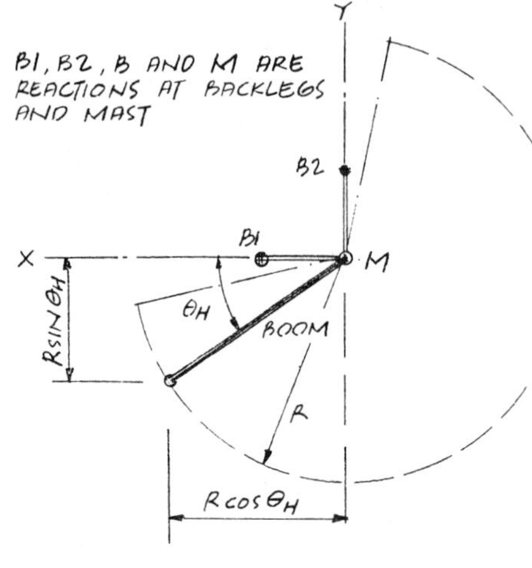

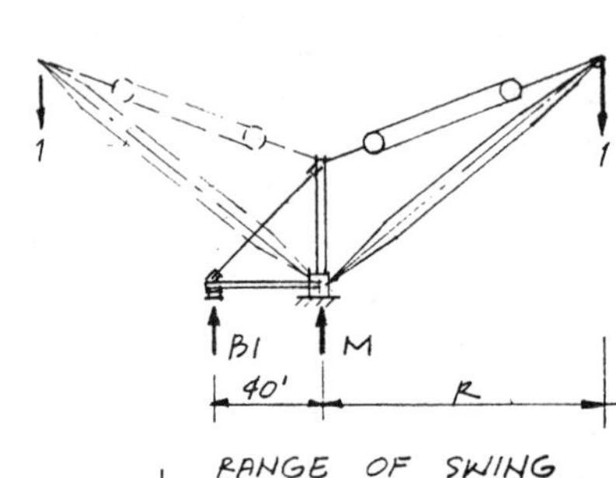

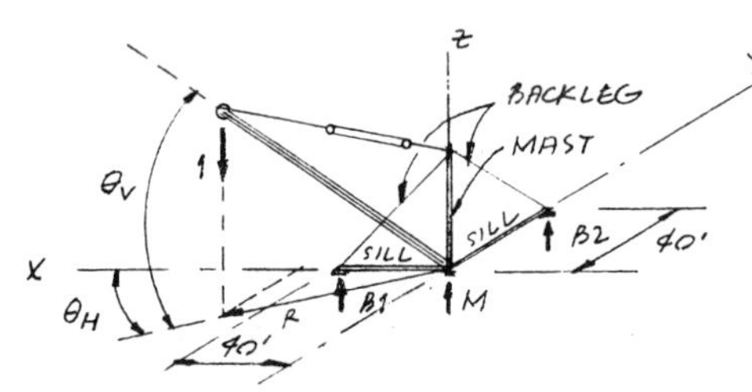

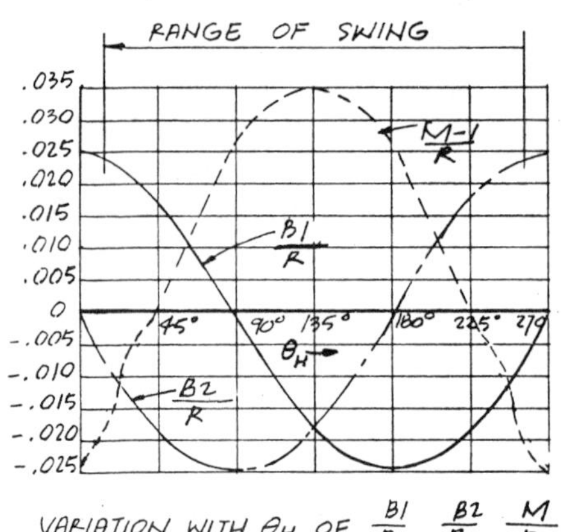

VARIATION WITH θ_H OF $\frac{B1}{R}$ $\frac{B2}{R}$ $\frac{M}{R}$

$$B1 = \frac{R\cos\theta_H}{40}\ ;\quad \frac{B1}{R} = \frac{\cos\theta_H}{40}$$

$$B2 = \frac{-R\sin\theta_H}{40}\ ;\quad \frac{B2}{R} = \frac{-\sin\theta_H}{40}$$

$$M = 1-(B1+B2);$$

$$\frac{M-1}{R} = \frac{\sin\theta_H - \cos\theta_H}{40}$$

FOR HOOK LOAD = P AT θ_H & R

$$B1 = \left(\frac{B1}{R}\right)\times R\times P$$

$$B2 = \left(\frac{B2}{R}\right)\times R\times P$$

$$M = P[1-(B1+B2)]$$

(↑ = + FOR REACTIONS)

Project : STIFFLEG DERRICK
Subject : UNIT LOAD AT HOOK - AXIAL FORCES - STIFFLEG FRAME
SHEET 14 OF

AXIAL FORCES - STIFFLEG FRAME - UNIT LOAD AT HOOK

$F_{BL} = -1.41 B1$

$F_{SILL} = +B1$

B1

$F_{SILL} = \frac{R}{40} \cos \theta_H$; VARIES, + .025R TO -.025R

$F_{BACKLEG} = \frac{-1.41}{40} R \cos \theta_H$; VARIES -.0354R TO +.0354R

(SEE 13 FOR VARIATION OF $\frac{B}{R}$ WITH θ_H)

F_{BM} = $(F_{BM})_H$, $(F_{BM})_V$

F_{MAST}, F_{BM}, F_{SILL}*, M, 100, H, R, $\theta_V = \tan^{-1} \frac{H}{R}$

REACH	F_{BM}	$(F_{BM})_V$
20	-2.5	-2.45↓
40		-2.29
60		-2.00
80		-1.50
90		-1.09
100		0

$F_{MAST} = F_{BM} \sin \theta_V - M = (F_{BM})_V - M$

FOR UNIT LOAD AT HOOK

$M = 1 + \frac{R}{40} (\sin \theta_H - \cos \theta_H)$; ↑ = +

R (FT)	$(F_{BM})_V$	θ_H (DEG)	$\sin \theta_H - \cos \theta_H$	M (+ = ↑)	F_{MAST} (+ = T)
20'	-2.45	0	-1.00	+0.50	+1.95
		30	-0.37	+0.82	+1.63
		60	+0.37	+1.18	+1.27
		90	+1.00	+1.50	+0.95
		120	+1.37	+1.69	+0.76
		135	+1.41	+1.71	+0.74
40'	-2.29	0		0	+2.29
		30		+0.63	+1.66
		60		+1.37	+0.92
		90		+2.00	+0.29
		120		+2.37	-0.08
		135		+2.41	-0.12
60'	-2.00	0		-0.50	+2.50
		30		+0.45	+1.55
		60		+1.56	+0.44
		90		+2.50	-0.50
		120		+3.06	-1.06
		135		+3.12	-1.12
80'	-1.50	0		-1.00	+2.50
		30		+0.25	+1.25
		60		+1.75	-0.25
		90		+3.00	-1.50
		120		+3.75	-2.25
		135		+3.82	-2.32
90'	-1.09	0		-1.25	+2.34
		30		+0.17	+0.92
		60		+1.83	-0.74
		90		+3.25	-2.16
		120		+4.08	-2.99
		135		+4.17	-3.08
100'	0	0		-1.5	+1.5
		30		-0.08	+0.08
		60		+1.93	-1.93
		90		+3.5	-3.50
		120		+4.43	-4.43
		135		+4.53	-4.53

* RESULTANT OF FORCES IN THE TWO SILL BEAMS

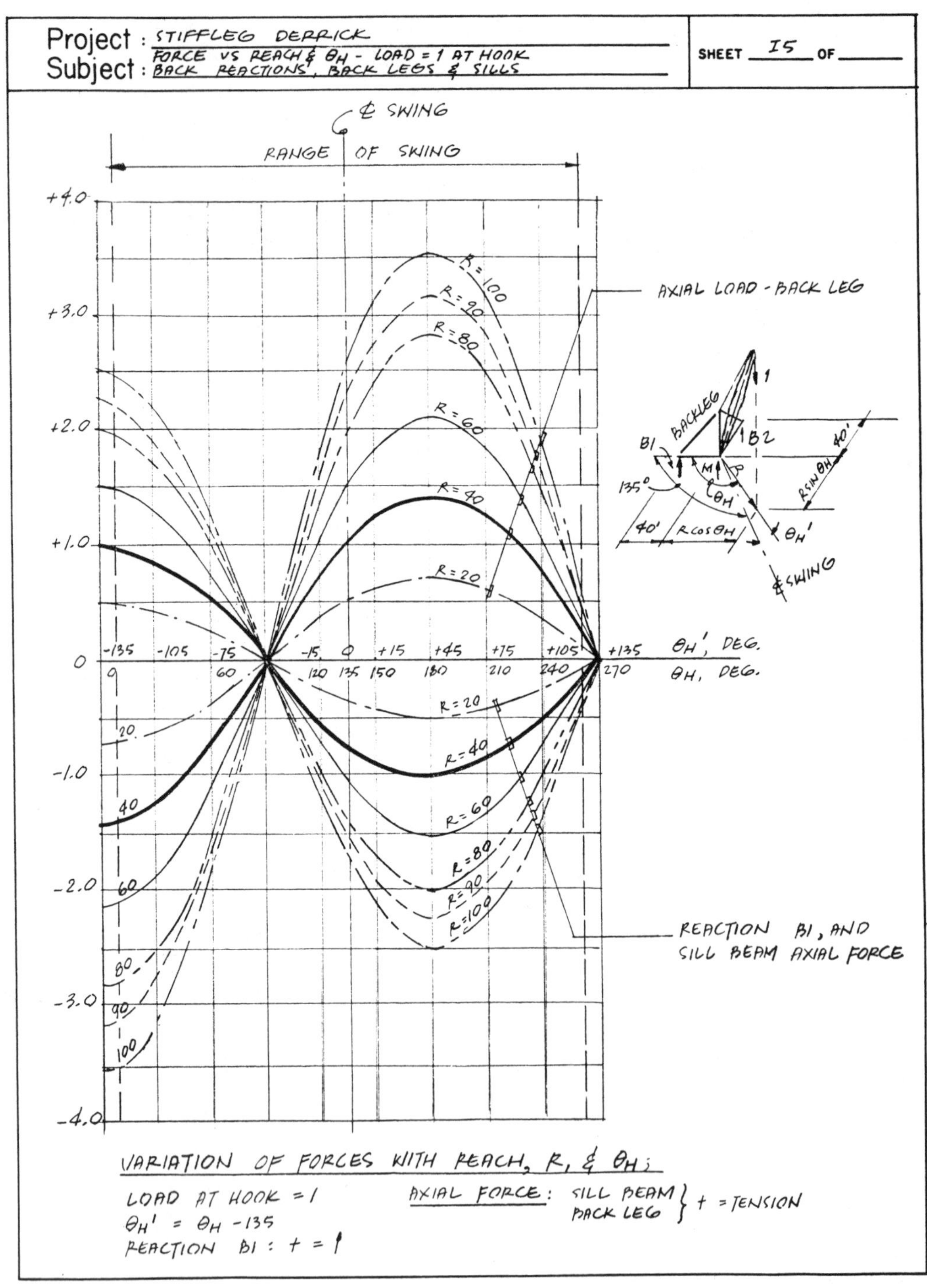
Project : STIFFLEG DERRICK
Subject : FORCE VS REACH & θH - LOAD = 1 AT HOOK
BACK REACTIONS, BACK LEGS & SILLS
SHEET I5 OF
℄ SWING
RANGE OF SWING
AXIAL LOAD - BACK LEG
REACTION B1, AND SILL BEAM AXIAL FORCE
R=100
R=90
R=80
R=60
R=40
R=20
+4.0
+3.0
+2.0
+1.0
0
-1.0
-2.0
-3.0
-4.0
-135 -105 -75 -15 0 +15 +45 +75 +105 +135
0 60 120 135 150 180 210 240 270
θH', DEG.
θH, DEG.
BACKLEG
B1
B2
M
R
135°
θH
40'
R cos θH
R sin θH
40'
θH'
℄ SWING
VARIATION OF FORCES WITH REACH, R, & θH;
LOAD AT HOOK = 1
θH' = θH - 135
REACTION B1 : + = ↑
AXIAL FORCE: SILL BEAM, BACK LEG } + = TENSION

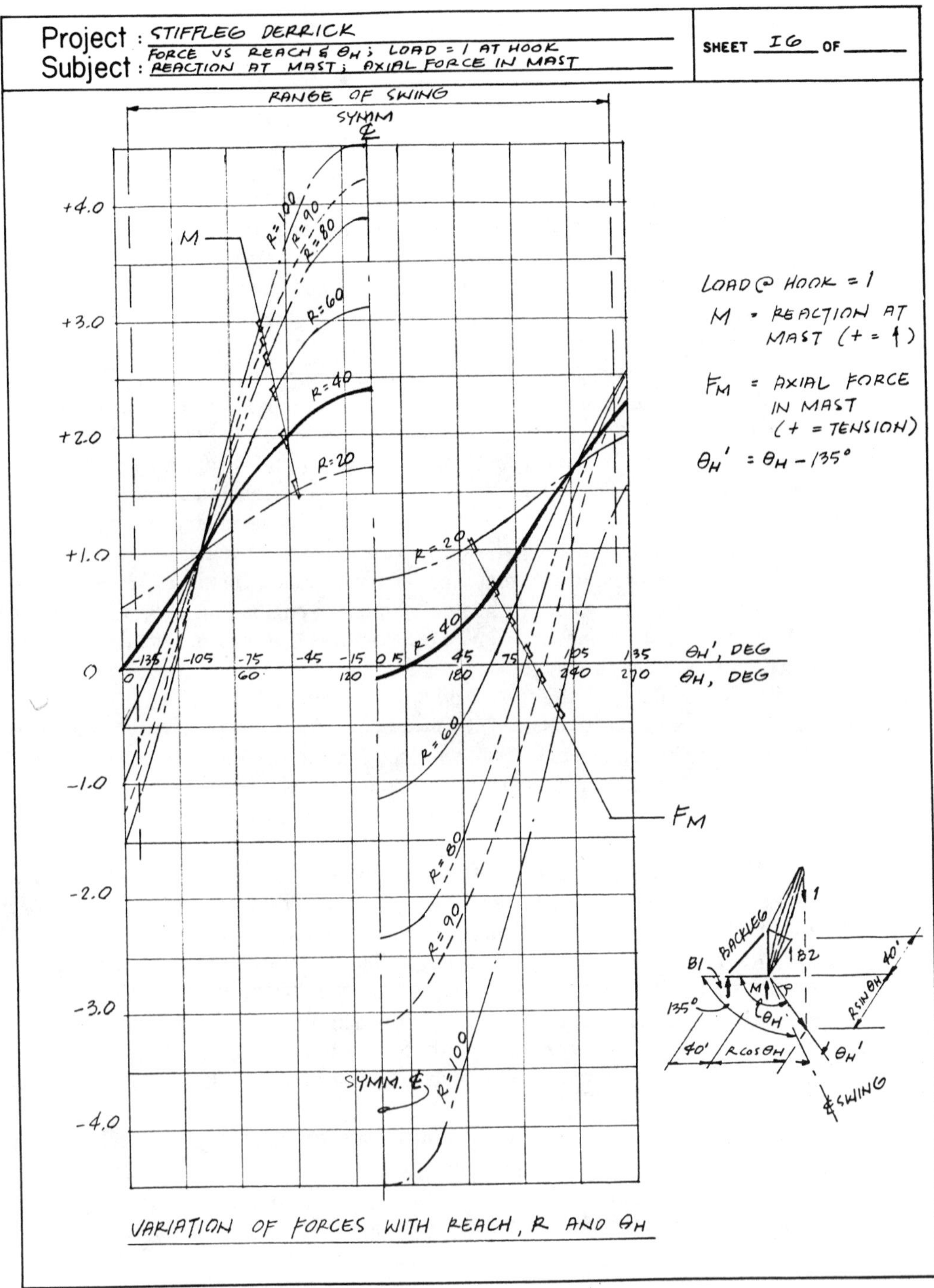
Project : STIFFLEG DERRICK
Subject : FORCE VS REACH & θ_H; LOAD = 1 AT HOOK
REACTION AT MAST; AXIAL FORCE IN MAST
SHEET 16 OF
RANGE OF SWING
SYMM ℄
M
R=100
R=90
R=80
R=60
R=40
R=20
LOAD @ HOOK = 1
M = REACTION AT MAST (+ = ↑)
F_M = AXIAL FORCE IN MAST (+ = TENSION)
$\theta_H' = \theta_H - 135°$
θ_H', DEG
θ_H, DEG
F_M
SYMM. ℄
BACKLEG
B1
B2
M
1
135°
θ_H
θ_H'
40'
R cos θ_H
R sin θ_H
℄ SWING
VARIATION OF FORCES WITH REACH, R AND θ_H

19.6 MEMBER SELECTION AND ANALYSIS

19.6.1 Boom

The boom design is a significant problem in itself; its study will also provide vital data for the rest of the system. The influence analysis has already supplied information on the compressive response of the boom to hook load. However, before selecting its parts, it is necessary to study in greater detail all the loads it will experience and how it will resist them. Sheets B1 through B5 do this.

Sheet B1 looks at the boom from three directions, examining the hook load and its conversion into topping line tension and boom compression, the lateral force, Q, applied at the tip, and its reactions at the boom heel and bottom of stays. It reveals some details required for connecting load and topping bails and leading the load line through the end of the boom on its passage to the load falls. We begin to see some of the sources of eccentricity at the connections—obvious at the boom heel, not yet so obvious at the head of the mast and point of boom. (The reader may wish to look ahead to sheets B3, B4, M3, and M4. The designer made rough early sketches to find out whether in fact there would be such eccentricity. The geometry of connections is often critical to design.) The cigar shape, with its parallel midsection and tapered ends, is an efficient way to use materials in compression members, discovered by sailing people centuries ago and validated by modern engineering theory. It will also provide compact end connections and help control clearance problems, which are indicated on this sheet as well as I1.

The boom will need to be *stayed* sideways for the action of slewing and/or wind forces (I_H) perpendicular to itself. The lower end of the stays must be pin connected on the axis of the boom heel pin. Any other location would demand changes in length, and therefore impossibly high tension as the boom pivots through its range of vertical angle. The bull wheel must be braced against the mast to ground the reaction from the stay.

The boom is conceived as a latticed compression member with parallel sides through the middle portion of its length. Its lower end tapers to provide a compact detail at the heel pin and to minimize interference with the mast when the boom is high or with the bull wheel when it reaches out flat. The upper end will taper to permit connection of the topping and load bails reasonably close to the boom axis. Lengths between splices are chosen for easy transport. The middle portion consists of two short, identical pieces. At times the operators may find it desirable to operate with a short boom. They may remove either one or both of the middle pieces provided all splices are identical.

The heel pin must be set forward of the mast to avoid physical interference. It is noted that the axis of the topping line will have some unknown and variable eccentricity with the intersection of mast and back leg. Both these facts will become important when the mast design is developed later.

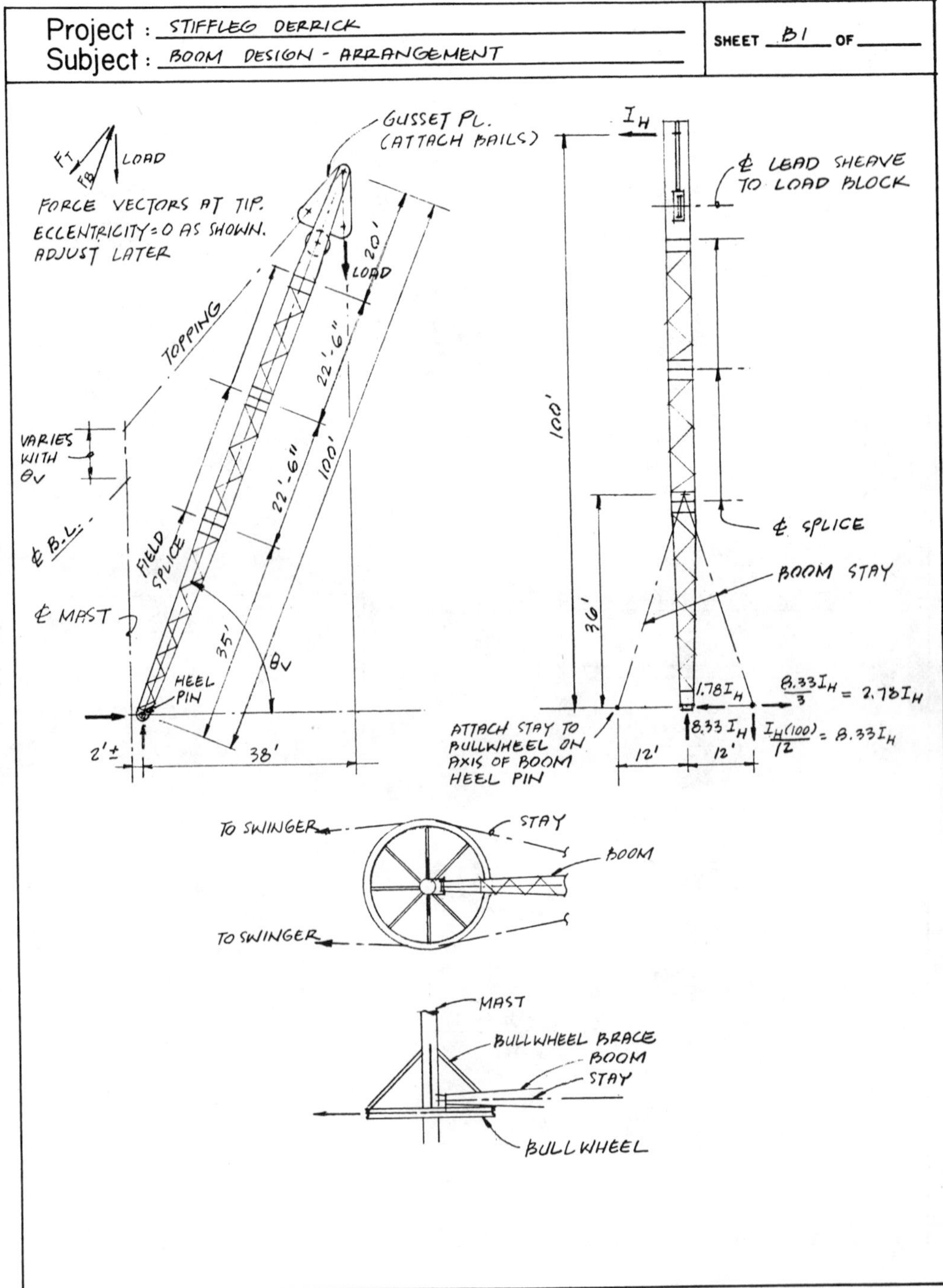
Project : STIFFLEG DERRICK
Subject : BOOM DESIGN - ARRANGEMENT
SHEET B1 OF
LOAD
FORCE VECTORS AT TIP.
ECCENTRICITY = 0 AS SHOWN.
ADJUST LATER
GUSSET PL.
(ATTACH BAILS)
LOAD
TOPPING
20'
22'-6"
22'-6"
100'
VARIES WITH θV
℄ B.L.
FIELD SPLICE
℄ MAST
35'
θV
HEEL PIN
2'±
38'
℄ LEAD SHEAVE TO LOAD BLOCK
100'
36'
℄ SPLICE
BOOM STAY
1.78IH
8.33IH/3 = 2.78IH
8.33 IH
IH(100)/12 = 8.33IH
ATTACH STAY TO BULLWHEEL ON AXIS OF BOOM HEEL PIN
12'
12'
TO SWINGER
STAY
BOOM
TO SWINGER
MAST
BULLWHEEL BRACE
BOOM
STAY
BULLWHEEL

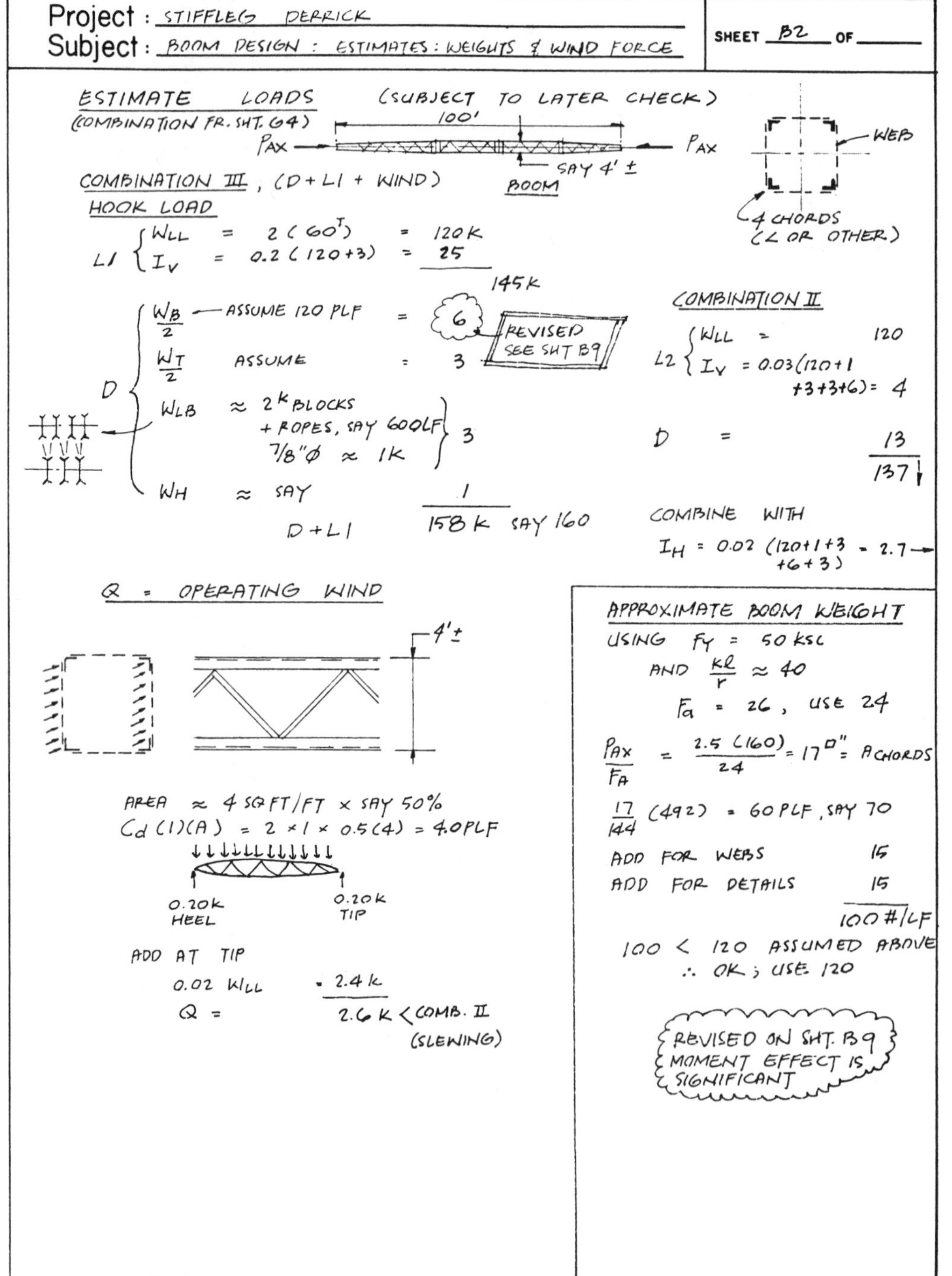

Project : STIFFLEG DERRICK
Subject : BOOM DESIGN : ESTIMATES : WEIGHTS & WIND FORCE
SHEET B2 OF
ESTIMATE LOADS (SUBJECT TO LATER CHECK)
(COMBINATION FR. SHT. G4)
100'
PAX
PAX
SAY 4' ±
BOOM
WEB
4 CHORDS (∠ OR OTHER)
COMBINATION III, (D + L1 + WIND)
HOOK LOAD
WLL = 2 (60T) = 120K
L1 IV = 0.2 (120+3) = 25
145K
WB/2 — ASSUME 120 PLF = 6
REVISED SEE SHT B9
WT/2 ASSUME = 3
D
WLB ≈ 2K BLOCKS + ROPES, SAY 600LF 7/8"Ø ≈ 1K 3
WH ≈ SAY 1
D + L1 158 K SAY 160
COMBINATION II
L2 WLL = 120
IV = 0.03(120+1+3+3+6) = 4
D = 13
137
COMBINE WITH
IH = 0.02 (120+1+3+6+3) = 2.7
Q = OPERATING WIND
4' ±
AREA ≈ 4 SQ FT/FT × SAY 50%
Cd (I)(A) = 2 × 1 × 0.5(4) = 4.0 PLF
0.20K HEEL
0.20K TIP
ADD AT TIP
0.02 WLL = 2.4 K
Q = 2.6 K < COMB. II (SLEWING)
APPROXIMATE BOOM WEIGHT
USING FY = 50 KSI
AND Kl/r ≈ 40
Fa = 26, USE 24
PAX/FA = 2.5 (160)/24 = 17□" = ACHORDS
17/144 (492) = 60 PLF, SAY 70
ADD FOR WEBS 15
ADD FOR DETAILS 15
100 #/LF
100 < 120 ASSUMED ABOVE ∴ OK ; USE 120
REVISED ON SHT. B9 MOMENT EFFECT IS SIGNIFICANT

On sheet B2, the load criteria established earlier on G4 are quantified for the condition of 60 tons of lifted load at 40-foot radius. To establish values for dead load, it is necessary to make some early assumptions, which will be reviewed from time to time and corrected as necessary. The reference to revision on sheet B9 indicates that some correction did become necessary.

In estimating the weights of suspended items (W_{LB}), the designer could refer to catalogs or consult with manufacturers' representatives. The consequence of error is small and becomes dangerous only if not corrected before the work is complete.

The boom supports its own weight as a simple span member in flexure, whose end reactions are vertical forces at the heel and tip each equal, approximately, to one half of the total weight. The tip reaction acts in the line of the hook and causes topping line and boom axial loads as do other hook loads. An approximation of the weight of the boom is made, based on a reasonable assumption of the value of Kl/r that is believed to be desirable. The value is chosen low enough to avoid the steep portion of the curve of F_a versus Kl/r (i.e., the region of high sensitivity of allowable load to changes in stiffness). Lifta uses AISC Table 3-50 to establish F_a.

Lateral wind pressure will also act on the boom as a flexural member. The mechanism for resisting the tip reaction will be explored on the next sheet; the quantity of the wind force is approximated here, based on earlier criteria on sheet G4. The wind exerts pressure on the area of actual surface that interrupts the wind stream, a small percentage of the gross area of the boom as seen in profile. Since a small angle between the boom transverse axis and the direction of flow exposes the far face as well as the near to the wind, a relatively high percentage is used in the calculation.

Load combinations II and III follow the criteria on sheet G4. It is not possible to predict which will control the design of the boom. Continuing the boom analysis on sheet B3, Lifta examines the effect of bending moments from the boom's dead weight. This bending is about the longitudinal x-x plane of the boom as seen in cross section. Moments vary with vertical angle of boom, or reach, acting otherwise as moments from uniformly distributed load on a simple span with maximum value at midspan. The table of maximum moments indicates that a flat boom will have over two and one half times as much dead weight bending moment as one reaching 40 feet. Recall that the boom is to be designed for a specified load at 40 feet and then the safe loads for further reaches determined. Boom bending can be expected to have significant effects as the boom reaches farther.

Lifta looks at the boom in two vertical projections to establish the effect of each type of applied load on the design loads and moments on the boom. P_B comes from the effective hook load. The heel reaction also includes the effect of the lower half of the boom. In addition, note, from the second view, that lateral forces at the tip are carried down the boom by cantilever bending; the boom in combination with the side stays ultimately converts the bending moment into a couple of longitudinal forces, one of which adds to boom axial compression.

A table summarizes, for load combinations II and III, the effects found to date. The biggest surprise is the magnitude of the bending moment about the y axis, which is

Project: STIFFLEG DERRICK
Subject: BOOM DESIGN LOADS
SHEET B3 OF

BENDING MOMENT FROM SELF WEIGHT

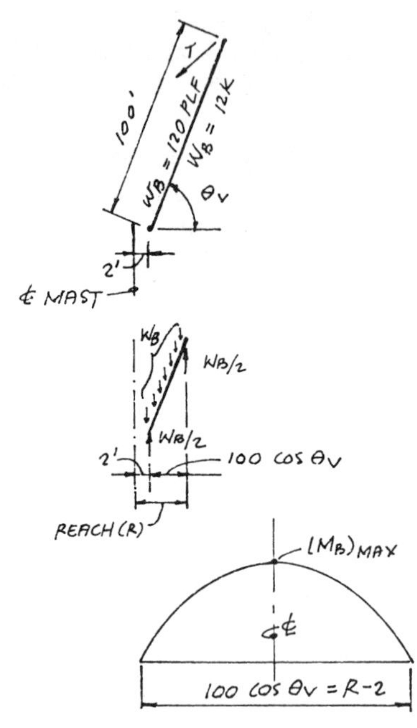

AT MID-LENGTH

$$(M_B)_{MAX} = \frac{W_B}{8}(R-2)$$

$$= 1.5\,(R-2)\ \text{FT.K}$$

R	$(M_B)_{MAX}$
20	27 'k
40	57
60	87
80	117
90	132
100	147
102	150

LOADS & MOMENTS - REACH = R = 40'

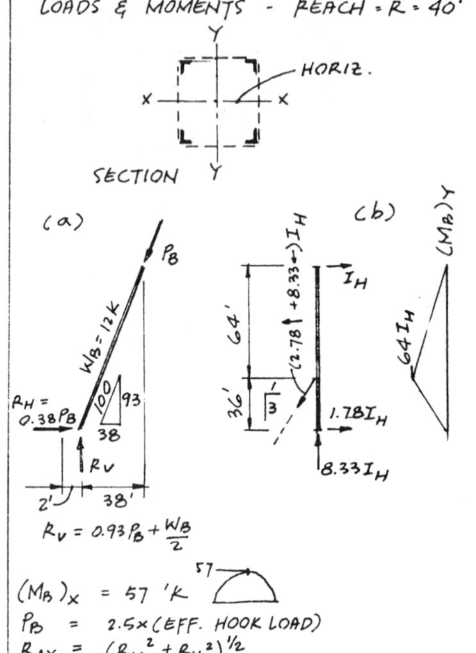

$R_V = 0.93P_B + \frac{W_B}{2}$

$(M_B)_X$ = 57 'K

P_B = 2.5×(EFF. HOOK LOAD)

$R_{AX} = (R_V^2 + R_H^2)^{1/2}$

NOTE: ECCENTRICITY OF HOOK LOAD AT BOOM TIP NOT INCLUDED HERE

	COMB I	COMB III	COMB II
EFFECTIVE HOOK LOAD	NOT CONTROLLING	160k	144k
P_B		400k	360k
P_H		2.6k	2.7k
W_B		12.0k	12.0k
R_V		378k	340k
R_H		152k	137k
R_{AX} (MAX. AX. FORCE)		407k	367k
M_{BX}		57'k	57'k
M_{BY}		166'k	173'k

SEE REVISIONS B9

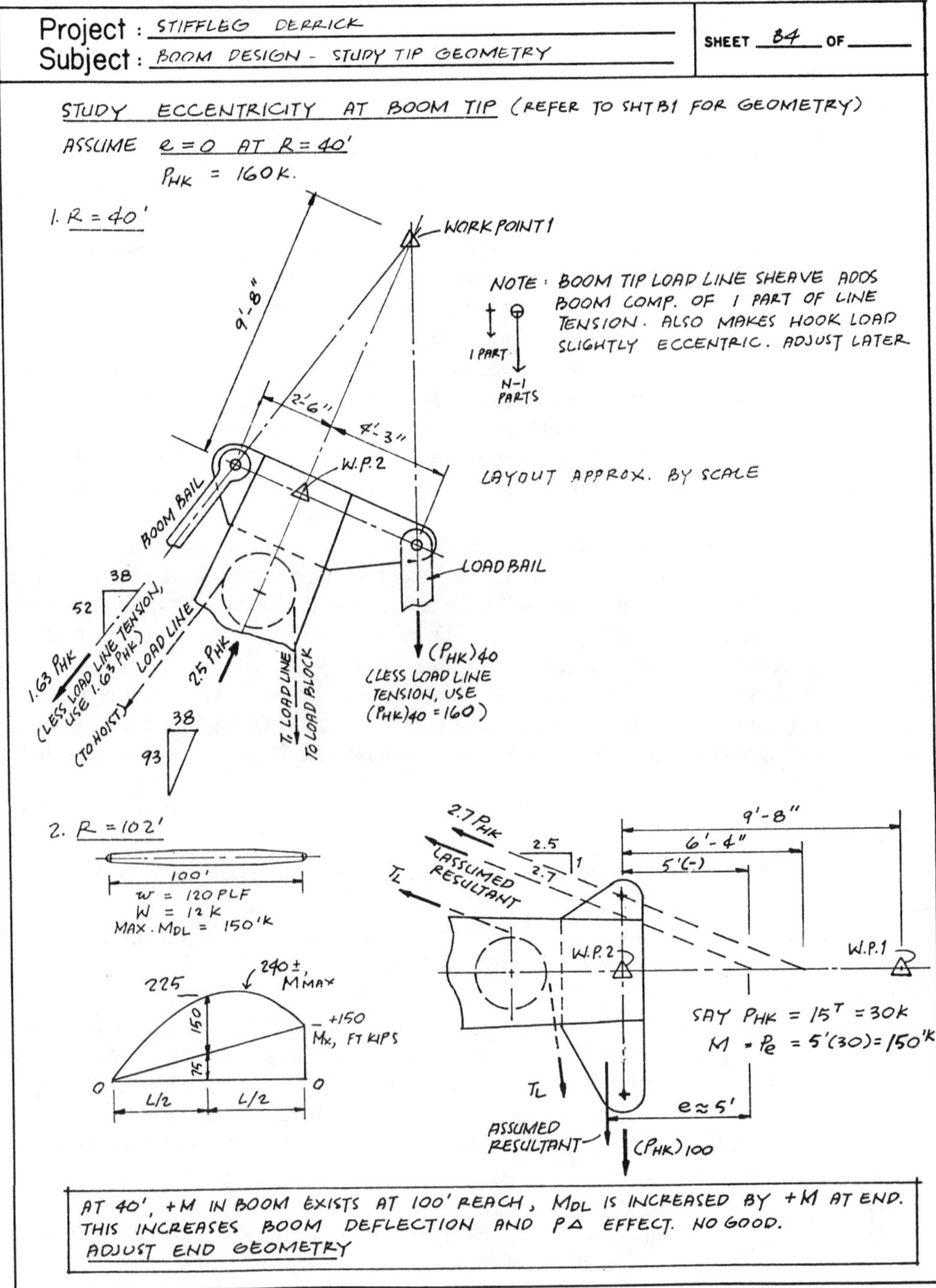
Project : STIFFLEG DERRICK
Subject : BOOM DESIGN - STUDY TIP GEOMETRY
SHEET 84 OF
STUDY ECCENTRICITY AT BOOM TIP (REFER TO SHT 81 FOR GEOMETRY)
ASSUME e = 0 AT R = 40'
PHK = 160K.
1. R = 40'
WORK POINT 1
9'-8"
NOTE: BOOM TIP LOAD LINE SHEAVE ADDS BOOM COMP. OF 1 PART OF LINE TENSION. ALSO MAKES HOOK LOAD SLIGHTLY ECCENTRIC. ADJUST LATER
1 PART
N-1 PARTS
2'-6"
4'-3"
W.P. 2
LAYOUT APPROX. BY SCALE
BOOM BAIL
LOAD BAIL
38
52
1.63 PHK
(LESS LOAD LINE TENSION, USE 1.63 PHK)
LOAD LINE
(TO HOIST)
2.5 PHK
TL LOAD LINE
TO LOAD BLOCK
(PHK)40
(LESS LOAD LINE TENSION, USE (PHK)40 = 160)
38
93
2. R = 102'
100'
w = 120 PLF
W = 12 K
MAX. MDL = 150'K
225
240±, MMAX
150
75
+150
Mx, FT KIPS
0
L/2
L/2
0
2.7 PHK
(ASSUMED RESULTANT
TL
2.5
2.7
1
9'-8"
6'-4"
5'(-)
W.P. 2
W.P. 1
SAY PHK = 15T = 30K
M = Pe = 5'(30) = 150'K
TL
e ≈ 5'
ASSUMED RESULTANT
(PHK)100
AT 40', +M IN BOOM EXISTS AT 100' REACH, MDL IS INCREASED BY +M AT END. THIS INCREASES BOOM DEFLECTION AND PΔ EFFECT. NO GOOD.
ADJUST END GEOMETRY

somewhat larger for the slewing force than for the wind. This emphasizes the importance of setting strict requirements on the acceleration characteristics of the hoist.

Knowing that the details at the boom tip are also a source of bending moment, Lifta next examines possible tip arrangements. Sheet B4 looks at one possible arrangement. Boom and load bails are located so that their force vectors intersect on the boom axis when the boom tip is at 40-foot reach. The load line is led through the boom over a sheave. It has a small effect on the total force system at the tip, and the system is essentially concurrent, causing little additional bending moment in the boom. However, when the boom rotates to the flat position, the forces are eccentric. Bending is caused in the boom, which is of the same sense as that caused by dead weight bending. This seems undesirable.

An alternative arrangement on sheet B5 introduces negative moment in both high and low boom positions. Although M_L at 40-foot reach looks high near the boom tip, it acts to reduce the midspan moment, which is the one most likely to cause boom buckling. It should be easy to reinforce the boom locally for the negative moment. With the boom flat, and for the very low value of hook load assumed, the effect of eccentricity is a small reduction in midspan moment.*

The designer is also concerned about boom deflection and the additional moment ($P\Delta$) that it may cause. This is approximately evaluated for the flat boom case. The formula used (see AISC Manual, page 2-114, case 1) accounts for flexural deflection and is usually sufficient for beams. For the open, trusslike configuration here, the designer estimates a possible addition for shear deflection; she decides, conservatively, to ignore the correcting live load moment. The result is a quite small $P\Delta$ moment and the conviction that boom deflection will not be a problem if the 4-foot assumed depth between chords is maintained. It is now time to start selecting members.

The process, starting on sheet B6, follows principles developed in Part II:

- For built-up latticed compression members with guidance from AISC Specification section 1.18;
- For combined axial load and bending with reference to a modified form of the interaction formulas in AISC Specification, 1.6.

The moment of inertia of the section is contributed by the chords only; the radius of gyration is almost exactly the distance from the axis to that of a chord. The effective moment of inertia must be reduced by the reduced stiffness in the regions of end tapers. Timoshenko and Gere[59] tabulate the effect of such tapers as a function of I_0/I_1 and h/l. The result in this case is an effective stiffness, measured by I, reduced

*On sheet M2, it will be found that the effective hook load for $R = 100$ feet is higher than what was assumed here. This reduces positive moment even further.

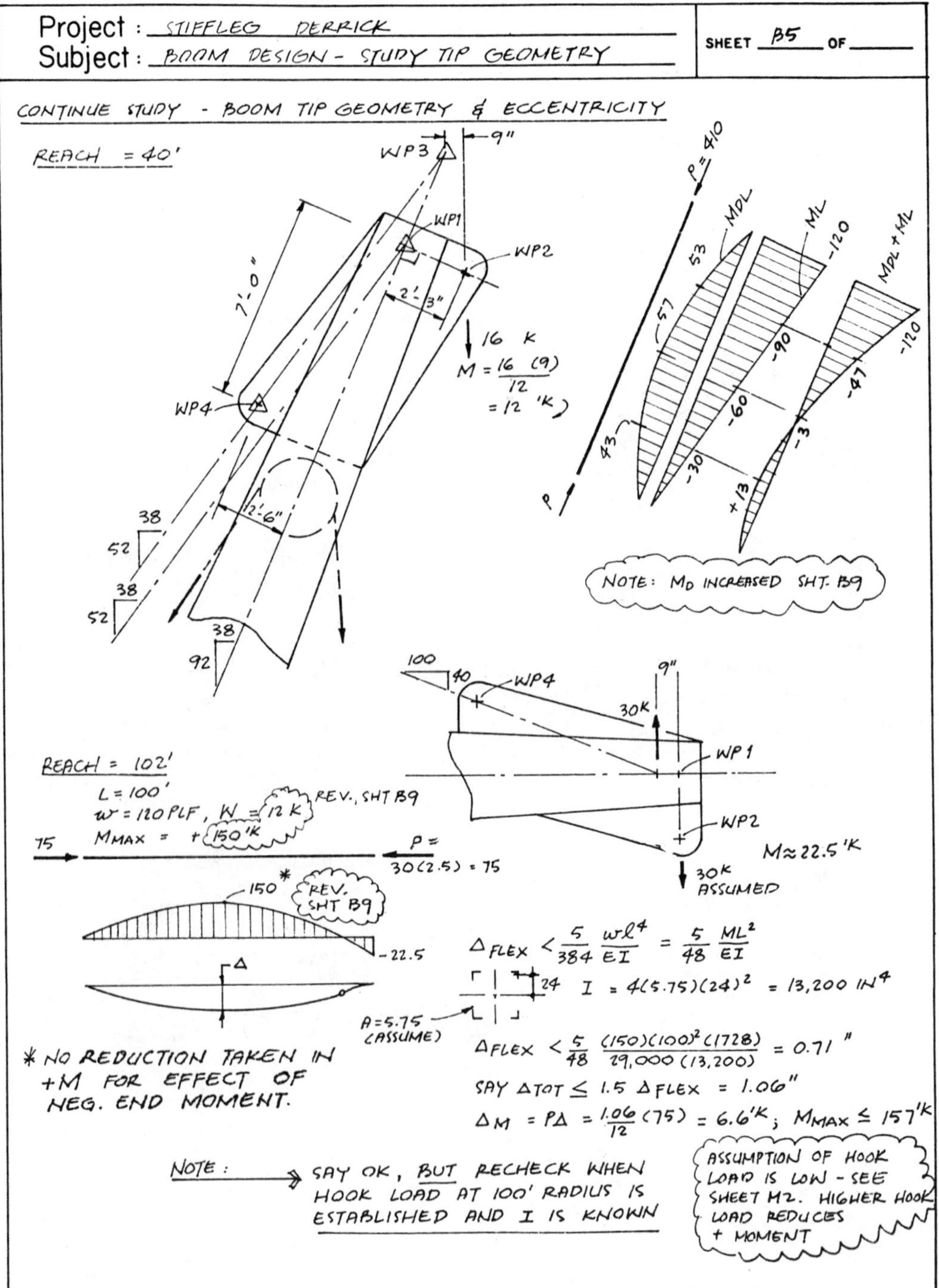

Project : STIFFLEG DERRICK
Subject : BOOM DESIGN - STUDY TIP GEOMETRY
SHEET B5 OF
CONTINUE STUDY - BOOM TIP GEOMETRY & ECCENTRICITY
REACH = 40'
WP3
9"
WP1
WP2
7'-0"
2'-3"
16 K
M = 16 (9) / 12 = 12 'K
WP4
2'-6"
38
52
38
52
38
92
P = 410
P
MDL
ML
MDL + ML
53
-120
57
43
-90
-60
-30
-47
-3
+13
-120
NOTE: MD INCREASED SHT. B9
100
40
WP4
9"
30K
WP1
WP2
M ≈ 22.5 'K
30K ASSUMED
REACH = 102'
L = 100'
w = 120 PLF, W = 12 K
REV., SHT B9
MMAX = + 150'K
75
P = 30(2.5) = 75
150*
REV. SHT B9
-22.5
Δ
ΔFLEX < 5/384 wl⁴/EI = 5/48 ML²/EI
24
I = 4(5.75)(24)² = 13,200 IN⁴
A = 5.75 (ASSUME)
ΔFLEX < 5/48 (150)(100)²(1728) / 29,000 (13,200) = 0.71"
SAY ΔTOT ≤ 1.5 ΔFLEX = 1.06"
ΔM = PΔ = 1.06/12 (75) = 6.6'K ; MMAX ≤ 157'K
* NO REDUCTION TAKEN IN +M FOR EFFECT OF NEG. END MOMENT.
NOTE : SAY OK, BUT RECHECK WHEN HOOK LOAD AT 100' RADIUS IS ESTABLISHED AND I IS KNOWN
ASSUMPTION OF HOOK LOAD IS LOW - SEE SHEET M2. HIGHER HOOK LOAD REDUCES + MOMENT

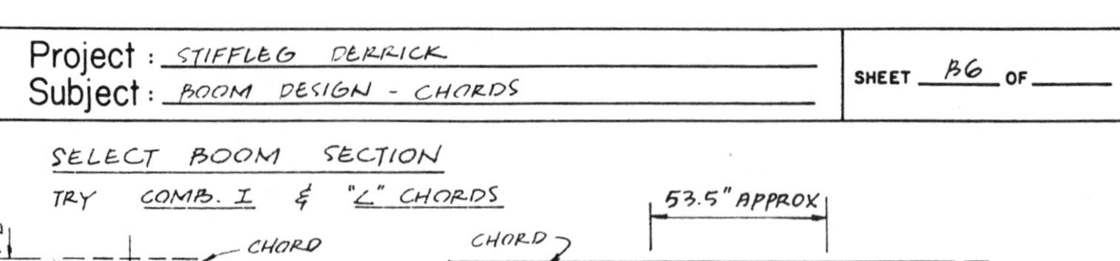

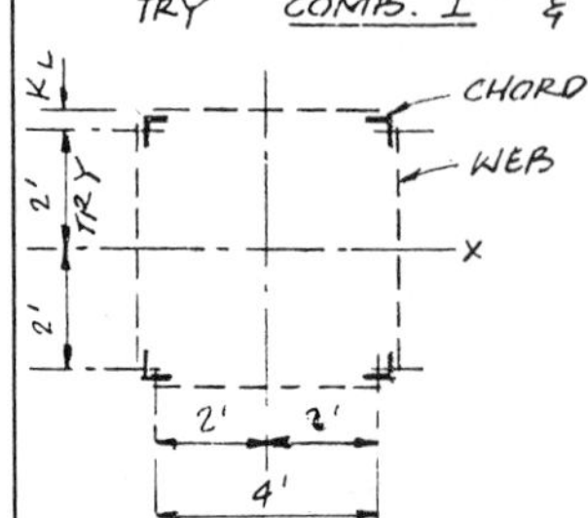

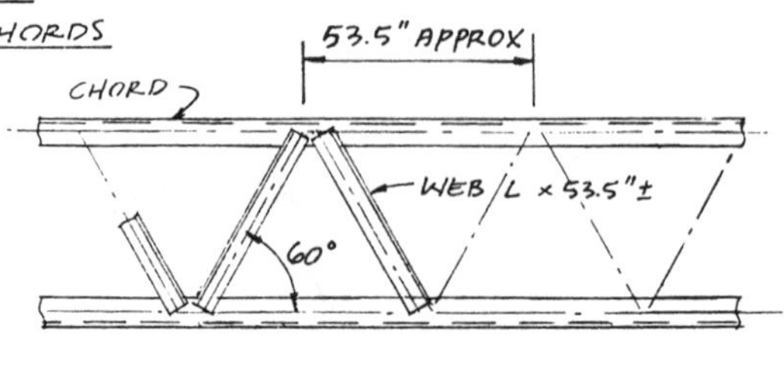

$A_{CH} = 4 \times A_L$ — SMALL - IGNORE

$(I_o)_{x\,or\,y} = 4\ A_L \times (24)^2 + 4 I_L$

$r_o = \sqrt{\frac{I}{A_{CH}}} = \sqrt{\frac{4A_L\,(24)^2}{4A_L}} = 24''$

$F_Y = 50$ ksi

$\frac{Kl}{r} = \frac{(1)(100)(12)}{24} = 50$

BUT

I VARIES — $100' = l$; 30, $50 = h$, 20; I_1, I_0

$I_1 \approx .2 I_0$; $\frac{h}{l} = .5$

$P_{CR} \approx \frac{9EI_o}{L^2}$; (COMPARE TO $\frac{\pi^2 EI}{L^2}$ FOR CONSTANT I)

$I_{EFF} \approx 0.91\ I_o{}^{*} = \frac{9}{\pi^2} I_o$

$r_{EFF} \approx 0.95\ r_o$

$\frac{Kl}{r} = 53$

$F_a = 23.9$ ksi (TABLE 3-50 AISC)

FOR AXIAL LOAD + BENDING IN X & Y DIRECTIONS, USE INTERACTION FORMULA, AISC 1.6-1a, BUT ADAPT BENDING TERMS USING f_a & F_a IN PLACE OF f_b & F_b

FROM BENDING,

$f_a = \pm \frac{M_{x\,or\,y}\,(12)}{2 A_L}$ — M, $f(2A_L)$, 48", $f(2A_L)$

FORMULA 1.6-1a BECOMES

$$\frac{f_{a1}}{F_a} \pm \frac{C_{mx} f_{a2}}{(1-\frac{f_{a1}}{F'_{ex}})F_a} \pm \frac{C_{my} f_{a3}}{(1-\frac{f_{a1}}{F'_{ey}})F_a} \le 1$$

WHERE :

$f_{a1} = \frac{P_B}{4A_L}$

$f_{a2} = \frac{M_x(12)}{2A_L(48)}$; $f_{a3} = \frac{M_Y(12)}{2A_L(48)}$

$F'_{ex} = F'_{ey} = 53.2$ (FOR $\frac{Kl}{r} = 53$ IN TABLE 9, AISC)

$C_{MX} = C_{MY} = 1$

$$\frac{1}{23.9 A_L}\left(\frac{P_B}{4} \pm \frac{M_x}{(1-\frac{f_{a1}}{53.2})8} \pm \frac{M_Y}{(1-\frac{f_{a1}}{53.2})8}\right) \le 1$$

IF $f_{a1} \approx 0.8 F_a$

$\frac{f_{a1}}{F'_{ex}} = \frac{f_{a1}}{F'_{ey}} = 0.36$

THEN USING POSITIVE VALUES IN BENDING TERMS,

$$\frac{1}{23.9}\left(\frac{P_B}{4} + \frac{M_x + M_Y}{8(.64)}\right) \le A_L$$

* TIMOSHENKO & GERE - THEORY OF ELASTIC STABILITY - 2ND EDITION - 1961 - McGRAW HILL p.113 et seq.

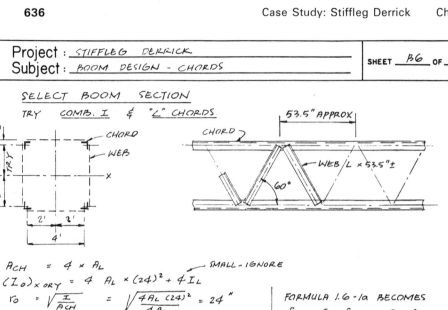

Project : STIFFLEG DERRICK
Subject : BOOM DESIGN – CHORDS
SHEET B7 OF ____

FROM SHTS B3 & B6

	CASE III	CASE II
P_B	400^K	360^K
M_X	$57'^K$	$57'^K$
M_Y	$166'^K$	$173'^K$
A_L (MIN)	6.01	5.65
$\frac{P_B}{4A_L F_a}$	0.70	0.67
f_{a1}	16.64	15.93
$1-\frac{f_{a1}}{F'_{ex}}$	0.69 > 0.64 OK	0.75 > 0.64 OK
USE A_L MIN	6.01	

ALT. 1

CHORDS : $\angle 6 \times 6 \times \frac{9}{16}$

A_L = 6.43"□

WT/FT, 4∠S = 87.6

b/t = $(5/9)(16)$ = 8.89 < 10.75 (AISC 1.9.1.2)

r_z OF ∠ = 1.18

$\ell/r = \frac{53.5}{1.18} = 45.3 < 53$ (AISC 1.18.2.4)

ALT. 2

CHORDS : □ 5×5×3/8 ⟸ PREFER (REVISED NEXT SHEET)

A (1 TUBE) = 6.58□"

WT/FT (4 TUBES) = 89.48

r OF TUBE = 1.86

$\ell/r = \frac{53.5}{1.86} = 28.8 \ll 53$

ALT2 IS SLIGHTLY HEAVIER THAN ALT. 1 BUT IS LESS VULNERABLE TO LOCAL DAMAGE.

CONNECTION OF, SAY, 4×3 TUBES FOR LACING, SEEMS SIMPLE AND AVOIDS ECCENTRICITY OF L LACING ON L CHORDS. DO NOT USE <4" TO AVOID BENDING IN FACE OF CHORD.

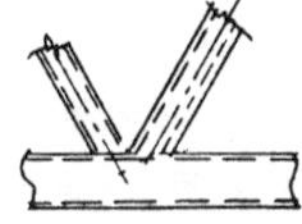

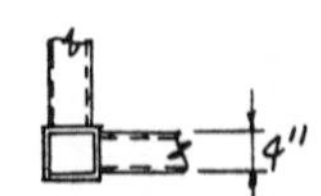

only 9% from what we would have if the midsection I were constant. This result would not be surprising to earlier generations of seafarers. The reduction in radius of gyration is only 5%.

In applying AISC formula 1.6-1a, the designer sees a need to adapt the bending terms. The usual bending moment formula,

$$f_b = \frac{My}{I},$$

must be converted to

$$f_{a2}\ (\text{or}\ f_{a3}) = \frac{M}{d \cdot A_{\text{chord}}}$$

since, in parallel chord trusses, moment effects are converted to axial loads in the chords. The factor $[1 - (f_a/F'_e)]$, which recognizes the effect of high axial load on flexural stability, is preserved in the usual form and will serve to increase the effect of bending in the interaction formula. This effect is estimated in the final formula for required chord area on this sheet. The estimate will turn out to be on the safe side when the calculation is made on sheet B7.

The calculation on B7 is done for both of load cases II and III. In searching for available sections, the designer finds an angle that fulfills all of the requirements. This was the type of section envisioned in earlier work, up to sheet B6, and is quite common in booms. However, she sees an opportunity to satisfy the same requirements using structural tubes with enhanced local stiffness indicator, r, and a chord less likely to twist or be damaged during use. The decision is made to follow this preference, and is maintained even though the actual tube selection is revised on the next sheet.

Sheet B8 is used to select web members, which must be designed for shear perpendicular to the boom axis. Calculated shear results from action of lateral loads at the boom tip and, for the vertical faces, from the dead weight of the boom. However, AISC 1.18.2.6 requires a minimum shear for lacing design of 2% of axial compression to guarantee the composite action of the built-up section.

Selection of the web member turns out to be controlled by the required stiffness, l/r. The cross-sectional area required to satisfy l/r is very much greater than that required to limit stress. Even the l/r requirement is overfulfilled by the lightest 4×3 tube available. The designer seeks a way to save weight of web members by using a smaller (3×2) tube, the lightest of which is still more than adequate. To permit this, she decides to change the chords from 5×5 tubes to thicker 4×4, with almost the same cross-sectional area and more than adequate stiffness. The reader may find it instructive to investigate the potential for bending in the chord face mentioned on sheet B7, and which leads to the designer's decision to control the depth of web members.

Project : STIFFLEG DERRICK
Subject : BOOM DESIGN - LACING

SHEET B8 OF ____

DESIGN SELECTION - LACING (REFER AISC 1.18)

FLAT FACES

□ 5×5
48"
53.5"
60°±
□ 5×5

SHEAR IN WEB - 2 FACES

@ 2% OF AXIAL COMP. = 400 (.02) = 8.0 k
OR P_H MAX (SHT. B3) = 3.0

DESIGN FOR 4.0 k/FACE

IN LACING, $P_{AX} = \frac{4.0}{\sin 60°} = 4.62$ k

4×3 TUBE, $r_y = 1.18$

$\frac{kl}{r_y} < \frac{53.5}{1.18} = 45.3 \ll 140$

$F_a = 25$

$A_R = 5.35/25 = 0.21$

LIGHTEST 4×3 TUBE ; A = 2.39 >> 0.21

USE 3×2 TUBE

BUT REVISE CHORDS TO:
- 4×4 × 1/2
- $A = 6.36\,\square'' > 6.01$
- W_T/ft (4 TUBES) = 86.5
- r = 1.39
- $\frac{l}{r} = \frac{53.5}{1.39} = 38.49 < 51$

SEE SHT. B7

3×2× 3/16 LACING

$A = 1.64\,\square''$; $W_T = 5.6$ PLF

$(\frac{b}{t})_{MAX} = \frac{2}{3/16} = 10.67 \ll \frac{238}{\sqrt{F_Y}}(1.92)$

$r_y = .771$

$\frac{kl}{r} = \frac{53.5}{.771} = 69.4 \ll 140$

$F_a = 21.1$ ksi

$f_a = \frac{5.35}{1.64} = 3.26 \ll 21.1$

NO SMALLER STIFF MEMBER ACCEPTABLE

USE 3×2× 3/16 □

VERTICAL FACES - 60° LACING

SHEAR IN WEB

@ 2% OF AXIAL COMP. = 8.0 k
ADD $\frac{W_B}{2}$ ≤ 6.0 k
14.0 k

DESIGN FOR $\frac{14.0}{2 \sin 60°} = 8.$ k

USING SAME TUBES AS IN FLAT FACE,

$F_a = 21.1$ ksi

$f_a = \frac{8.0}{1.66} = 4.82 \ll 21.1$

USE ▯ 3×2× 3/16

DETAILS

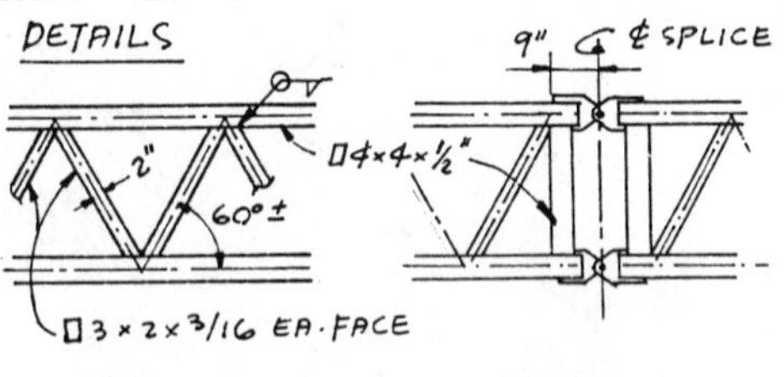

AT SPLICE OF CHORD

9"

END MOMENT/CORNER, DUE TO SHEAR

$\leq \frac{V}{4}(9'') = \frac{8.0}{4}(9) = 18.0''$k

$\frac{M}{S} = \frac{18.0}{6.13} = 2.94$ ksi

$\frac{f_b}{F_b} \leq \frac{2.94}{36} = 0.08$

FROM AXIAL FORCE (SHT B6, B7)

$\frac{\Sigma f_a}{F_a} = \frac{6.01}{6.36} = 0.94$

≤1.02 ≈ 1.0

SAY OK

BUT DO NOT ALLOW 9" DIMENSION TO INCREASE

Project : STIFFLEG DERRICK

Subject : BOOM DESIGN – WEBS

SHEET B9 OF ______

CHECK WEIGHT OF BOOM

4 @ 4 × 4 × ½ ☐ = 86.5 PLF × 100' = 8650 #

4 @ 3 × 2 × 3/16 ▯@60° = 2(5.6) × 4 × 100' = 4480 #

ADD OTHER DETAIL, SAY 4000 #

17,130 # – SAY 18,000 #

WAS ASSUMED @ 12K, SHT B2 !!

REVISE PREVIOUS CALCS.

	COMB. III WAS	COMB. III BECOMES	COMB II WAS	COMB II BECOMES
SHEET B2				
$\frac{W_B}{2}$	6	9	6	9
HOOK LOAD	160	163	144	147
SHEET B3				
P_B	400	408	360	368
P_H	2.6	2.6	2.7	2.9
W_B	12	18	12	18
R_V	378	388	340	351
R_H	152	155	137	140
R_{AX}	407	418	367	378
M_{BX}	57	86	57	86
M_{BY}	166	166	173	186

SHEET B5 : ADD 50% TO MOMENTS FROM BOOM DEAD WEIGHT

SHEET B7: RECALCULATION OF A_L REQ'D :

$$\frac{1}{23.9}\left(\frac{408}{4} + \frac{252}{8(.69)}\right) = 4.27 + 1.91 = 6.18^{\square}{}''$$

$$\frac{1}{23.9}\left(\frac{368}{4} + \frac{272}{8(.75)}\right) = 3.85 + 1.90 = 5.75^{\square}{}''$$

AVAILABLE : ▯ 4 × 4 × ½" ; A = 6.36 > 6.18

ALL MEMBER SIZES UNCHANGED.

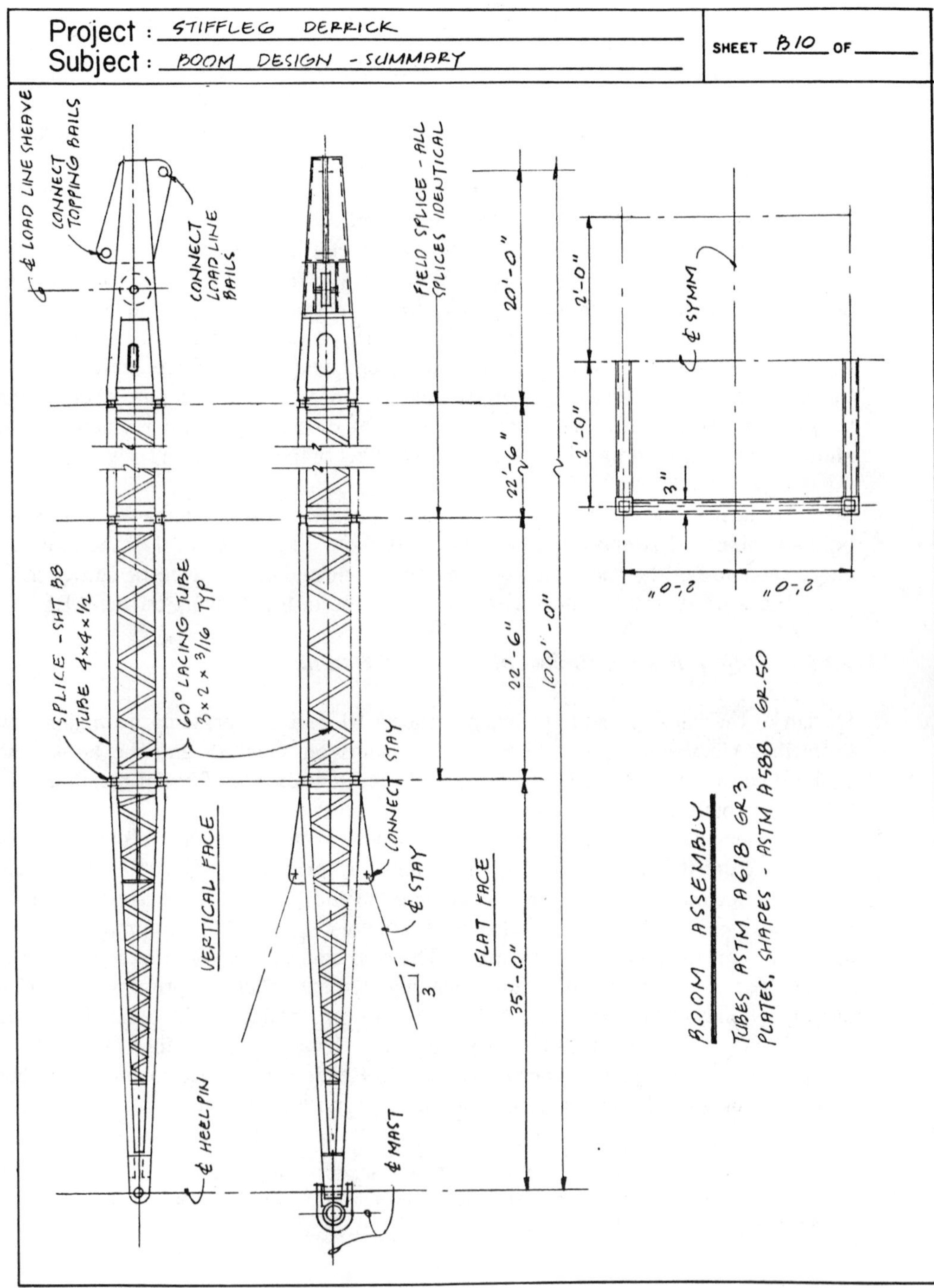

Project : STIFFLEG DERRICK
Subject : BOOM DESIGN - SUMMARY
SHEET B10 OF
℄ LOAD LINE SHEAVE
CONNECT TOPPING BAILS
CONNECT LOAD LINE BAILS
FIELD SPLICE - ALL SPLICES IDENTICAL
20'-0"
22'-6"
22'-6"
35'-0"
100'-0"
2'-0"
2'-0"
2'-0"
2'-0"
3"
℄ SYMM
SPLICE - SHT B8
TUBE 4×4×1/2
60° LACING TUBE
3×2×3/16 TYP
CONNECT STAY
℄ STAY
VERTICAL FACE
FLAT FACE
℄ HEEL PIN
℄ MAST
BOOM ASSEMBLY
TUBES ASTM A618 GR3
PLATES, SHAPES - ASTM A588 - GR.50

Before leaving this sheet, the designer sketches to scale a typical portion of the boom and a possible field splice. The splice detail is far from complete, but indicates a desire to use single pins at each corner, a boon to erectors who have to assemble and disassemble the boom whenever the derrick is moved. One result is that the lacing is interrupted at the splice, leaving the chords themselves to resist transverse shear through flexure. The moment effect is small, but enough to nudge the interaction formula slightly over the limit. Convinced that the shear and moment calculation is on the high side, she decides to accept the small indicated overstress.

A moment of truth is faced on sheet B9, when Lifta checks the weight of the boom against the earlier assumption. With an allowance for as yet undeveloped details, the result is 50% higher than the assumption that was included in all calculations up to this point. Rather than go back to redo all the earlier sheets, she decides to flag the problem on the early sheets and evaluate its effects at this point in the set. By this procedure, she preserves the process of development of design thinking and avoids the mess that results from crossing out and revising earlier numbers. More important, she will isolate the effect of the increase in boom weight on the design of the boom.

To do this, the effect on earlier results is presented in a comparison table for both combinations II and III. Combination III now assumes control of the required chord section, but the chord section previously selected turns out to be sufficient.

The work on the boom to date is summarized in the drawing on sheet B10.

19.6.2 Stiffleg Frame: Back Legs and Sill Beams

Design of the back leg and sill beams (sheets S1 and S2) presents no major new problems. Member loads come from coefficients on sheet I5 and the hook load established on sheet B9. Since compression and tension are of equal magnitude in both cases, both members are designed for compression.

The length of back leg is approximated from a sketch that examines the required reduction from the theoretical diagonal length to "practical" points for connecting pins.* An open, laced column is selected and the ends tapered for reasons similar to those considered for the boom. Field splices are not necessary. The common L chords seem quite reasonable for this member. The web angles, which are attached on the outside face of the chords, experience bending moment under compression, requiring an interaction check by criteria of AISC, 1.6. For the smallest readily available L, the utilization factors for axial load and bending are both very small and their sum is much less than 1. The designer decides to use that L. Readers might like to check whether double lacing with flat bars would turn out to be more economical.

*Sheet M3 will show, later, that at least one of these setback lengths, that at the top, is overestimated. Since Kl/r is small, the sensitivity of F_a to small changes in length is not great, so the design does not require change.

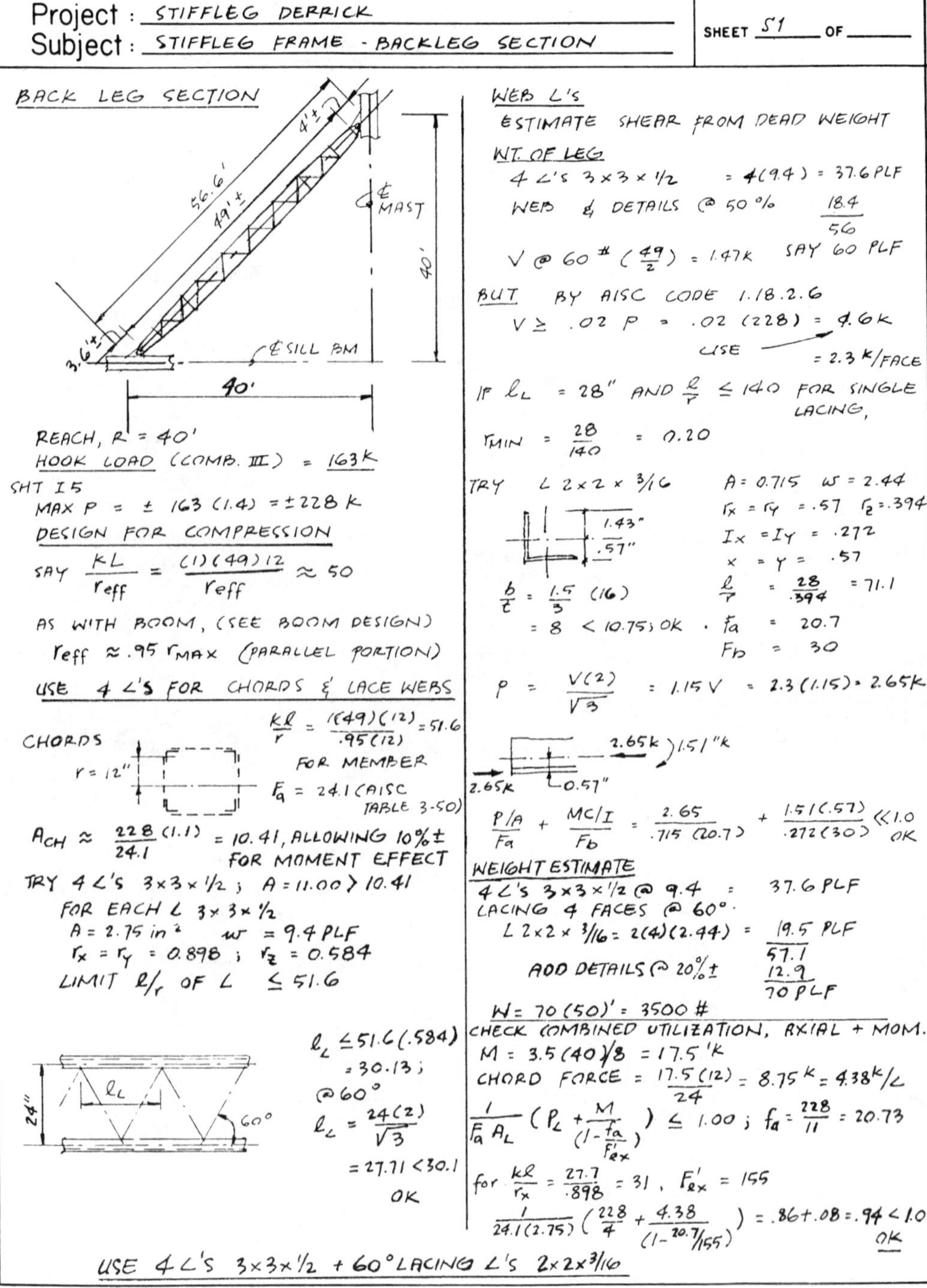

Project : STIFFLEG DERRICK

Subject : STIFFLEG FRAME - BACKLEG SECTION

SHEET 51 OF

BACK LEG SECTION

REACH, R = 40'

HOOK LOAD (COMB. III) = 163 K

SHT I5

MAX P = ± 163 (1.4) = ±228 K

DESIGN FOR COMPRESSION

SAY $\frac{KL}{r_{eff}} = \frac{(1)(49)12}{r_{eff}} \approx 50$

AS WITH BOOM, (SEE BOOM DESIGN)

$r_{eff} \approx .95\, r_{MAX}$ (PARALLEL PORTION)

USE 4 ∠'S FOR CHORDS & LACE WEBS

CHORDS

$\frac{K\ell}{r} = \frac{1(49)(12)}{.95(12)} = 51.6$ FOR MEMBER

r = 12"

F_a = 24.1 (AISC TABLE 3-50)

$A_{CH} \approx \frac{228}{24.1}(1.1) = 10.41$, ALLOWING 10% ± FOR MOMENT EFFECT

TRY 4 ∠'S 3×3×½ ; A = 11.00 > 10.41

FOR EACH ∠ 3×3×½

A = 2.75 in² w = 9.4 PLF

$r_x = r_y$ = 0.898 ; r_z = 0.584

LIMIT ℓ/r OF ∠ ≤ 51.6

$\ell_L \le 51.6(.584)$ = 30.13 ;

@ 60°

$\ell_L = \frac{24(2)}{\sqrt{3}}$ = 27.71 < 30.1 OK

WEB ∠'S

ESTIMATE SHEAR FROM DEAD WEIGHT

WT. OF LEG

4 ∠'S 3×3×½ = 4(9.4) = 37.6 PLF

WEB & DETAILS @ 50% 18.4

56

V @ 60# $(\frac{49}{2})$ = 1.47K SAY 60 PLF

BUT BY AISC CODE 1.18.2.6

V ≥ .02 P = .02 (228) = 4.6K

USE = 2.3 K/FACE

IF ℓ_L = 28" AND $\frac{\ell}{r} \le 140$ FOR SINGLE LACING,

$r_{MIN} = \frac{28}{140} = 0.20$

TRY ∠ 2×2×3/16 A = 0.715 w = 2.44

$r_x = r_y$ = .57 r_z = .394

$I_x = I_y$ = .272

x = y = .57

$\frac{b}{t} = \frac{1.5}{3}(16)$ = 8 < 10.75; OK

$\frac{\ell}{r} = \frac{28}{.394}$ = 71.1

F_a = 20.7

F_b = 30

$P = \frac{V(2)}{\sqrt{3}}$ = 1.15 V = 2.3 (1.15) = 2.65K

$\frac{P/A}{F_a} + \frac{MC/I}{F_b} = \frac{2.65}{.715(20.7)} + \frac{1.51(.57)}{.272(30)}$ << 1.0 OK

WEIGHT ESTIMATE

4 ∠'S 3×3×½ @ 9.4 = 37.6 PLF

LACING 4 FACES @ 60°.

∠ 2×2×3/16 = 2(4)(2.44) = 19.5 PLF

57.1

ADD DETAILS @ 20% ± 12.9

70 PLF

W = 70 (50)' = 3500 #

CHECK COMBINED UTILIZATION, AXIAL + MOM.

M = 3.5 (40)/8 = 17.5 'K

CHORD FORCE = $\frac{17.5(12)}{24}$ = 8.75 K = 4.38 K/∠

$\frac{1}{F_a A_L}\left(P_L + \frac{M}{(1-\frac{f_a}{F'_{ex}})}\right) \le 1.00$; $f_a = \frac{228}{11}$ = 20.73

for $\frac{k\ell}{r_x} = \frac{27.7}{.898}$ = 31 , F'_{ex} = 155

$\frac{1}{24.1(2.75)}\left(\frac{228}{4} + \frac{4.38}{(1-20.7/155)}\right)$ = .86 + .08 = .94 < 1.0 OK

USE 4 ∠'S 3×3×½ + 60° LACING ∠'S 2×2×3/16

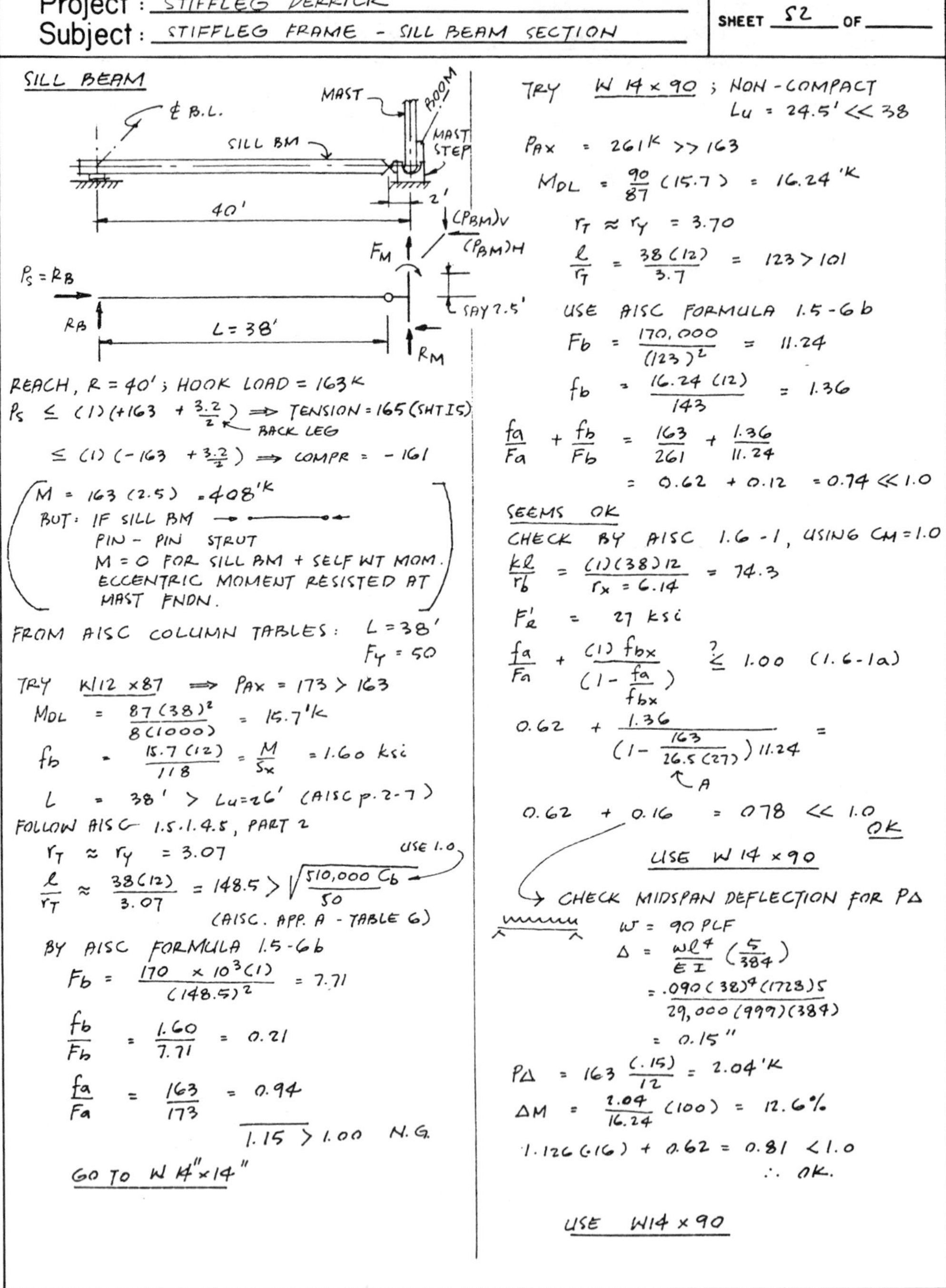

Project : STIFFLEG DERRICK
Subject : STIFFLEG FRAME - SILL BEAM SECTION
SHEET 52 OF

REACH, R = 40'; HOOK LOAD = 163^K

$P_S \leq (1)(+163 + \frac{3.2}{2}) \Rightarrow$ TENSION = 165 (SHT 15)
← BACK LEG

$\leq (1)(-163 + \frac{3.2}{2}) \Rightarrow$ COMPR = −161

(M = 163 (2.5) = 408'K
BUT: IF SILL BM •——• PIN − PIN STRUT
M = 0 FOR SILL BM + SELF WT MOM.
ECCENTRIC MOMENT RESISTED AT MAST FNDN.)

FROM AISC COLUMN TABLES: L = 38', F_Y = 50

TRY W12 × 87 ⇒ P_{AX} = 173 > 163

$M_{DL} = \frac{87(38)^2}{8(1000)} = 15.7'^K$

$f_b = \frac{15.7(12)}{118} = \frac{M}{S_x} = 1.60$ ksi

L = 38' > Lu = 26' (AISC p. 2-7)

FOLLOW AISC 1.5.1.4.5, PART 2

$r_T \approx r_y = 3.07$

$\frac{\ell}{r_T} \approx \frac{38(12)}{3.07} = 148.5 > \sqrt{\frac{510,000\,C_b}{50}}$ ← USE 1.0
(AISC. APP. A - TABLE 6)

BY AISC FORMULA 1.5-6b

$F_b = \frac{170 \times 10^3(1)}{(148.5)^2} = 7.71$

$\frac{f_b}{F_b} = \frac{1.60}{7.71} = 0.21$

$\frac{f_a}{F_a} = \frac{163}{173} = 0.94$

1.15 > 1.00 N.G.

GO TO W 14" × 14"

TRY W 14 × 90; NON-COMPACT
Lu = 24.5' << 38

P_{AX} = 261^K >> 163

$M_{DL} = \frac{90}{87}(15.7) = 16.24'^K$

$r_T \approx r_y = 3.70$

$\frac{\ell}{r_T} = \frac{38(12)}{3.7} = 123 > 101$

USE AISC FORMULA 1.5-6b

$F_b = \frac{170,000}{(123)^2} = 11.24$

$f_b = \frac{16.24(12)}{143} = 1.36$

$\frac{f_a}{F_a} + \frac{f_b}{F_b} = \frac{163}{261} + \frac{1.36}{11.24}$
= 0.62 + 0.12 = 0.74 << 1.0

SEEMS OK

CHECK BY AISC 1.6-1, USING C_M = 1.0

$\frac{K\ell}{r_b} = \frac{(1)(38)12}{r_x = 6.14} = 74.3$

F'_e = 27 ksi

$\frac{f_a}{F_a} + \frac{(1) f_{bx}}{(1 - \frac{f_a}{F'_{ex}})} \overset{?}{\leq} 1.00$ (1.6-1a)

$0.62 + \frac{1.36}{(1 - \frac{163}{26.5(27)})\,11.24} =$
↖ A

0.62 + 0.16 = 0.78 << 1.0 OK

USE W 14 × 90

→ CHECK MIDSPAN DEFLECTION FOR PΔ

w = 90 PLF

$\Delta = \frac{w\ell^4}{EI}\left(\frac{5}{384}\right)$

$= \frac{.090(38)^4(1728)5}{29,000(999)(384)}$

= 0.15"

$P\Delta = 163\,\frac{(.15)}{12} = 2.04'^K$

$\Delta M = \frac{2.04}{16.24}(100) = 12.6\%$

1.126 (.16) + 0.62 = 0.81 < 1.0
∴ OK.

USE W14 × 90

Following a weight check, the effect of dead weight moment is checked and found to be small.

For the sill beam, the designer decides to try a standard W, hoping to reduce fabrication costs and lured in part by the attractive walking surface it offers for workers. In this case, dead weight bending turns out to be significant, ruling out the first hopeful choice of a W12. The results of applying a simple interaction formula similar to AISC 1.6-1 show 97% utilization in compression plus an additional 21% in bending. The reasonable thing seems to be to add depth and lateral stiffness, the first to add section modulus, the second to increase F_a by reducing l/r. At the expense of only 3 pounds/foot, the new W14 selection accomplishes both goals, proving out, with much to spare, by the more precise interaction formulas of AISC 1.6-1. Noting a large l/d ratio of

$$\frac{38'(12)}{14} = 32.6$$

the designer decides to check dead weight deflection and the consequent $P\Delta$ moment. This turns out to be 12% of the dead weight moment, adding only

$$0.12(0.16) = 0.02$$

to the bending term of the interaction formula. Comfortable in this result, Lifta moves on to the mast.

19.6.3 Mast

Once again, the designer is faced with a complex member problem, perhaps even more complex than the boom. As part of the stiffleg frame, the mast must also rotate with the boom. From sheet I6, it is apparent that it must be both a tension and compression member. However, the significant compression occurs with the hook load, not at 40 feet, but at the farther reaches. The boom study (sheet B1) has already revealed an eccentric connection at the boom heel pin and a possible one at the masthead. It is decided to ease into the design problem by:

1. Studying the effect of the hook load through the full range of reach.
2. Examining the geometry at top and bottom.

This is done on sheets M1 through M4 and results in the design forces and moments on sheet M5.

Since it seems likely from the curves on sheet I6 that loads at long reach will control the design of the mast, Lifta decides that it is time to develop the curve of load versus reach. The general criteria are that:

- Lifted loads shall not exceed 60 tons at any reach.

- Members and connections chosen for 60 tons at 40 feet shall not be overstressed by any other combination of load and reach.
- Reactions established for 60 tons at 40 feet shall not be exceeded by other combinations of load versus reach.

It seems improbable that the second criterion can be applied to the mast design without making the rig almost useless in the outer reaches of its range. However, it can be applied to the other members.

Tables on sheet M1 assemble influence data from sheets I2, I5, and I6. In the table at the top of sheet M2, the range of coefficients for each member is compared to the coefficient for hook load at 40 feet. The effective hook load will vary inversely with the coefficient from a base of 167k at 40 feet (sheet B9). The permissible lifted load (*rated capacity*) is the effective hook load less the included constant dead weight effect, adjusted for the impact factor. The formulas are set out below the table.

Examining the table, it becomes obvious that the range for back leg, sill loads, and back leg reactions are the same and larger than that for boom and topping line. Control of capacity seems to be in the back legs and sills. From the formulas accompanying the table, curves are drawn for load versus reach. The lower curve, which has been adjusted down for dead load and impact, and will be converted from kip units to tons, can be published as a derrick capacity curve if no other limitations show up in later stages of design.

Note from the upper curve that the effective hook load at 100 feet is 67 kips. On sheet B5, it was assumed at 30. The change increases end negative moment within an acceptable range, and decreases positive moment, a favorable result.

Sheets M3 and M4 are geometric studies at mast head and foot, leading to evaluation of eccentricity of input loads at various reaches. On M3, the various problems of arrangement are addressed at the mast head. The mast section is assumed as 18 inches O.D. with unknown thickness.

1. It will be necessary to maintain a constant 45° angle between mast and back leg in the plane of the two. (The other slope shown is a projected slope when the boom and bail are in a plane 45° away from that of the back leg.)
2. To allow rotation, the mast must pass through a sleeve and rotate in a lubricated radial bearing.
3. To transfer the vertical component of back-leg reaction, both tension and compression, the sleeve must bear against a thrust bearing top and bottom.
4. Wings are provided on the sleeve to connect it to the back legs.
5. The lower boom bail will be connected by a pin to a point outside the mast.

Project : STIFFLEG DERRICK Subject : MAST	SHEET M1 OF

MAST — FROM F_M CURVES, SHEET I6, IT SEEMS CLEAR THAT MAST DESIGN WILL NOT BE CONTROLLED BY LOAD AT 40' REACH

COEFFICIENTS - F_M & M - SHT I-6

R, FT	θ_H'	COMP	θ_H'	TENSION
20	–	N.A.	135°	1.95
40	0°	-0.13	↓	2.25
60	↓	-1.15	↓	2.50
80	↓	-2.35	↓	2.50
90	↓	-3.05	↓	2.35
100	↓	-4.5	↓	1.5

ESTIMATE HOOK LOAD VS. REACH, BASED ON DESIGN CAPACITY OF VARIOUS MEMBERS

VARIATION OF MEMBER FORCES AND REACTIONS WITH EFFECTIVE HOOK LOAD
C = (LIFTED LOAD + BOOM TIP D.L.)(1 + IMPACT FACTOR)

R	COEFFICIENTS (SHTS I2, I5, I6)						
	MEMBERS (+ = TENSION)					REACTIONS (+ = THRUST; – = UPLIFT)	
	BOOM	TOPP'G	BACK LEG	SILL BEAM	MAST, F_M	M	B
20	-2.5	+1.55	+0.70 -0.70	-0.50 +0.50	+1.95	+1.7	-0.50 +0.50
40	↓	+1.63	+1.40 -1.40	-1.00 +1.00	-0.13 +2.25	0 +2.41	-1.00 +1.00
60	↓	+1.80	+2.15 -2.15	-1.50 +1.50	-1.15 +2.50	-0.5 +3.1	-1.50 +1.50
80	↓	+2.05	+2.80 -2.80	-2.00 +2.00	-2.35 +2.50	-1.00 +3.85	-2.00 +2.00
90	↓	+2.25	+3.20 -3.20	-2.25 +2.25	-3.05 +2.35	-1.25 +4.15	-2.25 +2.25
100	↓	+2.70	+3.50 -3.50	-2.50 +2.50	-4.5 +1.5	-1.50 +4.55	-2.50 +2.50

Project : STIFFLEG DERRICK	SHEET M2 OF
Subject : MAST	

WITH EFFECTIVE HOOK LOAD AT 40' REACH AS STANDARD, COMPARE VARIATION OF COEFFICIENTS FOR MEMBER FORCES AND REACTIONS

	RANGE	COMMENT
BOOM	CONSTANT	P_{HK} FOR FLAT BOOM LIMITED IN PART BY BOOM BENDING MOMENT
TOPPING	0.95 TO 1.66	NOT CONTROLLING FOR HOOK LOAD VS. REACH
BACK LEG	±0.5 TO ±2.5	CONTROLS HOOK LOAD VS. REACH (COEFF/1.4)
SILL BEAM	±0.5 TO ±2.5	— " — — " — — " — (COEFF/1.0) ← USE
MAST	−1.0 TO −3.62 +0.67 TO +1.11	COMP. + BENDING (TOP & BOTTOM) WILL CONTROL MAST DESIGN. STUDY GEOMETRY ASSURE TENSION CONTINUITY IN DETAILS
MAST STEP REACTION	+0.5 TO +1.90 $-\infty$ SINCE $R_{40} = 0$ BUT $\frac{-1.50}{+2.4} = -0.63$	REACTIONS DO NOT CONTROL MAST DESIGN
BACK LEG REACTIONS	±0.5 TO ±2.50	

ESTIMATE EFFECTIVE HOOK LOAD VS. REACH & LIFTED LOAD VS. REACH

AT ANY R, $P_{HK} = \frac{(P_{HK})_{R=40}}{\text{SILL BEAM COEFF}} \leq (P_{HK})_{R=40}$

$W_{LL} = \frac{P_{HK} - 18}{1.2}$, WHERE 18^K = DEAD LOAD EFFECT AT BOOM TIP; IMPACT = 20% OF W_{LL}

R	COEFF	P_{HK}	W_{LL}
20'	0.5	163	$120^K = 60^T$
40'	1.00	163	$120^K = 60^T$
60'	1.50	109	$76^K = 38^T$
80'	2.00	82	$53^K = 27^T$
90'	2.25	72	$45^K = 23^T$
100'	2.50	65	$39^K = 20^T$

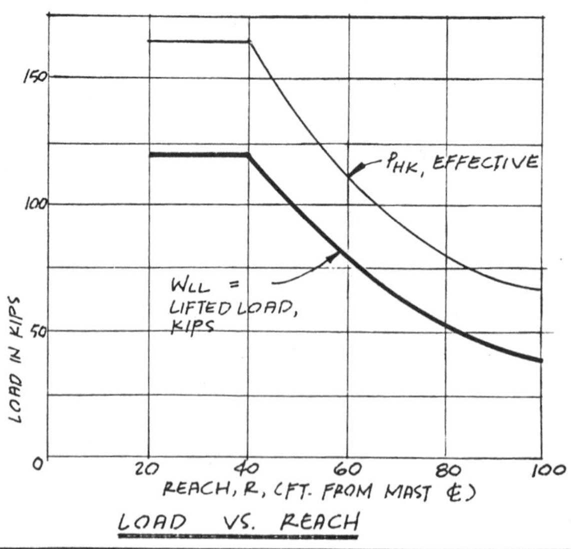

LOAD VS. REACH

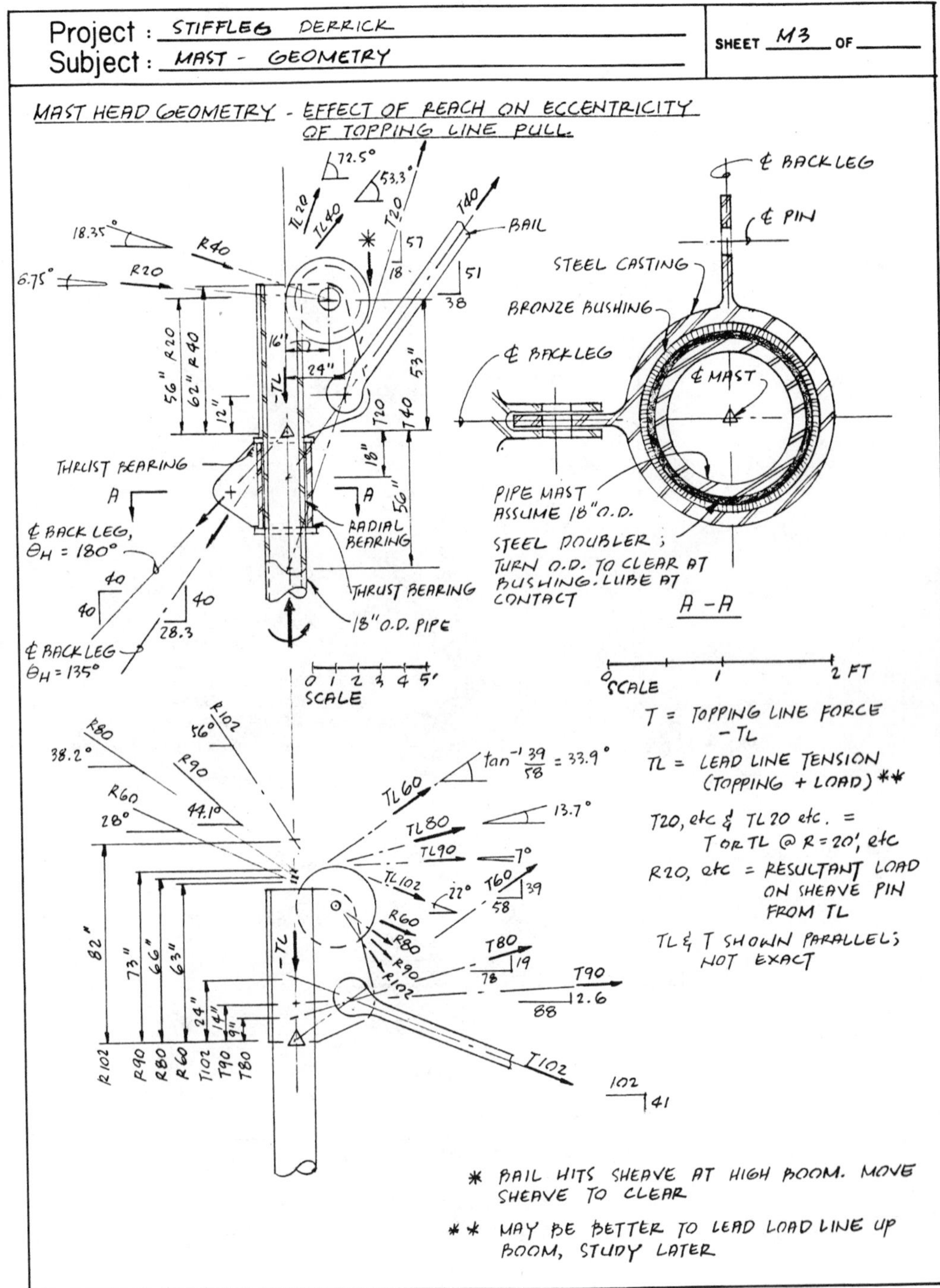
Project : STIFFLEG DERRICK
Subject : MAST - GEOMETRY
SHEET M3 OF
MAST HEAD GEOMETRY - EFFECT OF REACH ON ECCENTRICITY OF TOPPING LINE PULL.
¢ BACKLEG
¢ PIN
STEEL CASTING
BRONZE BUSHING
¢ BACKLEG
¢ MAST
BAIL
THRUST BEARING
RADIAL BEARING
THRUST BEARING
18" O.D. PIPE
PIPE MAST ASSUME 18" O.D.
STEEL DOUBLER ; TURN O.D. TO CLEAR AT BUSHING. LUBE AT CONTACT
A - A
¢ BACK LEG, θ_H = 180°
¢ BACKLEG θ_H = 135°
SCALE
2 FT
T = TOPPING LINE FORCE - TL
TL = LEAD LINE TENSION (TOPPING + LOAD) **
T20, etc & TL20 etc. = T OR TL @ R = 20', etc
R20, etc = RESULTANT LOAD ON SHEAVE PIN FROM TL
TL & T SHOWN PARALLEL; NOT EXACT
tan⁻¹ 39/58 = 33.9°
* BAIL HITS SHEAVE AT HIGH BOOM. MOVE SHEAVE TO CLEAR
** MAY BE BETTER TO LEAD LOAD LINE UP BOOM, STUDY LATER

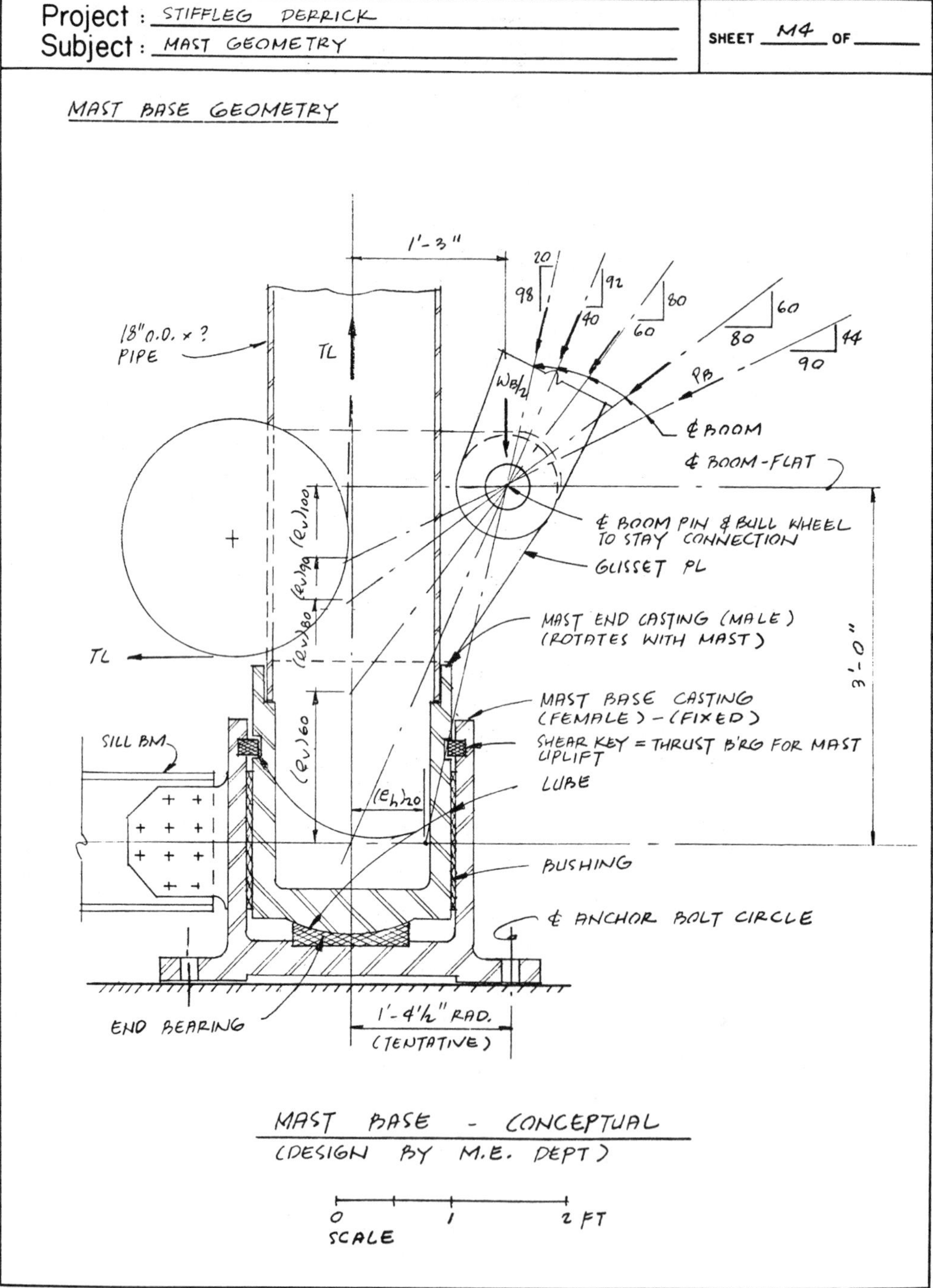
Project : STIFFLEG DERRICK
Subject : MAST GEOMETRY
SHEET M4 OF
MAST BASE GEOMETRY
1'-3"
18" O.D. × ? PIPE
TL
$W_B/2$
P_B
₵ BOOM
₵ BOOM-FLAT
₵ BOOM PIN & BULL WHEEL TO STAY CONNECTION
GUSSET PL
MAST END CASTING (MALE) (ROTATES WITH MAST)
MAST BASE CASTING (FEMALE) - (FIXED)
SHEAR KEY = THRUST B'RG FOR MAST UPLIFT
LUBE
BUSHING
₵ ANCHOR BOLT CIRCLE
SILL BM.
END BEARING
1'-4½" RAD. (TENTATIVE)
3'-0"
$(e_v)_{100}$
$(e_v)_{90}$
$(e_v)_{80}$
$(e_v)_{60}$
$(e_h)_{20}$
MAST BASE - CONCEPTUAL
(DESIGN BY M.E. DEPT)
0 1 2 FT
SCALE

Project : STIFFLEG DERRICK

Subject : MAST

SHEET M5 OF ____

FORCES IN MAST, $F_M = C_{FM} \times P_{HK}$ (SHT. I6 OR I4)

REACH	P_{HK} (SHT. M2)	F_M: $\theta'_H = 0$	F_M: $\theta'_H = 130°$
20	163	+121	+310
40	163	−70	+358
60	109	−122	+255
80	82	−190	+190
90	72	−221	+155
100	65	−295	+87

TL = TOPPING LEADLINE FORCE ≤ 50K

TL; θ/2; θ/2; e_{TL}; e_T; T; ℄ MAST

MOMENT AT TOP OF MAST (DUE TO ECCENTRICITY OF TOPPING)

REACH (FT)	T^* (KIPS)	T_H (KIPS)	TL (KIPS)	TL_H (KIPS)	e_T (INCHES)	e_{TL} (INCHES)	M_T (FT. KIPS)	M_{TL} (FT. KIPS)	$M_{T,MAX}$ (FT. KIPS)
20	253	76	≤50	15	−56	56	354	70	354
40	266	159	↓	30	−18	62	239	155	239
60	196	163	↓	42	0	63	0	221	221
80	168	163	↓	48	+9	66	122	264	350
90	162	161	↓	50	14	73	188	304	434
100	176	163	↓	46	24	82	326	314	548

$T_H = T\cos\theta_V$ WHERE θ_V = ∠'S ON SHT. M3

$M_T = T_H e_T/12$

$M_{TL} \leq TL_H e_{TL}/12$; $TL \leq 50K$

IF M_T AND M_{TL} ARE OF OPPOSITE SIGN, $(M_T)_{MAX} = M_T$

IF OF SAME SIGN, $(M_T)_{MAX} = M_{TL} + \left(\frac{T-TL}{T}\right) M_T$

$TL_H = 48$; 39; 30; 12; TL20; TL40; TL60; TL80; TL90; TL100; 50; TL; $TL_H = 43$; 50

T20, ETC, ARE PARALLEL TO TL20, ETC.

MOMENT AT BASE OF MAST (DUE TO ECCENTRICITY OF BOOM PIN)

REACH	e_V	e_H	P_{BM}	P_H	P_V	M_{BM} (FT. K.)	M_{TL} (FT. K.)	M_{MAX} (FT. K.)
20	-	8.5"	418K.	84K.	410K.	290	≤80	290
40	0	0	418	167	383	0	↓	80
60	15.5"	-	278	167	222	216	↓	296
80	25.0"	-	210	168	126	350	↓	430
90	28.5"	-	185	166	81	394	↓	474
100	36"	-	168	168	0	504	↓	584

* $T = P_{HK} \times$ COEFF (SHT I2); TL IS INCLUDED IN T, BUT NOT DEDUCTED IN CALCULATING T_H & M_T SINCE TL IS VARIABLE

6. Lead lines to both topping and load falls* will emerge from the mast at its top and be deflected by sheaves to head for the boom tip.
7. Both bail and sheave will transmit loads at varying eccentricity, causing bending moment in the mast varying as the product of load times eccentricity. In the calculations here, eccentricity is defined as vertical distance from the intersection of mast and back-leg axes. Moment is caused by the horizontal component of the bail and sheave pin forces. Slopes are found from sheet I2 (adjusted slightly). Eccentricities are established by scale.

Similar information is found on sheet M4 by studying the conditions at the base of the mast. The mast must rotate around the fixed mast step casting, which attaches to the sills and foundation, requiring a radial bearing. Thrust against the foundations must go through a lubricated thrust bearing. Net uplift requires shear keys between mast base and lower step casting. (Since the boom compression is introduced above the casting, the magnitude of uplift is the mast reaction, R_m on sheet I6, not the mast tension, F_m.) The boom has large force and the lead line sheaves relatively small forces. Both are attached eccentrically; in the case of the boom, eccentricity varies with reach. The boom pin is located with zero eccentricity for 40-foot reach, and eccentricities at other boom slopes are measured by scale.

Tables on sheet M5 are used to calculate the force and moment results to be considered in sizing the mast, which are then reassembled in a summary table on M6. The worst combination of compression and bending occurs at 100-ft reach. Unfortunately, the bending causes single curvature, which magnifies the buckling potential.

The designer here passes up an opportunity that may have occurred to the reader. Lowering the boom pin would decrease the base moment. It would increase moments at short reaches, but since axial loads in those cases are either tensile or small compressions, that could be tolerated. The significance of the unnecessarily large moment will be seen in the next two paragraphs.

An 18-inch O.D. pipe is checked for the design condition established previously. Wall thickness is assumed at 1 inch and strength and stiffness formulas evaluated. Results of interaction formulas (1.6-1a) and (1.6-1b) are unfavorable. Overutilization is a significant 39%.

The problem could be solved by increasing the wall thickness of the pipe about 25%, assuming 18-inch diameter × 1¼-inch pipe is available. However, noting that the highest utilization effect is in the bending term, the designer decides to increase the section modulus by increasing the diameter, with a small bonus in a higher value

*The wisdom of leading the load line over the mast head is debatable. It may foul the topping system on its way to the boom tip. A better path may be through the mast above the boom heel pin and up the upper face of the boom. This change is reserved for later study. It is not expected to result in a change in mast design. Error would be on the safe side.

Project : STIFFLEG DERRICK
Subject : MAST SECTION

SHEET M6 OF ____

MAST DESIGN

DESIGN FORCES & MOMENTS

REACH	FORCES (KIPS)				MAX. MOMENTS ('K)	
	P_{HK}	T_L	T_{MAX}	C_{MAX}	M_{MT}	M_{MB}
20	163	≤50	+310	–	354	290
40	163	≤50	+358	-70	239	80
60	109	≤50	+255	-122	221	296
80	82	≤50	+190	-190	350	430
90	72	≤50	+155	-221	434	474
100	65	≤50	+87	-295	548	584

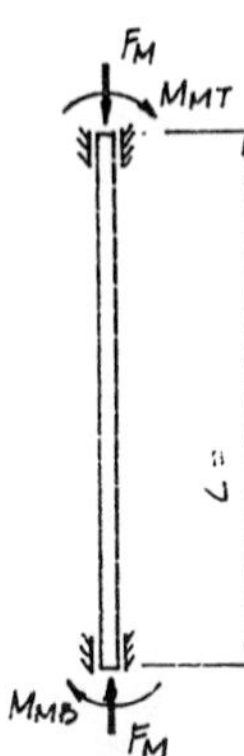

DESIGN FOR 100' REACH
(NOTE: LOWER FORCES INCLUDE EFFECTS ON MAST, BUT DO NOT GIVE REACTION. MOMENTS FROM BOOM FORCE INCLUDED, BUT NOT FORCE)

640'K, 295K, 22K, 22K, 295, 584 — LOAD — M: 548, 584

FOR 18" O.D. PIPE × 1" (MANUAL, p 2-66)

$I = 1936\ in^4 \quad A = 53.4\ in^2$

$r = 6.02\ in \quad S = 215\ in^3$

CHECK FOR COMPACT SECTION BY AISC 1.5.1.4.1

$\frac{d}{t} = \frac{18}{1} = 18 \ll \frac{3300}{F_y} = 66. \quad O.K.$

$F_b = .66\ F_y = 33\ ksi$

$f_b = 32.6\ ksi$

$\frac{KL}{r} \approx \frac{.8\,(40)\,12}{6.02} = 63.8$

$F_a = 22\ ksi$ (AISC TABLE 3-50)

$f_a = 5.52$

$F'_e = 36.5\ ksi$ (AISC TABLE 9)

$C_m = 0.6 + 0.4\,(548/584) = 0.98$

AISC FORMULA (1.6-1a), (1.6-1b)

(1.6-1a) $\frac{5.52}{22} + \frac{32.6\,(.98)}{33(1-\frac{5.52}{35.6})}$

$= 0.25 + 1.14 = 1.39 \gg 1.0$ N.G.

(1.6-1b) $\frac{5.52}{30} + \frac{32.6}{33} = 0.18 + 0.99 = 1.17 > 1.0$ N.G.

BENDING IS SOURCE OF PROBLEM

INCREASE S BY 35% APPROX W/ SOME INCR. IN AREA

TRY 22" O.D. × 7/8

22" O.D. PIPE × 7/8"

$A = 58.07\ ; \quad \frac{58.07}{53.4} = 1.09$

$S = 295\ ; \quad \frac{295}{215} = 1.37$

$I = 3245\ ; \quad r = 7.48$

$\frac{d}{t} = 25 \ll 3300/50\ ; \quad \therefore$ COMPACT

$\frac{KL}{r} \approx \frac{.8\,(40)\,12}{7.48} = 51.34$

$F_a = 24.15\ ksi\ ; \quad f_a = 5.08$

$F_b = 33\ ksi\ ; \quad f_b = 23.76$

$F'_e = 57.4$

$C_m = 0.98$

FORM. (1.6-1a)

$\frac{5.08}{24.15} + \frac{23.76\,(0.98)}{33(1-\frac{5.08}{57.4})} = 0.21 + 0.77 = 0.98$

FORM. (1.6-1b)

$\frac{5.08}{30} + \frac{23.76}{33} = 0.17 + 0.72 = 0.89$

OK

USE 22" O.D. × 7/8" PIPE

A 618, GR. 3

of F_a. The resulting 22-inch O.D. × ⅞-inch pipe reduces the combined utilization to 0.98, a 41% reduction achieved by a 9% addition to area. The temptation to go to an even larger diameter and thinner wall is resisted in order to avoid further eccentricity in the connections.

Mast design is dropped at this point, although there are clearly a number of details that will demand careful attention and may force partial reconsideration of the current decisions. That is left for another time outside the limits of this book.

19.6.4 Bails

Lifta moves on to consideration of the adjustable tension systems used to lift the load and the boom. The boom (topping line) bails, fixed-length extensions of the topping line falls, are examined first. In theory, the lower block could be connected to the mast and the upper to the boom, making the bails unnecessary. However, there are objections. Examination of the geometry on sheets B4, B5, and M3 will show that sheaves in the order of, say, 30-inch diameter would foul the lead (TL) sheave at a large reach, preventing the desired minimum reach of 20 feet. In addition, any additional length provided by running ropes must be multiplied by the number of parts of rope reeved over the blocks to hold large loads with small hoist line pull. The economy of fixed-length tension members is even more favorable when it is realized that they may have almost unlimited useful life, while running ropes have to be replaced periodically.

Diagrams on sheet I1 reveal that the topping line is shortest when the boom vertical angle is greatest. At that position, the designer will want the distance between upper and lower blocks to be the smallest acceptable to avoid chafing of the ropes on the sides of the sheaves as, in reeving, it moves from the plane of one sheave to that of another. This distance will be determined when the dimensions of the multisheave block are known, but it is not likely to be much less than the 20 feet shown by scale on sheet I1. The rest of the length goes to lower and upper bails.

The design of boom bails on sheet R1 follows the general approach to tension members discussed in Chapter 8 and special rules in the AISC Specification, section 1.14.5, for pin-connected eyebars. The rules are designed to guarantee gradual transition from the body of the eyebar to the enlarged head. The cross-sectional area of the body is sized for pure tension at the relatively low value of F_t permitted in AISC 1.5.1.1 for pin-connected members. The net area in tension is increased in the section across the head. This is intended to account for stress magnification at the hole, as well as the complex shear and bending effects in the region beyond the pin,which tend to cause local dishing or fracture.

The pin itself must be considered as a member in double shear and bending, and sufficient thickness of eyebar is necessary to reduce bearing stresses between pin and eyebar. Confident in the rules, the designer applies them like a cookbook, juggling all the dimensions to arrive at a consistent set satisfying all criteria. By luck, the right mix emerges from essentially the first try in the table, making it unnecessary to go further.

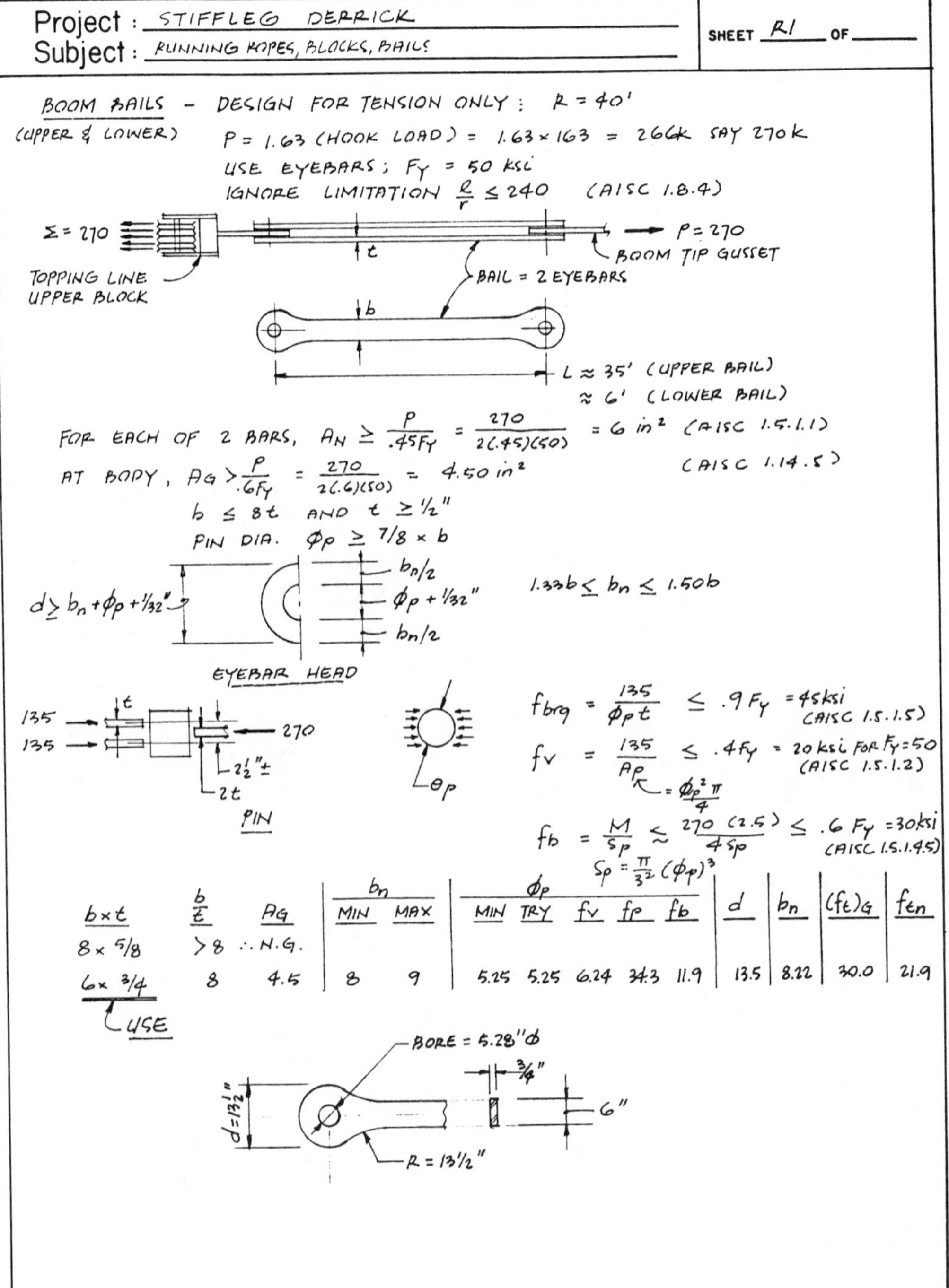

b×t	b/t	AG	bn MIN	bn MAX	φp MIN	φp TRY	fv	fp	fb	d	bn	(ft)G	ftn
8 × 5/8	> 8 ∴ N.G.												
6 × 3/4	8	4.5	8	9	5.25	5.25	6.24	34.3	11.9	13.5	8.22	30.0	21.9

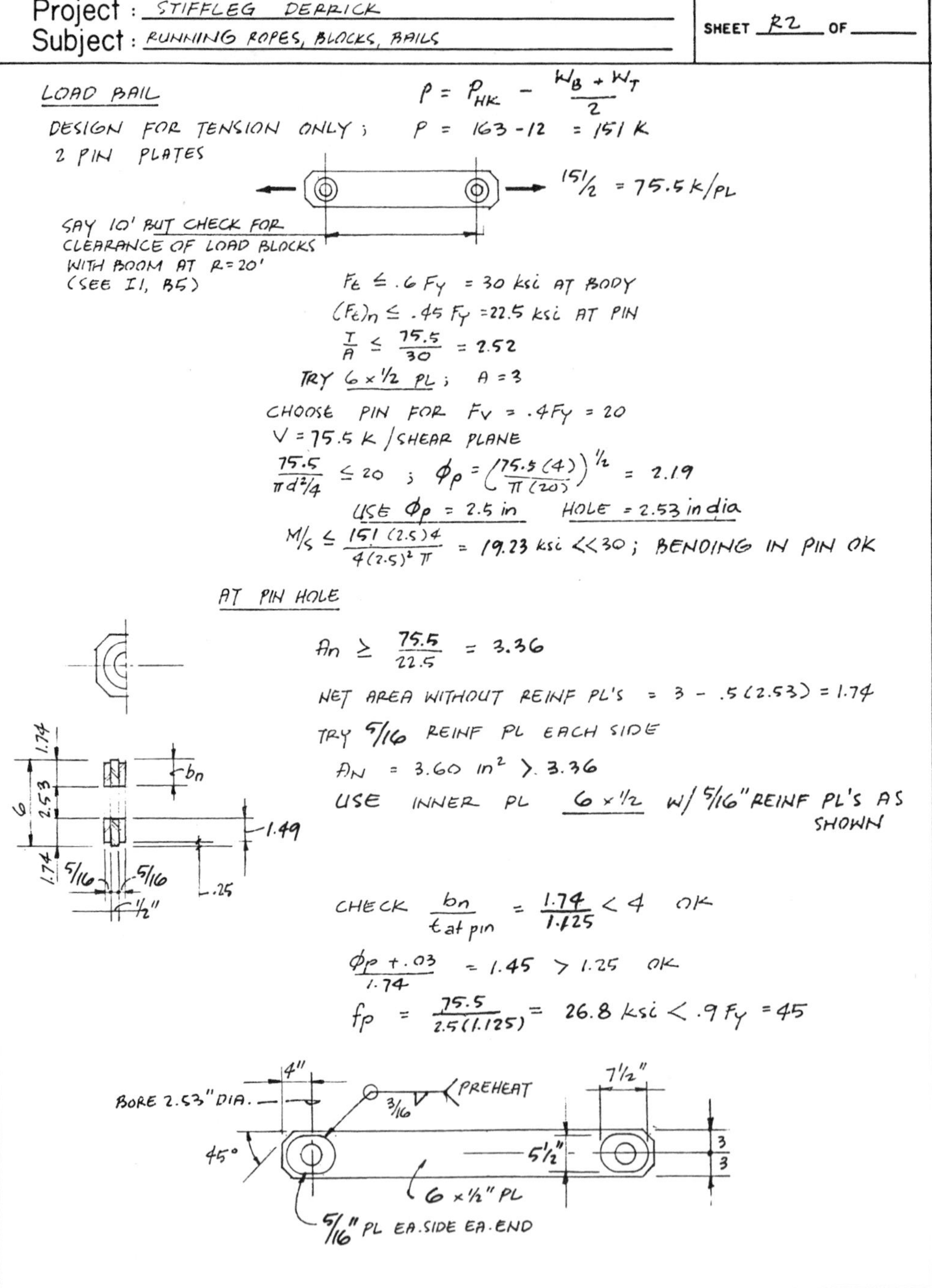

Project : STIFFLEG DERRICK
Subject : RUNNING ROPES, BLOCKS, BAILS
SHEET R2 OF

LOAD BAIL

$P = P_{HK} - \frac{W_B + W_T}{2}$

DESIGN FOR TENSION ONLY ; $P = 163 - 12 = 151$ K

2 PIN PLATES

$151/2 = 75.5$ K/PL

SAY 10' BUT CHECK FOR CLEARANCE OF LOAD BLOCKS WITH BOOM AT R = 20' (SEE I1, B5)

$F_t \le .6 F_y = 30$ ksi AT BODY

$(F_t)_n \le .45 F_y = 22.5$ ksi AT PIN

$\frac{T}{A} \le \frac{75.5}{30} = 2.52$

TRY 6 × ½ PL ; A = 3

CHOOSE PIN FOR $F_V = .4F_y = 20$

V = 75.5 K / SHEAR PLANE

$\frac{75.5}{\pi d^2/4} \le 20$; $\phi_p = \left(\frac{75.5(4)}{\pi(20)}\right)^{1/2} = 2.19$

USE $\phi_p = 2.5$ in HOLE = 2.53 in dia

$M/S \le \frac{151(2.5)4}{4(2.5)^2\pi} = 19.23$ ksi << 30 ; BENDING IN PIN OK

AT PIN HOLE

$A_n \ge \frac{75.5}{22.5} = 3.36$

NET AREA WITHOUT REINF PL'S = 3 − .5(2.53) = 1.74

TRY 5/16 REINF PL EACH SIDE

$A_N = 3.60$ in² > 3.36

USE INNER PL 6 × ½ W/ 5/16" REINF PL'S AS SHOWN

CHECK $\frac{b_n}{t_{at\ pin}} = \frac{1.74}{1.125} < 4$ OK

$\frac{\phi_p + .03}{1.74} = 1.45 > 1.25$ OK

$f_p = \frac{75.5}{2.5(1.125)} = 26.8$ ksi $< .9F_y = 45$

For the load bail (sheet R2), Lifta decides on a different form of pin-connected plate. Here it is desirable to use the shortest length that will avoid interference between the upper block and the boom. The expense of forming the eyebar head may not be justifiable in this case. AISC rules in Sections 1.14 and 1.5.1 cover this type of bar also. A required section is established for the body of the bar. Where a hole is to be drilled, the lost area is more than replaced by reinforcing plates. For reasons similar to those used for eyebars, dimensions in the region of the hole and beyond are controlled by the AISC rules.

19.6.5 Blocks and Ropes

On sheets R3 and R4, the designer studies the related problems of the ropes and blocks in the topping system. The use of wire ropes and safety factors for different wire rope uses are discussed in Chapter 8. The notes on safety factors on sheet R3 reflect considerable variance in practice regarding safety factors used in selection of wire rope. The safety factor

$$\frac{\text{strength}}{T_{\max}}$$

is based on minimum manufacturers' published breaking strength of new rope; in usual practice (see AISC Derrick Guide), the line pull is based on static load (i.e., without impact). The safety factor of 3 recommended by the AISC can be compared to values in Table 8.1 in this book, which range from 2.34 to 3.02 for the ductile structural steels. The designer is uncomfortable with the factor of 3 since the strength of running ropes can be expected to decline in service even before it becomes obvious to the user by observing individual broken wires. She decides to use a factor of 3.5 on the static pull for standing ropes. Raising or lowering the load or boom requires turning of sheaves by the running rope against friction at the bearings. Each successive part of line has more tension than the one before. In a falls of many parts, the maximum tension may be much higher than *tension force/number of parts*, the simple relation for an ideal, frictionless system. In addition, some force is necessary to bend the moving rope around each sheave against mechanical resistance in the rope. This force is a function of rope and sheave diameters. So the maximum tension for the moving rope is higher than for the stationary rope. The safety factor for this case is, prudently, increased also, since $T_{\max}$ is based on the load without impact. The designer chooses SF = 4, nominally 14% higher than for the stationary rope, but in operation close to the same.

Project : STIFFLEG DERRICK
Subject : RUNNING ROPES, BLOCKS, BAILS

SHEET R3 OF ______

TOPPING LINE ROPES & BLOCKS

HOOK LOAD (NOT INCLUDING IMPACT) = 138 ; P = 1.63 × 138 = 225 KIPS

SAFETY FACTOR ≥ 3.5* FOR STATIC PULL

≥ 4* FOR PULL ON MOST HEAVILY LOADED ROPE (ASSUME CONSTANT VELOCITY)

STATIC

$$\frac{(1.63)\,138\,(3.5)}{n\,(2)} = \text{STRENGTH} = \frac{394}{n}\;; \qquad n \geq \frac{394}{\text{STR}}$$

n = NUMBER OF PARTS
STRENGTH IS IN TONS OF 2000#
FROM USS TABLES

TO CALCULATE MAX. SINGLE LINE PULL - MOVING ROPE

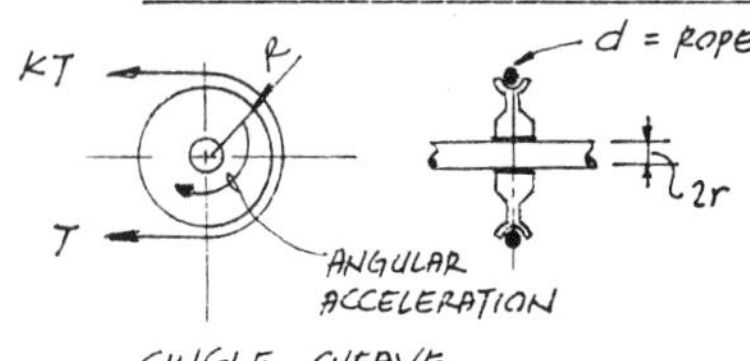

SINGLE SHEAVE

$$K = 1 + 0.3\frac{d^2}{R} + 2f\frac{r}{R}$$

f = FRICTION FACTOR ≈ 0.08 FOR BRONZE BUSHINGS
≈ 0.02 FOR ROLLER BEARINGS

FOR n PARTS OF ROPE

$$T_{MAX} = P\left(\frac{K^n(K-1)}{K^n-1}\right)K^2$$

(ADAPTED FR: WIRE ROPE ENG'G HANDBOOK" U.S. STEEL 1946)

WHERE:

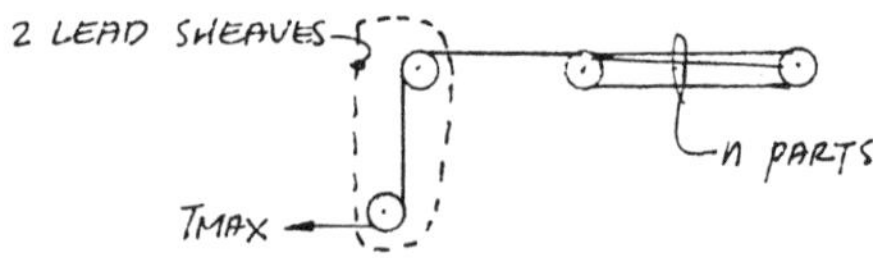

U.S. STEEL RECOMMENDS, MIN. TREAD DIAMETER

		D	USE R
3/4"φ	6×19 ;	= 22½"	= 12
7/8"	↓	26¼	14
1"		30	16

* AISC RECOMMENDS 3 AND 3.5 (IN "GUIDE FOR ANALYSIS OF GUY AND STIFFLEG DERRICKS, 1974, p.14) CITING ANSI B30.6. SEEMS LOW. SOME REGULATORY AGENCIES REQUIRE SAFETY FACTORS ≥ 5 FOR HOISTING ROPES. FINAL DECISION SHOULD BE MADE AFTER CHOICE OF SHEAVE TREAD DIAMETER AND BEARINGS.

SAFETY FACTOR IS BASED ON STRENGTH OF NEW WIRE ROPE AS PUBLISHED IN TABLES. IMPACT EFFECT INCLUDED IN SAFETY FACTOR FOR MOVING ROPE. ACTUAL SAFETY FACTOR DECLINES AS ROPE WEARS. DECLINE CAN BE LIMITED BY PROGRAM OF INSPECTION AND PERIODIC REPLACEMENT.

Project: STIFFLEG DERRICK
Subject: RUNNING ROPES, BLOCKS, BAILS, TOPPING LINES

SHEET R4 OF ____

USING 6 × 19 HOISTING ROPE ; IMPROVED PLOW STEEL

STRENGTHS PER U.S. STEEL CO. "MONITOR STEEL" AS TABULATED
HEMP CORE AS LISTED IN TONS OF 2000 #
I.W.R.C. = INDEPENDENT WIRE ROPE CORE = HEMP CORE × 1.075
BRIGHT STEEL (NOT GALVANIZED)

dia (in)	STRENGTH (TONS)		n	R	$3\frac{d^2}{R}$	IF r = 1.5" $2f\frac{r}{R}$		K	TMAX (TONS)		STR/TMAX	
	HEMP CORE	IWRC				f=.08	f=.02		f=.08	f=.02	f=.08	f=.02
3/4"	23.8		18									
		26.2	16	12	.014	.020		1.034	9.87		2.65	
							.005	1.019				
7/8"	32.2		13	14	.0164	.0171		1.0335	11.55			
							.0043	1.0207				
		35.4	13	14	.0164	.0171		1.0335	11.55		3.06	
							.0043	1.0207		10.38		3.41
1"	41.8		10	16	.0188	.0150		1.0338	14.37		2.91	
							.0038	1.0226		13.28		3.15
		46.0	12	16	.0188	.0150		1.0338	12.36		3.72	
							.0038	1.0226		11.31		4.07
		46.0	13	16	.0188	.0150		1.0338	11.58		3.97	
							.0038	1.0226		10.55		4.36
		46.0	14	16	—			1.0338	10.92		4.21	
								1.0226		9.90		4.65

USE: 1" dia - IMP. PLOW STEEL - I.W.R.C. - 12 PARTS
SHEAVE TREAD DIA. $\geq (16 - .5)2 = 31"$
ROLLER BEARINGS ON 3" DIA. SHAFT ; $f \leq .02$
(VERIFY SHAFT SIZE & "f" ON PURCHASE OF BLOCKS)
LEAD LINE PULL $= \frac{46(2)}{4.09} = 22.6$k.

REEVING

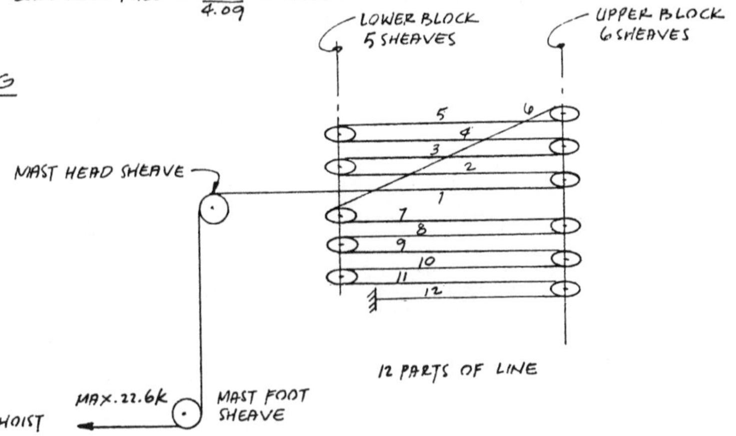

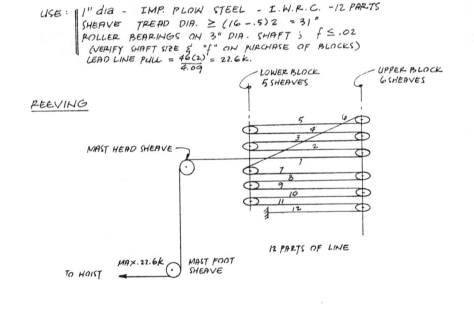

LOWER BLOCK 5 SHEAVES
UPPER BLOCK 6 SHEAVES
MAST HEAD SHEAVE
5 6 4 3 2 1 7 8 9 10 11 12
12 PARTS OF LINE
MAX. 22.6k
TO HOIST
MAST FOOT SHEAVE

Project : STIFFLEG DERRICK
Subject : LOAD LINE ROPES & BLOCKS

SHEET R5 OF

LOAD LINE ROPES & BLOCKS

HOOK LOAD = 138K - NO IMPACT (SHT. B2, B9)

LESS : $\frac{W_B}{2}$ = 9

$\frac{W_T}{2}$ = 12

126K = 63 TONS

USING 1" ϕ IMP. PLOW STEEL, I.W.R.C.

R = 16" ; f = .02 ; K = 1.0226 ; FOR T_{MAX}, SHT. R3

FROM SHT. R4

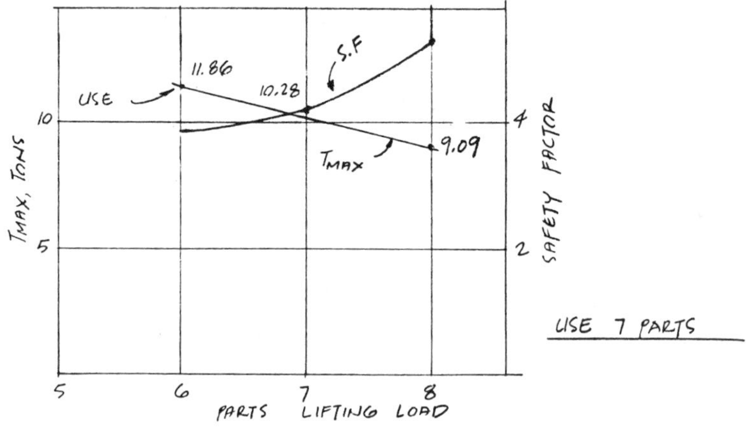

USE 7 PARTS

REEVING

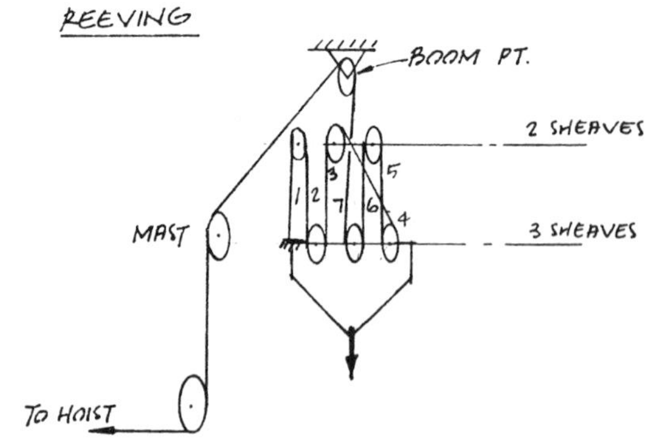

Some regulatory agencies responsible for public safety or public investment policy may insist on higher factors of safety, requiring either more parts of the same rope, stronger rope, or *downrating* the derrick's capacity.

The formula used for T_{max}, adapted from one published by U.S. Steel Company, accounts for both frictional and mechanical increments of tension. Programmed on a hand calculator, it quickly goes through a number of iterations to arrive at the combination of rope strength, friction factor, and number of parts necessary to achieve the safety factor sought.

The choice is made of 12 parts of rope with $f = 0.02$ rather than the alternative of 14 parts and $f = 0.08$. To get the lower friction factor will require more expensive bearings, but the saving of two sheaves can be expected to offset this, and the system will be lighter. This decision should be reviewed when prices are checked with manufacturers.

The designer studies a pattern of *reeving* for the 12 parts of line and extension to the hoist. The crossover at part 6 is an attempt to center the resultant tension; each part carries a different tension. The crossover is also the major source of chafing, which is limited by the distance between blocks and the consequent *fleet angle*. Manufacturers suggest a fleet angle no greater than 1½ degrees for good rope service.

The load falls is approached similarly on sheet R5, but with lighter load. Having established the desired sheaves, bearings, and rope size, the designer decides to use the same for the load live. The same formula results in a value of $n = 6$, which is then sketched for reeving to establish the necessary blocks. Note that the boom point lead sheave carries one of the necessary parts.

19.6.6 Miscellaneous Items

Sheet MS1 is added to establish loads for foundation design, which emerge easily from data already assembled. We are ready to drop the case study at this point, having illustrated the design process as intended and being limited by time and space. The designer cannot stop. Some of the remaining tasks are listed on sheets MS2 and MS3. Successful completion of these should bring the project to the end of Flow Chart 19.2 and transfer back to Flow Chart 19.1.

Project : STIFFLEG DERRICK

Subject : STIFFLEG BASE REACTIONS

SHEET MS1 OF ____

REACTIONS AT BASE OF STIFFLEG
(FOR FOUNDATION DESIGN & BASE DETAILS)

BACK LEGS $\theta_H = 0°$ OR $= 180°$

R_B R_B BOOM

$P_{HK} = 163\,K$ @ $R = 40'$

SHT. I5 ; $R_B = \pm 1.0(163) = \pm 163$ (+ = THRUST ; − = UPLIFT)

LESS $W_{SILL}/2 = -\ 2$

LESS $W_{BL}/2 = -\ 2$

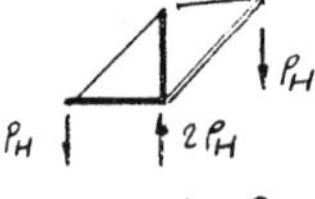

$\theta_H = 180°$

P_H P_H

$\theta_H = 0°$

USE 160K UPLIFT
168K THRUST

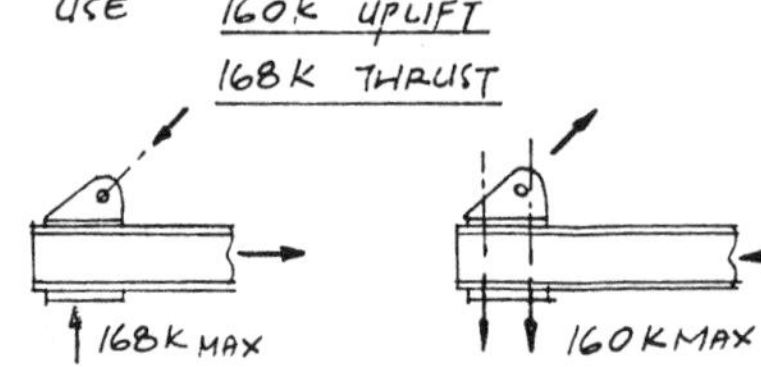

MAST STEP

$P_{HK} = 163$ @ $R = 40'$

SHT. I6 ; $R_M = -0.1(163) = 17\,K \downarrow$; $\theta_H = 0°$

$= +2.25(163) = 367\,K \uparrow$; $\theta_H = 180°$

	UPLIFT	THRUST
	−17	367
DEAD WEIGHTS		
$W_{BL}/2$	+2	+2
$W_{SILL}/2$	+2	+2
$W_{BOOM}/2$	+9	+9
$W_{TOP'G}/2$	+3	+3
	NET UPLIFT = 1 BUT SAY PROVIDE CAPACITY FOR 50K. MIN. UPLIFT	383 K

PROVIDE FOR SHEAR AT MAST FOUNDATION

SAY 3 LEAD LINES AT 24K + IMPACT = 90K

Project : STIFFLEG DERRICK
Subject : TASKS TO COMPLETE
SHEET MS2 OF ______

TASKS TO BE COMPLETED - (STRUCTURAL & COORDINATE WITH M.E., E.E.)

ADDITIONAL ELEMENTS OF MAIN SYSTEM

DESIGN BULL WHEEL WITH BRACING TO MAST, CONNECTION TO BOOM STAYS AND SWINGING ROPES.
DESIGN BOOM STAYS WITH END CONNECTIONS

IF JIB DESIRED:

1. DESIGN JIB AND ATTACHMENT TO BOOM CAPACITY, SAY 10^T MAX.
2. RECALCULATE EFFECT OF JIB WEIGHT ON MEMBER FORCES & REACTIONS
3. DECIDE IF CAPACITY AT MAIN HOOK TO BE REDUCED BY LOAD ON AUXILIARY HOOK.

OR 4. IF AUXILIARY HOOK LOAD TO BE ADDED TO MAIN HOOK LOAD, RECALCULATE SYSTEM AND STRENGTHEN AS NECESSARY.

DETAILS

BOOM TIP

LEAD SHEAVE DETAILS
GUSSETS
SEE SHT. B5 - BY ADJUSTING GEOMETRY : ATTEMPT TO REDUCE NEG. MOMENT FOR HIGH BOOM; PROVIDE FOR NEG. MOM. NEAR TIP; ATTEMPT TO INCREASE NEGATIVE MOMENT AT FLAT BOOM.
MISCELLANEOUS TIP DETAILS

BOOM HEEL

DESIGN DETAIL COMPATIBLE WITH MAST STEP CASTING

BOOM SPLICES - DESIGN

BOOM CONNECTION TO STAYS

MAST STEP CASTINGS

DESIGN & DETAIL
CONNECT TO MAST (FOR T & C)
CONNECT TO SILLS

MAST BASE

GUSSET TO BOOM HEEL PIN
CONNECT LEAD SHEAVES - PROVIDE FOR FAIR LEADING TO HOIST WITHOUT CHAFING FOR FULL RANGE OF SWING.

MAST HEAD

DESIGN HEAD CASTING - CONNECT MAST TO BACK LEGS
DETAIL GUSSET TO TOPPING LINE BAILS AND LEAD SHEAVES.

Project : STIFFLEG DERRICK
Subject : TASKS TO COMPLETE

SHEET MS3 OF ______

TASKS - (CONT'D)

BACK LEG.

DETAIL ENDS - COMPATIBLE WITH MAST HEAD AND SILL
COMPLETE DETAILS - FULL LENGTH

SILL BEAM

DETAIL OUTER ENDS TO MATCH BACK LEG DETAILS AND FOUNDATION REQMTS.
DETAIL INNER END, TO MATCH MAST STEP

TOPPING LINE - BLOCKS

PURCHASE TO SPECIFICATIONS IN CALC'S. IF NECESSARY TO REVISE SPECIFICATIONS, MAKE OTHER DETAILS TO SUIT
CHECK CLEARANCES

MAIN LOAD FALLS WITH HOOK

CALCULATE ROPE & BLOCK REQUIREMENTS - MATCH TO BAILS
PURCHASE TO SPECS.
CHECK CLEARANCES

SPECIFY HOIST

MISC DETAILS AS NECESSARY TO COMPLETE SYSTEM

REVIEW ALL CALCULATIONS FOR COMPATIBILITY, CORRECTNESS, COMPLETENESS
DRAW DETAILS TO SCALE
BUILD OPERATING MODEL (?)
ISSUE SPECIFICATIONS FOR MATERIALS & WORKMANSHIP
TEST PROTOTYPE (?)
STUDY LEAD OF LOAD LINE - OVER MAST? OR UP BOOM?
CHECK EFFECT OF OPERATING WITH SHORT BOOM, SETTING SPECIAL CAPACITY LIMITATIONS IF NECESSARY.

19.7 GLOSSARY: SPECIALIZED AND TRADE TERMS USED IN STEEL ERECTION AND DERRICK USAGE

These are common usages. Many of the terms have other definitions when used in different contexts. In many cases there are different terms that replace these, depending on locale or industry.

Backleg Diagonal member in a stiffleg frame.

Bail Steel plates, bars, or standing (nonrunning) wire ropes used in tension to extend the topping or load falls.

Blocks Sets of sheaves (or pulleys) on which running ropes are wound in order to increase mechanical advantage, i.e., to reduce the hoist line pull necessary to resist a load.

Boom A long, stiff member, used as part of a crane or derrick to reach from the base to the load position (in British usage, the term *jib* is usually equivalent to the American term *boom*).

Boom stays Tension rods used to stabilize a boom against lateral bending effects.

Bull wheel A beam bent to a large-diameter circle on which swinging ropes are wound; used to rotate the mast. Acts similarly to a pulley.

Derrick A term applied to a number of different pieces of hoisting equipment. Similar to a *Crane*, but in usual usage a derrick is a relatively fixed device and a crane a more mobile one.

Drilling derrick A fixed frame, usually pyramidal, used to raise, lower, and operate long earth or rock drills. Drill movement is on and around a vertical axis.

Falls A set of two blocks with ropes.

Fleet angle The angle of approach between a rope and the groove of a sheave. Fleet angles greater than one or two degrees may cause rapid wear due to chafing of the rope on the side of the groove.

Guy derrick A type of derrick utilizing a mast–boom–wire rope assembly stabilized by wire rope guys acting in tension.

Hoist A machine consisting of a source of rotative power and one or more drums on which running ropes are wound (in oil industry, called *drawworks*).

Impact Increase in the effects of moving gravity loads intended to account for vertical acceleration of the lifted load, as in starting and stopping a lift.

Jib In American usage, an extension of a boom used for light loads or a small boom of low capacity.

Load lines Wire ropes that raise and lower loads along a vertical line from the boom tip.

Luffing (or topping) The process of raising or lowering the boom.

Mast A vertical member. Usually, in a stiffleg or guy derrick, the axis of the mast is the axis of rotation.

Mobile cranes Lifting devices with booms, wire running ropes, and supporting frame mounted on a self-propelled mobile base. *Crawler cranes* move on caterpillar treads, *truck cranes* on truck wheels, and *gantry cranes* on special supporting frames often rolling on railroad wheels and tracks.

Parts of line In a continuous wire rope reeved over sets of blocks, the number of separate pieces that together share the load in the region between blocks.

Reeving Threading of ropes through various parts of a running rope system: hoist to snatch blocks to blocks to rope anchorage.

Running ropes Wire ropes (steel cables) that are wound in and out on drums in order to move parts of the derrick.

Sheave Pulley.

Sill beam Horizontal member in a stiffleg frame.

Slewing (or swinging) The process of rotating the mast and associated parts.

Slewing load A horizontal load, usually measured as a percentage of the weight of rotating parts, intended to account for angular acceleration in starting or stopping a swing.

Snatch block Often a single sheave block used to change the direction of a running rope.

Standing ropes Wire ropes (cables) of constant length, and not wound over sheaves or drums in use.

Stiffleg derrick A type of derrick utilizing a mast–boom–wire running rope assembly rotating around the mast axis, the assembly being supported and stabilized by a frame of stiff members in two planes anchored to a foundation.

Swinger Drum containing the rope used to rotate the mast.

Ton There are several definitions of a *ton*. The one used here is the usual one in current U.S. onshore construction practice, by which 1 ton = 2000 pounds = 2 kips. The reference is usually to weight, but it is also used in describing other forces or strength. The 2000-pound ton is sometimes referred to as a *short ton*, to distinguish it from the *long ton* of 2240-pounds used in the maritime industry. The *metric ton*, used in the MKS system to represent 1000 kilograms-force, is equal to 2204.6 pounds.

Topping lines Wire ropes that raise and lower the boom tip, rotating the boom around its heel.

Tower cranes Cranes mounted on top of a tall tower. A flat boom is counter-balanced by a counterweight to reduce overturning moment on the tower. A flat boom rotates around the mast. Hook and load falls move in and out along the boom. Common in modern building erection.

PROBLEMS

The problems suggested may be used for minor or major design investigations for students or others.

19.1. Do any or all of the tasks listed on Calculation Sheets MS2 and MS3.

19.2. If the stiffleg derrick of the case study is to be converted to *duty-cycle* use, lifting, swinging, and placing similar loads from a single position 40 times per hour for 25 years minimum:

- **(a)** What design changes would you make to improve the usefulness of the derrick for duty-cycle operation?
- **(b)** Prepare a capacity chart, load versus reach, for the derrick as designed. Consider fatigue and the nature of *impact* for the abruptly accelerating loads.
- **(c)** How would you change the specifications of the hoist?

19.3. Design the structure for a tower crane as commonly seen in urban building construction. Use as an erection crane for 10 tons maximum lift at 80-ft maximum reach, with rotation through 360°. (Manufacturers' catalogs describe the basic operations of these cranes.)

19.4. Investigate the similarities and differences between the stiffleg derrick designed here and other systems such as, for example, guy derrick, truck or caterpillar-mounted boom crane, or barge-mounted shear leg.

20
Case Study: Plate Girder Bridge, A Family of Designs

20.1 INTRODUCTION

20.1.1 General Remarks

The case study on design of plate girders for a highway bridge gives us the chance to consider a number of interesting ideas and opportunities in the design of steel structures.

- The arena for use of the design is in bridge building, a field that has traditionally been one of the most exciting for civil and structural engineers and the public.
- The national transportation network demands a large number of bridges, a high percentage of which are appropriately built of structural steel.
- Network demands include the need for many bridges designed to almost identical criteria with variations in span lengths only; for engineers, this presents good opportunities for design optimization.
- The span range being considered (60 feet $\leq$ span $\leq$ 120 feet) is one in which rolled sections become inadequate, yielding to the need for sections created by the designers and fabricated to their designs.

Viewed from below, a freeway crosses a San Francisco street on multiple parallel plate girders. Closeup of the same freeway crossing shows the web stiffening as well as both horizontal and vertical bracing. Two pipes are attached to the structure.

20.1.2 A Family of Designs

The problem is one that might be presented to the structural designers in a state highway or transportation department or to a consulting engineering company that designs transportation structures for such departments or foreign governments. If many structures are to be built over a period of years with similar design criteria and one physical variable, span, an engineering department will prepare one or several off-the-shelf families of similar designs that can be used as needed for the specific spans required in each highway project as they arise.

In our case, designs are to be prepared for two-lane simple-span bridges in the span range from 60 to 120 feet. Many such are needed as two-way bridges on highways and one-way bridges on freeways. The procedures followed are intended to seek maximum economy, taking advantage of opportunities to:

- Minimize steel weight to the extent consistent with economy.
- Minimize engineering costs by spreading them over a large number of similar projects.
- Minimize quality-control costs by the use of similar designs that present few special problems to quality-control personnel from project to project.
- Minimize construction costs by offering the construction industry a number of similar projects, increasing the feasibility of investing in cost-saving procedures.

Minimum weight design is sought by tailoring the needs of the plate girder cross section to the requirements of the moment and shear envelope diagrams, which vary continuously along the cross section. This becomes feasible as an approach to economy since standard rolled sections, limited to 36-inch maximum depth, are clearly not adequate, except, perhaps, in the shortest portions of the span range. However, it is not economical to create a continuously varying cross section. Economy is achieved by making decisions to change the cross section wherever the additional cost caused by the change is less than the saving resulting from the reduction in steel weight.

The case study as presented does not completely reflect current practice and current design opportunities in several important ways.

1. The selection of plate girders as it is done here ignores the participation of the concrete deck in the structural cross section. With the addition of inexpensive shear connectors to resist longitudinal shear flow at the plane where the steel girder interfaces with the concrete, the designers can achieve reliable composite action. This can be done with minimum added cost for shear connectors, no change in concrete requirements, and savings of structural steel. Composite design would often be the preferred choice over the design developed here, particularly in simple-span bridges. The advantages of composite design are not so persuasive in continuous girders where, at some parts of the span, the concrete deck is at the level of the tension flange. Most of the principles explored here apply equally to "all-steel" and composite designs, to simple and continuous spans. For special considerations applying to composite girders, the reader is referred to Chapter 14.
2. The calculations follow requirements of the AISC Design Specification, which is not identical to the specifications usually used in U.S. highway bridge design practice. The most commonly used specification in that design practice is the one published by the American Association of State Highway and Transportation Officials (AASHTO).[1] We have used the AISC Specification in detailed design since it is utilized in most of this book, and readers are already familiar with it. The design aids in the AISC Manual are based on it. Since the AISC Specification is based on authoritative opinion about safe design using structural steel, there is no objection, in principle, to its use. In actual highway bridge design, the engineers are usually required to follow the AASHTO Specification or some similar state specification. Resulting designs will differ in some, usually small, ways from those based on the AISC Specifications. We have followed the AASHTO Specification with regard to some issues specific to bridge requirements such as definition of live loads and limits of deflection.
3. Public demand for esthetically pleasing highway structures, particularly in urban areas, has led to the development of several box-girder cross sections.

Box girders may often be as economical as I-shaped plate girders, and, where torsional rigidity is desirable, offer superior structural properties. (Torsional rigidity is not always desirable.) In most respects, steel box girders are designed by procedures similar to those used for I-shaped girders.

The design of a family of related structures offers opportunities for computer-assisted design and optimization. This should become obvious particularly after studying the calculation sheets.

20.1.3 Design for a Single Case

The second half of the calculation set presents a specific plate girder design (90-foot span; interior girder), using hand and semiautomatic calculation procedures. These procedures help to:

1. Clarify for the designer the design issues that arise.
2. Clarify the nature of the analysis and the "art" of steel selection for students and novitiate engineers.

Once all the design issues have surfaced and the designers have a feel for the effects of various design choices, automatic procedures can be used to develop the entire family of designs. At the same time, it will become possible to explore areas of application of a number of alternative choices that exist. Since changing market conditions may change design choices, the pricing phase of a computer analysis should allow for changing cost data. Our study stops after the design of the 90-foot span interior girder.

20.1.4 Principles and Significant Issues

The principles of plate girder design are discussed in Chapter 14 of this book, as well as in many of the references. The specific AISC Specification requirements on plate girders are in Section 1.10; other sections also apply.

Some of the special issues that need to be understood are:

1. The search for strength and stiffness with minimum steel leads to deep sections. In the search for economical proportions of the web ($h_w \cdot t_w$), the thickness tends to become very small. Large ratios of h_w/t_w, desirable in deep girders to fully utilize the shear strength of the steel, may lead to shear instability in the web unless the web is stiffened.
2. Stiffeners used to prevent shear buckling must be stiff enough to do so effectively. Hence the principal control is the minimum I specified in AISC 1.10.5.4.

3. When *tension field action* is assumed (see Chapter 13), the assumption is that the compression diagonal of a plate element in shear is not effective. Web buckling is prevented by a tension field at an angle to the shearing vector, as in the web of a truss. Completing the trusslike action, the web stiffeners must act as compression members. Therefore, in addition to the stiffness criterion for stiffeners, we also have, in AISC 1.10.5.4., a minimum area requirement (Formula 1.10-3) to be applied if the designer takes advantage of the higher shear stresses permitted with tension field action.
4. Where such stiffeners are required, if they are not symmetrical with the centerline of the web, they act like eccentrically loaded columns, and therefore require additional area to account for the bending stress.
5. Bearing stiffeners transfer the end shear to the supports, acting together with a piece of the web in a manner similar to a column. The effective Kl/r of the "column" is reduced, since most of the web shear is transferred to the stiffeners progressively along the length of the stiffener. Since $M = 0$ at this point, the problem of tensile fatigue at the flange–stiffener weld does not arise. See calculation sheets 29 and 30 and AISC 1.10.5.1.
6. Repetitive loadings from passage of truck traffic (live load and impact) lead to the potential for fatigue failure. This arises in the tension region of the principal cross section and in the shear transfer at fillet welds. Stiffeners are stopped short of the tension flange in order to avoid notch effects that would result if they were welded to the flange. This is often misunderstood by zealous welders who feel they are doing a better job if they weld the ends to the flange. The detail we show in Section B-B, sheet 31 of the calculations, is intended to avoid such well-intended overzealousness.
7. It is not easy to assess the number of cycles of design live load to apply for use with AISC Specification, Appendix B. The standard AASHTO-HS20 design truck-trailer (see sheet A2) is a heavy one and does not represent most of the trucks passing over the bridge. Most are lighter; occasional trucks are heavier. The impact factor is a function of velocity, which is variable. Reference 63 reports statistical investigations of interest. We have assumed, conservatively, load condition 4 (maximum number of cycles) with the f_{sr} resulting from the HS20 truck-trailer.
8. The critical point for longitudinal tension fatigue turns out to be the bottom of the stiffeners. Extending the elastic stress diagram from the stress limit for fatigue at that point to the outer tension fiber results in a low limit on f_b. The ends of the stiffeners cannot be held back more than 6 web thicknesses (AISC 1.10.5.4.) without inviting web buckling. This raises the question of whether it would be more efficient to thicken the web and avoid the necessity for stiffeners. The price per pound of stiffeners is greater than the average price per pound of girder. This is not investigated in the calculations but should be in a complete design investigation.

9. The AISC Specification provides, in Sections 1.10.6 and 1.10.7, for adjusted values of allowable bending stress, F_b. Both provisions relate to potential loss of bending strength due to web instability and are triggered by large values of the ratio h_w/t_w. In our design, neither Section 1.10.6 nor 1.10.7 requires a reduction in effective F_b. The AISC Commentary sheds some light on these restrictions.

20.2 CALCULATIONS

20.2.1 Organization of Calculations

The calculations are presented in two groups of calculation sheets plus several sheets of appendix, organized as noted in the listing of Contents on sheet 1.

Sheets 2 through 11 are of general application to the design of all spans within the range being considered.

Sheets 12 through 20 are not used. This gives a designer the freedom to extend the general portion of the calculations at a later time.

Sheets 21 through 31 cover design calculations for one girder, an interior girder for 90-foot span using the first of two alternative arrangements of girders and bracing and one of the two different steels that are being considered.

An appendix to the calculations, found at the end of this chapter, includes special items pertinent to the design development.

1. Diagrams from the AASHTO Specification defining standard *truck* and *truck-trailer* loads and *lane loads* based on them. HS20 loading is used for the design, the most common choice today.
2. Table from AASHTO Specification listing the maximum shear and moment resulting from one lane of HS20 loading on simple spans of various lengths.

From item 2, we note that the moments and shears controlling the design for the span range we are interested in are all determined by the HS loads rather than the alternative lane loads, the latter being the control for longer spans. The reader may verify this with the help of the influence diagrams on sheet 11. The term *lane loads* should not be confused with the HS load, which is also considered to occupy one traffic lane. Where the HS20 load controls, there is one truck-trailer combination on the bridge in each lane. *Lane loading* represents a train of trucks on a long span simultaneously, converted for convenience to a uniformly distributed load plus a concentrated load.

Project : PLATE GIRDER BRIDGE
Subject : CONTENTS

SHEET 1 OF ____

CALCULATIONS FOR DESIGN OF
PLATE GIRDERS FOR 2 LANE HIGHWAY BRIDGE

DEVELOPMENT OF TYPICAL CROSS SECTIONS AND DESIGN BASIS FOR A FAMILY OF SIMILAR DESIGNS OF SIMPLE SPAN BRIDGES IN SPAN RANGE $60' \leq L \leq 120'$.

CALCULATIONS LEADING TO DESIGN OF INTERIOR PLATE GIRDER FOR 90' SPAN.

CONTENTS

20.2.2 Calculations: Part 1—Conditions Common to Range of Spans

General considerations. A general statement of the problem is given on sheet 2, and design conditions are established. The arrangements of roadway, plate girders, and bracing are developed on sheet 3, where two possible arrangements appear to be feasible, one with four parallel girders, the other with three. The better choice is not self-evident and will probably not be known until both schemes are tested for total cost. Roadway widths are slightly greater than would be minimally acceptable for two-lane bridges by AASHTO Standards.[1] The curb and offset from curb to guardrail would be consistent with AASHTO Standards, where minimum setback of rails is specified for safety reasons. Design of the rail itself is not developed in this set of calculations. They would be in a complete design. Design criteria for such rails exist in the AASHTO Specification. Criteria in the AISC Specification are insufficient since usual building design does not require such heavy guardrails.

Vertical planes of lateral (transverse) bracing are provided at 25-foot maximum spacing in accordance with AASHTO requirements for this type of deck system. Transverse wind or seismic loads could be resisted by horizontal bracing top and bottom similar to that shown in the upper plane. However, the vertical bracing helps to limit differential deflections between adjacent girders. This will also help preserve the integrity of the concrete deck.

The attempt will be made to make design controls for interior and exterior girders as nearly the same as possible, since it is desirable to have a uniform depth and possibly the same total design for all girders in a single bridge. In search of such balance, the deck, which will span between girders (i.e., in the transverse direction of the bridge), is made to cantilever beyond the exterior girders on each side. This attracts both live and dead load to the exterior girders, which would otherwise go to the interior girders. It also will reduce the deck span, and therefore its thickness, and, in our opinion, make the esthetic impact of the design more satisfactory.

Sheet 4 addresses some considerations to be weighed in comparison of alternatives 1 and 2; they do not lead to a clear choice. The ultimate choice may turn out to be different at the lower and upper ends of the range of spans and very likely unclear in midrange. The principal criterion for choice in this comparison is cost. Focus is on cost of a bridge rather than cost of individual parts. For comparison to other work, it is often desirable to convert the cost to cost per lineal foot or cost per square foot.

Assign loads to girders; approximate midspan moments. On sheet 5, we tentatively assign dead loads to interior and exterior spans of the two alternative arrangements. The choice of alternatives will determine the thickness of concrete slabs, important in the direct cost of the slab itself as well as its effect on the dead load supported by the girders. This would seem to speak for the four-girder alternative (number 2). However, that is not the only variable to consider in estimating the effect

Project : PLATE GIRDER BRIDGE - ALL SPANS
Subject : GENERAL CRITERIA

SHEET 2 OF

DESIGN OF PLATE GIRDERS FOR 2 LANE HIGHWAY BRIDGE
STATEMENT OF PROBLEM AND DESIGN CRITERIA

A FAMILY OF DESIGNS IS TO BE DEVELOPED FOR SIMPLE - SPAN TWO LANE HIGHWAY BRIDGES IN THE SPAN RANGE 60' ≤ SPAN ≤ 120'.
THE STRUCTURAL SYSTEM SHALL INCLUDE :

REINFORCED CONCRETE DECK, SPANNING PERPENDICULAR TO SPAN;
LONGITUDINAL PLATE GIRDERS;
LATERAL BRACING

COMPOSITE ACTION BETWEEN THE DECK AND GIRDERS SHALL NOT BE CONSIDERED.

THE RESULTING ASSEMBLED GIRDERS (OR POSSIBLY THE ENTIRE ASSEMBLED SUPERSTRUCTURE) SHOULD BE SUITABLE FOR PLACEMENT ON PREVIOUSLY PREPARED FOUNDATIONS, WHICH MAY BE PERMITTED TO HAVE MODERATE SETTLEMENT (CRITERIA TO BE SET LATER BASED ON SITE CONDITIONS)

DESIGN CONDITIONS

SPAN RANGE - 60' ≤ SPAN ≤ 120' IN INCREMENTS OF 10'

REINFORCED CONCRETE DECK - USING NORMAL WEIGHT CONCRETE AT 150#/FT^3 INCLUDING RESTEEL - AND CONFORMING TO AASHTO HIGHWAY BRIDGE SPECIFICATION (1977).

STRUCTURAL STEEL -

MATERIALS - A36 OR A588
PLATE GIRDER ASSEMBLIES WELDED; BRACING WELDED OR BOLTED.
DESIGN SPECIFICATIONS - AISC "SPECIFICATION FOR DESIGN, FABRICATION & ERECTION OF STRUCTURAL STEEL FOR BUILDINGS"* - 1978, EXCEPT - REQUIREMENTS FOR LIVE LOAD, IMPACT AND DEFLECTION BASED ON AASHTO SPECIFICATION (1977) - USING HS 20 LIVE LOADS.
FATIGUE CRITERIA - > 2,000,000 APPLICATIONS OF STANDARD HS 20 LOADS WITH IMPACT.

* NOT USUAL SPECIFICATIONS FOR DESIGN OF HIGHWAY BRIDGES

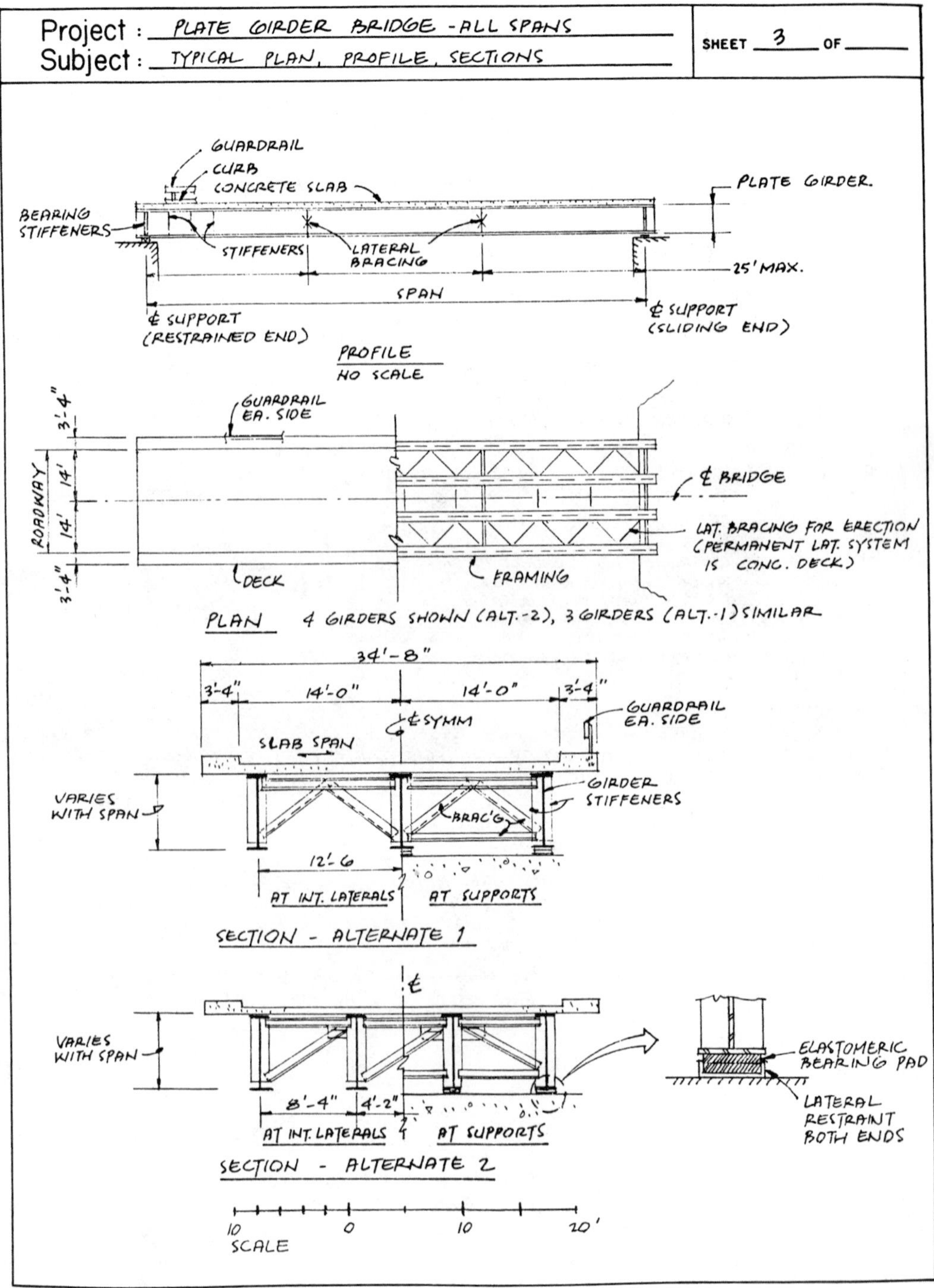
Project : PLATE GIRDER BRIDGE - ALL SPANS
Subject : TYPICAL PLAN, PROFILE, SECTIONS
SHEET 3 OF
GUARDRAIL
CURB
CONCRETE SLAB
PLATE GIRDER.
BEARING STIFFENERS
STIFFENERS
LATERAL BRACING
25' MAX.
SPAN
℄ SUPPORT (RESTRAINED END)
℄ SUPPORT (SLIDING END)
PROFILE
NO SCALE
GUARDRAIL EA. SIDE
3'-4"
14'
14'
3'-4"
ROADWAY
℄ BRIDGE
LAT. BRACING FOR ERECTION (PERMANENT LAT. SYSTEM IS CONC. DECK)
DECK
FRAMING
PLAN 4 GIRDERS SHOWN (ALT.-2), 3 GIRDERS (ALT.-1) SIMILAR
34'-8"
3'-4"
14'-0"
14'-0"
3'-4"
℄ SYMM
GUARDRAIL EA. SIDE
SLAB SPAN
VARIES WITH SPAN
GIRDER STIFFENERS
BRAC'G
12'-6
AT INT. LATERALS
AT SUPPORTS
SECTION - ALTERNATE 1
℄
VARIES WITH SPAN
8'-4"
4'-2"
AT INT. LATERALS
AT SUPPORTS
ELASTOMERIC BEARING PAD
LATERAL RESTRAINT BOTH ENDS
SECTION - ALTERNATE 2
10
0
10
20'
SCALE

Project : PLATE GIRDER BRIDGE - ALL SPANS
Subject : CONSIDERATION IN SELECTION - ALT 1 OR 2

SHEET 4 OF

CONSIDERATIONS IN SELECTION BETWEEN ALTERNATES # 1 & 2

WE ARE SEEKING MINIMUM COST FOR A COMBINATION OF REINFORCED CONCRETE AND STRUCTURAL STEEL.

COSTS OF THE REINFORCED CONCRETE DECK ARE MEASURED IN \$/C.Y.* OF REINFORCED CONCRETE; (OR, ALTERNATIVELY IN \$/ф OF DECK)
PRINCIPAL ELEMENTS IN COST OF REINF. CONCRETE (EACH TO BE CONVERTED TO \$/C.Y.)

a. \$/C.Y CONCRETE, PURCHASED AND PLACED
b. \$/lb RESTEEL, PURCHASED AND PLACED
(CONVERSION: \$/# × # RESTEEL/ф DECK × ф DECK/C.Y. CONC. = \$/C.Y.)
c. \$/SQ.FT SOFFIT FORMS, PLACE SUPPORT, REMOVE.
(CONVERSION: \$/SQ.FT × SQ.FT/C.Y. = \$/C.Y.)
d. \$/ф OF SURFACE - SCREED, COMPACT, FINISH
(CONVERSION: \$/SQ.FT. × SQ.FT./C.Y. = \$/C.Y.)

COST (\$/C.Y.) = Q_C(a) + Q_{ST}(b)(CONV.) + Q_S(c)(CONV.) + Q_S(d)(CONV.)
WHERE Q_C = QUANTITY OF CONC., C.Y.; Q_{ST} = QUANT. OF STEEL, #; Q_S = SQ. FT. OF SURFACE
OR COST, IN \$/ф = COST IN \$/C.Y. × C.Y./SQ.FT.

SINCE THE SLAB CONTRIBUTES TO DEAD LOAD ON GIRDER (DL. = WT. OF SLAB + WT. OF GIRDER), AND h_s WILL INCREASE WITH LARGER SPAN OF SLAB, THE LOADS AND ∴ THE REQUIRED STRENGTH OF GIRDER WILL INCREASE (WHEN MEASURED PER SQ.FT. OF BRIDGE) FOR ALT. 1 COMPARED TO ALT. 2

1" MIN.
1½"
2" CLR WEATHER PROTECT'N & WEARING SURF.
DIRECTION OF SLAB SPAN
h_s
1½" CLR
1½"
SOFFIT
TYP. REINF. CONC. DECK SLAB
$h_s \geq 7\frac{1}{2}$" & VARIES WITH SPAN
$w_s \geq (7.5/12)(150) = 93.8$ #/ф; SAY ≥100

COSTS OF THE GIRDERS WILL REPRESENT THE GREATEST PART OF THE COST OF STRUCTURAL STEEL, AND PROBABLY THE ONLY PART THAT WILL VARY BETWEEN ALT 1 & 2. THE BRACING SYSTEM WILL DEMAND A SMALL PERCENTAGE OF THE TOTAL STEEL. QUANTITY WILL BE VERY LITTLE AFFECTED BY THE CHOICE OF ALTERNATES. THE GUARDRAIL WILL BE INDEPENDENT OF THAT CHOICE.

IT IS NECESSARY TO EXAMINE THE EFFECT OF THE CHOICE OF ALTERNATE ON THE PLATE GIRDER WEIGHTS, AND THROUGH THE WEIGHTS, ON THEIR COSTS. OUR CONCERN HERE IS COST / BRIDGE OR COST / LIN. FT. OF BRIDGE, WHERE COST PER LIN. FT. :

FOR ALT. 1 = 3 × COST / LIN. FT. OF GIRDER ;
FOR ALT. 2 = 4 × COST / LIN. FT. OF GIRDER .

* \$/C.Y. = DOLLARS PER CUBIC YARD
\$/ф = \$/SQ.FT. = \$/FT² = DOLLARS PER SQUARE FOOT
\$/lb. = \$/# = DOLLARS PER POUND.

Project : PLATE GIRDER BRIDGE - ALL SPANS
Subject : DEAD LOADS - ESTIMATED
SHEET 5 OF

LOADS TO GIRDERS - DEAD LOADS/L.F. OF GIRDER

ALT. 1

3'-4" 14'-0" 14'-0" 3'-4"
9" = 112.5 #/□ SAY 10" = 125 #/□
4.8' 12.5' 12.5' 4.8'
3.33' ℄ SYMM. w_2 = 112 #/□ w_1 = 125 #/□
4.8' 12.5'

MOM. DISTR.

M_F	+2.61	-1.63	+1.63	-1.63	+1.63	-2.61
		-0.98	-0.49	+0.49	+0.98	
M	2.61	-2.61	+1.14	-1.14	+2.61	-2.61

ΔV FROM END MOM. 2.61 1.14 .12 .12

SHEAR, KIPS/FT .97 .78 .12 .90 .78 -.12 .66 .66

REACTIONS, K/FT	1.87	1.32
ADD ASSUMED GIRDER	0.30	0.30
w_D, K/FT	2.17	1.62

ALT. 2

9" = 112 #/□ 7.5" ±, SAY 100 #/□
4.8' 8.3' 8.3' 8.3' 4.8'
3.33' 3.33' w_2 = 112 #/□ w_1 = 100 #/□ w_2
℄ SYMM.

STIFFNESS DISTR., F: 0 | 1 .43 | .57 .57 | .43 1 | 0

M_F	2.32	-0.58	+.58	-.58	+.58	-.58	+.58	-2.32
		-1.74	-.87			+.87	+1.74	
			+.37	+.50	-.50	-.37		
				-.25	+.25			
			+.11	+.14	-.14	-.11		
				-.07	+.07			
			+.03	+.04	-.04	-.03		
				-.02	+.02			
			+.01	+.01	-.01	-.01		
M, K/FT	2.32	-2.32	+.23	-.23	+.23	-.23	2.32	-2.32

ΔV, (K/FT) 2.32 .23 .25 .25

V, K/FT .85 .41 .25 .66 .42 -.25 .17 .42

R, K/FT	1.51	0.59
ASSUMED WT OF GIRDER	.24	.24
w_D, K/FT	1.75	0.83

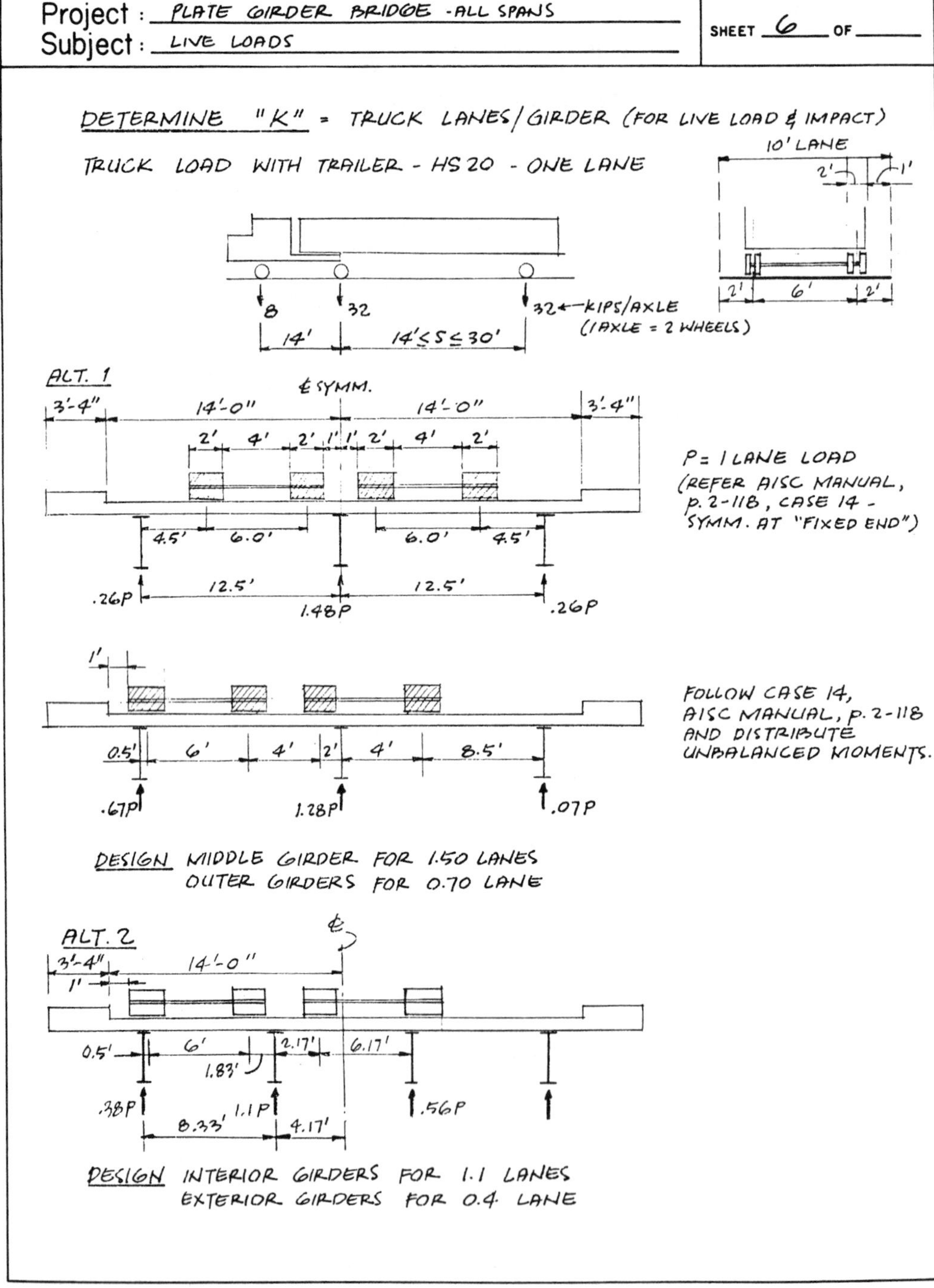
Project : PLATE GIRDER BRIDGE - ALL SPANS
Subject : LIVE LOADS
SHEET 6 OF
DETERMINE "K" = TRUCK LANES/GIRDER (FOR LIVE LOAD & IMPACT)
TRUCK LOAD WITH TRAILER - HS 20 - ONE LANE
10' LANE
8
32
32 ← KIPS/AXLE
(1 AXLE = 2 WHEELS)
14'
14' ≤ S ≤ 30'
ALT. 1
℄ SYMM.
P = 1 LANE LOAD
(REFER AISC MANUAL, p. 2-118, CASE 14 - SYMM. AT "FIXED END")
.26P
1.48P
.26P
FOLLOW CASE 14, AISC MANUAL, p. 2-118 AND DISTRIBUTE UNBALANCED MOMENTS.
.67P
1.28P
.07P
DESIGN MIDDLE GIRDER FOR 1.50 LANES
OUTER GIRDERS FOR 0.70 LANE
ALT. 2
.38P
1.1P
.56P
DESIGN INTERIOR GIRDERS FOR 1.1 LANES
EXTERIOR GIRDERS FOR 0.4 LANE

of the choice on total cost. The actual slab design is not done here. It would require discussion of reinforced concrete theory and the special considerations resulting from the heavy, moving concentrated loads from truck wheels on highway slabs. Although we avoid that byway here, a bridge designer would have to address it. The relative effects are approximated here as they contribute to the dead load on the girders.

A moment distribution calculation is used to establish the dead load contributed to each girder by the slab that spans transversely and is continuous over all girders as supports. This calculation assumes that the girders all deflect uniformly, so that, effectively, the slab's supports are unyielding. The vertical bracing discussed previously helps to provide this condition.

It is necessary at this stage to make assumptions for the weight of the girders themselves. These will be checked later.

To find the live load contribution to the load on the girders, we have to seek, *for each girder*, the position of live loads that will cause the greatest portion to go to it. This is done on sheet 6 for both alternative arrangements. The standard HS20 loading shown at the top represents the truck-trailer load (P) for one lane. For alternative 1, the maximum loads to the interior girder and the exterior girder result from different positions of the wheels. In alternative 2, one position gives both maxima.

It is interesting to note:

$$\text{Alternative 1: } [1.50 + 2(0.67)]P = 2.84P$$
$$\text{Alternative 2: } 2[1.1 + 0.38]P = 2.96P$$

The total live load resistance required for the group of girders is greater in alternative 2 than in alternative 1, an advantage for alternative 1. This effect is modified by the different distributions of dead load to the girders.

In a table on sheet 7, we look at both alternatives for interior and exterior girders at the extreme ends and middle of the span range. We will attempt to justify or adjust our original assumptions of girder dead weights, refining them somewhat and seeking the effect of span and alternative arrangements on that weight. This requires examination of the maximum moment that will occur in the midspan region, and which we expect to be the principal control on the cross section. The discussion below follows, roughly, the lines of the table.

Moments arise from dead load, live load, and impact. (As discussed later, a small but not controlling increase occurs from torsional effects on the bridge cross section resulting from wind or seismic loads.) Dead load moment comes from distributed load, $w_D = (w_D)_{\text{slab}} + (w_D)_{\text{steel}}$. The maximum live load moment per lane is found from the AASHTO table, sheet A4 (to be checked later) and modified by the factor K = live load lanes/girder, from sheet 6. The result is also multiplied by $(1 + I)$, where I (impact factor) is calculated by a formula from AASHTO relating the impact factor to span length. The impact factor derives from the dynamic vertical acceleration of the live load mass as the girder deflects during a passage of the vehicle over the bridge. Its maximum value, a function of vehicle velocity and girder stiffness,

Project : PLATE GIRDER BRIDGE - ALL SPANS
Subject : MIDSPAN MOMENTS; ADJUST DEAD LOADS

SHEET 7 OF ____

VERIFY ASSUMPTION OF DEAD WEIGHTS OF GIRDERS BY FINDING APPROXIMATE MOMENTS TO GIRDERS AT MIDSPAN (D + L + I)

	ALTERNATE 1						ALTERNATE 2					
	INNER GIRDER			OUTER GIRDER			INNER GIRDER			OUTER GIRDER		
① SPAN, FT (C-C)	60'	90'	120'	60'	90'	120'	60'	90'	120'	60'	90'	120'
② $(M_{LL})_{MAX}$ - 1 LANE (SHT. A4), FT. KIPS	807	1344	1883	807	1344	1883	807	1344	1883	807	1344	1883
③ IMPACT FACTOR AT 50/L + 125 (AASHTO)	.27	.23	.20	.27	.23	.20	.27	.23	.20	.27	.23	.20
④ $(M_I)_{MAX}$ - 1 LANE, FT-KIP	218	309	377	218	309	377	218	309	377	218	309	377
⑤ M_{L+I} = ② + ④ FT.KIP/LANE	1025	1653	2260	1025	1653	2260	1025	1653	2260	1025	1653	2260
⑥ K = LANE LOADS / GIRDER	1.50	1.50	1.50	0.70	0.70	0.70	1.10	1.10	1.10	0.40	0.40	0.40
⑦ $(M_{L+I})_{MAX}$ = ⑤ × ⑥ FT.KIP	1538	2480	3390	718	1157	1582	1128	1818	2486	410	661	904
⑧ w_D TO GIRDER												
a) = SLAB	1320	1320	1320	1870	1870	1870	600	600	600	1510	1510	1510
b) = GIRDER, SAY	300	300	300	300	300	300	240	240	240	240	240	240
c) = w_D, #/L.F.	1620	1620	1620	2170	2170	2170	840	840	840	1750	1750	1750
⑨ $(M_D)_{MAX}$ (FT.K.)	729	1640	2916	976	2197	3906	378	851	1512	788	1772	3150
⑩ $(M_{D+L+I})_{MAX}$ = ⑦ + ⑨, (FT.K)	2267	4120	6306	1694	3354	5488	1506	2669	3998	1198	2433	4054
⑪ h [I-section: A_F, A_W, A_F] h @ SAY, L/12 in	60	90	120	60	90	120	60	90	120	60	90	120
⑫ A_F @ $\frac{.9M(12)}{F_b h}$ & $F_b = 20$	20.4	24.7	28.4	15.3	20.1	24.7	13.6	16.0	18.0	10.8	14.6	18.2
⑬ SAY $A_W \approx A_F$; THEN $A_{TOT} \approx 3.3 A_F$ (WITH STIFFENERS)	67.3	81.5	93.7	50.5	66.3	81.5	44.7	52.9	59.4	35.6	48.2	60.2
⑭ WT. @ MIDSPAN, #/LF	230	279	320	172	226	277	153	181	203	122	165	206
⑮ USE ≈ .9 × ⑭ = w_D, #/LF	210	250	290	155	203	249	140	160	180	110	150	190
⑯ (8b × ⑮) × $L^2/8$	-40	-51	-18	-65	-98	-92	-39	-81	-144	-59	-91	-90
⑰ ADJUSTED M_{D+L+I} = ⑩ - ⑯, FT.K/FT	2237	4069	6288	1629	3256	5396	1467	2588	3854	1140	2342	3964
⑱ ADJUSTED DEAD LOAD = 8a + ⑮, #/LF, AVE	1530	1570	1610	2025	2073	2119	740	760	780	1620	1660	1700

APPROX. WEIGHT, STEEL / BRIDGE

ALT. 1 60' SPAN

2 OUTER, 1 INNER, 10% BRACING

[210 + 2(155)](1.1)(60) = 34,320 #

90' SPAN

[250 + 2(203)](1.1)(90) = 64,940 #

120' SPAN

[290 + 2(249)](1.1)(120) = 104,100 #

ALT. 2 60' SPAN

2 OUTER, 2 INNER, 10% BRACING

[140 + 110](2)(1.1)(60) = 33,000 #

90' SPAN

[160 + 150](2)(1.1)(90) = 61,380 #

120' SPAN

[180 + 190](2)(1.1)(170) = 97,680 #

ALT. 2 APPEARS TO BE SOMEWHAT LIGHTER WITH % OF DIFFERENCE GROWING WITH SPAN LENGTH. HOWEVER, TREND IS NOT STRONG AND BOTH ALTERNATES SHOULD BE CHECKED FOR FULL RANGE

→ ERROR IN WEIGHT OF GIRDER HAS SMALL EFFECT ON TOTAL MOMENT

is approximated by the AASHTO formula. This increase, intimately related to the live load, is expressed as a percentage addition to it. The sum of dead and live and impact loads must be resisted within the basic bending and shear stress limits in AISC Section 1.5. This differs from combinations with wind or seismic loads, where increases of allowable stress are permitted (Section 1.5.6).

For purposes of this table, an approximation of the ratio L/h is made at an assumed value of 12. The ratio of span to depth is critical to control of the maximum deflection from live load and impact, an important consideration in determining riding quality. The very low limit of deflection is set by the AASHTO specification at 1/800 × (span).

As an approach to this problem, assume, as in the table, that $F_b = 20$ ksi and

$$M_{L+I} = \tfrac{2}{3} M_{D+L+I}$$

Then

$$(f_b)_{L+I} = 13 \text{ ksi}$$

We make two reasonable approximations (see Fig. 20.1):

1. Our girders will have almost constant moment of interia.
2. The moment diagram from HS20 loading giving the maximum moment will give slightly less deflection than that from a uniformly distributed load with the same maximum moment.

The first approximation derived from the constant girder depth. For the second refer to moment diagrams in Fig. 20.1; deflection is a function of the area of M/EI diagrams.

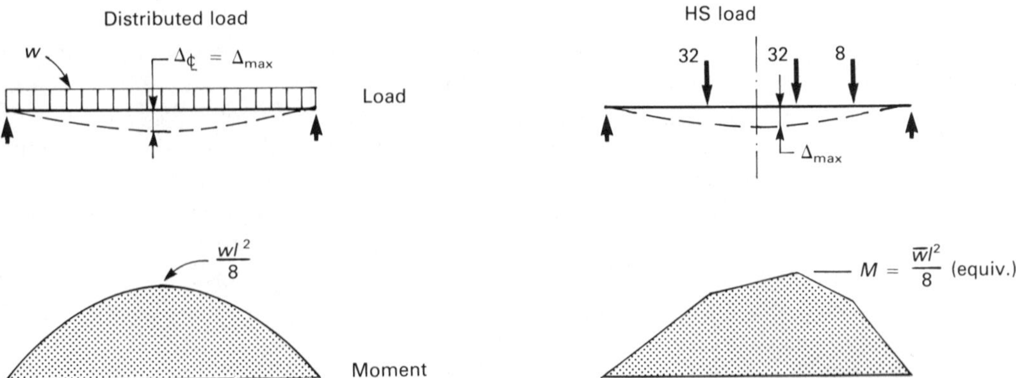

Figure 20.1 Maximum midspan moment; uniform load versus AASHTO load.

$$(M_{L+I})_{\max} \approx \frac{\bar{w}L^2}{8} = Sf_b,$$

with $\bar{w}$ = a fictitious equivalent intensity of distributed load.

$$\Delta_{L+I} = \frac{5}{384}\frac{\bar{w}L^4}{EI} = (M_{L+I})_{\max} \cdot \frac{5}{48}\frac{L^2}{EI} = Sf_b \times \frac{5}{48} \times \frac{L^2}{30{,}000} \times \frac{2}{Sh}$$

$$\frac{\Delta_{L+I}}{L} = \frac{10}{48(30{,}000)} \times \frac{L}{h} f_b \leq \frac{1}{800}$$

$$\frac{L}{h} \leq \frac{48(30{,}000)}{8000(13)} \leq 13.85; \quad \text{used } 12$$

Further assumptions are made that:

1. The flanges resist 90% of the total moment.
2. Web area ≈ area of one flange.
3. Stiffeners and bracing add 10% to the total weight.

None of these assumptions is completely accurate. In the case of a 90-foot span and for the section chosen on sheet 25, it turns out later that (see sheet 31):

1. The flanges provide approximately 80% of the total moment resistance.
2. The web weighs much more than one flange.
3. The stiffeners provide 5% of the total weight.

However, the purpose of the assumptions—to establish order of magnitude estimates of girder weight—was served. Since the girder weight contributes a small percentage of the total design moments and shears, the consequence of error in those assumptions is small. For the 90-foot span, the net error itself turns out to be quite small (again, see weight summary, sheet 31).

Calculated weights (pounds/linear foot) of steel on sheet 7 are somewhat different from that which had been assumed on sheet 5. This results in slightly revised total moments and adjusted dead loads. Before leaving this sheet, we try to use the calculated weights to see if we can detect a clear preference between the alternatives. There seems to be some preference, based on weight, for the four girder solution; but it is not clear enough to cause us to abandon the other alternative.

Choice of steel. The question addressed on sheets 8 and 9 is whether or not there is a clear preference between A36 and high-strength steels. Resolution of this

Project: PLATE GIRDER BRIDGE - ALL SPANS
Subject: CHOICE OF STEEL; F_b, F_{sr}

SHEET 8 OF

CHOICE OF STEEL AND F_b

CHOICE OF STEEL IS BETWEEN THE "BASIC" STRUCTURAL STEEL, A36 - (F_y = 36 ksi) AND HIGH STRENGTH LOW ALLOY STEELS ($42 \le F_y \le 65$ FOR PROBABLE RANGE OF THICKNESSES TO BE USED). FOR THE MATERIAL ALONE, THE RATIO OF COST/#/STRENGTH FAVORS THE HIGH STRENGTH STEELS*; i.e., WHERE DESIGN CRITERIA ARE LIMITED TO LEAST COST FOR STATIC LOAD CONDITIONS, A HIGH STRENGTH STEEL WOULD BE THE PROBABLE SELECTION. HOWEVER, WE NEED TO CHECK WHETHER OR NOT THE ADDITIONAL STRENGTH CAN BE UTILIZED IN THIS DESIGN - WHETHER IT ACTUALLY RESULTS IN HIGHER ALLOWABLE STRESSES. ISSUES ARE:

1. EFFECTS OF FATIGUE (AISC APP. B & SEC. 1.7)

NOT CHECKED HERE; TO BE CHECKED IN COMPLETE CALCULATIONS:

2. EFFECT OF DEFLECTION LIMITATIONS (AASHTO-1977 - PAR. 1.7.6)
3. EFFECT OF WEB DEPTH/THICKNESS RATIO ON F_b IN COMPRESSION (AISC - 1.10.6)
4. EFFECT OF COMBINATION OF HIGH TENSILE BENDING STRESS WITH HIGH SHEAR STRESS (AISC - 1.10.7)

1. FATIGUE $F_b = 0.60 F_y$ = 21.6 ksi (A36); VARIES 25.2 TO 39 ksi FOR H.S. STEELS

TO FIND F_{SR} (AISC APPENDIX B)

CYCLES OF DESIGN TRUCK LOAD > 2,000,000**

∴ LOADING CONDITION - (TABLE B1) = 4

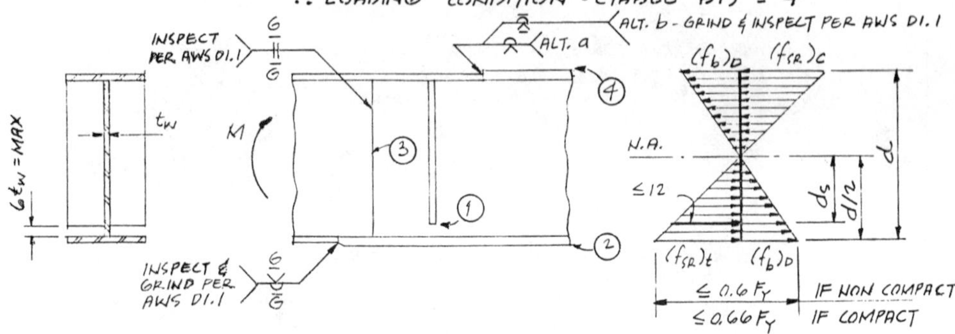

LOCATION	DET. (FIG. B1)	CATEGORY (TABLE B2)	F_{SR}, ksi (TABLE B3)	
1	7	C	12	$(F_{sr})_1 = 12$
2	11	B	16	$(f_{sr})_2 \le 12(\frac{d}{2d_s}) \le 16$
3	11	B	16	$(f_{sr})_3 \le 16$
4	ALWAYS	COMPRESSION;		NOT APPLICABLE

* SUBJECT TO CHANGES IN PRICING STRUCTURE

** SEE U.S. STEEL CO. "HIGHWAY STRUCTURES HANDBOOK" - CHAP. I/6 (NO DATE) FOR DISCUSSION OF FATIGUE IN BRIDGES.

Project : PLATE GIRDER BRIDGE – ALL SPANS

Subject : RATIOS, – M_D/M, M_L/M; $f_{b\,MAX}$ – SELECT STEEL

SHEET 9 OF

M_D & $M_{(L+I)}$, % OF TOTAL MOMENT

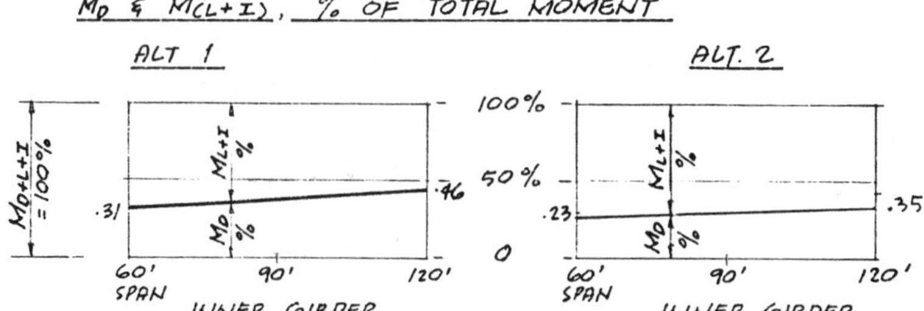

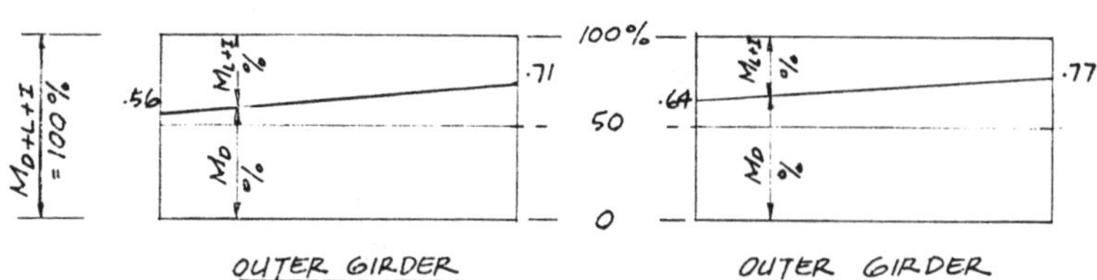

IF F_{sr} @ BOTTOM OF STIFF. = 12 ksi

AND $d/2 \approx .9 d_s$

THEN ALLOW FOR f_{sr} AT OUTER TENSION SURFACE $\leq \frac{1.0}{.9}(12) = 13.3$

LIMIT $(f_b)_{MAX}$ TO SMALLER OF $\frac{M_{(D+L+I)}}{M_{(L+I)}} \times 13.3$ ksi or $.6F_Y$

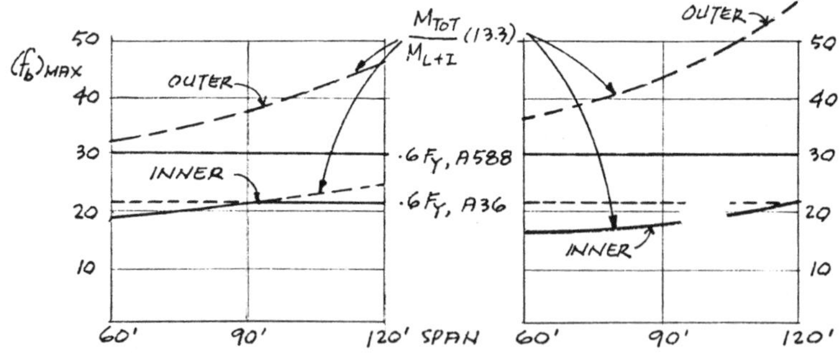

STEEL SELECTIONS*

OUTER GIRDER – A588
INNER GIRDER – NOT CLEAR CHOICE
MAY USE A36 FOR SPANS < 90'
A588 FOR SPANS > 90'

OUTER GIRDERS – A588
INNER GIRDERS – A36

* SUBJECT TO LATER REVIEW OF EFFECTS # 2 THRU # 4, PREVIOUS SHEET.
ULTIMATELY, SELECT ONE STEEL FOR ALL GIRDERS OF ANY BRIDGE

question may hang on whether the requirements set by fatigue (F_{sr}) will limit the effective maximum permissible tensile bending stress. The cycling portion of f_b is that caused by live load and impact; dead load stresses are constant. $(f_b)_{max}$ must be $\leq (f_b)_D + F_{sr} \leq F_b$. The smallest applicable value of F_{sr} is at the toe of stiffener welds where $F_{sr} = 12$ ksi. Extending $[(f_b)_D + 12]$ from the lower end of the stiffeners along the linear stress diagram to the outer tensile fibers leads to values of permissible $(f_b)_{max}$. These will vary depending on the ratio of dead load to (live load + impact), but are further limited to either $0.6F_y$ or, if compact, $0.66F_y$.

The table on sheet 7 is used to establish the ratios of dead load and live load plus impact moments to total moments. From that, it is possible to derive an approximation for $(f_b)_{max}$, which can then be compared to $0.6F_y$. The resulting curves provide interesting information. Fatigue becomes less of a control as, for the longer spans, the ratio of dead to live load becomes greater. For the outer girders, the extra strength available in A588 steel seems clearly useful. For the inner girders it may not be except for the longer spans. There is some advantage to using the same steel for all girders in a span. However, the decision for the inner girders is not clear and may be affected from time to time by changing relative costs of the steels.

There may also be some advantage to the higher-strength steel in web shear resistance. That is not addressed here, but can be seen in the calculations of the 90-foot span.

Lateral load effects. The effect of lateral loads from wind or earthquake is examined on sheet 10. Although the plate girders are a necessary part of the lateral resisting system, the effect of lateral loads is small enough not to be controlling in girder design. Wind or seismic effects on the lower half of the girder will be resisted by the lower flange as a horizontal beam between lateral braces. That will require some y-y bending strength, which will be easy to get if the flange width and proportions b_f/t_f are reasonable.

Influence lines: Shear and moment. Sheet 11 shows a family of influence lines for shear and moment. Influence lines are a classic and very powerful analysis tool. Their use is described in many texts, including Hsieh.[32] We have used them previously in the stiffleg derrick study. Each influence line shows, for a unit load at one position, an effect (in this case shear or moment) at another position. We have drawn the curves in dimensionless form, so they can be used for any span length.

Project : PLATE GIRDER BRIDGE - ALL SPANS	SHEET 10 OF ____
Subject : LATERAL LOAD EFFECTS	

CHECK EFFECT OF LATERAL FORCES ON PLATE GIRDERS.

EXAMINE LATERAL RESISTING SYSTEM:
CONSTRUCTION PHASE;
PERMANENT BRIDGE.

CONSTRUCTION PHASE

UPPER FLANGES ARE IN COMPRESSION DURING LIFTING AND WHEN GIRDERS ARE IN PLACE. PROVIDE LATERAL SYSTEM TO LIMIT L/r_T BETWEEN LATERAL SUPPORTS, ASSUMING TWO, MINIMUM, GIRDERS IN PLACE WITH LATERALS. ERECTOR MUST STABILIZE SINGLE GIRDERS UNTIL LATERAL SYSTEM IS EFFECTIVE.

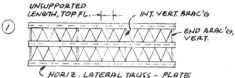

(1) UNSUPPORTED LENGTH, TOP FL. — INT. VERT. BRAC'G — END BRAC'G, VERT.

HORIZ. LATERAL TRUSS - PLATE GIRDER TOP FLANGES + DIAGONALS (ALT. 1 SHOWN, ALT. 2 SIMILAR)

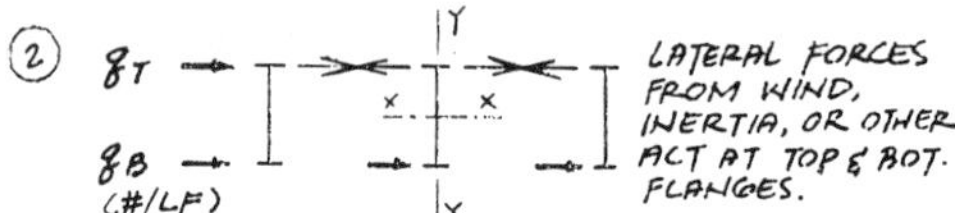

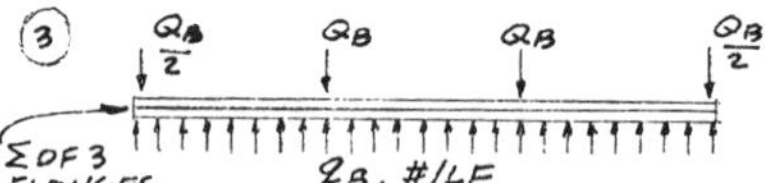

LOWER FLANGES ACT AS BEAM IN Y-Y DIRECTION, SPANNING BETWEEN VERTICAL BRACING

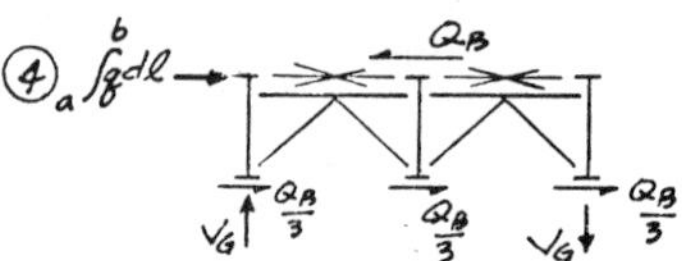

INT. LAT. BRACING TRANSFERS Q_B TO PLANE OF HORIZ. TRUSS. TORQUE RESISTED BY SMALL ADDED SHEARING LOAD IN OUTER GIRDERS: (<< (D+L+I).33)

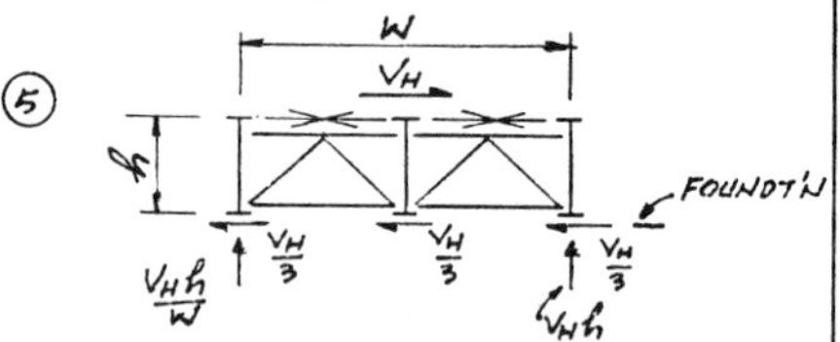

AT ENDS, REACTIONS OF HORIZ. TRUSS ARE TRANSFERRED TO FOUNDATION THRU END BRACING. OVERTURNING EFFECT CONVERTED TO VERTICAL COUPLE

PERMANENT BRIDGE LATERAL FORCES ARE FROM: WIND; EARTHQUAKE

RESISTING SYSTEM IS SAME EXCEPT HORIZONTAL TRUSS IS REPLACED BY THE MUCH STIFFER CONCRETE DECK. GIRDERS MAY BE EFFECTIVE AS FLANGES FOR THE HORIZONTAL GIRDER.

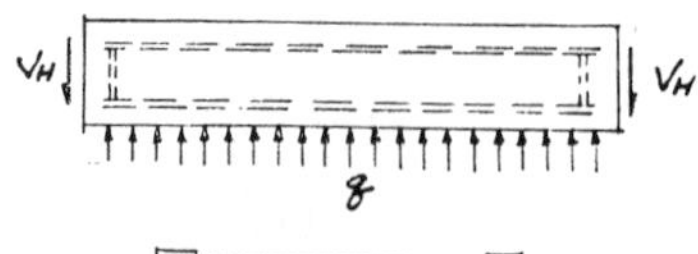

HORIZ. GIRDER - CROSS SECTION

PLATE GIRDERS ARE NECESSARY TO COMPLETE LATERAL SYSTEM. VERTICAL LOAD RESULTS FROM TORSIONAL EFFECT OF LOWER FLANGE HORIZONTAL FORCE. THIS IS SMALL. LOWER FLANGE MUST BE CHECKED FOR Y-Y BENDING BETWEEN INTERMEDIATE BRACING.

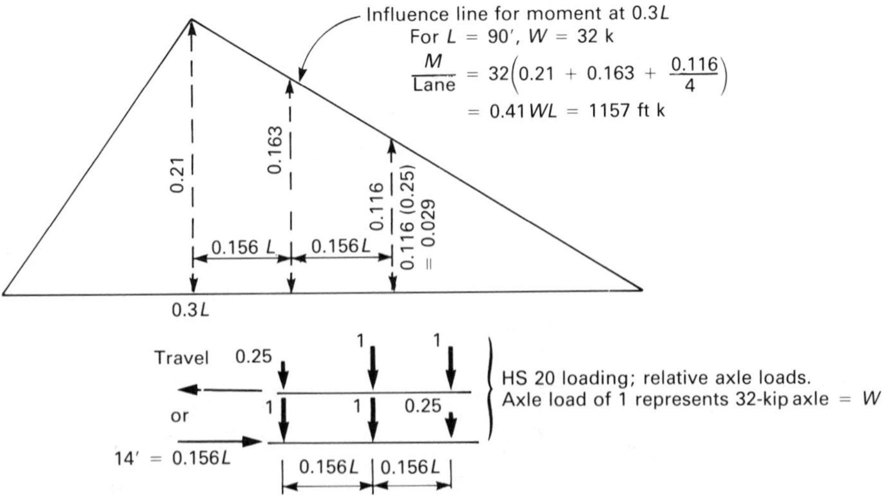

Figure 20.2 Unit AASHTO truck-trailer loads applied to influence line for moments at $0.3L$.

The influence lines can be used to create diagrams of maximum shear and moment. This is done later, on sheet 22, for a 90-foot span. They will be useful in tracing the effects of live load and impact. Since the standard truck-trailer load has three axle loads spaced at specified distances apart, a possible ordinate on the shear or moment diagram can be found as follows. The example in Fig. 20.2 is for moment at position $0.3L - 27$ feet where span $L = 90$ feet. The vehicle may move in either direction across the span.

If the load diagram is drawn to the same scale (for distance) as the influence diagram, an influence coefficient for the combination of loads is found. In this case, since the slope of the left portion of the influence line is much steeper than that of the right portion, the vehicle should be moved from left to right until the rear wheel is at the $0.3L$ point. The combined coefficient results:

$$0.21 + 0.163 + \frac{0.166}{4} = 0.41$$

Such coefficients, for each increment of $0.1L$, are used on sheet 22. When multiplied by the axle load of 32 kips and by K (discussed previously), a moment or shear ordinate results.

Some tentative general conclusions. The information common to all spans has now been assembled. Further work requires use of that information in designing individual spans. It would probably be sufficient to design at each 10-foot increment

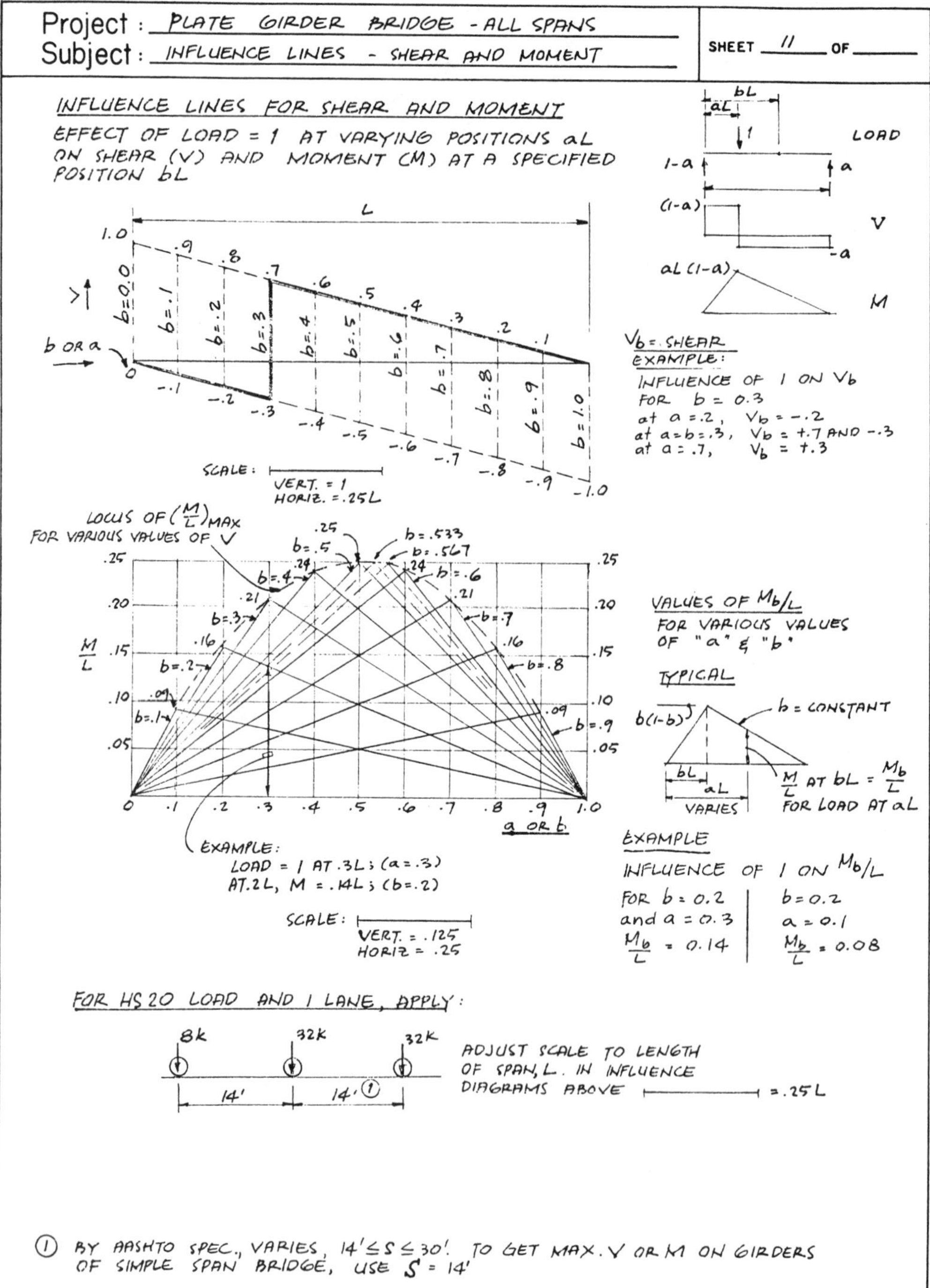
Project : PLATE GIRDER BRIDGE - ALL SPANS
Subject : INFLUENCE LINES - SHEAR AND MOMENT
SHEET 11 OF
INFLUENCE LINES FOR SHEAR AND MOMENT
EFFECT OF LOAD = 1 AT VARYING POSITIONS aL ON SHEAR (V) AND MOMENT (M) AT A SPECIFIED POSITION bL
LOAD
V
M
Vb = SHEAR
EXAMPLE:
INFLUENCE OF 1 ON Vb
FOR b = 0.3
at a = .2, Vb = -.2
at a = b = .3, Vb = +.7 AND -.3
at a = .7, Vb = +.3
b OR a
SCALE:
VERT. = 1
HORIZ. = .25L
LOCUS OF (M/L)MAX FOR VARIOUS VALUES OF V
VALUES OF Mb/L
FOR VARIOUS VALUES OF "a" & "b"
TYPICAL
b = CONSTANT
M/L AT bL = Mb/L FOR LOAD AT aL
VARIES
a OR b
EXAMPLE:
LOAD = 1 AT .3L; (a = .3)
AT .2L, M = .14L; (b = .2)
SCALE:
VERT. = .125
HORIZ = .25
EXAMPLE
INFLUENCE OF 1 ON Mb/L
FOR b = 0.2
and a = 0.3
Mb/L = 0.14
b = 0.2
a = 0.1
Mb/L = 0.08
FOR HS 20 LOAD AND 1 LANE, APPLY:
8k
32k
32k
14'
14' (1)
ADJUST SCALE TO LENGTH OF SPAN, L, IN INFLUENCE DIAGRAMS ABOVE = .25L
(1) BY AASHTO SPEC., VARIES, 14' ≤ S ≤ 30'. TO GET MAX. V OR M ON GIRDERS OF SIMPLE SPAN BRIDGE, USE S = 14'

of span (i.e., 60, 70, etc., through 120 feet), since the changes in significant design parameters with span are not rapid. Principal contributors to economy will be:

- Flange size
- Web size
- Stiffeners
- Total cost of web through sum of the weight of stiffeners and web, modified by extra cost of attaching stiffeners
- Choice between alternative steels
- Choice between alternative arrangements (i.e., number of girders)

Least-weight design seems to favor higher-strength steels, since the cost per pound for high-strength, low-alloy steels is not much more than for A36 steel, and

$$\frac{\$/\text{pound (A36)}}{\$/\text{pound (HS low alloy)}} > \frac{36}{F_y}$$

The cost ratio varies with time and market conditions. The preference for high-strength steel becomes stronger as the spans get longer, reflecting the increasing dominance of dead load over cycling loads. For shorter spans, it may still be desirable to consider A36 steel for inner girders, particularly if price ratios change or if the other constraints listed on sheet 8 turn out to affect the choice.

20.2.3 Calculations: Single Girder

Introduction. It is now time to examine a specific girder design in detail. This will test some of the assumptions we have made and provide further insights into a number of additional design selections that will affect the adequacy and efficiency of the system.

The procedure is developed, in sheets 21 through 30, and the resulting design shown on sheet 31. This is done for a 90-foot span interior girder of arrangement alternative 1 using steel of ASTM A588 ($F_y = 50$). The flow chart on sheet 21 describes a procedure similar to that used in Chapter 13. The resulting design is an "acceptable" result. It may not be the best result. Some ideas that arise in the course of this group of calculations merit further study and are discussed later.

The design calculations follow fairly closely the flow chart shown on sheet 21, which may be used at a later stage as the basis for computer programming. Several things are worth noting.

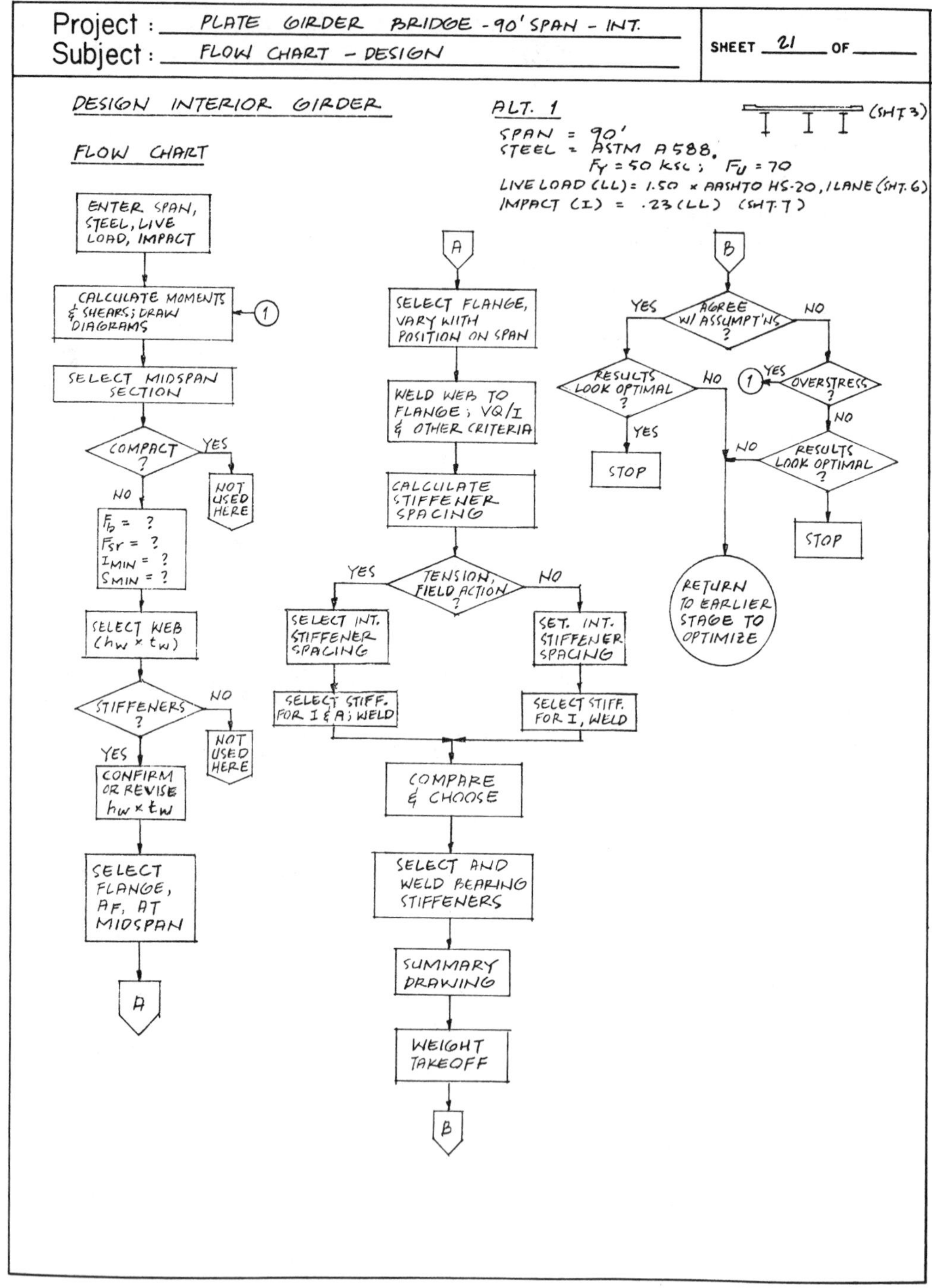

Project : PLATE GIRDER BRIDGE - 90' SPAN - INT.
Subject : FLOW CHART - DESIGN
SHEET 21 OF
DESIGN INTERIOR GIRDER
FLOW CHART
ALT. 1
(SHT 3)
SPAN = 90'
STEEL = ASTM A588.
Fy = 50 ksi ; Fu = 70
LIVE LOAD (LL) = 1.50 × AASHTO HS-20, 1 LANE (SHT. 6)
IMPACT (I) = .23 (LL) (SHT. 7)
ENTER SPAN, STEEL, LIVE LOAD, IMPACT
CALCULATE MOMENTS & SHEARS; DRAW DIAGRAMS
1
SELECT MIDSPAN SECTION
COMPACT ?
YES
NOT USED HERE
NO
Fb = ?
Fsr = ?
IMIN = ?
SMIN = ?
SELECT WEB (hw × tw)
STIFFENERS ?
NO
NOT USED HERE
YES
CONFIRM OR REVISE hw × tw
SELECT FLANGE, AF, AT MIDSPAN
A
A
SELECT FLANGE, VARY WITH POSITION ON SPAN
WELD WEB TO FLANGE; VQ/I & OTHER CRITERIA
CALCULATE STIFFENER SPACING
TENSION FIELD ACTION ?
YES
NO
SELECT INT. STIFFENER SPACING
SET. INT. STIFFENER SPACING
SELECT STIFF. FOR I & A; WELD
SELECT STIFF. FOR I, WELD
COMPARE & CHOOSE
SELECT AND WELD BEARING STIFFENERS
SUMMARY DRAWING
WEIGHT TAKEOFF
B
B
AGREE W/ ASSUMPT'NS ?
YES
NO
RESULTS LOOK OPTIMAL ?
NO
YES
STOP
1
YES
OVERSTRESS ?
NO
NO
RESULTS LOOK OPTIMAL ?
STOP
RETURN TO EARLIER STAGE TO OPTIMIZE

1. Design of compact sections turns out to be inappropriate. This conclusion is clear from sheet 23 for the present span. It is likely to be true for the entire family of designs as deflection limitations lead to deep, thin webs.
2. Web selection is made early, since web depth is the primary deflection control.
3. The requirement of a deep, thin web leads to considerations of shear stability; web stiffening becomes a major factor. The decision about whether or not to take advantage of the possible reserve shear strength in tension field action requires comparison of stiffener requirements with or without the stiffer stiffeners needed.
4. After a design emerges, it must be further checked for
   a. Conformance with early assumptions.
   b. Opportunities for optimization. The process of optimization may involve some considerations specific to the 90-foot span, some others general for the family of designs. Either is beyond the scope and purposes of the present case study.

Moments and shears. Design moments and shears are calculated on sheet 22. Coefficients for one lane of live load are derived from the influence lines on sheet 11 for each tenth point on the span. When multiplied by the axle load of 32 kips and, for moment, by a span of 90 feet, columns 5 and 6 of the table result. The maximum moment occurs several feet before midspan and can only be found when additional points are investigated between $0.4L$ and $0.5L$. Coefficients are found from the influence lines by scale, using procedures of the previous section. Maxima compare closely to those tabulated by AASHTO on sheet A4. Design values for the girder require application of the lanes/girder coefficient K from sheet 6 and the impact coefficient I from sheet 7. This results in columns 7 and 8. The resulting ordinates reflect, for each position, the maximum values of moment and shear from live load with impact. They do not all represent the same load position. Nor do the moment and shear ordinates at any point represent the results of the same position of load. The curves drawn from columns 7 and 8, M_{L+I} and V_{L+I} versus location, are *envelope* curves of maximum ordinates for each location.

Dead load, being constant, provides a single load case and single moment and shear diagrams, which are also envelope diagrams. The intensity of dead load is based on the results on sheet 7. Adding the dead load results to those for live load plus impact gives total ordinates in columns 11 and 12 and the summation curves. We resist the temptation to save time by producing only the combined results of dead, live, and impact loads. Relative values are significant in tracing the effects of span length on design. Separate values are necessary in considering fatigue and will be found useful in other ways later.

Project : PLATE GIRDER BRIDGE - 90' SPAN - INT.	SHEET 22 OF
Subject : MOMENT AND SHEAR ENVELOPES	

90' SPAN, INTERIOR GIRDER, ALT. 1 - MOMENT & SHEAR DIAGRAMS

STEEL A588; DEAD LOAD - SEE SHT. 7; LIVE LOAD ORDINATES BY SCALE FROM INFLUENCE DIAGRAMS, SHT. 11.

bL

L

b ①	bL (ft.) ②	LIVE LOAD ONE LANE				L.L. + I = L50(1+.23) ⑤ OR ⑥		D.L. @ 1.6 K/L.F. SEE SHT. 7		D + L + I	
		MOM. COEFF ③	SHEAR COEFF ④	MOM., (FT. K.) *⑤	SHEAR (KIPS) *⑥	MOM., (FT. K.) ⑦	SHEAR (K) ⑧	MOM. (FT. K.) ⑨	SHEAR (KIPS) ⑩	⑦+⑨, MOM., (FT. K.) ⑪	⑧+⑩ SHEAR, (KIPS) ⑫
0	0.0	0	2.012	0	64.4	0	119	0	72.0	0	191
0.100	9.0	.175	1.787	504	57.2	930	106	583	57.6	1513	164
0.200	18.0	.312	1.562	898	50.0	1657	92	1037	43.2	2694	135
0.300	27.0	.401	1.338	1155	42.8	2131	79	1360	28.8	3492	108
0.400	36.0	.448	1.113	1290	35.6	2380	66	1555	14.4	3935	80
0.433	39.0	.450	1.038	1296	33.2	2391	61	1574	9.6	3965	71
0.467	42.0	.460	0.963	1325	30.8	2445	57	1611	4.8	4056	62
0.500	45.0	.452	0.888	1302	28.4	2402	52	1620	0	4022	52

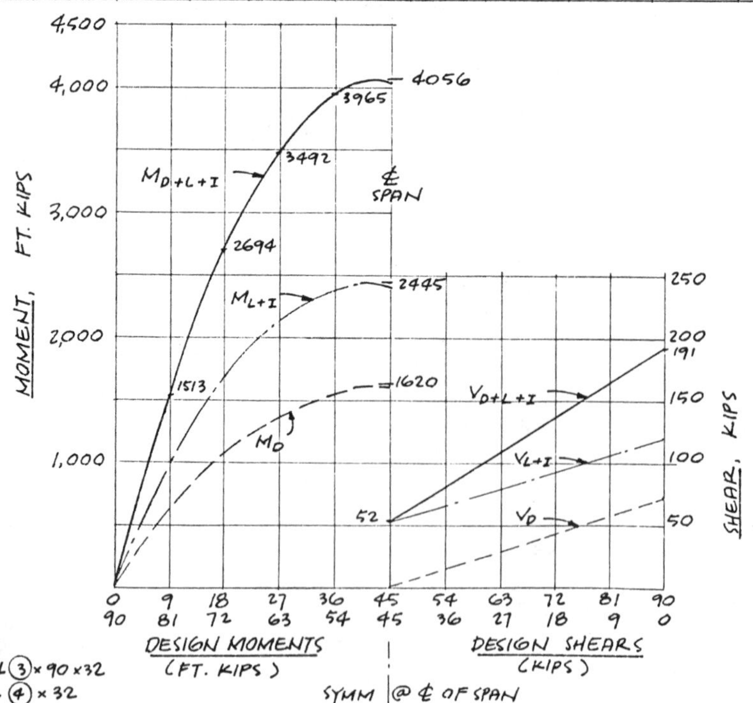

* COL ⑤ = COL ③ × 90 × 32

COL ⑥ = COL ④ × 32

Select midspan section. Sheets 23 through 25 are devoted to the search for the midspan cross section. They are dominated by consideration of deflection from live load plus impact. The very restrictive limit of $L/800$ set by the AASHTO Specification (Section 1.7.6) was adopted as part of our design criteria on sheet 2. Since deflection is controlled largely by the parameter M/EI, we are forced to seek a deep section, not necessarily the lightest one by stress criteria. Dead load deflection is a consideration. However, we are able to compensate for that by building in a crown (camber) during fabrication so that application of the dead load will cause the girder to deflect into a straight profile. The limit on deflection for the 90-foot span is calculated to be 1.35 inches.

Although we do not expect to be able to design an efficient section by the "compact section" criteria of AISC 1.5.1.4.1, we check the deflection that would result if the girder were proportioned by those criteria. The limit on web proportions, $d_w/t_w < 90.5$, would lead to a shallow section, requiring very heavy flanges by strength criteria, but one that deflects much more than permissible. The equation used for finding Δ_{L+I} assumes a constant moment of inertia, an approximation.

We are led to the use of noncompact sections, with deeper, but stiffened webs, and a limit on F_b set at $0.6F_y$. This is pursued on sheet 24. We seek a desirable value for depth by combining the required moment of inertia and the maximum permissible bending stress, $(f_b)_{max}$ (not necessarily equal to F_b). Assuming, as before, that the flanges provide 90% of the total moment resistance, we calculate a limit of 22 ksi on f_b, confirming the previous result on sheet 9. The calculated balance is found at $h \approx$ 80 inches. Our tentative choice for the web assumes $h_w = 80$, making the total depth slightly greater.

Two different limits of the ratio h_w/t_w are set by the AISC depending on the nature of web stiffening. These lead to two different values of minimum web thickness. Decimal results for thickness are increased to the smallest usual commercially available thickness equal to or greater than that required. A check of shear stress shows that, for the depth chosen, both minimum thicknesses result in maximum shear stresses (at end of girder), much less than the usual limit of $0.40F_y$. For insight into the stiffening problem, we consult AISC Tables 10-50 and 11-50, which are solutions to AISC formulas 1.10-1 and 1.10-2. We are able to ascertain quickly from Table 10-50 that, in the regions of high shear stress, the 3/8-inch web will have no difficulty resisting the shear with reasonable stiffener spacing. For a 5/16-inch web, we conclude that it would be necessary to take advantage of tension field action, which is likely to require heavier stiffeners.

A tentative decision is made to use the lighter web, and an approximate calculation of flange size is made. The formula used is discussed in Chapter 13. It leads to a section weighing very close to the weight approximated earlier on sheet 7. The flanges would resist 84% of the moment, the web 16%. This seems a reasonable solution, but it is decided to consider several other possibilities in search of possible improvement.

Project : PLATE GIRDER BRIDGE - 90' SPAN - INT.	SHEET 23 OF
Subject : SELECT MID-SPAN SECTION	

90' SPAN, INTERIOR GIRDER, ALT. 1
SELECT SECTION AT ℄ SPAN

$M_{D+(L+I)} = 1620 + 2445 = 4065\ k'$ (@ ℄ NEAR ℄)

$V_{D+(L+I)} = 52k$ @ ℄; $[(V_{MAX})_{D+(L+I)} = 72 + 119 = 191\ k$ AT ENDS$]$

LIMIT $\Delta_{L+I} \leq \frac{L}{800} = \frac{90(12)}{800} = 1.35''$ (AASHTO 1.7.6)

CAMBER FOR Δ_D

$F_Y = 50$ ksi

$F_b \leq .66\ F_Y = 33$ ksi IF COMPACT; $F_b \leq .6\ F_b = 30$ ksi IF NON-COMPACT

$F_{sr} = \begin{cases} 12 \text{ ksi AT STIFFENERS} \\ 16 \text{ ksi BELOW STIFFENERS} \end{cases}$

COMPACT OR NON-COMPACT ? (AISC 1.5.1.4.1)

CRITERIA FOR COMPACT

1. WELD CONT., FLG TO WEB ? DESIRABLE FOR FATIGUE & TO PREVENT CORROSION (EXTERIOR EXPOSURE)
2. $\frac{b}{2t}$ OF COMP. FLG $\leq \frac{65}{\sqrt{F_Y}} = 9.19$? NO APPARENT PROBLEM
3. $\frac{b}{t}$ FOR STIFFENED COMP. FLG. ? NOT APPLICABLE
4. $\frac{d_w}{t_w} \leq \frac{640}{\sqrt{F_Y}}\left(1 - 3.74\frac{f_a}{f_Y}\right) = 90.5$ for $f_a = 0$; USE 90

FOR $V_{MAX} = 191\ K$
AND $F_V = .4\ F_Y$ (AISC 1.5.1.2)
(OR LESS, DEPENDING ON STIFFENER SPACING.)

$A_W \approx 10\ IN^2 = d_w t_w$

$10 = 90\ t_w^2$

$t_w = .33$

$d_w \leq .33(90.5) = 30.0''$

$\frac{L}{h} \approx \frac{90(12)}{32} = 33.75$ (NG ?)

(AASHTO SPEC 1.7.4 SUGGESTS $\frac{L}{h} \leq 25$)

IF $d_w t_w = .33(30.0)$, AND

$M_{L+I} = 2445'^k = 29{,}340''^k \underset{\text{EQUIV.}}{\approx} \frac{w\ell^2}{8}$

$\Delta_{L+I} \approx \frac{5}{384}\frac{w\ell^4}{EI} = M_{L+I}\frac{5}{48}\frac{\ell^2}{EI}$

$= \frac{29{,}340(90)^2(144)}{30{,}000\ I}\left(\frac{5}{48}\right)$

$= 118{,}827/I$; SAY USE $120{,}000/I$

IF $A_F d F_b \approx .9\ M_{TOT.} = .9(4056)12$

$A_F \approx \frac{.9(4056)(12)}{34(24)} = 53.7$

$I \approx 2A_F(16)^2 + \frac{d_w^3 t_w}{12} = 28{,}236$

$\Delta_{L+I} = 4.25 \gg 1.35$ ∴ N.G.

BY AISC 1.10.2, LIMIT OF

$\frac{d_w}{t_w} = \frac{14{,}000}{\sqrt{F_Y(F_Y+16.5)}} = 243$

OR $\frac{2{,}000}{\sqrt{F_Y}} = 283$

FOR EFFICIENT USE OF WEB, $\frac{d_w}{t_w}$ SHOULD APPROACH 200 ±.
LIMITING AS IN COMPACT SECTIONS LEADS, BY WEB CRITERION, TO A BEAM DEPTH UNACCEPTABLE FOR DEFLECTION.

USE NON-COMPACT SECTION; LIMIT $F_b \leq .60 F_Y = 30$ ksi OR AS REDUCED BY OTHER CRITERIA.

Project: PLATE GIRDER BRIDGE - 90' SPAN - INT. SHEET 24 OF

Subject: SELECT MIDSPAN SECTION

90' SPAN, INTERIOR GIRDER; ALT. 1 - SELECT SECTION @ ℄ SPAN (CONT.)

CONSIDERING FATIGUE (SEE SHT. 9) EFFECTIVE LIMIT OF

$(f_b)_{MAX} \approx \frac{12}{.9}\frac{M_{TOT.}}{M_{L+I}} = \frac{12}{.9}\left(\frac{4056}{2445}\right) = 22.2$ ksi ; USE 22

FROM AASHTO 1.7.6 , $\Delta_{L+I} \leq 1.35'' \approx \frac{120{,}000}{I}$ ← SHT. 23

$I \geq 89{,}000$

$S \geq \frac{M}{(f_b)_{MAX}} \approx \frac{4065\,(12)}{22} = 2217$; $\frac{89{,}000}{2217} = 40.1 \approx \frac{h}{2}$

TRY $h_w = 80''$ $I \geq 89{,}000$ in^4

CHOOSE WEB

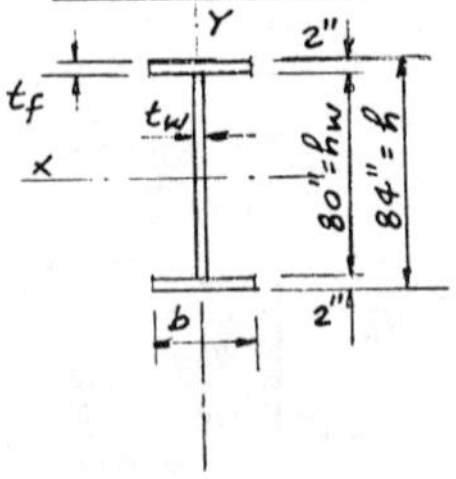

$\left(\frac{h_w}{t_w}\right) \leq \frac{14{,}000}{\sqrt{F_y(F_y+16.5)}} = 243$ (AISC 1.10.2 ; STIFFENERS NOT NECESSARILY REQUIRED)

$\left(\frac{h_w}{t_w}\right) \leq \frac{2{,}000}{\sqrt{F_Y}} = 283$ (AISC 1.10.2 ; STIFF. SPACING $a/h_w \leq 1.5$)

a) $\frac{h_w}{243} \leq t_w$; $t_w \geq 0.33''$ ⟹ 3/8" ℄ ; $t_w = .375''$; $\frac{h_w}{t_w} = 213$

b) $\frac{h_w}{283} \leq t_w$; $t_w \geq 0.28''$ ⟹ 5/16" ℄ ; $t_w = .3125''$; $\frac{h_w}{t_w} = 256$

c) MIN. DESIRABLE THICKNESS BY ENGR. DEP'T. POLICY ON BRIDGES = 1/4" < 5/16" ABOVE

CHECK FOR V_{MAX} = 191 k AT ENDS

3/8" WEB ; $f_v = \frac{191}{.375\,(80)} = 6.37$ ksi << $.4\,F_Y = 20$

5/16" WEB ; $f_v = 191/.3125(80) = 7.64$ ksi << 20

REFER TO AISC FORMULA 1.10-1 AND TABLES 10-50 AND 11-50
3/8" WEB SEEMS ACCEPTABLE WITHOUT TENSION FIELD ACTION.

TENTATIVELY USE : 5/16" WEB; WILL REQUIRE TENSION FIELD ACTION, BUT WIDELY SPACED STIFFENERS.

CHOOSE FLANGE (A_F & $b_F \times t_F$, LIMIT $b_F/2t_F$)

$S_x \approx \frac{th^2}{6} + A_F h$

$A_F \approx \frac{M}{(f_b)_{MAX}h} - \frac{t\,h_w}{6}$ ← NOTE SMALL WEB EFFECT

$= \frac{4065\,(12)}{22\,(84)} - \frac{.3125\,(80)}{6} = 26.4 - 4.17 = 22.23$ in^2

$\frac{b_F}{2t_F} \leq \frac{95}{\sqrt{F_Y}} = 13.44$ ksi (AISC 1.9.1.2 - COMP. FLG.)

FLANGE 18 × 1¼" ; $A_F = 24$ in^2; $\frac{b_F}{2t_F} = 7.20$

$w = [2(24) + .3125\,(80)]\,3.42 = 250$ #/LF + STIFF.

↖ COMPARE TO 279, SHT. 7

⟹ SEEMS OK, BUT COMPARE ALTERNATIVE CHOICES
MAINTAIN $I \geq 89{,}000$

Project: PLATE GIRDER BRIDGE - 90' SPAN, INT.
Subject: SELECT MIDSPAN SECTION
SHEET 25 OF

SELECT SECTION AT ℄ OF SPAN (CONT.)
COMPARE ALTERNATIVES

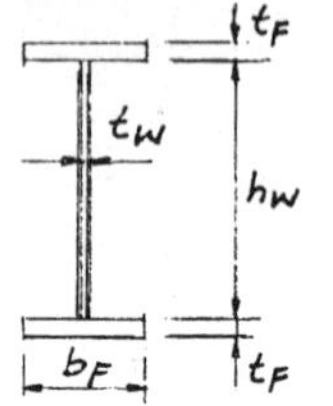

$I = 89{,}000\ IN^4$; $S \geq 2220$
t_W = VARIES (IN)
$h_W = C_1 t_W$ IN.
$I_W = \frac{C_1^3 t_W^4}{12}\ IN^4$
$I_F = 89{,}000 - I_W$
$A_F = b_F t_F = \frac{I_F}{2(\frac{h_W}{2}+1)^2}$, ASSUME $t_F = 2''$
$A = h_W t_W + 2A_F,\ IN^2$
$w = A(3.42)$, #/LF
$S = \frac{I}{(\frac{h_W}{2}+2)},\ IN^3$
(PROGRAMMED ON HAND CALCULATOR)

	$h_W = 280 t_W$			$h_W = 260 t_W$		$h_W = 240 t_W$		$h_W = 230 t_W$		$h_W = 220 t_W$	
t_W	5/16	3/8	7/16	5/16	3/8	5/16	3/8	5/16	3/8	5/16	3/8
h_W	87.5	105	122.5	81.25	97.5	75	90	71.88	86.25	68.75	82.5
I_W	17,445	36,175	67,020	13,968	28,964	10,986	22,781	9669	20,051	8,462	17,547
I_F	71,554	52,824	21,979	75,032	60,036	78,014	66,219	79,331	68,949	80,537	71,452
A_F	17.87	9.23	2.84	21.65	12.13	26.32	15.65	29.07	17.71	32.18	20.01
A	63.08	57.83	59.27	68.69	60.82	76.07	65.04	80.61	67.76	85.84	70.96
w	215.7	198	203	235	208	260	222	276	232	294	243
S	1945	1633	1407	2087	1753	2253	1893	2346	1972	2447	2057
	N.G.	N.G.	N.G.	N.G.	N.G.	✓	N.G.	✓	N.G.	✓	N.G.

USING 3/8" × 86" WEB AND ADJUSTING A_F FOR $I \geq I_{REQ.}$, $S \geq S_R$

$I_W = \frac{.375(86)^3}{12} = 19877$

$I_F \approx 2A_F(44)^2 = 3872\ A_F$

$I \approx 19877 + 3872\ A_F$

$S \approx I/45$

A_F	I_F	I	S	w, #/LF
16	61952	81829	1818	220
18	69696	89573	1990	233
20	77440	97317	2162	247
22	85184	105061	2335	261

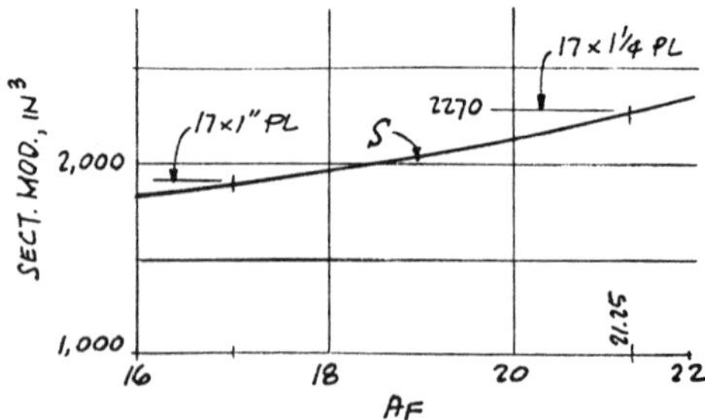

The set of equations at the top of sheet 25 lead to values of cross-section area, corresponding weight, and section modulus provided the ratio of h_w/t_w is known and a thickness of flange assumed, in this case at 2 inches. The results will be close for other flange thicknesses. The equations are easily programmable for a hand calculator, leading to results for any number of combinations of web thickness and depth. Displaying the results makes comparisons possible.

The solution of the previous sheet is almost exactly found by interpolating between the columns for 5/16-inch thickness and h_w/t_w of 260 and 240. However, we see that a ⅜ × 86.25-inch web leads to a girder of similar weight and only slightly deficient in section modulus. Increased section modulus can be found by adding a small amount of flange area. Reminded of the simpler web stiffening required for a 3/8-inch web, we decide to switch. The final table and curve relate flange area to I and S, for a 3/8 × 86-inch web. We have established our midspan section and are ready to look at ways to vary it as the moment envelope changes.

Modify section at reduced moment. One road to economy in plate girder design may be found through variation in the cross section along the length of the span. We travel this road partway on sheet 26. The envelope curve for moments is reproduced, smoothed out in the middle to eliminate the slight dip at midspan, which has little design significance. Required section modulus is calculated, assuming $F_b =$ 22 as previously established, and a curve of required S is drawn to scale. The strength requirement is that $S \leq S_R$ at every point. Since the midspan section was chosen for $(S_R)_{max}$, it is safe to reduce the section modulus everywhere else. Our previous deflection calculations assumed constant depth for the web, a common design basis for plate girder bridges of this type. The web thickness we used at midspan was initially selected to be sufficient for the high shear region at the end. The place for possible savings is in the flange. We elect to splice the flange plates at a point where the moment has reduced sufficiently to do so. A somewhat arbitrary decision is made to reduce the flange thickness to 3/4-inch at the splice, checking the ratio $b_f/2t_f$ to assure flange stability. The splice is made at the point where the new section modulus is sufficient. A full penetration groove weld is used to fully develop the 3/4-inch plate and then ground to avoid fatigue problems at the abrupt change in thickness. We will have a 48-foot piece at 1¼-inch thickness, plus short, thinner pieces at each end.

We do not choose to pursue this type of weight saving further. Additional splices are possible, but are not likely to save enough steel to pay for the extra welding that would be necessary.

Deflection check. Before finally accepting the design of web and flanges, some checks are necessary. They are made on sheet 27. First, deflection is checked by a graphic version of the moment area method, accounting this time for the change in moment of inertia that takes place beyond the splice. The solution is based on the moment envelope curve rather than the moment diagram for a specific load position.

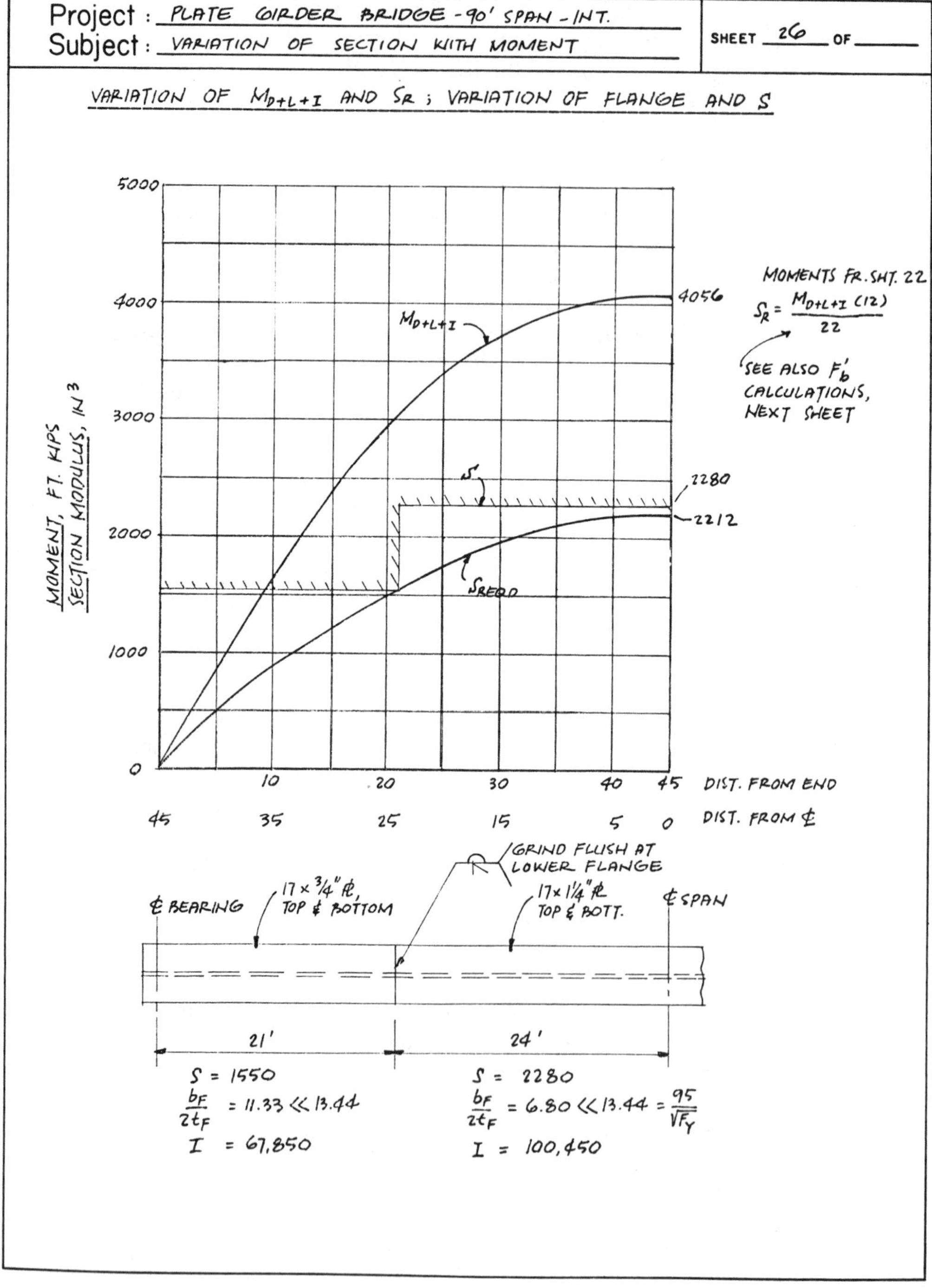

Project : PLATE GIRDER BRIDGE - 90' SPAN - INT.
Subject : VARIATION OF SECTION WITH MOMENT
SHEET 26 OF
VARIATION OF M D+L+I AND SR ; VARIATION OF FLANGE AND S
5000
4000
3000
2000
1000
0
MOMENT, FT. KIPS
SECTION MODULUS, IN 3
M D+L+I
4056
S
2280
2212
SREQD
MOMENTS FR. SHT. 22
SR = M D+L+I (12) / 22
SEE ALSO F'b CALCULATIONS, NEXT SHEET
10 20 30 40 45 DIST. FROM END
45 35 25 15 5 0 DIST. FROM ℄
GRIND FLUSH AT LOWER FLANGE
℄ BEARING
17 × 3/4" ℘, TOP & BOTTOM
17 × 1 1/4" ℘ TOP & BOTT.
℄ SPAN
21'
24'
S = 1550
bF/2tF = 11.33 << 13.44
I = 67,850
S = 2280
bF/2tF = 6.80 << 13.44 = 95/√FY
I = 100,450

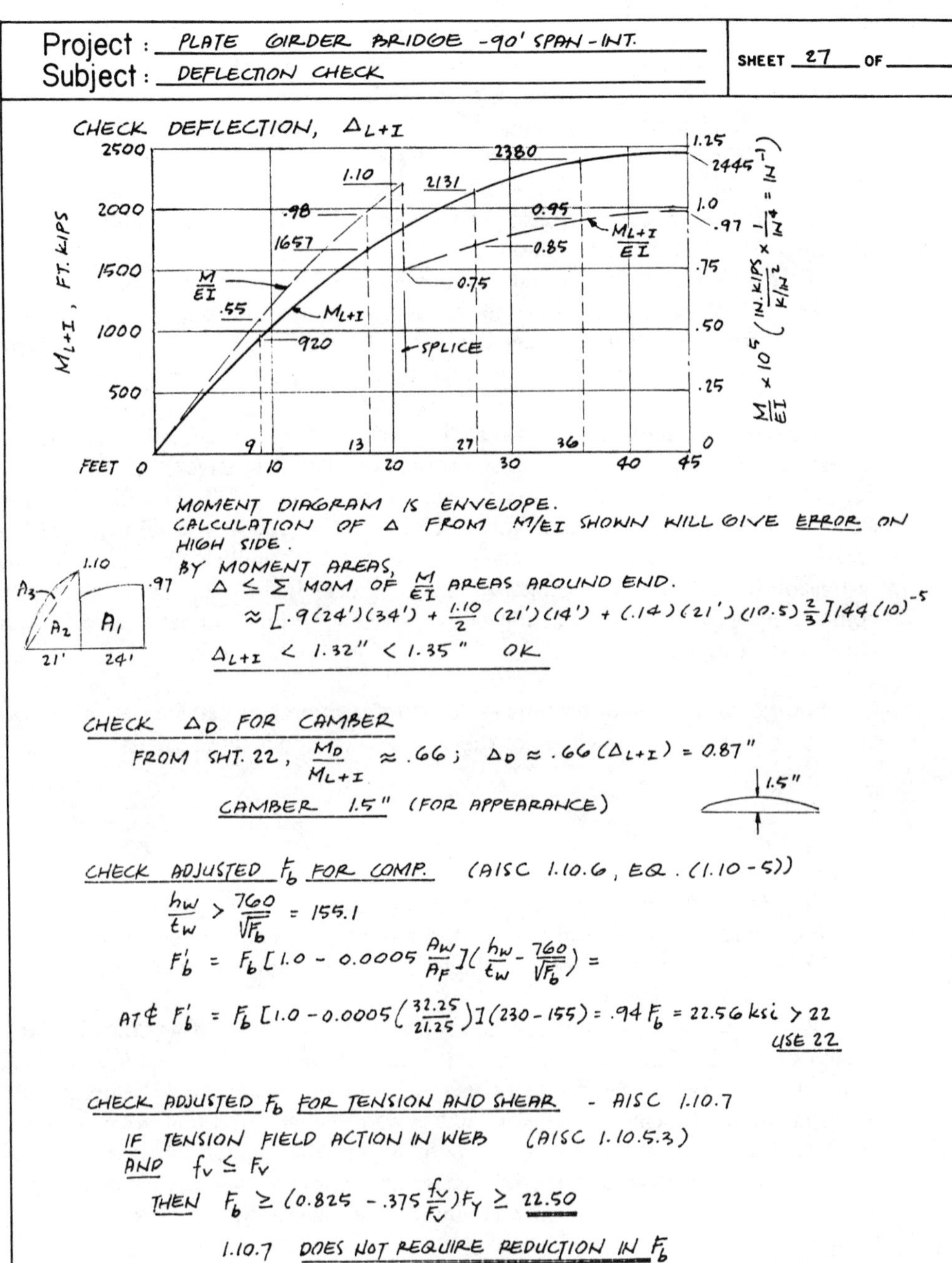

Project : PLATE GIRDER BRIDGE - 90' SPAN - INT.
Subject : DEFLECTION CHECK

SHEET 27 OF ____

CHECK DEFLECTION, Δ_{L+I}

MOMENT DIAGRAM IS ENVELOPE.
CALCULATION OF Δ FROM M/EI SHOWN WILL GIVE ERROR ON HIGH SIDE.

BY MOMENT AREAS,

$\Delta \leq \Sigma$ MOM OF $\frac{M}{EI}$ AREAS AROUND END.

$\approx [.9(24')(34') + \frac{1.10}{2}(21')(14') + (.14)(21')(10.5)\frac{2}{3}]144(10)^{-5}$

$\Delta_{L+I} < 1.32'' < 1.35''$ OK

CHECK Δ_D FOR CAMBER

FROM SHT. 22, $\frac{M_D}{M_{L+I}} \approx .66$; $\Delta_D \approx .66(\Delta_{L+I}) = 0.87''$

CAMBER 1.5" (FOR APPEARANCE)

CHECK ADJUSTED F_b FOR COMP. (AISC 1.10.6, EQ. (1.10-5))

$\frac{h_w}{t_w} > \frac{760}{\sqrt{F_b}} = 155.1$

$F'_b = F_b[1.0 - 0.0005\frac{A_w}{A_F}](\frac{h_w}{t_w} - \frac{760}{\sqrt{F_b}}) =$

AT ℄ $F'_b = F_b[1.0 - 0.0005(\frac{32.25}{21.25})](230 - 155) = .94F_b = 22.56$ ksi > 22

USE 22

CHECK ADJUSTED F_b FOR TENSION AND SHEAR - AISC 1.10.7

IF TENSION FIELD ACTION IN WEB (AISC 1.10.5.3)
AND $f_v \leq F_v$
THEN $F_b \geq (0.825 - .375\frac{f_v}{F_v})F_y \geq 22.50$

1.10.7 DOES NOT REQUIRE REDUCTION IN F_b

This introduces a small error on the high side, so the result will overestimate the actual deflection. M/EI areas and moments of the areas about the end of span are calculated by scale measurements on the M/EI diagram.

Dead load deflection is proportioned from that calculated for live load, the proportion being derived from the diagrams on sheet 22. The result is small, but would be distressing to an observer who might incorrectly interpret a sag in the span as a sign of weakness. The decision is made to camber the girders slightly more than the calculated dead load deflection.

Both of these deflection calculations ignore the participation of the concrete deck in flexural resistance. Although our design ignores the composite action of the deck with the girders, the structure will actually exhibit it. In both cases, the actual deflection can be expected to be less than that calculated.

Checks of allowable stress, F_b. The AISC Specification requires, in Sections 1.10.6 and 1.10.7, some adjustment of allowable stress, F_b. The check in 1.10.6 applies to the compression region and compensates for loss of longitudinal strength in the web from minor buckling. The potential reduction is a function of h_w/t_w and the relative areas of web and flange. Section 1.10.7 applies if we take advantage of tension field action. It accounts for the high principal tension resulting from the combination of tension field action due to shear with longitudinal tension. Neither check leads to a change in F_b.

Flange to web welding (sheet 28). Interaction between the flange and web requires connection for longitudinal shear flow. The applicable formula,

$$q = \frac{VQ}{I}$$

is familiar from texts in mechanics of solids. As used here, V is shear in kips, Q is the first moment about the neutral axis of the area beyond the plane being connected in in.^3, and I is the moment of inertia, in.^4. The resulting shear flow, q, has the dimensions kips per inch.

With a fillet weld on both sides, each resists half of the shear flow. This turns out to be much less than is permitted for the minimum-sized fillet weld allowed based on the thickness of the material. Although intermittent welding would be permitted by the calculation, reducing the total amount of welding, we reject that in favor of usual bridge practice of continuous welding. This has the virtues of superior fatigue properties and better corrosion resistance. In any case, continuous weld by automatic processes would very likely cost no more than intermittent welding if done by hand.

Web stiffeners (sheets 29 and 30). The web requires stiffening at the ends where it is to deliver a high concentrated force to the supports and along its length to

Project : PLATE GIRDER BRIDGE - 90' SPAN - INT. Subject : FLANGE TO WEB WELDING	SHEET 28 OF

90' SPAN, INTERIOR GIRDER; ALT. 1

WELD FLANGE TO WEB

WELD MUST BE CONTINUOUS FOR FATIGUE (ALSO CORROSION)
(REFER APP. B, FIG. B1, SKETCH 4 & 5)

CHECK SHEAR FLOW, $q = \frac{VQ}{I}$, (KIPS/IN) AT V_{MAX}, NEAR END

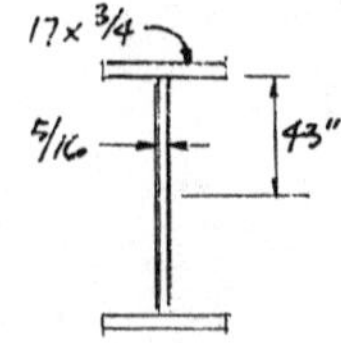

V = 191 AT ENDS
I = 67,850 (SHT. 26)
Q = 17 × .75 × 43.38 = 553
$q = \frac{VQ}{I} = 1.56$ K/IN

WITH FILLET WELD EA. SIDE
$\frac{1.56}{2} = q = .78$ K/IN/WELD

ALLOW ON THROAT, 8 ksi (LOAD COND. 4, CAT. F)
MIN. WELD = 5/16" (TABLE 1.17.2A) - FOR 1¼" FLANGE
5/16" FILLET; q_{ALL} = 5/16 (.707) 8 = 1.77 K/IN < .78
→ USE 5/16" FILLET WELD, EA SIDE (CONT.)

NOTE: AISC APP. B, TABLE B1, PERMITS INTERMITTENT WELDS UNDER CATEGORY F.

AWS WELDING CODE D.1.1, 1980, PAR. 9.19, REQUIRES CONTINUOUS WELDING BETWEEN PIECES OF BUILT-UP CROSS SECTION AT LONGITUDINAL JOINTS.

CONFIRMS USE OF CONTINUOUS WELDS AS ABOVE

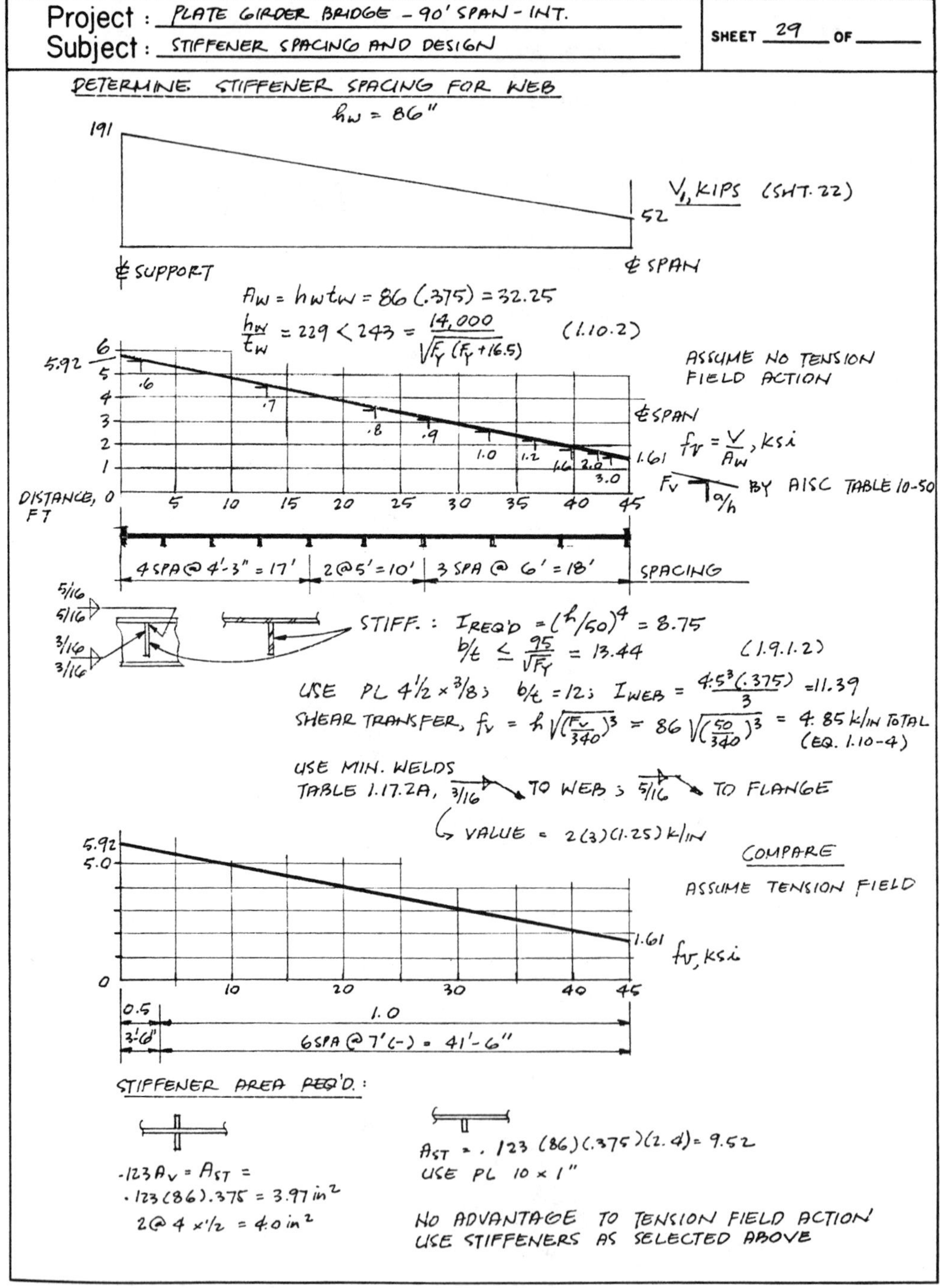
Project : PLATE GIRDER BRIDGE – 90' SPAN – INT.
Subject : STIFFENER SPACING AND DESIGN
SHEET 29 OF
DETERMINE STIFFENER SPACING FOR WEB
hw = 86"
191
V, KIPS (SHT. 22)
52
℄ SUPPORT
℄ SPAN
Aw = hwtw = 86 (.375) = 32.25
hw/tw = 229 < 243 = 14,000/√(Fy(Fy+16.5)) (1.10.2)
ASSUME NO TENSION FIELD ACTION
5.92
℄ SPAN
fv = V/Aw, ksi
1.61
Fv BY AISC TABLE 10-50
DISTANCE, FT
4 SPA @ 4'-3" = 17'
2@5' = 10'
3 SPA @ 6' = 18'
SPACING
5/16
3/16
STIFF.: IREQ'D = (h/50)^4 = 8.75
b/t ≤ 95/√Fy = 13.44 (1.9.1.2)
USE PL 4½ × 3/8; b/t = 12; IWEB = 4.5³(.375)/3 = 11.39
SHEAR TRANSFER, fv = h√(Fv/340)³ = 86√(50/340)³ = 4.85 k/in TOTAL (EQ. 1.10-4)
USE MIN. WELDS
TABLE 1.17.2A, 3/16 TO WEB; 5/16 TO FLANGE
VALUE = 2(3)(1.25) k/in
COMPARE
ASSUME TENSION FIELD
5.92
5.0
1.61
fv, ksi
0.5
1.0
3'-6"
6 SPA @ 7'(-) = 41'-6"
STIFFENER AREA REQ'D.:
.123 Av = AST =
.123 (86).375 = 3.97 in²
2@ 4 × ½ = 4.0 in²
AST = .123 (86)(.375)(2.4) = 9.52
USE PL 10 × 1"
NO ADVANTAGE TO TENSION FIELD ACTION
USE STIFFENERS AS SELECTED ABOVE

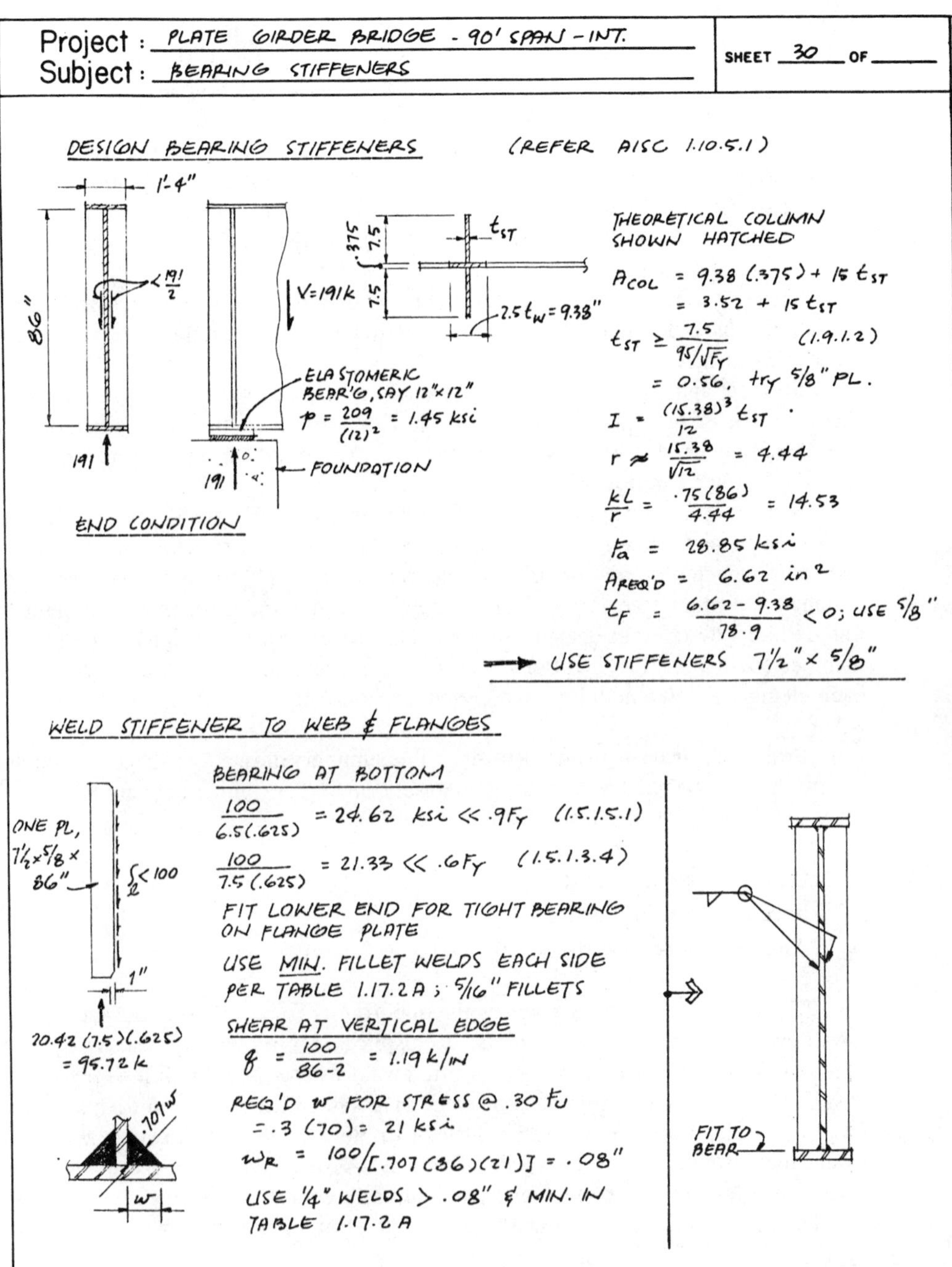
Project : PLATE GIRDER BRIDGE - 90' SPAN - INT.
Subject : BEARING STIFFENERS
SHEET 30 OF
DESIGN BEARING STIFFENERS (REFER AISC 1.10.5.1)
1'-4"
86"
191/2
191
V=191k
ELASTOMERIC BEAR'G, SAY 12"x12"
p = 209/(12)^2 = 1.45 ksi
191
FOUNDATION
END CONDITION
t_ST
.375
7.5
7.5
2.5 t_w = 9.38"
THEORETICAL COLUMN SHOWN HATCHED
A_COL = 9.38 (.375) + 15 t_ST
= 3.52 + 15 t_ST
t_ST ≥ 7.5 / 95/√F_Y (1.9.1.2)
= 0.56, try 5/8" PL.
I = (15.38)^3 t_ST / 12
r ≈ 15.38/√12 = 4.44
kL/r = .75(86)/4.44 = 14.53
F_a = 28.85 ksi
A_REQ'D = 6.62 in^2
t_F = (6.62 - 9.38)/78.9 < 0; USE 5/8"
USE STIFFENERS 7½" × 5/8"
WELD STIFFENER TO WEB & FLANGES
ONE PL, 7½ × 5/8 × 86"
q < 100
1"
20.42 (7.5)(.625) = 95.72 k
BEARING AT BOTTOM
100 / 6.5(.625) = 24.62 ksi << .9F_Y (1.5.1.5.1)
100 / 7.5(.625) = 21.33 << .6F_Y (1.5.1.3.4)
FIT LOWER END FOR TIGHT BEARING ON FLANGE PLATE
USE MIN. FILLET WELDS EACH SIDE PER TABLE 1.17.2A ; 5/16" FILLETS
SHEAR AT VERTICAL EDGE
q = 100 / 86-2 = 1.19 k/in
REQ'D w FOR STRESS @ .30 F_U
= .3 (70) = 21 ksi
w_R = 100/[.707(36)(21)] = .08"
USE 1/4" WELDS > .08" & MIN. IN TABLE 1.17.2A
.707w
w
FIT TO BEAR

protect the web against shear buckling. Intermediate stiffeners are addressed first, following the requirements of AISC 1.10.5.2. As mentioned earlier, Tables 10-50 and 11-50 are solutions to the formulas for F_v as a function of h_w/t_w and panel proportions a/h. The latter table accounts for tension field action.

The shear envelope is reproduced from sheet 22. Dividing by the web area, this is converted to an envelope of shear stresses for all points on the span. Ignoring possible tension field action, Table 10-50 provides regions where various ratios a/h are permissible. A practical spacing of stiffeners is devised in which spacing never exceeds that permitted.

The requirement for the stiffener itself is defined as a stiffness requirement, a minimum value of moment of inertia proportional to h^4. A small flat bar on one side of the web is sufficient.

Stiffener spacing with tension field action is checked separately with the help of Table 11-50. Fewer stiffeners are required. However, since they act as compression members, they need to be heavier than the stiffeners found earlier. The advantage lies with the simpler solution without tension field.

Bearing stiffeners at the ends act as columns. The web transfers most of its shear to these stiffeners through fillet welds, the small remaining part being transferred from web to flange to foundation in bearing. Similarly, the portion transferred to the stiffeners is further transferred to the flange in bearing and through the flange to the foundation. The column design follows usual procedures for design of compression members, using, for the conditions here, $K = 0.75$, and being sure that the b/t ratio of each element is low enough to avoid element buckling.

Summary drawing and take-off. The summary drawing on sheet 31 is made in a form that can be used in preparing contract drawings. All members are shown with their connections, with a few exceptions such as:

Connections to bracing
Base plates at foundations
Anchors connecting the top flange to the concrete slab

The process of drawing gives us the opportunity to be sure that our calculations have covered all the necessary points. Web splices, not previously discussed, do not require calculation since the full penetration weld will transfer the full strength of the web. It is necessary to inform a contractor that web splices are permitted; a 90-foot-long web without splices would be difficult to buy. If the web thickness varied, we would have to specify the locations of splices. As it is, we allow the contractor to select them for his own convenience, hoping the freedom will result in lower cost.

The camber diagram is shown as a series of straight chords between web splices, again to simplify fabrication.

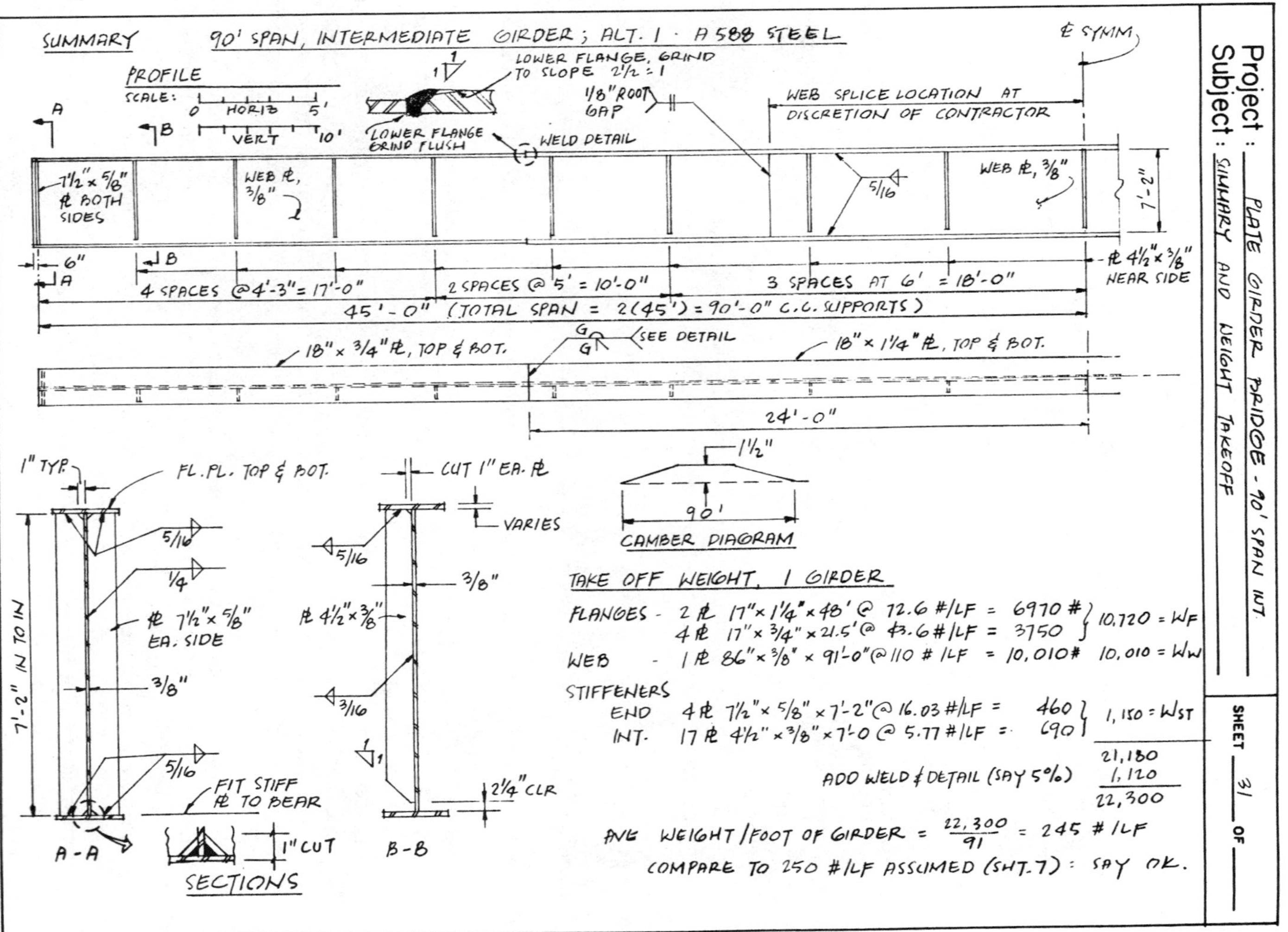

Project : PLATE GIRDER BRIDGE - 90' SPAN INT.
Subject : SUMMARY AND WEIGHT TAKEOFF
SHEET 31 OF
SUMMARY 90' SPAN, INTERMEDIATE GIRDER ; ALT. 1 · A 588 STEEL
PROFILE
SCALE: HORIZ 0 5' VERT 10'
℄ SYMM.
LOWER FLANGE, GRIND TO SLOPE 2½ : 1
LOWER FLANGE GRIND FLUSH
1/8" ROOT GAP
WEB SPLICE LOCATION AT DISCRETION OF CONTRACTOR
WELD DETAIL
7½" × 5/8" ℄ BOTH SIDES
WEB ℄, 3/8"
WEB ℄, 3/8"
7'-2"
6"
4 SPACES @ 4'-3" = 17'-0"
2 SPACES @ 5' = 10'-0"
3 SPACES AT 6' = 18'-0"
℄ 4½" × 3/8" NEAR SIDE
45'-0" (TOTAL SPAN = 2(45') = 90'-0" C.C. SUPPORTS)
18" × 3/4" ℄, TOP & BOT.
SEE DETAIL
18" × 1¼" ℄, TOP & BOT.
24'-0"
1" TYP.
FL. PL. TOP & BOT.
CUT 1" EA. ℄
VARIES
1½"
90'
CAMBER DIAGRAM
5/16
1/4
℄ 7½" × 5/8" EA. SIDE
3/8"
℄ 4½" × 3/8"
3/16
7'-2" IN TO IN
TAKE OFF WEIGHT, 1 GIRDER
FLANGES - 2 ℄ 17" × 1¼" × 48' @ 72.6 #/LF = 6970 #
4 ℄ 17" × 3/4" × 21.5' @ 43.6 #/LF = 3750 } 10,720 = WF
WEB - 1 ℄ 86" × 3/8" × 91'-0" @ 110 #/LF = 10,010 # 10,010 = WW
STIFFENERS
END 4 ℄ 7½" × 5/8" × 7'-2" @ 16.03 #/LF = 460
INT. 17 ℄ 4½" × 3/8" × 7'-0 @ 5.77 #/LF = 690 } 1,150 = WST
21,180
ADD WELD & DETAIL (SAY 5%) 1,120
22,300
FIT STIFF ℄ TO BEAR
2¼" CLR
1" CUT
A-A
B-B
SECTIONS
AVE WEIGHT / FOOT OF GIRDER = 22,300 / 91 = 245 #/LF
COMPARE TO 250 #/LF ASSUMED (SHT. 7) : SAY OK.

Once the drawing is complete, we are able to make a fairly accurate take-off of the weight of the girder, using, for density of steel,

$$\frac{492\text{ lb/c.f.}}{144\text{ in.}^2/\text{ft}^2} = 3.42\text{ lb/in.}^2\text{/ft of length}$$

The average weight of the girder, in pounds/linear foot, compares closely to that assumed on sheet 7, which was used in establishing design moments and shears. A substantial difference, particularly on the high side, would require:

1. A check to identify the source of the difference and to make sure there are no gross blunders.
2. Some adjustments in design to account for the increase in dead load.

Since the girder is only a small part of the dead load and an even smaller part of the total load, the required adjustment will be minor. In this case, none is necessary.

We have reached the end of that part of the calculations we planned to include in this case study. The work is not complete, but a direction charted. Before leaving the case study itself, we should attempt to envision the rest of the process to see what has arisen from the calculation process that should be looked at further, to make some judgments on whether we are proceeding on the right track.

20.3 RESULTS AND IMPLICATIONS OF CALCULATIONS

Recall our earlier stated purposes. We were to investigate the problem of plate girder design for a family of simple-span two-lane highway bridges within a given range. We were to develop, in one group of sheets, some general considerations common to the range. We were also to use these sheets to identify some choices that might differ in different parts of the range. We were then to select a particular girder to design. In the process of that design, we hoped to establish a design procedure that would use the information from the first set of calculations and lend itself to a high degree of automatic calculation. A designer could then turn out with minimum further effort a set of similar designs for different span lengths and both interior and exterior girders.

The fact that we were able to do the second part of the calculations from the data in the first part is an indication of some success in the initial definition of the problem. The fact that the weight of the girder we designed in detail turned out to be closer than we could have dared to hope to the weight predicted in the first part strengthens our confidence in our earlier evaluation of the problem. Similarly, our early conclusion that we could justify the use of high-strength steel seems to be validated to some degree. The usable flexural stress as limited by fatigue and other considerations is higher than could be provided by A36 steel, but not so much higher that a change in relative costs of the steels could not change our choice. The curves on sheet 9 show the effect of span length on that decision and also show that the advantages of high-strength steels can be expected to be much greater for the outer girders. Both differences are caused by the mix of dead and live load.

We have found that the use of a "compact section" by AISC criteria is not feasible and probably would not be for the entire range.

We have found it advantageous to use a thicker web than necessary for strictly shear stress considerations and that use of the higher allowable stresses permitted for tension field action requires additional cost of stiffening, which makes it disadvantageous. This conclusion is somewhat tentative and may not apply at different span lengths. We have not found any advantage, in this girder, in varying the thickness of web along the length of the girder. There may be such advantage for other spans. Since the web requires splicing in any case, the cost of the splice would not be an offsetting disadvantage.

We have found that considerable steel can be saved in the flanges by using smaller areas where the moment demand is less. We have also recognized a limit on the amount we can save in that manner, a limit set by the cost of welding splices. On the other hand, by assuming constant width of flange, we may have set an unnecessary constraint on our ability to reduce the total flange weight and cost. Relaxing this constraint may be beneficial, particularly for longer spans.

We have found that the minimum fillet weld sizes in AISC table 1.17.2A force us to use larger welds than can be justified by stress calculation. The form of the table is somewhat arbitrary, but we recognize that it arose from concern about the effectiveness of the welding procedure when small welds are made to heavy plates that dissipate heat rapidly. We speculate that proper control of preheat and interpass temperature may serve better than the arbitrary table and allow smaller welds. With the automatic processes that will no doubt be used in fabricating the girders, the cost of superior heat control may be justified by the saving in welding. Since we will be designing a large number of similar girders and many more are designed by others, there seems reason to undertake a search for pertinent existing research, and possibly to sponsor new research on the matter. If research supports alternatives to the arbitrary table, we might seek precedent in practice for deviating from the specification or propose a change in the specification.

The next stage of the process is to design the family of girders:

- For both alternative arrangements at, say, 10-foot increments of span.
- For steel of A558, Grade 50, for the entire range.
- For A36 steel for at least the shorter spans up to the limit where its use is clearly not justifiable.
- Searching for a possible span range where tension field action in the web is justifiable.
- Searching for further opportunities to vary flange and web areas along the span.

The design procedure lends itself to automatic calculation. The decision between alternative designs and the search for opportunities for improvement requires close scrutiny of the process and application of the designer's judgment.

Cost investigation should include not only average prices of steel, but the fabrication cost of the girder and, separately, of attachments such as stiffeners, gussets, and detail pieces. Recognition of existing fabrication techniques may lead to advantageous design adjustments. Conversely, there may be sufficient volume in the total work so that well thought out design may encourage new, more efficient fabrication processes.

In producing the final series of designs, the designer should keep in mind some of the secondary criteria set at the beginning. All designs are to be sufficiently similar so that fabricators will have incentives to develop efficient, repetitive processes, and quality-control problems will be minimized.

20.4 FURTHER CONSIDERATIONS

Some preconceptions went into the plate girder design that we should note. One has already been discussed. We have ignored composite action between the steel and concrete deck. This was done for the purposes of the book and is not necessarily recommended for practice. The reader may refer to Chapter 14 and investigate the design changes and cost implications that would result if composite design were used.

We also assumed, somewhat arbitrarily, that the depth of web would be constant for the entire length of span. To some extent, this constraint was set to simplify shop procedures. It is also consistent with some current thinking about the esthetics of the design of short-span bridges. Sometimes, as in the San Francisco BART System, designers have extended this concept of beauty by establishing a constant depth for all girders in a family of spans. When a series of varying length spans exists along a section of right of way, the eye follows one continuous line. Although both of these concepts may be justifiable and may, under the right circumstances, lead to least-cost design, they do not represent the least-weight solutions. The shear stresses we were able to use were far below the limit of $0.40F_y$ available from the inherent strength of the steel. The flanges are understressed through most of their length. A least-weight solution might emerge by varying the depth from a minimum set by shear requirements at the ends to a maximum set by moment requirements in the midspan region. The urgency of the least-weight solution would become greater for a country with a more limited primary steel industry than it appears to be for the United States. However, time, energy, and resource needs may change the present situation and demand new creative approaches by designers. In design of long-span plate girder bridges, the constant-depth solution is less common.

It is interesting to note, from our calculations, the increasing relative importance of dead load to live load with increasing span. This adds impetus to the concept of load factor design, which is gaining currency in U.S. bridge practice. In this approach a more consistent safety factor is sought by applying smaller "load factors" to the more predictable dead load than to the less predictable live, impact, and other loadings.

PROBLEMS

If the problems are done as class assignments, different students may be assigned different spans. The results may be integrated into a class discussion on the effects of span on the various elements of the design, for example:

Dead load versus total load and effect on the design.
Weight of web versus weight of flanges.
Span/depth ratios.
Tension field versus no tension field.

20.1. Do designs for interior plate girders for one or more spans in the range from 60 to 130 ft in increments of 10 ft. Use A36 steel and A588 and compare results. All decisions (e.g., same or varying web thickness? tension field action or no? number of flange splices?) may be different for different spans.

20.2. Do designs for exterior plate girders similar to Problem 20.1. If you do the same span(s) as in Problem 20.1, you can test the premise that the interior and exterior girders are close enough to being the same so that it would be best to make them identical.

20.3. Redesign the 90-ft-span girder (or other) as a composite girder. See Chapter 14. Compare results.

20.4. Write a computer program suitable for a personal office computer for designing plate girders for the family of designs that are the subject of this study.

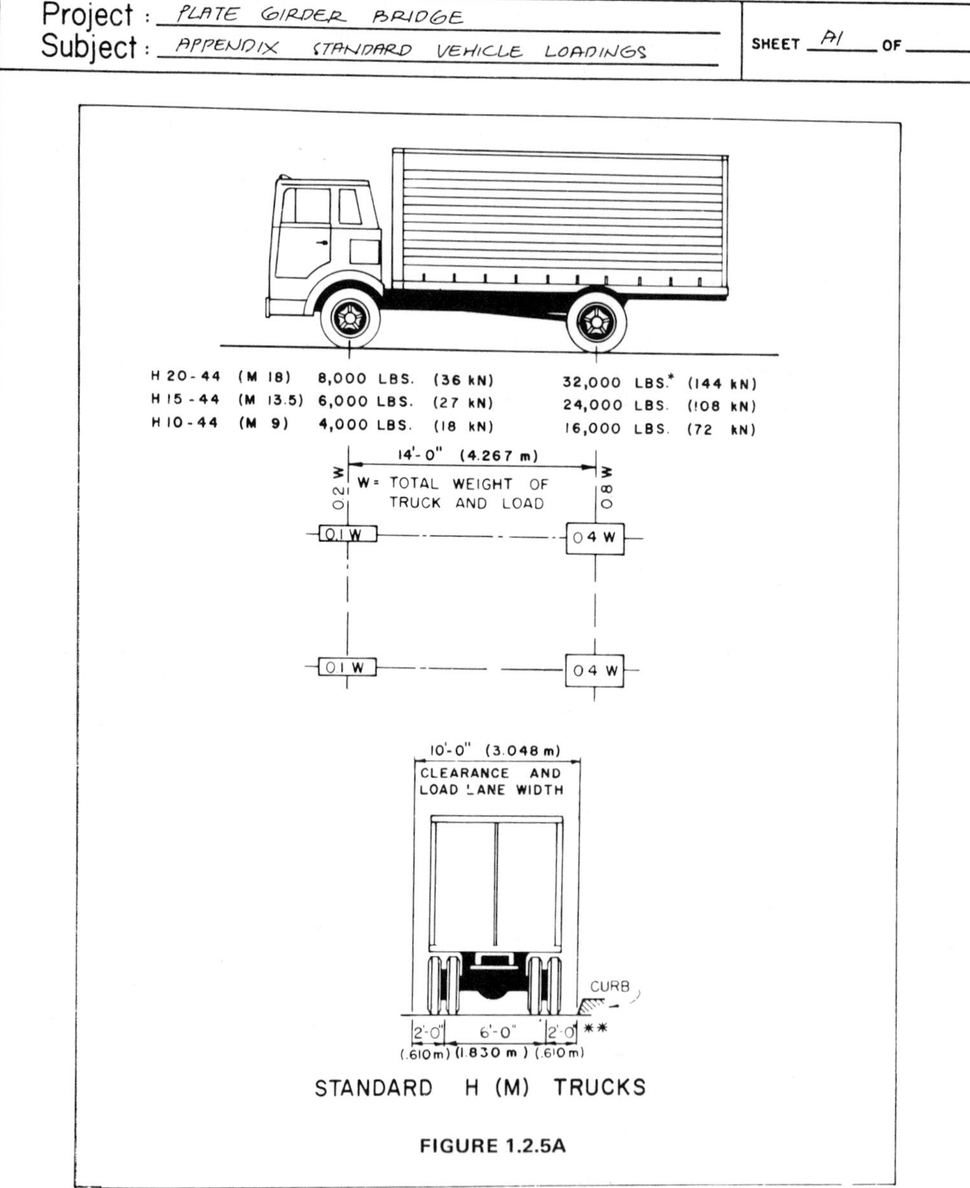

STANDARD H (M) TRUCKS

FIGURE 1.2.5A

*In the design of timberfloors and orthotropic steel decks (excluding transverse beams) for H20 (M 18) loading, one axle load of 24,000 pounds (108 kN) or two axle loads of 16,000 pounds (72 kN) each spaced 4 feet (1.219 m) apart may be used, whichever produces the greater stress, instead of the 32,000 pound (144 kN) axle shown.

**For slab design, the center line of wheels shall be assumed to be 1 foot (.305m) from face of curb. (See Art. 1.3.2(B))

Project : PLATE GIRDER BRIDGE
Subject : APPENDIX - STANDARD VEHICLE LOADINGS
SHEET A2 OF

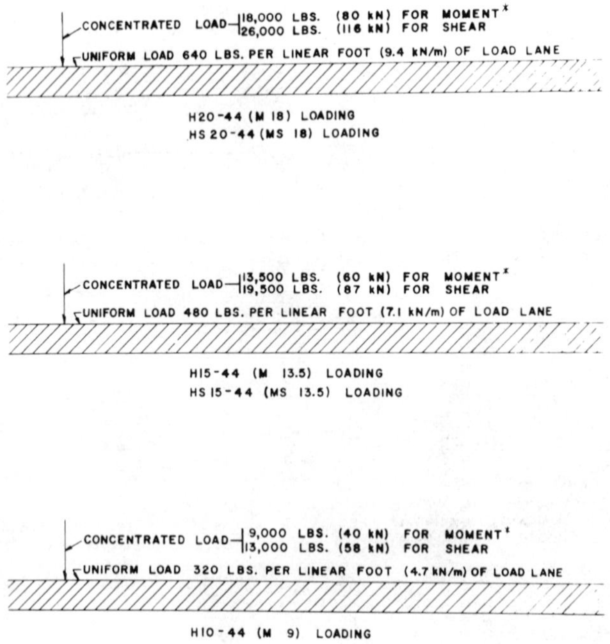

CONCENTRATED LOAD — 18,000 LBS. (80 kN) FOR MOMENT*
26,000 LBS. (116 kN) FOR SHEAR
UNIFORM LOAD 640 LBS. PER LINEAR FOOT (9.4 kN/m) OF LOAD LANE

H20-44 (M 18) LOADING
HS 20-44 (MS 18) LOADING

CONCENTRATED LOAD — 13,500 LBS. (60 kN) FOR MOMENT*
19,500 LBS. (87 kN) FOR SHEAR
UNIFORM LOAD 480 LBS. PER LINEAR FOOT (7.1 kN/m) OF LOAD LANE

H15-44 (M 13.5) LOADING
HS 15-44 (MS 13.5) LOADING

CONCENTRATED LOAD — 9,000 LBS. (40 kN) FOR MOMENT*
13,000 LBS. (58 kN) FOR SHEAR
UNIFORM LOAD 320 LBS. PER LINEAR FOOT (4.7 kN/m) OF LOAD LANE

H10-44 (M 9) LOADING

H (M) LANE AND HS (MS) LANE LOADINGS

FIGURE 1.2.5B

*For the loading of continuous spans involving lane loading refer to Article 1.2.8(C) which provides for an additional concentrated load.

Project : PLATE GIRDER BRIDGE Subject : APPENDIX - STANDARD VEHICLE LOADINGS	SHEET A3 OF ____

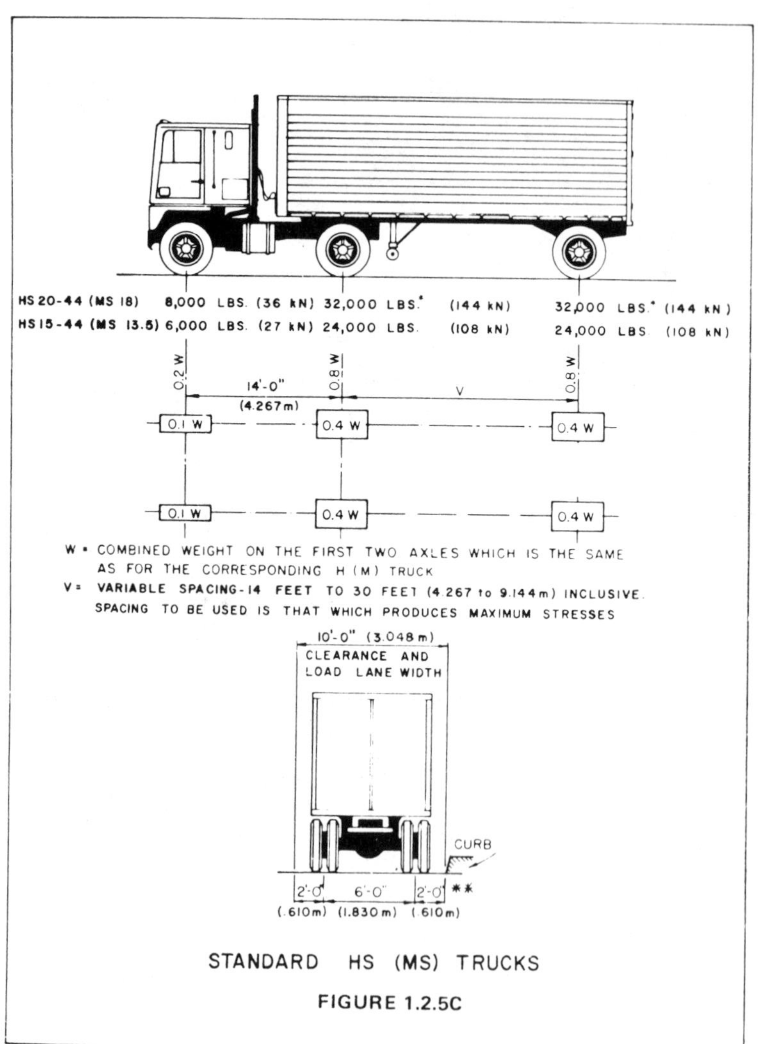

STANDARD HS (MS) TRUCKS

FIGURE 1.2.5C

*In the design of timber floors and orthotropic steel decks (excluding transverse beams) for HS20 (MS18) loading, one axle load of 24,000 pounds (108 kN) or two axle loads of 16,000 pounds (72 kN) each, spaced 4 feet (1.219 m) apart may be used, whichever produces the greater stress, instead of the 32,000 pound (144 kN) axle shown.

**For slab design, the center line of wheels shall be assumed to be 1 foot (.305 m) from face of curb. (See Art. 1.3.2 (B))

Project : PLATE GIRDER BRIDGE

Subject : APPENDIX - TABLE OF LIVE LOADS, SHEARS & MOMENTS, SIMPLE SPANS

SHEET A4 **OF** ______

452 HIGHWAY BRIDGES App. A

LOADING–HS 15-44(MS 13.5)

TABLE OF MAXIMUM MOMENTS, SHEARS AND REACTIONS.–SIMPLE SPANS, ONE LANE

Spans in feet (m); moments in thousands of foot-pounds (kNm); shears and reactions in thousands of pounds (kN).

These values are subject to specification reduction for loading of multiple lanes.

Impact not included.

Span	Moment	End shear and end reaction (a)	Span	Moment	End shear and end reaction (a)
1(.305)	6.0(b)(8.13)	24.0(b)(106.75)	42(12.802)	364.0(b)(493.51)	42.0(b)(186.82)
2(.610)	12.0(b)(16.27)	24.0(b)(106.75)	44(13.411)	390.7(b)(529.72)	42.5(b)(189.04)
3(.914)	18.0(b)(24.40)	24.0(b)(106.75)	46(14.021)	417.4(b)(565.92)	43.0(b)(191.26)
4(1.219)	24.0)b)(32.54)	24.0(b)(106.75)	48(14.630)	444.1(b)(602.12)	43.5(b)(193.49)
5(1.524)	30.0(b)(40.67)	24.0(b)(106.75)	50(15.240)	470.9(b)(638.45)	43.9(b)(195.27)
6(1.829)	36.0(b)(48.81)	24.0(b)(106.75)	52(15.850)	497.7(b)(674.79)	44.3(b)(197.05)
7(2.134)	42.0(b)(56.94)	24.0(b)(106.75)	54(16.459)	524.5(b)(711.13)	44.7(b)(198.83)
8(2.438)	48.0(b)(65.08)	24.0(b)(106.75)	56(17.069)	551.3(b)(747.46)	45.0(b)(200.16)
9(2.743)	54.0(b)(73.21)	24.0(b)(106.75)	58(17.678)	578.1(b)(783.80)	45.3(b)(201.49)
10(3.048)	60.0(b)(81.35)	24.0(b)(106.75)	60(18.288)	604.9(b)(820.13)	45.6(b)(202.83)
11(3.353)	66.0(b)(89.48)	24.0(b)(106.75)	62(18.898)	631.8(b)(856.61)	45.9(b)(204.16)
12(3.658)	72.0(b)(97.62)	24.0(b)(106.75)	64(19.507)	658.6(b)(892.94)	46.1(b)(205.05)
13(3.962)	78.0(b)(105.75)	24.0(b)(106.75)	66(20.117)	685.5(b)(929.42)	46.4(b)(206.39)
14(4.267)	84.0(b)(113.89)	24.0(b)(106.75)	68(20.726)	712.3(b)(965.75)	46.6(b)(207.28)
15(4.572)	90.0(b)(122.02)	25.6(b)(113.87)	70(21.336)	739.2(b)(1002.22)	46.8(b)(208.17)
16(4.877)	96.0(b)(130.16)	27.0(b)(120.10)	75(22.860)	806.3(b)(1093.19)	47.3(b)(210.39)
17(5.182)	102.0(b)(138.29)	28.2(b)(125.43)	80(24.384)	873.7(b)(1184.58)	47.7(b)(212.17)
18(5.486)	108.0(b)(146.43)	29.3(b)(130.33)	85(25.908)	941.0(b)(1275.83)	48.1(b)(213.95)
19(5.791)	114.0(b)(154.56)	30.3(b)(134.77)	90(27.432)	1,008.3(b)(1367.07)	48.4(b)(215.28)
20(6.096)	120.0(b)(162.70)	31.2(b)(138.78)	95(28.956)	1,074.9(b)(1457.37)	48.7(b)(216.62)
21(6.401)	126.0(b)(170.83)	32.0(b)(142.34)	100(30.480)	1,143.0(b)(1549.70)	49.0(b)(217.95)
22(6.706)	132.0(b)(178.97)	32.7(b)(145.45)	110(33.528)	1,277.7(b)(1732.33)	49.4(b)(219.73)
23(7.010)	138.0(b)(187.10)	33.4(b)(148.56)	120(36.576)	1,412.5(b)(1915.10)	49.8(b)(221.51)
24(7.315)	144.5(b)(195.91)	34.0(b)(151.23)	130(39.624)	1,547.3(b)(2097.86)	50.7 (225.51)
25(7.620)	155.5(b)(210.83)	34.6(b)(153.90)	140(42.672)	1,682.1(b)(2280.63)	53.1 (236.19)
26(7.925)	166.6(b)(225.88)	35.1(b)(156.12)	150(45.720)	1,856.3 (2516.81)	55.5 (246.86)
27(8.230)	177.8(b)(241.07)	35.6(b)(158.35)	160(48.768)	2,076.0 (2814.68)	57.9 (257.54)
28(8.534)	189.0(b)(256.25)	36.0(b)(160.13)	170(51.816)	2,307.8 (3128.96)	60.3 (268.21)
29(8.839)	200.3(b)(271.57)	36.6(b)(162.80)	180(54.864)	2,551.5 (3459.37)	62.7 (278.89)
30(9.144)	211.6(b)(286.89)	37.2(b)(165.47)	190(57.912)	2,807.3 (3806.19)	65.1 (289.56)
31(9.449)	223.0(b)(302.35)	37.7(b)(167.69)	200(60.960)	3,075.0 (4169.14)	67.5 (300.24)
32(9.754)	234.4(b)(317.81)	38.3(b)(170.36)	220(67.056)	3,646.5 (4943.98)	72.3 (321.59)
33(10.058)	245.8(b)(333.26)	38.7(b)(172.14)	240(73.152)	4,266.0 (5783.91)	77.1 (342.94)
34(10.363)	257.7(b)(349.39)	39.2(b)(174.36)	260(79.248)	4,933.5 (6688.93)	81.9 (364.29)
35(10.668)	270.9(b)(367.29)	39.6(b)(176.14)	280(85.344)	5,649.0 (7659.02)	86.7 (385.64)
36(10.973)	284.2(b)(385.32)	40.0(b)(177.92)	300(91.440)	6,412.5 (8694.18)	91.5 (406.99)
37(11.278)	297.5(b)(403.36)	40.4(b)(179.70)			
38(11.582)	310.7(b)(421.25)	40.7(b)(181.03)			
39(11.887)	324.0(b)(439.28)	41.1(b)(182.81)			
40(12.192)	337.4(b)(457.45)	41.4(b)(184.15)			

(a) Concentrated load is considered placed at the support. Loads used are those stipulated for shear.

(b) Maximum value determined by Standard Truck Loading. Otherwise the Standard Lane Loading governs.

App. A 453

LOADING–HS 20-44 (MS18)

TABLE OF MAXIMUM MOMENTS, SHEARS AND REACTIONS.–SIMPLE SPANS, ONE LANE

Spans in feet (m); moments in thousands of foot-pounds (kNm); shears and reactions in thousands of pounds (kN).

These values are subject to specification reduction for loading of multiple lanes.

Impact not included.

Span	Moment	End Shear and end reaction (a)	Span	Moment	End shear and end reaction (a)
1(.305)	8.0(b)(10.85)	32.0(b)(142.34)	42(12.802)	485.3(b)(657.99)	56.0(b)(249.09)
2(.610)	16.0(b)(21.70)	32.0(b)(142.34)	44(13.411)	520.9(b)(706.25)	56.7(b)(252.20)
3(.914)	24.0(b)(32.55)	32.0(b)(142.34)	46(14.021)	556.5(b)(754.51)	57.3(b)(254.87)
4(1.219)	32.0(b)(43,40)	32.0(b)(142.34)	48(14.630)	592.1(b)(802.78)	58.0(b)(257.98)
5(1.524)	40.0(b)(54.25)	32.0(b)(142.34)	50(15.240)	627.9(b)(851.32)	58.5(b)(260.21)
6(1.829)	48.0(b)(65.08)	32.0(b)(142.34)	52(15.850)	663.6(b)(889.72)	59.1(b)(262.88)
7(2.134)	56.0(b)(75.93)	32.0(b)(142.34)	54(16.459)	699.3(b)(948.12)	59.6(b)(265.10)
8(2.438)	64.0(b)(86.77)	32.0(b)(142.34)	56(17.069)	735.1(b)(996.67)	60.0(b)(266.88)
9(2.743)	72.0(b)(97.62)	32.0(b)(142.34)	58(17.678)	770.8(b)(1045.06)	60.4(b)(268.66)
10(3.048)	80.0(b)(108.47)	32.0(b)(142.34)	60(18.288)	806.5(b)(1093.46)	60.8(b)(270.44)
11(3.353)	88.0(b)(119.31)	32.0(b)(142.34)	62(18.898)	842.4(b)(1142.14)	61.2(b)(272.22)
12(3.658)	96.0(b)(130.16)	32.0(b)(142.34)	64(19.507)	878.1(b)(1190.54)	61.5(b)(273.55)
13(3.962)	104.0(b)(141.00)	32.0(b)(142.34)	66(20.117)	914.0(b)(1239.21)	61.9(b)(275.33)
14(4.267)	112.0(b)(151.85)	32.0(b)(142.34)	68(20.726)	949.7(b)(1287.62)	62.1(b)(276.22)
15(4.572)	120.0(b)(162.70)	34.1(b)(151.68)	70(21.336)	985.6(b)(1336.29)	62.4(b)(277.56)
16(4.877)	128.0(b)(173.55)	36.0(b)(160.13)	75(22.860)	1,075.1(b)(1457.65)	63.1(b)(280.67)
17(5.182)	136.0(b)(184.39)	37.7(b)(167.69)	80(24.384)	1,164.9(b)(1579.39)	63.6(b)(282.89)
18(5.486)	144.0(b)(195.23)	39.1(b)(173.92)	85(25.908)	1,254.7(b)(1701.14)	64.1(b)(285.12)
19(5.791)	152.0(b)(206.08)	40.4(b)(179.70)	90(27.432)	1,344.4(b)(1822.76)	64.5(b)(286.90)
20(6.096)	160.0(b)(216.93)	41.6(b)(185.04)	95(28.956)	1,434.1(b)(1944.38)	64.9(b)(288.68)
21(6.401)	168.0(b)(227.78)	42.7(b)(189.93)	100(30.480)	1,524.0(b)(2066.27)	65.3(b)(290.45)
22(6.706)	176.0(b)(238.62)	43.6(b)(193.93)	110(33.528)	1,703.6(b)(2309.77)	65.9(b)(293.12)
23(7.010)	184.0(b)(249.47)	44.5(b)(197.94)	120(36.576)	1,883.3(b)(2553.41)	66.4(b)(295.35)
24(7.315)	192.7(b)(261.27)	45.3(b)(201.49)	130(39.624)	2,063.1(b)(2797.20)	67.6 (300.68)
25(7.620)	207.4(b)(281.19)	46.1(b)(205.05)	140(42.672)	2,242.8(b)(3040.82)	70.8 (314.92)
26(7.925)	222.2(b)(301.26)	46.8(b)(208.17)	150(45.720)	2,475.1 (3355.79)	74.0 (329.15)
27(8.230)	237.0(b)(321.33)	47.4(b)(210.84)	160(48.768)	2,768.0 (3752.91)	77.2 (343.39)
28(8.534)	252.0(b)(341.66)	48.0(b)(213.50)	170(51.816)	3,077.1 (4171.99)	80.4 (357.62)
29(8.839)	267.0(b)(362.00)	48.8(b)(217.06)	180(54.864)	3,402.1 (4612.63)	83.6 (371.85)
30(9.144)	282.1(b)(382.48)	49.6(b)(220.62)	190(57.912)	3,743.1 (5074.97)	86.8 (386.09)
31(9.449)	297.3(b)(403.09)	50.3(b)(223.73)	200(60.960)	4,100.0 (5558.85)	90.0 (400.32)
32(9.754)	312.5(b)(423.69)	51.0(b)(226.85)	220(67.056)	4,862.0 (6591.99)	96.4 (428.79)
33(10.058)	327.8(b)(444.44)	51.6(b)(229.52)	240(73.152)	5,688.0 (7711.89)	102.8 (457.25)
34(10.363)	343.5(b)(465.73)	52.2(b)(232.19)	260(79.248)	6,578.0 (8918.57)	109.2 (485.72)
35(10.668)	361.2(b)(489.72)	52.8(b)(234.85)	280(85.344)	7,532.0 (10212.02)	115.6 (514.19)
36(10.973)	378.9(b)(513.72)	53.3(b)(237.08)	300(91.440)	8,550.0 (11592.24)	122.0 (542.66)
37(11.278)	396.6(b)(537.72)	53.8(b)(239.30)			
38(11.582)	414.3(b)(561.72)	54.3(b)(241.53)			
39(11.887)	432.1(b)(585.85)	54.8(b)(243.75)			
40(12.192)	449.8(b)(609.85)	55.2(b)(245.53)			

(a) Concentrated load is considered placed at the support. Loads used are those stipulated for shear.

(b) Maximum value determined by Standard Truck Loading. Otherwise the Standard Lane Loading governs.

APPENDIX A

Excerpt from Uniform Building Code

Chapter 23, *Uniform Building Code* (1982). Reproduced from the *Uniform Building Code*, 1982 edition, copyright 1982, with permission of the publisher, the International Conference of Building Officials.

Part V

ENGINEERING REGULATIONS—QUALITY AND DESIGN OF THE MATERIALS OF CONSTRUCTION

Chapter 23

GENERAL DESIGN REQUIREMENTS

Scope

Sec. 2301. This chapter prescribes general design requirements applicable to all structures regulated by this code.

Definitions

Sec. 2302. The following definitions give the meaning of certain terms used in this chapter:

DEAD LOAD is the vertical load due to the weight of all permanent structural and nonstructural components of a building, such as walls, floors, roofs and fixed service equipment.

LIVE LOAD is the load superimposed by the use and occupancy of the building not including the wind load, earthquake load or dead load.

LOAD DURATION is the period of continuous application of a given load, or the aggregate of periods of intermittent application of the same load.

Design Methods

Sec. 2303. (a) **General.** All buildings and portions thereof shall be designed and constructed to sustain, within the stress limitations specified in this code, all dead loads and all other loads specified in this chapter or elsewhere in this code. Impact loads shall be considered in the design of any structure where impact loads occur.

> **EXCEPTION:** Unless otherwise required by the building official, buildings or portions thereof which are constructed in accordance with the conventional framing requirements specified in Chapter 25 of this code shall be deemed to meet the requirements of this section.

(b) **Rationality.** Any system or method of construction to be used shall be based on a rational analysis in accordance with well-established principles of mechanics. Such analysis shall result in a system which provides a complete load path capable of transferring all loads and forces from their point of origin to the load-resisting elements. The analysis shall include but not be limited to the following:

1. **Distribution of horizontal shear.** The total lateral force shall be distributed to the various vertical elements of the lateral force-resisting system in proportion to their rigidities considering the rigidity of the horizontal bracing system or diaphragm. Rigid elements that are assumed not to be part of the lateral force-

resisting system may be incorporated into buildings, provided that their effect on the action of the system is considered and provided for in the design.

2. **Horizontal torsional moments.** Provision shall be made for the increased forces induced on resisting elements of the structural system resulting from torsion due to eccentricity between the center of application of the lateral forces and the center of rigidity of the lateral force-resisting system. Forces shall not be decreased due to torsional effects. For accidental torsion requirements for seismic design, see Section 2312 (e) 4.

3. **Stability against overturning.** Every building or structure shall be designed to resist the overturning effects caused by the lateral forces specified in this chapter. See Section 2311 (e) for wind and Section 2312 (f) for seismic.

4. **Anchorage.** Anchorage of the roof to walls and columns, and of walls and columns to foundations, shall be provided to resist the uplift and sliding forces which result from the application of the prescribed forces. For additional requirements for masonry or concrete walls, see Section 2310.

(c) **Critical Distribution of Live Loads.** Where structural members are arranged so as to create continuity, the loading conditions which would cause maximum shear and bending moments along the member shall be investigated.

(d) **Stress Increases.** All allowable stresses and soil-bearing values specified in this code for working stress design may be increased one-third when considering wind or earthquake forces either acting alone or when combined with vertical loads. No increase will be allowed for vertical loads acting alone.

(e) **Load Factors.** Load factors for ultimate strength design of concrete and plastic design of steel shall be as indicated in the appropriate chapters on the materials.

(f) **Load Combinations.** Every building component shall be provided with strength adequate to resist the most critical effect resulting from the following combination of loads (floor live load shall not be included where its inclusion results in lower stresses in the member under investigation):

1. Dead plus floor live plus roof live (or snow).[1]
2. Dead plus floor live plus wind (or seismic).
3. Dead plus floor live plus wind plus snow/2.[1]
4. Dead plus floor live plus snow plus wind/2.[1]
5. Dead plus floor live plus snow[2] plus seismic.

[1]Crane hook loads need not be combined with roof live load nor with more than three fourths of the snow load or one-half wind load.

[2]Snow loads over 30 psf may be reduced 75 percent upon approval of the building official, and snow loads 30 psf or less need not be combined with seismic.

Floor Design

Sec. 2304. (a) **General.** Floors shall be designed for the unit loads set forth in Table No. 23-A. These loads shall be taken as the minimum live loads in pounds per square foot of horizontal projection to be used in the design of buildings for the occupancies listed, and loads at least equal shall be assumed for uses not listed in this section but which create or accommodate similar loadings.

EXCEPTION: In designing floors to be used for industrial or commercial purposes, the actual live load caused by the use to which the building or part of the building is to be put shall be used in the design of such building or part thereof, and special provision shall be made for machine or apparatus loads when such machine or apparatus would cause a greater load than specified for such use.

(b) **Distribution of Uniform Floor Loads.** Where uniform floor loads are involved, consideration may be limited to full dead load on all spans in combination with full live load on adjacent spans and on alternate spans.

(c) **Concentrated Loads.** Provision shall be made in designing floors for a concentrated load as set forth in Table No. 23-A placed upon any space 2½ feet square, wherever this load upon an otherwise unloaded floor would produce stresses greater than those caused by the uniform load required therefor.

Provision shall be made in areas where vehicles are used or stored for concentrated loads consisting of two or more loads spaced 5 feet nominally on center without uniform live loads. Each load shall be 40 percent of the gross weight of the maximum size vehicle to be accommodated. The condition of concentrated or uniform live load producing the greater stresses shall govern. Garages for the storage of private pleasure cars shall have the floor system designed for a concentrated wheel load of not less than 2000 pounds without uniform live loads. The condition of concentrated or uniform live load producing the greater stresses shall govern.

Provision shall be made for special vertical and lateral loads as set forth in Table No. 23-B.

(d) **Partition Loads.** Floors in office buildings and in other buildings where partition locations are subject to change shall be designed to support, in addition to all other loads, a uniformly distributed dead load equal to 20 pounds per square foot.

(e) **Live Loads Posted.** The live loads for which each floor or part thereof of a commercial or industrial building is or has been designed shall have such designed live loads conspicuously posted by the owner in that part of each story in which they apply, using durable metal signs, and it shall be unlawful to remove or deface such notices. The occupant of the building shall be responsible for keeping the actual load below the allowable limits.

Roof Design

Sec. 2305. (a) **General.** Roofs shall sustain, within the stress limitations of this code, all "dead loads" plus unit "live loads" as set forth in Table No. 23-C. The live loads shall be assumed to act vertically upon the area projected upon a horizontal plane.

(b) **Distribution of Loads.** Where uniform roof loads are involved in the design of structural members arranged so as to create continuity, consideration may be limited to full dead loads on all spans in combination with full live loads on adjacent spans and on alternate spans.

EXCEPTION: Alternate span loading need not be considered where the uniform roof live load is 20 pounds per square foot or more and the provisions of Section 2305 (d) are met.

(c) **Unbalanced Loading.** Unbalanced loads shall be used where such loading will result in larger members or connections. Trusses and arches shall be designed to resist the stresses caused by unit live loads on one half of the span if such loading results in reverse stresses, or stresses greater in any portion than the stresses produced by the required unit live load upon the entire span. For roofs whose structure is composed of a stressed shell, framed or solid, wherein stresses caused by any point loading are distributed throughout the area of the shell, the requirements for unbalanced unit live load design may be reduced 50 percent.

(d) **Snow Loads.** Snow loads full or unbalanced shall be considered in place of loads set forth in Table No. 23-C, where such loading will result in larger members or connections.

Potential accumulation of snow at valleys, parapets, roof structures and offsets in roofs of uneven configuration shall be considered. Where snow loads occur, the snow loads shall be determined by the building official.

Snow loads in excess of 20 pounds per square foot may be reduced for each degree of pitch over 20 degrees by R_s as determined by the following formula:

$$R_s = \frac{S}{40} - \frac{1}{2}$$

WHERE:

R_s = Snow load reduction in pounds per square foot per degree of pitch over 20 degrees.

S = Total snow load in pounds per square foot.

(e) **Special-purpose Roofs.** Roofs to be used for special purposes shall be designed for appropriate loads as approved by the building official.

Greenhouse roof bars, purlins and rafters shall be designed to carry a 100-pound-minimum concentrated load in addition to the live load.

(f) **Water Accumulation.** All roofs shall be designed with sufficient slope or camber to assure adequate drainage after the long-time deflection from dead load or shall be designed to support maximum loads including possible ponding of water due to deflection. See Section 2307 for deflection criteria.

Reduction of Live Loads

Sec. 2306. The design live load determined using the unit live loads as set forth in Table No. 23-A for floors and Table No. 23-C, Method 2, for roofs may be reduced on any member supporting more than 150 square feet, including flat slabs, except for floors in places of public assembly and for live loads greater than 100 pounds per square foot, in accordance with the following formula:

$$R = r\,(A - 150) \quad \ldots\ldots\ldots\ldots\ldots\ldots\ldots (6\text{-}1)$$

The reduction shall not exceed 40 percent for members receiving load from one level only, 60 percent for other members, nor R as determined by the following formula:

$$R = 23.1\,(1 + D/L) \quad \ldots\ldots\ldots\ldots\ldots\ldots\ldots (6\text{-}2)$$

WHERE:

R = Reduction in percent.

r = Rate of reduction equal to .08 percent for floors. See Table No. 23-C for roofs.

A = Area of floor or roof supported by the member.

D = Dead load per square foot of area supported by the member.

L = Unit live load per square foot of area supported by the member.

For storage live loads exceeding 100 pounds per square foot, no reduction shall be made, except that design live loads on columns may be reduced 20 percent.

The live load reduction shall not exceed 40 percent in garages for the storage of private pleasure cars having a capacity of not more than nine passengers per vehicle.

Deflection

Sec. 2307. The deflection of any structural members shall not exceed the values set forth in Table No. 23-D, based upon the factors set forth in Table No. 23-E. The deflection criteria representing the most restrictive condition shall apply. Deflection criteria for materials not specified shall be developed in a manner consistent with the provisions of this section. See Section 2305 (f) for camber requirements. Span tables for light wood frame construction as specified in Sections 2517 (d) and 2517 (h) 2 shall conform to the design criteria contained therein, except that where the dead load exceeds 50 percent of the live load, Table No. 23-D shall govern. (For aluminum, see Section 2803.)

Special Design

Sec. 2308. (a) **General.** In addition to the design loads specified in this chapter, the design of all structures shall consider the special loads set forth in Table No. 23-B and in this section.

(b) **Retaining Walls.** Retaining walls shall be designed to resist the lateral pressure of the retained material in accordance with accepted engineering practice. Walls retaining drained earth may be designed for pressure equivalent to that exerted by a fluid weighing not less than 30 pounds per cubic foot and having a depth equal to that of the retained earth. Any surcharge shall be in addition to the equivalent fluid pressure.

(c) **Heliport and Helistop Landing Areas.** In addition to other design requirements of this chapter, heliport and helistop landing or touchdown areas shall be designed for the maximum stress induced by the following:

1. Dead load plus actual weight of the helicopter.
2. Dead load plus a single concentrated impact load covering 1 square foot of 0.75 times the fully loaded weight of the helicopter if it is equipped with hydraulic-type shock absorbers, or 1.5 times the fully loaded weight of the helicopter if it is equipped with a rigid or skid-type landing gear.
3. The dead load plus a uniform live load of 100 pounds per square foot. The required live load may be reduced in accordance with the formula in Section 2306.

Walls and Structural Framing

Sec. 2309. (a) **General.** Walls and structural framing shall be erected true and plumb in accordance with the design.

(b) **Interior Walls.** Interior walls, permanent partitions, and temporary partitions which exceed 6 feet in height shall be designed to resist all loads to which they are subjected but not less than a force of 5 pounds per square foot applied perpendicular to the walls. The deflection of such walls under a load of 5 pounds per square foot shall not exceed 1/240 of the span for walls with brittle finishes and 1/120 of the span for walls with flexible finishes. See Table No. 23-J for earthquake design requirements where such requirements are more restrictive.

> **EXCEPTION:** Flexible, folding or portable partitions are not required to meet the load and deflection criteria but must be anchored to the supporting structure to meet the provisions of this code.

Anchorage of Concrete or Masonry Walls

Sec. 2310. Concrete or masonry walls shall be anchored to all floors and roofs which provide lateral support for the wall. Such anchorage shall provide a positive direct connection capable of resisting the horizontal forces specified in this chapter or a minimum force of 200 pounds per lineal foot of wall, whichever is greater. Walls shall be designed to resist bending between anchors where the anchor spacing exceeds 4 feet. Required anchors in masonry walls of hollow units or cavity walls shall be embedded in a reinforced grouted structural element of the wall. See Sections 2312 (j) 2 C and 2312 (j) 3 A.

Wind Design

Sec. 2311. (a) **General.** Every building or structure and every portion thereof shall be designed and constructed to resist the wind effects determined in accordance with the requirements of this section. Wind shall be assumed to come from any horizontal direction. No reduction in wind pressure shall be taken for the shielding effect of adjacent structures.

Structures sensitive to dynamic effects, such as buildings with a height-width ratio greater than five, structures sensitive to wind-excited oscillations, such as vortex shedding or icing, and buildings over 400 feet in height, shall be, and any structure may be, designed in accordance with approved national standards.

(b) **Basic Wind Speed.** The minimum basic wind speed for determining design wind pressure shall be taken from Figure No. 4, these higher values shall be the minimum basic wind speeds.

(c) **Exposure.** An exposure shall be assigned at each site for which a building or structure is to be designed. Exposure C represents the most severe exposure and has terrain which is flat and generally open, extending one-half mile or more from the site. Exposure B has terrain which has buildings, forest or surface irregularities 20 feet or more in height covering at least 20 percent of the area extending one mile or more from the site.

(d) **Design Wind Pressures.** Design wind pressures for structures or elements of structures shall be determined for any height in accordance with the following formula:

$$p = C_e C_q q_s I \qquad \text{(11-1)}$$

WHERE:

p = Design wind pressure.

C_e = Combined height, exposure and gust factor coefficient as given in Table No. 23-G.

C_q = Pressure coefficient for the structure or portion of structure under consideration as given in Table No. 23-H.

q_s = Wind stagnation pressure at the standard height of 30 feet as set forth in Table No. 23-F.

I = Importance factor as set forth in Section 2311 (h).

(e) **Primary Frames and Systems.** The primary frames or load-resisting system of every structure shall be designed for the pressures calculated using Formula (11-1) and the pressure coefficients, C_q, of either Method 1 or Method 2. In addition, design of the overall structure and its primary load-resisting system shall conform to Section 2303.

The base overturning moment for the entire structure, or for any one of its individual primary lateral resisting elements, shall not exceed two thirds of the dead-load-resisting moment. The weight of earth superimposed over footings may be used to calculate the dead-load-resisting moment.

1. **Method 1 (Normal Force Method).** Method 1 shall be used for the design of gabled rigid frames and may be used for any structure. In the Normal Force Method, the wind pressures shall be assumed to act simultaneously normal to all exterior surfaces. For pressures on leeward walls, C_e shall be evaluated at the mean roof height.

2. **Method 2 (Projected Area Method).** Method 2 may be used for any structure less than 200 feet in height except those using gabled frames. This method may be used in stability determinations for any structure less than 200 feet high. In the Projected Area Method, horizontal pressures shall be assumed to act upon the full vertical projected area of the structure, and the vertical pressures shall be assumed to act simultaneously upon the full horizontal projected area.

(f) **Elements and Components of Structures.** Design wind pressures for each element or component of a structure shall be determined from Formula (11-1) and C_q values from Table No. 23-H, and shall be applied perpendicular to the surface. For outward acting forces the value of C_e shall be obtained from Table No. 23-G based on the mean roof height and applied for the entire height of the structure. Each element or component shall be designed for the more severe of the following loadings:

1. The pressures determined using C_q values for elements and components acting over the entire tributary area of the element.

2. The pressures determined using C_q values for local areas at discontinuities such as corners, ridges and eaves. These local pressures shall be applied over a distance from a discontinuity of 10 feet or 0.1 times the least width of the structure, whichever is less.

The wind pressures from Subsections (e) and (f) need not be combined.

(g) **Miscellaneous Structures.** Greenhouses, lath houses, agricultural buildings or fences 12 feet or less in height shall be designed in accordance with Section 2311. However, three fourths of q_s, but not less than 10 pounds per square foot, may be substituted for q_s in Formula (11-1). Pressures on local areas at discontinuities need not be considered.

(h) **Importance Factor.** A factor of 1.15 shall be used for essential facilities which must be safe and usable for emergency purposes after a windstorm in order to preserve the health and safety of the general public. Such facilities shall include:

1. Hospitals and other medical facilities having surgery or emergency treatment areas.
2. Fire and police stations.
3. Municipal government disaster operation and communication centers deemed to be vital in emergencies.
4. Buildings where the primary occupancy is for assembly use for more than 300 people.

A factor of 1.0 shall be used for all other buildings.

Earthquake Regulations

Sec. 2312. (a) **General.** Every building or structure and every portion thereof shall be designed and constructed to resist stresses produced by lateral forces as provided in this section. Stresses shall be calculated as the effect of a force applied horizontally at each floor or roof level above the base. The force shall be assumed to come from any horizontal direction.

Structural concepts other than set forth in this section may be approved by the building official when evidence is submitted showing that equivalent ductility and energy absorption are provided.

Where prescribed wind loads produce higher stresses, such loads shall be used in lieu of the loads resulting from earthquake forces.

(b) **Definitions.** The following definitions apply only to the provisions of this section:

BASE is the level at which the earthquake motions are considered to be imparted to the structure or the level at which the structure as a dynamic vibrator is supported.

BOX SYSTEM is a structural system without a complete vertical load-carrying space frame. In this system the required lateral forces are resisted by shear walls or braced frames as hereinafter defined.

BRACED FRAME is a truss system or its equivalent which is provided to resist lateral forces in the frame system and in which the members are subjected primarily to axial stresses.

DUCTILE MOMENT-RESISTING SPACE FRAME is a moment-resisting space frame complying with the requirements for a ductile moment-resisting space frame as given in Section 2312 (j).

ESSENTIAL FACILITIES—See Section 2312 (k).

LATERAL FORCE-RESISTING SYSTEM is that part of the structural system assigned to resist the lateral forces prescribed in Section 2312 (d).

MOMENT-RESISTING SPACE FRAME is a vertical load-carrying space frame in which the members and joints are capable of resisting forces primarily by flexure.

SHEAR WALL is a wall designed to resist lateral forces parallel to the wall.

SPACE FRAME is a three-dimensional structural system without bearing walls, composed of interconnected members laterally supported so as to function as a complete self-contained unit with or without the aid of horizontal diaphragms or floor-bracing systems.

VERTICAL LOAD-CARRYING SPACE FRAME is a space frame designed to carry all vertical loads.

(c) **Symbols and Notations.** The following symbols and notations apply only to the provisions of this section:

C = Numerical coefficient as specified in Section 2312 (d).

C_p = Numerical coefficient as specified in Section 2312 (g) and as set forth in Table No. 23-J.

D = The dimension of the structure, in feet, in a direction parallel to the applied forces.

δ_i = Deflection at level i relative to the base, due to applied lateral forces, Σf_i, for use in Formula (12-3).

$F_i F_n F_x$ = Laterial force applied to level i, n or x, respectively.

F_p = Lateral forces on a part of the structure and in the direction under consideration.

F_t = That portion of V considered concentrated at the top of the structure in addition to F_n.

f_i = Distributed portion of a total lateral force at level i for use in Formula (12-3).

g = Acceleration due to gravity.

$h_i h_n h_x$ = Height in feet above the base to level i, n or x respectively.

I = Occupancy Importance Factor as set forth in Table No. 23-K.

K = Numerical coefficient as set forth in Table No. 23-I.

Level i

I = Level of the structure referred to by the subscript i.

i = 1 designates the first level above the base.

Level n = That level which is uppermost in the main portion of the structure.

Level x = That level which is under design consideration.

x = 1 designates the first level above the base.

N = The total number of stories above the base to level n.

S = Numerical coefficient for site-structure resonance.

T = Fundamental elastic period of vibration of the building or structure in seconds in the direction under consideration.

T_s = Characteristic site period.

V = The total lateral force or shear at the base.

W = The total dead load as defined in Section 2302 including the partition loading specified in Section 2304 (d) where applicable.

EXCEPTION: *W* shall be equal to the total dead load plus 25 percent of the floor live load in storage and warehouse occupancies. Where the design snow load is 30 psf or less, no part need be included in the value of *W*. Where the snow load is greater than 30 psf, the snow load shall be included; however, where the snow load duration warrants, the building official may allow the snow load to be reduced up to 75 percent.

$w_i w_x$ = That portion of W which is located at or is assigned to level i or x respectively.

W_p = The weight of a portion of a structure or nonstructural component.

Z = Numerical coefficient dependent upon the zone as determined by Figures No. 1, No. 2 and No. 3 in this chapter. For locations in Zone No. 1, $Z = 3/16$. For locations in Zone No. 2, $Z = 3/8$. For locations in Zone No. 3, $Z = 3/4$. For locations in Zone No. 4, $Z = 1$.

(d) **Minimum Earthquake Forces for Structures.** Except as provided in Section 2312 (g) and (i), every structure shall be designed and constructed to resist minimum total lateral seismic forces assumed to act nonconcurrently in the direction of each of the main axes of the structure in accordance with the following formula:

$$V = ZIKCSW \qquad \text{(12-1)}$$

The value of K shall be not less than that set forth in Table No. 23-I. The value of C and S are as indicated hereafter except that the product of CS need not exceed 0.14.

The value of C shall be determined in accordance with the following formula:

$$C = \frac{1}{15\sqrt{T}} \qquad \text{(12-2)}$$

The value of C need not exceed 0.12.

The period T shall be established using the structural properties and deformational characteristics of the resisting elements in a properly substantiated analysis such as the following formula:

$$T = 2\pi\sqrt{\left(\sum_{i=1}^{n} w_i\delta_i^2\right) \div \left(g\sum_{i=1}^{n} f_i\delta_i\right)} \qquad \text{(12-3)}$$

where the values of f_i represent any lateral force distributed approximately in accordance with the principles of Formulas (12-5), (12-6) and (12-7) or any other rational distribution. The elastic deflections, δ_i, shall be calculated using the applied lateral forces, f_i.

In the absence of a determination as indicated above, the value of T for buildings may be determined by the following formula:

$$T = \frac{0.05h_n}{\sqrt{D}} \qquad \text{(12-3A)}$$

Or in buildings in which the lateral force-resisting system consists of ductile moment-resisting space frames capable of resisting 100 percent of the required lateral forces and such system is not enclosed by or adjoined by more rigid elements tending to prevent the frame from resisting lateral forces:

$$T = 0.10N \qquad \text{(12-3B)}$$

The value of S shall be determined by the following formulas, but shall be not less than 1.0:

$$\text{for } T/T_s = 1.0 \text{ or less} \quad S = 1.0 + \frac{T}{T_s} - 0.5\left[\frac{T}{T_s}\right]^2 \qquad \text{(12-4)}$$

$$\text{for } T/T_s \text{ greater than 1.0 or less} \quad S = 1.2 + 0.6\frac{T}{T_s} - 0.3\left[\frac{T}{T_s}\right]^2 \qquad \text{(12-4A)}$$

WHERE:

T in Formulas (12-4) and (12-4A) shall be established by a properly substantiated analysis but T shall be not less than 0.3 second.

The range of values of T_s may be established from properly substantiated geotechnical data, in accordance with U.B.C. Standard No. 23-1, except that T_s shall not be taken as less than 0.5 second nor more than 2.5 seconds. T_s shall be that value within the range of site periods, as determined above, that is nearest to T.

When T_s is not properly established, the value of S shall be 1.5.

EXCEPTION: Where T has been established by a properly substantiated analysis and exceeds 2.5 seconds, the value of S may be determined by assuming a value of 2.5 seconds for T_s.

(e) **Distribution of Lateral Forces.** 1. **Structures having regular shapes or framing systems.** The total lateral force V shall be distributed over the height of the structure in accordance with Formulas (12-5), (12-6) and (12-7).

$$V = F_t + \sum_{i=1}^{n} F_i \qquad \text{(12-5)}$$

The concentrated force at the top shall be determined according to the following formula:

$$F_t = 0.07TV \qquad \text{(12-6)}$$

F_t need not exceed $0.25V$ and may be considered as 0 where T is 0.7 second or less. The remaining portion of the total base shear V shall be distributed over the height of the structure including level n according to the following formula:

$$F_x = \frac{(V - F_t)\, w_x h_x}{\sum_{i=1}^{n} w_i h_i} \qquad \text{(12-7)}$$

At each level designated as x, the force F_x shall be applied over the area of the building in accordance with the mass distribution on that level.

2. **Setbacks.** Buildings having setbacks wherein the plan dimension of the tower in each direction is at least 75 percent of the corresponding plan dimension of the lower part may be considered as uniform buildings without setbacks, provided other irregularities as defined in this section do not exist.

3. **Structures having irregular shapes or framing systems.** The distribution of the lateral forces in structures which have highly irregular shapes, large differences in lateral resistance or stiffness between adjacent stories, or other unusual structural features, shall be determined considering the dynamic characteristics of the structure.

4. **Accidental torsion.** In addition to the requirements of Section 2303 (b) 2, where the vertical resisting elements depend on diaphragm action for shear distribution at any level, the shear-resisting elements shall be capable of resisting a torsional moment assumed to be equivalent to the story shear acting with an eccentricity of not less than 5 percent of the maximum building dimension at that level.

(f) **Overturning.** At any level the incremental changes of the design overturning moment, in the story under consideration, shall be distributed to the various resisting elements in the same proportion as the distribution of the shears in the resisting system. Where other vertical members are provided which are capable of partially resisting the overturning moments, a redistribution may be made to these members if framing members of sufficient strength and stiffness to transmit the required loads are provided.

Where a vertical resisting element is discontinuous, the overturning moment carried by the lowest story of that element shall be carried down as loads to the foundation.

(g) **Lateral Force on Elements of Structures and Nonstructural Components.** Parts or portions of structures, nonstructural components and their anchorage to the main structural system shall be designed for lateral forces in accordance with the following formula:

$$F_p = ZIC_pW_p \qquad \text{(12-8)}$$

The values of C_p are set forth in Table No. 23-J. The value of the I coefficient shall be the value used for the building.

EXCEPTIONS: 1. The value of I for panel connectors shall be as given in Section 2312 (j) 3 C.

2. The value of I for anchorage of machinery and equipment required for life safety systems shall be 1.5.

The distribution of these forces shall be according to the gravity loads pertaining thereto.

For applicable forces on diaphragms and connections for exterior panels, refer to Sections 2312 (j) 2 C and 2312 (j) 3 C.

(h) **Drift and Building Separations.** Lateral deflections or drift of a story relative to its adjacent stories shall not exceed 0.005 times the story height unless it can be demonstrated that greater drift can be tolerated. The displacement calculated from the application of the required lateral forces shall be multiplied by $(1.0/K)$ to obtain the drift. The ratio $(1.0/K)$ shall be not less than 1.0.

All portions of structures shall be designed and constructed to act as an integral unit in resisting horizontal forces unless separated structurally by a distance sufficient to avoid contact under deflection from seismic action or wind forces.

(i) **Alternate Determination and Distribution of Seismic Forces.** Nothing in Section 2312 shall be deemed to prohibit the submission of properly substantiated technical data for establishing the lateral forces and distribution by dynamic analyses. In such analyses the dynamic characteristics of the structure must be considered.

(j) **Structural Systems.** 1. **Ductility requirements.** A. All buildings designed with a horizontal force factor $K = 0.67$ or 0.80 shall have ductile moment-resisting space frames.

B. Buildings more than 160 feet in height shall have ductile moment-resisting space frames capable of resisting not less than 25 percent of the required seismic forces for the structure as a whole.

EXCEPTION: Buildings more than 160 feet in height in Seismic Zones Nos. 1 and 2 may have concrete shear walls designed in accordance with Section 2627 or braced frames designed in conformance with Section 2312 (j) 1 G of this code in lieu of a ductile moment-resisting space frame, provided a K value of 1.00 or 1.33 is utilized in the design.

C. In Seismic Zones No. 2, No. 3 and No. 4 all concrete space frames required by design to be part of the lateral force-resisting system and all concrete frames located in the perimeter line of vertical support shall be ductile moment-resisting space frames.

EXCEPTION: Frames in the perimeter line of the vertical support of buildings designed with shear walls taking 100 percent of the design lateral forces need only conform with Section 2312 (j) 1 D.

D. In Seismic Zones No. 2, No. 3 and No. 4 all framing elements not required by design to be part of the lateral force-resisting system shall be investigated and shown to be adequate for vertical load-carrying capacity and induced moment due to $3/K$ times the distortions resulting from the code-required lateral forces. The rigidity of other elements shall be considered in accordance with Section 2303 (b) 1.

E. Moment-resisting space frames and ductile moment-resisting space frames may be enclosed by or adjoined by more rigid elements which would tend to prevent the space frame from resisting lateral forces where it can be shown that the action or failure of the more rigid elements will not impair the vertical and lateral load-resisting ability of the space frame.

F. Necessary ductility for a ductile moment-resisting space frame shall be provided by a frame of structural steel with moment-resisting connections (complying with Section 2722 for buildings in Seismic Zones No. 3 and No. 4 or Section 2723 for buildings in Seismic Zones No. 1 and No. 2) or by a reinforced concrete frame (complying with Section 2625 for buildings in Seismic Zones No. 3 and No. 4 or Section 2626 for buildings in Seismic Zones No. 1 and No. 2).

EXCEPTION: Buildings with ductile moment-resisting space frames in Seismic Zones No. 1 and No. 2 having an importance factor I greater than 1.0 shall comply with Section 2625 or 2722.

G. In Seismic Zones No. 3 and No. 4 and for buildings having an importance factor I greater than 1.0 located in Seismic Zone No. 2, all members in braced frames shall be designed for 1.25 times the force determined in accordance with Section 2312 (d). Connections shall be designed to develop the full capacity of the members or shall be based on the above forces without the one-third increase usually permitted for stresses resulting from earthquake forces.

Braced frames in buildings shall be composed of axially loaded bracing members of A36, A441, A500 Grades B and C, A501, A572 (Grades 42, 45, 50 and 55) or A588 structural steel, or reinforced concrete members conforming to the requirements of Section 2627.

H. Reinforced concrete shear walls for all buildings shall conform to the requirements of Section 2627.

I. In structures where $K = 0.67$ and $K = 0.80$, the special ductility requirements for structural steel or reinforced concrete specified in Section 2312 (j) 1 F, shall apply to all structural elements below the base which are required to transmit to the foundation the forces resulting from lateral loads.

2. **Design requirements.** A. **Minor alterations.** Minor structural alterations may be made in existing buildings and other structures, but the resistance to lateral forces shall be not less than before such alterations were made, unless the building as altered meets the requirements of this section.

B. **Reinforced masonry or concrete.** All elements within structures located in Seismic Zones No. 2, No. 3 and No. 4 which are of masonry or concrete shall be reinforced so as to qualify as reinforced masonry or concrete under the provisions of Chapters 24 and 26. Principal reinforcement in masonry shall be spaced 2 feet maximum on center in buildings using a moment-resisting space frame.

C. **Diaphragms.** Floor and roof diaphragms and collectors shall be designed to resist the forces determined in accordance with the following formula:

$$F_{px} = \frac{\sum_{i=x}^{n} F_i}{\sum_{i=x}^{n} w_i} w_{px} \qquad \ldots\ldots\ldots\ldots\ldots\ldots (12\text{-}9)$$

WHERE:

F_l = the lateral force applied to level l.

w_l = the portion of W at level l.

w_{px} = the weight of the diaphragm and the elements tributary thereto at level x, including 25 percent of the floor live load in storage and warehouse occupancies.

The force F_{px} determined from Formula (12-9) need not exceed $0.30ZIw_{px}$.

When the diaphragm is required to transfer lateral forces from the vertical resisting elements above the diaphragm to other vertical resisting elements below the diaphragm due to offsets in the placement of the elements or to changes in stiffness in the vertical elements, these forces shall be added to those determined from Formula (12-9).

However, in no case shall lateral force on the diaphragm be less than $0.14ZIw_{px}$.

Diaphragms supporting concrete or masonry walls shall have continuous ties between diaphragm chords to distribute, into the diaphragm, the anchorage forces specified in this chapter. Added chords may be used to form subdiaphragms to transmit the anchorage forces to the main cross ties. Diaphragm deformations shall be considered in the design of the supported walls. See Section 2312 (j) 3 A for special anchorage requirements of wood diaphragms.

3. **Special requirements.** A. **Wood diaphragms providing lateral support for concrete or masonry walls.** Where wood diaphragms are used to laterally support concrete or masonry walls the anchorage shall conform to Section 2310. In Zones No. 2, No. 3 and No. 4 anchorage shall not be accomplished by use of toenails or nails subjected to withdrawal; nor shall wood framing be used in cross-grain bending or cross-grain tension.

B. **Pile caps and caissons.** Individual pile caps and caissons of every building or structure shall be interconnected by ties, each of which can carry by tension and compression a minimum horizontal force equal to 10 percent of the larger pile cap or caisson loading, unless it can be demonstrated that equivalent restraint can be provided by other approved methods.

C. **Exterior elements.** Precast or prefabricated nonbearing, nonshear wall panels or similar elements which are attached to or enclose the exterior shall be designed to resist the forces determined from Formula (12-8) and shall accommodate movements of the structure resulting from lateral forces or temperature changes. The concrete panels or other similar elements shall be supported by means of cast-in-place concrete or mechanical connections and fasteners in accordance with the following provisions:

Connections and panel joints shall allow for a relative movement between stories of not less than two times story drift caused by wind or $(3.0/K)$ times the calculated elastic story displacement caused by required seismic forces, or ½ inch, whichever is greater. Connections to permit movement in the plane of the panel for story drift shall be properly designed sliding connections using slotted or oversized holes or may be connections which permit movement by bending of steel or other connections providing equivalent sliding and ductility capacity.

Bodies of connectors shall have sufficient ductility and rotation capacity so as to preclude fracture of the concrete or brittle failures at or near welds.

The body of the connector shall be designed for one and one-third times the force determined by Formula (12-8). Fasteners attaching the connector to the panel or the structure such as bolts, inserts, welds, dowels, etc., shall be designed to ensure ductile behavior of the connector or shall be designed for four times the load determined from Formula (12-8).

Fasteners embedded in concrete shall be attached to or hooked around reinforcing steel or otherwise terminated so as to effectively transfer forces to the reinforcing steel.

The value of the coefficient I shall be 1.0 for the entire connector assembly in Formula (12-8).

(k) **Essential Facilities.** Essential facilities are those structures or buildings which must be safe and usable for emergency purposes after an earthquake in order to preserve the health and safety of the general public. Such facilities shall include but not be limited to:

1. Hospitals and other medical facilities having surgery or emergency treatment areas.
2. Fire and police stations.
3. Municipal government disaster operation and communication centers deemed to be vital in emergencies.

The design and detailing of equipment which must remain in place and be functional following a major earthquake shall be based upon the requirements of Section 2312 (g) and Table No. 23-J. In addition, their design and detailing shall consider effects induced by structure drifts of not less than $(2.0/K)$ times the story drift caused by required seismic forces nor less than the story drift caused by wind. Special consideration shall also be given to relative movements at separation joints.

(l) **Earthquake-recording Instrumentations.** For earthquake-recording instrumentations see Appendix, Section 2312 (l).

TABLE NO. 23-A—UNIFORM AND CONCENTRATED LOADS

USE OR OCCUPANCY		UNIFORM LOAD[1]	CONCENTRATED LOAD
CATEGORY	DESCRIPTION		
1. Armories		150	0
2. Assembly areas[4] and auditoriums and balconies therewith	Fixed seating areas	50	0
	Movable seating and other areas	100	0
	Stage areas and enclosed platforms	125	0
3. Cornices, marquees and residential balconies		60	0
4. Exit facilities[5]		100	0[8]
5. Garages	General storage and/or repair	100	[3]
	Private pleasure car storage	50	[3]
6. Hospitals	Wards and rooms	40	1000[2]
7. Libraries	Reading rooms	60	1000[2]
	Stack rooms	125	1500[2]
8. Manufacturing	Light	75	2000[2]
	Heavy	125	3000[2]
9. Offices		50	2000[2]
10. Printing plants	Press rooms	150	2500[2]
	Composing and linotype rooms	100	2000[2]
11. Residential[6]		40	0[8]
12. Rest rooms[7]			
13. Reviewing stands, grandstands and bleachers		100	0
14. Roof deck	Same as area served or for the type of occupancy accommodated		
15. Schools	Classrooms	40	1000[2]
16. Sidewalks and driveways	Public access	250	[3]
17. Storage	Light	125	
	Heavy	250	
18. Stores	Retail	75	2000[2]
	Wholesale	100	3000[2]

[1]See Section 2306 for live load reductions.
[2]See Section 2304 (c), first paragraph, for area of load application.
[3]See Section 2304 (c), second paragraph, for concentrated loads.
[4]Assembly areas include such occupancies as dance halls, drill rooms, gymnasiums, play-

grounds, plazas, terraces and similar occupancies which are generally accessible to the public.

5Exit facilities shall include such uses as corridors serving an occupant load of 10 or more persons, exterior exit balconies, stairways, fire escapes and similar uses.

6Residential occupancies include private dwellings, apartments and hotel guest rooms.

7Rest room loads shall be not less than the load for the occupancy with which they are associated, but need not exceed 50 pounds per square foot.

8Individual stair treads shall be designed to support a 300-pound concentrated load placed in a position which would cause maximum stress. Stair stringers may be designed for the uniform load set forth in the table.

TABLE NO. 23-B—SPECIAL LOADS[1]

USE		VERTICAL LOAD	LATERAL LOAD
CATEGORY	DESCRIPTION	(Pounds per Square Foot Unless Otherwise Noted)	
1. Construction, public access at site (live load)	Walkway See Sec. 4406	150	
	Canopy See Sec. 4407	150	
2. Grandstands, reviewing stands and bleachers (live load)	Seats and footboards	120[2]	See Footnote 3
3. Stage accessories, see Sec. 3902 (live load)	Gridirons and fly galleries	75	
	Loft block wells[4]	250	250
	Head block wells and sheave beams[4]	250	250
4. Ceiling framing (live load)	Over stages	20	
	All uses except over stages	10[5]	
5. Partitions and interior walls, see Sec. 2309 (live load)			5
6. Elevators and dumbwaiters (dead and live load)		2 x Total loads[6]	
7. Mechanical and electrical equipment (dead load)		Total loads	
8. Cranes (dead and live load)[7]	Total load including impact increase	1.25 x Total load[7]	0.10 x Total load[8]
9. Balcony railings, guard rails and handrails	Exit facilities serving an occupant load greater than 50		50[9]
	Other		20[9]
10. Storage racks	Over 8 feet high	Total loads[10]	See Table No. 23-J

(Footnotes on following page)

FOOTNOTES FOR TABLE NO. 23-B

[1]The tabulated loads are minimum loads. Where other vertical loads required by this code or required by the design would cause greater stresses they shall be used.

[2]Pounds per lineal foot.

[3]Lateral sway bracing loads of 24 pounds per foot parallel and 10 pounds per foot perpendicular to seat and footboards.

[4]All loads are in pounds per lineal foot. Head block wells and sheave beams shall be designed for all loft block well loads tributary thereto. Sheave blocks shall be designed with a factor of safety of five.

[5]Does not apply to ceilings which have sufficient total access from below, such that access is not required within the space above the ceiling. Does not apply to ceilings if the attic areas above the ceiling are not provided with access. This live load need not be considered acting simultaneously with other live loads imposed upon the ceiling framing or its supporting structure.

[6]Where Appendix Chapter 51 has been adopted, see reference standard cited therein for additional design requirements.

[7]The impact factors included are for cranes with steel wheels riding on steel rails. They may be modified if substantiating technical data acceptable to the building official is submitted. Live loads on crane support girders and their connections shall be taken as the maximum crane wheel loads. For pendant-operated traveling crane support girders and their connections, the impact factors shall be 1.10.

[8]This applies in the direction parallel to the runway rails (longitudinal). The factor for forces perpendicular to the rail is 0.20 × the transverse traveling loads (trolley, cab, hooks and lifted loads). Forces shall be applied at top of rail and may be distributed among rails of multiple rail cranes and shall be distributed with due regard for lateral stiffness of the structures supporting these rails.

[9]A load per lineal foot to be applied horizontally at right angles to the top rail.

[10]Vertical members of storage racks shall be protected from impact forces of operating equipment or racks shall be designed so that failure of one vertical member will not cause collapse of more than the bay or bays directly supported by that member.

TABLE NO. 23-C—MINIMUM ROOF LIVE LOADS[1]

	METHOD 1			METHOD 2		
	TRIBUTARY LOADED AREA IN SQUARE FEET FOR ANY STRUCTURAL MEMBER					
ROOF SLOPE	0 to 200	201 to 600	Over 600	UNIFORM LOAD[2]	RATE OF REDUC-TION r (Percent)	MAXIMUM REDUC-TION R (Percent)
1. Flat or rise less than 4 inches per foot. Arch or dome with rise less than one eighth of span	20	16	12	20	.08	40
2. Rise 4 inches per foot to less than 12 inches per foot. Arch or dome with rise one eighth of span to less than three eighths of span	16	14	12	16	.06	25
3. Rise 12 inches per foot and greater. Arch or dome with rise three eighths of span or greater	12	12	12	12	No Reductions Permitted	
4. Awnings except cloth covered[3]	5	5	5	5		
5. Greenhouses, lath houses and agricultural buildings[4]	10	10	10	10		

[1]Where snow loads occur, the roof structure shall be designed for such loads as determined by the building official. See Section 2305 (d). For special purpose roofs, see Section 2305 (e).

[2]See Section 2306 for live load reductions. The rate of reduction r in Section 2306 Formula (6-1) shall be as indicated in the table. The maximum reduction R shall not exceed the value indicated in the table.

[3]As defined in Section 4506.

[4]See Section 2305 (e) for concentrated load requirements for greenhouse roof members.

TABLE NO. 23-D—MAXIMUM ALLOWABLE DEFLECTION FOR STRUCTURAL MEMBERS[1]

TYPE OF MEMBER	MEMBER LOADED WITH LIVE LOAD ONLY (L.L.)	MEMBER LOADED WITH LIVE LOAD PLUS DEAD LOAD (L.L. + K D.L.)
Roof Member Supporting Plaster or Floor Member	$L/360$	$L/240$

[1]Sufficient slope or camber shall be provided for flat roofs in accordance with Section 2305 (f).

L.L. = Live load

D.L. = Dead load

K = Factor as determined by Table No. 23-E

L = Length of member in same units as deflection

TABLE NO. 23-E—VALUE OF "K"

WOOD		REINFORCED CONCRETE[2]	STEEL
Unseasoned	Seasoned[1]		
1.0	0.5	$[2 - 1.2\,(A'_s/A_s)] \geqq 0.6$	0

[1]Seasoned lumber is lumber having a moisture content of less than 16 percent at time of installation and used under dry conditions of use such as in covered structures.

[2]See also Section 2609.

A'_s = Area of compression reinforcement.

A_s = Area of nonprestressed tension reinforcement.

TABLE NO. 23-F—WIND STAGNATION PRESSURE (q_s) AT STANDARD HEIGHT OF 30 FEET

Basic wind speed (mph)[1]	70	80	90	100	110	120	130
Pressure q_s (psf)	13	17	21	26	31	37	44

[1]Wind speed from Section 2311 (b).

TABLE NO. 23-G—COMBINED HEIGHT, EXPOSURE AND GUST FACTOR COEFFICIENT (C_e)

HEIGHT ABOVE AVERAGE LEVEL OF ADJOINING GROUND, IN FEET	EXPOSURE C	EXPOSURE B
0- 20	1.2	0.7
20- 40	1.3	0.8
40- 60	1.5	1.0
60-100	1.6	1.1
100-150	1.8	1.3
150-200	1.9	1.4
200-300	2.1	1.6
300-400	2.2	1.8

TABLE NO. 23-H—PRESSURE COEFFICIENTS (C_q)

STRUCTURE OR PART THEREOF	DESCRIPTION	C_q FACTOR
Primary frames and systems	**Method 1** (Normal Force Method)	
	Windward wall	0.8 inward
	Leeward wall	0.5 outward
	Leeward roof or flat roof	0.7 outward
	Windward roof	
	Slope<9:12	0.7 outward
	Slope 9:12 to 12:12	0.4 inward
	Slope>12:12	0.7 inward
	Wind parallel to ridge	
	Enclosed structures	0.7 outward
	Open structures[1]	1.2 outward
	Method 2 (Projected Area Method)	
	On vertical projected area	
	Structures 40 feet or less in height	1.3 horizontal any direction
	Structures over 40 feet in height	1.4 horizontal any direction
	On horizontal projected area	
	Enclosed structure	0.7 upward
	Open structure[1]	1.2 upward
Elements and components	Wall elements	
	All structures	1.2 inward
	Enclosed structures	1.1 outward
	Open structures	1.6 outward
	Parapets	1.3 inward or outward
	Roof elements	
	Enclosed structures	
	Slope<9:12	1.1 outward
	Slope 9:12 to 12:12	1.1 outward or 0.8 inward
	Slope>12:12	1.1 outward or inward
	Open structures[1]	
	Slope<9:12	1.6 outward
	Slope 9:12 to 12:12	1.6 outward or 0.8 inward
	Slope>12:12	1.6 outward or 1.1 inward
Local areas at discontinuities[2]	Wall corners	2.0 outward
	Canopies or overhangs at eaves or rakes	2.8 upward
	Roof ridges at ends of buildings or eaves and roof edges at building corners	3.0 upward

(Continued)

TABLE NO. 23-H—PRESSURE COEFFICIENTS (C_q)—(Continued)

STRUCTURE OR PART THEREOF	DESCRIPTION	C_q FACTOR
	Eaves or rakes without overhangs away from building corners and ridges away from ends of building	2.0 upward
	Cladding connections Add 0.5 to outward or upward C_q for appropriate location	
Chimneys, tanks and solid towers	Square or rectangular	1.4 any direction
	Hexagonal or octagonal	1.1 any direction
	Round or elliptical	0.8 any direction
Open-frame towers[3] [4]		2.0 any direction
Signs, flagpoles, lightpoles, minor structures		1.4 any direction

[1]A structure with more than 30 percent of any one side open shall be considered an open structure. Nonimpact-resistant glazing shall be considered as an opening.

[2]Local pressures shall apply over a distance from the discontinuity of 10 feet or 0.1 times the least width of the structure, whichever is smaller.

[3]The area to which the design pressure shall be applied shall be the projected area of all elements other than those in planes parallel to the direction of application.

[4]For radio and transmission towers, the area shall be the projected area of the members on one face multiplied by 2.0 for rectangular towers and 1.8 for triangular towers.

TABLE NO. 23-I—HORIZONTAL FORCE FACTOR *K* FOR BUILDINGS OR OTHER STRUCTURES[1]

TYPE OR ARRANGEMENT OF RESISTING ELEMENTS	VALUE[2] OF *K*
1. All building framing systems except as hereinafter classified	1.00
2. Buildings with a box system as specified in Section 2312 (b) **EXCEPTION:** Buildings not more than three stories in height with stud wall framing and using plywood horizontal diaphragms and plywood vertical shear panels for the lateral force system may use $K = 1.0$.	1.33
3. Buildings with a dual bracing system consisting of a ductile moment-resisting space frame and shear walls or braced frames using the following design criteria: a. The frames and shear walls or braced frames shall resist the total lateral force in accordance with their relative rigidities considering the interaction of the shear walls and frames b. The shear walls or braced frames acting independently of the ductile moment-resisting portions of the space frame shall resist the total required lateral forces c. The ductile moment-resisting space frame shall have the capacity to resist not less than 25 percent of the required lateral force	0.80
4. Buildings with a ductile moment-resisting space frame designed in accordance with the following criteria: The ductile moment-resisting space frame shall have the capacity to resist the total required lateral force	0.67
5. Elevated tanks plus full contents, on four or more cross-braced legs and not supported by a building	2.5[3]
6. Structures other than buildings and other than those set forth in Table No. 23-J	2.00

[1]Where wind load as specified in Section 2311 would produce higher stresses, this load shall be used in lieu of the loads resulting from earthquake forces.

[2]See Figures Nos. 1, 2 and 3 in this chapter and definition of Z as specified in Section 2312 (c).

[3]The minimum value of KC shall be 0.12 and the maximum value of KC need not exceed 0.25.

The tower shall be designed for an accidential torsion of 5 percent as specified in Section 2312 (e) 4. Elevated tanks which are supported by buildings or do not conform to type or arrangement of supporting elements as described above shall be designed in accordance with Section 2312 (g) using $C_p = .3$.

TABLE NO. 23-J—HORIZONTAL FORCE FACTOR C_p FOR ELEMENTS OF STRUCTURES AND NONSTRUCTURAL COMPONENTS

PART OR PORTION OF BUILDINGS	DIRECTION OF HORIZONTAL FORCE	VALUE OF C_p[1]
1. Exterior bearing and nonbearing walls, interior bearing walls and partitions, interior nonbearing walls and partitions—see also Section 2312 (j) 3 C. Masonry or concrete fences over 6 feet high	Normal to flat surface	0.3[6]
2. Cantilever elements: a. Parapets	Normal to flat surfaces	0.8
b. Chimneys or stacks	Any direction	
3. Exterior and interior ornamentations and appendages	Any direction	0.8
4. When connected to, part of, or housed within a building: a. Penthouses, anchorage and supports for chimneys, stacks and tanks, including contents b. Storage racks with upper storage level at more than 8 feet in height, plus contents c. All equipment or machinery	Any direction	0.3[2 3]
5. Suspended ceiling framing systems (applies to Seismic Zones Nos. 2, 3 and 4 only)—see also Section 4701 (e)	Any direction	0.3[4 7]
6. Connections for prefabricated structural elements other than walls, with force applied at center of gravity of assembly	Any direction	0.3[5]

[1] C_p for elements laterally self-supported only at the ground level may be two thirds of value shown.

[2] W_p for storage racks shall be the weight of the racks plus contents. The value of C_p for racks over two storage support levels in height shall be 0.24 for the levels below the top two levels. In lieu of the tabulated values steel storage racks may be designed in accordance with U.B.C. Standard No. 27-11.

Where a number of storage rack units are interconnected so that there are a minimum of four vertical elements in each direction on each column line designed to resist horizontal forces, the design coefficients may be as for a building with K values from Table No. 23-I, $CS = 0.2$ for use in the formula $V = ZIKCSW$ and W equal to the total dead load plus 50 percent of the rack-rated capacity. Where the design and rack configurations are in accordance with this paragraph, the design provisions in U.B.C. Standard No. 27-11 do not apply.

[3] For flexible and flexibly mounted equipment and machinery, the appropriate values of C_p shall be determined with consideration given to both the dynamic properties of the equipment and machinery and to the building or structure in which it is placed but shall be not less than the listed values. The design of the equipment and machinery and their anchorage is an integral part of the design and specification of such equipment and masonry.

For essential facilities and life safety systems, the design and detailing of equipment which must remain in place and be functional following a major earthquake shall consider drifts in accordance with Section 2312 (k).

[4]Ceiling weight shall include all light fixtures and other equipment which is laterally supported by the ceiling. For purposes of determining the lateral force, a ceiling weight of not less than 4 pounds per square foot shall be used.

[5]The force shall be resisted by positive anchorage and not by friction.

[6]See also Section 2309 (b) for minimum load and deflection criteria for interior partitions.

[7]Does not apply to ceilings constructed of lath and plaster or gypsum board screw or nail attached to suspended members that support a ceiling at one level extending from wall to wall.

TABLE NO. 23-K—VALUES FOR OCCUPANCY IMPORTANCE FACTOR I

TYPE OF OCCUPANCY	I
Essential facilities[1]	1.5
Any building where the primary occupancy is for assembly use for more than 300 persons (in one room)	1.25
All others	1.0

[1]See Section 2312 (k) for definition and additional requirements for essential facilities.

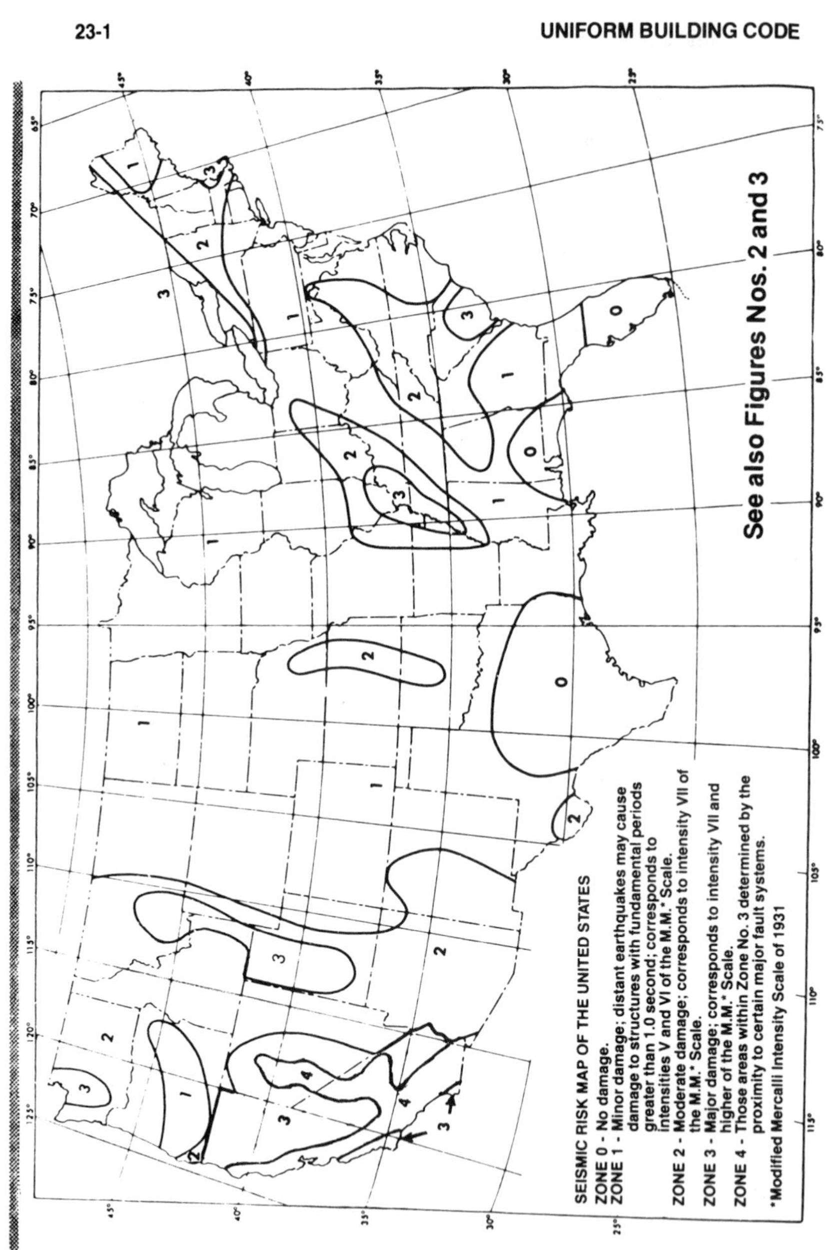
See also Figures Nos. 2 and 3
SEISMIC RISK MAP OF THE UNITED STATES
ZONE 0 - No damage.
ZONE 1 - Minor damage; distant earthquakes may cause damage to structures with fundamental periods greater than 1.0 second; corresponds to intensities V and VI of the M.M.* Scale.
ZONE 2 - Moderate damage; corresponds to intensity VII of the M.M.* Scale.
ZONE 3 - Major damage; corresponds to intensity VII and higher of the M.M.* Scale.
ZONE 4 - Those areas within Zone No. 3 determined by the proximity to certain major fault systems.
*Modified Mercalli Intensity Scale of 1931

FIGURE NO. 1—SEISMIC ZONE MAP OF THE UNITED STATES
For areas outside of the United States, see Appendix Chapter 23

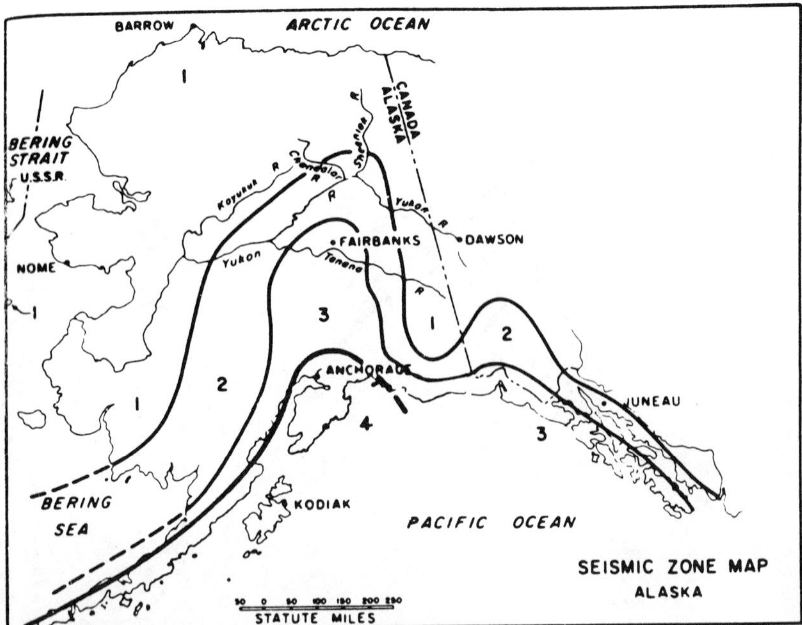

FIGURE NO. 2

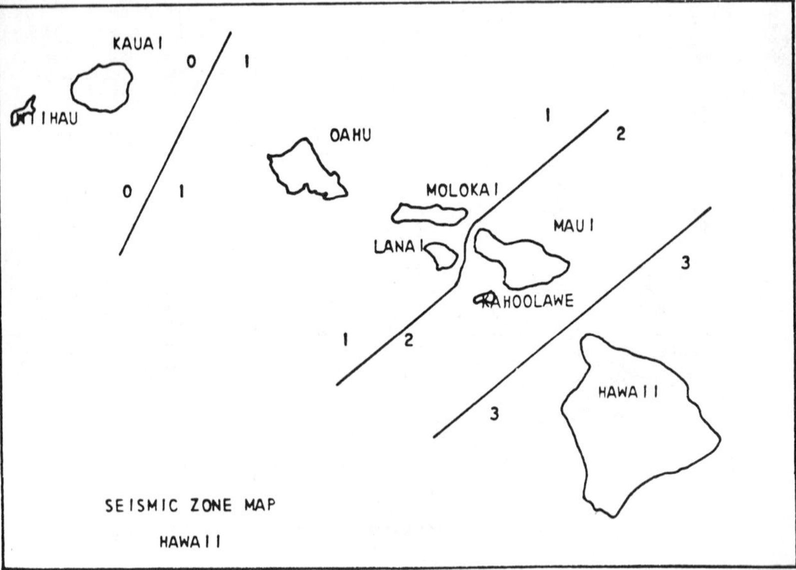

FIGURE NO. 3

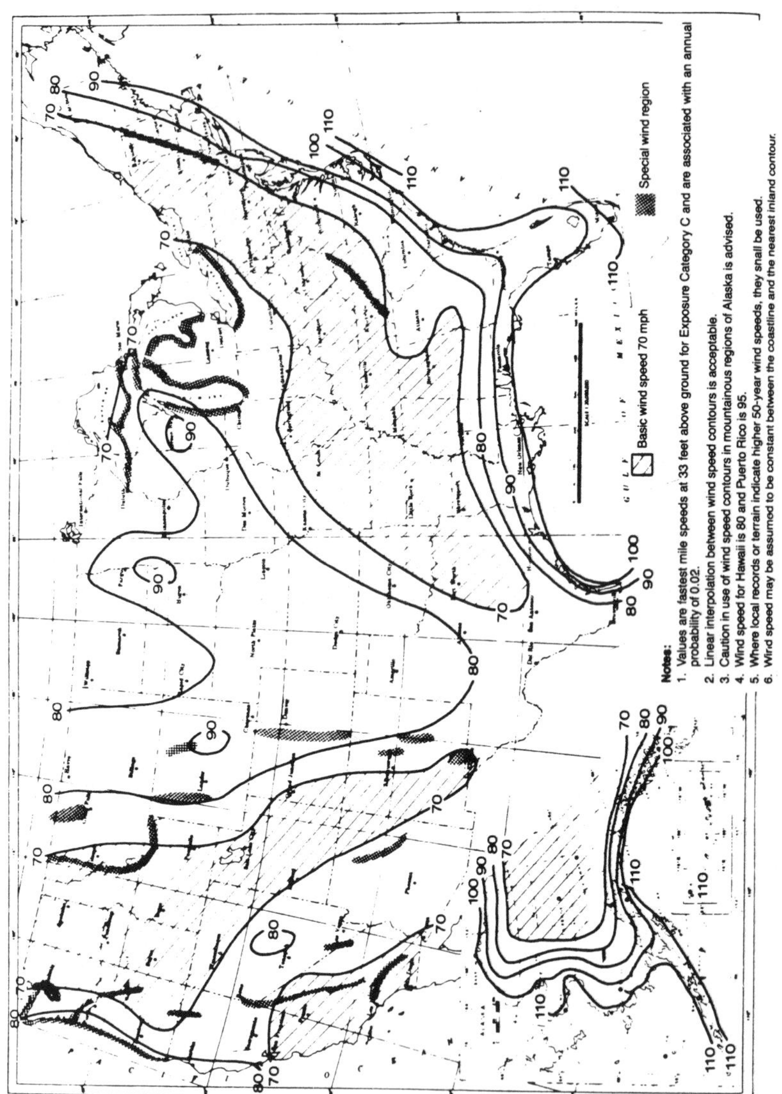

Figure No. 4—Basic Wind Speeds in Miles Per Hour

APPENDIX B

Nomenclature, Symbols, Glossaries

NOMENCLATURE

A Gross area of an axially loaded compression member, Part 2 (square inches)

A_b Nominal body area of a fastener (square inches); area of an upset rod based upon the major diameter of its threads, i.e., the diameter of a coaxial cylinder which would bound the crests of the upset threads (square inches)

A_c Actual area of effective concrete flange in composite design (square inches)

A_e Effective net area of an axially loaded tension member (square inches)

A_f Area of compression flange (square inches)

A_n Net area of an axially loaded tension member (square inches)

A_s Area of steel beam in composite design (square inches)

A_s' Area of compressive reinforcing steel (square inches)

A_{sr} Area of reinforcing steel providing composite action at point of negative moment (square inches)

A_{st} Cross-sectional area of stiffener or pair of stiffeners (square inches)

A_w Area of girder web (square inches)

Definitions of letter symbols listed are part of the Nomenclature listing of the AISC Specification, 1978, reprinted here with permission. When the same symbols are given different meanings in the text, they are defined where used.

C_b Bending coefficient dependent upon moment gradient

C_m Coefficient applied to bending term in interaction formula for prismatic members and dependent upon column curvature caused by applied moments

C'_m Coefficient applied to bending term in interaction formula for tapered members and dependent upon axial stress at the small end of the member

C_t Reduction coefficient in computing effective net area of an axially loaded tension member

C_v Ratio of "critical" web stress, according to the linear buckling theory, to the shear yield stress of web material

C_2 Increment used in computing minimum edge distance for oversized and slotted holes

D Factor depending upon type of transverse stiffeners; outside diameter of tubular member (inches)

E Modulus of elasticity of steel (29,000 kips per square inch)

E_c Modulus of elasticity of concrete (kips per square inch)

F_a Axial compressive stress permitted in a prismatic member in the absence of bending moment (kips per square inch)

F_{as} Axial compressive stress permitted in the absence of bending moment, for bracing and other secondary members (kips per square inch)

F_b Bending stress permitted in a prismatic member in the absence of axial force (kips per square inch)

F'_b Allowable bending stress in compression flange of plate girders as reduced for hybrid girders or because of large web depth-to-thickness ratio (kips per square inch)

F'_e Euler stress for a prismatic member divided by factor of safety (kips per square inch)

$F'_{e\gamma}$ Euler stress for a tapered member divided by factor of safety (kips per square inch)

F_p Allowable bearing stress (kips per square inch)

F_{sr} Stress range (kips per square inch)

F_t Allowable axial tensile stress (kips per square inch)

F_u Specified minimum tensile strength of the type of steel or fastener being used (kips per square inch)

F_v Allowable shear stress (kips per square inch)

F_y Specified minimum yield stress of the type of steel being used (kips per square inch)

I_{tr} Moment of inertia of transformed composite section (inches^4)

K Effective length factor for a prismatic member

M Moment, Part 1 (kip-feet); factored bending moment, Part 2 (kip-feet)

M_1 Smaller moment at end of unbraced length of beam-column

$\boldsymbol{M_2}$ Larger moment at end of unbraced length of beam-column

$\boldsymbol{M_D}$ Moment produced by dead load

$\boldsymbol{M_L}$ Moment produced by live load

$\boldsymbol{M_p}$ Plastic moment (kip-feet)

$\boldsymbol{N}$ Length of bearing of applied load (inches)

$\boldsymbol{P_{cr}}$ Maximum strength of an axially loaded compression member or beam, Part 2 (kips)

$\boldsymbol{P_e}$ Euler buckling load, Part 2 (kips)

$\boldsymbol{S_s}$ Section modulus of steel beam used in composite design, referred to the bottom flange (inches^3)

$\boldsymbol{S_{tr}}$ Section modulus of transformed composite cross section, referred to the bottom flange; based upon maximum permitted effective width of concrete flange (inches^3)

$\boldsymbol{T_b}$ Specified pretension of a high-strength bolt (kips)

$\boldsymbol{V_h}$ Total horizontal shear to be resisted by connectors under full composite action (kips)

$\boldsymbol{V'_h}$ Total horizontal shear provided by the connectors in providing partial composite action (kips)

$\boldsymbol{V_u}$ Statical shear produced by "ultimate" load in plastic design (kips)

$\boldsymbol{Y}$ Ratio of yield stress of web steel to yield stress of stiffener steel

$\boldsymbol{Z}$ Plastic section modulus (inches^3)

$\boldsymbol{a}$ Clear distance between transverse stiffeners (inches); dimension parallel to the direction of stress, Sect. B1, Table B2 (inches)

$\boldsymbol{b}$ Actual width of stiffened and unstiffened compression elements (inches); dimension normal to the direction of stress, Sect. B1, Table B2 (inches)

$\boldsymbol{b_e}$ Effective width of stiffened compression element (inches)

$\boldsymbol{b_f}$ Flange width of rolled beam or plate girder (inches)

$\boldsymbol{d}$ Depth of beam or girder (inches); diameter of a roller or rocker bearing (inches); nominal diameter of a fastener (inches)

$\boldsymbol{f_a}$ Computed axial stress (kips per square inch)

$\boldsymbol{f_b}$ Computed bending stress (kips per square inch)

$\boldsymbol{f'_c}$ Specified compression strength of concrete (kips per square inch)

$\boldsymbol{f_t}$ Computed tensile stress (kips per square inch)

$\boldsymbol{f_v}$ Computed shear stress (kips per square inch)

$\boldsymbol{f_{vs}}$ Shear between girder web and transverse stiffeners (kips per linear inch of single stiffener or pair of stiffeners)

$\boldsymbol{g}$ Transverse spacing between fastener gage lines (inches)

$\boldsymbol{h}$ Clear distance between flanges of a beam or girder at the section under investigation (inches)

k Coefficient relating linear buckling strength of a plate to its dimensions and condition of edge support; distance from outer face of flange to web toe of fillet of rolled shape or equivalent distance on welded section (inches)

l For beams, distance between cross sections braced against twist or lateral displacement of the compression flange (inches); for columns, actual unbraced length of member (inches); unsupported length of a lacing bar (inches)

l_b Actual unbraced length in plane of bending (inches)

l_{cr} Critical unbraced length adjacent to plastic hinge (inches)

n Modular ratio (E/E_c)

q Allowable horizontal shear to be resisted by a shear connector (kips)

r Governing radius of gyration (inches)

r_b Radius of gyration about axis of concurrent bending (inches)

r_T Radius of gyration of a section comprising the compression flange plus one-third of the compression web area, taken about an axis in the plane of the web (inches)

s Longitudinal center-to-center spacing (pitch) of any two consecutive holes (inches)

t Girder, beam, or column web thickness (inches); thickness of a connected part (inches); wall thickness of a tubular member (inches)

t_b Thickness of beam flange or moment connection plate at rigid beam-to-column connection (inches)

t_f Flange thickness (inches)

x Subscript relating symbol to strong axis bending

y Subscript relating symbol to weak axis bending

z Distance from the smaller end of a tapered member (inches)

α Ratio of hybrid girder web yield stress to flange yield stress

β Ratio S_{tr}/S_s or S_{eff}/S_s

γ Tapering ratio of a tapered member or unbraced segment of a tapered member; subscript relating symbol to tapered members

Δ Displacement of the neutral axis of a loaded member from its position when the member is not loaded (inches)

Welding symbols are found in Figure 12.13. Glossaries of specialized terms are found in Chapter 12 for welding, Chapter 13 for mechanical fasteners, and Chapter 19 for steel erection and derrick usage. A limited list of definitions of terms is found in Section 2.3.

References

1. American Association of State Highway and Transportation Officials (AASHTO), *Standard Specification for Highway Bridges*. Washington, D.C.: The Association, 1977 (with more recent supplements).
2. American Concrete Institute, *ACI Standard Building Code Requirements for Reinforced Concrete*. Detroit: The Institute, 1983.
3. American Institute of Steel Construction, *Code of Standard Practice for Steel Buildings and Bridges* (in reference 7).
4. American Institute of Steel Construction, "Commentary on Highly Restrained Welded Connections," *AISC Engineering Journal*, 3rd quarter, 1973.
5. American Institute of Steel Construction, *Design Manual for Orthotropic Steel Plate Deck Bridges*. Chicago: The Institute, 1963.
6. American Institute of Steel Construction, *Guide to the Analysis of Guy and Stiffleg Derricks*. Chicago: The Institute, 1974.
7. American Institute of Steel Construction, *Manual of Steel Construction* (8th ed.). Chicago: The Institute, 1980.
8. American Institute of Steel Construction, *Plastic Design in Steel*. Chicago: The Institute, 1959.
9. American Institute of Steel Construction, *Iron and Steel Beams, 1873–1952*. Compiled and edited by Herbert W. Ferris. Chicago: The Institute, 1953.
10. American Institute of Steel Construction, *Specification for Design, Fabrication and Erection of Steel Buildings* (in reference 7).
11. American Iron and Steel Institute, *Plastic Design of Multi-Story Steel Frames* Washington, D.C.: The Institute, 1968.

12. American Iron and Steel Institute, *Specification for the Design of Cold-Formed Steel Structural Members*. Washington, D.C.: The Institute, 1980.
13. American National Standards Institute, *B58.1—Building Code Requirements for Minimum Design Loads in Buildings and Other Structures*. Washington, D.C.: The Institute (updated periodically).
14. American National Standards Institute, *B30.6—Safety Code for Derricks*. Washington, D.C.: The Institute (updated periodically).
15. American Society for Metals, *The Metals Handbook*, vol. 2. (8th ed.). Metals Park, Ohio: The Society, 1964.
16. American Society of Civil Engineers, *Consulting Engineering—A Guide for the Engagement of Engineering Services*, Manual 45. New York: The Society, 1975.
17. American Welding Society, *Structural Welding Code—Steel*, AWS D.1.1-80. Miami: The Society, 1980.
18. Barrie, D. S., and B. C. Paulson, Jr., *Professional Construction Management*. New York: McGraw-Hill, 1978.
19. Beaufait, F. W., *Basic Concepts of Structural Analysis*. Englewood Cliffs, N.J.: Prentice-Hall, 1977.
20. Bleich, Friedrich, and others, *Mathematical Theory of Vibration in Suspension Bridges*. Washington, D.C.: U.S. Bureau of Public Roads, 1950.
21. Bleich, Friedrich, *Buckling Strength of Metal Structures*. New York: McGraw-Hill, 1952.
22. Blodgett, Omer W., *Design of Welded Structures*. Cleveland: James F. Lincoln Arc Welding Foundation, 1966.
23. Blume, J. A., N. M. Newmark, and L. H. Corning, *Design of Reinforced Concrete Buildings for Earthquake Motions*. Chicago: The Portland Cement Association, 1961.
24. Brockenbrough, R. L., and B. G. Johnston, *Steel Design Manual*. Pittsburgh: U.S. Steel Corp., 1981.
25. Bronowski, Jacob, *The Ascent of Man*. Boston: Little, Brown, 1973.
26. Construction Specification Institute, *Uniform Construction Index*. Washington, D.C.: The Institute.
27. Dieter, George E., *Mechanical Metallurgy* (2nd ed.). New York: McGraw-Hill, 1961.
28. Galambos, Theodore V., "Load and Resistance Factor Design," *AISC Engineering Journal*, 3rd quarter, 1981.
29. Gaylord, E. H., Jr., and C. N. Gaylord, *Design of Steel Structures* (2nd ed.). New York: McGraw-Hill, 1972.
30. Glahn, Else, "Chinese Building Standards in the 12th Century," *Scientific American*, May 1981.
31. Holesapple, J. C., "The Fabricator/Designer Connection," *Civil Engineering, ASCE*, Nov. 1982.
32. Hsieh, Yuah-Yu, *Elementary Theory of Structures*. Englewood Cliffs, N.J.: Prentice-Hall, 1970.
33. International Conference of Building Officials, *Uniform Building Code (UBC)*. Whittier, Calif.: The Conference, 1982.
34. Johnson, J. B., C. W. Bryan, and F. E. Turneaure, *Modern Framed Structures* (10th ed.). New York: Wiley, 1944.
Johnston, B. G., F. J. Lin, and T. V. Galambos, *Basic Steel Design* (2nd ed.). Englewood Cliffs, N.J.: Prentice-Hall, 1980.

751

36. Johnston, Bruce G., ed., *Structural Stability Research Council, Guide to Stability Design Criteria for Metal Structures* (3rd ed.). New York: Wiley, 1976.
37. Lin, T. Y., and D. S. Stotesbury, *Structural Concepts and Systems for Architects and Engineers*. New York: Wiley, 1981.
38. Marcus, S. H., *Basics of Structural Steel Design*. Reston, Va.: Reston, 1977.
39. McCormac, J. C., *Structural Steel Design* (3rd ed.). New York: Harper & Row, 1981.
40. Means, R. S., *Building Construction Cost Data—1983*. Kingston, Mass.: The Robert Snow Means Co., Inc., 1983.
41. Means, R. S., *Means Systems Costs, 1983*. Kingston, Mass.: The Robert Snow Means Co., Inc., 1983.
42. Metal Building Manufacturers' Association, *Metal Building Systems Manual*. Cleveland: The Association, 1981.
43. Nervi, P. L., *Structures*. English translation by G. Salvadori and M. Salvadori. New York: McGraw-Hill, 1956.
44. Pfrang, E. O., and R. Marshall, "Collapse of the Kansas City Hyatt Regency Walkways—The NBS Report," *Civil Engineering, ASCE*, July 1982.
45. Popov, Egor P., *Mechanics of Materials* (2nd ed.). Englewood Cliffs, N.J.: Prentice-Hall, 1976.
46. Popov, E. P., and C. W. Roeder, "Design of Eccentrically Braced Frames," *Engineering Journal, AISC*, 2nd quarter, 1978.
47. Preece, F. Robert, *Structural Steel in the 80's—Materials, Fastening and Testing*. 171 Second St., San Francisco: The Steel Committee of California, 1981.
48. Roark, R. J., and W. C. Young, *Formulas for Stress and Strain* (5th ed.). New York: McGraw-Hill, 1976.
49. Roeder, C. W., and E. P. Popov, "Eccentrically Braced Steel Frames for Earthquakes," *Journal of the Structural Division, ASCE*, Vol. 104, No. ST3, March 1978.
50. Rolfe, S. T., "Fracture and Fatigue Control in Steel Structures," *Engineering Journal, AISC*, Sept. 1980.
51. Salvadori, Mario, *Building—The Fight against Gravity*. New York: Atheneum, 1979.
52. Salvadori, M., and M. Levy, *Structure in Architecture* (2nd ed.). Englewood Cliffs, N.J.: Prentice-Hall, 1981.
53. Seaburg, Paul A., "The ABC's (and q) of Cold-Formed Steel Design," *Civil Engineering, ASCE*, Jan. 1981.
54. Shank, M. E., ed., *Control of Steel Construction to Avoid Brittle Failure*. New York: The Welding Research Council, 1957.
55. Steel Joist Institute, *Standard Specifications, Load Tables and Weight Tables for Steel Joists and Joist Girders*. Richmond, Va.: The Institute (updated periodically).
56. Tall, Lambert, ed., *Structural Steel Design*. New York: Ronald Press, 1964.
57. Teal, Edward J., *Practical Design of Eccentric Braced Frames to Resist Seismic Forces*. San Francisco: The Structural Steel Educational Council, 1979 (contains bibliography).
58. Timoshenko, S., *Strength of Materials*, 2 vol. (3rd ed.). New York: Van Nostrand Reinhold, 1956 (reprinted, R. E. Krieger, Huntington, N.Y., 1976).
59. Timoshenko, S., and J. M. Gere, *Theory of Elastic Stability* (2nd ed.). New York: McGraw-Hill, 1961.
60. Timoshenko, S., and J. N. Goodier, *Theory of Elasticity* (2nd ed.). New York: McGraw-Hill, 1951.

61. Torroja, Eduardo, *Philosophy of Structures*. English version by J. J. Polivka and M. Polivka. Berkeley: University of California Press, 1958.
62. Troitsky, M. S., *Orthotropic Bridges—Theory and Design*. Cleveland: James F. Lincoln Arc Welding Foundation, 1967.
63. United States Steel Co., *Highway Structures Design Handbook*, 2 vol. Pittsburgh: The Company (chapters of varying dates).
64. United States Steel Supply Co., *Wire Rope Engineering Handbook*. Pittsburgh: The Company (to 1982).
65. University of Washington Experiment Station, *Aerodynamic Theory of Suspension Bridges*, Bulletin No. 116, 3 parts. Seattle: University of Washington Press, 1952.
66. Van Vlaak, Lawrence H., *Materials for Engineering*. Reading, Mass.: Addison-Wesley, 1982.
67. American Institute of Steel Construction, *Engineering for Steel Construction*. Chicago: The Institute, 1984.

Index